Plant and Vegetation

Volume 12

Series editor
M.J.A. Werger, Utrecht, The Netherlands

Plant and Vegetation is a new Springer series comprising a series of books that present current knowledge and new perspectives on world vegetation. Examining the ecology of plants and vegetation at all scales – from plant to landscape – and covering key issues such as globalization, invasive species, climate change and the dynamics of plant biodiversity, this book series draws together a wide range of material of interest to plant ecologists, vegetation scientists, and geographers around the world. The series provides a valuable resource for both graduate students and researchers in environmental and biological sciences, as well as for landscape planners and policy makers involved in land-use and restoration projects at local, regional and international levels.

More information about this series at http://www.springer.com/series/7549

Javier Loidi
Editor

The Vegetation of the Iberian Peninsula

Volume 1

Springer

Editor
Javier Loidi
Department of Plant Biology and Ecology
University of the Basque Country (UPV/EHU)
Bilbao, Spain

Additional material to this book can be downloaded from http://extras.springer.com.

ISSN 1875-1318 ISSN 1875-1326 (electronic)
Plant and Vegetation
ISBN 978-3-319-54782-4 ISBN 978-3-319-54784-8 (eBook)
DOI 10.1007/978-3-319-54784-8

Library of Congress Control Number: 2017941283

Printed on acid-free paper

This Springer imprint is published by Springer Nature
The registered company is Springer International Publishing AG
The registered company address is: Gewerbestrasse 11, 6330 Cham, Switzerland

Preface

The Introduction of Geobotany in the Iberian Countries

A book about the vegetation of the Iberian Peninsula is of great interest and importance to all of Europe, due to the importance of this territory in terms of its plants and habitat diversity. When the opportunity of writing such a book presented itself, instead of writing it alone, using the abundant bibliographic sources at hand and my own experience, I decided to make it a collaborative project in which a large number of skilled and experienced colleagues from Spain and Portugal, representing the majority of the vegetation scientists currently active in these countries, could participate. This book tries to summarise the knowledge and experience that a complete generation of Spanish and Portuguese geobotanists have accumulated during their lives in their research done during the last quarter of the twentieth century and the dawn of the twenty-first. The book is divided into three main parts: the first deals with general issues which influence vegetation distribution, such as the relief and the geology of the area, as well as its climate. This part includes one chapter about the Iberian flora and another on the biogeographical division of the Iberian Peninsula. In another chapter, the conceptual framework, supported by the general theory of vegetation dynamics and by the theory of dynamic-catenal phytosociology which has been developed under the basic concept of potential natural vegetation (i.e. vegetation series, geoseries and geopermaseries), is commented upon in order to better interpret the landscape from an ecological and dynamic point of view. The second part of the book consists of the systematic description of the vegetation of Iberia and the Balearic Islands. For that the territory has been divided into 14 regions following geographic-biogeographic criteria (Fig. 1). Each of these regions is described by authors having extensive experience in the area; the phytosociological system (or Braun-Blanquet approach) is applied to give structure to the description. The third part is formed by a number of chapters dealing with specific aspects of Iberian vegetation which deserve in-depth treatment, because some of them deal with particular habitat

Fig. 1 The Iberian Peninsula and Balearic Islands divided into 14 regions described in the chapters of this book: (*1*) the lowlands and midlands of northwestern Atlantic Iberia; (*2*) the high mountain area of northwestern Spain, the Cantabrian range, the Galician-Leonese mountains and the Bierzo trench; (*3*) the Pyrenees; (*4*) Trás-os-Montes and Beira Alta; (*5*) the Duero Basin; (*6*) the Iberian ranges and highlands; (*7*) the Ebro Basin; (*8*) the Sistema Central (Central Range); (*9*) the Coastal Levantine area; (*10*) the Balearic Islands; (*11*) Lusitania; (*12*) La Mancha; (*13*) Bética and Southwest Andalusia; and (*14*) the arid southeast

groups (coasts, wetlands, high mountains or gypsum and dolomite vegetation) or with particular issues related to management (forests) or to biodiversity (alien flora). The final chapter about vegetation-plot databanks might be useful for completing the information provided.

In many of the chapters, and particularly in the descriptive part of the book, the phytosociological classification has been taken as the common system to formalise the information concerning plant communities. This is due to the common usage of the Braun-Blanquet approach in the Iberian countries and the fact that most of the authors have been trained in this approach. Hence, by using the same units for communities, as well as for bioclimatic and for biogeographic terms, the highest degree of coherence within the different chapters and parts of the book has been achieved. This hopefully will bestow a high consistency to the book as it will make the different chapters and parts easy to understand, making it also easily applicable to the adopted habitat typology of the EU for conservation policy. The typology and authorship adopted is that of Rivas-Martínez et al. (2011) for the Iberian Peninsula, enabling to refer authorship citation to that work and avoiding the explicit mentioning of authorship.

The information contained in this book is meant to be useful for vegetation scientists and ecologists throughout the world who want to have a source of updated and accurate information on Iberian ecosystems. Travellers and visitors of the country will find useful information for interpretation of the landscapes and the vegetation formations they find on their trips.

The Initial Steps of Geobotany in the Iberian Peninsula

Surveys on the vegetation and landscape of Iberia have been continuous throughout the twentieth century, but they were initiated by important earlier contributions, some by foreign researchers, that set the cornerstones of the subsequent development in this field in the Iberian Peninsula.

The first clear geobotanical observations done in the area date from the nineteenth century and originate from the Swiss botanist Pierre Edmond Boissier in his famous *Voyage Botanique dans le Midi de l'Espagne pendant l'année 1837* (1839–1845). He included a chapter describing the landscape (*géographie botanique*) and a famous diagram with the altitudinal belts of the Sierra Nevada and surrounding mountains (Fig. 2). Another remarkable attempt to improve the geobotanical

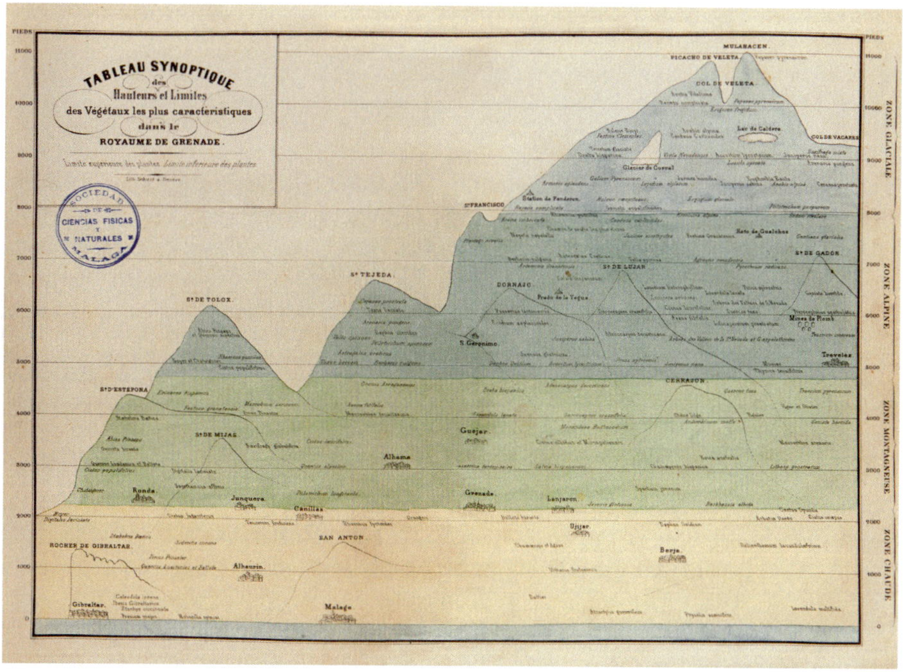

Fig. 2 Diagram by E. Boissier (1839) of the altitudinal belts of the Sierra Nevada and surrounding mountains

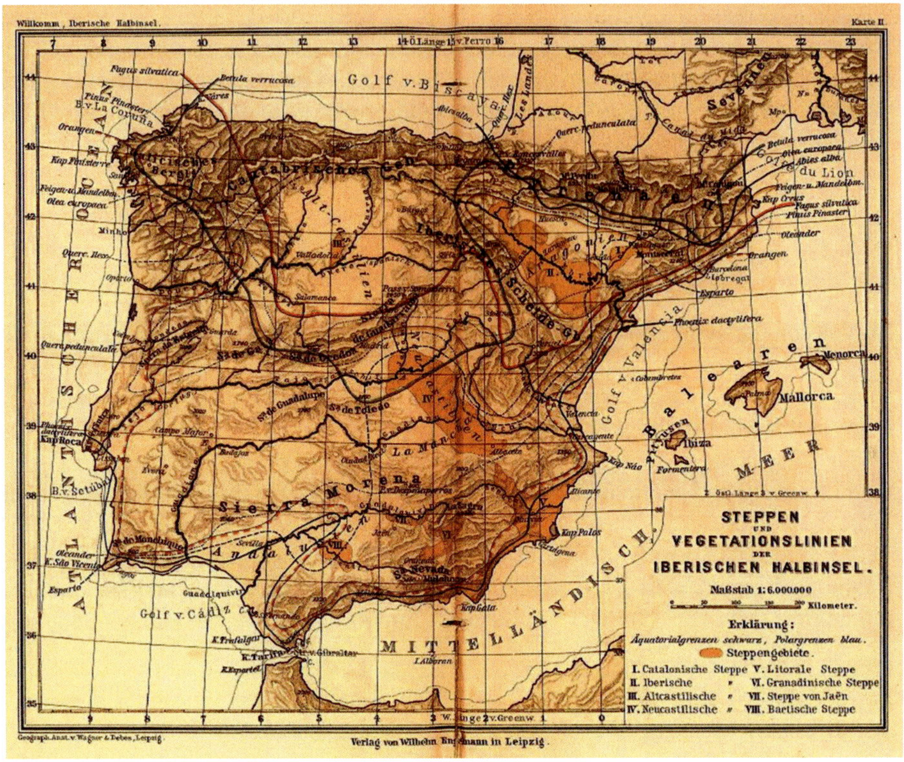

Fig. 3 Map by M. Willkomm (1852) of the steppe areas of the Iberian Peninsula

knowledge on the Andalusian mountains was done by Simón de Rojas Clemente (1864) describing the species composition of lichens in the vegetation belts of the Baetic mountains. These initial contributions were focused in the Sierra Nevada area of southern Spain, which is a very attractive territory for botanists due to its singularity and high concentration of endemic taxa, but these studies had limited influence on the development of geobotany in Spain.

A much more influential author in the subsequent development of the basic geobotanical conceptual framework for Iberia was the German botanist Mauritius Willkomm, who wrote two works on this topic, one (1852) about the supposed Iberian steppes (Fig. 3) and another with a more general treatment of the peninsular vegetation, which was published 1 year after his death (1896). In both works, an extensive description of the vegetation of the Iberian Peninsula is provided, and they can be considered the starting point of the development of geobotany at the scale of the entire peninsula. For a considerable period of time, Willkomm's description and ideas were entirely accepted by the local scientists and his influence endured for a long time after these publications. Nonetheless, in the early twentieth century, a remarkable author appeared: Emilio Huguet del Villar, a geographer and naturalist who was deeply influenced by Clements, was critical of the Willkommian

Fig. 4 The SIGMA excursion in Catalonia, Easter 1934. Among other participants, there are J. Susplugas, R. Tüxen, W. Koch, M. Klika, J. Cuatrecasas, R. Molinier, J. Braun-Blanquet, P. Font Quer and W. Rothmaler

tradition and denied the existence of true steppes in Iberia, an idea that had rooted deeply in the thinking of many naturalists in Spain as well as in the Germanic tradition until recently (Jäger 1971). Huguet worked extensively in edaphology and wrote a noteworthy textbook entitled *Geobotánica* (1929).

A turning point in the Iberian geobotanical history was the excursion of the SIGMA (Station Internationale de Géobotanique Méditerranéenne-Alpine) led by Braun-Blanquet at Easter of 1934 through Catalonia (Díaz González 2004). It was organised by a prominent Catalan botanist, Pius Font Quer, who had earlier contacted Braun-Blanquet and was interested in the new discipline of phytosociology. The excursion was attended by many Spanish and European colleagues and was the first important demonstration in the field of the methods and procedures of this school in the Iberian Peninsula (Fig. 4). As a result of those contacts, a Spanish student, J González-Albo, was sent to the SIGMA in Montpellier to be trained by Braun-Blanquet and soon started his research in the area of Madrid. Unfortunately, he was prevented from completing his research as a consequence of the Spanish Civil War, but still wrote two meritorious works (1934, 1940).

After the war, contacts were re-established in two ways. One was by means of Braun-Blanquet himself, who had been working intensively in the eastern Pyrenees and published an extraordinary monograph about its vegetation in 1948. Shortly after that, Font Quer recruited a young botanist in Barcelona, Oriol de Bolòs, who was committed to be trained by him by making a survey of a substantial area. The selected territory was the Ebro Valley and several field campaigns took place in the early 1950s (Fig. 5). The monograph appeared in 1958, and it is one of the most important contributions ever done on the Iberian vegetation, being still constantly

Fig. 5 Seeing the Ebro steppe: J. Braun-Blanquet, Prof. P. Font Quer, Mrs. Braun-Blanquet and P. Montserrat

consulted and cited. This is the starting point of the Catalan school of vegetation science which has been working mainly in the eastern Iberian Peninsula and the Balearic Islands and currently is mostly located in Barcelona.

Another important event that happened in the 1950s was the contact established at the International Botanical Congress in Stockholm by Salvador Rivas Goday, a professor of the University of Madrid, who attended that meeting and attracted the attention of Braun-Blanquet and Tüxen. As a result of this, the 10th IPE (Internationale Pflanzengeographische Exkursion) excursion in the summer of 1953 across a large part of Spain (Fig. 6) was organised, with the participation of several prominent scientists from the Germanic area, particularly Tüxen, Oberdorfer, Lüdi, Gams, Kubiena, etc., and local organisers Rivas Goday, Bolós and Fernández Galiano. The results of this excursion were published in two volumes at the Institute Stiftung Rübel in Zürich in 1956 and 1958, and they are, as the aforesaid monographs by Braun-Blanquet, milestones in the subsequent development of research on vegetation. These presented some of the first synthetic summaries for the Iberian vegetation with a vegetation map by Rivas Goday (1956) (Fig. 7) and the large monograph by Tüxen and Oberdorfer (1958) establishing the basic units and patterns of the temperate Iberian vegetation.

In Portugal the beginnings were also led by Braun-Blanquet, who made contact with a skilled and enthusiastic engineer, António Rodrigo Pinto da Silva, who, together with Arnaldo Rozeira, established a working group which explored the entire country in three long excursions in the 1950s (Fig. 8). The results were published in four papers over several years (Braun-Blanquet et al. 1952, 1956, 1965 and 1972). This dedication of Braun-Blanquet to Iberia was completed with a later monograph on the Basque Country (1967) and reveals his commitment and devotion to the introduction of his method in the Iberian countries.

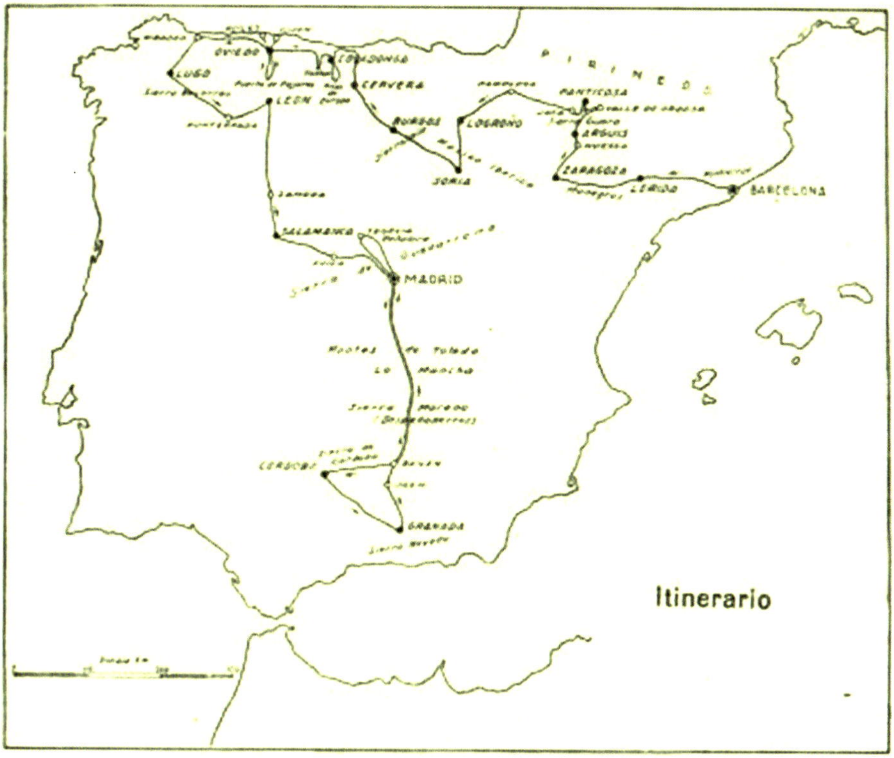

Fig. 6 Map of the itinerary of the IPE excursion through Spain in 1953 (This figure is largely commented in the text and is from the report: Rivas Goday 1956)

Undoubtedly, such efforts bore fruit. From the original centres of Barcelona, Madrid and Lisbon, the exploration of Iberia has grown intensively from the 1960s until the beginning of this century, resulting in a huge amount of published data (over 145,000 relevés in SIVIM, Font et al. 2009) in a countless number of publications. Two persons have been particularly relevant in this development, as they have guided most of the scientists who have done the fieldwork and analyses: Salvador Rivas-Martínez and Oriol de Bolòs. Rivas-Martínez was intensively trained in learning the Iberian flora by his father during his early childhood and youth and later, when he was a student, spent up to three summer stays with Tüxen in Stolzenau (Zentralstelle für Vegetationskartierung), but without having also direct contact with Braun-Blanquet (Loidi 1996). For that reason Rivas-Martínez was intensely influenced by Tüxen, while Bolòs was purely a Braun-Blanquet pupil. We safely can say that the current state of knowledge of the Iberian vegetation is due to their influence and constant supervision. In the following generation, the development was explosive, initially in Spain but somewhat later also in Portugal. Led by these authorities, a group of people, such as Manuel Costa, Jesús Izco,

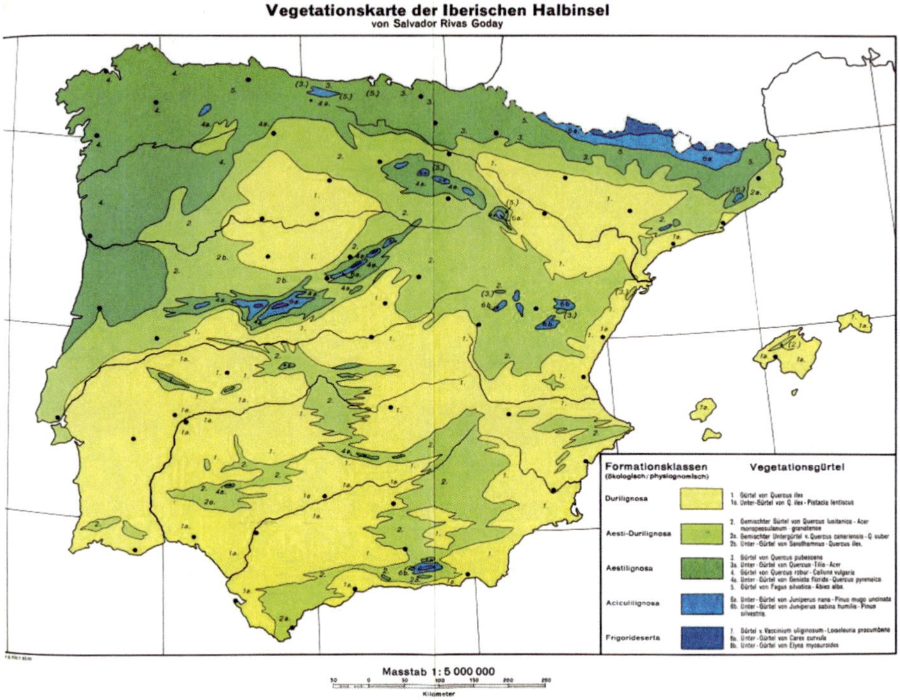

Fig. 7 Rivas Goday's map of the Iberian vegetation of 1956 (This figure is also commented in the text and is from Rivas Goday 1956)

Photo 23. De droite à gauche: Mme et M. A. R. PINTO DA SILVA, Prof. A. ROZEIRA, BR.-BL., chauffeur, assistant M. FONTES, Mme BR.-BL., J. MALATO-BELIZ dans la pineraie près de Tomar.

Fig. 8 Braun-Blanquet in Portugal. In a pine woodland near Tomar, from *right to left*: M. A. R. Pinto da Silva, Prof. A. Rozeira, J. Braun-Blanquet, M. Fontes, Mrs. Braun-Blanquet and J. Malato Beliz, among other persons

Miguel Ladero and Wolfredo Wildpret, started to work intensely in phytosociology under the leadership of the Rivas family in Madrid, while Josep Vigo was trained in Barcelona. In Portugal, the flame was temporarily put out as the original masters got many duties in the forest and agronomic service of their country, but a new and enthusiastic researcher, Mario Fernandes Lousã, restarted the Lusitanian tradition and developed the current flourishing Portuguese vegetation scientist group.

The development during the last decades has been documented sufficiently by other authors (Rivas-Martínez 1996) and it is unnecessary to repeat it in detail. In any case, it can be said that after the intensive, descriptive activity of the period between the 1970s and the 2010s, many researchers have tended to diverge into different fields in which their expertise has been advantageously used, particularly in the field of conservation biology, with the inventory and management of terrestrial habitats, endangered populations and species as well as community ecology.

Bilbao, Spain Javier Loidi

References

Boissier PE (1839–1845) *Voyage Botanique dans le Midi de l'Espagne pendant l'année 1837*. Paris

Braun-Blanquet J (1948) *La végétation alpine des Pyrénées Orientales*. Monografías de la Estación de Estudios Pirenaicos y del Instituto Español de Edafología, Ecología y Fisiología Vegetal, 9 (Bot. 1). (Commun. Stat. Int. Géobot. Médit. Alpine 98). Barcelona, p 306

Braun-Blanquet J (1967) Vegetationsskizzen aus dem Baskenland mit Ausblicken auf das weitere Ibero-Atlantikum. II Teil. Vegetatio 14(1–4):1–126

Braun-Blanquet J, de Bolòs O (1958) Les groupements végétaux du bassin moyen de l'Ebre et leur dynamisme. Anales Estación Experimental Aula Dei 5(1–4):1–266

Braun-Blanquet J, Pinto da Silva AR, Rozeira A, Fontes F (1952) Résultats de deux excursions Géobotaniques à travers le Portugal septentrional et moyen. I. Une incursion dans la Serra da Estrela. Agron Lusit 4(4):303–323

Braun-Blanquet J, Pinto da Silva AR, Rozeira A (1956) Résultats de deux excursions Géobotaniques à travers le Portugal septentrional et Moyen. II. Chênaies à feuilles caduques (Quercion occidentale) et Chênaies à feuilles persistantes (Quercion fagineae) au Portugal. Agron Lusit 18(3):167–234

Braun-Blanquet J, Pinto da Silva AR, Rozeira A (1965) Résultats de trois excursions Géobotaniques à travers le Portugal septentrional et moyen, III. Landes à Cistes et Ericácees (Cisto-Lavanduletea et Calluno-Ulicetea). Agron Lusit 23(4):229–213

Braun-Blanquet J, Braun-Blanquet G, Rozeira A, Pinto da Silva AR (1972) Résultats de trois excursions Géobotaniques à travers le Portugal septentrional et moyen, IV. Equisse sur la végétation dunale. Agron Lusit 33(1–4):217–234

de Rojas Clemente S (1864) Tentativa sobre la liquenología geográfica de Andalucía. Revista Progr Ci Exact 14(1):39–58. Madrid

Díaz González TE (2004) Pasado, presente y futuro de la Fitosociología española. Lazaroa 25:3–13

Font X, Rodríguez-Rojo MP, Acedo C, Biurrun I, Fernández-González F, Lence C, Loidi J, Ninot JM (2009) SIVIM: an on-line database of Iberian Macaronesian vegetation. Waldökologie, Landschaftsforschung und Naturschutz 9:15–22

González-Albo J (1934) Introducción al estudio de la sociología vegetal por el método de Braun-Blanquet. Reseñas Científicas de la Soc Esp Hist Nat 9:81–100

González-Albo J (1940) Datos sobre la flora y fitosociología de la provincia de Madrid. Bol Real Soc Esp Hist Nat 38:9–18

Huguet del Villar E (1929) *Geobotánica*. Ed. Labor. Barcelona. p 339

Jäger E J (1971) Die Pflanzengeographische Stellung der "Steppen" der Iberischen Halbinsel. Flora 160:217–256

Loidi J (1996) *Laudatio*: la obra científica del Profesor Salvador Rivas-Martínez. In: Loidi J (ed) Avances en Fitosociología 143–148. Ed. UPV/EHU

Rivas Goday S (1956) Übersicht über die Vegetationsgürtel der Iberischen Halbinsel: Kennzeichnende Arten und Gesellschaften. Die Pflanzenwelt Spaniens I. Veröff Ber Geobot Inst ETH Stiftung Rübel 31:32–69

Rivas-Martínez S (1996) La fitosociología en España. In: Loidi J (ed) Avances en Fitosociología 149–174. Ed. UPV/EHU

Rivas-Martínez S et al (2011) Mapa de series, geoseries y geopermaseries de vegetación de España (Memoria del Mapa de vegetación de España) Parte II. Itinera Geobotanica 18:5–800

Tüxen R, Oberdorfer E (1958) Die Pflanzenwelt Spaniens II. Eurosiberische Phanerogamen-Gesellschaften Spaniens. Veröff Ber Geobot Inst ETH Stiftung Rübel 32:1–328

Willkomm M (1852) Die Strand- und Steppengebiete der Iberischen Halbinsel und deren Vegetation. Friedrich Fleischer Verlag, p 275

Willkomm M (1896) Grundzüge der Pflanzenverbreitung auf der Iberischen Halbinsel. In: Engler A, Drude O (eds) *Die Vegetation der Erde*. Wilhelm Engelmann Verlag. Leipzig, p 395

Contents

Contributors

Carlos Aedo Real Jardín Botánico-CSIC, Madrid, Spain

Carlos Aguiar CIMO-Centro de Investigação de Montanha, Escola Superior Agrária de Bragança, Bragança, Portugal

Juan Luis Aguirre Department of Life Sciences (Plant Biology), University of Alcalá, Madrid, Spain

Paulo Alves Universidade de Porto, Porto, Portugal

Javier Amigo Universidade de Santiago de Compostela, Santiago, Spain

Alejandro Aparicio Department of Life Sciences (Plant Biology), University of Alcalá, Madrid, Spain

Antoni Buira Real Jardín Botánico-CSIC, Madrid, Spain

Paloma Cantó Department of Plant Biology II, Faculty of Pharmacy, Complutense University of Madrid, Madrid, Spain

Empar Carrillo Institute for Research on Biodiversity (IRBio) & Department of Evolutionary Biology, Ecology and Environmental Sciences, University of Barcelona, Barcelona, Spain

José Carlos Costa Botanical Center, Technical University of Lisbon, Lisbon, Portugal

Manuel Costa ICBiBE-Botanical Garden, Universitat de València, València, Spain

Vicenta de la Fuente Department of Biology, Faculty of Sciences, Universidad Autónoma de Madrid, Madrid, Spain

Sara Del Río Department of Biodiversity and Environmental Management (Area of Botany), Mountain Livestock Farming Institute (Join Center CSIC-ULE), Faculty of Biological and Environmental Sciences, University of León, León, Spain

Tomás Emilio Díaz González Department of Organisms and Systems, Biologia Faculty, University of Oviedo, Oviedo, Spain

Marta Fernández-Albert Real Jardín Botánico-CSIC, Madrid, Spain

Albert Ferré Institute for Research on Biodiversity (IRBio) & Department of Evolutionary Biology, Ecology and Environmental Sciences, University of Barcelona, Barcelona, Spain

Rosario G. Gavilán Department of Plant Biology II, Faculty of Pharmacy, Universidad Complutense, Madrid, Spain

Luis Herrero Department of Biodiversity and Environmental Management (Area of Botany), Mountain Livestock Farming Institute (Join Center CSIC-ULE), Faculty of Biological and Environmental Sciences, University of León, León, Spain

Miguel Ladero Department of Botany, Faculty of Pharmacy, University of Salamanca (Spain), Salamanca, Spain

Javier Loidi Department of Plant Biology and Ecology, University of the Basque Country (UPV/EHU), Bilbao, Spain

Leopoldo Medina Real Jardín Botánico-CSIC, Madrid, Spain

Joaquín Molero Department of Botany, Faculty of Pharmacy, Campus de la Cartuja, University of Granada, Granada, Spain

Josep M. Ninot Institute for Research on Biodiversity (IRBio) & Department of Evolutionary Biology, Ecology and Environmental Sciences, University of Barcelona, Barcelona, Spain

Manuel Peinado Department of Life Sciences (Plant Biology), University of Alcalá, Madrid, Spain

Ángel Penas Department of Biodiversity and Environmental Management (Area of Botany), Mountain Livestock Farming Institute (Join Center CSIC-ULE), Faculty of Biological and Environmental Sciences, University of León, León, Spain

João Jose Pradinho Honrado Universidade de Porto, Porto, Portugal

Salvador Rivas-Martínez Phytosociological Research Center (CIF), Madrid, Spain

Salvador Rivas-Sáenz Phytosociological Research Center (CIF), Madrid, Spain

Manuel Antonio Rodríguez-Guitián Universidade de Santiago de Compostela, Santiago, Spain

Daniel Sánchez-Mata Department of Plant Biology II, Faculty of Pharmacy, Universidad Complutense, Ciudad Universitaria, Madrid, Spain

Pilar Soriano ICBiBE-Botanical Garden, Universitat de València, València, Spain

Carlos Vila-Viçosa CIBIO – Centro de Investigação em Biodiversidade e Recursos Genéticos/InBIO Laboratório Associado, Universidade do Porto, Vairão, Portugal

Part I
General Conditions

Chapter 1
Introduction to the Iberian Peninsula, General Features: Geography, Geology, Name, Brief History, Land Use and Conservation

Javier Loidi

Abstract A brief description of the physical structure of the Iberian Peninsula is given: its position, size and main structural and lithologic entities. Also some general explanations about the main relief units ordered around the central core of the Peninsula, or *Meseta*, are offered. The etymology of the names of Iberia and Hispania is commented upon and a concise report of the human history in Antiquity is given. Traditional land-use by humans is considered important to correctly interpret the current landscape in adaptation to the conditions of the different parts of the territory. A special comment about irrigation and exploitation of the freshwater resources is made as regards its importance in the Mediterranean part of the Peninsula. Finally, some issues influencing conservation of terrestrial ecosystems are treated: the impact of the protected areas resulting from the policy of the administrations in the last decades, the landscape changes as a result of the rural abandonment (the "ecology of abandonment"), the urban development with the entailed artificialization of the land and the impact of the modern technicized forestry.

1.1 Physical Structure

1.1.1 Position and Size

The Iberian Peninsula or Iberia (Fig. 1.1) has a total extent of 583,832 km^2 encompassing continental Spain and Portugal, the Principality of Andorra, the British colony of Gibraltar and some areas of France in the Pyrenees. To this area, we add the 4992 km^2 of the Balearic Islands, which are related to the Iberian Peninsula as they form part of the same continental platform, entirely belonging to one of its structural units, and have been several times connected to the Peninsula

J. Loidi (✉)
Department of Plant Biology and Ecology, University of the Basque Country (UPV/EHU), Bilbao, Spain
e-mail: javier.loidi@ehu.eus

J. Loidi (ed.), *The Vegetation of the Iberian Peninsula*, Plant and Vegetation 12,
DOI 10.1007/978-3-319-54784-8_1

Fig. 1.1 The Iberian Peninsula with its main structural units

during the ice ages of the Pleistocene. Iberia is a prominent large sized peninsula in the southwestern end of Europe, reaching the westernmost (9° 29′ W in Cape Rocha, Portugal) and southernmost (36° 00′ N in Tarifa) points of Europe, as only small areas in western Ireland and southern Kriti are beyond these coordinates in the European context. This peninsula has a massive pentagonal shape with large areas of medium to high elevation in its centre, viz. the huge central plateau or *Meseta*; the average elevation of Iberia is ca. 600 m asl. The straight line distance between peripheral opposite points of the peninsula oscillates between the extreme values of 1170 and 860 km. This bulky and prominent appendix of the European continent is welded to it by means of a wide isthmus of 435 km length, along the Pyrenees, one of the most prominent mountain ranges of the region. This type of connection, unifying but also separating, reveals the intense relationships of Iberia with Europe but also its independence and isolation from the rest of the continent, which is clearly translated into its history and culture, economy and biological content. It is an almost-island located between Europe and Africa and between the Atlantic and the Mediterranean, and its idiosyncrasy is determined by this double dualism of its geographical surroundings.

1.1.2 Geological Outline of the Iberian Peninsula

Iberia is a highly diverse territory from the geological point of view, in terms of rock types and relief models, with materials of a long array of ages. Essentially, Iberia was constructed around a primary old core to which successive portions were added in the course of time. This evolution can be divided into several periods, some of them corresponding to the build up of reliefs due to tectonic efforts (orogenesis), and others to the destruction of such reliefs due to erosion (erosive and sedimentary periods). This geological history is basically conditioned by its position between the African and the Eurasian plates in the middle of the western Tethys Sea and there are two momentous events: The Hercynian Orogeny, at the end of the Palaeozoic, and the Alpine Orogeny during the Tertiary. As a consequence, we can divide the Peninsula into two main sectors: One belonging to the old or Hercynian cycle, the old core of Iberia, and a second one related with the modern Alpine and post alpine stages. According to this, in the scheme by Terán and Solé Sabarís (1978), three main units have been recognized for the Iberian Peninsula.

1.1.2.1 The Hercynian Basement or Hesperian Shield

The Hercynian orogenic period took place at the end of the Paleozoic as a result of the collision between Laurasia and Gondwana that created the large supercontinent of Pangea. With this folding large amounts of the deposited materials in the seas which covered the area of the Peninsula so far were raised and emerged, forming the basis of the primitive Iberia. This primitive basement has been traditionally called the Hesperian Shield. These materials are the oldest of Iberia (pre-Cambrian and Paleozoic) and, depending on the conditions in which they were formed, are plutonic rocks, such as granites, formed after solidification of previously melted rocks, together with a wide variation of metamorphic rocks, including gneiss, quartzite, slate and schist. Most of these materials are siliceous and they constitute the old structural basement of Iberia, around which the other structural elements were later added. This Hesperian Shield underlies the basement of the central areas of Iberia, partially buried below more recent materials, but also emerging extensively in the western half of the Peninsula and in the core of several of the main mountain ranges, such as the Pyrenees, the Sierra de Demanda, the Central Range and the Sierra Nevada. These old siliceous materials, folded and wrapped during the Hercynian orogeny into mountain systems, eroded during the Mesozoic and were transformed into peneplains. The erosion material was deposited in the surrounding seas, such as the Thetys Sea, and mixed with biogenic lime. This process constituted the rocks which built the subsequent set of structures resulting from the Alpine Orogeny. This orogeny resulted in the current existence of an area of high elevation, or plateau, in the central section of the Iberian Peninsula, called the *Meseta*. It constitutes the basic structural element of modern Iberia, around which the rest of the structures are attached.

1.1.2.2 The Alpine Relief

After the dismantling by erosion of the Hercynian relief during the long Secondary era and its deposition into the sedimentary basins surrounding the Hesperian Shield, different phases of the Alpine Orogeny affected Iberia in the Tertiary, being responsible of the uplift of the main mountain reliefs that are currently found. This orogeny was generated by the African plate approaching the Eurasian plate and lifting most of the Iberian ranges. After the compressive phases, a generalized distension happened in the last part of the Tertiary and some tectonic fosses opened. They, as well as the peripheral basins such as those of the Ebro and Guadalquivir, were filled by the materials eroded from the high new reliefs formed in the Alpine orogeny. Among the mountains lifted in this orogeny there are three main groups: those resulting from the deformations of the previously existing Hesperian Shield, the so-called Mesetan Massifs; those formed in the borders of the Meseta; and those newly arisen from the bottom of the surrounding sedimentary basins, the true Alpine Massifs. The latter are basically formed from sedimentary materials, mostly limestone, sandstone or marl, such as the Pyrenees, the Iberian Ranges, the Catalonian Coastal Ranges and the Baetic Ranges. These Alpine massifs often show an emerging old Hercynian core too. The Mesetan Massifs resulted from a renewing of ancient Hercynian faults and the uplift of parts of the primitive Hesperian Shield, as e.g., the Central Range and the Montes de Toledo. They are formed mostly from siliceous rocks, such as granite, slate, gneiss, quartzite and schist, although there are areas with sandstone and limestone. Along the borders of the Meseta there are other groups of massifs, such as the Cantabrian Range, the Sierra Morena and the Iberian System.

1.1.2.3 The Great Depressions

Embedded between these ranges, there are a number of wide depressions which are filled by Tertiary materials and drained by the main Iberian rivers. There are four depressions or main valleys: Ebro, Duero, Tagus-Guadiana and Guadalquivir, and they can be divided into two categories: those which lie over the Meseta (upper Duero and Tagus-Guadiana) and those which are enclosed at the periphery of the Meseta (Ebro, Guadalquivir and the lower Douro and Tagus-Sado). Another division can be made based on its recent geologic history: depressions which were disconnected from the open sea during the Tertiary and were filled with sediments of the surrounding land areas, with abundant evaporitic materials (Ebro, upper Duero and Tagus-Guadiana), or depressions which never lost their connection with the open sea (Guadalquivir, lower Tagus-Sado). Each of these big valleys is covered by recent deposits (marl, sand) which are favourable for agriculture. For that reason those areas comprise a major part of the agricultural wealth of the Peninsula as well as of its human population. Huge irrigation infrastructures have been developed in many of these territories since several centuries, transforming the agrarian economy of many districts.

1.1.3 *Relief Units of the Iberian Peninsula*

In this part some more details are given about the main structural units in which Iberia can be divided (Loidi 1999). They are ordered according to their relation to the Meseta or Hesperian Shield, the basement of the Iberian Peninsula (Figs. 1.1 and 1.2).

1.1.3.1 Units Within the Meseta

The Basins Atop the Meseta

This large high-elevation area (211.000 km^2) occupies the centre of the Iberian Peninsula and is divided by the Central Range into two halves called submesetas, the northern submeseta or Duero Basin, and the southern submeseta or upper

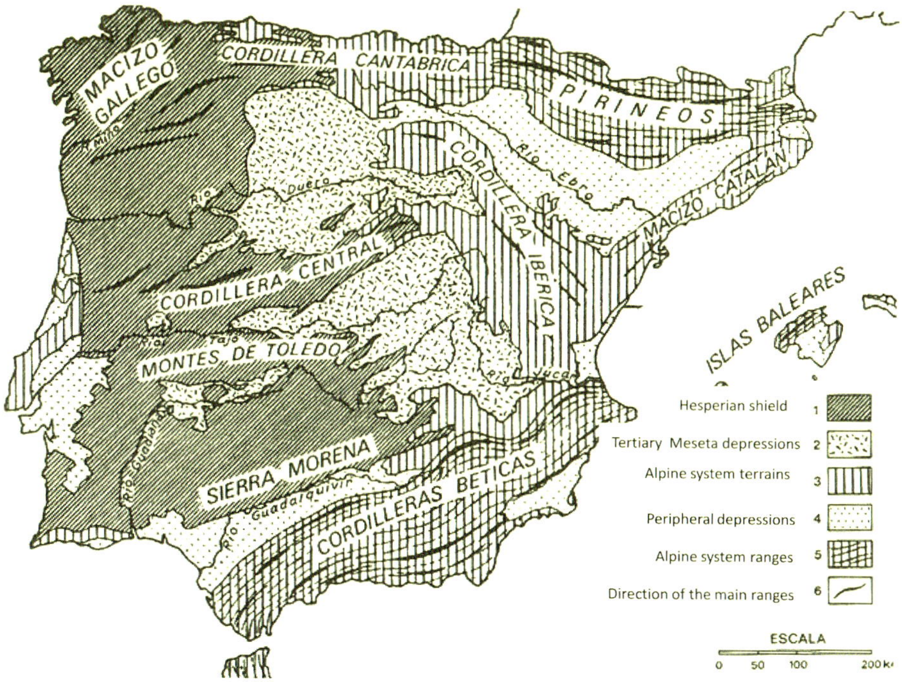

Fig. 1.2 Simplified geological structure of the Iberian Peninsula after Terán and Solé Sabarís (1978)

Tagus-Guadiana basin: La Mancha. Both basins fringe at their western border on the old siliceous terrains of the Hesperian Shield in the peneplains of Extremadura and Zamora, and these depressions are filled with sediments of Tertiary age, deposited in inland seas and subsequently not folded because they are of more recent age than the Alpine ontogeny. These materials often constitute evaporitic rocks, often gypsaceous, together with marl and limestone. As the Meseta is slightly tilted to the west, the drainage is basically towards the Atlantic, and is done by the big rivers flowing westwards, such as the Duero, Tagus and Guadiana; only the Júcar flows eastwards towards the Mediterranean Sea, draining the southeastern part of La Mancha.

The Upper Duero Basin

This is a perfectly tray-shaped basin concealed by the Mountains of León and the Cantabrian Range in the northwest and north, the northern Iberian System in the east, and the Central Range in the south. Its western limit is formed by the escarpments of the Meseta, with the lower Duero in its last stretch towards the Atlantic Ocean crossing northern Portugal (Douro). It covers about 55,000 km^2 occupying most of the northern Submeseta and an elevation ranging 700–800 m asl.

The Upper Tagus-Guadiana Basin: La Mancha

This basin extends over the southern Submeseta at an elevation between 600 and 700 m Its limits are defined by the Central range in the north, the Iberian System in the east and the Sierra Morena in the south. Its western limits are somewhat undefined as this basin grades into the Montes de Toledo and the Extremadurean peneplain. In the Guadiana headwaters the terrain is so flat that the streams possess little energy and speed and a large part of the water sinks into a huge aquifer system which, when it emerges at some points, forms lakes (ojos) or swamp areas (tablas).

The Central Range

This is the huge dividing range which crosses the Meseta in a southwest-northeast direction for 440 km, dividing it into two halves. It ranges from the Serra da Estrêla at its western end to the Sierra de Ayllón at the opposite end, encompassing a number of sierras such as Gata, Peña de Francia, Candelario, Gredos and Guadarrama. Many of these reach more than 2000 m, with the highest peak, the Pico Almanzor (Gredos) reaching 2592 m. This range is formed by blocks of the old Hesperian Massifs lifted by the Alpine Orogeny and separated by valleys corresponding to tectonic fosses. The bedrock is siliceous and old and includes granite, quartzite and slate. Glaciers have left their footprints in the main massifs of the Central Range.

The Montes de Toledo

These are a group of mountains of modest elevation (Corocho de Rocigalgo 1440 m; Sierra de Guadalupe, Las Villuercas 1603 m) dividing the southern submeseta into two watersheds, that of the Tagus and that of the Guadiana rivers. They have a similar origin as the Central Range and share many of its lithological and geomorphological characteristics, occupying an important area in western La Mancha and eastern Extremadura.

1.1.3.2 Units Bordering the Meseta

The Lower Douro Basin

This area has a sizeable extent in northern Portugal comprising the regions of Douro Litoral, Alto Douro and Tras os Montes. The lower stretch of the Duero river (Douro) crosses northern Portugal carving, together with its tributaries, deep and narrow valleys in the siliceous rocks of the Meseta basement.

Mountains of Galicia and León

In the northwestern part of the Iberian Peninsula, including Galicia and the neighbouring parts of León and Zamora, the Hesperian Shield folds and fragments due to the forces of the Alpine Orogeny and results in a series of reliefs of uneven elevation and morphology. Rocks are basically siliceous: quartzite, gneiss, slate, with outcrops of granite. The Sil depression (El Bierzo and Valdeorras) profoundly divides this unit, leaving its highest massifs on its southern side, with important mountains above 2000 m, such as the Teleno (2188 m). Glacial footprints are visible in the summital areas of the higher mountains.

The Cantabrian Range

The northen side of the Meseta is bordered by a high range of mountains of over 220 km long, parallel to the Atlantic coastal shoreline. This area is called Cantabria, a name that covers both the range and to the narrow fringe of low land between the mountain chain and the coast. The Cantabrian Range contains several massifs with numerous peaks over 2000 m, with its highest summit in Torre Cerredo (2648 m) in the Picos de Europa. The northern slope is very steep, the distance between the high summits (1800–2600 m) and the sea being only 40–60 km. This leaves the Cantabrian Fringe as a narrow stretch of land separating the Cantabrian Range and the sea (Gulf of Biscay) and being crossed by a high number of fast-running rivers coming from the high elevations towards the sea and excavating a steep relief with deep valleys. Rock materials are Paleozoic, as all the Hesperian ones, the siliceous ones (slate, quartzite, sandstone and conglomerates) being more abundant, but in the eastern sectors, as in the Picos de Europa or Peña Ubiña, Carboniferous limestone is also important (Caliza de Montaña). These calcareous massifs are often submitted to karstification and the water has excavated narrow gorges, such as that of the Cares. The southern slope is less steep because the northern Meseta plain lies at 800 m. Glacial activity has been important during ice ages in the high-altitude areas.

The Iberian System

This system limits the eastern side of the Meseta, spanning about 400 km from the Sierra de la Demanda, in its northwestern end, to the Maestrazgo reliefs, close to the Mediterranean shore in the Valencia region, where it contacts with the Catalan Coastal Range and the Baetic Ranges. It is a vast complex of mountain chains and high plateaus (páramos), separated by several internal depressions, which form an uneven system of elevated lands separating the Meseta from the Ebro Basin. Summits often reach over 2000 m, attaining their maximal height in the northern sector (Moncayo, 2315 m) and being more modest in the south (Javalambre, 2020 m). Regarding the origin and age of its rock types, the Iberian System can be considered as a mixed cordillera. It is built of a combination of emerged materials, some belonging to the old core (Hesperian Shield), as is the case for the Sierra de Demanda or Moncayo, with Pre-Cambrian and Paleozoic siliceous materials (quartzite, siliceous conglomerates or slate), and others belonging to the sedimentary cover (or burden), mostly of the Triassic, Jurassic and Cretaceous periods, such as limestone, abundantly represented in most of the mountains and páramos. Some tectonic depressions intermingle with the high ranges and were filled with eroded soft materials such as marls, and form important corridors as that of Calatayud-Teruel.

Sierra Morena

The Meseta ends at its southern border in a vast dissected area called Sierra Morena, which really is an aggregate of low elevation ranges occupying a space of almost 500 km long, from the Serra de Monchique (Algarve, Portugal) to the Sierra de Relumbrar at the eastern end, and a maximal width of 120 km along the north-south transect across western Andalusia and southern Extremadura. It is the southern border of the Hesperian Shield which was faulted during the Alpine Orogeny and that resulted in a sort of doorstep, clearly distinguishable, towards the Guadalquivir Depression. Rocks are mostly siliceous and of old age: quartzite, slate and sometimes limestone, alternating with granitic intrusions. Elevations are modest (highest elevation in the Sierra Madrona, Bañuela summit 1323 m) and the landscape is dominated by hills of rounded summits in which husbandry has been traditionally the most important land use, hosting a great part of the wooded pastures (or dehesa) landscapes existing today.

1.1.3.3 Relief Units Outside the Meseta

Basque-Cantabrian Mountains

They form a system of west-east oriented reliefs connecting the Cantabrian Range and the Pyrenees, parallel to the shores of the Gulf of Biscay. They are formed by three parallel mountain alignments of elevations of less than 2000 m, the northern one being the highest, with peaks such as Castro Valnera 1718, Aizkorri 1549 and Ortzainzurieta 1567 m. They occupy a wide fringe of land in the headwaters of the Ebro river and in those mountain ranges, mostly limestone or sandstone intermingle with marly valleys. Materials are overwhelmingly Cretaceous and were lifted during the Alpine Orogeny. The southernmost alignment confronts the Ebro Basin and the northernmost one marks the border of the Cantabrian Fringe in its eastern stretch.

Pyrenees

This important range stretches from northeastern Navarre, near to the Gulf of Biscay, to northeastern Catalonia, close to the Mediterranean shore, stretching 400 km long and 150 km wide in its central sector. East-west disposed along the connecting isthmus between Iberia and the rest of Europe, it forms a formidable wall with most of its summits over 2000 but with several peaks in the central part even over 3000 m (Aneto 3404 m). At the same time, this connecting element is also an isolating one. The Pyrenees have an axial range, which is mostly formed by the old Paleozoic core, with siliceous materials such as quartzite, slate, schist, greywacke and conglomerates combined with granite and granodiorite outcrops, and the burden of younger rocks (Mesozoic and Tertiary limestone and marl) surrounding it. Structurally, the Pyrenees are divided into an array of three parallel ranges: the Axial Range (basically corresponding to the core), the Inner Range and the Outer Range. Between the latter two ranges there is the Inner Depression (Canal de Berdún, Graus, Conca de Tremp). The Axial Pyrenees show footprints of intense glacial activity and are enveloped by the Pre-Pyrenees, a wide fringe of mountains and valleys basically formed by the Inner Depressions and the Outer Ranges.

Catalan Coastal Ranges

Two parallel range systems, uplifted during the Alpine Orogeny, stretch along the Mediterranean coast of Catalonia. They span 180 km from the Empordá area in the north, where they connect with the eastern Pyrenees, to the Ports de Beseit, where they connect with the eastern Iberian System. Elevations are relatively modest (Montseny 1712 m) and materials are diverse, being either siliceous (granite, slate) or limestone depending on localities. The mountain alignments are disposed in a southwest-northeast direction and are interrupted by the river Ebro, which carved narrow canyons in its last stretch towards the Mediterranean.

The Baetic Ranges

These are one of the most important structural units of Iberia as they stretch from the Gibraltar area (Campo de Gibraltar) at the southwestern end, to the island of Minorca, encompassing the numerous southeast-northeast oriented massifs of southern Andalusia to the southern part of the Valencian region (Dianic Ranges) and the entire Balearic Islands. In the Peninsula *sensu stricto* the Baetic Ranges are some 600 km long and in the Balearic archipelago they cover another 180 km. The width of this complex is also noteworthy as it attains 150 km at its widest stretch. Its peninsular area occupies about a half of the Andalusian region, the whole of Murcia and southern Alicante. It has numerous peaks over 2000 m and the Sierra Nevada holds the highest elevations of the Baetics and the highest peak of the Iberian Peninsula (Mulhacén 3481 m). In this highest massif the southernmost glacier footprints of Iberia are found. Rock types are very diverse, and are dominated by base-rich materials such as dolomite, limestone and marl. In some areas the old core of siliceous materials emerges, mostly in the Penibetic sector, being either metamorphic or mafic: slate, schists, marble and peridotites. Between some of these massifs, some depressions became filled with sediments of soft marly materials and formed corridors important for human communications such as the Vega de Antequera, the Vega de Granada, the Hoya de Guadix and Baza.

Great Sedimentary Basins External to the Meseta

Ebro Basin This basin is concealed in the triangle formed by the Pyrenees, the Iberian System and the Catalan Coastal Ranges, thus forming a closed-to-the-sea basin. It occupies a vast area in the northeastern part of the Peninsula and sediments are lacustrine continental largely due to the circumstance that this basin became endorrheic during a long time between the Oligocene and the Pliocene when the connection with the sea was interrupted and an inland sea developed in the basin. This explains the abundance of evaporitic rocks such as gypsum-rich marls and the salinity phenomena in its depressions.

Guadalquivir Basin Between the Sierra Morena and the Baetic Ranges there is this sedimentary basin that is open to the Atlantic; it has been filled with recent marine deposits, rich in clay, eroded from the surrounding massifs. Due to their high content of expansible clay, these terrains support an intensive and highly productive cereal agriculture.

Sado-Lower Tagus Basin At the western end of the Iberian Peninsula a particular sedimentary basin, open to the Atlantic, lies between the westernmost reliefs of both the Central Range and the Sierra Morena ensemble. The northern limits are the calcareous massifs of the Serras de Montejunto and Candeeiros, the western part ends at the Serra de São Mamede, while the Serras de Ossa, Monfurado and Grandola conceal it in the south. Materials are recently deposited and the catchment area is drained by the Sado and Tagus (Tejo) rivers. It is an important economic area in Portugal, submitted to intensive agriculture and with a high human population density.

Types of Substrates

As a consequence of this basic structure, we distinguish three types of substrate-conditioned landscape types in Iberia, depending on the dominance of the types of rocks: siliceous, limestone and marly (Fig. 1.3) which strongly determine the geomorphological conditions (Terán and Solé Sabarís 1978).

– **The siliceous areas** are spread mostly in the western halve of the Peninsula, where the Hesperian Shield emerges with its old Paleozoic rocks (slate, granite, gneiss, quartzite and so on). Other siliceous areas are those of the emerging

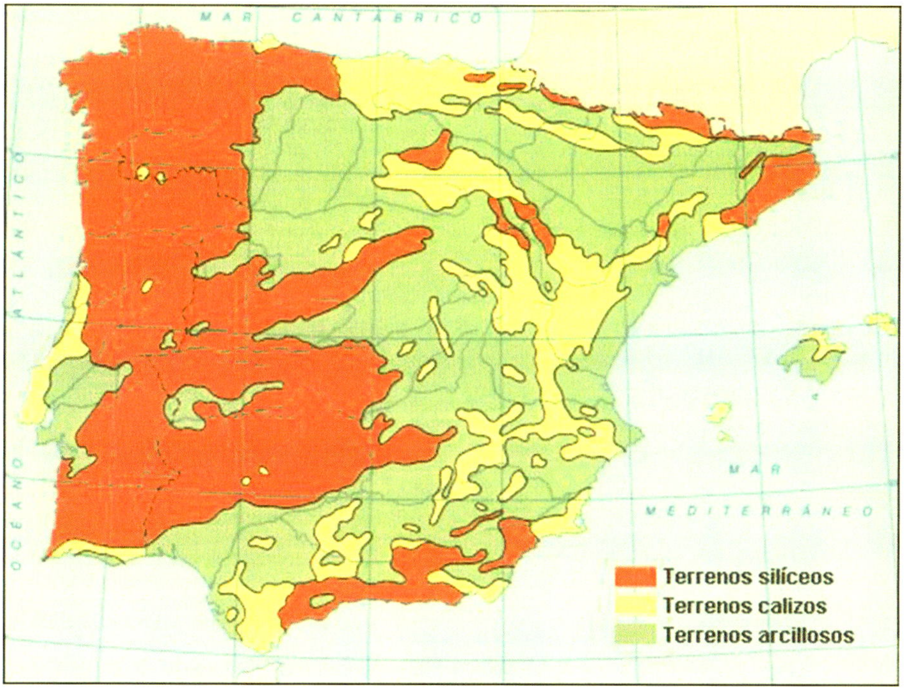

Fig. 1.3 Dominant rock types: siliceous (silíceos) granite, gneiss, slate, quartzite, schists, sandstone, etc.; limestone (calizos); limestone, dolomite; and clayey (arcillosos): marl, flysch, argillite, etc., after Terán and Solé Sabarís (1978)

Paleozoic cores of the main mountain ranges such as the Pyrenees, the Baetic Ranges (Sierra Nevada and Filabres), the Iberian System (Sierra de Demanda, Moncayo) and others. Areas of younger sedimentary rocks such as sandstone, are also included in this category, many in the Iberian System (Neila-Urbión-Cebollera, Espadá). Reliefs are smooth and undulating and only where ice age glaciers have been active it is possible to find abrupt steep areas. These areas usually are poor for agriculture and are prone to be exploited for husbandry.

- **The limestone areas** are characterized by dissected terrains due to the chemical erosion of the limestone and dolomite. They consist of Mesozoic or early Tertiary materials, common in the eastern Cantabrian Range, the Pyrenees, the Catalonian Coastal Range, the Iberian System and the Baetic Ranges. Often these areas are submitted to karstic erosion.
- **The clayey areas** greatly coincide with the great depressions: the Castilian Meseta plains of the upper Duero and Tagus-Guadiana catchment areas, the Ebro, Guadalquivir and lower Tagus-Sado, filled with younger soft materials. Strata remain horizontal as they have been formed after the last folding period of the Alpine Orogeny.

1.2 Iberia, Name and History

The name *Iberia* is of Greek origin (Ιβερια): the name the ancient Greeks used for the westernmost land they knew in the Mediterranean (a "far west" for ancient Mediterranean civilizations). It probably is related to the names used for the river that flowed there (river *Iber* or modern Ebro) and for the people who populated those Mediterranean shores and adjacent area, the Iberians. Another name applied for the same territory was *Hispania*, used by the Romans. This word probably derived from the Phoenicean *i-sch-phannim*, which means island or land of rabbits. The contacts of the Romans with the Carthaginians probably facilitated their adoption of this Punic denomination for the peninsula, and actually, this peninsula became famous among the Romans due to its abundance of olive oil and rabbits, as is evident from coins and other documents. Another theory is that *Hispania* is a corruption of an older Greek name, *Hesperia* (Χεςπερια), given to a mythic country at the western end of the world, largely unknown and attributed to have fabulous wealth. For the Greeks of the archaic period, *Hesperia* was applied first to Italy and later to Iberia while they broadened their geographical horizon. In any case, the use of the name Iberia gradually declined as it became more generally replaced by Hispania, although for a long time both terms were synonymous, as both encompassed this southern and westernmost peninsula of Europe. Since the Middle Ages, the term Hispania has been corrupted to España and has been applied only to the country which emerged from the fusion of the central and eastern medieval kingdoms: Spain. Portugal, a name derived from *Portus Cale*, the Roman name for Oporto, finally became excluded from the geographical unit of Spain from the seventeenth century onwards. The Roman province of Lusitania, one of the provinces of the *Diocesis Hispaniarum* in the lower empire, encompassed a great part of modern Portugal but also vast areas in western Spain. There is a tendency, not strongly supported by historical data, to consider Lusitania synonymous with Portugal. But the current frontiers between Spain and Portugal do not coincide either with Hispania nor with Lustania, and thus, when the entire peninsula is meant, it has been generally accepted to use the name Iberia or Iberian Peninsula. The Arabs, who dominated great parts Iberia for centuries during Middle Ages, called this country Al-Andalus (الأندلس), a term which was later applied only to the part dominated by them.

Old descriptions of the Iberian Peninsula are offered by some Greek and Roman authors, such as Herodotus, Polybius, Strabo and Appian of Alexandria. Their texts focus in the Roman history but also indicate some features of this territory. This concerned its size, which was considered large at that time, and it described the peninsula as wild, with a small population and covered with forests, woodlands and poor soils, and having unevenly irrigated territories. Strabo, however, gives an extensive description as he devotes the entire volume III of his Geography to the description of this land. In it he establishes one of the topical descriptive ideas about the shape of the Iberian Peninsula, saying that it looks like a bull skin: "Iberia ... resembles the skin of a bull, laid in the sense of its length from the west to the east and in the sense of its width from north to south, in a way that the front is towards the east". This sentence of the Hellenic geographer has been responsible for the commonly used sobriquet of the Iberian Peninsula as the "bull skin" (Fig. 1.4).

Strabo also indicates the existence of fertile valleys, densely populated and devoted to intense agricultural production. This implies the existence in Iberia of an agrarian society that was already well developed during the Roman domination and surely profoundly transformed large parts of the territory. After the period of the Barbarian invasions and the Visigoth kingdom, the Arabs appeared in the history of this part of the world when, in 711, they crossed the strait, which so far had been called the Pillars of Hercules but thereafter was called the strait of Gibraltar (from *Yabal Tāriq*, جبل طارق , or Tarik's Mountain). They seized the country and occupied the major part of the territory in the first centuries of their domination period. After a few centuries, their domination area was progressively reduced until their presence in Iberia was ended. This happened almost eight centuries after they entered in the Peninsula; it ended when the last Muslim kingdom of Granada was conquered by Castile in 1492. This Arabic period, lasting almost 800 years, was extremely important for the cultural and economic acquis of the country as the Arabs left a relevant and still visible heritage in some areas, particularly in the field of agriculture.

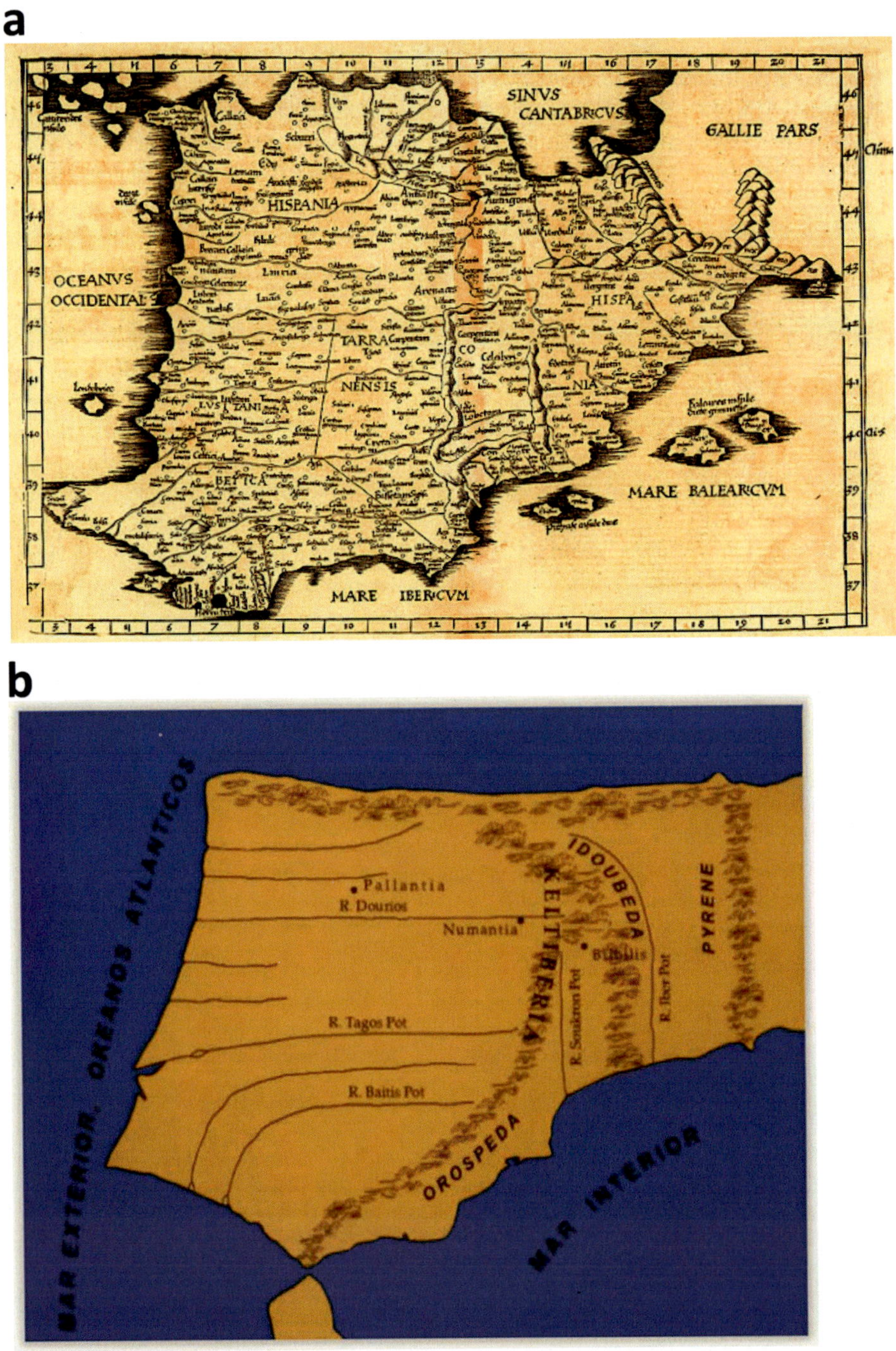

Fig. 1.4 Cartographic reconstruction of the Iberian Peninsula after the descriptions of ancient Greek authors: Ptolomeus (**a**) and Strabo (**b**). Notice that the Pyrenees are North-South oriented in both cases (The maps are from ancient classic authors (Ptolomeus and Strabo))

1.3 Land Use by Humans

In terms of land use models, Iberia is essentially divided into two major parts, coincident with the two main climatic types (Mediterranean and Temperate) and the two biogeographical regions (Mediterranean and Eurosiberian). The Temperate or Eurosiberian part covers a continuous area in the northern and northwestern territories of the Peninsula, as well as discrete isolated areas in the mountains of central Iberia. Here, the land use system is comparable to that of the rest of Atlantic Europe, where the traditional agrarian economy has been based in cattle breeding, either for milk or for meat, combined with small fields farming of crops such as maize, potatoes and other vegetables, together with more ancient cultures of apples and chestnuts. This produces a sort of subsistence farming system conditioned in large part by the dissected topography dominant in most of this northern area. The resultant landscape has been a typical *bocage*, a mosaic of small fields, woodlands and pastures separated by hedgerows, a structurally complex landscape of high aesthetical and diversity value. That bocage is combined with larger areas of heathlands or forests and grasslands in the upper elevation areas in the mountains.

Mediterranean Iberia is dominated by a totally different landscape resulting from the traditionally adapted land-use to the intrinsic climatic summer drought. Agriculture is important and in many areas the overwhelming land-use. Annual crops, particularly cereals, are dominant in the central Iberian areas where the flat topography is suited for this use, i.e. the basins of the big rivers: Ebro, Duero, Tagus, Guadiana (La Mancha) and Guadalquivir. In some of these areas and in many others where climatic conditions are not too cold, extensive Mediterranean fruit cultures are also found: vineyards, olive tree groves, almonds, carob trees, etc., which cover huge areas, mostly in the southern half of the Peninsula. These Mediterranean fruit cultures need no irrigation, as in the case of cereals, because these crops are derived from Mediterranean floristic ancestors thus are basically pre-adapted to such a climate; for that reason they are considered *secano* crops. Complementarily, the wet areas of river banks and humid soils are submitted to other exploitation systems with other sets of crops (see below).

All this cultivation combines with the breeding of domestic animals; in the predominantly agricultural areas husbandry is subsidiary and occupies a subordinate position, exploiting the leftovers the agriculture cannot take advantage of: grazing marginal areas on steep slopes, fallow fields, cereal fields after being harvested to eat the fallen grains, the straw and the emerging weed seedlings, etc. Nonetheless, in other areas where agriculture is not profitable (poor soils) or even possible (steepness of the soils, rockiness), husbandry replaces agriculture as the main activity and produces a different landscape dominated by scrub, heathland and grassland, where crops are cultivated marginally and only for local subsistence. These areas are in the rugged and steep mountainous territories, such as the Sierra Morena, Montes de Toledo, Iberian System, Baetic Mountains, etc. There, often, grazed woodlands, as a traditional, combined exploitation system, survive and occupy large areas: they are the *dehesas* or *montados*.

1.3.1 The Management of the Water Resources: Irrigation

Sometimes it is said, that for Spaniards water is not only an issue of economic importance, but an obsession penetrated in the collective mentality of the society. Possibly, this mentality is influenced by the so called "regenerationism", a political and social movement, led by Joaquín Costa at the dawn of the twentieth century, which tried to regenerate decadent Spain and proposed, among other things, the expansion of irrigation in the country. In any case, and besides the influence of the political doctrines, this is connected with the intrinsic characteristics on the climate of the country, which leads to the use of the surface and subterraneous freshwater resources for irrigation of crop fields. This is quite a common feature in many parts of the world where precipitation is scarce or has a seasonal distribution mismatching the water requirements of crop species. Of course, Mediterranean countries belong to this category and they have historically developed techniques and constructions to supply water to the fields in order to improve the productivity of their crops. Such activities were already developed in Iberia before the Roman period, and it is certain that during the Roman domination irrigation developed technologically and to an extent that matched the expansion of agriculture and trade. This improvement decreased during the Visigoth domination, a period which came to an end when the Arabs invaded the Peninsula as from 711 onwards.

The Arabs, the new lords of the country, were well acquainted with exploiting the irrigated areas intensively and efficiently. Together with new crops, such as rice, sugar cane or oranges, they introduced new irrigation techniques and procedures and built important hydraulic works. The irrigation system established by them was very advanced for its time and agriculture under irrigation was greatly developed during the Al-Andalus period. Their heritage is imprinted in the current irrigation agriculture practised in many traditionally irrigated areas of the Iberian Peninsula (Valencia, Murcia, Aragón). As a sign of this heritage and tradition, there are a number of words of Arabic origin in the currently spoken Spanish language, which have to do with irrigation and water:

Acequia. Irrigation ditch.
Albañal. Sewer, channel for dirty residual water.
Alberca. Pool of water for irrigation.
Albufera. Pond.
Alcantarilla. Sewer, drain for pluvial or residual water.
Alfaguara. Spring of abundant water which emerges with force.
Alfarda. Fee paid for the irrigation water use.
Aljibe. Cistern, usually below ground, to collect the water from the rain.
Aljofaina (or *jofaina*). Washbowl.
Azud. Noria, but also weir or dam to deviate the water in a stream, raise its level upstream or regulate its flow.
Marjal. Marsh.
Noria. Noria, a treadmill or wheel to extract water.
Rambla. Sandy riverbed which drains the water during floods.

Tanda. Irrigation turn.
Zahorí. Dowser, a person who has the ability to discover hidden things, particularly water in the underground.

Traditionally, before the twentieth century, agriculture under irrigation was basically confined to river floodplains and some areas gained from the lower terraces due to works in earlier times. During the past 100 years the irrigated area has expanded enormously, enabled by the modern, technologically advanced works, which include big channels and dams upstream in the rivers, which provided an abundant water supply to the river network from the surrounding mountain ranges. Typically, in all Mediterranean territories, the most valuable arable land almost always corresponds to the flood plains of the rivers: deep, moist and fertile soils which can be kept in production during the dry and hot summer. It is estimated that currently, in economic terms, the productivity of an irrigated area is seven times higher than a non-irrigated one. This has induced the Spanish terms of *secano* (dry, non-irrigated land) and *regadío* (irrigated land), which very eloquently indicate the human appreciation of each of these categories. Since centuries this has lead to the practically complete cultivation and human occupation of the perifluvial area, having destroyed or altered its natural forests almost totally and transforming practically all the area originally occupied by the fluvial and perifluvial ecosystems into agricultural land and urban areas.

Currently, the irrigated surface of Spain is 3.7 million hectares, which means ca. 7% of the total national area and almost a 20% of the total arable land (Fig. 1.5). Portugal has 617,000 ha of irrigated land (6.7% of its continental territory). These figures are comparable to those from other Mediterranean countries. They illustrate the importance of agriculture under irrigation and justify the enormous resources allocated to maintain the hydraulic infrastructure. Of the irrigated area in Spain, about one million hectares are considered as "historical" as they were in service before 1950. They include the traditional irrigation districts of Valencia and Murcia, well documented by historians to have been active since antiquity. Another million hectares were put in service during the government of General Franco and the rest is due to more recent implementation. This surface is served by a set of 1225 dams (plus ca. 200 in Portugal) which have a total reservoir capacity of 56 km^3, which is much more than the amount of surface water available to be regulated, particularly in the watersheds of the southern part of the Peninsula. Most of these dams have been built recently but some of them surmount to centuries ago. The oldest in service dates from Roman times: Proserpina dam (near Mérida) from the first century BC. Besides, a huge network of irrigation channels has been built to transport the water to the consumption points. The freshwater managed in this manner is used mostly for agriculture (ca. 80% in Spain), and the rest is for other uses (industrial, domestic), making the discussion of the water use basically an agrarian question.

Fig. 1.5 Irrigated areas in Spain (The map is from the Ministry of Agriculture, Fisheries and Environment of Spain (http://www.mapama.gob.es/es/), http://www.mapama.gob.es/es/desarrollo-rural/temas/gestion-sostenible-regadios/mapas.aspx. It was acceded on 12 September 2016)

1.4 Some Issues About Conservation

This section pays attention to some processes and actions that took place recently and that profoundly affect changes in the landscape and biodiversity of Iberia. They have to do with initiatives of the administrations and with social and economic phenomena having an influence in large parts or in the whole territory. They result from the economic and demographic developments over the past 40 years and from the different policies applied during the last decades.

1.4.1 The Areas with a Protection Status

In spite of the efforts and investments made by the different administrations in recent decades in order to create a huge network of protected areas, the current conservation situation of the vegetation in Iberia is not entirely in accordance with those efforts. With the implementation of the Natura 2000 network and the additional number of areas dependent of regional administrations, an important part of the total area of the Iberian Peninsula has obtained a certain protection status: with 137,365 km^2, which means 27.2% of its territory, Spain has the largest total extent of protected areas in the EU. To this we should add the contribution of Portugal with

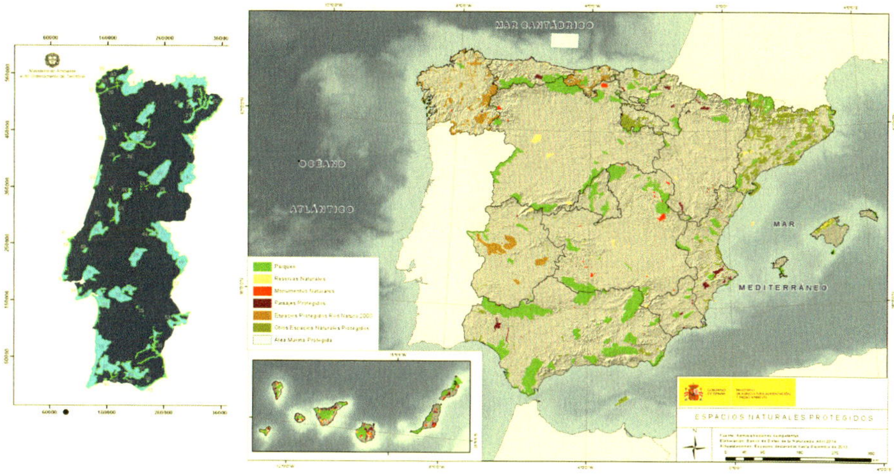

Fig. 1.6 Protected areas in Spain and Portugal (The maps are from the Ministry of Agriculture, Fisheries and Environment of Spain (http://www.mapama.gob.es/es/) and the Instituto da Conservaçao da Natureza e Florestas (http://www.icnf.pt/portal/ap/areas-protegidas), http://www.mapama.gob.es/es/biodiversidad/servicios/banco-datos-naturaleza/informacion-disponible/ENP.aspx. It was acceded on 05 February 2015, http://www.icnf.pt/portal/ap/rnap. It was acceded on 27 March 2015)

19,010 km^2, which means 20.7% of its territory (European Environmental Agency). This is a considerable effort which has been repeatedly praised at innumerable occasions and I consider it being of high value also (Fig. 1.6).

1. The conserved areas, to different degrees, have enhanced the successional trends promoting the naturalization of ecosystems. Human pressure has released, enhanced by rural abandonment and the introduction of the use of fossil fuels for domestic uses (see below), and a general tendency towards more natural, or less human influenced, vegetation types is evident: forests, shrublands, etc. Because of this, ecosystems which are dependent on a certain human-induced disturbance regime clearly retreat and this poses an important question in conservation priorities and in the criteria for management.
2. The managing criteria in these conservation areas are far from being fixed; often there are no recipes of what to do and managers have to take decisions with little or no scientific background. There are scarcely general rules of what to do to manage hunting, fishing, husbandry, forestry, etc. and this represents a source of uncertainty.
3. For some administrations, not very fervent in issues related with biodiversity and ecosystems conservation, a conserved areas network can be considered as an alibi to do nothing outside the conserved areas, forgetting that conservation has to be done not only inside the network but on the entire territory.
4. Around these protected areas (National Parks, natural parks, etc.), certain economic activities have been developed, such as ecologic or nature-loving tourism. This has become an important option for the spare time use of people living in urban areas, often complemented with the trade in several local products such as

cheese, wine, olive oil, honey, scented plants, handicraft, pottery, etc. These activities tend to offer some jobs to the local population, encouraging them to stick to the territory and to conserve also the mores, culture and traditions of the area concerned. The success of this has been low so far.

5. In general terms the scientific use of this huge network has been modest in comparison with the possibilities glimpsed. If a more abundant resources allocation would have been provided, important large scale projects could have been established, for instance large scale surveys on certain ecosystems, or projects aimed at the monitoring of terrestrial ecosystems in relation to global change, or dealing with the management for preserving diversity as a result of planned actions.

1.4.2 The Urban Development, Artificialization of the Land

Human societies have experienced profound changes in their functioning by virtue of the economic development taken place in the last decades, as a result of the associated phenomena of globalization and adoption of new technologies. This has affected the distribution patterns of the human population in most parts of the world. Also in Iberia this has changed the model and severity of impacts upon natural and semi-natural ecosystems. One of the most evident phenomena is urban expansion. This affects the occupancy of the territory by human settlements: dwellings, factories, roads, channels, railroads and all types of infrastructure typical of our present industrial and post-industrial modern societies. This sort of urban or peri-urban occupancy of the terrain is an inevitable outcome of economic and demographic growth. It has been called *artificialization of the soil* (as this concept excludes the agricultural land and the intensively grazed areas). Within the past six or seven decades this process has been laxly controlled or proceeded temporarily even unrestrictedly and resulted in a huge extent of land being artificialized. This took place during the times of explosive economic development and of expansion of touristic urban areas. More recently, some regulations have been implemented in order to control it but the building pressure is so strong that the process has continued ahead nonetheless. The area of land affected by this artificialization becomes mostly covered by concrete or asphalt and is irreversibly sealed, becoming useless for any other purposes. Only in Spain, in 2006, the extent of the area under such conditions amounted to 1,017,359 hectares, or 2.1% of the total area of the country. A high proportion of this artificialized area has been created in the last decades: the growth rate in the period 2000–2006 has been 2 ha per hour and since 1987 artificialization has transformed 1/3rd of all the surface area that has been artificialized in Spain during all previous centuries together. The coastal areas have suffered more intensely from this process and here the proportion reaches 1/2. This littoral concentration was concentrated on the Mediterranean shores; a full 34% of the first kilometer inland is already artificialized (Jiménez Herrero 2008). Big urban areas, such as Madrid, Barcelona, Lisbon or other large cities, as well as the coastal fringes submitted to the touristic pressure, are the most affected areas. Mountainous areas are not alien to this phenomenon because the urban expansion occurs in valley bottoms sealing the most fertile soils and it will perform a sort of agrarian suicide if food shortage would occur.

Another matter concerns the town model that is being implemented with these developments. Current urban development stimulates a change in town model from a compact city into a scattered urban area. The former is consistent with the urban tradition in Latin countries, where cities and towns were built densely, often with buildings of several stories and without any or little green areas. Such a town provides lower quality dwellings to its inhabitants and they have to live with a certain degree of overcrowding, with few opportunities to enjoy a little garden or a clear and bright surrounding. However, this model creates a city that is bustling and full of life, with walking distances, or distances that are easy to endow with public transport. The scattered model complies with the northern European tradition of living in small one-family dwellings surrounded by small gardens and grouped in low density towns. A green matrix of trees and parks intermingles with the houses and preserves a more nature-friendly appearance. However, distances are long and cars become necessary for going to work, for shopping, etc., making public transport less efficient and more expensive. This latter type is replacing the traditional Latin town model and this entails a much higher demand of energy and soil for building. It would be very convenient to return to the Latin urban tradition of the compact city, which is much less energy- and land-demanding and more sustainable.

1.4.3 The Rural Abandonment

As in large areas elsewhere in Europe, Iberia is severely affected by the phenomenon of the rural abandonment. People move from rural to urban areas and concentrate in cities, leaving large territories empty of people, particularly those of lower agricultural value. This process has intensely affected many interior areas of Iberia, where there are hundreds of entirely abandoned villages and many more inhabited by only a few old persons. The process encompasses a profound change in land use which can be summarized as follows:

(a) Areas with strong relief are released from the intense ancestral pressure of sheepherding and, alternatively, also of the self-sustaining agriculture performed by the traditional human population. These are the most abandoned areas and they experience a strong recovery of all the forms of natural ecosystems, in a process of re-naturalization of the territory.
(b) Agriculturally productive areas, usually flat and with fertile soils, are more intensely exploited with the use of machinery and chemicals (fertilizers and biocides). This entails a loss in important biodiversity elements such as weed communities, hedges, small woodlands, etc., which were constituent of the traditional agrarian landscape.

(c) Urban areas expand vigorously, encroaching upon large portions of land. Residential areas, industrial settlements, infrastructures of all types are being built as a result of the economic changes.

The most important territorial effect resulting from the new population model is the first one, in which land use changes in favour of lower pressure upon mountain areas trigger secondary succession towards more natural vegetation forms. Pastures are transformed into shrublands, trees encroach in scrub initiating their transformation into woodlands, forests formerly used for wood extraction are no longer pruned which results in increasing ages and sizes of the trees, and forest regeneration is promoted by the establishment of young cohorts of the tree species. All these processes convey to a homogenisation of the landscape with the blurring of the diversity between the different tesellas in which the territory was divided. This also leads to the gradual disappearance of many vegetation types which were bound to a particular exploitation regime by humans, such as mowing, manuring, grazing, etc., activities that now get in disuse, or are tied to some particular exploitation systems, such as transhumance of sheep flocks, that are now abandoned. This can be summarized under the term "abandonment ecology" which covers a group of phenomena that are consequences of the release of human pressure upon the territory due to the abandonment of traditional land-use practices.

1.4.4 Forestry

Travelling through Iberia brings the surprise of finding many areas covered by pines and/or eucalypts. For a while, it could be thought that the traveller is crossing a boreal country full of conifer forests or is visiting Australia with its widespread eucalypt woodlands. But for sure this is not the case; these are plantations done by the hand of humans. These plantations are the result of a policy of the governments addressing the problem of the historic shortage of wood in the Iberian countries. Such a policy was initiated in the early twentieth century and was boosted in periods of economic autarchy linked to specific political and economic difficulties during the last century. The idea was that these countries should approach self-sufficiency in forest products as a way to offset the imports of such commodities. The basic criteria for the selection of species to be planted was to use the highest-productive species among the climatically adapted tree species worldwide, a sort of agriculture of trees, that was begun in the Prussian forestry of the late nineteenth century and readily accepted early on by the Iberian technicians and politicians. The result has been a huge increase in the surface area planted by pines (or other conifers) and eucalypts, especially in marginal areas with a low agricultural potential, such as mountainous regions or with sandy substrates. These tree plantations became so extensive, that they became the dominant element in the landscapes in many areas of the Peninsula. As shown in the Table 1.1, the proportion of tree plantations in

Table 1.1 Comparison of the areas occupied by the main planted tree species and native tree species groups in peninsular Portugal and Spain

	Portugal		Spain	
Species	ha	% of the forested area of the country	ha	% of the forested area of the country
Eucalypts	811,943	26	633,000	3.5
Pinus pinaster	714,445	23	1,373,000	7.5
Pinus pinea	145,742	6	390,000	2.1
Pinus halepensis			1,926,000	10.5
Pinus nigra s.l.			625,000	3.4
Pinus radiata			295,000	1.6
Other conifers	73,217	2	432,000	2.4
Castanea sativa	41,410	1	228,000	1.2
Other planted broadleaved trees (Populus etc.)			109,000	0.6
Plantations	**1,749,488**	**58**	**6,011,000**	**47.6**
Quercus ilex s.l.	331,179	11	2,792,000	15.3
Quercus suber	736,775	23	301,000	1.6
Other broadleaved trees	177,176	6	702,000	3.9
Quercus pyrenaica			1,034,000	5.7
Quercus robur (+ petraea)			459,000	2.5
Quercus faginea (+ pubescens)			334,000	1.8
Other Quercus spp.	67,116	2		
Fagus sylvatica			486,000	2.7
Pinus uncinata			97,000	0.5
Juniperus (thurifera + phoenicea)			391,000	2.1
Abies (alba + pinsapo)			20,000	0.1
Natural forests	**1,312,246**	**42**	**6,616,000**	**52.4**
Total	**3,061,734**	**100**	**12,627,000**	**100.0**

Figures are in hectares (Inventario Florestal Nacional-6 2013 for Portugal and Montero and Serrada 2013 for Spain)

comparison with the natural and semi-natural forests in both countries is enormous, and even higher in Portugal than in Spain.

The results of such a gigantic effort during one century have been regrettably disappointing. The pretended total autarchy was never achieved because the average productivity was low. With the exception of the rainy areas of the north and the west, the rest of the territory is submitted to a Mediterranean climate with dry summers, severely limiting tree growth in that season. It is illusory to hope for high production values under such conditions. On other occasions the plantations were established claiming a protective target, particularly aiming at soil erosion. In such cases, benefits were diminished because the introduced exotic species produced an acidic litter negatively affecting the edaphic processes. Moreover, these conifer and eucalypt species are full of highly flammable resins and exudates which increase the risk of wildfires.

Those plantations sometimes have replaced former natural forests, but more often they have been planted in grazing areas covered by open scrub or dry grasslands that sustained the traditionally large livestock patrimony that existed historically in Iberia. The decrease of that livestock paralleled the increase of the plantations, producing a profound landscape transformation and a deep socio-demographic transformation in the rural areas: from open treeless landscapes to pine-planted hills and sierras. From this point of view, tree plantations became 'tree waste lands' that only supported poor, extensive grazing, satisfying the extensive, widespread feeling in the society that the country was lacking trees. The landscape produced by these plantations is clearly artificial, comprising a mosaic of geometric tessellas covering the land with each tessella a plantation of evenly-aged trees in regular lines, as if they were soldiers in a military training excercise.

If we consider the impact of these exotic tree plantations on biodiversity and naturalness, we cannot bestow the same category on them as on natural autochthonous forests. Natural forests are the result of a long process of joint evolution of organisms which live and interact together for a long time. The introduction of alien species alters this close-fitting framework of mutual interactions built during evolutionary time. Arbitrarily changing the tree species composition of a forest entails severe distortion in ecosystem functions and only impoverishment can be expected. For that reason, exotic trees plantations cannot be considered true forests because they entail a fundamental alteration of the natural order, not only in ecosystem functioning, but also in a biogeographical and evolutionary sense, which implies that in each specific habitat there has to be one tree species (or group of species), and no other ones, to form the natural forest. Planting and cultivating exotic trees makes no forests; it is a sort of agriculture or gardening of trees.

Another important point on this issue is that of the methods applied in modern forestry. Technology has also arrived in this activity and now heavy machinery and chemicals for different purposes are used: fertilizers and biocides, as in any conventional agriculture. The use of machinery for tilling the terrain and to open-up roads has become regular, dramatically increasing soil erosion in areas with steep slopes. This is a very serious point because it entails a severe loss of fertility. Soil cannot be considered a renewable resource as it takes centuries to millennia to be formed; degrading and/or eroding it is an irresponsible bet.

It should be pointed out, however, that these conifer and eucalypt plantations must not be confounded with the natural conifer forests existing in several parts of Iberia, often in high-elevation areas, on mountains and high plateaus, or in xeric areas under severe Mediterranean conditions.

References

Inventario Florestal Nacional-6 (2013) Áreas dos usos do solo e das espécies florestais de Portugal continental. Resultados preliminares. [pdf]. Instituto da Conservação da Natureza e das Florestas (ICNF). Lisboa, p 34

Jiménez Herrero LM (2008) Ocupación del suelo en España y cambio global: artificialización, urbanización y litorización. Cambio Global en España 2020: 94–99. Fundación CONAMA. Madrid

Loidi J (1999) General description of the Iberian Peninsula: substrate and relief. In: Rivas-Martínez S, Loidi J, Costa M, Díaz TE, Penas A (eds) Iter Ibericum A.D. MIM. Itinera Geobotanica 13, pp 17–40

Montero G, Serrada R (eds) (2013) La situación de los bosques y el sector forestal en España-ISFE. Sociedad Española de Ciencias Forestales. Lourizán, Pontevedra, p 253

Terán M, Solé Sabarís L (1978) Geografía general de España Ed. Aeriel, Barcelona, p 549

Web Pages

European Environemntal Agency (http://www.eea.europa.eu)

Chapter 2
Bioclimatology of the Iberian Peninsula and the Balearic Islands

Salvador Rivas-Martínez, Ángel Penas, Sara del Río, Tomás Emilio Díaz González, and Salvador Rivas-Sáenz

Abstract According to criteria and proposals of S. Rivas-Martínez, S. Rivas-Sáenz and A. Penas (Glob Geobot 1(1):1–634, 2011a) with which they established a world bioclimatic classification system, two macrobioclimates (Temperate and Mediterranean), eight bioclimates (temperate hyperoceanic, temperate oceanic, temperate xeric, mediterranean pluviseasonal-oceanic, mediterranean pluviseasonal-continental, mediterranean xeric-oceanic and mediterranean desertic-oceanic), two bioclimatic variants (steppic and submediterranean), 11 thermotypes (thermotemperate, mesotemperate, supratemperate, orotemperate, cryorotemperate, inframediterraean, thermomediterranean, mesomediterranean, supramediterranean, oromediterranean and cryoromediterraean) and seven ombrotypes (arid, semiarid, dry, subhumid, humid, hyperhumid and ultrahyperhumid) are recognized in the Iberian Peninsula and Balearic Islands. In addition to this, six types of continentality (euhyperoceanic, subhyperoceanic, semihyperoceanic, euoceanic, semicontinental and subcontinental) are also recognized. Furthermore, relationships between potential natural vegetation (sigmeta, geosigmeta, permasigmata, minorisigmeta and geopermasigmeta) and the bioclimatic units existing in the Iberian Peninsula and Balearic Islands are set.

S. Rivas-Martínez (✉) • S. Rivas-Sáenz
Phytosociological Research Center (CIF), E-28409 Los Negrales, Madrid, Spain
e-mail: rivas-martinez.cif@tsai.es

Á. Penas • S. del Río
Department of Biodiversity and Environmental Management (Area of Botany), Mountain Livestock Farming Institute (Joint Center CSIC-ULE), Faculty of Biological and Environmental Sciences, University of León, Campus of Vegazana, s/n, León E-24071, Spain

T.E. Díaz González
Department of Organisms and Systems, Biologia Faculty, University of Oviedo, E-33071 Oviedo, Spain

J. Loidi (ed.), *The Vegetation of the Iberian Peninsula*, Plant and Vegetation 12,
DOI 10.1007/978-3-319-54784-8_2

2.1 Notions on Bioclimatology

Bioclimatology is a scientific field related to Geobotany that studies the relationships between the climate and the distribution of living beings and their communities on Earth. This discipline, coincident with what has been called Phytoclimatology, began to be structured by correlating numeric climate values (mostly temperature and rainfall data) with the distribution areas of plants, formations, plant communities and ecosystems. At a later stage, information from dynamic-catenal phytosociology, (vegetation series, geoseries and geopermaseries) has been incorporated.

During the past 25 years, a worldwide bioclimatic classification (del Río et al. 2011a) has been largely developed. The aim is to yield an easily quantifiable bioclimatic typology showing a close relationship between vegetational components and climate parameters. Considering the high predictive value of the bioclimatic units, they can be used in other sciences, in studies and conservation programmes for biodiversity, as well as in agricultural and forestry resources management. Our increasingly detailed knowledge on the distribution of vegetation all over the World, as well as the modifications caused in the potential natural vegetation and its substitution stages by means of climatic, edaphic, geographic and anthropic factors, is improving the recognition of bioclimatic and vegetational frontiers with higher accuracy and objectivity. Once the limiting boundaries of the vegetation series, geoseries and geopermaseries are known, and also the distribution areas of the bioindicator species are mapped, it is possible to calculate the numerical bioclimatic threshold values that envelope them. The spaces corresponding to the bioclimatic units (bioclimates, thermotypes, ombrotypes and continentality) have been progressively delimited and adjusted. The obtained biophysical models have demonstrated a high level of reciprocity in the climate-vegetation relationship patterns, and this makes it possible to create significantly more precise bioclimatic and biogeographic maps of the world. This is achieved on basis of the highly reciprocal and predictive relationships between patterns in climate, defined by bioclimatic data, and patterns in vegetation types and the bioindicators. An important number of authors have contributed to the task (Gaussen 1947, 1954, 1955a, b; Bagnouls and Gaussen 1953, 1957; Trochain 1952; Emberger 1955, 1959; Quézel 1957, 1965, 1999; Meher-Homji 1963; Troll and Paffen 1964; Hueck and Seibert 1972; Sánchez Egea 1975; Eddy 1976; Daget 1977; Donadieu 1977; Box 1981; Rivas-Martínez 1981, 1982a, b, 1983, 1984, 1985, 1987a, b, 1988, 1991, 1996, 1997, 2001, 2004, 2005a, b, 2006, 2007, 2009; Montero de Burgos and González Rebollar 1987; Thornthwaite 1984; Takhtajan 1986; Le Houerou 1986, 1995; Pons and Reille 1988; Bolòs 1989; Ladero et al. 1987, 1994; Rivas-Martínez et al. 1991, 1999a, b, 2007, 2011a, b; Duplessy et al. (eds.) 1991; Quézel and Barbero 1993; Orlóci 1994; Peñalba 1994; Moreno and Oechel 1995; Grabherr et al. 1994; Blasi 1996; Fernández-González 1997; Copons and Bordonau 1997; Gavilán and Fernández-González 1997; Amigo and Ramírez 1998; Rivas-Martínez and Costa 1998; Gavilán et al. 1998, 2001; Pott 1998, 2001;

Herráez 1999; Ruíz-Zapata 1999; Galán et al. 1999; Menzel and Fabian 1999; Creus and Saz 1999; Moreno et al. 1990; Almarza 2000; Budong et al. 2000; Jalut et al. 2000; Manrique and Fernández-Cancio 2000; Murphy 2000; González-Hidalgo et al. 2001; Ramil-Rego et al. 2001; Neq et al. 2001; Barber et al. 2001; Oturbay and Loidi 2001; Goddard et al. 2001; Rodríguez et al. 2001; Navarro and Maldonado 2002a, b; Sparks et al. 2002; Sanz-Elorza et al. 2003; Davis et al. 2003; Mesquita et al. 2004; Lousã 2004; Solanki et al. 2004; Camarero and Gutiérrez 2004; Mesquita 2005; Fernández-González et al. 2005; del Río 2005; del Río et al. 2005a, b, 2007, 2009, 2011a, b; del Río and Penas 2006a, b; Navarro and Ferreira 2007; Moreno 2007; Rivas-Martínez and Rivas Sáenz 2009).

The global bioclimatic classification we use is based on the axiomatic principle that main climatic factors influencing vegetation and plant life, i.e. photoperiod, continentality, seasonality of the precipitation, mediterraneity, aridity, orobioclimates (mountain bioclimates), can be used to establish entities significantly coincident with biodiversity distribution patterns (Rivas-Martínez et al. 2011a, b). Our classification recognizes five macrobioclimates, 28 bioclimates and eight bioclimatic variants for the whole World. The **macrobioclimate** is the highest typological unit in the bioclimatic classification system; it is an eclectic biophysical model, delimited by specific climatic and vegetational values, and covering a wide territorial range and which is connected with the main types of climates, biomes and biogeographical regions to day accepted on Earth. There are five macrobioclimates: tropical, mediterranean, temperate, boreal and polar. Each of them, as well as their 28 subordinate units or **bioclimates** (Fig. 2.2), are represented by a group of vegetation types, biocoenoses, ecoregions, formations and plant communities. Within each of the bioclimates, there is climatic variability which can be expressed by means of the bioclimatic belts – **thermotypes** (Fig. 2.3) and **ombrotypes** (Fig. 2.4) – as well as by a certain number of other units which have been identified according to the seasonal temperature and yearly rainfall periods.

2.1.1 Bioclimatic Classification System

To classify the bioclimatic models that we use, i.e. macrobioclimates, bioclimates, as well as bioclimatic belts and variants, it is essential to use bioclimatic keys and a synoptical table of a worldwide bioclimatic classification system according to developed computer programs (originally Luengo, Penas and Rivas-Martínez, 1995, unpublished), and recently adapted to the computation requirements of the last bioclimatic world classification by Rivas Sáenz (http://www.globalbioclimatics.org) and Rivas-Martínez et al. 2011a.

2.1.2 Parameters and Bioclimatic Indices

The most common bioclimatic parameters and indices used in our bioclimatic classification are detailed below (a more detailed list in Rivas-Martínez et al. 2011a). The list starts with precipitation parameters (rainfall expressed in mm), followed by temperature parameters (with averages in degrees Celsius and positive temperature of indices in tenths of degrees Celsius), seasonality, ending with the bioclimatic indices which are simple arithmetic formulas that include parameters.

Precipitation Parameters

P	average annual precipitation in millimetres or litres per square metre.
P_i	average monthly precipitation, where i: 1 = January, . . ., 12 = December.
Pcm_1	precipitation of the warmest four consecutive months of the year.
Pcm_2	precipitation of the four consecutive months following the warmest of the year.
Pcm_3	precipitation of the four consecutive months before the warmest of the year.
Pd	precipitation of the driest quarter of the year.
Pp	positive annual precipitation of the months with Ti higher than 0 °C, ΣPpi 1–12 > 0 °C.
Ps	precipitation for the summer quarter.
Pss	precipitation for the warmest six consecutive months of the year.
Psw	precipitation for the coldest six consecutive months of the year.
Pw	precipitation for the winter quarter.

Temperature Parameters

T	mean annual temperature in centigrade.
Ti	mean monthly temperature in centigrades, where i: 1 = January, ..., 12 = December.
Tmin	average temperature of the coldest month of the year in C°.
Tp	positive annual temperature. Total in tenths of degrees Celsius of the average monthly temperatures higher than 0°, $\Sigma Ti_{1-12} > 0$ °C.
Tpi	positive monthly temperature, where i: 1 = January, ..., 12 = December, in tenths of degrees Celsius.
Tps	positive temperature of the summer quarter, in tenths of degrees Celsius, Σ Tps, Ti > 0 °C.
Tpw	positive temperature of the coldest quarter, in tenths of degrees Celsius, Σ Tpw, Ti > 0 °C.
Ts	average temperature of the summer quarter, N: T 6–8, S T 12–2.
M	average temperature of the maximums of the coldest month.
m	average temperature of the minimums of the coldest month.
m_i	average monthly temperature of the minimums, where i: 1 = January, ..., 12 = December.

Seasonality Parameters

Tr_1 quarter corresponding to the winter solstice (winter, N: 12–2, S: 6–8).
Tr_2 quarter corresponding to the spring equinox (spring, N: 3–5, S: 9–11).
Tr_3 quarter corresponding to the summer solstice (summer, N: 6–8, S; 12–2).
Tr_4 quarter corresponding to the autumn equinox (autumn, N: 9–11, S: 3–5).
Cm_1 warmest consecutive 4 months of the year.
Cm_2 four consecutive months after the warmest 4 months of the year.
Cm_3 four consecutive months before the warmest 4 months of the year.

Bioclimatic Indexes

Ic simple continentality index or annual thermic interval (Tmax-Tmin in degrees Celsius).
Id diurnality index or daily thermic interval (Tcmax-Tcmin in degrees Celsius).
Io annual ombrothermic index (Pp/Tp) 10.
Iom monthly ombrothermic index (Pi/Tpi) 10, where i: 1 = January, ..., 12 = December
Ios_1 ombrothermic index of the hottest month of the summer quarter (Tr_3).
Ios_2 ombrothermic index of the hottest 2 months of the summer quarter (Tr_3).
Ios_3 ombrothermic index of the summer quarter (Tr_3).
Ios_4 ombrothermic index of the 4-month period resulting from adding the summer quarter (Tr_3) and the month immediately preceding it.
It Thermicity index $(T + M + m)\ 10 \cong (T + Tmin \times 2)\ 10$.
Itc compensated thermicity index.

2.1.2.1 *Continentality Indices* (Ic) *and Types*

Continentality indices reflect the amplitude of the annual oscillation of temperature. The degree of continentality is directly proportional to this amplitude; the opposite concept is oceanity. Unfrozen seas, lakes and oceans tend to absorb the temperature contrast, while the opposite occurs inland, further away from the coast. The most widely used indices to express continentality/oceanity can be grouped into simple and compensated ones. Simple indices are those that only reflect the difference between extreme monthly mean temperatures (Ic), while the compensated ones are those that correct that amplitude based on altitude or latitude.

Amongst the compensated indices proposed in order to offset the effect of the increase of seasonal thermic amplitude with the increase of latitude, there is Gorezynski's index [c = (1.7Ic/sen lat.)−20.4]. Another one is Conrad's index which attempts to generalise the previous one [c = (1.7Ic/sen lat. +10°) – 14] by adjusting on a decimal scale the zero value (extremely oceanic, for the Faroe Islands) and the 100 value (extremely continental, for Verchojansk, in northeastern Siberia). Currey's continentality quotient offers a good correlation between continentality and vegetation in the northern territories of the northern hemisphere, which is obtained by dividing the annual thermal amplitude (Ic) by a third of the latitude plus one [Cc = Ic/ (1 + 1/3lat.)]. Based on this quotient, Currey considered territories with values lower than 0.6 as hyperoceanic, between 0.6 and 1.1 as oceanic, between 1.1 and 1.7 as subcontinental, between 1.7 and 2.3 as continental, and over 2.3 as hypercontinental.

As the effects of the increase in continentality on the vegetation are very evident outside of the eutropical latitudinal zone (0°–23° N & S), it has not been necessary to use the compensated continentality indices within the tropical latitudes. The simplicity of the indices used in our bioclimatic classification has been one of the objectives we have proposed, although in some cases this has meant a small loss of accuracy, which we believe is compensated by the easiness and accessibility of the data.

Because of its simplicity, availability of data and excellent global correlation with vegetation distribution, in the bioclimatic classification of the Earth we have used the simple continentality index (Ic), whose origins lie in the first oceanity map of the Earth by Supan. This continentality index expresses in degrees Celsius the difference between the average temperature of the hottest month (Tmax) and that of the coldest month of the year (Tmin). Ic = Tmax−Tmin. The types, subtypes and levels of continentality that are recognised are shown in Table 2.1 and Fig. 2.1. By averaging the value of the subtypes, we obtain strong and weak levels, which are allocated in each case depending on their greater (strong) or lesser (weak) proximity to the extreme values in the table (0 and 66).

Apart from the continentality categories recognised using the simple continentality index [Ic = Tmax − Tmin], namely 3 types, 9 subtypes and 18 levels [Itinera Geobot. 17: 16–17, tb. 2. 2007], sometimes it is useful to use the terms from another scale using nine continentality categories, especially tailored to correlate with certain types of vegetation and bioclimates (Table 2.2).

2.1.2.2 Compensation of the Summer Ombrothermic Indices

According to its definition, the Mediterranean macrobioclimate is an extratropical type (> 23° N & S) in which the summer (the hottest season of the year) coincides with a drought period of at least two consecutive months in which the precipitation is lower than or equal to twice the temperature (P = 2T). In contrast, a territory is not Mediterranean if the ombrothermic index of the two hottest months of the summer quarter Ios_2 is higher than 2 ($Ios_2 > 2$). If Ios_2 is less than or equal to 2.0 ($Ios_2 \leq 2.0$), the territory may be Mediterranean or not. Hence, if the availability of water in the soil can be compensated with the precipitation of the previous month, i.e. if P (June + July + August)/T (June + July + August) in the northern hemisphere, or otherwise P (December + January + February)/T (December + January + February) in the southern hemisphere, is higher than 2.0 ($Ios_3 > 2.0$) then the territories are not Mediterranean. If the Ios_3 is higher than or equal to 2.0 ($Ios_3 \leq 2.0$), the territory may be Mediterranean or not, because with a deficient Ios_3 compensation may occur with the precipitation from the previous month (May or November

Table 2.1 Types, subtypes and levels of simple continentality (Ic) recognized on earth

Types	Subtypes	Levels	Values
1. Hyperoceanic (0–11)	Ultrahyperoceanic	1.1a. strong	0–2.0
		1.1b. weak	2.0–4.0
	Euhyperoceanic	1.2a. strong	4.0–6.0
		1.2b. weak	6.0–8.0
	Subhyperoceanic	1.3a. strong	8.0–10.0
		1.3b. weak	10.0–11.0
2. Oceanic (11–21)	Semihyperoceanic	2.1a. strong	11.0–12.0
		2.1b. weak	12.0–14.0
	Euoceanic	2.2a. strong	14.0–15.0
		2.2b. weak	15.0–17.0
	Semicontinental	2.3a. weak	17.0–19.0
		2.3b. strong	19.0–21.0
3. Continental (21–66)	Subcontinental	3.1a. weak	21.0–24.0
		3.1b. strong	24.0–28.0
	Eucontinental	3.2a. weak	28.0–37.0
		3.2b. strong	37.0–46.0
	Hypercontinental	3.3a. weak	46.0–56.0
		3.3b. strong	56.0–66.0

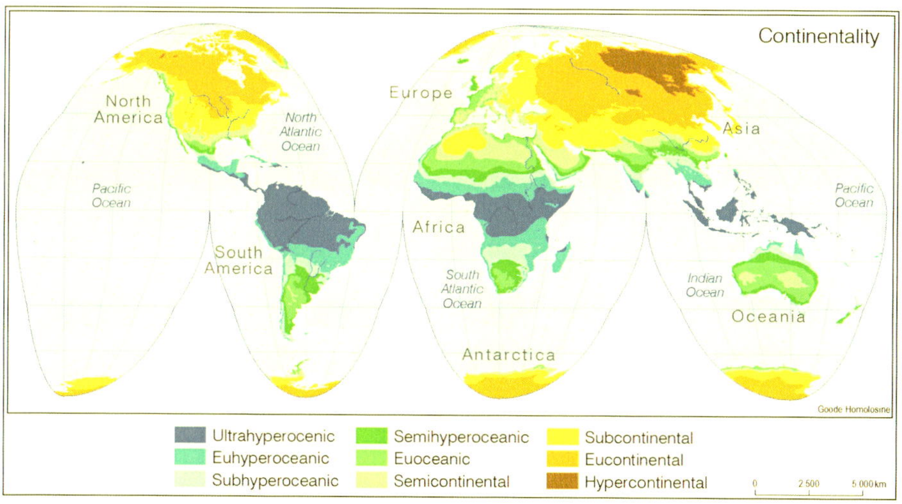

Fig. 2.1 Continentality map of the Earth (From Rivas-Martínez et al. 2011a)

respectively), i.e. if P (May + June + July + August)/T (May + June + July + August) in the northern hemisphere, or P (November + December + January +February)/T (November +December + January + February) in the southern hemisphere is higher than 2.0 ($Ios_4 > 2.0$), the territories are not bioclimatically mediterranean; in the

Table 2.2 Relationship between the terms and subtypes of continentality with the values of the simple continentality index (Ic)

Expressions	Ic values	Subtypes	Ic values
Extreme hyperoceanic	0.0–8.0	Ultrahyperoceanic	0.0–4.0
		Euhyperoceanic	4.0–8.0
Moderate hyperoceanic	8.0–14.0	Subhyperoceanic	8.0–11.0
		Semihyperoceanic	11.0–14.0
Balanced oceanic	14.0–17.0	Euoceanic	14.0–17.0
Moderate continental	17.0–28.0	Semicontinental	17.0–21.0
		Subcontinental	21.0–28.0
Extreme continental	28.0–66.0	Eucontinental	28.0–46.0
		Hypercontinental	46.0–66.0

Table 2.3 Compensation table modified. Intervals of the values of the annual ombrothermic indices (Io) which, depending on the values of the summer ombrothermic values (Ios_2, Ios_3, Ios_4), can be compensated and change from the Mediterranean macrobioclimate to the temperate one (submediterranean variant)

Io	Ios2	Ios3	Ios4
2.0–2.8	≥1.9	≥2.0	≥2.0
2.8–3.6	≥1.8	≥1.9	≥2.0
3.6–4.8	≥1.8	≥1.9	≥2.0
4.8–5.5	≥1.7	≥1.9	≥2.0
5.5–6.2	≥1.6	≥1.9	≥2.0
6.2–7.0	≥1.5	≥1.8	≥2.0
7.0–8.0	≥1.4	≥1.8	≥2.0
8.0–9.0	≥1.3	≥1.8	≥2.0
9.0–10.0	≥1.2	≥1.7	≥2.0
10.0–11.0	≥1.1	≥1.7	≥2.0
11.0–12.0	≥1.0	≥1.6	≥2.0
12.0–13.0	≥0.9	≥1.6	≥2.0
13.0–14.0	≥0.8	≥1.5	≥2.0
14.0–16.0	≥0.7	≥1.5	≥2.0
16.0–18.0	≥0.6	≥1.4	≥2.0
18.0–21.0	≥0.5	≥1.2	≥2.0
21.0–24.0	≥0.4	≥1.0	≥2.0
≥24.0	≤0.4	≤1.0	≥2.0

opposite case, they are definitely mediterranean. Ombrothermic values have a highly discriminatory value on the frontier between mediterranean-temperate territories and mediterranean-boreal territories. The compensable values of the summer ombrothermic indices are shown in Table 2.3.

Table 2.4 Levels and threshold values for submediterraneity corresponding to the submediterranean bioclimatic variant on Earth

Submediterraneity levels	Isbm
Extremely weak submediterranean	1–30
Highly weak submediterranean	30–80
Weak submediterranean	80–180
Strong submediterranean	180–320
Highly strong submediterranean	320–460
Extremely strong submediterranean	460–580

2.1.2.3 *Submediterraneity Index* (Isbm)

A locality or territory with a temperate, boreal or polar macrobiocolimate can be considered as submediterranean, i.e. it belongs to the submediterranean bioclimatic variable [Sbm], if at least in 1 month of the summer quarter its average precipitation in millimetres is 2.8 tenths less the average temperature in degrees Celsius [Te, Bo, Po, 23°-90° N & S, Iosi: $P < 2{,}8\ T$]. The submediterraneity value is calculated as follows: Vsbmn =280−(100 Iosi). Similarly, the submediterraneity index is calculated from the sum of the submediterraneity values of the summer monthly ombrothermic indices Ios1, Ios2, Ios3, Ios4, providing they are positive, without taking into account the submediterraneity values if they are negative. On Earth, depending on the amount of summer drought or aridity, measured in hundreds of ombrothermic units, six levels of submediterraneity are distinguished (Table 2.4).

2.1.2.4 *Thermicity indices* (It, Itc)

This is the sum, in tenths of a degree, of T (average annual temperature), m (average temperature of the minimals of the coldest month) and M (average temperature of the maximals of the coldest monthly period) $It = (T + m + M) \times 10$. It is therefore an index that considers the intensity of the cold in the coldest month, a limiting factor for many plants and plant communities, together with the average annual temperature. The correlation between the values of this index and the vegetation is quite satisfactory in warm and temperate climates. In cold climates with It or Itc values of less than 120, and in continental climates with an $Ic > 21$, it is more significant and precise to use the value of the positive annual temperature (Tp).

In the extratropical zones of the world (northwards and southwards of the 23° N & S parallels), the thermicity index (It) can be modified into the compensated thermicity index (Itc) to be applied in the areas where the 'excess' of cold or warmth occurring during the cold season in territories with a highly continental or a highly hyperoceanic tendency, respectively, should be compensated. It is a modification to make such highly continental or highly oceanic territories comparable with the rest. The way of doing this is as follows: if the simple

Table 2.5 Calculation of compensation values according to the value of the continentality index, in order to obtain the compensated thermicity index (Itc)

Ic	fi	Ci	Max. value
$Ic \leq 8$	$f_0 = 10$	$C_l = C_0$; $C_0 = f_0\ (8-Ic)$	$C_0 = -\ 80$
$18 < Ic \leq 21$	$f_1 = 5$	$C_i = C_1$; $C_1 = f_1\ (Ic-18)$	$C_1 = +\ 15$
$21 < Ic \leq 28$	$f_2 = 15$	$C_i = C_1 + C_2$; $C_1 = f_1\ (21-18) = 15$; $C_2 = f_2\ (Ic-21)$	$C_2 = +\ 105$
$28 < Ic \leq 46$	$f_3 = 25$	$C_i = C_1 + C_2 + C_3$; $C_1 = 15$; $C_2 = f_2\ (28–21) = 105$; $C_3 = f_3\ (Ic-28)$	$C_3 = +\ 450$
$46 < Ic \leq 65$	$f_4 = 30$	$C_i = C_1 + C_2 + C_3 + C_4$; $C_1 = 15$; $C_2 = 105$; $C_3 = f_3\ (46–28) = 450$; $C_4 = f_4\ (Ic-46)$	$C_4 = +\ 570$

continentality index (Ic) is between 8 and 18, Itc varies just as Ic (It = Itc). But if the continentality index does not reach or exceed these values, the thermicity value has to be compensated by adding or subtracting a compensation value (Ci). Itc = It ± Ci.

In strongly hyperoceanic extratropical zones (Ic < 8.0), the compensation value (C_0) is calculated by multiplying by ten the difference between 8.0 and the Ic of the location: $C_0 = (8.0-Ic) \times 10$. This value (C_0) is subtracted from the thermicity index: $Itc = It-C_0$.

In continental or semicontinental extratropical climates (Ic > 18.0), a compensation value (Ci) is added to the thermicity index: $Itc = It + C_i$. These compensation values are calculated according to the figure for the simple continentality index (Ic). So, when the continentality is moderate ($18.0 < Ic \leq 21.0$), the compensation value (C_1) is obtained by multiplying by f_1 ($f_1 = 5$) the result of the subtraction between the Ic of the location and 18. When the continentality is higher (Ic > 21.0), the compensation is done by adding values (C_1, C_2, C_3, C_4) which increase proportionally to the increase in the multiplying factor (fi) as a function of the increase in continentality. Therefore: $Itc = It + (C_1 + C_2 + C_3 + C_4)$. The compensation values based on the simple continentality index (Ic) and the progressive corrective factor for continentality (fi) are obtained in the way shown in Table 2.5.

In practice, for the extratropical territories on Earth (>23° N & S) with a continentality index ≥21, as well as for all of the locations with an It or Itc < 120, it is equivalent to use the positive temperature value (Tp) to calculate the thermotype.

2.1.3 Keys for Bioclimatic Classification

For an easy determination of bioclimatic types, simple dichotomous keys have been developed using latitudinal zones, bands and thermic zones (Table 2.6) as well as

Table 2.6 Thermic and latitudinal zones and latitudinal bands on earth

Thermic zones	Latitudinal zones	Latitudinal bands	Latitude
A. WARM	1. EUTROPICAL	1a. equatorial	7° N–7° S
(tropical)	0°–23° N & S	1b. low eutropical	7°–15° N & S
0°–36° N & S		1c. high eutropical	15°–23° N & S
	2. SUBTROPICAL	2a. low subtropical	23°–29° N & S
	23°–36° N & S	2b. high subtropical	29°–36° N & S
B. TEMPERATE	3. EUTEMPERATE	3a. low eutemperate	36°–43° N & S
36°–66° N & S	36°–51° N & S	3b. high eutemperate	43°–51° N & S
	4. SUBTEMPERATE	4a. low subtemperate	51°–59° N & S
	51°–66° N & S	4b. high subtemperate	59°–66° N & S
C. POLAR	5. POLAR	5a. low polar	66°–78° N & S
66°–90° N & S	66°–90° N & S	5b. high polar	78°–90° N & S

bioclimatic parameters and indices. Such keys have been published initially by Rivas-Martínez et al. (2011a, b), and have been updated in March 2016 in the framework of a "Synoptical worldwide bioclimatic classification system", which is summarized in Table 2.8.

Key for Macrobioclimates

1. Intertropical territories between parallels 23° N and S (eutropical latitudinal zone) TROPICAL
1. Territories north or south of these latitudes 2
2. Territories between parallels 23° and 36° N & S (subtropical latitudinal zone) 3
2. Territories to the N & S of parallel 36° (eutemperate, subtemperate, and polar zones) 11
3. Eurasia: latitude 26°–36° N, longitude 70°–120° E, altitude ≥2.000 m 10
3. Does not meet the conditions 4
4. In subtropical territories (parallels 23–36° N & S) at least two of the thermicity parameters or indices must have the following values: average annual temperature $T \geq 25°$, average temperature of the mini-mums of the coldest month $m \geq 10°$, thermicity index It, $Itc \geq 580$ TROPICAL
4. Does not meet the conditions 5
5. In territories with altitudes of less than 200 m (•) at least two of the thermicity parameters or indices must have the following values: average annual temperature $T > 21°$, average temperature of the maximums of the coldest month $M > 18°$, thermicity index, It, $Itc \geq 470$ 6
5. Does not meet the conditions 8
6. Subtropical zone (parallels 23° a 36° N & S): precipitation for the warmest 6-month period of the year should be higher than that of the coldest 6-month period, $Pss > Psw$ TROPICAL
6. Does not meet the conditions 7

7. Summer without hydric deficit: ombrothermic index of the hottest 2 months of the summer quarter $Ios_2 > 2$ or ombrothermic summer index resulting from the compensation Iosc >2	Tropical
7. Does not meet the conditions	9
8. Summer without hydric deficit; ombrothermic index of the hottest consecutive 2 months of the summer quarter $Ios_2 > 2$ or ombrothermic summer index resulting from the compensation Iosc >2	Temperate
8. Does not meet the conditions	Mediterranean
9. Maximum precipitation during the warmest 4 months of the year, Pcm3 < Pcm1 > Pcm2	Tropical
9. Does not meet the conditions	Mediterranean
10. Summer without hydric deficit; ombrothermic index of the hottest 2 months of the summer quarter $Ios_2 > 2$ or ombrothermic summer index resulting from the compensation Iosc >2	Temperate
10. Does not meet the conditions	Mediterranean
11. In territories of northern latitude >72° or southern latitude >56° and altitudes of less than 200 m (•) positive annual temperature Tp < 380.	Polar
11. In territories with altitudes of less than 200 m (•): positive annual temperature Tp > 380	12
12. In territories with altitudes less than 200 m (•), continentality index Ic < 11, positive temperature of the summer quarter Tps >320, average annual temperature <6°, positive annual temperature 320–720	Boreal
12. Does not meet the conditions	13
13. In territories with altitudes of less than 200 m (•). depending on the continentality index Ic, the values for the average annual temperature T and positive annual temperature Tp, must be for: Ic < 21, T < 5.3° and Tp < 720; for: Ic = 21–28, T < 4.8° to nd Tp < 740; for: Ic > 28, T < 3.8° and Tp < 800	Boreal
13. Does not meet the conditions	14
14. Summer without hydric deficit; ombrothermic index of the hottest 2 months of the summer quarter $Ios_2 > 2$ or ombrothermic summer index resulting from the compensation $Iosc_4 > 2$	Temperate
14. Does not meet the conditions	Mediterranean

(•) If the locality has an altitude over 200 m, the temperature values must be assessed: increase of T by 0.6°, and of M by 0.5°, and It or Itc by 13 units for every 100 m in excess of this altitude. If it is situated to the north of parallel 48° N or south of 51° S, the values of the average annual temperature and positive annual temperature Tp must be assessed, increasing T by 0.4° and Tp by 12 units for each

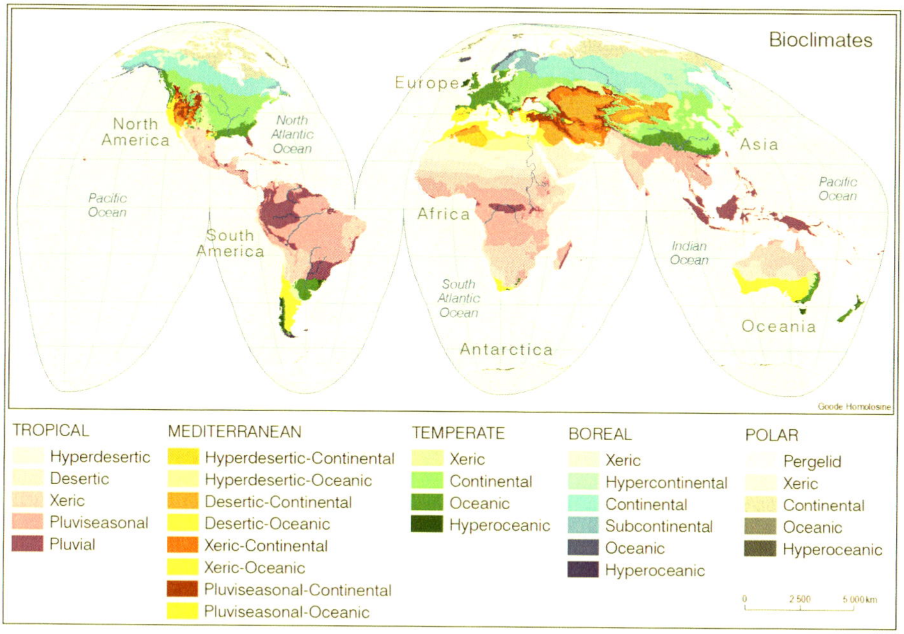

Fig. 2.2 Bioclimates map of the Earth (From Rivas-Martínez et al. 2011a)

Table 2.7 Distribution of the bioclimatic variants in the macrobioclimates of the earth

Bioclimatic variants	Tr	Me	Te	Bo	Po
Steppic (stp) (*)	–	•	•	•	•
Submediterranean(sbm) (**)	–	–	•	•	•

Tr tropical, *Me* Mediterranean, *Te* temperate, *Bo* Boreal, *Po* polar

100 m in excess this altitude. When Ic $\geq$ 21 (continental) or when It or Itc < 120 the thermotype must be calculated based on the positive annual temperature (Tp), and the values of Tp at altitudes over 200 m are modified by increasing by 55 units for each 100 m in increase from this altitude (Fig. 2.2).

2.1.4 *Bioclimatic Variants*

The bioclimatic variants are very informative and interesting typological units which can be recognised within the corresponding specific bioclimates, enabling to identify climatic peculiarities of ombric and occasionally thermic nature. Out of the eight bioclimatic variants distinguished for the entire wold, in the Iberian Peninsula there can be recognized two: the **steppic** and the **submediterranean** (Table 2.7).

(*) *Steppic* (stp). This bioclimatic variant is found in the mediterranean, temperate, boreal and polar macrobioclimates, having at least a low semicontinental character ($Ic > 17$). 'Steppic' means that the rainfall in the winter quarter is lower than in the summer quartrer [$Ps \geq Pw$]. In consequence, the annual ombrothermic index ranges between lower hyperarid and upper subhumid: [$6.0 \geq Io > 0.2$] and at least during 1 month in the summer (Ps_1), the rainfall in mm is lower than threefold the temperature in degrees Celsius [Ps_i: $P < 3T$].

The steppic nature can be recognized in very different continental or semicontinental plant communities and formations, due to the appearance of xerophytic types of vegetation and the fragility of the forests as a result of the limited amount of water available in periods associated with both solstices. The most characteristic plant formations of the Earth that correspond to this bioclimatic variable are the steppes and steppic forests of Eurasia, the large wooded or unwooded prairies of North America, the steppic deserts of Central Asia, and the Mediterranean steppic xerophytic holarctic forests, shrubland and grasslands. The steppic tundra and taiga formations found in the boreal and polar macrobioclimates are restricted to areas with low summer and winter rainfall in Asia and North America. In summary, it can be said that the steppic bioclimatic variant mainly corresponds to continental or semicontinental bioclimatic types with low summer mediterraneity and with low rainfall also during the winter solstice.

(**) *Submediterranean* (sbm). This bioclimatic variant is common in the temperate macrobioclimate, and scarce in the boreal and polar ones. In it at least in 1 month during the summer, the average of rainfall in millimetres is 2.8. times lower than the average temperature in degrees Celsius [Ios_i: $P < 2.8T$].

The most characteristic temperate submediterranean plant formations are those representing a transition or ecotone between the temperate bioclimates without summer aridity, and those which are genuinely mediterranean, where the summer drought ($P < 2T$) occurs during 2 months or more. In the Holarctic kingdom, the most representative plant formations are usually those whose mature stage consists of sclerophyllous or deciduous marcescent dry forests, as well as xerophytic conifer forests.

2.1.5 Bioclimatic Belts

The bioclimatic belts are types of climatic conditions (climatic envelopes) that occur in areas following an altitudinal or latitudinal clinosequence. They are defined according to the thermoclimatic and ombroclimatic factors (It, Itc, Tp, Io). Each bioclimatic belt has specific plant formations and communities: the vegetation belts, which are an expression of the universal phenomenon of zonation The thresholds between the ombroclimatic categories (defined by means of Io) are

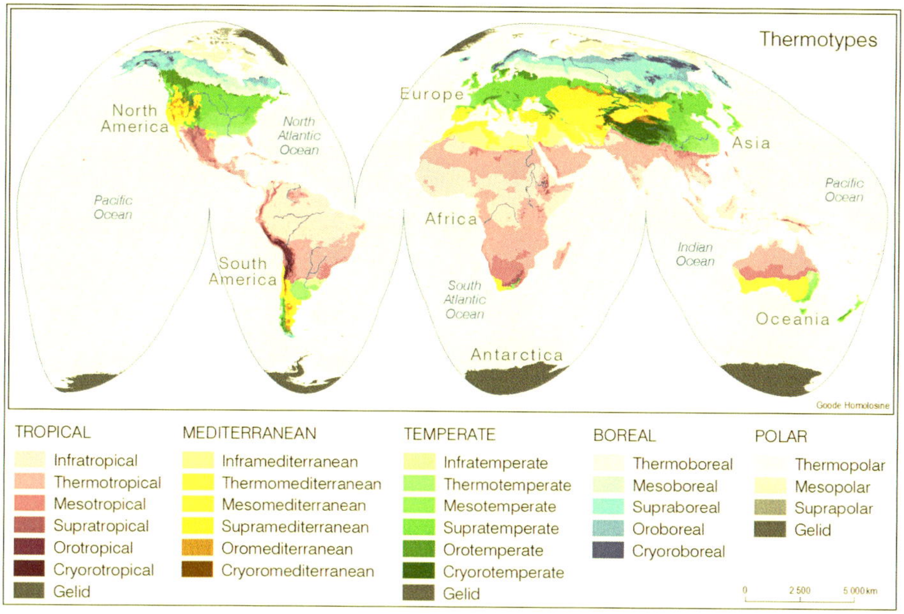

Fig. 2.3 Thermotypes map of the Earth (From Rivas-Martínez et al. 2011a)

the same all over the world, the thermoclimatic categories (defined after It, Itc and Tp) differ slightly across the majority of macrobioclimates.

Thermotypes The thermoclimatic types or thermotypes are calculated by using the thermic parameters described (It, Itc, Tp). For the sake of global convenience, due to their thermoclimatic and vegetational peculiarities, a particular altitudinal or latitudinal sequence of thermotypes is recognised for each of the Earth's macrobioclimates: Tropical (infra-, thermo-, meso-, supra-, oro-, cryoro- and gelid); Mediterranean (infra-, thermo-, meso-, supra-, oro-, cryoro- and gelid); Temperate (infra-, thermo, meso-, supra-, oro-, cryoro- and gelid) – which in this case, correspond respectively to the classic altitudinal belts and horizons: infracoline, thermocoline, eucoline, submontane, montane, high montane, subalpine, alpine, subnival, nival and gelid-; Boreal (thermo-, meso-, supra-, oro-, cryoro- and gelid); Polar (thermo-, meso-, supra- and gelid). In order to achieve closer concordance with the variability in vegetation patterns, at times it is necessary to distinguish within the bioclimatic belts between an upper and lower half of their thermic and ombric intervals, to which we refer to as thermotypic and ombrotypic bioclimatic horizons. If it were necessary to create an even more precise subdivision, these intervals could be further divided into upper and lower levels (Fig. 2.3).

Ombrotypes These types try to express the water availability conditions of the climate and are defined by means of a quotient between the rainfall values and the evaporation values, which are fundamentally related to the temperature. Thus their

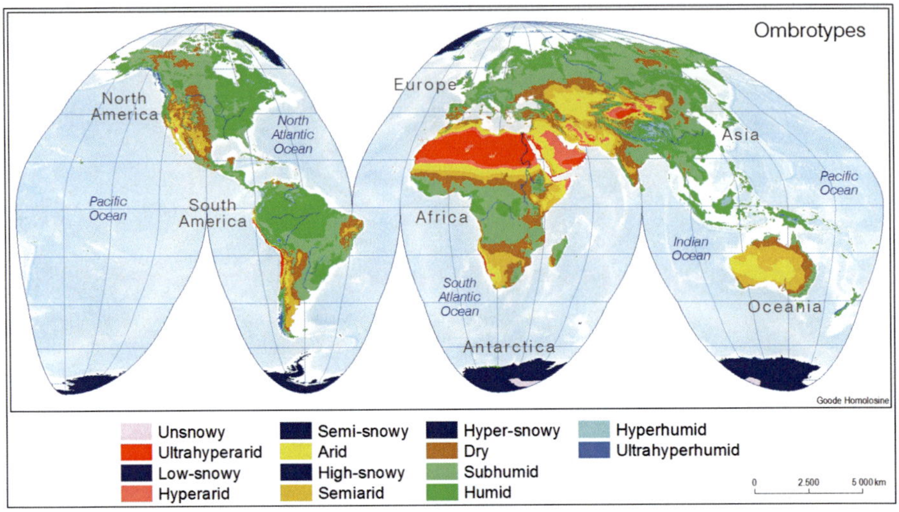

Fig. 2.4 Ombrotypes map of the Earth (From Rivas-Martínez et al. 2011a)

definition is based on precipitation/temperature-quotients, specifically in those between the average rainfall in millimetres and the temperature sum in degrees Celsius of the periods of the year whose average monthly temperatures are higher than zero degrees Celsius. Amongst others, these quotients include the annual ombrothermic index (Io = (Pp/Tp) 10), the quarterly ombrothermic index (Iotr = (Pptri/Tptri) 10) and the monthly ombrothermic index (Iom = (Pi/Tpi) 10). Due to the high predictive value and correlation between the annual ombrothermic values and the structures of the potential climatophillous vegetation all over the world, we have used them for more than a decade in our bioclimatic classification in order to establish the ombric types and horizons, replacing the scales with the simple average rainfall values (Fig. 2.4, Table 2.8).

2.2 Bioclimatic Units of the Iberian Peninsula and Balearic Islands

According to the proposal of Rivas-Martínez et al. (2011) two Macrobioclimates, Temperate and Mediterranean, are recognized in the Iberian Peninsula and Balearic islands (Fig. 2.5).

2.2.1 *Temperate Macrobioclimate*

The Temperate macrobioclimate applies to all the extratropical territories on Earth, at any altitude and continentality value between the latitudes 23° to 66° N and 23° to 55° S, in which there is not a period of two or more consecutive months with aridity during the summer or the hottest season of the year, or if so, they are not compensated (see compensationable summer ombrothermic indices). This means that the value in millimetres of the average rainfall for the hottest 2-months period of the summer quarter is more than twice the average temperature in degrees Celsius of the hottest 2-months period in the summer quarter ($Ps_2 > 2Ts_2$).

The Temperate macrobioclimate (small territory is indicated by (*)] is found in the following biogeographical regions (by continents): **Africa**; Capensis (*); **Australia and Polynesia**: Australian Temperate and Neozelandic; **Eurasia**: Sino-Japanese, Eurosiberian, East Siberian (*), and Tibetan-Himalayic; **North America**: Atlantic North American and Rocky Mountains; **South America**: Pampean and Valdivian-Magellanian. Four bioclimates are recognised within the temperate macrobioclimate, depending on the continentality and ombrothermic index: Temperate hyperoceanic (teho), Temperate oceanic (teoc), Temperate continental (teco) and Temperate xeric (texe). The Temperate macrobioclimate represents 20.35% of the total area of the Iberian Peninsula and Balearic Islands. Its borders are the Cantabrian Sea to the north, the whole of the Galician coast to the vicinity of Porto to the west, the Solsonés, Plana de Vic, Conflent, Montseny-Guilleries regions are its limit in the east and in the south the beginning of the first foothills of the Pyrenean and Cantabrian ranges and their continuity into the Galician-Leonese Mountains, continuing towards the coast crossing inner Portugal across the Montemuro and Caramulo ranges and the northwestern Serra da Estrêla, as it can be seen in the map of Fig. 2.5. We can also see interior bioclimatic temperate islands, in the high mountains of the Teleno and Aquilianos mountains, Queixa and Cabrera ranges, Iberic and Central ranges as well as in the Gúdar and Javalambre massifs. The rest of the Iberian Peninsula has a Mediterranean macrobioclimate. The Temperate macrobioclimate is diversified into three bioclimates, hyperoceanic (12.60%), oceanic (86.63%), xeric (0.77%) (Fig. 2.5).

2.2.2 *Mediterranean Macrobioclimate*

The Mediterranean macrobioclimate applies to all extratropical regions on Earth at any altitude and with any continentality value between the latitudes 23° and 52° N & S, in which there are at least two consecutive months with aridity during the warmest period of the year This means that the value in millimetres of the average rainfall of the hottest 2 months of the summer quarter is lower than twice the average temperature of the hottest 2 months of the summer quarter expressed in degrees Celsius ($Ps_2 < 2Ts_2$).

Table 2.8 Synoptical table of worlwide bioclimatic classification (III.2016, updated)

Macrobioclimates[a]	Bioclimates[e]	Abbr.	Bioclimatic ranges				Thermoclimatic types			Abbr.	Ombroclimatic types		Sigla
Tropical			Ic	Io	Iod2			It (Itc)	Tp[b]			Io	
Zone warm: equatorial, eutropical and subtropical (0°–36° N & S). In subtropical (23°36° N & S) at <200 m two values: $T \geq 25°$, $m \geq 10°$, $Itc \geq 580$. If $Pcm_2 < Pcm_1 > Pcm_3$ y $Pss > Psw$, two values: $T \geq 21°$, $M \geq 18°$, It ≥470. Eurasia and Africa: 25° to 36° N > 2000 m is not tropical.	**Tr. Pluvial**	trpl	–	≥3.6	>2.5	–	1. Infratropical	710–890	>2900	itr	1. Ultrahyperarid	<0.2	uha
	Tr. Pluviseasonal	trps	–	≥3.6	≤2.5	–	2. Thermotropical	490–710	>2300	ttr	2. Hyperarid	0.2–0.4	har
	Tr. Xeric	trxe	–	1.0–3.6	–	–	3. Mesotropical	320–490	>1700	mtr	3. Arid	0.4–1.0	ari
	Tr. Desertic	trde	–	0.2–1.0	–	–	4. Supratropical	160–320	> 950	str	4. Semiarid	1.0–2.0	sar
	Tr. Hyperdesertic	trhd	–	<0.2	–	–	5. Orotropical	<160	450–950	otr	5. Dry	2.0–3.6	dry
							6. Cryorotropical	–	1–450	ctr	6. Subhumid	3.6–6.0	shu
							7. Gelid[c]	–	0	gtr	7. Humid	6.0–12.0	hum
											8. Hyperhumid	12.0–24.0	hhu
											9. Ultrahyperhumid	≥24.0	uhh
Mediterranean			Ic	Io				It (Itc)	Tp[b]			Io	
Zone warm: subtropical and temperate eutemperate (23°–52° N & S), with aridity $P < 2\,T$, at least 2 months in summer: $Ios_2 \leq 2$, $Iosc_4 \leq 2$. In subtropical (23°–35° N & S) at least two values: $T < 25°$, $m < 10°$, $Itc < 580$.	**Me. Pluviseasonal Oceanic**	mepo	≤21	> 2.0	–	–	1. Inframediterranean	450–580	>2400	ime	1. Ultrahyperarid	<0.2	uha
	Me. Pluviseasonal Continental	mepc	>21	> 2.0	–	–	2. Thermomediterranean	350–450	>2100	tme	2. Hyperarid	0.2–0.4	har
	Me. Xeric Oceanic	mexo	≤21	1.0–2.0	–	–	3. Mesomediterranean	220–350	>1500	mme	3. Arido	0.4–1.0	ari
	Me. Xeric Continental	mexc	>21	1.0–2.0	–	–	4. Supramediterranean	<220	>900	sme	4. Semiarid	1.0–2.0	sar
	Me. Desertic Oceanic	medo	≤21	0.2–1.0	–	–	5. Oromediterranean	–	450–900	ome	5. Dry	2.0–3.6	dry
	Me. Desertic Continental	medc	>21	0.2–1.0	–	–	6. Cryoromediterranean	–	1–450	cme	6. Subhumid	3.6–6.0	shu
	Me. Hyperdesertic Oceanic	meho	≤21	< 0.2	–	–	7. Gelid[c]	–	0	gme	7. Humid	6.0–12.0	hum
	Me. Hyperdesertic Continental	mehc	>21	< 0.2	–	–					8. Hyperhumid	12.0–24.0	hhu
											9 Ultrahyperhumid	≥ 24.0	uhh
Temperate			Ic	Io				It (Itc)	Tp (b and d)			Io	
Zone warm: subtropical and temperate (23°–66° N & 23°–54° S). From 23°–35° N & S, at <200 m, at least two values: $T < 21°$, $M < 18°$, $Itc < 470$. $Ios_2 > 2$, $Iosc_4 > 2$.	**Te. Hyperoceanic**	teho	≤11	>3.6	–	–	1. Infratemperate	>410	>2350	ite	4. Semiarid	<2.0	sar
	Te. Oceanic	teoc	11–21	>3.6	–	–	2. Thermotemperate	290–410	>2000	tte	5. Dry	2.0–3.6	dry
	Te. Continental	teco	>21	>3.6	–	–	3. Mesotemperate	190–290	>1400	mte	6. Subhumid	3.6–6.0	shu
	Te. Xeric	texe	≥4	≤3.6	–	–	4. Supratemperate[d]	<190	>800	ste	7. Humid	6.0–12.0	hum
							5. Orotemperate[d]	–	380–800	ote	8. Hyperhumid	12.0–24.0	hhu
							6. Cryorotemperate	–	1–380	cte	9. Ultrahyperhumid	≥ 24.0	uhh
							7. Gelid[c]	–	0	gte			

Boreal			Ic	Io	Tp	T			Tp			Io	
Zones temperate and cold (42°–72° N, 49°–56° S). At <200 m: Ic ≤ 11:T ≤ 6°, Tp = 380–720; Tps >320; Ic = 11–21: T ≤ 5.3°, Tp = 380–720; Ic = 21–28: T ≤ 4.8°, Tp = 380–740; Ic = 28–45: T ≤ 4.3°, Tp = 380–800; Ic ≥ 45: T ≤ 0°, Tp = 380–800.	**Bo. Hyperoceanic**	boho	≤11	>3.6	≤720	<6.0°	1. Thermoboreal	–	>680	tbo	4. Semiarid	<2.0	sar
	Bo. Oceanic	booc	11–21	>3.6	≤720	≤5.3°	2. Mesoboreal	–	580–680	mbo	5. Dry	2.0–3.6	sec
	Bo. Subcontinental	bosc	21–28	>3.6	≤740	≤4.8°	3. Supraboreal	–	480–580	sbo	6. Subhumid	3.6–6.0	shu
	Bo. Continental	boco	28–46	>3.6	≤800	≤3.8°	4. Oroboreal	–	380–480	obo	7. Humid	6.0–12.0	hum
	Bo. Hypercontinental	bohc	>46	–	≤800	≤0.0°	5. Cryoroboreal	–	1–380	cbo	8. Hyperhumid	12.0–24.0	hhu
	Bo. Xeric	boxe	<46	≤3.6	≤800	≤3.8°	6. Gelid[c]	–	0	gbo	9. Ultrahyperhumid	≥ 24.0	uhh
Polar			Ic	Io	Tp				Tp			Io	
Zones temperate and cold (51°–90° N & S). At <100 m: Tp < 380.	**Po. Hyperoceanic**	poho	≤11	>3.6	>0	–	1. Infrapolar[f]		380–600	ipo	4. Semiarid	<2.0	sar
	Po. Oceanic	pooc	11–21	>3.6	>0	–	2. Thermopolar	–	280–380	tpo	5. Dry	2.0–3.6	dry
	Po. Continental	poco	>21	>3.6	>0	–	3. Mesopolar	–	100–280	mpo	6. Subhumid	3.6–6.0	shu
	Po. Xeric	poxe	≥4	≤3.6	>0	–	4. Suprapolar[c]	–	1–100	spo	7. Humid	6.0–12.0	hum
	Po. Pergelid	pope	–	–	−0	–	5. Gelid[c, g]	–	0	gpo	8. Hyperhumid	12.0–24.0	hhu
											9. Ultrahyperhumid	≥24.0	uhh

[a]North and south of equatorial and eutropical latitudinal belt (23° N & 23° S), if the locality is at 200 m altitude or higher, the thermical values at this altitude must be calculated, increasing T by 0.6°, M by 0.5°, and It or Itc by 13 units, for every 100 m higher than 200 m. But if the loaclity is north of 48° N or south of 51° S, the increases are T by 0.4° and Tp by 12 units, for every 100 m higher than 200 m. [b]If Ic ≥ 21 (continental) or It or Itc < 120 the thermotype must by calculated through Tp values, and the theoric values of Tp at 200 m increase by 55 units for every 100 m exceding that altitude. [c]In the pergelid bioclimate (polar), the upper suprapolar and the gelid thermotype we recognize the following ombrotypes (chionotypes): un-snowy (<50 mm), scanty-snowy (50–200 mm), low-snowy (200–400 mm), medium-snowy (400–600 mm), high-snowy (600–1000 mm), super-snowy (1000–2000 mm), ultra-snowy (>2000 mm). [d]The hemiboreal thermotypes (hbo) are used inside the temperate macrobioclimate, north of 45° N and south of 49° S, with the following values: Ic < 21, altitude <400 m, tp 720–900; Ic 21–28, altitude <600 m, tp 740–900; Ic > 28, alt. < 1000 m, tp 800–900. [e]Bioclimatic variants: steppic (stp), submediterranean (sbm), bixeric (bix), antitropical (ant), seropluvial (spl), polar semiboreal (posebo), semipolar subantarctic (seposa), tropical semimediterranean desertic (trsmd), polar euhyperoceanic (poeuhy), boreal subantarctic (bosuba) and temperte subantartic (tesuba). [f]The infrapolar thermotype (Tp: 380–600) only corresponds to the semipolar antarctic insular bioclimatic variant on the coast [values. Alt < 100 m, T < 7.5°, Tp < 600, Tps < 280, Tmax < 10°, Ic < 8, Io > 10]; the polar semiboreal bioclimatic variant corresponds to semicontinental-hyperoceanic hyperhumid oroboreal treeless arctic tundra territories [values: Tp 380–480, Ic < 28, Tmax ≤11°, Tps ≤ 320]. [g]In the polar pergelid bioclimate we identify three gelid thermotypes in the Antarctic Region and two in the Circumarctic Region: hypogelid (T – 10° to –25°, Tp 0, Tpmax < 30, Tpamax < 500, Twmax < –7°); hypergelid (T – 25° to –45°, Tp 0, Tpmax 0, Tpamax 300, Twmax –7° to –22°); ultragelid (T < –45°, Tp 0, Tpmax 0, Tpamax 0, Twmax < –22°) (only in Antarctica)

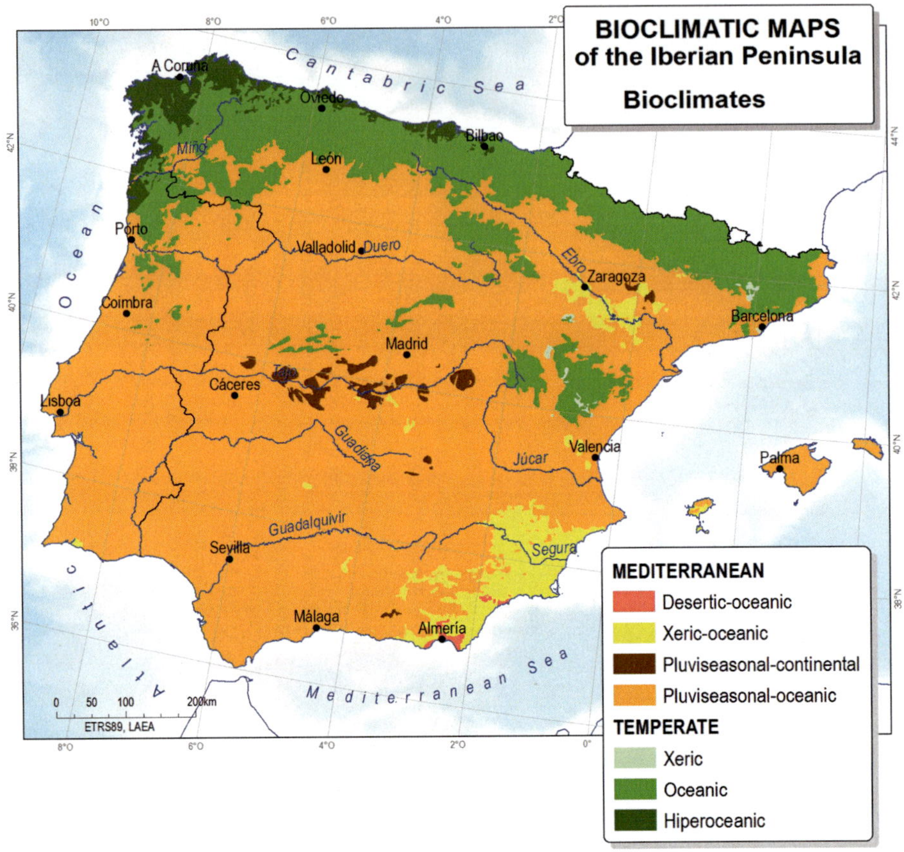

Fig. 2.5 Macrobioclimates and bioclimates in the Iberian Peninsula and Baleric Islands

The Mediterranean macrobioclimate is mainly represented in the central and western part of all of the continents, except Antarctica. The Mediterranean macrobioclimate is found in the nine following biogeographical regions (by continents); **Africa**: Capensis, Mediterranean and Saharan-North Arabian; **Australia and Polynesia**: Mediterranean Australian; **Eurasia**: Irano-Turanian, Saharan-North Arabian and Mediterranean; **North America**: Californian, Great Basin and Rocky Mountains; **South America**: Middle Chilean-Patagonian. In the Mediterranean macrobioclimate, depending on the continentality and ombrothermic index, we can distinguish eight bioclimates: Mediterranean pluviseasonal oceanic (mepo), Mediterranean pluviseasonal continental (mepc), Mediterranean xeric oceanic (mexo), Mediterranean xeric continental (mexc), Mediterranean desertic oceanic (medo), Mediterranean desertic continental (medc), Mediterranean hyperdesertic oceanic (meho) and Mediterranean hyperdesertic continental (mehc).

This macrobioclimate represents 79.65% of the total area of the Iberian Peninsuls and the entire Balearic Islands. Five bioclimates are recognized in this macrobioclimate: mediterranean pluviseasonal oceanic (74.19%), mediterraean pluviseasonal continental (1.06%), mediterranean xeric oceanic (4.22%), mediterranean xeric continental (0.01%) and mediterranean desertic oceanic (0.17%). (Fig. 2.5).

2.2.3 *Bioclimatic Variants*

Two bioclimatic variants are recognized in the Iberian Peninsula and Balearic Islands: Submediterranean and Steppic (Figs. 2.6, 2.7 and 2.8).

The submediterranean variant represents 65.64% of the total area of the Temperate macrobioclimate (Temperate hyperocanic 14.83%; Temperate oceanic 84.65%; Temperate xeric 0.52%).

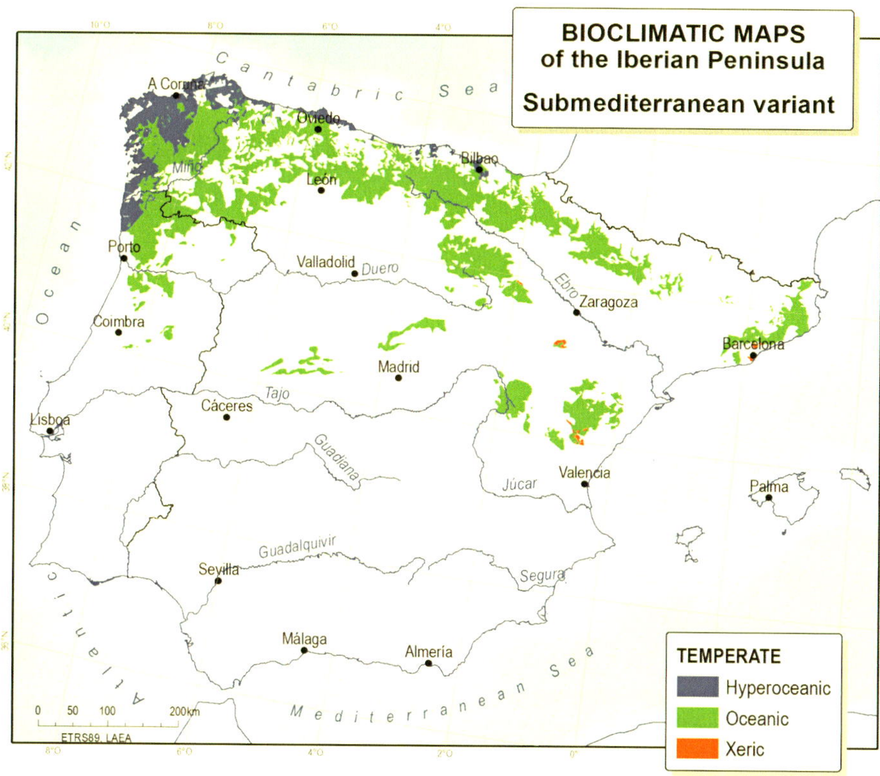

Fig. 2.6 Submediterranean variant in the Iberian Peninsula and Baleric Islands

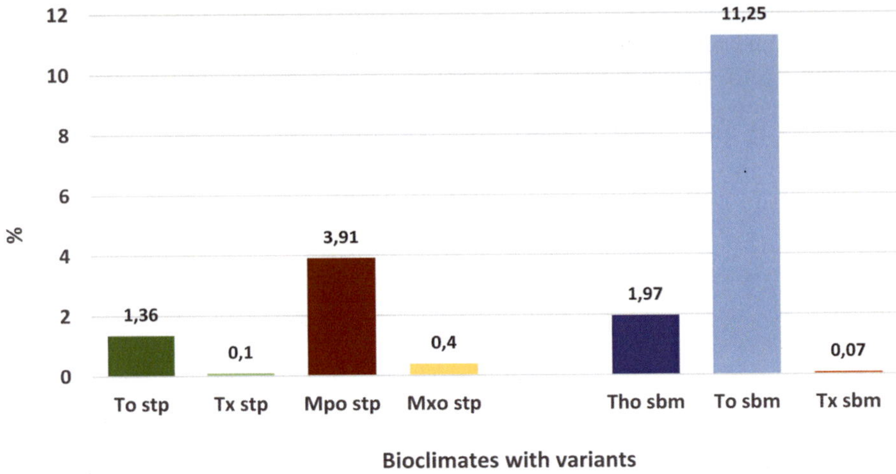

Fig. 2.7 Proportional area covered by the steppic and submediterranean bioclimatic variants compared to the total area of the Iberian Peninsula and Balearic Islands. *Tho* temperate hyperoceanic, *To* temperate oceanic, *Tx* temperate xeric, *Mpo* Mediterranean pluviseasonal oceanic, *Mxo* Mediterranean xeric oceanic, *stp* steppic, *sbm* submediterranean

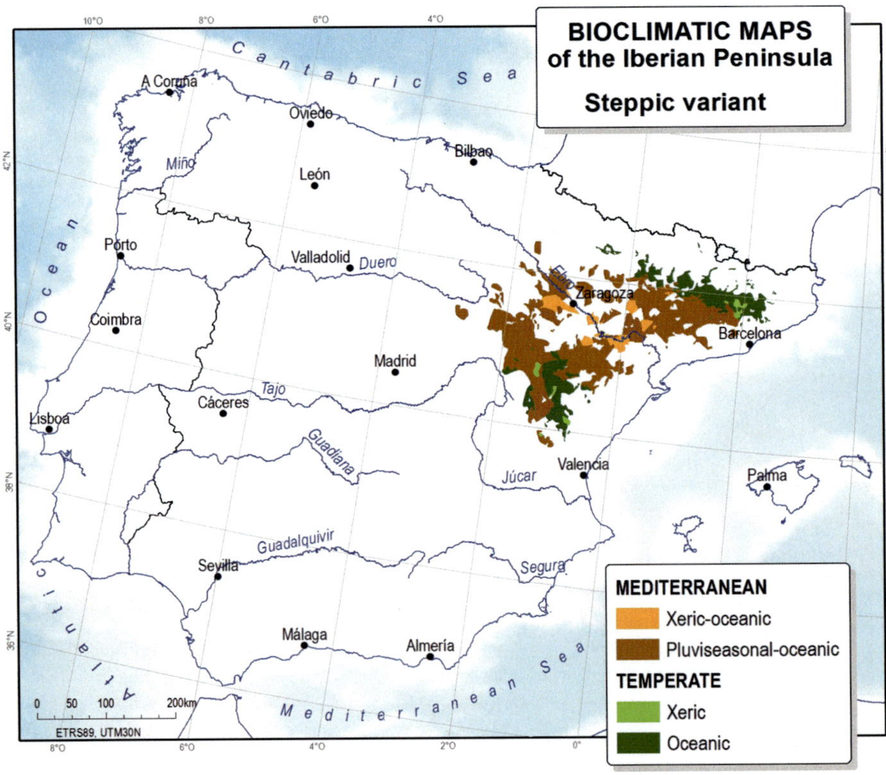

Fig. 2.8 Steppic variant in the Iberian Peninsula and Baleric Islands

The steppic variant represents 7.20% of the total area of the Temperate macrobioclimate (Temperate oceanic 93.14%; Temperate xeric 6.86%) and 5.41% of the total area of the Mediterranean macrobioclimate (Mediterranean pluviseasonal oceanic 90.77%; Mediterranean xeric oceanic 9.23%).

2.2.4 *Thermotypes*

In the Iberian Peninsula and Balearic Islands the following temperate thermotypes are recognized: thermo- (1.38%), meso- (7.29%), supra- (9.91%), oro- (1.26%) and cryorotemperate (0.03%). (Table 2.9, Fig. 2.9).

In the Mediterranean macrobioclimate in the Iberian Peninsula and Balearic Islands the following thermotypes are recognized: infra- (0.17%), thermo- (11.34%), meso- (46.19%), supra- (22.17%), oro- (0,.2%) and cryoromediterranean (0.03%). (Table 2.10, Fig. 2.9).

2.2.5 *Ombrotypes*

The ombrotypes recognized in the Iberian Peninsula and Baleric Islands are: arid (0.19%), semiarid (4.16%), dry (44.62%), subhumid (28.39%), humid (17.92%), hyperhumid (4.55%) and ultrahyperhumid (0.17%). The ombrotypes of the Temperate macrobioclimate are shown in Figs. 2.10, 2.11 and Table 2.11.

The ombrotypes of the Mediterranean macrobioclimate are shown in Figs. 2.10 and 2.11, and in Table 2.12.

Table 2.9 Bioclimates and thermotypes in the temperate macrobioclimate

Bioclimates/thermotypes temperate	tte	mte	ste	ote	cte
Temperate hyperoceanic	•	•	•	–	–
Temperate hyperoceanic submediterranean	•	•	•	–	–
Temperate oceanic	•	•	•	•	•
Temperate oceanic submediterranean	•	•	•	•	–
Temperate oceanic steppic	–	•	•	–	–
Temperate xeric submediterranean	–	•	•	–	–
Temperate xeric steppic	–	–	•	–	–

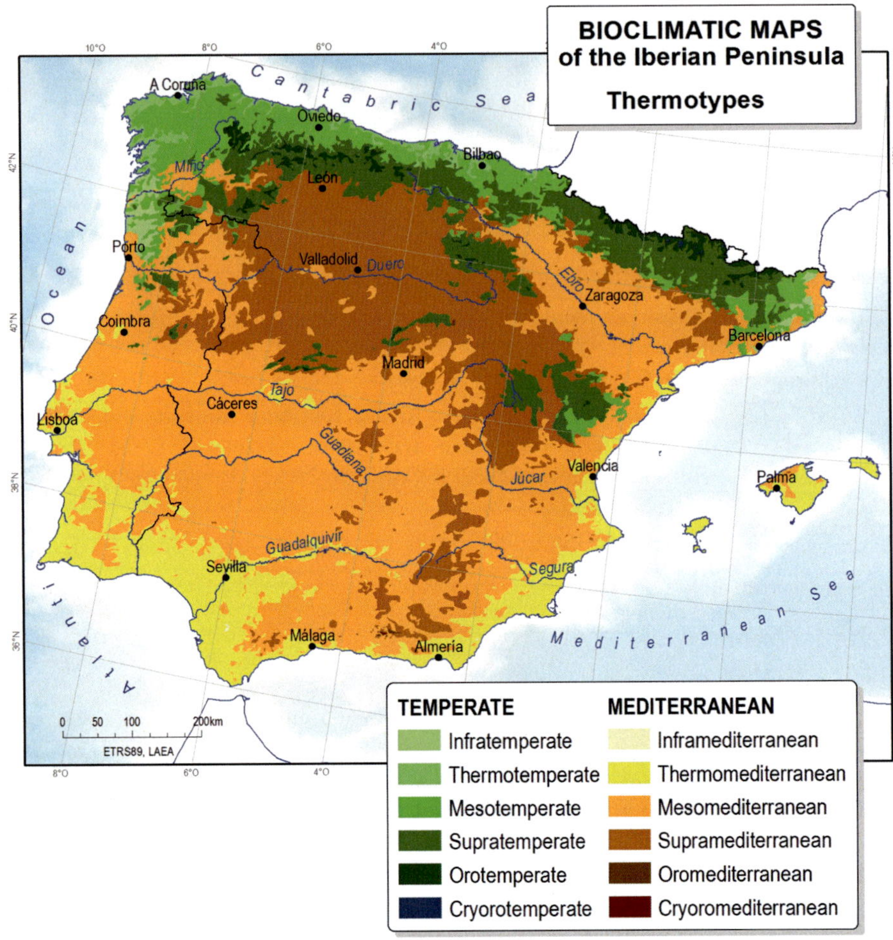

Fig. 2.9 Thermotypes in the Iberian Peninsula and Baleric Islands

Table 2.10 Bioclimates and thermotypes in the Mediterranean macrobioclimate

Bioclimates/thermotypes	ime	tme	mme	sme	ome	cme
Mediterraean pluviseasonal oceanic	•	•	•	•	•	•
Mediterraean pluviseasonal oceanic steppic	–	–	•	•	–	–
Mediterraean pluviseasonal continental	–	•	•	•	•	•
Mediterraean pluviseasonal continental steppic	–	–	•	–	–	–
Mediterraean xeric oceanic	•	•	•	•	–	–
Mediterraean xeric oceanic steppic	–	–	•	•	–	–
Mediterraean xeric continental	–	–	•	–	–	–
Mediterraean desertic oceanic	•	•	–	–	–	–

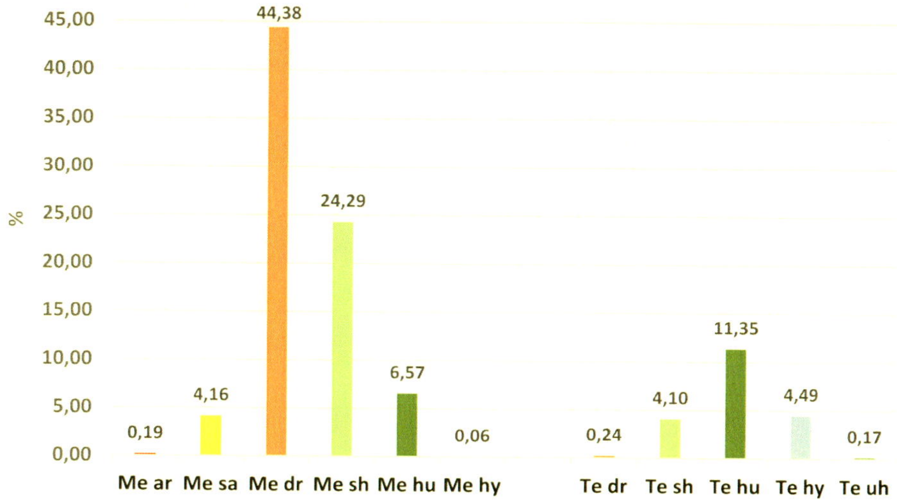

Fig. 2.10 Proportional area covered by the ombrotypes in the temperate and Mediterranean macrobioclimates compared to the total area of the Iberian Peninsula and Balearic Islands. *Me* Mediterranean, *Te* temperate, *ar* arid, *sa* semiarid, *dr* dry, *sh* subhumid, *hu* humid, *hy* hyperhumid, *uh* ultrahyperhumid

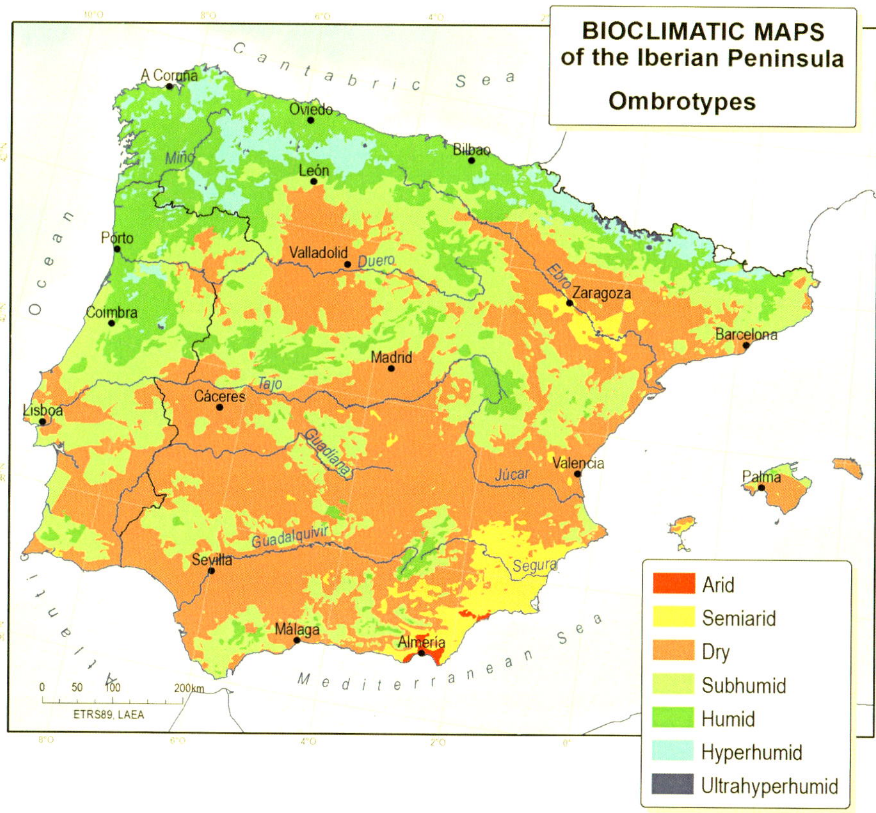

Fig. 2.11 Ombrotypes in the Iberian Peninsula and Baleric Islands

Table 2.11 Bioclimates and ombrotypes in the temperate macrobioclimate

Bioclimates/ombrotypes	dry	suh	hum	hhu	uhu
Temperate hyperoceanic	–	•	•	•	–
Temperate hyperoceanic submediterranean	–	•	•	•	–
Temperate oceanic	–	•	•	•	•
Temperate oceanic submediterranean	•	•	•	•	–
Temperate oceanic steppic	–	•	–	–	–
Temperate xeric submediterranean	•	–	–	–	–
Temperate xeric steppic	•	–	–	–	–

Table 2.12 Bioclimates and ombrotypes in the Mediterraean macrobioclimate

Bioclimates/ombrotypes	ari	sar	dry	suh	hum	hhu
Mediterraean pluviseasonal-oceanic	•	•	•	•	•	•
Mediterraean pluviseasonal-oceanic steppic	–	•	•	•	–	–
Mediterraean pluviseasonal-continental	–	–	•	•	•	–
Mediterraean pluviseasonal-continental steppic	–	–	•	–	–	–
Mediterraean xeric-oceanic	•	•	•	•	–	–
Mediterraean xeric-oceanic steppic	–	•	•	–	–	–
Mediterraean xeric-continental	–	•	–	–	–	–
Mediterraean desertic-oceanic	•	•	–	–	–	–

2.2.6 Continentality

The Continentality in the Iberian Peninsula covers the range from hyperoceanic to subcontinental subtypes (Table 2.13, Fig. 2.12).

2.3 Relationships Between the Potential Natural Vegetation and Bioclimatology

The relationships between the potential natural vegetation and bioclimatology are shown in this section. To that effect, tables including the vegetation complex units or sigmataxa: sigmeta (S.), minorisigmeta (Ms), permasigmeta (Ps), geosigmeta(Gs) and geopermasigmeta (Gps) existing in each bioclimate and bioclimatic variant

Table 2.13 Continentaly types in the Iberian Peninsula and Balearic Islands

Types	Subtypes	Values	Km^2	%
Hyperoceanic	Euhyperoceanic	4.0–8.0	508.35	0.09
	Subhyperoceanic	8.0–11.0	22,805.82	3.88
Oceanic	Semihyperoceanic	11.0–14.0	73,923.78	12.58
	Euoceanic	14.0–17.0	186,35.30	31.77
	Semicontinental	17.0–21.0	297,426.68	50.60
Continental	Subcontinental	21.0–28.0	6415.02	1.09

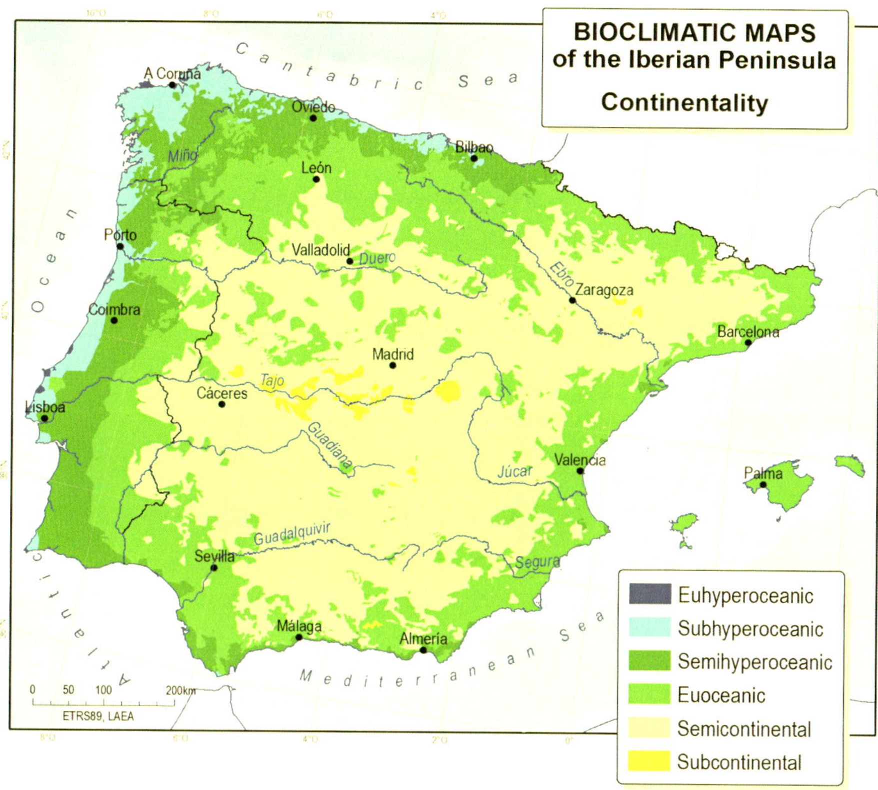

Fig. 2.12 Continentality in the Iberian Peninsula and Baleric Islands

(submediterranean or steppic) with representation in the Iberian Peninsula are shown. The vegetation series and geoseries (sigmeta and geosigmeta) have been split into the following groups in each bioclimatic unit: climatophilous, edaphoxerophilous, climato-temporihygrophilous and hygrophilous. Geopermasigmeta, permasigmeta and minorisigmeta have been independently considered. Some features (preferably soil characteristics) of each sigmataxon well as the thermotypes and ombrotypes in which they can grow also have been indicated, in brackets, according to the Rivas-Martínez et al. (2011) approach.

Appendices

A. Temperate Macrobioclimate

A.1. Potential Natural Vegetation Units (Sigmataxa) in the Temperate Hyperoceanic Bioclimate

Temperate hyperoceanic	Thermotypes			Ombrotypes		
Sigmataxa	1	2	3	4	5	6
Climatophilous sigmeta						
Blechno spicant-Querco roboris S. (acidophilous)	•	•	•	–	•	•
Hyperico pulchri-Querco roboris S. (acidophilous)	•	•	•	–	•	•
Saxifrago spathularis-Fago sylvaticae S. (acidophilous)	–	•	•	–	–	•
Saxifrago hirsutae-Fago sylvaticae S. (neutro-acidophilous)	–	•	•	–	–	•
Climato-temporihygrophilous sigmeta						
Polysticho setiferi-Fraxino excelsioris S. (neutro-acidophilous)	•	•	–	•	•	–
Hygrophilous geosigmeta						
Hyperico androsaemi-Ulmo glabrae Gs. (neutro-acidophilous)	–	•	–	–	•	•
Permasigmeta						
Spergulario rupicolae-Armerio depilatae Ps. (haloanemogenous rock littoral)	•	–	–	•	•	–
Otantho maritimi-Ammophilo australis Ps. (coastal dune)	•	–	–	•	•	

1 = thermotemperate, 2 = mesotemperate, 3 = supratemperate, 4 = subhumid, 5 = humid, 6 = hyperhumid

A.2. Potential Natural Vegetation Temperate Units (Sigmataxa) in the Hyperoceanic Bioclimate Submediterranean Variant

Temperate hyperoceanic submediterranean	Thermotypes			Ombrotypes		
Sigmataxa	1	2	3	4	5	6
Climatophilous sigmeta						
Viburno tini-Querco roboris S. (acidophilous)	•	–	–	–	•	•
Rusco aculeati-Querco roboris S. (acidophilous)	•	•	•	–	•	•
Edaphoxerophilous sigmeta						
Lauro nobilis-Querco ilicis S. (relict)	•	•	–	•	•	–
Climato-temporihygrophilous sigmeta						
Polysticho setiferi-Fraxino excelsioris S. (neutro-acidophilous)	•	•		•	•	–
Hyperico androsaemi-Querco roboris S. (acidophilous)		•	•	•	•	–
Hygrophilous geosigmeta						
Carici lusitanicae-Alno glutinosae Gs. (dystrophic lentic freshwater)	•	•	–	•	•	–
Senecioni bayonensis-Alno glutinosae Gs. (soft freshwater)	•	•	–	•	•	–
Hyperico androsaemi-Alno glutinosae Gs. (hard freshwater)	•	•	–	•	•	•
Carici reuterianae-Betulo celtibericae Gs. (soft freshwater)	–	•	•	•	•	•

(continued)

Temperate hyperoceanic submediterranean	Thermotypes			Ombrotypes		
Sigmataxa	1	2	3	4	5	6
Salici angustifolio-albae Gs. (hard freshwater)	–	–	•	•	•	•
Geopermasigmeta						
Crithmo maritimi-Armerio pubigerae Gps. (haloanemogenous rock littoral)	•	–	–	•	•	–
Leucanthemo crassifolii-Festuco pruinosae Gps. (haloanemogenous rock littoral)	•	–	–	•	•	–
Puccinellio maritimae-Sarcocornio perennis Gps. (halophilous mareal)	•	•	–	•	•	

1 = thermotemperate, 2 = mesotemperate, 3 = supratemperate, 4 = subhumid, 5 = humid, 6 = hyperhumid

A.3. Potential Natural Vegetation Units (Sigmataxa) in the Temperate Oceanic Bioclimate

Temperate oceanic	Thermotypes					Ombrotypes			
Sigmataxa	1	2	3	4	5	6	7	8	9
Climatophilous sigmeta									
Blechno spicant-Querco roboris S. (acidophilous)	•	•	•	–	–	–	•	•	–
Hyperico pulchri-Querco roboris S. (acidophilous)	•	•	•	–	–	–	•	•	•
Pteridio aquilini-Querco pubescentis S. (neutro-acidophilous)	–	•	•	–	–	•	•	–	–
Helleboro occidentalis-Fago sylvaticae S. (neutro-acidophilous)	–	•	•	–	–	–	•	•	–
Blechno spicant-Fago sylvaticae S. (acidophilous)	–	•	•	–	–	–	•	•	–
Carici sylvaticae-Fago sylvaticae S. (neutro-acidophilous)	–	•	•	–	–	–	•	•	–
Geranio nodosi-Fago sylvaticae S. (neutro-acidophilous)	–	•	•	–	–	–	•	•	–
Pulmonario longifoliae-Querco petraeae S. (acidophilous)	–	•	•	–	–	–	•	•	–
Saxifrago hirsutae-Fago sylvaticae S. (acidophilous)	–	•	•	–	–	–	•	•	–
Echinosparto horridi-Pino pyrenaicae S. (basophilous)	–	–	•	–	–	•	•	–	–
Veronico officinalis-Pino pyrenaicae S. (acidophilous)	–	–	•	–	–	–	•	•	–
Polygalo calcareae-Pino catalaunicae S. (basophilous)	–	–	•	–	–	•	•	•	–
Pino uncinato-pyrenaicae S. (acidophilous)	–	–	•	–	–	–	•	•	–
Galio rotundifolii-Pino pyrenaicae S. (acidophilous)	–	–	•	–	–	•	•	•	–
Goodyero repentis-Pino pyrenaicae S. (basophilous)	–	–	•	–	–	–	•	•	–
Teucrio pyrenaici-Pino pyrenaicae S. (basophylous)	–	–	•	–	–	•	•	•	–
Buxo sempervirentis-Abieti albae S. (neutro-acidophilous)	–	–	•	–	–	–	•	•	–
Pulmonario affinis-Abieti albae S. (neutro-acidophilous)	–	–	•	–	–	–	•	•	•
Scillo liliohyacinthi-Fago sylvaticae S. (neutro-acidophilous)	–	–	•	–	–	–	•	•	–
Luzulo niveae-Fago sylvaticae S. (acidophilous)	–	–	•	–	–	–	•	•	–

(continued)

Temperate oceanic	Thermotypes					Ombrotypes			
Sigmataxa	1	2	3	4	5	6	7	8	9
Prunello hastifoliae-Querco petraeae S. (acidophilous)	–	–	•	–	–	–	•	•	–
Festuco altissimae-Abieti albae S. (neutro-acidophilous)	–	–	•	–	–	–	–	•	–
Goodyero repentis-Abieti albae S. (acidophilous)	–	–	•	–	–	–	–	•	•
Lysimachio nemorum-Fago sylvaticae S. (neutro-acidophilous)	–	–	•	–	–	–	–	•	•
Carici caudatae-Fago sylvaticae S. (neutrophilous)	–	–	•	•	–	–	•	•	–
Sorbo aucupariae-Fago sylvaticae S. (neutro-acidophilous)	–	–	–	•	–	–	–	•	–
Avenello ibericae-Fago sylvaticae S. (acidophilous)	–	–	–	•	–	–	–	•	•
Vaccinio microphylli-Junipero alpinae S. (acidophilous)	–	–	–	•	–	–	–	•	•
Daphno cantabricae-Arctostaphylo uvaeursi S. (basophilous)	–	–	–	•	–	–	–	•	–
Roso pendulinae-Fago sylvaticae S. (neutro-acidophilous)	–	–	–	•	–	–	–	•	•
Rhododendro ferruginei-Pino uncinatae S. (acidophilous)	–	–	–	•	–	–	–	•	•
Rhododendro ferruginei-Abieti albae S. (acidophilous)	–	–	–	•	–	–	–	•	•
Arctostaphylo uvaeursi-Pino uncinatae S. (acidophilous)	–	–	–	•	–	–	–	•	•
Climato-temporihygrophilous sigmeta									
Luzulo henriquesii-Pruno lusitanicae S.	–	•	–	–	–	–	•	•	–
Fraxino angustifoliae-Ulmo glabrae S. (acidophilous)	–	•	•	–	–	–	–	•	–
Hyperico androsaemi-Ulmo glabrae S.	–	•	–	–	–	–	–	•	–
Isopyro thalictroidis-Querco roboris S. (neutrophilous)	–	•	–	–	–	–	•	–	–
Aceri opali-Querco petraeae S. (acidophilous)	–	–	•	–	–	–	•	•	–
Roso pendulinae-Acero pseudoplatani S. (colluvial)	–	–	•	•	–	–	•	•	–
Hygrophilous geosigmetum and minorisigmeta									
Equiseto hyemalis-Alno glutinosae Gs. (hard freshwater)	–	•	•	–	–	•	•	–	–
Euphorbio hybernae-Fraxino excelsioris Gs. (hard freshwater)	–	–	•	–	–	–	•	•	–
Salici atrocinereo-daphnoidis Ms. (hard freshwater)	–	–	•	–	–	–	•	•	–
Agrostio stoloniferae-Myricario germanicae Ms. (hard freshwater)	–	–	•	–	–	–	•	•	–
Geopermasigmeta									
Teesdaliopsio confertae-Festuco eskiae Gps. (acidophilous)	–	–	–	•	–	–	–	•	–
Oxytropido neglecto-halleri Gps. (basophilous)	–	–	–	•	•	–	•	•	–
Saxifrago bryoidis-Minuartio sedoidis Gps. (acidophilous)	–	–	–	–	•	–	•	•	–
Oreochloo blankae-Carici curvulae Gps. (acidophilous)	–	–	–	–	•	–	–	•	•
Hieracio breviscapi-Festuco airoidis Gps. (acidophilous)	–	–	–	–	•	–	•	•	–
Junco trifidi-Oreochloo blankae Gps. (acidophilous)	–	–	–	–	•	–	–	•	–
Oxytropido halleri-Kobresio myosuroidis Gps. (basophilous)	–	–	–	–	•	–	–	•	–
Oxytropido foucaudii-Kobresio myosuroidis Gps. (basophilous)	–	–	–	–	•	–	–	•	•
Oxytropido neglectae-Kobresio myosuroidis Gps. (basophilous)	–	–	–	–	•	–	–	•	–

(continued)

Temperate oceanic	Thermotypes					Ombrotypes			
Sigmataxa	1	2	3	4	5	6	7	8	9
Minuartio cerastiifoliae-Androsaco ciliatae Gps. (basophilous)	–	–	–	–	•	–	–	•	•
Minuartio sedoidis-Androsaco ciliatae Gps. (acidophilous)	–	–	–	–	•	–	–	•	•

1 = thermotemperate, 2 = mesotemperate, 3 = supratemperate, 4 = orotemperate, 5 = cryorotemperate, 6 = subhumid, 7 = humid, 8 = hyperhumid, 9 = ultrahyperhumid

A.4. Potential Natural Vegetation Units (Sigmataxa) in the Temperate Oceanic Bioclimate with Submediterranean Variant

Temperate oceanic submediterranean	Thermotypes			Ombrotypes		
Sigmataxa	1	2	3	4	5	6
Climatophilous sigmeta						
Violo willkommii-Querco fagineae S. (basophilous)	•	•	–	•	–	–
Carici depauperatae-Querco pubescentis S. (acidophilous)	•	•	–	•	•	–
Melampyro pratensis-Querco pyrenaicae S. (acidophilous)	•	•	–	•	•	•
Buxo sempervirentis-Querco pubescentis S. (basophilous)	•	•	–	•	•	–
Pulmonario longifoliae-Querco fagineae S. (acidophilous)	•	•	–	•	•	–
Lonicero periclymeni-Querco pyrenaicae S. (acidophilous)	•	•	–	–	•	–
Buxo sempervirentis-Querco subpyrenaicae S. (basophilous)	•	•	–	•	•	–
Holco mollis-Querco pyrenaicae S. (acidophilous)	•	•	–	–	•	•
Rusco aculeati-Querco roboris S. (acidophilous)	•	•	–	–	•	•
Linario triornithophorae-Querco pyrenaicae S. (acidophilous)	•	•	–	–	•	–
Roso arvensis-Querco pubescentis S. (basophilous)	•	•	–	–	•	–
Emero majoris-Abieti albae S. (neutrophilous)	•	•	–	–	•	•
Buxo sempervirentis-Fago sylvaticae S. (neutrophilous)	•	•	–	–	•	•
Epipactido helleborines-Fago sylvaticae S. (neutrophilous)	•	•	–	–	•	–
Lathyro linifolii-Querco petraeae S. (acidophilous)	•	•	–	–	•	•
Vaccinio myrtilli-Querco roboris S. (acidophilous)	•	•	–	–	•	•
Saniculo europaeae-Ilici aquifolii S. (neutrophilous)	•	•	–	–	•	–
Sileno melliferae-Querco faginae S. (basophilous)	–	•	–	•	•	–
Ononido aragonensis-Pino ibericae S. (climatophilous)	–	•	–	•	•	–
Junipero sabino-orocantabricae S. (basophilous)	–	•	–	•	•	–
Pteridio aquilini-Pino ibericae S. (acidophilous)	–	•	–	•	•	•
Cephalanthero rubrae-Querco pyrenaicae S. (acidophilous)	–	•	–	•	•	–
Pulmonario longifoliae-Querco pyrenaicae S. (neutrophilous)	–	•	–	•	•	–
Telino patentis-Querco fagineae S. (basophilous)	–	•	–	•	•	–
Genisto falcatae-Querco pyrenaicae S. (acidophilous)	–	•	–	•	•	–
Linario triornithophorae-Querco petraeae S. (acidophilous)	–	•	–	–	•	•
Luzulo henriquesii-Querco petraeae S. (acidophilous)	–	•	–	–	•	•
Galio odorati-Querco petraeae S. (acidophilous)	–	•	–	–	•	•

(continued)

Temperate oceanic submediterranean	Thermotypes			Ombrotypes		
Sigmataxa	1	2	3	4	5	6
Omphalodo nitidae-Fago sylvaticae S. (neutron-acidophilous)	–	•	–	–	•	•
Galio rotundifolii-Fago sylvaticae S. (acidophilous)	–	•	•	–	•	•
Eryngio juresiani-Betulo celtibericae S. (acidophilous)	–	•	–	–	•	•
Primulo acaulis-Fago sylvaticae S. (neutrophilous)	–	•	–	–	•	–
Calluno vulgaris-Pino ibericae S. (acidophilous)	–	•	•	–	•	–
Junipero sabinae-Pino ibericae S. (basophilous)	–	•	•	•	•	–
Ephedro nebrodensis-Junipero sabinae S. (basophilous)	–	•	•	–	•	–
Lithodoro diffusae-Junipero sabinae S. (basophilous)	–	•	•	–	•	•
Avenello ibericae-Querco orocantabricae S. (acidophilous)	–	•	•	–	–	•
Luzulo henriquesii-Betulo celtibericae S. (acidophilous, chionophilous)	–	•	•	–	–	•
Junipero sabinae-Pino uncinatae S. (basophilous)	–	–	•	•	•	–
Pulsatillo fontqueri-Pino uncinatae S. (basophilous)	–	–	•	–	•	•
Avenello ibericae-Junipero alpinae S. (acidophilous)	–	–	•	–	•	•
Vaccinio myrtilli-Junipero alpinae S. (acidophilous)	–	–	•	–	•	•
Genisto sanabrensis-Junipero alpinae S. (acidophilous)	–	–	•	–	•	•
Avenello ibericae-Pino ibericae S. (acidophilous)	–	–	•	–	•	•
Vaccinio myrtilli-Pino ibericae S. (acidophilous, relict)	–	–	•	–	•	•
Lycopodio clavati-Junipero alpinae S. (acidophilous)	–	–	•	–	–	•
Avenello ibericae-Pino uncinatae S. (acidophilous)	–	–	•	–	•	•
Climatophilous and edaphoxerophilous sigmeta						
Physospermo cornubiensis-Querco suberis S. (acidophilous)	•	–	–	•	•	–
Carici basilaris-Querco suberis S. (acidophilous)	•	–	–	•	•	–
Genisto falcatae-Querco rotundifoliae S. (basophilous)	•	•	–	•	•	–
Cephalanthero longifoliae-Querco rotundifoliae S. (basophilous, relict)	•	•	–	•	•	–
Asplenio onopteridis-Querco ilicis S. (acidophilous, mafic)	•	•	–	•	•	–
Junipero hemisphaerico-thuriferae S. (basophilous)	–	•	–	•	–	–
Hedero helicis-Querco rotundifoliae S. (basophilous)	–	•	–	•	•	–
Junipero thuriferae-Pino latisquamae S. (basophilous)	–	•	–	•	–	–
Spiraeo obovatae-Querco rotundifoliae S. (basophilous)	–	•	–	•	•	–
Teucrio scorodoniae-Querco rotundifoliae S. (acidophilous)	–	•	–	•	•	–
Berberido cantabricae-Querco fagineae S. (basophilous)	–	•	–	•	•	–
Lauro nobilis-Querco ilicis S. (relict)	•	–	–	–	•	–
Teucrio salviastri-Querco suberis S. (acidophilous, relict)	•	–	–	–	•	–
Edaphoxerophilous sigmetum						
Asplenio adiantinigri-Querco rotundifoliae S. (acidophilous)	•	•	–	•	•	–
Daphno laureolae-Querco ilicis S. (basophilous)	•	•	–	•	•	–
Buxo sempervirentis-Junipero phoeniceae S. (basophilous)	•	•	–	•	•	–
Teucrio salviastri-Querco rotundifoliae S. (acidophilous, relict)	•	•	–	–	•	•
Arctostaphylo crassifoliae-Pino catalaunicae S. (silicolous)	–	•	–	•	–	–
Lonicero xylostei-Pino salzmannii S. (calcodolomiticolous)	–	•	–	•	–	–
Telino patentis-Pino salzmannii S. (calcodolomiticolous)	–	•	–	•	•	–

(continued)

Temperate oceanic submediterranean	Thermotypes			Ombrotypes		
Sigmataxa	1	2	3	4	5	6
Buxo sempervirentis-Pino catalaunicae S. (calcodomiticolous)	–	•	–	•	•	–
Lithodoro diffusae-Querco rotundifoliae S. (basophilous)	–	•	–	•	•	–
Climato-temporihygrophilous sigmeta						
Carici depressae-Querco canariensis S. (neutro-acidophilous)	•	–	–	•	•	–
Polysticho setiferi-Fraxino excelsioris S. (acidophilous)	•	–	–	•	•	–
Frangulo alni-Pruno lusitanicae S. (acidophilous)	•	–	–	•	•	•
Omphalodo nitidae-Fraxino angustifoliae S. (basophylous)	•	–	–	–	•	•
Crataego laevigatae-Querco roboris S. (neutro-acidophilous)	•	•	–	•	•	–
Doronico pardalianchis-Fraxino excelsioris S. (neutro-acidophilous)	•	•	–	•	•	–
Poo nemoralis-Tilio platyphylli S. (colluvial)	•	•	–	•	•	–
Buxo sempervirentis-Querco rotundifoliae S. (acidophilous)	•	•	–	•	•	–
Hyperico androsaemi-Querco roboris S. (acidophilous)	•	•	–	–	•	–
Helleboro occidentalis-Tilio cordatae S. (basophilous)	•	•	–	–	•	–
Brachypodio sylvatici-Fraxino excelsioris S. (neutrophilous)	–	•	–	•	•	–
Luzulo henriquesii-Acero pseudoplatani S. (acidophilous)	–	•	–	–	•	–
Violo mirabilis-Ulmo glabrae S. (colluvial)	–	•	–	–	•	–
Melico uniflorae-Betulo celtibericae S. (acidophilous)	–	•	•	–	•	•
Hygrophilous geosigmetum and minorisigmeta						
Lamio flexuosi-Alno glutinosae Gs. (soft freshwater)	•	–	–	•	•	–
Hedero hibernicae-Fraxino angustifoliae Gs. (soft freshwater)	•	–	–	•	•	–
Rusco aculeati-Fraxino angustifoliae Gs. (hard freshwater)	•	–	–	•	–	–
Lithospermo purpureocaerulei-Ulmo minoris Gs. (hard freshwater)	•	–	–	•	•	–
Senecioni bayonensis-Alno glutinosae Gs. (soft freshwater)	•	–	–	•	•	–
Hyperico androsaemi-Alno glutinosae Gs. (hard freshwater)	•	–	–		•	–
Salici cantabricae Gs. (hard freshwater)	•	•	–	•	•	•
Lonicero xylostei-Alno glutinosae Gs. (hard freshwater)	•	•	–	•	•	–
Festuco giganteae-Fraxino excelsioris Gs (hard freshwater)	•	•	–	•	•	–
Lathraeo clandestinae-Populo nigrae Gs. (hard freshwater)	•	•	–	•	•	–
Valeriano pyrenaicae-Alno glutinosae Gs. (soft freshwater)	•	•	–	•	•	–
Carici pendulae-Alno glutinosae Gs. (soft freshwater)	•	•	–	•	•	–
Viburno lantanae-Ulmo minoris Gs. (hard freshwater)	•	•	–	•	•	–
Salici lambertiano-albae Gs. (hard freshwater)	•	•	–	•	•	–
Betulo meridionalis-Salici albae Gs. (hard freshwater)	•	•	–	•	•	–
Salici salviifoliae Gs. (soft freshwater)	•	•	–	•	•	–
Carici reuterianae-Betulo celtibericae Gs. (soft freshwater)	•	•	–		•	–
Salici angustifolio-albae Gs. (hard freshwater)	–	•	–	•	•	–
Salici lambertiano-angustifoliae Ms. (hard freshwater)	–	•	–	•	•	–
Rubo lainzii-Salici atrocinereae Gs. (soft freshwater)	–	•	–	•	•	–
Salici lambertiano-salviifoliae Ms. (soft freshwater)	–	•	–	•	•	–
Aceri campestris Fraxino excelsioris S. (neutro-acidophilous)	–	•	–	–	•	–
Salici cantabrico-albae Gs. (hard freshwater)	–	•	–	–	•	–

(continued)

Temperate oceanic submediterranean	Thermotypes			Ombrotypes		
Sigmataxa	1	2	3	4	5	6
Salici cantabrico-bicoloris Gs. (hard freshwater, chionophilou)	–	–	•	–	–	•
Veratro albi-Salici bicoloris Ms. (soft freshwater, chionophilous)	–	–	•	–	–	•
Geopermasigmeta						
Hieracio myriadeni-Festuco carpetanae Gps. (acidophilous)	–	•	•	–	•	•
Teesdaliopsio confertae-Festuco summilusitanae Gps. (acidophilous)	–	•	•	–	•	•
Armerio microcephalae-Festuco aragonensis Gps. (acidophilous)	–	•	•	–	•	•
Antennario dioicae-Festuco carpetanae Gps. (acidophilous)	–	•	•	–	•	•
Agrostio rupestris-Armerio bigerrensis Gps. (acidophilous)	–	–	•	–	•	•

1 = mesotemperate, 2 = supratemperate, 3 = orotemperate, 4 = subhumid, 5 = humid, 6 = hyperhumid

A.5. Potential Natural Vegetation Units (Sigmataxa) in the Temperate Xeric Bioclimate with Submediterranean Variant

Temperate xeric submediterraean			
Sigmataxa	mte	ste	shu
Climatophilous sigmeta			
Buxo sempervirentis-Querco subpyrenaicae S. (basophilous)	•	•	•

mte mesotemperate, ste supratemperate, shu subhumid

B. Mediterranean Macrobioclimate

B.1. Potential Natural Vegetation Units (Sigmataxa) in the Mediterranean Pluviseasonal Oceanic Bioclimate

Mediterranean pluviseasonal-oceanic	Thermotypes						Ombrotypes					
Sigmataxa	1	2	3	4	5	6	7	8	9	10	11	12
Climatophilous sigmeta												
Cneoro tricocci-Ceratonio siliquae S. (basophilous)	–	•	–	–	–	–	–	•	•	–	–	–
Oleo sylvestris-Querco suberis S. (acidophilous)	–	•	–	–	–	–	–	–	•	•	–	–
Asparago aphylli-Querco suberis S. (acidophilous)	–	•	–	–	–	–	–	–	•	•	–	–
Viburno tini-Querco rivas-martinezii S. (basophilous)	–	•	–	–	–	–	–	–	–	•	–	–
Viburno tini-Oleo sylvetris S. (basophilous)	–	•	–	–	–	–	–	–	–	•	–	–

(continued)

Mediterranean pluviseasonal-oceanic	Thermotypes						Ombrotypes					
Sigmataxa	1	2	3	4	5	6	7	8	9	10	11	12
Myrto communis-Querco suberis S. (acidophilous)	–	•	–	–	–	–	–	–	–	•	•	–
Querco alpestris-broteroi S. (basophilous)	–	•	–	–	–	–	–	–	–	•	•	–
Lavandulo viridis-Querco suberis S. (acidophilous)	–	•	–	–	–	–	–	–	–	•	•	–
Rhamno fontquerani-Querco rotundifoliae S. (basophilous, dolomitícolous, ultramafic)	–	•	•	–	–	–	–	–	•	•	–	–
Asplenio onopteridis-Querco suberis S. (acidophilous)	–	•	•	–	–	–	–	–	•	•	–	–
Aro neglecti-Oleo sylvestris S. (verticola)	–	•	•	–	–	–	–	–	•	•	–	–
Smilaco asperae-Querco suberis S. (acidophilous)	–	•	•	–	–	–	–	–	•	•	–	–
Viburno tini-Fraxino orni S. (basophilous)	–	•	•	–	–	–	–	–	–	•	•	–
Sanguisorbo hybridae-Querco broteroi S. (acidophilous)	–	•	•	–	–	–	–	–	–	•	•	–
Arisaro simorrhini-Querco pyrenaicae S. (acidophilous)	–	•	•	–	–	–	–	–	–	•	•	–
Arisaro simorrhini-Querco broteroi S. (basophilous)	–	•	•	–	–	–	–	–	–	•	•	–
Teucrio baetici-Querco suberis S. (acidophilous)	–	•	•	–	–	–	–	–	–	•	•	–
Asparago acutifolii-Querco rotundifoliae S. (basophilous)	–	–	•	–	–	–	–	–	•	•	–	–
Viburno tini-Querco ilicis S. (basophilous)	–	–	•	–	–	–	–	–	•	•	–	–
Hedero hybernicae-Querco fagineae S. (basophilous)	–	–	•	–	–	–	–	–	–	•	•	–
Doronico plantaginei-Querco canariensis S. (acidophilous)	–	–	•	–	–	–	–	–	–	•	•	–
Arbuto unedonis-Querco pyrenaicae S. (acidophilous)	–	–	•	–	–	–	–	–	–	•	•	–
Fraxino angustifoliae-Acero monspessulani S. (silicolous)	–	–	•	–	–	–	–	–	–	–	•	–
Holco mollis-Querco pyrenaicae S. (acidophilous)	–	–	•	–	–	–	–	–	–	–	•	•
*Luzulo baeticae-Querco pyrenaicae*S. (acidophilous)	–	–	•	–	–	–	–	–	–	–	•	•
Pistacio terebinthi-Querco broteroi S. (acidophilous)	–	–	•	–	–	–	–	–	–	•	–	–
Adenocarpo decorticantis-Querco rotundifoliae S. (acidophilous)	–	–	•	•	–	–	–	–	•	•	–	–
Cephalanthero rubrae-Querco fagineae S. (basophilous)	–	–	•	•	–	–	–	–	•	•	–	–
Fraxino orni-Querco fagineae S. (basophilous)	–	–	•	•	–	–	–	–	–	•	–	–
Adenocarpo decorticantis-Querco suberis S. (acidophilous)	–	–	•	•	–	–	–	–	–	•	–	–

(continued)

Mediterranean pluviseasonal-oceanic	Thermotypes						Ombrotypes					
Sigmataxa	1	2	3	4	5	6	7	8	9	10	11	12
Junipero lagunae-Querco suberis S. (acidophilous)	–	–	•	•	–	–	–	–	–	•	•	–
Spiraeo obovatae-Querco rotundifoliae S. (basophilous)	–	–	•	•	–	–	–	–	•	•	•	–
Genisto falcatae-Querco pyrenaicae S. (acidophilous)	–	–	•	•	–	–	–	–	–	•	•	•
Violo willkommii-Querco fagineae S. (basophilous)	–	–	•	•	–	–	–	–	–	•	–	–
Lonicero periclymeni-Querco pyrenaicae S. (acidophilous)	–	–	•	•	–	–	–	–	–	–	•	•
Adenocarpo decorticantis-Querco pyrenaicae S. (acidophilous)	–	–	–	•	–	–	–	–	–	•	•	–
Sileno melliferae-Querco faginae S. (basophilous)	–	–	–	•	–	–	–	–	–	•	•	–
Pulmonario longifoliae-Querco pyrenaicae S. (acidophilous)	–	–	–	•	–	–	–	–	–	•	•	–
Telino patentis-Querco fagineae S. (basophilous)	–	–	–	•	–	–	–	–	–	•	•	–
Sorbo torminalis-Querco pyrenaicae S. (acidophilous)	–	–	–	•	–	–	–	–	–	•	•	–
Berberido hispanicae-Querco pyrenaicae S. (climatophilous)	–	–	–	•	–	–	–	–	–	•	•	–
Pyro bourgaeanae-Querco pyrenaicae S. (acidophilous)	–	–	–	•	–	–	–	–	–	•	•	–
Berberido hispanicae-Querco alpestris S. (basophilous, calcodolomiticolous)	–	–	–	•	–	–	–	–	–	•	•	–
Pteridio aquilini-Pino ibericae S. (acidophilous)	–	–	–	•	–	–	–	–	–	•	•	•
Luzulo forsteri-Querco pyrenaicae S. (acidophilous)	–	–	–	•	–	–	–	–	–	•	•	•
Primulo balearicae-Aceri granatensis S. (basophilous)	–	–	–	•	–	–	–	–	–	–	•	–
Festuco merinoi-Querco pyrenaicae S. (acidophilous)	–	–	–	•	–	–	–	–	–	–	•	•
Junipero sabinae-Pino latisquamae S. (basophilous)	–	–	–	•	•	–	–	–	–	•	•	–
Climatophilous and edaphoxerophilous sigmeta and minorisigmeta												
Prasio majoris-Oleo sylvestris S. (basophilous)	–	•	–	–	–	–	–	–	•	–	–	–
Myrto communis-Querco rotundifoliae S. (acidophilous)	–	•	–	–	–	–	–	–	•	•	–	–
Aristolochio baeticae-Junipero turbinatae Ms. (basophilous)	–	•	–	–	–	–	–	–	•	•	–	–
Rhamno oleoidis-Querco rotundifoliae S. (basophilous, calcodolomiticolous)	–	•	–	–	–	–	–	–	•	•	•	–
Pistacio lentisci-Pino halepensis S. (basophilous, calcodolomiticolous)	–	•	•	–	–	–	–	•	•	–	–	–

(continued)

Mediterranean pluviseasonal-oceanic	Thermotypes						Ombrotypes					
Sigmataxa	1	2	3	4	5	6	7	8	9	10	11	12
Ephedro fragilis-Pino halepensis S. (basophilous)	–	•	•	–	–	–	–	•	•	–	–	–
Rhamno almeriensis-Pino halepensis S. (calcodolomiticolous)	–	•	•	–	–	–	–	•	•	•	–	–
Carici bracteosae-Querco ilicis S. (acidophilous, basophilous and dolomiticolous)	–	•	•	–	–	–	–	–	•	–	–	–
Clematido cirrhosae-Querco rotundifoliae S. (basophilous and dolomiticolous)	–	•	•	–	–	–	–	–	•	•	–	–
Rubio longifoliae-Querco rotundifoliae S. (basophilous)	–	•	•	–	–	–	–	–	•	•	–	–
Querco cocciferae-Pino acutisquamae S. (ultramafic)	–	•	•	–	–	–	–	–	–	•	•	–
Pino acutisquamae S. (dolareniticolous)	–	•	•	•	–	–	–	–	•	•	•	–
Cyclamini balearici-Querco ilicis S. (basophilous, dolomiticolous)	–	•	•	•	–	–	–	–	–	•	•	–
Bunio macucae-Abieti pinsapo S. (ultramafic)	–	•	•	•	–	–	–	–	–	–	•	•
Paeonio broteri-Abieti pinsapo S. (calcodolomiticolous)	–	•	•	•	–	–	–	–	–	–	•	•
Rhamno lycioidis-Querco cocciferae Ms. (basophilous)	–	–	•	–	–	–	–	•	•	–	–	–
Querco cocciferae-Pino halepensis S. (basophilous)	–	–	•	–	–	–	–	•	•	–	–	–
Querco rotundifoliae S. (basophilous)	–	–	•	–	–	–	–	–	•	•	–	–
Pyro bourgaeanae-Querco rotundifoliae S. (acidophilous)	–	–	•	–	–	–	–	–	•	•	–	–
Paeonio coriaceae-Querco rotundifoliae S. (basophilous)	–	–	•	–	–	–	–	–	•	•	•	–
Physospermo cornubiensis-Querco suberis S. (acidophilous)	–	–	•	–	–	–	–	–	–	•	•	–
Hedero helicis-Querco rotundifoliae S. (basophilous)	–	–	•	•	–	–	–	–	•	•	–	–
Junipero lagunae-Querco rotundifoliae S. (acidophilous)	–	–	•	•	–	–	–	–	•	•	–	–
Genisto hystricis-Querco rotundifoliae S. (acidophilous)	–	–	•	•	–	–	–	–	•	•	•	–
Spiraeo obovatae-Querco fagineae S. (basophilous)	–	–	•	•	–	–	–	–	–	•	•	–
Genisto falcatae-Querco rotundifoliae S. (basophilous)	–	–	•	•	–	–	–	–	–	•	•	–
Carici basilaris-Querco suberis S. (acidophilous)	–	–	•	•	–	–	–	–	–	•	•	–
Buxo sempervirentis-Querco rotundifoliae S. (basophilous)	–	–	–	•	–	–	–	–	•	•	•	–
Berberido hispanicae-Junipero thuriferae S. (basophilous, dolomiticolous)	–	–	–	•	–	–	–	–	•	•	–	–

(continued)

Mediterranean pluviseasonal-oceanic	Thermotypes						Ombrotypes					
Sigmataxa	1	2	3	4	5	6	7	8	9	10	11	12
Junipero thuriferae-Querco rotundifoliae S. (basophilous)	–	–	–	•	–	–	–	–	•	•	–	–
Junipero hemisphaerico-thuriferae S. (basophilous)	–	–	–	•	–	–	–	–	•	•	–	–
Berberido hispanicae-Querco rotundifoliae S. (basophilous, calcodolomiticolous)	–	–	–	•	–	–	–	–	•	•	•	–
Junipero thuriferae-Pino latisquamae S. (basophilous)	–	–	–	•	–	–	–	–	–	•	–	–
Teucrio scorodoniae-Querco rotundifoliae S. (acidophilous)	–	–	–	•	–	–	–	–	–	•	•	–
Junipero phoeniceae-Pino latisquamae S. (calcodolomiticolous)	–	–	–	•	–	–	–	–	–	•	•	–
Rhamno infectorii-Junipero sabinae S. (basophilous, calcodolomiticolous)	–	–	–	•	•	–	–	–	–	–	•	•
Daphno hispanicae-Pino nevadensis S. (basophilous, calcodolomiticolous)	–	–	–	•	•	–	–	–	–	•	–	–
Edaphoxerophilous sigmetum and minorisigmeta												
Junipero turbinatae-Pino halepensis S. (basophilous)	–	•	–	–	–	–	–	•	•	–	–	–
Chamaeropo humilis-Junipero phoeniceae Ms. (basophilous)	–	•	–	–	–	–	–	•	•	–	–	–
Astragalo salvatoris-Junipero macrocarpae Ms. (littoral windy)	–	•	–	–	–	–	–	–	•	–	–	–
Phillyreo angustifoliae-Rhamno angustifolii Ms. (dune)	–	•	–	–	–	–	–	–	•	–	–	–
Osyrio quadripartitae-Junipero turbinatae Ms. (coastal dune)	–	•	–	–	–	–	–	–	•	–	–	–
Phlomido purpureae-Junipero turbinatae Ms. (mafic)	–	•	–	–	–	–	–	–	•	–	–	–
Clematido balearicae-Junipero turbinatae Ms. (coastal dune)	–	•	–	–	–	–	–	–	•	•	–	–
Querco cocciferae-Junipero turbinatae Ms. (basophilous)	–	•	–	–	–	–	–	–	•	•	–	–
Vinco difformis-Ceratonio siliquae S. (basophilous)	–	•	–	–	–	–	–	–	•	•	–	–
Edaphoxerophilous sigmetum and minorisigmetum												
Daphno gnidi-Junipero navicularis S. (acidophilous, relict)	–	•	–	–	–	–	–	–	•	•	–	–
Aro sagittifolii-Phillyreo rodriguezii S. (littoral)	–	•	–	–	–	–	–	–	•	•	–	–
Cneoro tricocci-Buxo balearicae Ms. (calcodolomiticolous)	–	•	–	–	–	–	–	–	•	•	–	–
Asparago horridi-Junipero turbinatae Ms. (calcodolomiticolous)	–	•	–	–	–	–	–	–	•	•	•	–
Ulici argentei-Querco rotundifoliae S. (acidophilous)	–	•	–	–	–	–	–	–	–	•	–	–

(continued)

Mediterranean pluviseasonal-oceanic	Thermotypes						Ombrotypes					
Sigmataxa	1	2	3	4	5	6	7	8	9	10	11	12
Rubio longifoliae-Junipero macrocarpae Ms. (coastal dune)	–	•	–	–	–	–	–	–	–	•	–	–
Junipero turbinatae Ms. (coastal dune)	–	•	•	–	–	–	–	–	•	–	–	–
Genisto majoricae-Buxo balearicae Ms. (calcodolomiticolous)	–	•	•	–	–	–	–	–	•	•	–	–
Rhamno ludovicisalvatoris-Junipero turbinatae Ms. (dolomiticolous)	–	•	•	–	–	–	–	–	•	•	–	–
Asparago albi-Oleo sylvestris S. (acidophilous)	–	•	•	–	–	–	–	–	•	•	–	–
Cytiso eriocarpi-Junipero lagunae Ms. (acidophilous, relict)	–	•	•	–	–	–	–	–	•	•	–	–
Rusco aculeati-Junipero lagunae Ms. (acidophilous, relict)	–	–	•	–	–	–	–	–	•	•	–	–
Lonicero implexae-Querco rotundifoliae S. (basophilous)	–	–	•	–	–	–	–	–	–	•	•	–
Vinco difformis-Lauro nobilis S. (basophilous)	–	–	•	–	–	–	–	–	–	–	•	•
Teucrio salviastri-Querco suberis S. (acidophilous)	–	–	•	–	–	–	–	–	–	•	•	
Rhamno myrtifolii-Junipero turninatae Ms. (dolomiticolous)	–	•	•	•	–	–	–	–	•	•	•	–
Rhamno lycioidis-Junipero phoeniceae S. (basophilous)	–	–	•	•	–	–	–	–	•	•	–	–
Rhamno lycioidis-Pino halepensis S. (calcodolomiticolous)	–	–	•	•	–	–	–	–	•	•	–	–
Buxo sempervirentis-Junipero phoeniceae Ms. (basophilous)	–	–	•	•	–	–	–	–	•	•	•	–
Tucrio salviastri-Querco rotundifoliae S. (acidophilous, relict)	–	–	•	•	–	–	–	–	–	–	•	•
Festuco merinoi-Junipero lagunae Ms. (acidophilous)	–	–	–	•	–	–	–	–	–	–	•	–
Rhamno infectorii-Junipero phoeniceae Ms. (basophilous)	–	–	–	•	–	–	–	–	•	•	–	–
Berberido hispanicae-Junipero phoeniceae S. (basophilous, calcodolomiticolous)	–	–	–	•	–	–	–	–	•	•	–	–
Climato-temporihygrophilous and edaphoxerophilous sigmeta												
Aro neglecti-Querco suberis Gs. (paleodune)	–	•	–	–	–	–	–	–	•	•	–	–
Climato-temporihygrophilous sigmeta												
Rusco hypophylli-Querco canariensis S. (acidophilous)	–	•	•	–	–	–	–	–	–	•	•	–
Euphorbio monchiquensis-Querco canariensis S. (acidophilous)	–	•	•	–	–	–	–	–	–	•	•	–
Oenantho crocatae-Querco pyrenaicae S. (acidophilous)	–	•	•	–	–	–	–	–	–	•	•	–

(continued)

Mediterranean pluviseasonal-oceanic	Thermotypes						Ombrotypes					
Sigmataxa	1	2	3	4	5	6	7	8	9	10	11	12
Ulici welwitschiani-Querco broteroi S. (basophilous)	–	•	•	–	–	–	–	–	•	•	–	–
Vinco difformis-Ulmo minoris S.	–	•	•	–	–	–	–	–	•	•	•	–
Oleo sylvestris-Querco alpestris S. (neutro-basophilous)	–	•	•	–	–	–	–	–	•	•	•	–
Clematido campaniflorae-Celtido australis S.	–	–	•	–	–	–	–	–	•	•	•	–
Carici depressae-Querco canariensis S. (neutro-acidophilous)	–	–	•	–	–	–	–	–	–	•	•	–
Frangulo alni-Pruno lusitanicae S. (acidophilous)	–	–	•	–	–	–	–	–	–	•	•	–
Pyro bourgaeanae-Querco broteroi S. (acidophilous)	–	–	•	•	–	–	–	–	•	•	•	–
Daphno latifoliae-Acero granatensis S. (basophilous)	–	–	–	•	–	–	–	–	–	•	•	–
Hygrophilous geosigmeta and minorisigmeta												
Inulo crithmoidis-Tamarici boveanae Ms. (halophilous)	•	•	–	–	–	–	•	•	–	–	–	–
Viti sylvestris-Salici atrocinereae Ms. (dystrophic lentic freshwater)	–	•	–	–	–	–	–	–	•	•	–	–
Erico terminalis-Nerio oleandri Ms. (hard freshwater, serpentinícolous)	–	•	–	–	–	–	–	–	•	•	–	–
Arisaro proboscidei-Alno glutinosae Gs. (soft freshwater)	–	•	–	–	–	–	–	–	•	•	–	–
Equiseto telmateiae-Salici pedicellatae Gs. (hard freshwater)	–	•	–	–	–	–	–	–	•	•	•	–
Irido foetidissimae-Fraxino angustifoliae Gs. (soft freshwater)	–	•	•	–	–	–	–	–	•	•	–	–
Clematido campaniflorae-Salici neotrichae Gs. (soft freshwater)	–	•	•	–	–	–	–	–	•	•	•	–
Saccharo ravennae-Tamarici canariensis Ms. (hard freshwater)	–	•	•	–	–	–	–	•	•	•	–	–
Nerio oleandri-Populo albae Gs. (hard freshwater)	–	•	•	–	–	–	–	–	•	–	–	–
Salici pedicellatae-Populo albae Gs. (hard freshwater)	–	•	•	–	–	–	–	–	•	•	–	–
Vinco difformis-Populo albae Gs. (hard freshwater)	–	•	•	–	–	–	–	–	•	•	–	–
Ranunculo ficariae-Fraxino angustifoliae Gs. (soft freshwater)	–	•	•	–	–	–	–	–	•	•	–	–
Hedero helicis-Ulmo minoris Gs. (hard freshwater)	–	•	•	–	–	–	–	–	•	•	–	–
Vinco difformis-Fraxino angustifoliae Gs. (hard freshwater)	–	•	•	–	–	–	–	–	•	•	–	–
Oenantho crocatae-Nerio oleandri Gs. (soft freshwater)	–	•	•	–	–	–	–	–	•	•	–	–

(continued)

Mediterranean pluviseasonal-oceanic	Thermotypes						Ombrotypes					
Sigmataxa	1	2	3	4	5	6	7	8	9	10	11	12
Coriario myrtifoliae-Salici angustifoliae Ms. (hard freshwater)	–	•	•	–	–	–	–	–	•	•	–	–
Erico terminalis-Salici angustifoliae Ms. (hard freshwater)	–	•	•	–	–	–	–	–	•	•	–	–
Salici atrocinereo-australis Ms. (soft freshwater)	–	•	•	–	–	–	–	–	•	•	–	–
Dorycnio recti-Salici pedicellatae Gs. (hard freshwater)	–	•	•	–	–	–	–	–	•	•	–	–
Pyro bourgaeanae-Flueggeo tinctoriae Ms. (hard freshwater)	–	•	•	–	–	–	–	–	•	•	–	–
Polygono equisetiformis-Tamarici africanae Ms. (hard freshwater)	–	•	•	–	–	–	–	–	•	•	–	–
Nerio oleandri-Salici pedicellatae Gs. (hard freshwater)	–	•	•	–	–	–	–	–	–	•	–	–
Crataego brevispinae-Populo albae Gs. (hard freshwater)	–	•	•	–	–	–	–	–	–	•	–	–
Galio viridiflori-Salici pedicellatae Gs. (hard freshwater)	–	•	•	–	–	–	–	–	–	•	•	–
Carici lusitanicae-Alno glutinosae Gs. (dystrophic lentic freshwater)	–	•	•	–	–	–	–	–	•	•	•	–
Campanulo primulifoliae-Alno glutinosae Gs. (soft freshwater)	–	•	•	–	–	–	–	–	–	–	•	•
Campanulo primuliflorae-Rhododendro pontici Ms. (neutro-acidophilous)	–	•	•	–	–	–	–	–	–	–	•	•
Biaro carratracensis-Ulmo minoris Gs. (hard freshwater)	–	•	•	•	–	–	–	–	•	•	–	–
Suaedo braunblanquetii-Tamarici boveanae Ms. (halophilous)	–	–	•	–	–	–	–	•	•	–	–	–
Saponario officinalis-Salici lambertianae Ms. (hard freshwater)	–	–	•	–	–	–	–	–	•	•	–	–
Populo albae Gs. (hard freshwater)	–	–	•	–	–	–	–	–	•	•	–	–
Rusco aculeati-Fraxino angustifoliae Gs. (hard freshwater)	–	–	•	–	–	–	–	–	•	•	–	–
Salici atrocinereae-Populo albae Gs. (hard freshwater)	–	–	•	–	–	–	–	–	•	•	–	–
Holoschoeno australis-Salici arenariae Ms.	–	–	•	–	–	–	–	–	–	•	•	–
Scrophulario scorodoniae-Alno glutinosae Gs. (soft freshwater)	–	–	•	–	–	–	–	–	•	•	•	–
Carici pendulae-Salici atrocinereae Ms. (soft freshwater)	–	–	•	–	–	–	–	–	–	•	–	–
Lithospermo purpureocaerulei-Ulmo minoris Gs. (hard freshwater)	–	–	•	–	–	–	–	–	–	•	•	–
Senecioni bayonensis-Alno glutinosae Gs. (soft freshwater)	–	–	•	–	–	–	–	–	–	•	•	–
Lamio flexuosi-Alno glutinosae Gs. (soft freshwater)	–	–	•	–	–	–	–	–	–	•	•	–

(continued)

Mediterranean pluviseasonal-oceanic	Thermotypes						Ombrotypes					
Sigmataxa	1	2	3	4	5	6	7	8	9	10	11	12
Hedero hibernicae-Fraxino angustifoliae Gs. (soft freshwater)	–	–	•	–	–	–	–	–	–	•	•	–
Rhododendro pontici-Alno glutinosae Gs. (soft freshwater)	–	–	•	–	–	–	–	–	–	–	•	•
Rubio tinctorum-Populo albae Gs. (hard freshwater)	–	–	•	•	–	–	–	•	•	–	–	–
Salici neotrichae Gs. (hard freshwater)	–	–	•	•	–	–	–	•	•	–	–	–
Tamarici canariensis Ms. (hard freshwater)	–	–	•	•	–	–	–	•	•	–	–	–
Crataego granatensis-Salici neotrichae Gs. (hard freshwater)	–	–	•	•	–	–	–	•	•	•	–	–
Suaedo braunblanquetii-Tamarici canariensis Ms. (halophilous)	–	–	•	•	–	–	–	–	•	–	–	–
Opopanaco chironium-Ulmo minoris Gs. (hard freshwater)	–	–	•	•	–	–	–	–	•	•	–	–
Galio broteriani-Betulo parvibracteatae Gs. (soft freshwater)	–	–	•	•	–	–	–	–	•	•	–	–
Humulo lupuli-Alno glutinosae Gs. (hard freshwater)	–	–	•	•	–	–	–	–	•	•	–	–
Salici salviifoliae Ms. (soft freshwater)	–	–	•	•	–	–	–	–	–	•	•	–
Salici lambertiano-salviifoliae Ms. (soft freshwater)	–	–	•	•	–	–	–	–	–	•	•	–
Rubo ulmifolii-Nerio oleandri Ms. (hard freshwater)	–	•	•	–	–	–	–	•	•	–	–	–
Populo nigrae-Salici neotrichae Gs. (hard freshwater)	–	–	–	•	–	–	–	–	•	•	–	–
Salici discoloro-angustifoliae Ms. (hard freshwater)	–	–	–	•	–	–	–	–	•	•	–	–
Galio broteriani-Alno glutinosae Gs. (soft freshwater)	–	–	–	•	–	–	–	–	•	•	•	–
Querco pyrenaicae-Fraxino angustifoliae Gs. (soft freshwater)	–	–	–	•	–	–	–	–	•	•	•	–
Aceri granatensis-Fraxino angustifoliae Gs. (soft freshwater)	–	–	–	•	–	–	–	–	–	•	–	–
Viburno lantanae-Ulmo minoris Gs. (hard freshwater)	–	–	–	•	–	–	–	–	–	•	•	–
Salici lambertiano-albae Gs. (hard freshwater)	–	–	–	•	–	–	–	–	–	•	•	–
Betulo meridionalis-Salici albae Gs. (hard freshwater)	–	–	–	•	–	–	–	–		•	•	–
Agrostio stoloniferae-Myricario germanicae Ms. (hard freshwater)	–	–	–	•	–	–	–	–		•	•	–
Carici camposii-Salici atrocinereae Ms. (soft freshwater)	–	–	–	•	–	–	–	–		•	–	–
Rubo lainzii-Salici atrocinereae Ms. (soft freshwater)	–	–	–	•	•	–	–	–		•	•	–

(continued)

Mediterranean pluviseasonal-oceanic	Thermotypes						Ombrotypes					
Sigmataxa	1	2	3	4	5	6	7	8	9	10	11	12
Geopermasigmeta and permasigmeta												
Cistancho phelypaeae-Sarcocornio fruticosae Gps. (halophilous littoral)	•	•	–	–	–	–	•	•	–	–	–	–
Crithmo maritimi-Limonio malacitani Gps. (haloanemogamous rock littoral)	•	•	–	–	–	–	–	–	•	–	–	–
Limonio pseudodictyoclado-carregadorensis Gps. (haloanemogamous rock littoral)	–	•	–	–	–	–	–	•	–	–	–	–
Loto cretici-Ammophilo australis Gps. (coastal dune)	–	•	–	–	–	–	–	•	•	–	–	–
Crithmo maritimi-Limonio rigualii Gps. (haloanemogamous rock littoral)	–	•	–	–	–	–	–	–	•	–	–	–
Crithmo maritimi-Limonio dufourii Gps. (haloanemogamous rock littoral)	–	•	–	–	–	–	–	–	•	–	–	–
Limonio artruchio-minuti Gps. (haloanemogamous rock littoral)	–	•	–	–	–	–	–	–	•	–	–	–
Crithmo maritimi-Limonio girardiani Gps. (haloanemogamous rock littoral)	–	•	–	–	–	–	–	–	•	–	–	–
Crithmo maritimi-Limonio gibertii Gps. (haloanemogamous rock littoral)	–	•	–	–	–	–	–	–	•	–	–	–
Crithmo maritimi-Limonio balearici Gps. (haloanemogamous rock littoral)	–	•	–	–	–	–	–	–	•	•	–	–
Limonio emarginati Gps. (haloanemogamous rock littoral)	–	•	–	–	–	–	–	–	–	•	–	–
Puccinellio ibericae-Sarcocornio perennis Gps. (halophilous mareal)	–	•	•	–	–	–	–	–	•	–	–	–
Limonio bellidifolii-Sarcocornio fruticosae Gps. (halophilous)	–	•	•	–	–	–	–	–	•	–	–	–
Medicagini marinae-Ammophilo australis Gps. (coastal dune)	–	•	•	–	–	–	–	–	•	•	–	–
Limonio minuto-fontqueri Gps. (haloanemogamous rock littoral)	–	•	•	–	–	–	–	–	•	•	–	–
Crithmo maritimi-Dauco hispanici Gps. (haloanemogamous rock littoral)	–	•	•	–	–	–	–	–	•	•	–	–
Armerio ruscinonensis Gps. (haloanemogamous rock littoral)	–	–	•	–	–	–	–	–	–	•	–	–
Suaedo braunblanquetii Gps. (halophilous inland)	–	–	•	–	–	–	–	•	•	–	–	–
Otantho maritimi-Ammophilo australis Gps. (coastal dune)	–	–	•	–	–	–	–	–	–	•	•	–
Puccinellio maritimae-Sarcocornio perennis Gps. (halophilous mareal)	–	–	•	–	–	–	–	–	–	•	•	–
Puccinellio lagascanae Ps. (halophilous inland)	–	–	•	•	–	–	–	•	•	–	–	–
Puccinellio caespitosae-Suaedo braunblanquetii Ps. (halophilous inland)	–	–	•	•	–	–	–	•	•	–	–	–

(continued)

Mediterranean pluviseasonal-oceanic	Thermotypes						Ombrotypes					
Sigmataxa	1	2	3	4	5	6	7	8	9	10	11	12
Crithmo maritimi-Armerio pubigerae Gps. (haloanemogamous rock littoral)	–	–	•	–	–	–	–	–	–	•	•	–

1 = inframediterranean, 2 = thermomediterranean, 3 = mesomediterranean, 4 = supramediterranean, 5 = oromediterranean, 6 = cryoromediterraean, 7 = arid, 8 = semiarid, 9 = dry, 10 = subhumid, 11 = humid, 12 = hyperhumid

B.2. Potential Natural Vegetation Units (Sigmataxa) in the Mediterranean Pluviseasonal Oceanic Bioclimate with Steppic Variant

Mediterraean pluviseasonal-oceanic steppic	Thermotypes		Ombrotypes	
Sigmataxa	1	2	3	4
Climatophilousand edaphoxerophilous sigmeta and minorisigmeta				
Junipero phoeniceo-thuriferae S., Ms. (vertic, gypsicolous))	•	–	•	•
Arbuto unedonis-Pino halepensis S. (basophilous)	•	–	•	•
Permasigmetum				
Artemisio gargantae-Pucciniello pungentis Ps. (halophilous inland)	–	•	–	•

1 = mesomediterranean, 2 = supramediterranean, 3 = semiarid, 4 = dry.

B.3. Potential Natural Vegetation Units (Sigmataxa) in the Mediterranean Pluviseasonal Continental Bioclimate with Steppic Variant

Mediterraean pluviseasonal-continental	Thermotypes				Ombrotypes		
Sigmataxa	1	2	3	4	5	6	7
Climatophilous and edaphoxerophilous sigmeta							
Genisto versicoloris-Junipero hemisphaericae S. (acidophilous, mafic)	–	–	•	–	–	•	•
Edaphohygrophilous minorisigmeta							
Pyro bourgaeanae-Flueggeo tinctoriae Ms. (soft freshwater)	•	•	–	–	•	•	–
Geopermasigmeta							
Erigeronto frigidi-Festuco clementei Gps. (acidophilous, mafic)	–	–	–	•	•	–	–

1 = thermomediterranean, 2 = mesomediterranean, 3 = oromediterranean, 4 = cryoromediterraean, 5 = dry, 6 = subhumid, 7 = humid

B.4. Potential Natural Vegetation Units (Sigmataxa) in the Mediterranean Xeric Oceanic Bioclimate

Mediterranean xeric-oceanic	Thermotypes				Ombrotypes			
Sigmataxa	1	2	3	4	5	6	7	8
Climatophilous sigmeta and minorisigmeta								
Maytenoeuropaei-Periploco angustifoliae S. (mafic)	•	•	–	–	•	•	–	–
Zizipho loti Ms. (dune)	•	•	–	–	•	•	–	–
Zizipho loti-Mayteno europaei S. (psamophilous)	•	•	–	–	•	•	–	–
Rhamno rivagodayani-Periploco angustifoliae S. (basophilous)	•	•	–	–	–	•	–	–
Calicotomo intermediate-Mayteno intermediate Ms. (basophilous)	•	•	–	–	–	•	•	–
Cneoro tricocci-Ceratonio siliquae S. (basophilous)	–	•	–	–	–	•	•	–
Cneoro tricocci-Pistacio lentisci Ms. (basophilous)	–	•	–	–	–	•	–	–
Arisaro simorrhini-Tetraclinido articulatae S. (basophilous)	–	•	–	–	–	•	–	–
Chamaeropo humilis-Rhamno lycioidis Ms.	–	•	–	–	–	•	–	–
Mayteno europaei-Oleo sylvestris S. (basophilous, mafic)	–	•	–	–	–	•	–	–
Aristolochio baeticae-Junipero turbinatae Ms. (basophilous)	–	•	–	–	–	•	•	•
Pistacio lentisci-Pino halepensis S. (basophilous, calcodolomiticolous)	–	•	•	–	–	•	•	–
Rhamno lycioidis-Querco cocciferae S. (basophilous)	–	–	•	–	–	•	•	–
Querco cocciferae-Pino halepensis S. (basophilous)	–	–	•	–	–	•	•	–
Ephedro fragilis-Pino halepensis S.	–	–	•	–	–	•	•	–
Rhamno almeriensis-Pino halepensis S. (calcodolomiticolous)	–	–	•	–	–	•	•	•
Edaphoxerophilous sigmeta and minorisigmeta								
Rhamno angustifolii-Junipero turbinatae Ms. (coastal dune)	•	•	–	–	•	•	–	–
Corematо albi-Junipero macrocarpae Ms. (coastal dune)	–	•	–	–	–	•	–	–
Junipero turbinatae-Pino halepensis S. (basophilous)	–	•	–	–	–	•	•	–
Asparago horridi-Junipero turbinatae S. (calcodolomiticolous)	–	•	–	–	–	–	•	•
Hygrophilous geosigmeta and sigmeta								
Zizipho loti-Nerio oleandri S. (very hard freshwater)	•	•	–	–	•	•	–	–
Inulo crithmoidis-Tamarici boveanae Gs. (halophilous)	•	•	–	–	•	•	–	–
Lonicero biflorae-Populo albae Gs. (hard freshwater)	–	•	–	–	–	•	–	–
Rubio longifoliae-Nerio oleandri S. (hard freshwater)	–	•	•	–	–	•	–	–
Saccharo ravennae-Tamarici canariensis Gs. (hard freshwater)	–	•	•	–	–	•	•	–
Rubo ulmifolii-Nerio oleandri S. (hard freshwater)	–	•	•	–	–	–	•	•
Limonio delicatuli-Nerio oleandri S. (hard freshwater)	–	–	•	–	–	•	–	–
Suaedo braunblanquetii-Tamarici boveanae Gs. (watercourse and lacustrine, halophilous)	–	–	•	–	–	•	•	–
Rubio tinctorum-Populo albae Gs. (hard freshwater)	–	–	•	–	–	•	•	–
Crataego granatensis-Salici neotrichae Gs. (hard freshwater)	–	–	•	•	–	•	•	•
Suaedo braunblanquetii-Tamarici canariensis Ms. (lacustrine, halophilous)	–	–	•	•	–	–	•	–
Geopermasigmeta and permasigmatea								
Cistancho phelypaeae-Sarcocornio fruticosae Gps. (halophilous littoral)	•	•	–	–	•	•	–	–

(continued)

Mediterranean xeric-oceanic	Thermotypes				Ombrotypes			
Sigmataxa	1	2	3	4	5	6	7	8
Frankenio corymbosae-Arthrocnemo macrostachyi Ps. (halophilous littoral)	•	•	–	–	•	•	–	–
Limonio cossoniani-Lycio intricati Ps. (haloanemogenous rock littoral)	•	•	–	–	–	•	–	–
Limonio pseudebusitani Gps. (haloanemogenous rock littoral)	–	•	–	–	–	•	–	–
Limonio caprariensis Gps. (haloanemogenous rock littoral)	–	•	–	–	–	•	–	–
Loto cretici-Ammophilo australis GPS (coastal dune)	–	•	–	–	–	•	•	–
*Suaedo braunblanquetii*Ps. (halophilous inland)	–	–	•	–	–	•	•	–
Puccinellio caespitosae-Suaedo braunblanquetii Gps. (halophilous inland)	–	–	•	•	–	•	•	–
Puccinellio lagascanae Ps. (halophilous inland)	–	–	•	•	–	•	•	–

1 = inframediterranean, 2 = thermomediterranean, 3 = mesomediterranean, 4 = supramediterranean, 5 = arid, 6 = semiarid, 7 = dry, 8 = subhumid

B.5. Potential Natural Vegetation Units (Sigmataxa) in the Mediterranean Xeric Bioclimate with Steppic Variant

Mediterranean xeric-oceanic steppic			
Sigmataxa	mme	sar	dry
Climatophilous sigmeta			
Junipero phoeniceo-thuriferae S. (vertic, giypsicola)	•	•	•
Climatophilous and xerophilous sigmeta			
Arbuto unedonis-Pino halepensis S. (basophilous)	•	•	•

mme mesomediterranean, *sar* semiarid, *dry* dry

B.6. Potential Natural Vegetation Units (Sigmataxa) in the Mediterranean Desertic Oceanic Bioclimate

Mediterraean desertic-oceanic				
Sigmataxa	ime	tme	ari	sar
Climatophilous sigmeta				
Mayteno europaei-Periploco angustifoliae S. (basophilous)	•	•	•	–
Zizipho loti-Mayteno europaei S. (basophilous)	•	•	•	•
Minorisigmeta				
Zizipho loti Ms. (dune)	•	•	•	–
Rhamno angustifolii-Junipero turbinatae Ms. (dune)	•	•	•	•
Hygrophilous geosigmeta				
Inulo crithmoidis-Tamarici boveanae Gs. (halophilous)	•	•	•	•
Zizipho loti-Nerio oleandri Gs. (very hard freshwater)	•	•	•	•

(continued)

Mediterraean desertic-oceanic				
Sigmataxa	ime	tme	ari	sar
Geopermasigmeta and permasigmeta				
Cistancho phelypaeae-Sarcocornio fruticosae Gps. (halophilous littoral)	•	•	•	•
Frankenio corymbosae-Arthrocnemo macrostachyi Pss. (halophilous littoral)	•	•	•	•

ime inframediterranean, *tme* thermomediterranean, *ari* arid, *sar* semiarid

References

Almarza C (2000) Variaciones climáticas en España. Época instrumental. In: Balarión L (ed) El cambio climático. El Campo de las Ciencias y de las Artes. Servicio de Estudios del BBVA, Madrid, pp 69–86

Amigo J, Ramírez C (1998) A bioclimatic classification of Chile: woodland communities in the temperate zone. Plant Ecol 136:9–26

Bagnouls F, Gaussen H (1953) Saison sèche et régime xérothermique. Documents pour les cartes des productions végétales, t. III, vol. I, art. 8, p 47. Toulouse

Bagnouls F, Gaussen H (1957) Les climats biologiques et leur classification. Ann. Geogr. Paris.

Barber A, Tun J, Crespo MB (2001) A new approach of the bioclimatology and potential vegetation of the Yucatán Peninsula (México). Phytocoenologia 31(1):1–31

Blasi C (1996) Un approccio fitoclimatico allo studio dei cambiamenti climatici in Italia. S It E Atti 17:39–43

Box EO (1981) Macroclimate and plant forms: an introduction to predictive modeling in Phytogeography. Tasks for vegetation science, vol 1. Junk Publ, La Haya, pp 1–258

Budong Q, Xu H, Corte Real J (2000) Spatial temporal structures of quasi-periodic oscillations in precipitation over Europe. Int J Climatol 20:1583–1598

Camarero JJ, Gutiérrez E (2004) Pace and pattern of recent treeline dynamics: response of ecotones to climatic variability in the Spanish pyrenees. Climate Change 63:181–200

Copons R, Bordonau J (1997) El registro glacial correspondiente a la pequeña Edad del Hielo en la Península Ibérica. In: Ibáñez JJ, Valero BL, Mahado C (eds) El paisaje mediterráneo a través del espacio y del tiempo. Implicaciones en la desertificación. Geoforma Ediciones, Logroño, pp 295–310

Creus J, Saz MA (1999) Estudio de la variabilidad climática del último milenio a partir de series de temperatura y precipitación reconstruidas en el NE español. In: Raso Nadal JM, Martín-Vide J (eds) La Climatología en los albores del siglo XXI. Asociación Española de Climatología (AEC), Barcelona, pp 155–164

Daget P (1977) Le bioclimat méditerranéen: caractères généraux, modes de caractérisation. Vegetatio 34(1):1–20

Davis BAS et al (2003) The temperature of Europe during the Holocene reconstructed from pollen data. Quat Sci Rev 22(15–17):1701–1716

de Bolòs O (1989) Bioclimatologia i geografia botànica. Mem Real Acad Ci Barcelona 48 (9):423–444

del Río S (2005) El cambio climático y su influencia en la vegetación de Castilla y León (España). Itinera Geobot 16:5–535

del Río S, Penas A (2006a) Potential areas of evergreen forests in Castile and Leon (Spain) according to future climate change. Phytocoenologia 36(1):45–66. doi:10.1127/0340-269X/ 2006 /0036-0045

del Río S, Penas A (2006b) Potential distribution of semi-deciduous forests in Castile and Leon (Spain) in relation to climatic variations. Plant Ecol 185:269–282. doi:10.1007/s11258-006-9103-x
del Río S, Penas A, Fraile R (2005a) Analysis of recent climatic variations in Castile and Leon (Spain). Atmos Res 73(1–2):69–85. doi:10.1016/j.atmosres.2004-06-005
del Río S, Penas A, Pérez-Romero R (2005b) Potential areas of deciduous forests in Castile and Leon (Spain) according to future climate change. Plant Biosyst 139(2):222–233. doi:10.1080/11263500500158835
del Río S, Fraile R, Herrero L, Penas A (2007) Analysis of recent trends in mean maximum and minimum temperatures in a region of the NW of Spain (Castilla y León). Theor Appl Climatol 90(1–2):1–12. doi:10.1007/s00704-006-0278-9
del Río S, Herrero L, Penas A (2009) Recent climatic trends in Castilla and León (Spain) and its possible influence on the potential natural vegetation. Acta Bot Gallica 156(4):625–636
del Río S, Herrero L, Pinto-Gomes C, Penas A (2011a) Spatial analysis of mean temperature trends in Spain over the period (1961–2006). Glob Planet Chang 78(1–2):65–75. doi:10.1016/j.gloplacha.2011.05.012
del Río S, Herrero L, Fraile R, Penas A (2011b) Spatial distribution of recent rainfall trends in Spain (1961–2006). Int J Climatol 31(5):656–667. doi:10.1002/ joc.2111
Donadieu P (1977) Contribution à une synthèse bioclimatique et phytogéographique au Maroc. Inst. Agron. et Vétérinaire Hassan II, Maroc, pp 1–155
Duplessy JC, Pons A, Fantechi R (eds) (1991) Climate and global change. Commission of the European Communities, Luxembourg
Eddy JA (1976) The Maunder minimum. Science 192:1189–1202
Emberger L (1955) Une classification biogéographique des climats. Rec des travaux des Labor de Bot Géol et Zool de la fac des Sci de l'Univ de Montpellier Sér Bot 7:3–43
Emberger L 1959 Rapport sur la carte écologique de la zone méditerranéenne. Annexe 2. F.A.O./U.N.E.S.C.O. Study Group on the ecological map of the Mediterranean Region.
Fernández-González F (1997) Bioclimatología. In: Izco J et al (eds) Botanica. McGraw-Hill, Madrid, pp 607–682
Fernández-González F, Loidi J, Moreno JC (2005) Impactos sobre la biodiversidad vegetal, 5. In: Moreno JM (ed) Evaluación preliminar de los impactos en España por efecto del cambio climático. MMA. Secretaría General Técnica. Centro de Publicaciones, Madrid, pp 183–248
Galán E, Basilla D, Fernández F, Cervera B (1999) Evolución de las precipitaciones anuales en la meseta meridional durante el siglo XX. In: Raso Nadal JM, Martín-Vide J (eds) La Climatología en los albores del siglo XXI. Asociación Española de Climatología (AEC), Barcelona, pp 169–180
Gaussen H (1947) La carte botanique du monde au 1/1000000. C R Acad Sci 224:589
Gaussen H (1954) Expression du milieu par des formules écologiques. Leur représentation cartographique. Les divisions écologiques du monde. Colloques internationaux du C.N.R.S. Paris et Ann. Biol. 31(5–6): 257–267. Paris
Gaussen H (1955a) Détermination des climats par la méthode des courbes ombrothermiques. Compt Rend Hebd Séances Acad Sci 240:642–644
Gaussen H (1955b) Les climats analogues à l'échelle du monde, vol 41. C. R. Acad. Agr., Paris
Gavilán R, Fernández-González F (1997) Climatic discrimination of Mediterranean broadleaved sclerophyllous and deciduous forests in Central Spain. J Veg Sci 8:377–386
Gavilán R et al (1998) Climatic classification and ordination of the Spanish Sistema central: relationships with potential vegetation. Plant Ecol 139:1–11
Gavilán R, Fernández-González F, Rivas-Martínez S (2001) Variaciones bioclimáticas en Madrid: un estudio sobre el cambio climático local. In: Gómez Mercado F, Mota Poveda JF (eds) Vegetación y Cambios Climáticos. Servicio de Publicaciones, Universidad de Almería, Almeria, pp 243–256
Goddard L, Mason SJ, Zebiak SE, Ropelewski CF, Bascher R, Cane MA (2001) Current approaches to seasonal-to-interanual climate predictions. Int J Climatol 21:1111–1152
González-Hidalgo JC, de Luis M, Reventós J, Sánchez JR (2001) Spatial distribution of seasonal rainfall trends in a Western Mediterranean area. Int J Climatol 21:843–860

Grabherr G et al (1994) Climate effects on mountain plants. Nature 369:448
Herráez I 1999 El cambio climático: fuentes de información. In: Ruiz Zapata B, Dorado M, Gil MJ, Valdeolmillos A (eds) Efectos del cambio climático en la Región Mediterránea durante los últimos 3000 años. Madrid, pp 40–53
Hueck K, Seibert P (1972) Vegetationskarte von Südamerika. G. Fischer-Verlag, Stuttgart
Jalut G et al (2000) Holocene climatic changes in the western Mediterranean, from South-East France to South-East Spain. Palaeogeogr Palaeoclimatol Palaeoecol 160:255–290
Ladero M, Díaz TE, Penas A, Rivas-Martínez S, Valle C (1987) Datos sobre la vegetación de las Cordilleras Central y Cantábrica (II Excursión Internacional de Fitosociología). Itinera Geobot 1:3–147
Ladero M, Valle CJ, Santos MT, González MT, García PA, García-Baquero G, Hernández P (1994) Mapas ombroclimáticos de las provincias de Salamanca y Zamora. Anais do Instituto Superior de Agronomía 44:733–753
Le Houerou N (1986) The desert and arid zones of Northern Africa. In: Hot deserts and arid shrublands. Elsevier Publications, Amsterdam, pp 101–147
Le Houerou N (1995) Bioclimatologie et biogeography des steppes arides du Nord de l'afrique. Options Méditerr Série B 10:1–396. Montpellier
Lousã M (2004) Bioclimatologia e series de vegetação de Portugal. Lazaroa 25:83–86
Luengo MA, Penas A, Rivas-Martínez S (inéd.). Biocli. Programa Informático para el procesamiento de datos climáticos. www.globalbioclimatics.org
Manrique E, Fernández-Cancio A (2000) Extreme climatic events in dendroclimatic reconstructions from Spain. Clim Chang 44:123–138
Meher-Homji VM (1963) Les bioclimats du Sub-continent Indien el leurs types analogues dans le Monde. Doc Cartes des Prod Veget 4(1):3–254
Menzel A, Fabian P (1999) Growing season extended in Europe. Nature 397:659
Mesquita S (2005) Modelaçao Bioclimatica de Portugal continental. Instituto Superior Técnico, Universidade Técnica de Lisboa, Lisboa
Mesquita S, Capelo J, de Sousa J (2004) Bioclimatologia da Ilha da Madeira: abordagem numérica. Quercetea 6:47–59
Montero de Burgos JL, González Rebollar JL (1987) Diagramas bioclimáticos. In: Rivas-Martínez S (ed) Memoria del mapa de series de vegetación de España, ICONA, Serie Técnica. Madrid. ICONA, Madrid, pp 227–268
Moreno JM (2007) El cambio climático antropogénico es un asunto de ayer, no de mañana: la acción no puede esperar. Lección inaugural del Curso Académico 2007/2008 de la Universidad de Castilla-La Mancha. Ed. Servicio de Publicaciones de la Universidad de Castilla-La Mancha. ISBN 978-84-8427-583-1
Moreno JM, Oechel W (1995) Global chance and Mediterranean-type ecosystems. Ecological studies, vol 117. Springer-Verlag
Moreno JM, Díaz Pineda F, Rivas-Martínez S (1990) Climate and vegetation at the Eurosiberian-Mediterranean boundary in the Iberian Peninsula. J Veg Sci 1: 233–244.
Murphy J (2000) Predictions of climate change over Europe using statistical and dynamical downscaling techniques. Int J Climatol 20:489–501
Navarro G, Ferreira W (2007) Mapa de vegetación de Bolivia, escala 1: 250.000. CD-Rom. The Nature Conservancy. ISBN:978-99954-0-168-9
Navarro G, Maldonado M (2002a) Geografía ecológica de Bolivia. Vegetación y ambientes acuáticos. Editorial: Centro de Ecología Simón I. Patiño-Departamento de Difusión Cochabamba Bolivia ISBN:99905-0-225-0
Navarro G Maldonado M (2002b) Geografía Ecológica de Bolivia. Ed. Centro de Ecología Simón. Patiño, 719 pp. ISBN: 99905-0-225-0. Cochabamba, Bolivia
Neq M, Todd M, Hulme M, Jones P (2001) Precipitation measurement and trends in the twentieth century. Int J Climatol 21:1889–1922
Orlóci L (1994) Global warming. The process and its anticipated phytoclimatic effects in temperate and cold zone. Coenoses 9(2):69–74

Oturbay A, Loidi J (2001) Cambio climático: predicción de su influencia en la distribución de especies arbóreas en el País Vasco. In: Gómez Mercado F, Mota Poveda JF (eds) Vegetación y Cambios Climáticos. Servicio de Publicaciones Universidad de Almería, Almería, pp 165–176

Peñalba MC (1994) The history of the Holocene vegetation in northern Spain from pollen analysis. J Ecol 82:815–832

Pons A, Reille M (1988) The Holocene-Pleistocene and upper-Pleistocene pollen record from Padul (Granada, Spain) – a new study. Palaeogeogr Palaeoclimatol Palaeoecol 66:243–263

Pott R (1998) La evolución postglaciar de los bosques frondosos caducifolios en Europa. Itinera Geobot 11:32–65

Pott R (2001) The alpine timber-line as a mirror of climatic fluctuations. In: Gómez Mercado F, Mota Poveda JF (eds) Vegetación y Cambios Climáticos. Servicio de Publicacíones Universidad de Almería, Almería, pp 119–138

Quézel P (1957) Peuplement vegetal des hautes montagnes de l'Afrique du Nord. Lechevalier, Paris, p 463

Quézel P (1965) La végétation du Sahara. Fischer Verlag, Stuttgart, p 333

Quézel P (1999) Les grandes structures de végétation en region méditerranéenne: facteurs déterminants dans leur mise en place post-glaciaire. Geobios 32:19–32

Quézel P, Barbero M (1993) Variations climatiques au Sahara et en Afrique sèche depuis le Pliocéne. Bull d'Ecologie 24(2–4):191–202

Ramil-Rego P, Muñoz C, Iriarte MJ, Gómez-Orellana L, Rodríguez MA (2001) Vegetación y cambio climático en los territorios del norte de la Península Ibérica durante los últimos 18000 años. In: Gómez Mercado F, Mota JF (eds) Vegetación y Cambios Climáticos. Servicio de Publicacíones Universidad de Almería, Almería, pp 139–149

Rivas-Martínez S (1981) Les étages bioclimatiques de la végétation de la Péninsule Ibérique. Anales JardBot Madrid 37(2):251–268

Rivas-Martínez S (1982a) Les étages bioclimatiques, secteurs chorologiques et series de végétation de l'Espagne mediterranén ne. Ecologia Mediterr 8(1–2):275–288

Rivas-Martínez S (1982b) Mapa de las series de vegetación de la provincia de Madrid. Publ. Serv. Forestal del Medio Ambiente y Contra Incendios, Diputac. Provinc, Madrid

Rivas-Martínez S (1983) Series de vegetación de la región eurosiberiana de la Península Ibérica. Lazaroa 4:155–166

Rivas-Martínez S (1984) Pisos bioclimáticos de España. Lazaroa 5:33–43

Rivas-Martínez S (1985) Biogeografía y Vegetación. Discurso de ingreso como Académico de Número. Real Academia de Ciencias Exactas, Físicas y Naturales, Madrid, p 86

Rivas-Martínez S (1987a) Mapa de series de vegetación de España. ICONA, Serie Técnica. 268 p. + 30 mapas, Madrid

Rivas-Martínez S (1987b) Nociones sobre Fitosociología, Biogeografía y Bioclimatología. In: Peinado M, Rivas-Martínez S (eds.) La vegetación de España. Servicio de Publicaciones. Universidad de Álcala de Henares, Madrid, pp 19–45. ISSN: 84-600-5146-3

Rivas-Martínez S (1988) Bioclimatología, Biogeografía y Series de Vegetación de Andalucía occidental. Lagascalia 15(extra):91–119

Rivas-Martínez S (1991) Bioclimatic belts of West Europe (relations between bioclimate and plant ecosystems). Proc. Eur. School Climate Nat. Hazards Course (Arles, 1990). pp 225–246. Strasbourg

Rivas-Martínez S (1996) Geobotánica y Climatología. Discurso investidura Dr. 'honoris causa' Universidad de Granada. Serv. Publ. Universidad de Granada, Granada, p 98

Rivas-Martínez S (1997) Syntaxonomical synopsis of the potential natural plant communities of North America, I. Itinera Geobot 10:5–148

Rivas-Martínez S (2001) Biogeographic map of South America: scale 1:22 mill. Cartographic Service, University of León, Spain (30.04.2001)

Rivas-Martínez S (2004) Sinopsis biogeográfica, bioclimática y vegetacional de América del Norte. Fitosociología 41(1), suppl. 2:19–52

Rivas-Martínez S (2005a) Avances en Geobotánica. Discurso apertura curso 2005. Real Acad. Farmacia, Madrid, p 142

Rivas-Martínez S (2005b) Notions on dynamic-catenal phytosociology as a basis of landscape science. Plant Biosyst 139:135–144

Rivas-Martínez S (2006) Series, geoseries y geopermaseries de vegetación de España. Acto Acad. Invest. Doctores Honoris Causa: 55–145. 84-9773-253-7. Publ Univ León

Rivas-Martínez S (2007) Globalbioclimatics, internet: http://www.globalbioclimatics.org

Rivas-Martínez S (2009) Ensayo geobotánico global sobre la Macaronesia. In: Beltrán Tejera E, Afonso Carrrillo J, García Gallo A, Rodríguez Delgado O (eds) Homenaje al Profesor Dr. Wolfredo Wildpret de la Torre. Instituto de Estudios Canarios, La Laguna (Tenerife. Islas Canarias). Monografía 78:255–269. ISBN:978-84-88366-82-5

Rivas-Martínez S, Costa M (1998) Datos sobre la vegetación y el bioclima del Valle de Arán. Acta Bot Barcinon 45:473–499

Rivas-Martínez S, Rivas Sáenz S (2009) Synoptical worldwide bioclimatic classification system (summarized table). Phytosociological Research Center. Web site at http://www.globalbioclimatics.org/book/namerica2/table.htm

Rivas-Martínez S, Báscones JC, Díaz González TE, Fernández-González F, Loidi J (1991) La vegetación del Pirineo Occidental y Navarra. Itinera Geobot 5:5–456

Rivas-Martínez S, Sánchez-Mata D, Costa M (1999a) North American Boreal and western temperate vegetation. Itinera Geobot 12:5–316

Rivas-Martínez S, Cantó P, Fernández-González F, Molina JA, Pizarro JM, Sánchez-Mata D (1999b) Synopsis of the Sierra de Guadarrama vegetation. Itinera Geobot 13:189–206

Rivas-Martínez, coauthors: Andalucía (Alfredo Asensi, Blanca Díez-Garretas, Joaquín Molero, Francisco Valle & Eusebio Cano), Aragón (Manuel Costa & María Luisa López), Asturias (Tomás E. Díaz & José Antonio F. Prieto), Baleares (Leonardo Llorens), Canarias (Marcelino del Arco, W. Wildpret, P.L. Pérez de Paz, O. Rodríguez, J.R. Acebes, A. García, V. E. Martín, J. A. Reyes Betancort, M. Salas, M.A. Díaz, J.A. Bermejo, R. González, M.V. Cabrera y S. García), Cantabria (José Antonio F. Prieto & Tomás E. Díaz), Castilla-La Mancha (Federico Fernández & Daniel Sánchez-Mata), Castilla y León (Ángel Penas), Cataluña (Ramón Masalles & Manuel Costa), Ceuta y Melilla (Alfredo Asensi & Blanca Díez-Garretas), Extremadura (Miguel Ladero & Ángel Amor), Galicia (Jesús Izco & Javier Amigo), La Rioja (Javier Loidi, José Antonio Molina & Gonzalo Navarro), Madrid (Paloma Cantó), Murcia (Francisco Alcaraz), Navarra (Javier Loidi & Juan Carlos Báscones), País Valenciano (Manuel Costa & Pilar Soriano), País Vasco (Javier Loidi) (2007) Mapa de series, geoseries y geopermaseries de vegetación de España (Memoria del mapa de vegetación potencial de España,). Parte I. Itinera Geobot 17:5–436

Rivas-Martínez S, Rivas Sáenz S, Penas A (2011a) Worldwide bioclimatic classification system. Glob Geobot 1(1):1–634. + 4 maps

Rivas-Martínez, coauthors: Andalucía (Alfredo Asensi, Blanca Díez-Garretas, Joaquín Molero, Francisco Valle & Eusebio Cano), Aragón (Manuel Costa & Luis Villar), Asturias (Tomás E. Díaz & José Antonio F. Prieto), Baleares (Leonardo Llorens), Canarias (Marcelino del Arco, W. Wildpret, P.L. Pérez de Paz, O. Rodríguez, J.R. Acebes, A. García, V. E. Martín, J. A. Reyes Betancort, M. Salas, M.A. Díaz, J.A. Bermejo, R. González, M.V. Cabrera y S. García), Cantabria (José Antonio F. Prieto & Tomás E. Díaz), Castilla-La Mancha (Federico Fernández & Daniel Sánchez-Mata), Castilla y León (Ángel Penas, Luis Herrero & Sara del Río), Cataluña (Ramón Masalles & Manuel Costa), Ceuta y Melilla (Alfredo Asensi & Blanca Díez-Garretas), Extremadura (Miguel Ladero & Ángel Amor), Galicia (Jesús Izco & Javier Amigo), La Rioja (Javier Loidi & Gonzalo Navarro), Madrid (Paloma Cantó), Murcia (Francisco Alcaraz), Navarra (Javier Loidi & Juan Carlos Báscones), País Valenciano (Manuel Costa & Pilar Soriano), País Vasco (Javier Loidi) (2011b) Mapa de series, geoseries y geopermaseries de vegetación de España (Memoria del mapa de vegetación potencial de España, 2011). Parte II. Itinera Geobot 18(1):5–424

Rodríguez MA, Muñoz C, Ramil-Rego P (2001) Variaciones espaciales en la distribución de la vegetación como respuesta a los cambios climáticos globales, a partir del último máximo glaciar, en la Serra dos Ancares (NW Ibérico). In: Gómez Mercado F, Mota JF (eds) Vegetación y Cambios Climáticos. Servicio de Publicaciones Universidad de Almería, Almería, pp 153–164

Ruíz-Zapata B (1999) El cambio climático en la Península ibérica. In: Ruíz-Zapata B, Dorado M, Gil MJ, Valdeolmillos A (eds) Efectos del cambio climático en la Región Mediterránea durante los últimos 3000 años. Madrid, pp 86–95

Sánchez Egea J (1975) El Clima. Los dominios climácicos y los pisos de vegetación de las provincias de Madrid. Ávila y Segovia: ensayo de un modelo fitoclimático Anal Jard Bot 32 (2):1039–1078

Sanz-Elorza M et al (2003) Changes in the high-mountain vegetation of the Central Iberian Peninsula as a probable sign of global warming. Ann Bot 92:273–280

Solanki SK, Usoskin IG, Kromer B, Schüssler M, Beer J (2004) Unusual activity of the sun during recent decades compared to the previous 11.000 years. Nature 431:1084–1087

Sparks TH, Menzel A (2002) Observed changes in seasons: an overview. Int J Climatol 22:1715–1725

Takhtajan A (1986) In: Cronquist A (ed) Floristic regions of the world. University of California Press, Berkeley, p 522

Thornthwaite CW (1984) An approach towards a rational classification of climate. Geogr Rev 38:55–94

Trochain J (1952) Les territoires phytogéographiques de l'Afrique noire francaise d'après leur pluviométrie. Recueil de trav. de la Fac. des Sc. de Montpellier. fasc. 5

Troll C, Paffen K (1964) Die Jahreszeitenklimate der Erde. (The seasonal climates of the Earth). Erdkunde 18:5–28. + map

Chapter 3
Dynamism in Vegetation. Vegetation Changes on a Short Time Scale

Javier Loidi

Abstract The study of dynamic processes is one of the main issues in vegetation ecology. This chapter is an attempt to summarize the basic concepts to be applied when studying the dynamics of vegetation. As succession is the central process in vegetation dynamics, some relevant theories and models are explained, especially those concerning inhibition, facilitation and tolerance. Plant strategies related to dynamism, particularly those of r- and K-selection and Grime's CRS model, are commented upon. Several types of succession are distinguished, and the concept of Vegetation Series (Sigmetum) is introduced, with comments on the diverse types, i.e. the climatophilous, edaphoxerophilous and edaphohygrophilous vegetation series. Also the sigmeta under extreme conditions are explained (permasigmeta or permaseries). Different disturbance regimes result in specific seral vegetation types and the persistence of these disturbance regimes shape the landscapes, particularly in heavily populated territories. The extant landscapes are a cultural construction in combination with the natural biotic element. In order to formalize a dynamic-catenal vegetation science, the concept of geoseries or geosigmetum is explained, separated into topographic geoseries (across an eco-topographic gradient) and cliserial geoseries (across an altitudinal gradient in mountains). Finally the zonations of permaseries (geopermasigmeta or geopermaseries) are discussed.

3.1 Introduction

The idea that vegetation is not static and changes occur over time is old, but it was Clements, originally in his publications of 1916 and 1928, who formulated a "theory of succession of the plant communities". Shifting environmental factors are responsible for vegetation change over time, and these can include climatic factors as well as the hydrologic regime of the soil, the effects of substrate erosion, etc. However, the temporal scale of a noticeable effect of those variations is longer than the brevity of the life span of most of plants. This makes it possible to separate

J. Loidi (✉)
Department of Plant Biology and Ecology, University of the Basque Country (UPV/EHU), Bilbao, Spain
e-mail: javier.loidi@ehu.eus

J. Loidi (ed.), *The Vegetation of the Iberian Peninsula*, Plant and Vegetation 12,
DOI 10.1007/978-3-319-54784-8_3

the changes driven by environmentally long-span variations from those due to the internal dynamics of genetic variation. The population and community dynamics that are mainly governed by the disturbance regime occurring in the ecosystem, can be considered independent from the environmental temporal variation happening at a slower pace. Thus we can consider the changes happening in plant populations and communities at a brief time scale within the framework of what we will call ***dynamics***, and it can be defined *as a set of mechanisms which drive **succession**, and which correspond to a change or evolution in plant communities at a determined site over time.*

Succession integrates variations in the relationships among coexisting plant populations that together constitute the plant community. It integrates the history of the competition among them with the integration of their life cycles in an unstable balance. This instability results in change and tends to promote the progressive dominance of the most competitive species. In general terms, succession starts with pioneer stages, is followed by the intermediate ones and finishes with the mature stage. This progressive tendency towards dominance of the most competitive species leads to a progressive slowing down of the speed of change until a certain stability or "steady state" is reached. This stable, or mature, phase (Climax by Clements, or Potential Natural Vegetation by Tüxen) will be in balance with the environmental conditions and will be altered only when these environmental conditions change or when a disturbance episode happens. The different stages of succession can be formally defined as associations, so that each succession can be regarded as series of associations replacing one another over time. Thus, we can formally define a determined succession course by knowing the associations composing that successional course and comparing it with others. In this way we can develop a body of organized knowledge about succession.

3.2 Some Theories About Models and Mechanisms of Succession

It is evident that during succession a multitude of phenomena interact to influence the performance of the different plant species within the community. For instance, the availability of seeds and other propagules (seed rain, etc.), herbivory of selected species, capability of micro-environmental changes, and other factors (Glenn-Lewin and van der Maarel 1992). In spite of the complexity of such related phenomena, there have been some interesting proposals to model succession.

One of the best known theories is that of Connell and Slatyer (1977) in which three models for succession can be distinguished: facilitation, inhibition and tolerance.

Facilitation This postulates that the preceding community creates the conditions suitable for establishing the subsequent stage. In this model the early stages will prepare, by means of changing the local conditions (soil, micro-climate, shelter, etc.), the establishment of more advanced stages. This responds to the Clementsian (Clements 1928) concept of succession in vegetation.

Inhibition This opposes the facilitation model, and it postulates that the early stages create or influence conditions that impede the establishment of the subsequent ones. Only the die off of plants of the early stages will allow the restart of succession towards mature stages.

Tolerance In this model succession will be determined by the competitive capacities and life span of the different species in such a way that the most long-lived and competitive will ultimately survive longer and dominate in the mature stages. The established species do not inhibit nor favor the germination and establishment of new species; each of the populations will depend on their own abilities to survive. This presumes neutrality in the interaction of species as regards recruitment.

These three models are not mutually exclusive and can be combined in a complete successional course.

Another attempt to model succession is that of Egler (1954), developed for abandoned fields ('old fields'); it allows for two hypotheses:

Relay floristics. There are successive groups of species which are replaced over time as a result of changes in the environmental conditions occurring at the site: a sequence of changes in species composition occurs. This model harmonizes with the Clementsian conceptions.

Initial floristic composition. This model considers that most of the species initially colonize the site together: the species of the pioneer stage, the intermediate stages and the mature stages, and they will dominate in their respective stages. First all the species are present but only those of the initial stages develop, until this phase is over and they are replaced by the emerging ones of subsequent stages. In this model, succession only depends on the developmental dynamics of the species initially established and not on the changes in the environmental conditions induced by previous stages.

3.3 Plant Strategies in Relation to Succession

As plant strategy we understand as the set of morphological and functional adaptations of plants that ensure the success in establishment and occupation of a certain site and an optimal exploitation of its resources. Thus, plants will have an adaptive syndrome developed to surmount the difficulties of survival, in interaction with the environment and in competition with other plants. The significance of those strategies for the dynamics of vegetation is very high as they are characteristic of the different stages of succession, from initial to mature.

Some authors proposed a model of two extreme basic strategies. The most common is that of r- and K-selection by MacArthur and Wilson (1967) which proposes two extreme possibilities for plants:

The **r-selected** plants are short-lived (often therophytes), have a fast growth, low stature, low proportion of roots to total plant weight, high production of small seeds that are easily dispersed over long distances and able to keep their germination capacity for a long time. Accordingly, they contribute substantially to the soil seed bank of the surrounding areas. As a result of their low competitive abilities these species are typical in the pioneer stages of succession, disappearing as soon as other species develop.

The **K-selected** plants possess the opposite syndrome: they are long-lived, usually big-sized, slow-growing, have a low seed production without efficient dispersal mechanisms and the seeds do not retain their germination capability longer than a few weeks, making them unsuitable to contribute significantly to a seed bank. However, these species are highly competitive in terms of exploiting the local resources of water, nutrients and light, ultimately outcompeting other plants and becoming dominant in the mature phases of succession.

Besides this dual-strategy model, there are models based in three extreme strategies, such as the triangular model by Grime (1979, 1985). Each of the primary strategies is represented by one vertex of a triangle and any plant species can be scored along these three axes according to the proportion of these primary strategies it possesses; the combination of these scores determines the plant species' position within the triangle (Fig. 3.1).

Ruderal species (R) are pioneer or opportunistic species and coincide with the r-selected plants explained above. These plants are strongly related to the occurrence of frequent disturbance.

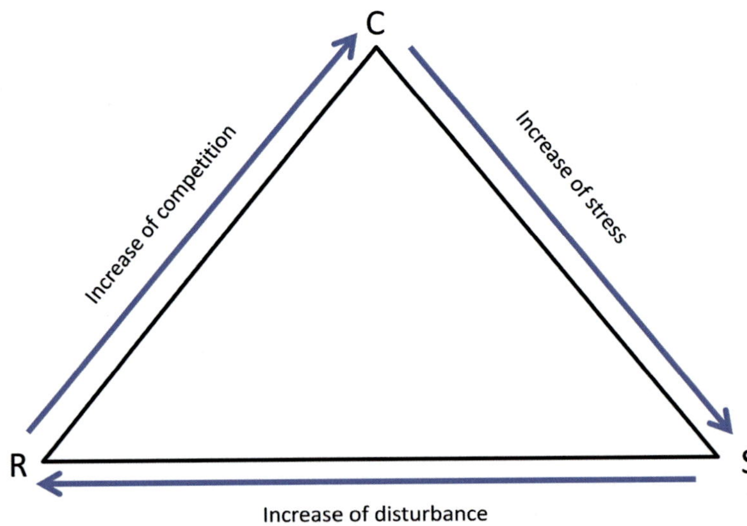

Fig. 3.1 Grime's three plant strategies model. *C* competitive, *S* stress tolerant, and *R* ruderal strategy

Stress-tolerant species (S) can endure adverse environmental conditions (stress) because of their reduced size, slow growth, possession of several defensive morphological traits such as spines, thick cuticles (sclerophylly), succulency, reduced leaves (microphylly), resinous glands, abundant hairs, and other phenological and physiological adaptations. They produce relatively few flowers and hence, fruits and seeds are scarce. They inhabit stressfull biotopes such as exceptionally dry or exceptionally wet, or rocky places, saline soils, heavy metal substrates, etc.

Competitor species (C) partially correspond to the category of K-selected plants. They grow vigorously, with much foliage along their stems, allowing them to out-shade their neighbours, with an efficient root absorption capacity and a high growth rate. They thrive in habitats with a good nutrient and water availability in which disturbance and environmental stress are light or absent and the main factor influencing community dynamics is competition. Many trees are included in this category, and also perennial grasses and megaforbs.

In this three-type model it is assumed that the intensity of competition for available resources increases as disturbance and environmental stress diminish. Since all plant species in a community will have a position in the triangle depending on their traits, i.e. its scores as regards stress resistance, disturbance tolerance and competitive vigour, a plant community will also show the balance of these three phenomena resulting from the averaging of the constituent species. Thus, the structure and floristic composition of a community will reflect the balance between these three phenomena at the community level.

3.4 Types of Succession

Succession phenomena, besides being intrinsically complex at a functional level, are also diverse and can be classified into several types based on several criteria (Fig. 3.2) (Dierschke 1994).

- According to the factors driving succession:

  Exogenous (ecogenous or allogenous). This succession is driven by changes in environmental conditions, either human or naturally induced. It can be both progressive or regressive. Examples are climate change, changes in salinity or soil humidity, etc.

  Endogenous (phytogenous or endogenous). In this case the changes consist of the appearance and expansion of new species at the site and the disappearance of others, due to the intrinsic evolution of the community's biota. This is the case when shade-tolerant plants become established after the tree canopy develops, or when litter steadily accumulates to a closed organic layer on top of the soil.

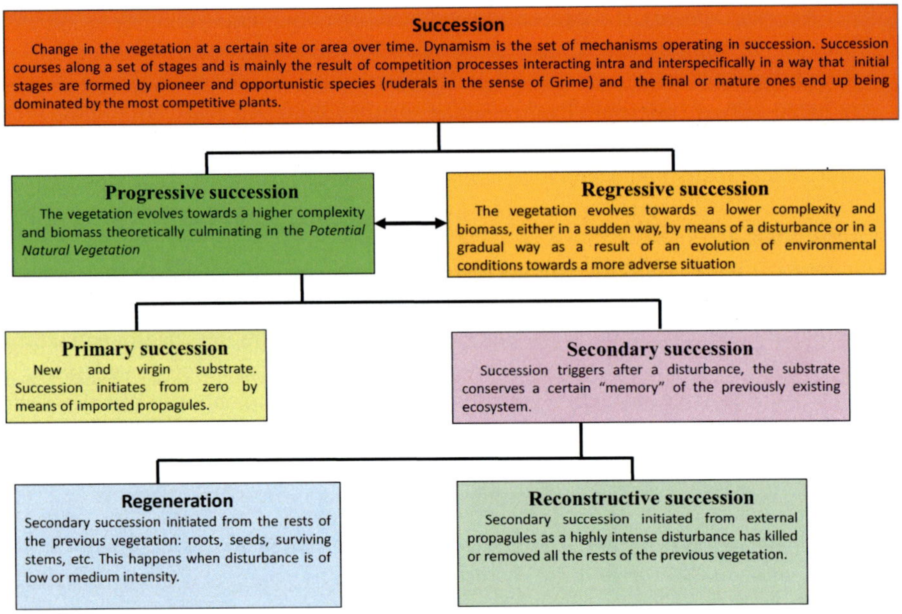

Fig. 3.2 Various types of succession

Endo-exogenous (mixed). Such a succession is a combination of the previous ones.

- Depending on the direction of the succession:

Progressive. The succession evolves towards the mature stage. It can be:

Primary succession. This is the succession developing on an entirely new, bare substrate which is colonized for the first time, e.g. in case of new volcanic material, or new soil becoming free of ice after a glacier thaws, etc. Thus, it is a primary colonization by plants on a substrate without any "memory" (rests of soil, roots, seed bank, or other elements left by a previous vegetation) from the past.

Secondary succession. This is the succession that unfolds after a disturbance has removed part of the previously existing ecosystem biomass. That previous vegetation leaves always a certain "heritage" in the form of a seed bank, soil organic matter, subterraneanously living rests (roots, etc.), which influence the subsequent succession. It is by far the most frequent form of succession we can currently observe due to the massive impact of human-induced disturbances practically all over the world.

Regressive. This succession evolves from mature stages back towards the initial stages. It can happen suddenly, as in disturbance events (floods, fires, etc.), or can be slow, as in the degenerative processes.

Cyclic. In this succession progressive and regressive phases alternate over time, e.g. in deforestation-afforestation cycles of forested areas in which wildfires or strong winds occur regularly, destroying part of the forest canopy and triggering the subsequent gap-filling processes.

- Depending on the spatial scale:

Micro-succession. This typically happens at a very local scale in small plots in the context of a larger community, such as succession in a gap created by the fall of a tree in the forest.

Local succession. This develops in a middle-sized area that a person can view from a single vantage point. It is the most commonly considered succession or succession *sensu stricto*.

Regional succession. This is succession at a landscape scale.

- According to the time span:

Current succession. This is the succession that happens within a relatively short time span, from a few months or years to several centuries. (Like local succession, it is a succession *sensu stricto*).

Secular succession. This the succession that happens over a historical time scale. It is studied by means of indirect methods (palynology, anthracology, etc.)

3.5 Some Basic Concepts in Succession

Tessella This word is used differently in different contexts, and has a specific meaning in the conceptual framework of succession. A tessella is an ecologically homogeneous area in which only one type of Potential Natural Vegetation can be distinguished (Bolòs 1963; Rivas-Martínez 1976, 1987, 2007; Loidi 1990). In Bolòs' words it is "any portion of land ecologically homogeneous and able to sustain a certain stable community and its seral associations". A tessella will be an ecologically equipotential area which has the same potential in all its parts. In a primeval non-anthropic landscape, each tessella will be occupied by one characteristic vegetation type that can be classified at the association level; often several somewhat similar tessellas can sustain a unique association of Potential Natural Vegetation which, in such a case, could diversify into variants.

This reciprocal relationship is established as follows:

Tessella ↔ association of Potential Natural Vegetation

In the modern landscape it is difficult to recognize a tessella, because all territory is profoundly influenced and transformed by human action. Thus, natural or nearly-natural vegetation is rare and often completely lacking from many areas, while seral stages are much more abundant. These seral stages cover several tessellas as they have a wider ecological range, thus masking the recognition of the different tessellas to be identified.

3.5.1 Potential Natural Vegetation and Climax

The stable or mature phase of a successional series was called *climax* by Clements as he considered it was basically related with the climatic conditions of the area. Later, Tüxen (1956) enriched this conceptual framework by considering that each mature stage was dependent not only on the climate but also on soil conditions, such as bedrock etc. He coined the concept of *Potential Natural Vegetation* (PNV), which had also a more theoretical component as it should be inferred in any surveyed territory by studying the existing plant communities. The PNV can be determined even in areas without remnants of the mature stages, just by acquiring sufficient knowledge of the successional course of the communities replacing one another over time. Each PNV is preceded by several earlier states of vegetation which can be formalized as previous stages or previous plant communities (Fig. 3.2). These communities will also replace the PNV when it is destroyed by a disturbance and in that case they will have a seral character as they are substitution communities. These seral, or substitution, communities can be defined after their species composition and/or their dominance or physiognomy. The concept of Potential Natural Vegetation (PNV) is still under discussion (Loidi et al. 2010; Loidi and Fernández-González 2012). It has often been interpreted as the vegetation which potentially would grow at a particular place if no disturbance would happen during a sufficiently long span of time. This is more a theoretical assertion than a real one as probably there are no places in the world free from any disturbance events. Another wrong idea is that PNV is the same as the primeval vegetation that existed before humans altered the ecosystems and transformed ecosystems in the present, secondary vegetation types existing almost everywhere. Certainly the PNV idea is a theoretical assertion because it is emphasized that it corresponds to current (*heutige*) conditions not to past ones, and it reflects how the vegetation would correspond in a non-disturbed situation under a given set of environmental conditions. It is the most competitive vegetation type to be found in an ecologically homogeneous area (tessella).

Within the several PNVs existing in a given territory, some will correspond to situations in which the condition of the soil preponderates over others (mostly climate) often because the substrate entails a certain stress for the vegetation growing on it. This is the case with hydromorphic soils with permanent or regular flooding, salinity, tidal flood, steep slopes, stony soil, high wind exposure, abundance of heavy metals, etc. Such situations determine the existence of an adapted vegetation and hence a particular PNV type. These communities are basically related to such edaphic and topographic conditions and not so much to the climatic conditions. Such communities, not related to the climate and adapted to the corresponding edaphic stress, have been traditionally called *permanent communities*. The Clementsian climax concept entails a climatic determinism of the final

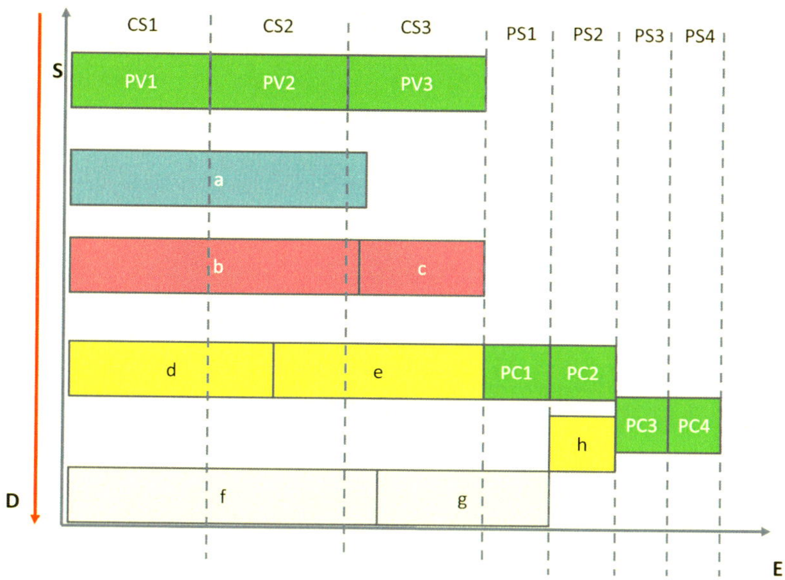

Fig. 3.3 Diagram representing the positions of the vegetation stages or plant communities which can be found in the landscape along a successional trend. (S), reversed by disturbance (D), and across the ecological gradient (E). CS1 to CS3 represent three climatophilous vegetation series (forest in many cases) and PS1 to PS4 represent four permaseries, all of them ordered along a gradient which becomes extreme and stressful towards the right end. PV1 to PV3 are the Potential Vegetation of the climatophilous series and PC1 to PC4 are the permanent communities of the permaseries; a to h are seral communities or substitution stages: *grey* ones are pioneer stages, *yellow* and *lilac* are seral scrub and grasslands, *blue* are secondary forests or hedges. Frequently seral stages span a broader ecological range than that of the climatophilous vegetation series, producing mismatchings and overlappings. The permaseries have less seral stages because ecological factors are more stressful, until the extreme situation in which there are no seral stages (PS3 and PS4). The resulting diagnosis for the vegetation series will be:
CS1 = PV1 + a + b + d + f;
CS2 = PV2 + a + b + d + e + f;
CS3 = PV3 + a + c + e + f + g;
PS1 = PC1 + g;
PS2 = PC2 + h;
PS3 = PC3; PS4 = PC4.
The collection of the climatophilous sigmeta (from CS1 to CS3) is the Geosigmetum.
The collection of the Permasigmeta (from PS1 to PS4) is the Geopermasigmetum

community, it mirrors the climate of the area and all the vegetation of a climatic territory converges, by means of succession, towards a unique type of climax regardless of its bedrock type. This is the *monoclimax* or unifying climax concept. This conception was contradicted by the *polyclimax* concept in which each edaphic situation will have its own *edaphic climax* (or PNV) type. The substrate diversity cannot be reduced to an unifying entity by the combined action of climate and vegetation succession; it is conveyed to a set of different climaxes, one for each case, and this diversity is also irreducible (Fig. 3.3).

Therefore, in any territory several tessellas will be found and they can be grouped into three main categories: those which basically mirror the climate (*climatophile*), those which are deviant from the moisture condition emerging from the regional climate: those which are drier (*edaphoxerophile*), such as steep slopes, summits and ridges, and those which are moister (*edaphohygrophile*), such as riverbanks, depressions and any places which are wet due to gravitational input of water. In addition to these, in some areas there will be other special biotopes (sand dunes, salt marshes, ponds, mires, rocky cliffs and screes, etc.) in which other azonal permanent communities will be found.

3.5.2 Primeval Vegetation

This is the vegetation that occurs without the interference of man, i.e. the vegetation that dominated before the time humans started to severely influence terrestrial ecosystems by means of agriculture and animal husbandry at the dawn of the Neolithic. In this primary situation without human transformation, each tesella should be occupied by its PNV, with the exception of the minor areas that are recovering from a natural disturbance and are occupied temporally by a seral community. Human-induced severe disturbances occurred later in history, initially in Europe and Asia, and later in most of the rest of the world, and these have been so extreme in terms of extent, intensity and recurrence, that in many areas the substrate conditions have changed and are now probably different from the original ones (because of soil erosion, changes in topography, etc.). On top of that, after several millennia of human civilization, climate has also changed to such a degree that it will not be possible to restore the original vegetation because it was very probably adapted to climatic conditions than differed from the current ones. Therefore, the terrestrial surface of our days is probably conditioned by a different climate than that which existed a couple of millennia ago, before the Neolithic. Thus, human induced and natural changes have tailored a new scenario in large parts of the world, making it an impossible task to match primeval vegetation with the modern PNV.

3.5.3 Primary Succession

Primary succession starts at an empty substrate that has never carried any previous vegetation or ecosystem (Fig. 3.2). On this "virgin" substrate, consisting solely of mineral material without soil and seed bank, succession starts initially with pioneer stages of vegetation. This initial phase is due to the arrival, germination and establishment of propagules coming from elsewhere, produced by plants living in the surroundings or in places within reach of their dispersal units. The initial stages of succession start to inhabit a bare substrate and start to form the soil. The soil will develop parallel to the course of the succession. Primary succession is much less common than secondary succession, just because a new substrate is necessary to start it. However, there are several situations which typically support primary succession:

1. Bare rock after severe erosion which has totally removed previous vegetation and soil, leaving the bedrock completely uncovered. Some substrates are particularly prone to be easily eroded, such as marls on steep slopes, which become easily bared as a result of torrential rains and other events. This is also the case on steep slopes suffering landslides after earthquakes, causing the exposure of bare rock.
2. Some quick sedimentary processes such as those of coastal or inland sand dunes when affected by strong storms moving huge amounts of sand, or on river beaches which appear in new places after a big flood, etc.
3. Lava flows and volcanic ash deposits resulting from eruptions.
4. Islands emerging from the sea, as result of volcanic activity (as in the previous case) or by the slow emergence of a part of the continental platform.
5. Self decay of weak bedrock materials on cliffs or steep slopes (similar to 1)
6. Bare substrate left behind after the withdrawal of the ice of glaciers. This phenomenon has been extremely important at the end of each Pleistocenic ice age, particularly in Europe and North America.
7. Stone screes accumulated at the foot of cliffs, mostly due to gelifraction, in the steep areas in the mountains. The upper part of the stone accumulation receives a constant input of stone fragments and is the unstable top of the scree, while in the lower part the stone masses tend to stabilize and colonization by vascular plants takes place.
8. Surfaces which have been flooded for a long time and become uncovered after drainage.

Human intervention is usually absent in primary succession, though in some cases it is present, e.g. in abandoned quarries and mines, in debris accumulations, on buildings and any type of human constructions which can be colonized by plants when abandoned for a long time, etc.

3.5.4 Secondary Succession. Disturbances

The course of any succession process can be interrupted by an event, called disturbance (Fig. 3.4), which can stop it or even move it backwards, partially or totally, destroying the vegetation previously existing. This disturbance determines the restart of succession from the state in which the ecosystem has been left: this is called *secondary succession*. Secondary succession, initiated after a disturbance, means that there is a sort of "heritage" of the previous vegetation, in the form of a seed bank, root bank or any surviving parts of the species which lived there before, and that conditions to a certain extent the course and outcome of the whole process.

The destroying events, or *disturbances*, can be natural or human-induced. The disturbance can also be a change in the site conditions which interferes with the normal functioning of a biological system. Among natural disturbances there are physical events such as spontaneous wildfires, hurricanes, heavy rains, floods,

Disturbance

Event limiting biomass by causing its total or partial destruction or a change in environmental conditions interfering with the normal functioning of a biological system.

Type of disturbance/Intensity	Removal of aerial parts	Changes in soil conditions: nutrients, water and erosion of fine materials.	Turning the soil over
Fire	+++	+	
Logging (forest)	+++	+	
Snowslide	+++	+	
Husbandry (influence of big herbivores: consumption of vegetal material, trampling and nitrification)	++	+	
Agriculture	+++	+++	+++
Treefall (forest)	+		
Flooding	++	+	
Plague	++		
Pollution	+ - ++	+	

+++ intense, ++ medium, + low

Fig. 3.4 Types and severity of disturbances

earthquakes, snow avalanches, etc., but also biotic disturbances such as plagues and herbivory can be considered natural if they are not induced by humans.

Anthropic disturbances can be very diverse and presently these are much more common and important than natural ones, particularly in areas where the human population density is high. Man influences vegetation not only by means of activities which could be qualified as rural, such as fire, tilling, grazing by herds, watering, firewood extraction, logging, etc., but also by means of typically urban activities such as road paving, urban development, water course channelling, etc. There is almost no human activity not entailing a certain disturbance in natural vegetation.

Among the most frequent disturbances are fire, damage by wind, water or ice, abrasion by sand or other small particles blown by wind, herbivory, pathogen attacks, soil erosion, flooding, logging, mowing of herbaceous vegetation, trampling by big animals, etc. (Fig. 3.5). In many of these cases, plants do not disappear completely. Some disturbances favour certain species, e.g. those adapted to heavy herbivory and trampling can become dominant at sites that are heavily and frequently grazed; they are better adapted than their competitors in the regeneration (secondary succession) taking place after disturbance (Burrows 1990). In any case, any disturbance is an instant withdrawal episode interrupting progressive succession that restarts as soon as that episode is over. Succession and disturbance are linked phenomena.

Secondary succession after a disturbance episode will be faster the more *resilient* the ecosystem is. The concept of resilience (Grime 1979) can be used to indicate the speed of recovery to the initial state of a disturbed system.

As a result of the human-induced disturbance regimes, extant landscapes on most of the earth's land surface are profoundly transformed and show a very different image from what they would have if only natural disturbances were operating (Fig. 3.6). In the case of the Iberian Peninsula, it is astonishing to see

Disturbance

Frequencies of the types of disturbances.

Frequency of the disturbance	Episodic Low frequency, irregular	Recurrent Medium frequency	Systematic High frequency, regular
Fire	+	+	
Logging	+	+	
Snowslide	+	+	
Husbandry		+	+
Agriculture		+	+
Treefall	+		
Flooding	+	+	+
Plague	+		
Pollution	+	+	+

Fig. 3.5 Frequencies of the types of disturbances

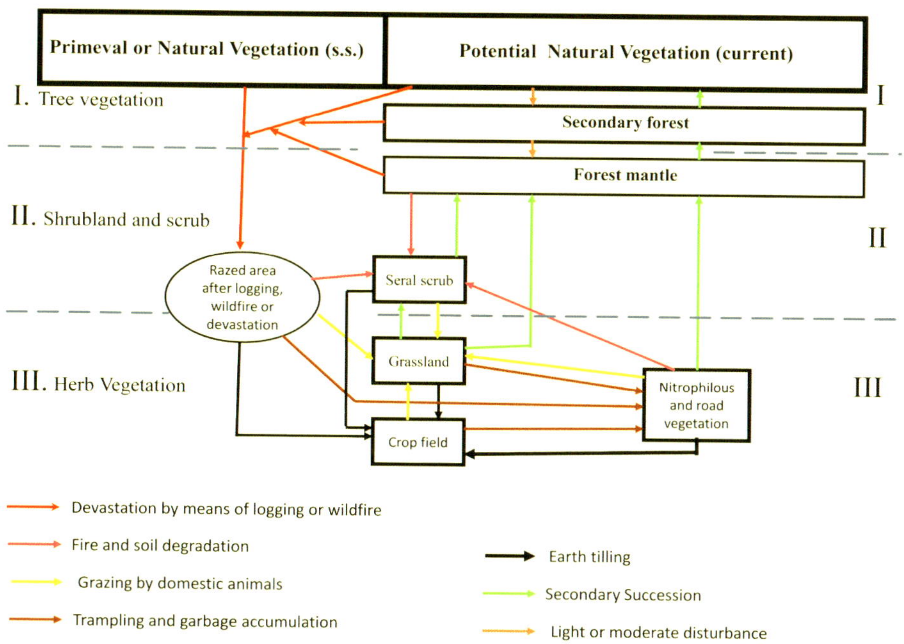

Fig. 3.6 Diagram representing a model of the dynamic relationships of the different stages of vegetation resulting from the main disturbances in a temperate or Mediterranean territory

the huge land surface covered by low scrub (tomillar or matorral), particularly in hilly areas, complemented with extensive grasslands in other mountainous districts. This is by no means a natural vegetation but an anthropic one related to the traditional exploitation system. Such vegetation types have been qualified as semi-natural as they are not artificial in the strict sense but are evidently related to a human activity. The old Strabo-attributed story of the squirrel which could cross the entire Peninsula from the Pyrenees to the southern extreme jumping from branch to branch without touching the ground has been strongly influential in the idealistic image of the ancient landscape and ecosystems that existed in the peninsula before the Roman domination. Seemingly this story is entirely false as Strabo only said that "Hispania was a great extension of forests and mountains."; no mention of the squirrel has been found in his Geography (first century AD). Nevertheless, it is very reasonable to think that forests and mature vegetation types covered a larger extent in Iberia at that time than they do today. It is common knowledge and likely true that people are guilty of causing the current degraded state of the land. Logging, wood cutting and induced fires, combined with very aggressive forms of husbandry with sheep and goats, have been the main disturbances which have driven the current landscape to have so many treeless hills and mountains.

3.6 Vegetation Series or Sigmetum

The set of vegetation types which, by virtue of succession replace each other at a particular site has been called a *series*. Already Clements (1916) used this term with this meaning but nowadays it has been used more profusely and with a more accurate conceptual meaning (Alcaraz 1996; Géhu and Rivas-Martínez 1981; Loidi and Báscones 1995; Rivas-Martínez 1976, 1978, 1987, 1994; Rivas-Martínez et al. 2007). We can also define a *Vegetation Series* as the ordered set of plant communities which can substitute each other over time at a certain place (Bolòs 1962) or also the particular sequence of stages occurring inside the ecologically homogeneous space called *tessella*, or the set of associations which can occupy the same tessella depending on its degree of degradation (Bolòs 1962, 1963).

Summarizing, succession within the limits of a tessella can be formed by stages replacing each other over time, but we distinguish three categories in these stages:

Pioneer stages (or communities). In the initial phases of succession plant communities present a variable floristic composition, species number and cover. The plants are often allochthonous and possess the r-selection syndrome: short life span, high growth rate, small size, low proportion of biomass of the underground parts in relation to the total biomass, high seed production, small seeds dispersing over long distances and viable for a long time, etc.

Intermediate stages (or communities). The plants belonging to these stages are longer-lived (hemicryptophytes, chamaephytes, nanophanerophytes) and their communities have a higher stability in structure and composition. The majority of the grasslands, shrublands and secondary forests in many parts of the world are such intermediate stages. They are usually the communities appearing after disturbances such as fire, grazing, logging and other very frequently human-induced disturbances. For that reason, these intermediate stages (also called *seral stages* or communities) dominate in the cultural landscapes (those profoundly modelled by humans).

Final stages (or communities). They represent the maximally possible complexity and biomass in the vegetation for each tessella. These stages can also be considered as the potential vegetation of a site which is reached after sufficient time has passed without disturbance. This can be confounded with the Potential Natural Vegetation (PNV), a concept that concerns the current potentiality, without waiting sufficiently long to reach the real potential vegetation. For that reason the PNV is used to define and name each Vegetation Series. In the final, or mature, stages long-lived and large-sized species (phanerophytes, often big trees) dominate, which have a slow growth and seeds with a low dispersal ability, etc.; it resembles the K-selection syndrome or the competitive plant type. These final stages also contain species that are not competitive but stress tolerant as they need to adapt to the resource-poor niches that are not entirely exploited by the big and strong competitors, such as shade tolerant species, lianas, epiphytes, etc.

A Vegetation Series is also called a *sigmetum*, being the total of all communities which constitute the series and which are related by dynamic links. Each Vegetation Series or sigmetum occupies one tessella (or small group of similar tessellas) and is named by its specific PNV association. In other words, each tessella contains its Vegetation Series or sigmetum (Fig. 3.7). There is variability within the sigmeta as not all of them reflect the same type of environmental conditions. There are those which occupy large areas of land and are conditioned by non-exceptional conditions, i.e. basically by the regional climate and the regional bedrock type. These are the *climatophilous sigmeta* only fed with rainwater. In the general topographic model of ridge-slope-piedmont-valley bottom, there is room for four types of sigmeta: *edaphoxerophilous*, *climatophilous*, *temporihygrophilous* and *edaphohygrophilous*, the latter two having additional water contribution due to slope runoff (Fig. 3.7).

Others are conditioned by extremely stressful conditions, such as permanent floods, steep slopes, rocky substrates with shallow soils, sandy mobile substrates, salinity, etc., and they will comprise specialized plant communities. These are the *permasigmeta*. In the permasigmeta the succession course is shortened, sometimes to only one stage, as a result of extremely limiting conditions, and for that they have been traditionally called *permanent communities*. Depending on the general stress conditions we can group permasigmeta into several categories, thus we can have *cryopermasigmeta* (cold), *saxipermasigmeta* (rocky substrate), *hygropermasigmeta* (flood) and *halopermasigmeta* (salinity). An intermediate category of sigmetum between the highly stressed permasigmetum and the climatophilous sigmetum is the *minorisigmetum* (Rivas-Martínez et al. 2011), in which the stressing conditions are not so extreme enabling the existence of perennial seral stages but preventing from reaching the potential natural vegetation corresponding to the climate and substrate of the non stressed surrounding areas.

Any terrestrial part of the world can be split up into a mosaic of tesellas resembling a puzzle, and each of them will be occupied by its specific sigmetum. The spatial arrangement of the sigmeta is the target of landscape vegetation science and will be treated in the following section. Sigmeta can contain a certain internal heterogeneity due to geographic or to ecological diversity, and that heterogeneity is addressed by the term *faciation*. Any analysis of the landscape has to take into

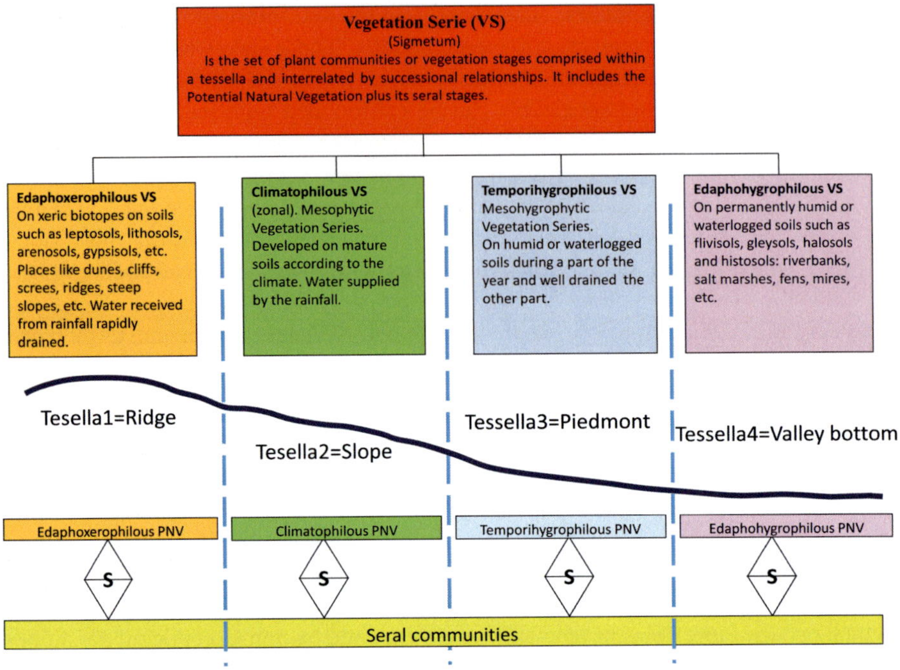

Fig. 3.7 Vegetation series or sigmeta and their main types arranged in the zonation along the ridge-slope-piedmont-valley bottom model with succession (S) operating within each of the sigmeta

account the ecological gradients and the successional situation superimposed on it (Fig. 3.7). When doing so, the accuracy and predictive capacity of such an analysis is long-reaching, but the initial step is to recognize the spatial delimitation of the ecologically homogeneous spaces, which are the tessellas.

3.7 Vegetation Geoseries or Geosigmetum

According to the conceptual framework of the dynamic-catenal phytosociology (Rivas-Martínez et al. 2007), the ordered arrangement of vegetation units across a gradient is named with the prefix geo-. Vegetation zonation is one of the oldest targets of study in vegetation science and every descriptive study of the plant communities of any territory encompasses the description of the zonations observed. The spatial arrangement of the sigmeta is the aim of landscape vegetation science. It is focused on the study of the zonation of the sigmeta. The nature of the zonations depends on the gradient type, thus large scale gradients (climatic) will produce zonations of broadly defined vegetation units, while fine scale gradients (soil moisture or salinity, microtopography) will determine zonations of fine scale vegetation units. Accordingly, we can coarsely separate two types of zonations:

Topographic zonations (Topographic Geoseries or Topogeosigmeta). These occur when the gradient manifests itself at a topographic level, i.e. coastal sand dunes gradient, salt marsh gradient, soil moisture gradient across the edge of a lake, a gradient in a mire, etc.

Cliserial zonations (Cliserial Geoseries or Oreogeosigmeta). These occur when the gradient is a climatic one manifesting itself across an altitudinal transect. The zonation is typically that of the vegetation belts of a mountain. It can encompass all the Vegetation Series arranged along the altitude range of a mountain or steep territory and can be of medium to large size (at least two bioclimatic belts). If the mountain is lithologically homogeneous and the bedrock is the same over the whole mountain, then we can have an Iso-oreogeosigmetum. If it is geologically diverse and several types of bedrock occur, then we have a Hetero-oreogeosigmetum (Fig. 3.8).

In highly stressed areas under extremely limiting conditions where permasigmeta occur, we can also find zonations along the gradients of these factors along which several permasigmeta are ordered contiguously, each one of them occupying a permatessella, called *geopermasigmeta* or *geopermaseries* (Fig. 3.9). A similar discussion can be made to explain the *geominorisigmetum* or *geominoriseries.*

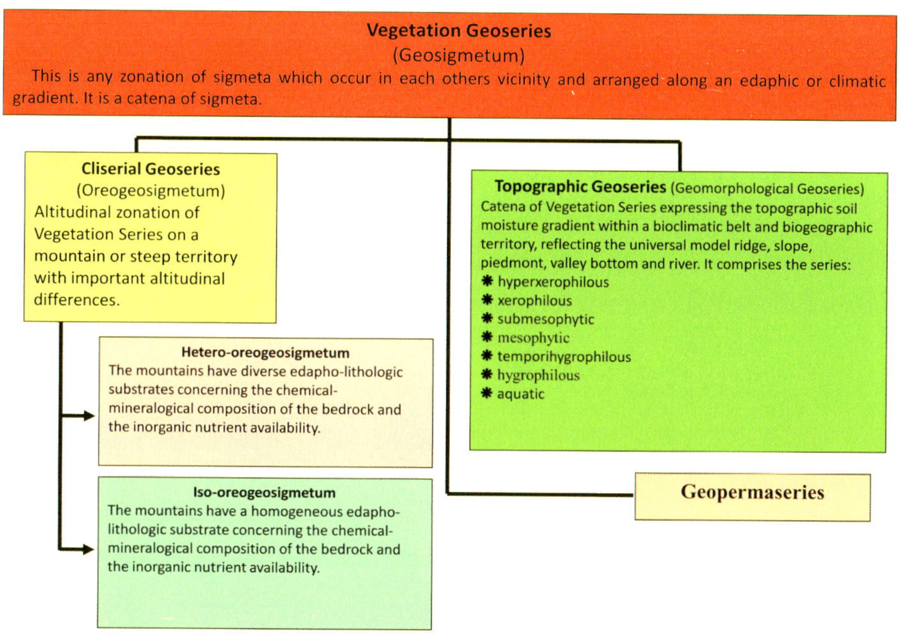

Fig. 3.8 Diagram explaining Vegetation Geoseries or Geosigmeta

Vegetation Geopermaseries
(Geopermasigmetum)

It is a catena of permaseries or permasigmeta which are subject to extreme and stressing conditions due to climate (high mountains and polar zones) or exceptional microtopographic or edaphic situations (rocky places, cliffs, salt marshes; etc.), which originate a high number of contiguous ecological residences populated by diverse specialized perennial plant communities which lack other perennial seral stages having seemingly reached a certain balance with such conditions.

Cryogeopermaseries

Polar (tundral) geopermaseries

Corresponds to areas under cold climate (polar), where low temperatures are the limiting factor in the ecosystems. Sedges, small chamaephytes, bryophytes and lichens are the most common plant types.

High mountain geopermaseries

In the high mountain (cryoro belt) cold also is the limiting factor, similarly as in the previous type. Alpine grasslands or dwarf shrublands are the most frequent types of vegetation.

Saxigeopermaseries

Rocky cliffs geopermaseries

The vertical cliff in a rocky mountain has a number of crevices in which some very specialized stress tolerant species live. Oozing rocks represent a variant as well as those cliffs with additional nitrogen nutrient inputs due to animals (bird nests) or humans.

Scree geopermaseries

Screes formed by stone accumulation at the foot of cliffs, usually due to gelifraction, are an unstable substrate which only specialized plants can colonize.

Chionophilous (snow) geopermaseries

Zonation is determined by permanence of snow, which in turn is dependant on its degree of accumulation. Hollows and corners where it gets accumulated are called snowfields and are found in snowy climates such as high-mountain, boreal or polar climates

Hygrogeopermaseries

Freshwater inland geopermaseries

Ponds, lakes and swampy areas with still freshwater have a clear hydric gradient, from the moist to the permanently flooded soil, in a transition of increasing anoxic stress.

Mires geopermaseries

Depressions which fill up with water in territories under cold climate accumulate organic matter and decomposition is slowed down or stopped, accumulating in the pond and forming peat deposits. Such environments are difficult for many vascular plants, being dominated by bryophytes; the most frequent group of vascular plants are the Cyperaceae. The influence of stagnant water soaked up by the peat mosses establishes the gradient which drives the plant communities zonation.

Halogeopermaseries

Halophilous tidal geopermaseries

Coastal salt marshes influenced by tides and sheltered from waves are populated by vegetation adapted to the high sea water salinity. The gradient is given by the tidal floods influence. Vascular vegetation is dominated by succulent Chenopodioideae in the Extratropical countries and by mangroves in the Tropical ones with warm waters.

Halophilous inland geopermaseries

In the inland endorrheic depressions or basins, soluble salts accumulate in the soil under certain climatic and geohydrologic conditions, reproducing partially the conditions of the coastal salt marshes. The gradient is given basically by the saline concentration in the soil water.

Coastal dunes geopermaseries

Between the beach and the fixed dune, the coastal Sandy deposits are modelled by the wind and the salinity brought by the sea spray, creating a gradient of mobility and salinity which clearly results in the zonation of the foredunes, white or mobile dunes and grey dunes typical of this geopermaseries all over the world. The substrate mobility and coarseness (sand) entail dry edaphic conditions (almost immediate percolation and no water retention capacity) which is added to the more or less intense salinity brought by the sea spray (at least in the fringes nearest to the sea). The wind itself has a strong evaporation effect and a mechanical abrasive action on the plants.

Coastal halo-anemogenous rupicolous geopermaseries

The cliff coasts are influenced by the sea wind which imposes its mechanical and desiccating action and brings small drops of sea water spreading marine salt on the cliff. This effect is added to the rockiness of this habitat and all together determine the environmental gradient of this highly stressed geopermaseries.

Fig. 3.9 Diagram summarizing the vegetation geopermaseries or geopermasigmeta

References

Alcaraz F (1996) Fitosociología integrada, paisaje y biogeografía. In: Loidi J (ed) Avances en Fitosociología. Ed. UPV/EH, Bilbao, pp 59–94

Burrows CJ (1990) Processes of vegetation change. Chapman & Hall, London, p 551

Clements FE (1916) Plant succession: an analysis of the development of the vegetation. Publ. Carnegie Institute, Washington DC, p 515

Clements FE (1928) Plant succession and indicators. Wilson, New York

Connel JH, Slatyer RO (1977) Mechanisms of succession in natural communities and their role in community stability and organization. Am Nat 111:1119–1144

de Bolòs O (1962) El paisaje vegetal barcelonés. Cátedra Ciudad de Barcelona, Excmo. Ayuntamiento, Barcelona

de Bolòs O (1963) Botánica y geografía. Mem Real Acad Ci Artes Barcelona 34:443–491

Dierschke H (1994) Pflanzensoziologie. UTB. Verlag Eugen Ulmer, Stuttgart, p 683

Egler FE (1954) Vegetation science concepts. I. Initial floristic composition, a factor in old-field vegetation development. Vegetatio 4:412–417

Géhu JM, Rivas-Martínez S (1981) Notions fondamentales de Phytosociologie. In: Dierscke H (ed) Sytaxonomie. Berlin International symposium IVV. Ed. Cramer, Vaduz, pp 5–53

Glenn-Lewin DC, van der Maarel E (1992) Patterns and processes of vegetation dynamics. In: Glenn-Lewin DC, Peet RK, Veblen TT (eds) Plant succession. Theory and prediction. Chapman & Hall, London, pp 11–59

Grime JP (1979) Plant strategies and vegetation processes. Wiley, London, p 222

Grime JP (1985) Towards a functional description of vegetation. In: White J (ed) The population structure of vegetation, Handbook of vegetation science, vol 3, pp 503–514

Loidi J (1990) Vegetation series: its use for small scale geobotanical mapping. Phytocoenosis 3:119–122

Loidi J, Báscones JC (1995) Memoria del mapa de series de vegetación de Navarra. Ed. Gobierno de Navarra, Pamplona, p 99

Loidi J, del Arco M, Pérez de Paz PL, Asensi A, Díez Garretas B, Costa M, Díaz González TE, Izco J, Penas A, Rivas-Martínez S, Sánchez-Mata D (2010) Understanding properly the 'potential natural vegetation' concept. J Biogeogr 37:2209–2211

Loidi J, Fernández-González F (2012) Potential natural vegetation: reburying or reboring. J Veg Sci 23:596–604

Mac Arthur RH, Wilson ED (1967) The theory of island biogeography. Princeton University Press, Princeton

Odum EP (1969) The strategy of ecosystem development. Science 164:262–270

Rivas-Martínez S (1976) Sinfitosociología, una nueva metodología para el estudio del paisaje vegetal. Anales Inst Bot Cavanilles 33:179–188

Rivas-Martínez S (1978) Sobre las sinasociaciones de la Sierra de Guadarrama. In: Tüxen R (ed) *Assoziationskomplexe (Sigmeten)*. Berlin international symposium IVV. Ed. Cramer, Vaduz, pp 189–212

Rivas-Martínez S (1987) Memoria del mapa de series de vegetación de España. ICONA, Madrid

Rivas-Martínez S (1994) Dinamic-zonal phytosociology as landscape science. Phytocoenologia 24:23–25

Rivas-Martínez S et al (2007) Mapa de series, geoseries y geopermaseries de vegetación de España [Memoria del mapa de vegetación potencial de España] Parte 1. Itinera Geobotanica 17:5–436

Rivas-Martínez S et al (2011) Mapa de series, geoseries y geopermaseries de vegetación de España [Memoria del mapa de vegetación potencial de España] Parte 1. Itinera Geobotanica 18 (1, 2):5–800

Tüxen R (1956) Die heutige potentielle natürliche Vegetation als Gegenstand der Vegetationskartierung. Angew. Pflanzensoziologie 13:5–42. Stolzenau/Weser

Chapter 4
The Iberian Vascular Flora: Richness, Endemicity and Distribution Patterns

Carlos Aedo, Antoni Buira, Leopoldo Medina, and Marta Fernández-Albert

Abstract An historical review of the Iberian flora is presented here to give a summary of the main milestones from the fourteenth century until today. More than 400 years of floristic work has been briefly analysed from Clusius' time until today, with special attention on the oldest and less known aspects. The Iberian flora comprises 6276 species including 739 non-native species. 1258 species (being 22.8% of the native species) are endemic. This shows that the Iberian flora is rather special; even so, the richness and endemism levels are comparable to other Mediterranean basin countries such as Greece or Morocco, and lower than Turkey. The rate of generic endemism, however, is considerably higher in the Iberian flora. The species richness and the endemicity patterns are analysed on basis of the native and endemic species observed in 50 × 50 km squares. The highest observed species richness areas generally overlap with the richest in endemic species. They are located in the main Iberian Mountain Ranges. The Baetic System shows the highest rate of endemism, followed by the Cantabrian Mountains, the Central System and the Pyrenees. The composition of the Iberian flora is examined in large groups. The Compositae is the largest family at species and endemism levels, and also provides the largest number of alien species. *Limonium*, *Centaurea* and *Carex* are the largest genera in the Iberian flora; the two former have their diversification centres in the Mediterranean Basin. More than one third of the Iberian species has been classified in Spain in one or other special category and included in Red Data Lists or Books. At least 20 taxa can be considered as extinct in the territory of the Iberian flora.

C. Aedo (✉) • A. Buira • L. Medina • M. Fernández-Albert
Real Jardín Botánico-CSIC, Plaza de Murillo 2, 28014 Madrid, Spain
e-mail: aedo@rjb.csic.es

J. Loidi (ed.), *The Vegetation of the Iberian Peninsula*, Plant and Vegetation 12,
DOI 10.1007/978-3-319-54784-8_4

4.1 Historic Review

Towards the late Renaissance in Europe, botany began to be separated from medicine. Plant books then gradually started to become more similar to current floras or catalogues that show an interest in the plants themselves. At this time, the Flemish physician Carolus Clusius (1526–1609) travelled to Spain and Portugal. Between 1564 and 1565 he explored parts of Andalusia, Aragón, the Castilian regions, Valencia and the central region of Portugal during 8 months, collecting plants to be dried and seeds and sent them to Flanders. In 1576 he published his *Rariorum aliquot stirpium per Hispanias observatarum historia*, gathering information about 300 plants (225 from Spain and Portugal) and including 233 original pictures (Fig. 4.1). Later part of those plants was sent to Caspar Bauhin, and are now located at the herbarium at Basel University (BAS), and to Felix Platter, now located at the herbarium at Berne University (Bern). In these herbaria, specimens do not have accurate labels and are difficult to assign to authors or localities. These plants are probably the oldest collected in Iberia and preserved in accordance with scientific purposes, although even without detailed labels. Clusius' work proved of key importance as a source of knowledge of the Iberian flora. Many of the elements that later formed a regular part of floras were already glimpsed. Although the plants are organized in a first book dedicated to trees and shrubs and a second one dedicated to the herbaceous plants, certain species are already recognized and grouped by morphological affinities (*i.e.*: eight different *Erica* and five *Narcissus* species are mentioned), often as we still do it today. The plants are described according to their main organs and habitat, specific localities, and flowering period and vernacular names. Ramón-Laca and Morales (2005) have highlighted the relevance of this work and have considered it rightly as an Iberian "protoflora".

Colmeiro (1858: 155) refers to some collections of the late 16th and early seventeenth centuries. These concern the plants that Agustín León and Francisco Micó (c. 1528-c. 1583) sent to Jacques Daléchamps (1513–1588), which were published in his *Historia generalis plantarum* (1586). Unfortunately, the French herbarium of this author has been lost. Furthermore, the Spanish physician Melchor Villena (1564–1655) also collected a herbarium which was preserved in Valencia until the end of the eighteenth century, but has probably disappeared today (Colmeiro 1858: 158).

The Saxon physician and botanist Joachim Burser (1583–1639) made extensive journeys throughout Europe in search of plant specimens. Though he did not publish any book, he gathered his plants in a bound herbarium 'book' in 25 volumes (*Hortus siccus*), which is kept at Uppsala University (UPS). The significance of this herbarium is well known because it allowed Linné to inspect many of the plants that Bauhin included in his *Pinax Theatri botanici* (1623). During his trip Burser studied the southern slopes of the Pyrenees, as in at least one dozen plants the location "In montibus Pyrenaeis Hispaniam versus" is mentioned (Juel 1936). Since apparently none of the collections mentioned before persists till today, these plants of Burser

Fig. 4.1 *Dracaena drago* L. One of the first drawings of this Macaronesian species, observed by Clusius at the Nossa Senhora da Graça Covent, in Lisbon (Photo by Ramón-Laca at Kraków)

would be the oldest ones from Iberia collected and preserved until today (Fig. 4.2). The date of these collections has not been determined with precision yet but can be estimated at between 1600 and 1620, when they were already mentioned by Bauhin in the *Prodromos theatri botanici* (1620). This book also mentions 17 plants which Bauhin received from Jacobo Albinus, a physician from Hamburg, who travelled through Spain in the early seventeenth century (between 1614 and 1620). Perhaps some of these plants are preserved in the Basel University Herbarium (BAS), but so far this has not been verified.

Fig. 4.2 *Arenaria tetraquetra*. Conserved at Burser's herbarium (UPS), from a southern slope of the Pyrenees. It is one of the earliest plants collected from Iberia (probably before 1620) and preserved until today (Photo by Mats Hjertson, courtesy of the UPS)

Jacques Barrelier (1606–1673) travelled through Spain (Barcelona, Valencia, Córdoba, etc.) and collected plants in the middle of the seventeenth century. In his work *Plantae per Galliam, Hispaniam et Italiam observatae*, which was published posthumously in 1714 by Antoine de Jussieu (1686–1758), plant observations are included in a very similar style as they were in Clusius' work, also with lots of icons. Unfortunately there is no evidence whether his collections are still preserved today. The same applies to the *Viridarium Lusitanicum* from Gabriel Grisley (1749), which lists 1618 plants collected mainly in the surroundings of Lisbon (Fernandes 1987), but no further information on these collections has been found.

The family Salvador was a pharmacist's dynasty from Barcelona that built up a Natural History Cabinet in the seventeenth and eighteenth centuries. Its founder was Joan Salvador i Boscà (1598–1681), who collected plants around 1626 on the outskirts of Barcelona in collaboration with the Dutchman Guillem (Willem) Boel. From these collections some plants are still preserved (Ibáñez 2006: 33). His son, Jaume Salvador i Pedrol (1649–1740), travelled in 1687 with Joseph Pitton de Tournefort (1656–1708) through Catalonia and Valencia, where they collected many specimens that are located in his own herbarium as well as in Tournefort's, preserved in the Museum of Natural History in Paris (P-TRF). The son of Jaume, Joan Salvador i Riera (1683–1725), collected plants in Majorca and Minorca and sent duplicates to Hans Sloane and James Petiver (BM). He worked on *Botanomasticon catalonicum* (Bolòs 1946) that never was published. He travelled with Antoine de Jussieu between 1716–1717 across Spain and Portugal collecting specimens which are preserved in the Museum of Natural History in Paris (P-JU and P-LA) and in the Salvador's Herbarium. This collection is very important for the knowledge of the Iberian flora for the same reasons that are going to be indicated for Tournefort's. Joan's brother, Josep Salvador i Riera (1690–1761), collected plants in 1725 on Minorca. The subsequent generations preserved the collections but they did not maintain significant collecting activities. At the beginning of the twentieth century Pius Font i Quer (1888–1964) rescued the collections; they now are located in the Botanic Institute of Barcelona (BC). This is the oldest herbarium established in Spain that has reached our days. The Salvador's collection consists of 4025 specimens, mostly Iberian (Ibáñez 2006: 40). Its importance is enhanced because of the presence of materials sent by significant botanists already mentioned above (Jussieu and Tournefort), and also Herman Boerhaave (1668–1739), Pierre Magnol (1638–1715), James Petiver (1658–1718) and Sébastien Vaillant (1669–1722), as a result of the excellent relationships between the Salvador family and the European naturalists at that time.

At that time European travellers were exploring the Iberian Peninsula. One of the most important explorations are Tournefort's, who visited the eastern coast of the Peninsula in 1687 and a wider area between October 1688 and March 1689 (Fig. 4.3). Henriques (1890, 1898) made a list of plants collected and localities visited by Tournefort from manuscripts discovered in the library of the Botanical Garden of Coimbra. Tournefort's herbarium is preserved at the Museum of Natural History in Paris (P-TRF). Many labels only indicate the name of the plant. However, sometimes they include a precise locality that allows to assign the specimen to a place he visited on his travels through the Peninsula. Tournefort's is undoubtedly the largest and most valuable antique plant collection from Iberia. Furthermore, its value redoubles due to its location in a major institution.

In the seventeenth century botanical knowledge made great progress. While earlier only 500–600 plant species were known, Bauhin in his *Pinax* (1623) recognized 6000, and made the first global effort to systemize the names and to

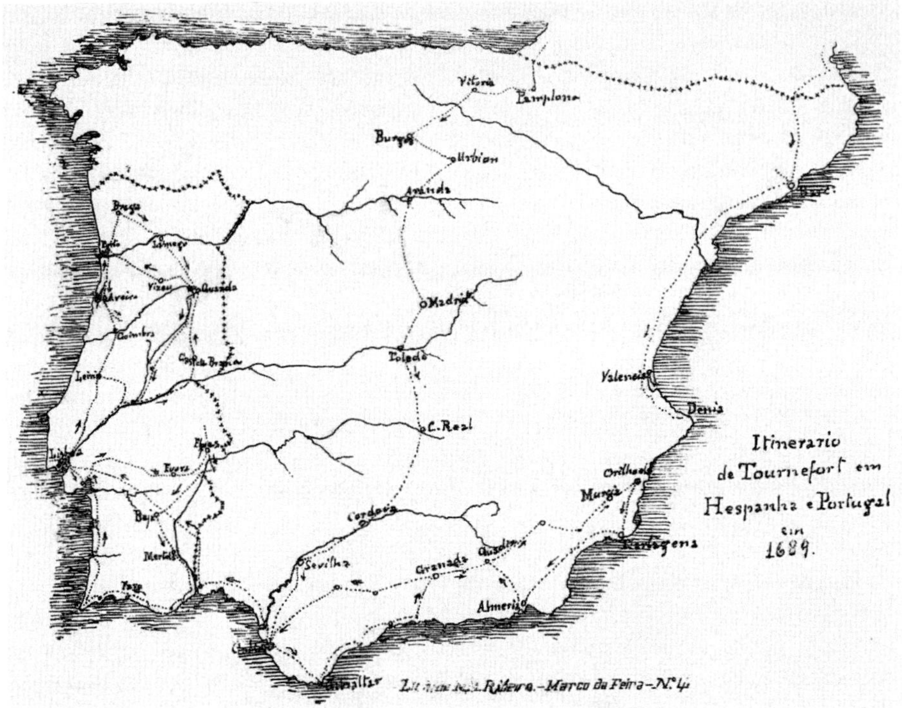

Fig. 4.3 Tournefort's trips in Spain and Portugal in 1687 and 1688 (Picture from Henriques 1890)

establish a complete synonymy, while John Ray (1627–1705) distinguished some 8000 species. As a result of this increase in species numbers recognized a general classification of plants was needed: examples are the ones proposed, with different artificial criteria, by Robert Morison (1620–1683), John Ray, Pierre Magnol and J. Tournefort. The genus concept (which we owe to Tournefort) was defined and strengthened, and genera were grouped in families (Pierre Magnol and later Michel Adanson (1727–1806)) (Dayrat 2003). This built the foundations for a future evolutionary interpretation. The highlight for the eighteenth century is the sexual system of Carl von Linné (1707–1778) which meant an important step forward as regards artificial systems and which had great success due to its simplicity. However, Linné's most important accomplishment is his system of binomial nomenclature (although it already had some precedents) which, linked with an extensive synonymy, facilitates a common framework of knowledge. Bernard de Jussieu (1699–1777) and Antoine-Laurent de Jussieu (1748–1836) proposed a natural method of classification based on shared character sets, not only for genera but also for families. This classification was consolidated by other authors and many aspects have been maintained even today.

In this scientific context, Joseph Quer (1685–1764) published his *Flora Española* in 1762–1764, though the last two volumes were published posthumously by Gómez Ortega (1784). It describes about 2050 species (Colmeiro 1885: CXX), sorted alphabetically. It is based on collections by the author himself, now preserved in Geneva (G). Unfortunately, the specific locations mentioned in these books are not reflected on his herbarium labels. Quer's work undoubtedly has great merit, but its weakness was its explicit rejection of the Linnean nomenclature, and this was partially corrected by Gómez Ortega in the two last volumes.

In Spain the Royal Botanic Garden (*Real Jardín Botánico*) was founded in Madrid in 1755 but its current location dates back to 1781. Joseph Quer was the first botanist responsible for it. Its consolidation as a scientific institution occurred in 1801, when Antonio José de Cavanilles (1745–1804) took the reins and founded the herbarium MA that kept about 12,000 plants at his death (Colmeiro 1858: 175). This author also mentioned the existence of 1507 ancient plants in the herbarium, of which there is hardly any trace today (Fig. 4.4). MA currently is the main herbarium of the Iberian Peninsula, with about 1,050,000 sheets. Cavanilles' activity focused on descriptions of plants coming from the New World. That is why his work had relatively little direct impact on the elaboration of the Iberian flora.

Fig. 4.4 *Viola cornuta* collected in Asturias by Esteban de Prado around 1762–1783, one of the oldest specimens in the MA herbarium (Photo by C. Aedo)

The Botanical Garden of Ajuda (*Jardim Botânico da Ajuda*) also was founded in the mid-eighteenth century, in 1768, near Lisbon, followed in 1771 by the Botanical Garden of Coimbra (*Jardim Botânico de Coimbra*) under the impetus of the Italian botanist Domenico Vandelli (1735–1816). When Felix de Avellar Brotero (1744–1828) was in charge of the Coimbra institution he gathered a herbarium with the intention of preparing his *Flora lusitanica* (1804). Specimens belonging to this herbarium nowadays have vanished (Pereira Coutinho 1916). This first Portuguese flora described about 1900 species, using Linné sexual system and, more important, incorporating his binomial nomenclature, aligning this work with its age in contrast with the Spanish flora of Quer. A few years later Johann Centurius Hoffmannsegg (1766–1849) and Johann Heinrich Link (1767–1851) published the *Flore portugaise* (1809–1840), which consolidated Portugal as a country with a much better known flora than Spain. The plants of these German botanists are preserved in different European herbaria.

During much of the nineteenth century the peninsular botany suffered a severe crisis which is not surprising given the social and political upheavals of the time. For Spain, after the failed attempt of Mariano Lagasca (1776–1839), who apparently lost the materials he had been preparing for a Spanish flora around 1823 (González and Rodríguez 1996: 623), we mention the compilations carried out by Miguel Colmeiro (1816–1909) between 1885 and 1889, and by Mariano del Amo (1809–1896) in the *Flora fanerogámica de la Península Ibérica* between 1871 and 1873, both stockpiling herbarium specimens and bibliographic information but without a critical analysis. Only towards the end of the century the situation improved with the publication of the *Prodromus florae Hispanicae* (1861–1880) by Heinrich Moritz Willkomm (1821–1895) and Johan Lange (1818–1898). This work listed and described the vascular plants of mainland Spain, excluding the Balearic Islands and Portugal, although both territories were mentioned occasionally in taxa distribution notes. According to the taxa numbers provided at the end of each group, the *Prodromus* described 1030 genera and 5092 species, as well as a considerable number of infraspecific taxa. These infraspecific taxa, usually varieties, have been accepted in many cases by later authors as species or subspecies. Willkomm (1893) published subsequently the *Supplementum*, increasing taxa numbers to 1048 genera and 5570 species. Lange's plants are preserved in the Natural History Museum of Denmark (C) and Willkomm's at the Botanical Garden of Coimbra (COI). This meticulous and careful work, based on the critical study of the materials collected by the authors for years, can be considered as the starting point of the modern knowledge on the flora of Spain's mainland. In Portugal, where little of Brotero's work remained at the end of the nineteenth century, botanical activities were reactivated by Júlio Henriques (1838–1928), who collected throughout the country and consolidated the current herbarium of Coimbra. The impulse he gave to floristic studies was reflected in the *Flora de Portugal* published in 1913 by Antonio Xavier Pereira Coutinho (1851–1939). Spanish botanists of that time did not respond with works of the same level; they continued to produce more or less complete compilations, such as Blas Lázaro é Ibiza (1858–1921) with his *Compendio de la flora española* (1920–1921), of which the third edition includes 5531 species of vascular plants; and Arturo Caballero (1877–1950) with his *Flora Analítica de España* (1940). It is noted, however, that around this time a new approach, equivalent to that in other more advanced countries, was breaking through. Two good examples of this were Carlos Pau (1857–1937) and, later, Pius Font i Quer. These botanists conducted a major work in collecting and describing critical species and significantly improving herbaria and libraries. Unfortunately these initiatives did not result in works of synthesis and finally they were truncated as a result of the Spanish Civil War.

Between 1964 and 1980 *Flora Europaea*, the reference work until now for Europe, was published. Its five volumes, with effective keys and concise descriptions, is considered a great revolution in the knowledge of European floristics and also an example of cooperation among the 187 participating authors. Finally, there was a modern work that allowed botanist to compare Iberian plants with those that grew in neighbouring countries, and to identify them easily, despite the lack of illustrations. Also, as always happens with this kind of publications, the shortcomings inherent in all syntheses were showing up. This encouraged the publication of numerous new species and claims of other, forgotten species that became apparent thanks to this powerful tool of comparison.

On the occasion of the centenary of the completion of the work of Willkomm and Lange, Bertram E. Smythies (1912–1999) published in 1984 an updated catalogue, *Flora of Spain and the Balearic Islands*, which takes its taxonomic basis from the *Flora Europaea*. This author compiles in addition some taxa proposed by Pau, Sennen and other botanists that had been omitted in the *Flora Europaea*. In total Smythies recognised 5323 species of which 910 would be endemic to the mainland of Spain and the Balearic Islands.

The increase in floristic studies in the last third of the twentieth century has improved significantly the knowledge of our flora and has crystallized into an important number of catalogues and regional floras. Towards the end of the twentieth century it became evident that, in Europe, only Spain and some Balkan countries, such as Greece, lacked of a modern flora. In response to this gap, in Spain different plans were drawn to settle eventually in the *Flora iberica* project at the beginning of the 1980s and under the impulse of Santiago Castroviejo (1946–2009) (Fig. 4.5).

This project has generated assets worthy of being exploited. Herbaria and botanical libraries have been improved and strengthened in quality. New taxonomists have been formed and developed within the project. At the present time, many of the main international projects, such as *Flora of China* or *Flora of North America*, are requiring the participation of our specialists, which means a recognition for the project that has educated them. However, the state of affairs within our borders is completely different. The curricular assessment of floristic and taxonomic publications is very low. In practice it is a mere ornament to the curriculum of the authors. This is being a general problem that is happening because of the unspecific evaluation of the curricula and it is affecting seriously certain disciplines where the publication of books and monographs is really

Fig. 4.5 Santiago Castroviejo (1946–2009), main driver and coordinator of the *Flora iberica* project since its initial years. He was a great defender of plant collecting, and always endeavoured to convey this interest to his disciples. In the picture taken in Madeira, he shows the correct way to collect and prepare a *Carex* specimen in the field (Photo by L. Medina)

important. If nothing changes, these disciplines, even with the greatest international recognition, will disappear from our universities and research centers. In addition, another more specific problem of the natural sciences, and particularly in botany, is arising here. The generation that launched *Flora iberica* in the 1980s were, at the same time that they were developing their work, encouraging the training of researchers in various laboratory techniques to improve our knowledge of plants at all levels. Part of these researchers who were promoted thanks to the open-mindedness of their advisors have now positions of responsibility and have failed to rise to the occasion. They maintain an ungenerous position with respect to the floristic studies, instead of promoting a harmonious development of all botanical specialtizations. All of this is seriously threatening the future of this discipline in Spain. While other countries around us are devoted to the study of tropical floras, ours has hardly taken steps in that direction. The opportunity that we have high-potential personnel and the advantage of a common language with many countries in the Neotropics seems to be hopelessly squandered. Something so easy in appearance like completing the job started with the *Flora iberica* seems today an unreachable desire as regards a *Flora Macaronesica*. Plants from the Azores, Canary Islands and Madeira will have to wait for a better opportunity.

Although none of this is inevitable it seems that the current great moment of the floristic studies will be a mere shooting star.

4.2 Geographical Delimitation and Source of Data

All data provided in this chapter refer to the Iberian Peninsula and the Balearic Islands. The Iberian Peninsula consists of Andorra and mainland Spain and Portugal. Likewise some small islands along the Iberian coast have been included as part of the mainland (*e.g.* Columbretes, Cíes, Ons, Alborán and Berlengas islands). The Balearic Islands include Majorca, Minorca, Ibiza, Formentera and some other smaller islands. The flora of these territories is hereinafter referred to as *Iberian flora*.

The data of taxa have been extracted from the *Flora iberica* (Castroviejo 1986–2016). For the two major subfamilies of the Compositae: Cichorioideae and Asteroideae and the whole Gramineae family, which remain unfinished in the *Flora iberica*, we have used the Med-Checklist (Greuter et al. 2008) and the *Flora Europaea* (Tutin et al. 1964–1980) respectively. Nevertheless, if published reviews or genus drafts of the *Flora iberica* were available, these were preferred over other works. We have also considered the new taxa described after the publication of their corresponding genera in the *Flora iberica*, which are compiled in the project website (www.floraiberica.es). The total number of species slightly changes according to the updating of the work. The numbers indicated here slightly differ from Aedo et al. (2013) since the volumes XVI(I) and XX of the *Flora iberica* have been published.

The circumscription of families used by the *Flora iberica* is based on Stebbins (1974) for flowering plants and Pichi Sermolli (1977) for ferns and fern allies. Although these systems may seem obsolete nowadays, they were considered "modern" in 1986 when the first volume of this work was published. Thus, the number of families would significantly be altered if other classification systems such APG III were used. We mainly have used the species taxonomical category to carry out the numerical data analyses. When data refer to number of taxa, this includes the species and subspecies but not the varieties or lower categories. Hybrids have been omitted.

The chorological data to draw the maps of richness of native, endemic and alien species have been taken from the Anthos Project (www.anthos.es), a project developed to gather and record the results of the ongoing work within the *Flora iberica* framework. That database compiles more than 1.5 million of occurrence data from bibliographic sources and revised herbarium material. Additionally, we have used data from the Flora-on Project (www.flora-on.pt) to complement the information on Portugal. We have considered as endemic all the species that exclusively occur in the territory of the Iberian flora, as well as those whose distributions slightly overflow the Iberian limits on the northern side of the Pyrenees.

4.3 Species Richness

The Iberian flora is comprised of 189 families, 1278 genera and 6276 species (7099 taxa). From these, 19 families (10%), 146 genera (11.4%) and 739 species (11.8%) are considered non-native. Mainland Spain has the highest species richness (5987 species, 6740 taxa), representing 95.4% of the total species number. That is consistent with the fact that mainland Spain has the largest area and the highest diversity of habitats and biogeographic territories. Mainland Portugal comprises 2844 species (3042 taxa), representing 45.3% of the total. Only 125 species (2%) are exclusively found in mainland Portugal. Finally, the Balearic Islands are home to 1532 species (1602 taxa), representing 24.4% of the total. From these, 164 species (2.6% of the total) only occur in the Balearics, and thus within the territory of the Iberian flora.

According to Tutin et al. (1964–1980) the European flora would consist of 11,557 species. This number is probably obsolete today, although it can be useful to broadly compare the plant species richness of different countries. The Iberian flora contains about 54% of the European plant species. Other Mediterranean floras are equally rich: the Italian flora with 6711 species (Conti et al. 2005), contains 58% of the European species; Greece with 5752 species (Dimopoulos et al. 2013) comprises around 50% of the total. In Central Europe the number of species is considerably smaller. Germany, with 3349 species (Haeupler and Muer 2001), comprises 29% of the total number of species in Europe, and Great Britain and Ireland, with around 2049 species, 18%. The Mediterranean African countries have somewhat lesser rich floras, possibly due to their larger extensions of desert as well as a lower level of exploration. According to Médail and Quézel (1997) there would be about 4200 species in Morocco and 3150 in Algeria. The Balearic Islands are noticeably less rich than the other large Mediterranean Islands to the east, e.g. Corsica (2313 taxa) (Jeanmonod and Gamisans 2007), Sardinia (2407 taxa) and Sicily (3010 taxa) (Conti et al. 2005). The smaller size of the Balearics and the lower elevation of its mountains would explain this fact.

The pattern of species richness within the Iberian Peninsula is shown in Fig. 4.6. It is based on the number of native species observed in 50 × 50 km squares. Probably these data show a considerable bias because the sampling effort over the entire territory is rather uneven. Because of this, almost 40% of the squares are under-sampled (<4000 observations) and show very low richness values (number of species <800). But it is clear that the areas with the highest observed species richness (number of species >1600) are located in the Pyrenees and Cantabrian mountains in the north, and in Sierra Nevada in the south. Some parts of Central System (Sierra de Guadarrama) and Baetic System (Almijara and Serranía de Ronda) are almost equally rich.

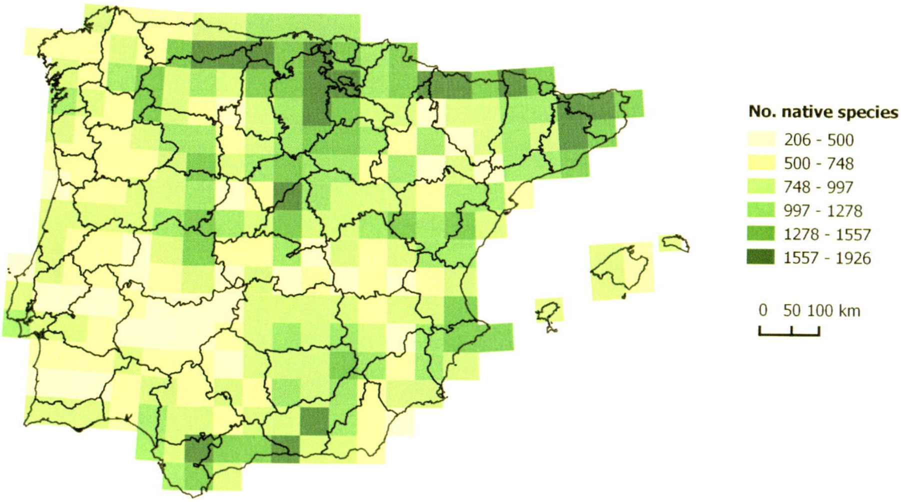

Fig. 4.6 Observed richness pattern in native plant species in the Iberian flora

4.4 Composition in Large Groups

As regards the composition of the Iberian flora in large groups, there is a predominance of dicots (80.1% of species), followed by monocots (17.5%). Gymnosperms and pteridophytes are marginally represented with only 0.4% and 1.9% of the species, respectively. The ten largest families comprise 57.6% (3615 species) of the total Iberian vascular flora. The Compositae is the largest family with 850 species (12.9%), followed by the Leguminosae with 530 species (8.7%) and the Gramineae with 463 species (7.3%). There are slight differences in composition and relative importance of the ten richest families over mainland Spain, Portugal and the Balearic Islands. One of the most significant differences is the lower representation of Rosaceae in the Balearic Islands and mainland Portugal. This is owing to the absence of the genus *Alchemilla* in those territories. A very analytic treatment was adopted for *Alchemilla* in the *Flora iberica*, resulting in many species gathered in the high northern mountains and consequently increasing the weight of the Rosaceae in mainland Spain.

Regarding the contribution of the ten largest families in genera, we can observe that the Gramineae are proportionally much richer in genera than in species. Although some of its genera, such as *Festuca*, *Bromus* or *Agrostis*, are very large in the Iberian flora (61, 23 and 21 species respectively), there are 61 genera with just one species. The Umbelliferae show a similar pattern comprising 6.9% of total number of genera but only 3.6% of total number of species. The Plumbaginaceae is the most disproportionate family, consisting of only four genera (2 of them with only 1 species) but with 162 species (mostly endemic) in the Iberian flora (Fig. 4.7).

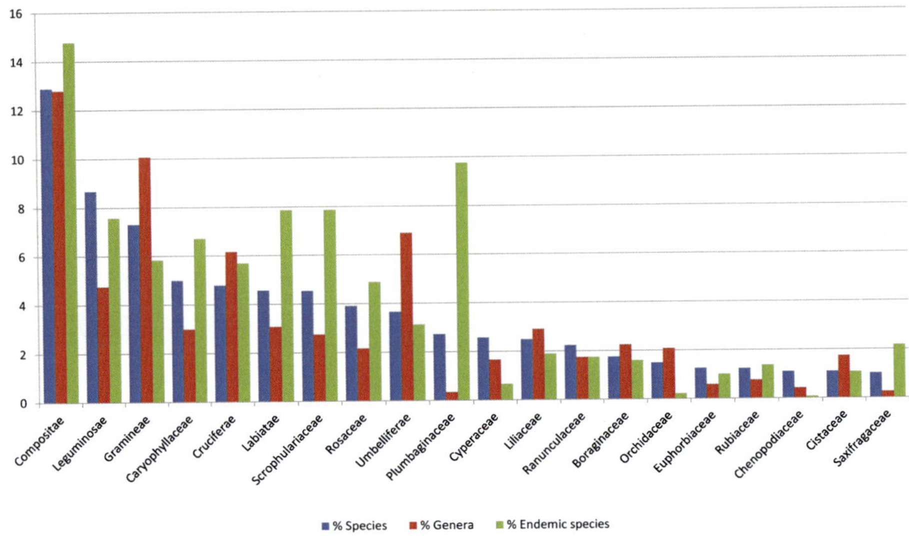

Fig. 4.7 Percentage of total numbers of species, genera and endemics in the 20 largest families of the Iberian flora. These 20 families represent 74.4% of the total number of species, 66.4% of total number of genera and 85.9% of total number of endemic species

The 20 largest genera in the Iberian flora contribute more than 20% of the total species number. The most diverse are *Limonium* (109 species), *Centaurea* (94 species), *Carex* (92 species) and *Alchemilla* (83 species) (Fig. 4.8). A detailed analysis reveals considerable differences between the territories. *Centaurea, Carex* and *Alchemilla*, but also *Festuca, Teucrium* and *Saxifraga*, are very diversified in the main Iberian mountain ranges (mostly within mainland Spain), and contain many endemic species. All these genera are much less represented in mainland Portugal, mostly due to the lack of big mountain chains. In contrast, two leguminous genera *Trifolium* and *Vicia*, which include many widely distributed species but no endemics, gain importance in terms of species contribution. The most diverse genus found in the Balearic Islands is *Limonium* (46 species), which ranks fifth in mainland Spain and is much less significant in Portugal. This genus is considered to have two centres of diversification: the coasts of the western Mediterranean region and the Asian steppes (Erben 1993).

The information available on the relative weight of the major taxonomic groups is highly heterogeneous and sometimes hardly comparable. Floras of other Mediterranean regions have also low rates of pteridophytes and gymnosperms: 0.9 and 0.2% for Turkey (Davis et al. 1988), 1.7 and 0.2% for California (Baldwin et al. 2012), and 1.2 and 0.1% for the Cape Floristic Region (CFR) (Goldblatt and Manning 2002). It is well known that the relative weight of the pteridophytes is much higher in tropical floras, for instance it ranges from 17.3 and 22.8% in the Gulf of Guinea islands (Exell 1944) and it is more than 8.5% in Ecuador. The percentage of monocots in the Iberian flora (17.5%) is very similar as in the whole

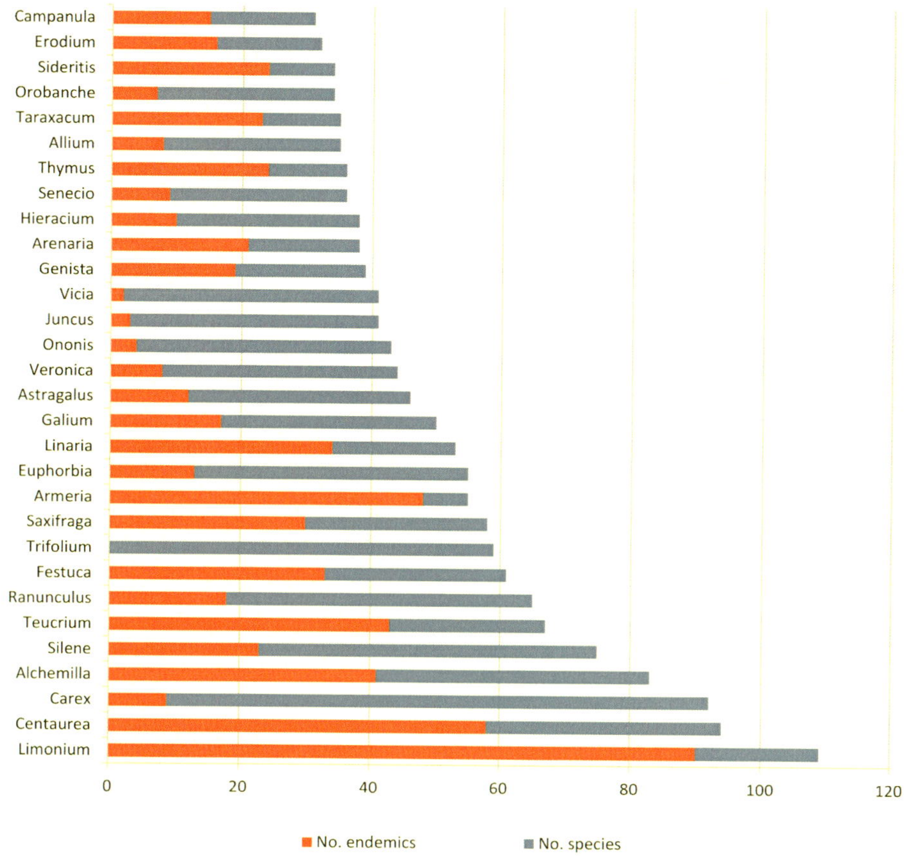

Fig. 4.8 Number of species and endemics in the 30 largest genera of the Iberian flora

of Europe (18%), Turkey (16.4%, Davis et al. 1988), or California (17.8%, Baldwin et al. 2012), but lower than in the CFR (24.5%, Goldblatt and Manning 2002). The ratio of species per genus, 5.3 in the Iberian flora (Table 4.1), is also similar to the Italian and Californian rates but low compared to Turkey and the CFR.

Concerning the 15 largest families from other Mediterranean regions, we observe some similarities. The Iberian list of 15 largest families corresponds for 100% (but they differ in order) with that of Italy (Conti et al. 2005), 87% of Turkey (Davis et al. 1988), 73.3% of California (Jepson Flora Project 2013) and 47% of the Cape Region (Goldblatt and Manning 2002). The largest family for the five regions is the Compositae; Leguminosae is the second one for all, except for Italy where it is third. Compositae is usually the largest family in floras of arid to semi-arid regions and Leguminosae are well developed in most parts of the world (Goldblatt and Manning 2002). With regard to the genera, 11 of the 15 largest genera of the Italian flora correspond with the Iberian ones. Some of the dissimilarities are due to different taxonomic criteria. For instance, *Hieracium* is the largest genus in the

Table 4.1 Attributes of Mediterranean floras (considering native species only) of the Cape Floristic Region, California, Turkey, the Iberian Peninsula and Italy

	Cape Floristic Region (CFR)	California	Turkey	Iberian Peninsula	Italy
Area (km^2)	90,000	423,970	783,562	583,832	301,338
No. of families	172	162	163	159	152
Endemic families	4	?	0	0	0
No. of genera	992	991	1146	1053	1049
Endemic genera	162 (16.3%)	25 (2.2%)	15 (1.3%)	27 (2.3%)	7 (0.7%)
No. of species	9087	5271	8575	5537	5933[a]
Endemic species	6226 (68.5%)	1580 (30%)[a]	2651 (30.9%)	1258 (22.7%)	799 (13.5%)[a]
Species/genera	9.2	5.3	7.5	5.3	5.7
Non-native species	?	1086	217	739	778
% of the total flora in the					
15 largest families	71%	59%	76.4%	69.9%	70.5%
10 largest genera	21,40%	16%	18.2%	12.6%	11.6%

[a]Estimated values from the original data (Sources: CFR: Goldblatt et al. (2005); California: Jepson Flora Project (2013); Turkey: Davis et al. (1988); Italy: Conti et al. (2005), endemic genera in Peruzzi et al. (2014), percentage of flora in Beard et al. (2008))

Italian flora, whereas it is not among the top 15 largest genera of the Iberian flora. *Hieracium* is a very polymorphic group which also shows a high variability in the Iberian Peninsula, but its taxonomic treatment in *Flora iberica* was more synthetic and resulted in few accepted species.

4.5 Floristic Affinities

As mentioned above, the 95.5% of the species from mainland Portugal are shared with mainland Spain, just because no biogeographic barriers exist between these territories. From the 125 non-shared species, 43 are non-natives, 64 are endemic and 18 native non-endemic. Some of the latter are aquatic plants (e.g. *Damasonium alisma* and *Utricularia subulata*), which normally have erratic distributions related to habitat availability and their dispersal systems. A small number of them are shared with North Africa, such as *Volutaria cuprinoides*, *Echium tuberculatum*, *Hyacinthoides mauritanica* and *Asplenium hemionitis* (also found in the Macaronesian islands). From the 164 non-shared species between mainland Spain and the Balearic Islands, five are non-natives, 103 are endemic and 56 native non-endemic. Some belonging to the last group, such as *Allium ebusitanum* and *Scilla numidica*, are from the northwestern coast of Africa; the closest point to the Algerian coast is only 235 km. There are also some Tyrrhenian species, shared with Sardinia or Corsica, such as *Brimeura fastigiata*, *Serapias nurrica* and *Helicodiceros muscivorus*.

We do not aim to analyse in depth the similarities between the Iberian flora and the surrounding regions, but some uncommon species distributions are mentioned below (Fig. 4.9). The relationship between the south of the Iberian Peninsula and North Africa has been well known for a long time. A great number of species from northwestern Africa occur in southern Spain since an obvious connection via the Strait of Gibraltar exists. For instance, over 70 North African plant species occur in the sub-desertic areas of southeastern Spain, such as *Tetraclinis articulata*, *Polygala balansae*, *Anthemis chrysanta*, *Enneapogon persicus*, *Mauranthemum decipiens*, *Salsola webbii* and *Narcissus pachybolbus*. It is interesting to note that some species, such as *Arenaria pungens*, *Epilobium atlanticum*, *Eryngium glaciale*, *Euphrasia willkommii*, *Saxifraga trabutiana* and *Sedum melanantherum*, are found both in the high Baetic Mountains and in the great Moroccan Mountain Chains (Atlas and Rif).

Many species occurring in the Euro-Siberian region of the Peninsula are distributed across central and northern Europe. More unusual are the distributions of some plants growing in the northwest of the Peninsula, such as *Saxifraga spathularis* and *Erica mackaiana*, which are otherwise only found on the coast of Ireland. Others, such as *Omphalodes littoralis* and *Cochlearia aestuaria*, are restricted to the northern coast of Spain and the northwestern coast of France. Many boreal-alpine plants occur in the high mountains of the Pyrenees. However, a small number of plants from the Alps, such as *Androsace lactea*, *Primula pedemontana* and *Salix breviserrata*, are only known from the Cantabrian Mountains. Similarly, the Euro-Siberian plants *Corydalis intermedia* and *Gentianella amarella*, occur in the Iberian System but not in the Pyrenees. These disjunction distributions could be explained by long distance dispersal or local extinction processes.

Other singular distributions are those of *Astragalus oxyglottis* and *Krascheninnikovia ceratoides*. Within the Iberian Peninsula, they occur in the arid regions of the Ebro Depression and Baza-Guadix, but the nearest populations are in the steppes of the Eastern Mediterranean. Similarly, *Commicarpus africanus* occurs from East Africa to the Middle East, but a few small populations are known from southeastern Spain. Finally, *Rhododendron ponticum* is a relic of evergreen forests that were widely distributed throughout Europe during the Tertiary, whereas it has, at present, a reduced disjunct distribution including the southwestern Iberian Peninsula, the Black Sea coast and Lebanon (Cross 1975).

Fig. 4.9 Some species with uncommon distribution ranges: (**a**) *Erica mackaiana*; (**b**) *Bellium bellidioides*; (**c**) *Anthemis chrysanta*; (**d**) *Helicodiceros muscivorus*; (**e**) *Polygala balansae*; (**f**) *Krascheninnikovia ceratoides*; (**g**) *Commicarpus africanus*; (**h**) *Tetraclinis articulata*; (**i**) *Gentianella amarella* (Photos (**a-e**) and (**g-h**) by C. Aedo, (**f**) by A. Buira, (**i**) by L.M. Ferrero)

4.6 Endemicity

One of the best indicators for determining the originality of a flora is endemicity. There is not any Iberian endemic family, but there are some endemic genera. Twenty-one endemic genera (Fig. 4.10) and six subendemics have been recorded (Table 4.2). We consider subendemic genera those whose distributions slightly overflow the Iberian limits at the northern side of the Pyrenees: *Borderea*, *Dethawia*, *Endressia*, *Petrocoptis* and *Xatardia*. Similarly we have taken into account *Naufraga*, known from the north of Majorca and considered extinct from the island of Corsica. All Iberian endemic genera are paucispecific, with only one or two species, except *Petrocoptis* which has nine. The rate of endemic genera in the Iberian flora is significantly higher than that of other countries in the Mediterranean Basin such as Italy (7) (Peruzzi et al. 2014), Greece (8) (Dimopoulos et al. 2013) or Turkey (15) (Davis et al. 1988). It is important to note that these rates are strongly influenced by analytical or synthetic taxonomic trends. For instance, the genus *Femeniasia*, which was considered endemic in the Balearics, has been recently transferred to *Carduncellus*.

The number of endemic species in the Iberian flora has been estimated at 1258, which is 22.7% of the total native species number. Some examples of Iberian endemic plant species are shown in Fig. 4.11. Analysing each territory separately, mainland Spain has 1116 Iberian endemic species (21.3% of the total native species number of mainland Spain), 822 of which (15.7%) are unique for the country. The Balearic Islands contain 125 (9.1%) Iberian endemic species, 103 (7.5%) of these being exclusive for this archipelago. The very nature of islands contributes to isolation, which normally results in high levels of endemism. However the closeness of the Balearics to the mainland (86 km) and the relatively low altitude of their mountains make its flora moderately rich in endemism. 336 Iberian endemic species (13.7% of the total native species number) are known from mainland Portugal, but only 65 (2.7%) are exclusive of that territory. Most of the Iberian endemics from Portugal also occur in Spain since there is no biogeographic separation. As we have discussed above, the lack of big mountain chains in Portugal compared with Spain is likely the cause of its lower proportion of endemic species.

In fact, a further examination of the Peninsula shows that the areas richest in endemic species fully coincide with the main Iberian mountain ranges (Fig. 4.12). The Picos de Europa in the Cantabrian Mountains, and Sierra de Almijara, Sierra Nevada and Sierras de Cazorla and Segura in the Baetic System gather the greatest number of endemic species, with densities over 150 endemics per 1000 km^2. Some 50 × 50 km squares of the Andalusian Baetic Mountains reach up to 16% of endemism. According to Médail and Quézel (1997), this mountain range is one of the richest areas of plant endemicity in the Mediterranean Basin. In particular, the Sierra Nevada is the most important massif for the endemic flora of the Baetic range (Blanca 2001). The relict confinement of Arctic-Alpine taxa, the diversity of habitats and the height gradient of this mountainous area produced intense speciation processes (Blanca 2001; Mendoza-Fernández et al. 2015). Although the

Fig. 4.10 Representatives of some endemic Iberian genera: (**a**) *Hymenostemma pseudoanthemis*; (**b**) *Euzomodendron bourgaeanum*; (**c**) *Gyrocaryum oppositifolium*; (**d**) *Teesdaliopsis conferta*; (**e**) *Pseudomisopates rivasmartinezii*; (**f**) *Xatardia scabra*; (**g**) *Petrocoptis pyrenaica subsp. glaucifolia*; (**h**) *Odontitella virgata*; (**i**) *Lycocarpus fugax* (Photos by C. Aedo)

Table 4.2 Endemic and subendemic genera of the Iberian Flora

Genera	Family	Distribution
Avellara Blanca & C. Díaz	Compositae	S Peninsula
Boleum Desv.	Cruciferae	Ebro valley
[a]*Borderea* Miég.	Dioscoriaceae	Pyrenees
Castrilanthemum Vogt & Oberpr.	Compositae	Sierra de Cazorla
[a]*Dethawia* Endl.	Umbelliferae	Pyrenees and Cantabrian Mountains
Distichoselinum García Martín & Silvestre	Umbelliferae	S Peninsula
[a]*Endressia* J. Gay	Umbelliferae	N Peninsula
Euzomodendron Coss.	Cruciferae	Almería
Guillonea Coss.	Umbelliferae	E Peninsula
Guiraoa Coss.	Cruciferae	SE Peninsula
Gyrocaryum B. Valdés	Boraginaceae	W Peninsula
Hispidella Lam.	Compositae	W Peninsula
Hymenostemma Kunze ex Willk.	Compositae	S Peninsula
Lepidophorum Neck.	Compositae	W Peninsula
Lycocarpus O.E. Schulz	Cruciferae	SE Peninsula
Naufraga Constance & Cannon	Umbelliferae	Majorca (apparently extinct in Corsica)
Odontitella Rothm.	Scrophulariaceae	W Peninsula
Ortegia L.	Caryophyllaceae	NW Peninsula
Periballia Trin.	Gramineae	C-W Peninsula
[a]*Petrocoptis* Endl.	Caryophyllaceae	Pyrenees and Cantabrian Mountains
Phalacrocarpum (DC.) Willk.	Compositae	NW Peninsula
Pseudomisopates Güemes	Scrophulariaceae	Sierra de Gredos
Prolongoa Boiss.	Compositae	C-E Peninsula
Pterocephalidium G. López	Dipsacaceae	C-W Peninsula
Rothmaleria Font Quer	Compositae	S Baetic Mountains
Teesdaliopsis (Willk.) Gand.	Cruciferae	Cantabrian Mountains and Serra da Estrela
[a]*Xatardia* Meisn. & Zeyh.	Umbelliferae	Pyrenees

[a]Subendemic genera whose distribution slightly overflows the Iberian limits on the northern slope of the Pyrenees

Pyrenees have the highest species richness, they are poorer in endemic species than the Cantabrian Mountains, the Central System and the Sierra Nevada. The Pyrenees form a geographical barrier which contributes to the isolation of the Peninsula. However their mountains are fairly connected with the high southern European mountains, which would explain its lower number of Iberian endemic species. As regards Portugal, the richest area occurs in the Serra da Estrêla, the highest mountain of the country.

The ten largest families contain over 74% of the total endemic species. It is usual for endemism to reach high values in the largest plant families (Beard et al. 2008). The Compositae show the highest levels of endemism (203 species, 14.8% of total

Fig. 4.11 Some examples of endemic Iberian species: (**a**) *Omphalodes nitida*; (**b**) *Allium pyrenaicum*; (**c**) *Senecio rodriguezii*; (**d**) *Narcissus cyclamineus*; (**e**) *Silene stockenii*; (**f**) *Klasea algarbiensis*; (**g**) *Limonium daveaui*; (**h**) *Stauracanthus genistoides*; (**i**) *Linaria hirta* (Photos (**a–e**) and (**g-i**) by C. Aedo, (**f**) by A. Buira)

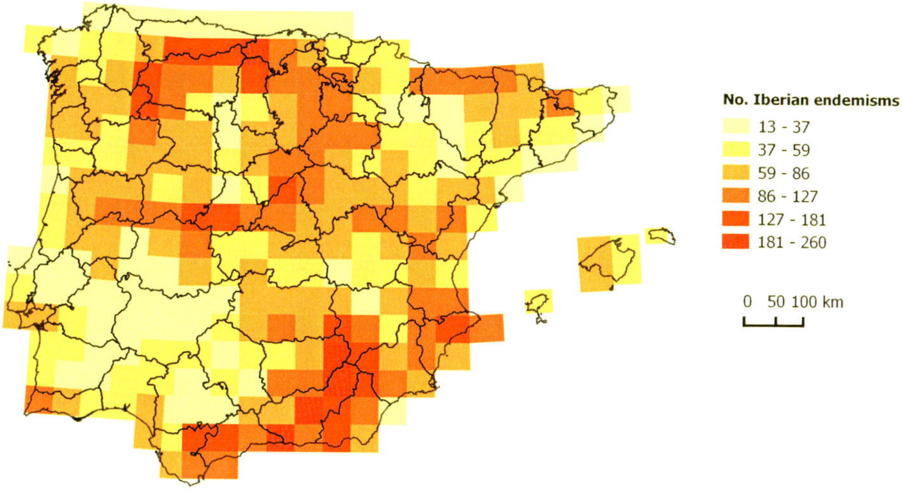

Fig. 4.12 Pattern of richness in endemic species in the Iberian flora

endemic species), which is consistent with the predominance of the family in the Iberian flora (Fig. 4.7). *Centaurea* is the largest genus within the Compositae, both in endemic species and total flora. Plumbaginaceae is the second family in terms of endemic species, but the number of genera within the family in the Iberian Peninsula is only four (Fig. 4.7). Both *Armeria* and *Limonium* are very diverse genera in Mediterranean Basin, and most of the Iberian species are endemic. Two other families with many endemic species are Labiatae and Scrophulariaceae. Some genera belonging to these families, such as *Teucrium, Thymus, Sideritis, Linaria* and *Antirrhinum*, have the highest species diversity in the Mediterranean region. The most diverse genera in the Iberian flora are generally those that also provide many endemic species but there are exceptions. *Carex, Trifolium, Ononis* and *Vicia* are very large genera; all of them have more than 40 species (*Carex* has 92), but only few, if any, are endemic (Fig. 4.8).

The Iberian flora has a high degree of originality (22.7% endemism); however, it is comparable to other countries in the Mediterranean Basin. Turkey appears to be significantly richer both at species (Table 4.1) and endemism levels (34%). Greece maintains a rate of endemism around 22% (Dimopoulos et al. 2013); in Italy it would be 13.5% according to Conti et al. (2005) or 18.9% according to Peruzzi et al. (2014); and it is around 21.4% in Morocco (Médail and Quézel 1997). According to these authors, the whole Mediterranean Basin scores an endemism rate of around 50%, but when analysing the endemism rates by countries the values are, on average, much lower. On the other hand, it should be taken into consideration that these countries differ in their proportion that enjoys a Mediterranean climate. The floras of Central European countries are poorer at species and endemism levels. As an example, Haeupler and Muer (2001) assessed the rate of endemism of the German flora at 0.7%. The values for other Mediterranean-type biogeographic regions show also high rates of endemism (Table 4.1): 30% in

California, 52.5% in the Southwest of Australia (Beard et al. 2008) and up to 68.5% in the Cape Floristic Region (CFR). The CFR is by far much richer at the species level than the others, as well as in rates of endemism at family and generic levels. The CFR is, indeed, one of the world's richest areas of plant species diversity (Goldblatt and Manning 2002).

4.7 Extinct and Threatened Species

Over the past 30 years, attention to the conservation problems of the Iberian flora (Bañares et al. 2004 and references herein) has increased. At the same time our knowledge of threats and risks to our vascular flora has increased considerably. As a result, 2258 species (36.6% of the total) have been catalogued in Spain (excluding Canary Islands) under some categories and included in Red Data Lists or Books, 1009 in the National Red List (Bañares et al. 2008) and 1979 in several regional Red Data Books and Lists.

A total of 1743 species (28.3% of the total) have been legally protected in Spain: 188 species under EU laws, 158 under national laws and 1681 under regional laws. Portugal has no specific legislation; nevertheless 110 taxa are included in the European Habitats Directive. Even though a detailed analysis has not been carried out, a trend to protect rare and endemic species regardless of the habitat they occupy can be observed. Many endemic species form scarce populations and grow in mountain habitats; they are not subject to specific threats except for the fact that they are rare. On the other hand, some non-endemic species associated with vulnerable coastal habitats and wetlands are not catalogued. In any case, protecting about one third of all species seems somewhat excessive and certainly impracticable.

An interesting aspect of species conservation is the extinction analysis. This topic has been analyzed in Aedo et al. (2014); it concludes that 20 taxa, two of them endemic, can be considered extinct in the territory of the Iberian flora. We have excluded *Nolletia chrysocomoides* because it was wrongly identified (Aedo 2014). Only one species, *Armeria arcuate*, may be extinct in Portugal; but its taxonomic status remains uncertain (Nieto Feliner 1990).

The highest number of extinct species occurred in wetlands, followed by dunes and ravines. High mountain plants only account for a minor part of the extinct species. It should be noted that more than half of the species live in coastal areas. Caution should be used when considering the dates referring to the last citing of a species. Except in very specific and recent cases, there have not been monitoring studies which would allow us to identify with accuracy the time of disappearance. References to 5 species stopped in the late eighteenth century. The last reference for about half the supposedly extinct species dates from the nineteenth century, with a peak towards the end of that century followed by a decrease during the twentieth and early twenty-first centuries. Among the probable causes of plant extinction the most important is habitat loss and overgrazing, affecting two thirds of the taxa. In only one instance displacement by an alien and invasive species is designated as the cause of extinction. In just over a quarter of the species the cause of disappearance is completely unknown.

Over-collection has been put forward at times as one of the factors causing the disappearance of species (Barreno et al. 1984: 49; Norton et al. 1994; Lorite et al. 2003: 15). Data referring to the Spanish flora do not support this hypothesis. For most of the extinct species fewer than three herbarium specimens per species have been found and this can hardly have affected their survival. Regarding the species designated as extinct at one time or another in red lists or books, it is striking that the majority subsequently has been removed due to the appearance of new wild populations. This indicates that exploration and collection work has had a very positive result on the knowledge of species threat categories (Aedo et al. 2014).

Greuter (1994) discussed the extinction of species in the Mediterranean biomes and established that the documented extinction rate in the Mediterranean Basin (0.11%) is very low in comparison with other Mediterranean-type areas of the world such as the 0.30% of the Cape floristic region, 0.40% of California or 0.66% of Western Australia. The extinction rate of 0.31% obtained for the Iberian flora is more like those of the Cape floristic region or California than the rest of the Mediterranean floras. For the European area, data of IUCN (Bilz et al. 2011) show very low extinction rates ranging from 0.024 to 0.028%, but only considering policy plant species. Some neighbouring European countries have calculated similar extinction rates. For example, Cheffings and Farrel (2005) compiled for Great Britain 13 extinct species out of a total of 2951 species (Preston et al. 2002) which results in an extinction rate of 0.44%. Italy lists 21 extinct taxa which represent a calculated extinction rate of 0.31% from 6711 species (Abbate et al. 2007) and the values for Metropolitan France show an extinction rate of 0.53%, from a total of c. 5000 species (Olivier et al. 1995) and 24 extinct species (Monocorps and Siblet 2012).

The extinction rate calculated for the Iberian flora might seem relatively low and close to a natural extinction rate. It should not be forgotten, however, that this rate of extinction has occurred in a very short period of time of about 200 years. In any case a cursory reading of these data would suggest that the conservation status of the Iberian flora is relatively good. Unfortunately we have records of hundreds of species that have very few populations and that could be on their way to extinction (Bañares et al. 2008). If the management of these populations is inadequate, the extinction rate could increase dramatically in the coming years (Hanski and Ovaskainen 2002).

4.8 Alien Species

The number of non-native plants is growing at present and usually provides controversial data. Some authors prefer to include all alien species in the floras and checklist. Others, however, prefer to stick to those that are fully naturalized,

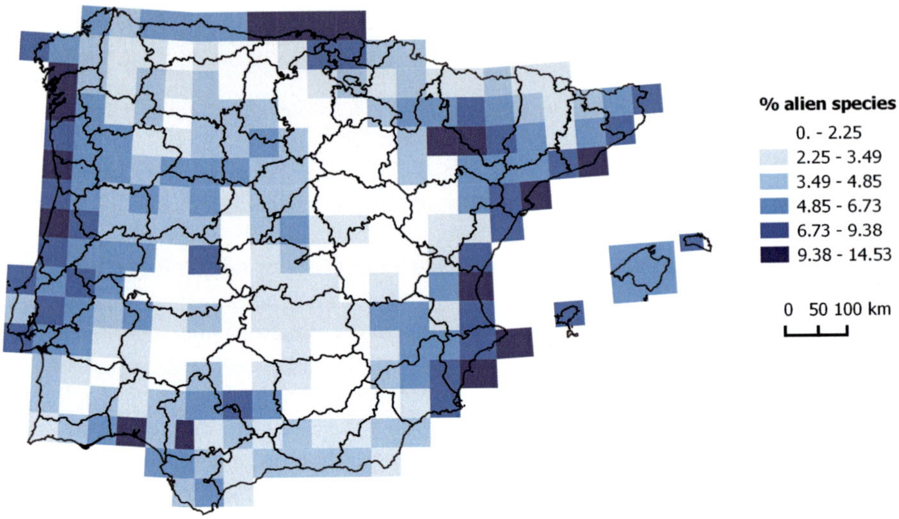

Fig. 4.13 Pattern of the percentage of alien species in the Iberian flora

excluding those that are merely occasional. Therefore, available data are not homogeneous even in the same flora, due to different author's criteria. In the Iberian flora we have classified 739 species as non-native, representing 11.8% of the total flora. This number includes both naturalized and casual species with respect to invasion status, and archaeophytes and neophytes with respect to their residence time. The percentage of these species is very similar in mainland Spain and the Balearics (about 10%); however it is noticeably higher in Portugal (13.9%).

By analysing the percentage of alien species throughout the Iberian Peninsula (Fig. 4.13), a trend to gather in the coastal areas can be observed. According to Gassó et al. (2009), high anthropogenic disturbance, low altitude, short distance to the coastline and a dry, hot climate are directly correlated to alien invader richness in Spain (Gassó et al. 2009). However, it is important to note that predominantly agricultural squares from the centre and south of the Peninsula would probably have higher rates of alien species if they were properly sampled. Thus, for instance, some agricultural regions from the Ebro Depression which have been quite well sampled show high rates of alien species.

The alien species in the Iberian flora belong to about 85 families, dominated by the three largest families (Compositae, Gramineae and Leguminosae) (Fig. 4.14). According to Lambdon et al. (2008), these three large worldwide occurring plant families, which have a weedy tendency and have undergone major radiations in temperate regions, are also the largest in the alien flora of Europe. Other families providing several non-native species and thoroughly spread in the Iberian Peninsula are the Solanaceae, Amaranthaceae and Oxalidaceae. The commonest genera, with the highest number of occurrence records, are *Amaranthus, Oxalis, Conyza, Bidens, Paspalum, Datura* and *Chamaesyce*. They are all neophytes mostly from the Neotropics and North America, comprising mainly urban and agricultural weeds.

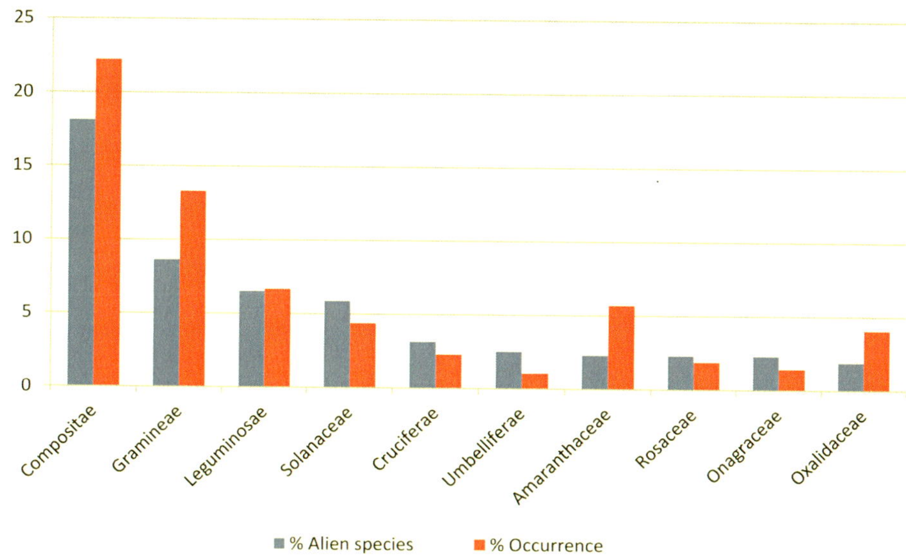

Fig. 4.14 The ten largest families of alien flora existing in the Iberian Peninsula: percentage of alien species in the total alien flora and percentage of records of alien species. These ten families represent 53.7% of the alien species and 63% of alien species records in the Iberian flora

Our estimation is very similar to the one proposed by Sanz Elorza et al. (2004) for Spain, who indicated a rate of 10–13.4%. These rates are very close to others estimated for nearby areas, such as 16.4% for Corsica (Jeanmonod and Gamisans 2007), 11% for France (Heywood 1989) and 10.5% for Italy (Conti et al. 2005). Baldwin et al. (2012) recorded 19.9% for California and Weeb et al. (1988) up to 50% for New Zealand. The Mediterranean Basin, with a long history of migrations and trade, may have better resisted the recent invasion of non-native floras than oceanic islands or 'climatically isolated' regions.

Acknowledgments The authors wish to acknowledge the efforts of those professionals and amateur botanists who regularly contribute to the knowledge of the Spanish flora depositing their collections in public herbaria. We thank N. Fumeaux, J. Güemes, M. Hjertson, N. Ibáñez, G. López, R. Morales, F. Muñoz Garmendia, M. Porto and L. Ramón-Laca for providing bibliographic or herbaria data. We are grateful to Helen Álvaro for correcting the English version of the manuscript. L.M. Ferrero generously provided the picture of *Gentianella amarella*. This work has been supported and funded by the Spanish Government through the *Flora iberica* project (CGL2014-52787-C3-0P) and by the Anthos project (Fundación Biodiversidad-CSIC).

References

Abbate G, Alessandrini A, Conti F (2007) Vascular plants. In: Blasi C, Boitani L, La Posta S, Manes F, Marchetti M (eds) Biodiversity in Italy. Palombi Editori, Roma, pp 149–161

Aedo C (2014) *Nolletia chrysocomoides* (Desf.) Less. (Compositae), especie a excluir de la flora europea. Acta Bot Malac 39:320–321

Aedo C, Medina L, Fernández-Albert M (2013) Species richness and endemicity in the Spanish vascular flora. Nord J Bot 31:478–488. doi:10.1111/j.1756-1051.2012.00009.x
Aedo C, Medina L, Barberá P, Fernández-Albert M (2014) Extinctions of vascular plants in Spain. Nord J Bot 33:83–100. doi:10.1111/njb.00575
Amo M (1871–1873) *Flora fanerogámica de la Península Ibérica.* vol 2. Impr. de D. Indalecio Ventura, Granada
Anthos (2015) *Anthos. Sistema de Información de las Plantas de España.* http://www.anthos.es. Accessed 18 Feb 2015
Baldwin BG, Goldman D, Keil DJ, Patterson R, Rosatti ThJ (eds) (2012) The Jepson manual. Vascular plants of California, 2nd edn. Thoroughly Revised and Expanded. University of California Press, Berkeley
Bañares Á, Blanca G, Güemes J, Moreno JC, Ortiz S (eds) (2004) Atlas y libro rojo de la flora vascular amenazada. Dirección General de Conservación de la Naturaleza & Sociedad Española de Biología de la Conservación de Plantas, Madrid, p 1072
Bañares Á, Blanca G, Güemes J, Moreno JC, Ortiz S (eds) (2008) *Lista Roja 2008 de la flora vascular española.* DG Medio Natural y Política Forestal (MAGRAMA) y SEBICOP, Madrid, p 86
Barrelier J (1714) Plantae per Galliam, Hispaniam et Italiam observatae. Stephanum Ganeau, Paris
Barreno E, Bramwell D, Cabezudo B, Cardona MA, Costa M, Fernández Casas FJ, Fernández-Galiano E, Fernández Prieto JA, Gómez Campo C, Hernández Bermejo E, Heywood VH, Izco J, Llorens L, Molero Mesa J, Montserrat P, Rivas Martínez S, Sáenz Laín C, Santos Guerra A, Valdés Castrillón B, Wildpret de la Torre W (1984) Listado de plantas endémicas, raras o amenazadas de España. Información Ambiental 3:49–72
Bauhin C (1620) Prodromos theatri botanici. Typis Pauli Iacobi, Frankfurt
Bauhin C (1623) Pinax theatri botanici. Sumptibus & typis Ludovici Regis, Basileae
Beard JS, Chapman AR, Giota P (2008) Species richness and endemism in the Western Australian flora. J Biogeogr 27:1257–1268. doi:10.1046/j.1365-2699.2000.00509.x
Bilz M, Kell SP, Maxted N, Lansdonw RV (2011) European red list of vascular plants. Publication Office of the European Union-IUCM, Luxembourg. http://ec.europa.eu/environment/nature/conservation/species/redlist/downloads/European_vascular_plants.pdf
Blanca G (2001) Flora amenazada y endémica de Sierra Nevada. Universidad de Granada, Granada
Bolòs A (1946) El herbario Salvador. Collect Bot (Barcelona) 1:1–8
Brotero FA (1804) Flora Lusitanica. Tipographia Regia, Lisboa
Caballero A (1940) Flora analítica de España. Sociedad Anónima Española de Traductores y Autores, Madrid
Castroviejo S (ed gen) (1986–2016) *Flora iberica.* 19 vols. Real Jardín Botánico-CSIC, Madrid [detailed list of volume editors and genera authors at www.floraiberica.es]
Cheffings CM, Farrell L (eds) (2005) The vascular plant red data list for great Britain. Species Status 7:1–116. http://jncc.defra.gov.uk/pdf/pub05_speciesstatusvpredlist3_web.pdf
Clusius C (1576) *Rariorum aliquot stirpium per Hispanias observatarum historia.* Antwerpen
Colmeiro M (1858) La botánica y los botánicos de la Península Hispano-Lusitana. Estudios bibliográficos y biográficos. Impr. M. de Rivadeneyra, Madrid
Colmeiro M (1885) *Enumeración y revisión de las plantas de la Península Hispano-Lusitana é islas Baleares*, vol 1. Impr. Fuentenebro, Madrid
Conti F, Abbate G, Alessandrini A, Blasi C (2005) An annotated checklist of the Italian Vascular Flora. Palombi Editore, Roma
Cross JR (1975) Biological flora of the British Isles: Rhododendron ponticum. J Ecol 63:345–364
Daléchamps J (1586) Historia generalis plantarum. Lugduni
Davis PH, Mill RR, Tan K (1988) Flora of Turkey and the East Aegean Islands, vol 10. Edinburgh University Press, Edinburg, p 590
Dayrat B (2003) Les botanistes et la flore de France. Muséum Nationale d'Histoire Naturelle, París
Dimopoulos P, Raus T, Bergmeier E, Constantinidis T, Iatrou G, Kokkini S, Strid A, Tzanoudakis D (2013) Vascular plants of Greece: an annotated checklist. Englera 31:1–372

Erben M (1993) Limonium mill. In: Castroviejo S et al (eds) Flora Iberica, vol 3, Plumbaginaceae (partim)-Capparaceae. Real Jardín Botánico-CSIC, Madrid, pp 2–143
Exell AW (1944) Catalogue of the vascular plants of S. Tomé (with Principe and Annobon). British Museum, London
Fernandes A (1987) História da botânica em Portugal até finais do século XIX. História e desenvolvimiento da ciencia em Portugal. Academia de Ciencias de Lisboa, Lisboa
Flora-On: Flora de Portugal Interactiva 2015 Sociedade Portuguesa de Botânica. http://www.flora-on.pt. Accessed 18 Feb 2015
Gassó N, Sol D, Pino J, Dana E, Lloret F, Sanz-Elorza M, Sobrino E, Vilà M (2009) Exploring species attributes and site characteristics to assess plant invasions in Spain. Divers Distrib 15:50–58. doi:10.1111/j.1472-4642.2008.00501.x
Goldblatt P, Manning J (2002) Plant diversity of the Cape region of Southern Africa. Ann Mo Bot Gard 89:281–302
Goldblatt P, Manning JC, Snijman D (2005) Cape plants: corrections and additions to the flora. 1. Bothalia 35:35–46. doi:10.4102/abc.v35i1.367
Gómez Ortega C (1784) *Continuacion de la flora Española, ó historia de las plantas de España, que escribia Don Joseph Quer, cirujano consultor del Exército.* 2 vols. Impr. Joachin Ibarra, Madrid
González A, Rodríguez R (1996) En torno a la 'Flora española': dos proyectos fallidos en los años centrales del siglo XIX. Anales Jard Bot Madrid 54:622–626
Greuter W (1994) Extinctions in Mediterranean areas. Philos Trans R Soc Lond B 344(41):46
Greuter W et al (1984–2008) Med-checklist. A critical inventory of vascular plants of the circum-mediterranean countries, vol 4. Conservatoire et Jardin Botaniques, Genève
Grisley G (1749) Viridarium Lusitanicum. Antonii Craesbeeck, Verona
Haeupler H, Muer T (2001) Bildatlas der Farn- und Blütenpflanzen Deutschlands. Verlag Eugen Ulmer, Stuttgart
Hanski I, Ovaskainen O (2002) Extinction debt at extinction threshold. Conserv Biol 16:666–673
Henriques J (1890) Explorações botânicas em Portugal por Tournefort. Bol Soc Brot 8:191–247
Henriques J (1898) Explorações botânicas em Hespanha por Tournefort. Bol Soc Brot 15:108–174
Heywood VH (1989) Patterns, extents and modes of invasions by terrestrial plants. In: Drake JA, Mooney H (eds) Biological invasions, a global perspective. SCOPE. John Wiley & Sons, London, pp 31–55
Hoffmannsegg JC, Link JHF (1809–1840) *Flore portugaise.* Impr. Charles Fréderic Amelang, Berlin
Ibáñez N (2006) *Estudis sobre cinc herbaris històrics de l'Institut Botànic de Barcelona.* PhD thesis. Universidad de Barcelona, Barcelona
Jeanmonod D, Gamisans J (2007) Flora Corsica. Édisud, Aix-en-Provence
Jepson Flora Project (eds) (2013) *Jepson eFlora.* http://ucjeps.berkeley.edu/IJM.html. Accessed 18 Feb 2015
Juel HO (1936) Joachim Burser's Hortus siccus. Lundequistska, Uppsala
Lambdon Ph W, Pyšek P, Basnou C, Hejda M, Arianoutsou M, Essl F, Jarošík V, Pergl J, Winter M, Anastasiu P, Andriopoulos I, Brundu G, Celesti-Grapow L, Chassot P, Delipetrou P, Josefsson M, Kark S, Klotz S, Kokkoris Y, Kühn I, Marchante H, Perglová I, Pino J, Vil M, Zikos A, Roy D, Hulme PE (2008) Alien flora of Europe: species diversity, temporal trends, geographical patterns and research needs. Preslia 80:101–149
Lázaro é Ibiza B (1920–1921) *Compendio de la flora española.* Tercera edición corregida y aumentada. Impr. Clásica Española, Madrid
Lorite J, López RM, Ruiz Girela M (2003) Conservación de la flora amenazada de Sierra Nevada. Conserv Veg 8:15–16
Médail F, Quézel P (1997) Hot-spots analysis for conservation of plant biodiversity in the Mediterranean basin. Ann Mo Bot Gard 84:112–127
Mendoza-Fernández AJ, Pérez-García FJ, Martínez-Hernández F, Salmerón-Sánchez F, Medina-Cazorla JM, Garrido-Becerra JA, Martínez-Nieto MI, Merlo ME, Mota JF (2015) Areas of

endemism and threatened flora in a Mediterranean hotspot: southern Spain. J Nat Conserv 23:35–44. doi:10.1016/j.jnc.2014.08.001

Monocorps S, Siblet J-Ph (2012) La Liste rouge des espèces menacées en France - Chapitre Flore vasculaire de France métropolitaine: premiers résultats pour 1 000 espèces, sous-espèces et variétés. IUCN, FCBN, MNHN. http://inpn.mnhn.fr/docs/LR_FCE/Dossier_presse_Liste_rouge_Flore_vasculaire_de_metropole_5_Nov_2012.pdf

Nieto Feliner G (1990) Armeria Willd. In: Castroviejo S et al (eds) Flora Iberica, vol. 2, - Platanaceae-Plumbaginaceae (partim). Real Jardín Botánico-CSIC, Madrid, pp 642–721

Norton DA, Lord JM, Given DR, De Lange PJ (1994) Over-collecting: an overlooked factor in the decline of plant taxa. Taxon 43:181–185

Olivier L, Galland JP, Maurin H (1995) Livre rouge de la flore menacée de France, Tome 1: Espèces prioritaires. MNHN SPN, Collection Patrimoine Naturel 20, Paris. http://inpn.mnhn.fr/docs/LR_FCE/livrerouge1.pdf

Pereira Coutinho AX (1913) A Flora de Portugal. A. Alves & Cia, Paris

Pereira Coutinho AX (1916) Plantas portuguesas dos Herbarios de Brotero e de Valorado existentes na Universidade de Lisboa. Arq Univ Lisboa 3:333–379

Peruzzi L, Conti F, Bartolucci F (2014) An inventory of vascular plants endemic to Italy. Phytotaxa 168:1–75. http://dx.doi.org/10.11646/phytotaxa.168.1.1

Pichi Sermolli REG (1977) Tentamen Pteridophytorum genera in taxonomicum ordinem redigendi. Webbia 31:313–512

Preston CD, Pearman DA, Dines TD (2002) New atlas of the British and Irish flora. University Press, Oxford

Quer J (1762–1764) *Flora española, ó historia de las plantas, que se crian en España*. 4 vols. Impr. Joachin Ibarra, Madrid

Ramón-Laca L, Morales R (eds) (2005) Descripción de algunas plantas raras encontradas en España y Portugal. Charles de L'Ecluse de Arras. Junta de Castilla y León, Valladolid

Sanz Elorza M, Dana Sánchez ED, Sobrino Vesperinas E (2004) Atlas de las plantas alóctonas invasoras en España. Ministerio de Medio Ambiente, Madrid

Smythies BE (1984) Flora of Spain and the Balearic Islands. Englera 3:1–881

Stebbins GL (1974) Flowering plants. Evolution above the species level. Belknap Press, Cambridge

Tutin TG et al (1964–1980) Flora europaea, vol 5. Cambridge University Press, Edinburg

Weeb CJ, Sykes WR, Garnock-Jones PJ (1988) Flora of New Zealand, vol 4, Naturalised Pteridophytes, Gymnosperms, Dicotyledons. Botany Division D.S.I.R., Christchurch

Willkomm M (1893) *Supplementum prodromi florae Hispanícae sive enumeratio et descriptio omnium plantarum inde ab anno 1862 usque ad annum 1893 in Hispania detectarum quae innotuerunt auctori, adjectis locis novis specierum jam nolarum, auctore*. Sumtibus E. Schweizerbart, Stuttgart

Willkomm M, Lange J (1861–1880) *Prodromus florae hispanicae seu synopsis methodica omnium plantarum in Hispania sponte nascentium vel frequentius cultarum quae innotuerunt, auctoribus*. 3 vols. Sumtibus E. Schweizerbart, Stuttgart

Chapter 5
Biogeographic Units of the Iberian Peninsula and Baelaric Islands to District Level. A Concise Synopsis

Salvador Rivas-Martínez, Ángel Penas, Tomás Emilio Díaz González, Paloma Cantó, Sara del Río, José Carlos Costa, Luis Herrero, and Joaquín Molero

Abstract The following biogeographic units for the Iberian Peninsula and Balearic Islands have been established according to the conceptual geobotanical proposals of Rivas-Martínez et al. (Parte I. Itinera Geobotanica 17:5–436, 2007; Parte II. Itinera Geobotanica 18 (1):5–424, 2011a; Glob Geobot 1(1):1–634, 2011b; Int J Geobot Res 1(1):21–40, 2011c and Int J Geobot Res 4(1):1–64, 2014): one kingdom (Holartic), two regions (Eurosiberian, Mediterranean), eight provinces, 16 subprovinces, 49 sectors and 264 districts. The potential natural vegetation: climatophilous, climato-temporihygrophilous, xerophilous, hygrophilous sigmetum or geopermasigmetum (series and geopermaseries) are shown for each province and sector. We also point out some of their specific features. Biogeographic maps up to district level are shown.

S. Rivas-Martínez (✉)
Phytosociological Research Center (CIF), E-28409 Los Negrales, Madrid, Spain
e-mail: rivas-martinez.cif@tsai.es

Á. Penas • S. del Río • L. Herrero
Department of Biodiversity and Environmental Management (Area of Botany). Mountain Livestock Farming Institute (Joint Center CSIC-ULE), Faculty of Biological and Environmental Sciences, University of León, Campus of Vegazana, s/n, León E-24071, Spain

T.E. Díaz González
Department of Organisms and Systems, Biology Faculty, University of Oviedo, E-33071 Oviedo, Spain

P. Cantó
Department of Plant Biology II, Faculty of Pharmacy, Complutense University of Madrid, Madrid E-28071, Spain

J.C. Costa
Botanical Center, Technical University of Lisbon, Tapada da Ajuda, Lisbon 1349-017, Portugal

J. Molero
Department of Botany, Faculty of Pharmacy, Campus de la Cartuja, University of Granada, Granada E-18071, Spain

J. Loidi (ed.), *The Vegetation of the Iberian Peninsula*, Plant and Vegetation 12,
DOI 10.1007/978-3-319-54784-8_5

5.1 Notions on Biogeography

Biogeography is the science which studies the distribution of species, communities, habitats, biocoenoses and natural ecosystems on Earth, as well as the relationships between them. It takes into account the distribution areas of taxa and syntaxa (chorology), in addition to information from other natural sciences (geography, botany, zoology, soil science, bioclimatology, geology, etc.), attempting to establish a hierarchical biogeographic classification of the territories on the planet. The main typological units in decreasing rank are: kingdom, region, province, sector, district, country, landscape cell and tesella (Rivas-Martínez et al. 2007, 2011a, b, c, 2014). Terrestrial biogeography has been twinned with phytogeography due to the value of vascular plant species and communities in its definition and delimitation on Earth. Oceanic biogeography should be studied with oceanological research methods.

The elementary biogeographic terrestrial unit of the lowest rank is the tessella, defined as a geographic space of greater or lesser extension, that is ecologically homogeneous, which means that it has only a single type of potential natural vegetation (climatophilous, edaphoxerophilous or edaphohygrophilous) and, as a consequence, only one successional trend of natural substitution communities (see Chap. 3). Within the framework of dynamic-catenal phytosociology, when zonation takes place in areas under extreme conditions, such as polar, fluvial, lake and marine landscapes, deserts, high mountain summits, dunes, rock formations, coastal cliffs, etc., the specialized vegetation growing in each of the elementary spaces is not replaced by perennial non-nitrified seral communities; in such cases the ecological homogeneous spaces or tesella are considered permatesella. Both, the tesella and the permatessella, are the only biogeographic units which can be repeated disjointed. In a bottom-to-top progression, we will define the various territorial categories used in biogeography:

- The landscape cells, such as horsts, peneplains, river valleys, lake systems, marshes, high mountain summits, etc., are constituted by a mosaic of tessellas or permatessellas with their corresponding complexes of plant communities, assembled by networks of geosigmeta and geopermasigmeta based on the geomorphology and the soils of the territory.
- The biogeographical country must be an extensive and clearly delimited geographic territory which possesses an abundant group of landscape cells, species, associations, and above all, its own topographical geosigmeta.
- The district is a group of biogeographical countries, characterized by the existence of a high number of differential species and even endemic taxa, especially in the coastal, oreadic (high elevation areas) or interior halophilous zones, which permit their distinction from the adjacent territories; it also comprises as characteristic units some associations, series, geoseries and geoclinoseries (cliserial zonations, see Chap. 3) which are absent from nearby districts.
- The sector is a set of biogeographical districts with a large-scale geographic entity, which possesses its own endemic taxa, associations and vegetation series,

as well as original topographical and geoclinosequential (altitudinal zonations) geoseries which are generally due to the existence of exclusive geoclinoseries, climatophilous series, permanent and subserial communities, as well as paleo-climatic evidences and former migratory routes.

- The province is a vast geographic territory which, as well as possessing a large number of endemisms and differential species (its own subelement), has particular macroseries. It is also characteristic of each biogeographical province to hold geomacroseries and a particular altitudinal vegetation zonation or exclusive geoclinoseries.
- The region is a very extensive territory, formed by a group of biogeographical provinces which has a flora or regional floristic element with endemic species, genera or even families; in addition it has its own particular megaseries, geomegaseries and geomegapermaseries and in consequence, its own bioclimatic and vegetation belts (Rivas-Martínez 2005).
- Finally, the kingdom is the supreme unit of biogeography, generally pluricontinental and multinsular, which in addition to taxonomic and ecosystematic considerations, addresses the origins of the flora and fauna, as well as the origin of the great continents, orogenies and particular macrobioclimates.

As it is by now traditional in this science, the denominations of the biogeographic units –both the primary and the auxiliary units (from the subregion to the area)– are given based on known geographical, orographic or historical designations which are considered to be more or less coincident with the area they are intending to represent. Orthographically, all the units are considered to be proper names identifying the zone. The names of high ranks (provinces or higher) are formed, if necessary, by two geographical nouns, joined by means of a hyphen; on the contrary, the lower ranks (sector or less) are joined by the conjunction 'and'; all the double biogeographical names maintain the initial capital in both and conserve their condition of a proper noun.

It must be emphasized that the biogeographical units can only be accurately delimited through their diagnosis and through the corresponding maps. All the territories –except the tessella and permatessella– must be contiguous by land, lake or sea routes, and include all the small orographic accidents and lithological diversity which exist in the area. Sometimes, in the biogeographical territories as a whole, there are introgressions by other adjacent territories, and these "islands" frequently occur in regions with a varied lithology or in areas near regional or provincial boundaries. Their possible typological independence, always of a lower rank than the area into which they introgress, depends on their originality, floristic richness and phytocoenotics, as well as on their surface area.

One of the criteria traditionally used for recognizing and delimiting biogeographic units as well as determinin their entity, is to incorporate the information of the geographic distribution of those taxa (families, genera, species and subspecies) which are narrowly restricted to a particular area up to the biogeographical province rank. These taxa are termed endemic taxa or endemisms. Endemisms have been successfully used to define and delimit the chorological or biogeographical units (provinces, sectors and districts), and they form part of the phytogeographical subelement which characterize them. Moreover, it is better to name the endemisms (taxa or syntaxa) which occupy a greater area or are regional, and those which for migratory reasons are dispersed across various biogeographical regions, as phytogeographical elements or geoelements of those higher units.

5.2 Concepts on Vegetation Series and Lanscape Phytosociology

Nowadays, the development of dynamic-catenal phytosociology and the syntaxonomic knowledge of broad territories of the Earth, as well as the cartographic delimitation of vegetation series, geoseries and geopermaseries, when available, have become the essential criteria for defining biogeographical units, in addition to suitably compiled and mapped bioclimatic and soil factors.

The vegetation series, also termed sigmetum (in honor of S.I.G.M.A.), expresses the whole set of plant communities or stages which can be found within similar tessellar spaces as a result of the succession process, and includes both the representative association of the mature stage, series head or potential natural vegetation, which is used as a nomenclatural reference, and the initial or subseral associations that may replace it. Based on this concept, the vegetation series or sigmetum represents the basic unit or essential model of dynamic phytosociology. Distinctions can also be made between climatophilous, xerophilous, temporihygrophilous and hygrophilous series. Climatophilous or zonal series are located on mature soils according to the mesoclimate, and only receive rainwater: mesophytic, submesophytic and subxerophytic; the temporihygrophilous series, included among the climatophilous, are those which have additional water contribution due to their topographical circumstances, and they thus develop on flooded or very wet soils throughout part of the year, and –at least during the summer or dry period– the soil horizons are well-drained and aerated. Finally the xerophilous series are found in particularly dry or xerophytic soils or biotopes such as lithosols, arenosols, very windy sites, steep slopes, crests, ledges, etc.; and the hygrophilous series grow on particularly wet soils and biotopes such as fluvisols, halosols, histosols, etc., and are found in river beds, marsh areas, salt flats, peat bogs, etc.

The vegetation geoseries or geosigmetum is the basic unit of dynamic-catenal phytosociology. It corresponds to a catena of vegetation series which is found around a given bioclimatic belt and biogeographic territory in the heart of the universal crest-slope-valley model. This topographic framework makes it possible to distinguish the three geomorphological aspects of any complete catena where the vegetation series constituting the geosigmetum are located in zones; the xerophilous series and geoseries (hyperxerophilic and xerophilic) are located in the driest sites (crests, escarpments, lithosols, etc.); the climatophilous and temporihygrophilous series and geoseries are located on slopes and foothills where greater humidity is contributed by rainfall and run-on; and the hygrophilous series and geoseries are found in the valleys and watercourses (fluvial, lake and water-courses), among which the river fractogeosigmetum (partial geosigmetum) is of great importance to plant landscape science due to its extrazonality, and also, in combination with the xerophilous and climatophilous sigmeta and geosigmeta, to the definition and structuring of regional and global biogeography.

The vegetation geopermaseries, also known as geopermasigmeta, is the catenal expression of a set of neighboring permaseries or permasigmeta, delimited by changing topographic or soil situations. These are determined by conditions of extreme climate (high mountains and polar areas) and exceptional microtopographic and soil conditions (walls, rock formations, marine cliffs, salt flats, etc.) which give rise to a large number of neighboring ecological residences populated by diverse permanent perennial plant communities (continuous vegetation permaseries) with absence of non-nitrophilous seral perennial communities which appear to have reached their equilibrium. The most favorable sites for the existence of geopermaseries or geopermasigmeta, in adition to sites corresponding to permanent types of vegetation in extreme high-mountain and polar region bioclimates, are ledges, rock crevasses, cliffs and coastal rock formations bathed by sea waters, peat bogs, wind drifts, mobile sand dunes, lake shores, streams etc. (Rivas-Martínez et al. 2011b).

5.3 Biogeographic Units of the Iberian Peninsula and Balearic Islands to District Level

According to the concepts previously defined, we show the hierachical typology of the biogeographic units recognized in the the Iberian Peninsula and Balearic Islands at the present time. In some cases we also comment on some exclusive units from France. In this case, we indicate it as [France]. The name of biogeographic units are in English, followed by its Spanish name in brackets and in the case of districts its extension is given in square kilometers, also in brackets. To achieve this synthesis we have considered various proposals from other authors, such as the most general approaches of Quézel (1985), Takhtajan (1986), Costa (1997), Rivas-Martínez et al. (2011b) to more specific or wider territories such as the proposals of Rivas-Martínez (1969, 1973, 1985, 1987, 1988), Pons and Quézel (1985), Alcaraz (1996), and also the proposals concerning limited territories such as those by Braun-Blanquet and Bolòs (1958), Rivas Goday and Borja (1961), Rivas Goday and Rivas-Martínez (1968), Alcaraz (1984), Peinado and Martínez Parras (1987), Navarro Andrés and Valle Gutiérrez (1987), Rivas-Martínez and Pizarro (1988), Loidi and Báscones (1995), Molina et al. (1993), Rivas-Martínez (1963, 1981), Rivas Goday (1964), Rigual (1972), Esteve (1973), Rivas-Martínez et al. (1984, 1987, 1990a, b, c, 1991a, b, 1997), Folch (1986), Bolòs (1967, 1987), Costa (1987),

Izco (1987), Ladero et al. (1987), Martínez-Parras and Peinado (1987), Asensi and Díaz Garretas (1987), Pérez Raya et al. (1990), Molero Mesa and Pérez Raya (1987), Díaz González and Fernández Prieto, (1988, 1994), Alcaraz et al. (1989, 1991), Peinado et al. (1992), Navarro, G. (1989), Berastegui et al. (1997), Loidi et al. (1997), Villa et al. (1997), Cantó P. (2007). We also referred to the most recent approaches established by Rivas-Martínez and Loidi (1999), Cantó (2007) and Rivas-Martínez et al. (2014).

According to our studies the territory of the Iberian Peninsula and Balearic Islands share two biogeographic regions: the Eurosiberian and Mediterranean, both included in the Holartic Kingdom. The numeric synthesis is shown in the next table:

Biogeographic unit	Eurosiberian region	Mediterranean region	Total
Subregions	2	1	3
Provinces	2	6	8
Subprovinces	6	10	16
Sectors	12	37	49
Districts	64	182	264

The listing below shows the biogeographic typology pointing out its hierarchical structure up to district level and also the approximate extension in square kilometers. Its geographical distribution can be seen in the corresponding maps (Figs. 5.1, 5.2 and 5.3).

I. EUROSIBERIAN Region (*Región EUROSIBERIANA*)

IA. ALPINE-CAUCASIAN Subregion (*Subregión ALPINA-CAUCÁSICA*)

Ia. PYRENEAN Province (*Provincia PIRENAICA*)

Iaa. EAST PYRENEAN Subprovince (*Subprovincia PIRENAICA ORIENTAL*)

1. EAST PYRENEAN Sector (*Sector PIRENAICO ORIENTAL*)

1a. High Ampurdán Pyrenean District (*Distrito Pirenaico Altoampurdanés*) (944 km^2)
1b. Conflent and Ripollés District (*Distrito Conflentino-Ripollés*) (2069 km^2)
1c. Andorra and Cerdaña District (*Distrito Andorrano-Cerdañés*) (1916 km^2)
1d. High Pallars East Pyrenean District (*Distrito Pirenaico Oriental Altopallarés*) (2466 km^2)
1e. Ariège East Pyrenean District (*Distrito Pirenaico Oriental Ariegense*) [France]
1f. Cadí Sierran District (*Distrito Serrano Cadiés*) (863 km^2)
1g. South Bergueda and Solsona District (*Distrito Surberguedano-Solsonés*) (2109 km^2)
1h. Montseny District (*Distrito Montsignático*) (852 km^2)
1i. Osona and Olot District (*Distrito Ausetano-Olotense*) (2638 km^2)

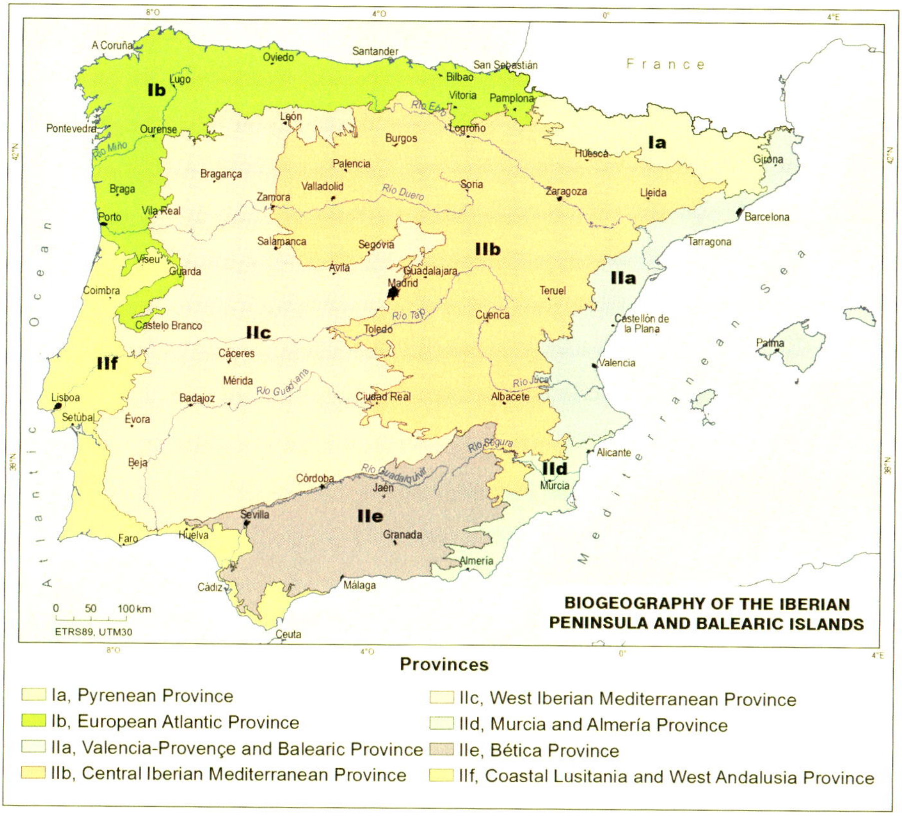

Fig. 5.1 Biogeographic map of the Iberian Peninsula and Balearic Islands at province level

Iab. Central Pyrenean Subprovince (*Subprovincia PIRENAICA CENTRAL*)

2. CENTRAL PYRENEAN Sector (*Sector PIRENAICO CENTRAL*)

   2a. Central High Pyrenean District (*Distrito Altopirenaico Central*) (1928 km^2)

   2b. Bigorra Central Pyrenean District (*Distrito Pirenaico Central Bigorrés*) [France]

   2c. Arán Central Pyrenean District (*Distrito Pirenaico Central Aranés*) (346 km^2)

3. WEST PYRENEAN Sector (*Sector PIRENAICO OCCIDENTAL*)

   3a. Aragón West Pyrenean District (*Distrito Pirenaico Occidental Aragonés*) (474 km^2)

   3b. Navarra West Pyrenean District (*Distrito Pirenaico Occidental Navarro*) (1298 km^2)

   3c. French West Pyrenean District (*Distrito Pirenaico Occidental Francés*) [France]

Fig. 5.2 Biogeographic map of the Iberian Peninsula and Balearic Islands at subprovince level

Iac. Aragón Prepyrenean Subprovince (*Subprovincia Prepirenaica Aragonesa*)

4. Aragón Prepyrenean Sector (*Sector Prepirenaico Aragonés*)

4a. Low Pallars District (*Distrito Bajopallarés*) (2175 km^2)
4b. Low Ribagorza and Sobrarbe District (*Distrito Bajorribagorzano-Sobrarbés*) (2634 km^2)
4c. Guara Sierran District (*Distrito Serrano Guarense*) (1222 km^2)
4d. Jacetania District (*Distrito Jacetano*) (1936 km^2)

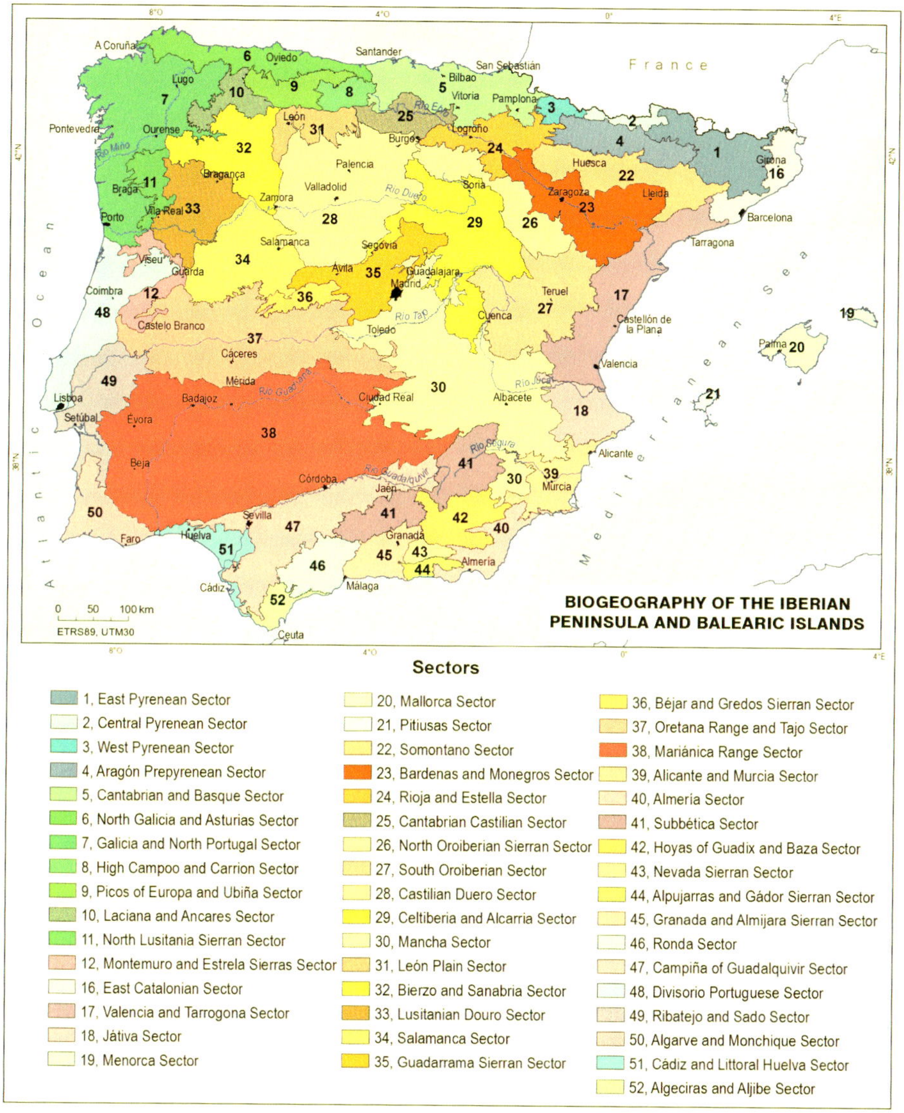

Fig. 5.3 Biogeographic map of the Iberian Peninsula and Balearic Islands at sector level

IB. ATLANTIC-CENTRAL EUROPEAN Subregion (*Subregión* ATLÁNTICA-CENTROEUROPEA)

Ib. EUROPEAN ATLANTIC Province (*Provincia* ATLÁNTICA EUROPEA)

Iba. CANTABRIAN ATLANTIC Subprovince (*Subprovincia* CANTABROATLÁNTICA)

5. CANTABRIAN AND BASQUE Sector (*Sector* CÁNTABRO-VASCÓNICO)

5a. Labourd and Baztán District (Distrito *Labortano-Baztanés*) (933 km^2)
5b. Guipúzcoa District (Distrito *Guipuzcoano*) (2046 km^2)
5c. Biscay District (Distrito *Vizcaíno*) (2965 km^2)
5d. Valnera and Trasmiera District (Distrito *Valnerano-Trasmierano*) (948 km^2)
5e. North Cantabrian District (Distrito *Cántabro Septentrional*) (1907 km^2)
5f. South Cantabrian District (Distrito *Cántabro Meridional*) (1390 km^2)
5g. Pamplona District (Distrito *Pamplonés*) (1785 km^2)
5h. Urbasa Sierran District (Distrito *Serrano Urbaseño*) (1038 km^2)
5i. Álava District (Distrito *Alavés*) (1637 km^2)
5j. Losa and Omecillo District (Distrito *Losino-Omecillés*) (762 km^2)

6. North galicia and Asturias Sector (*Sector Galaico septentrional-Asturiano*)

6a. Cuera and Sueve Sierras District (Distrito *Serrano Cuerano-Suevense*) (218 km^2)
6b. Oviedo District (Distrito *Ovetense*) (4058 km^2)
6c. Northwest Asturias District (Distrito *Asturiano Noroccidental*) (2798 km^2)
6d. North Galician District (Distrito *Galaico Septentrional*) (2620 km^2)

7. Galicia and north Portugal Sector (*Sector Galaico-Portugués septentrional*)

7a. Lugo District (Distrito *Lucense*) (4456 km^2)
7b. Brigantium District (Distrito *Brigantino*) (2327 km^2)
7c. Compostela District (Distrito *Compostelano*) (6361 km^2)
7d. Cíes and Ons Islands District (Distrito *Insular de Cíes y Ons*) (11 km^2)
7e. Low Miño and Pontevedra Sierran District (Distrito *Bajomiñense-Serrano Pontevedrés*) (5509, km^2)
7f. Braga District (Distrito *Bracarense*) (4225 km^2)
7g. Porto and Low Douro District (Distrito *Portueño-Bajoduriense*) (2384 km^2)
7h. Valdeorras District (Distrito *Valdeorrense)* (448 km^2)
7i. Orense and Lemos District (Distrito *Orensano-Lemosano*) (1638 km^2)
7j. Navia District (Distrito *Naviano*) (1521 km^2)

Ibb. Orocantabrian Subprovince (*Subprovincia Orocantábrica*)

8. High Campoo and Carrión Sector (*Sector Altocampurriano-Carrionés*)

8a. High Campoo District (Distrito *Altocampurriano*) (919 km^2)
8b. Lieébana District (Distrito *Lebaniego*) (529 km^2)
8c. High Esla and Carrión District (Distrito *Altoeslano-Altocarrionés*) (1150 km^2)
8d. Espigüete Sierran District (Distrito *Serrano Espigüeteño*) (386 km^2)

9. Picos of europa and Ubiña Sector (*Sector Picoeuropeano-Ubiñense*)

9a. Picos of Europa District (Distrito *Picoeuropeano*) (710 km^2)
9b. Redes District (Distrito *Redesano*) (643 km^2)
9c. Somiedo and Ubiña District (Distrito *Somedano-Ubiñense*) (1168 km^2)
9d. Central Orocantabrian District (Distrito *Orocantábrico Central*) (1028 km^2)
9e. Mampodre Sierran District (Distrito *Serrano Mampodrense*) (281 km^2)
9 f. Babia and Torío District (Distrito *Babiano-Toriano*) (1119 km^2)

10. Laciana and Ancares Sector (*Sector Lacianiego-Ancarense*)

10a. Laciana District (Distrito *Lacianiego*) (931 km^2)
10b. High Narcea District (Distrito *Altonarceense*) (1143 km^2)
10c. Omaña District (Distrito *Omañés*) (536 km^2)
10d. Ancares Sierran District (Distrito *Serrano Ancarense*) (756 km^2)
10e. Caurel Sierran District (Distrito *Serrano Caureliano*) (1006 km^2)

Ibc. Atlantic Orolusitania Subprovince (*Subprovincia Orolusitana Atlántica*)

11. North lusitania Sierran Sector (*Sector Serrano Norlusitano*)

11a. Peneda and Xurés Sierras District (Distrito *Serrano Penedano-Juresiano*) (2113 km^2)
11b. Barroso and Cabreira Sierras District (Distrito *Serrano Barroseño-Cabreirés*) (892 km^2)
11c. Alvão and Marão Sierras District (Distrito *Serrano Alvão-Marão*) (995 km^2)

12. Montemuro and Estrela Sierras Sector (*Sector Serrano Montemuro-Estrelense*)

12a. Montemuro and Caramulo Sierras District (Distrito *Serrano Montemuro-Caramulo*) (1955 km^2)
12b. Estrela Sierran District (Distrito *Serrano Estrelense*) (2694 km^2)
12c. Guarda District (Distrito *Guardense*) (567 km^2)

II. MEDITERRANEAN Region (*Región MEDITERRÁNEA*)

IIA. WEST MEDITERRANEAN Subregion (*Subregión MEDITERRÁNEA OCCIDENTAL*)

IIa. VALENCIA-PROVENÇE AND BALEARIC Province (*Provincia VALENCIANA-PROVENZAL-BALEAR*)

IIaa. CATALONIAN AND PROVENÇE Subprovince (*Subprovincia CATALANA-PROVENZAL*)

16. EAST CATALONIAN Sector (*Sector CATALÁN ORIENTAL*)

16a. Rosellón District (Distrito *Rosellonés*) [France]
16b. Selva and Ampurdán District (Distrito *Selvatano-Ampurdanés*) (3447 km^2)
16c. Vallés District (Distrito *Vallesano*) (1404 km^2)
16d. High Penedés and Montserrat District (Distrito *Altopenedesano-Montserratino*) (951 km^2)

IIab. VALENCIA Subprovince (*Subprovincia VALENCIANA*)

17. VALENCIA AND TARRAGONA Sector (*Sector VALENCIANO-TARRACONENSE*)

17a. Low Penedés and Alt Camp District (Distrito *Bajopenedesano-Altocampino*) (1857 km^2)
17b. Igualada and Prades District (Distrito *Igualadino-Pradesano*) (1429 km^2)
17c. Low Ebro District (Distrito *Bajoebrense*) (1450 km^2)
17d. Gandesa and Priorato District (Distrito *Gandesano-Priorateño*) (1288 km^2)
17e. Beceite and Morella District (Distrito *Puertobeceitano-Morellano*) (2133 km^2)
17f. Maestrat District (Distrito *Maestrazguero*) (3115 km^2)
17g. Espadán Sierran and Castellón District (Distrito *Serrano Espadano-Castellonense*) (2782 km^2)
17h. Huerta of Valencia an Túria District (Distrito *Huertano Valenciano-Turiano*) (4611 km^2)

18. JÁTIVA Sector (*Sector SETABENSE*)

18a. Játiva District (Distrito *Setabense*) (2494 km^2)
18b. Alcoy and Denia District (Distrito *Alcoyano-Dianense*) (2658 km^2)
18c. Allora and Cofrentes District (Distrito *Allorano-Cofrentino*) (1122 km^2)
18d. Yecla and Villena District (Distrito *Yeclano-Villenense*) (1332 km^2)

IIac. BALEARIC ISLANDS Subprovince (*Subprovincia BALEAR*)

19. MENORCA Sector (*Sector MENORQUÍN*)

19a. North Menorca District (Distrito *Menorquín Septentrional*) (324 km^2)
19b. South Menorca District (Distrito *Menorquín Meridional*) (377 km^2)

20. MALLORCA Sector (*Sector MALLORQUÍN*)

20a. Mallorca East Sierran District (Distrito *Serrano Levantino Mallorquín*) (540 km^2)
20b. Mallorca Central Plain District (Distrito *Llano Central Mallorquín*) (2143 km^2)
20c. Mallorca Tramuntana Sierran District (Distrito *Serrano Tramuntano Mallorquín*) (964 km^2)
20d. Cabrera Islands District (Distrito *Insular de Cabrera*) (13 km^2)

21. PITIUSAS Sector (*Sector PITIÚSICO*)

21a. Ibiza Island District (Distrito *Insular Ebusitano*) (574 km^2)
21b. Formentera Island District (Distrito *Insular Formenterano*) (83 km^2)

IIb. CENTRAL IBERIAN MEDITERRANEAN Province (*Provincia MEDITERRÁNEA IBÉRICA CENTRAL*)

IIba. LOW ARAGÓN AND HIGH EBRO Subprovince (*Subprovincia BAJOARAGONESA-ALTOEBRENSE*)

22. SOMONTANO Sector (*Sector SOMONTANO*)

22a. Manresa and Segarra District (Distrito *Manresano-Segárrico*) (2536 km^2)
22b. Noguera District (Distrito *Noguerano*) (2932 km^2)
22c. Aragón Somontano District (Distrito *Somontano Aragonés*) (4888 km^2)
22d. Cinco Villas District (Distrito *Cincovillés*) (2038 km^2)

23. BARDENAS AND MONEGROS Sector (*Sector BARDENERO-MONEGRINO*)

23a. Alcañiz District (Distrito *Alcañizano*) (3219 km^2)
23b. Low Cinca and Segriá District (Distrito *Bajocincano-Segriano*) (3260 km^2)
23c. Monegros District (Distrito *Monegrino*) (2638 km^2)
23d. Belchite and Híjar District (Distrito *Belchitano-Hijarense*) (1738 km^2)
23e. Bardenas District (Distrito *Bardenero*) (1751 km^2)
23f. Zaragoza Steppe District (Distrito *Zaragozano Estepario*) (4083 km^2)

24. Rioja and Estella Sector (*Sector Riojano-Estellés*)

24a. Low Rioja and Tafalla District (Distrito *Bajorriojano-Tafallés*) (4677 km^2)
24b. High Rioja and Estella District (Distrito *Altorriojano-Estellés*) (2552 km^2)
24c. Low Irati and Sangüesa District (Distrito *Bajoiratiano-Sangüesino*) (1128 km^2)

25. Cantabrian Castilian Sector (*Sector Castellano Cantábrico*)

25a. Bureba District (Distrito *Burebano*) (1078 km^2)
25b. Burgos Páramo District (Distrito *Parameño Burgalés*) (1208 km^2)
25c. High Burgos Merindades District (Distrito *Merindano Altoburgalés*) (1225 km^2)
25d. Miranda and Treviño District (Distrito *Mirandés-Treviñés*) (665 km^2)
25e. Palencia Low Campoo District (Distrito *Bajocampurriano Palentino*) (880 km^2)

IIbb. Oroiberian Subprovince (*Subprovincia Oroibérica*)

26. North Oroiberian Sector (*Sector Oroibérico Septentrional*)

26a. Moncayo Sierran District (Distrito *Serrano Moncayense*) (1947 km^2)
26b. Bilbilis and Cucalón Sierran District (Distrito *Bilbilitano-Serrano Cucalonense*) (4841 km^2)
26c. Cameros Sierran District (Distrito *Serrano Camerano*) (1476 km^2)
26d. Urbión Sierran District (Distrito *Serrano Urbionense*) (1197 km^2)
26e. Demanda Sierran District (Distrito Serrano Demandés) (1347 km^2)

27. South Oroiberian Sector (*Sector Oroibérico Meridional*)

27a. High Jiloca District (Distrito *Altojiloquense*) (1642 km^2)
27b. Gúdar Sierran District (Distrito *Serrano Gudárico*) (1607 km^2)
27c. Javalambre Sierran District (Distrito *Serrano Javalambrense*) (2173 km^2)
27d. Ademuz and Teruel District (Distrito *Ademuceño-Turolense*) (1662 km^2)
27e. Aliaga and Montalbán District (Distrito *Aliagueño-Montalbanés*) (2244 km^2)
27f. Albarracín District (Distrito *Albarracinense*) (1678 km^2)
27g. North Cuenca Sierran District (Distrito *Serrano Conquense Septentrional*) (3784 km^2)

27h. South Cuenca Sierran District (Distrito *Serrano Conquense Meridional*) (3819 km^2)

IIbc. CASTILIAN Subprovince (*Subprovincia CASTELLANA*)

28. CASTILIAN DUERO Sector (*Sector CASTELLANO DURIENSE*)

28a. Castilian Ribaduero District (Distrito *Riberoduriense Castellano*) (4880 km^2)
28b. Low Arlanza and Cerrato District (Distrito *Bajoarlanzano-Cerrateño*) (2560 km^2)
28c. Tierra of Campos District (Distrito *Terracampino*) (11,795 km^2)
28d. Tierras of Medina and Armuña District (Distrito *Medinense-Armuñés*) (2274 km^2)
28e. Burgos District (Distrito *Burgalés*) (1923 km^2)
28f. Tierra of Pinares and Low Adaja District (Distrito *Terrapinariego-Bajoadajense*) (2750 km^2)
28g. Tierra of Arévalo and Moraña District (Distrito *Arevalense-Morañés*) (4491 km^2)

29. CELTIBERIA AND ALCARRIA Sector (*Sector CELTIBÉRICO-ALCARREÑO*)

29a. Soria District (Distrito *Soriano*) (7009 km^2)
29b. High Arlanza and Covarrubias District (Distrito *Altoarlanzano-Covarrubiense*) (777 km^2)
29c. Segovia District (Distrito *Segoviano*) (1698 km^2)
29d. High Jalón District (Distrito *Altojalonés*) (2019 km^2)
29e. Molina District (Distrito *Molinés*) (2845 km^2)
29f. High Alcarria District (Distrito *Altoalcarreño*) (5655 km^2)
29g. Obispalía District (Distrito *Obispaleño*) (2966 km^2)

30. MANCHA Sector (*Sector MANCHEGO*)

30a. Low Madrid District (Distrito *Bajomatritense*) (6201 km^2)
30b. Sagra District (Distrito *Sagreño*) (4782 km^2)
30c. Sanjuán Mancha District (Distrito *Manchego Sanjuanés*) (11,028 km^2)
30d. Calatrava District (Distrito *Calatraveño*) (3209 km^2)
30e. Montiel District (Distrito *Montielense*) (2501 km^2)
30f. Cuenca Mancha District (Distrito *Manchego Conquense*) (4526 km^2)
30g. Júcar Mancha District (Distrito *Manchego Sucronense*) (7213 km^2)
30h. Jumilla and Hellín District (Distrito *Jumillano-Hellinense*) (3567 km^2)
30i. Espuña Sierran District (Distrito *Serrano Espuñense*) (2762 km^2)

IIc. West Iberian Mediterranean Province (*Provincia Mediterránea Ibérica Occidental*)

IIca. Carpetania and León Subprovince (*Subprovincia Carpetana-Leonesa*)

31. León plain Sector (*Sector Planileonés*)

31a. Valdavia District (Distrito *Valdaviés*) (1189 km^2)
31b. León Páramo District (Distrito *Parameño Leonés*) (4177 km^2)

32. Bierzo and Sanabria Sector (*Sector Berciano-Sanabrés*)

32a. Bierzo District (Distrito *Berciano*) (1451 km^2)
32b. Cabrera and Montes of León District (Distrito *Cabreireño-Monteleonés*) (1414 km^2)
32c. Zamora and Sanabria District (Distrito *Zamorano-Sanabrés*) (5825 km^2)
32d. Queixa Sierran District (Distrito *Serrano Queixense*) (2339 km^2)
32e. Maragatería District (Distrito *Maragato*) (2450 km^2)

33. Lusitanian Douro Sector (*Sector Lusitano Duriense*)

33a. Braganza District (Distrito *Braganzano*) (4041 km^2)
33b. Chaves and Verín District (Distrito *Chavesano-Verinense*) (3149 km^2)
33c. Terraquente District (Distrito *Terraquentino*) (1178 km^2)
33d. Vila Real District (Distrito *Vilarrealeño*) (1798 km^2)
33e. North Beira District (Distrito *Beirense Septentrional*) (1490 km^2)

34. Salamanca Sector (*Sector Salmantino*)

34a. Low Salamanca District (Distrito *Bajosalmantino*) (2984 km^2)
34b. Tormes District (Distrito *Tormesino*) (1982 km^2)
34c. High Salamanca District (Distrito *Altosalmantino*) (9787 km^2)
34d. Batuecas and Malcata Sierran District (Distrito *Batueco-Serrano Malcateño*) (1050 km^2)

35. Guadarrama Sierran Sector (*Sector Serrano Guadarrámico*)

35a. Ayllón Sierran District (Distrito *Serrano Ayllonense*) (2223 km^2)
35b. Riaza District (Distrito *Riazano*) (437 km^2)
35c. Ávila District (Distrito *Abulense*) (917 km^2)
35d. Corneja and Amblés District (Distrito *Cornejano-Amblense*) (1325 km^2)
35e. High Guadarrama District (Distrito *Altoguadarrámico*) (3274 km^2)
35f. High Madrid District (Distrito *Altomatritense*) (2313 km^2)
35g. Central Alberche District (Distrito *Alberchense Central*) (1100 km^2)

36. Bejar and Gredos Sierran Sector (*Sector Serrano Bejarano-Gredense*)

36a. Ávila Paramera Sierran District (Distrito *Serrano Parameño Abulense*) (391 km^2)
36b. Serrota Sierran District (Distrito *Serrano Serrotense*) (576 km^2)
36c. East Gredos Sierran District (Distrito *Serrano Gredense Oriental*) (597 km^2)
36d. High Gredos Sierran District (Distrito *Serrano Altogredense*) (441 km^2)
36e. Tormantos Sierran District (Distrito *Serrano Tormantino*) (617 km^2)
36f. Béjar Sierran District (Distrito *Serrano Bejarano*) (412 km^2)

IIcb. Lusitania And Extremadura Subprovince (*Subprovincia Lusa-Extremadurense*)

37. Oretana Range and Tajo Sector (*Sector Cordillerano Oretano-Tagano*)

37a. Talavera District (Distrito *Talaverano*) (4600 km^2)
37b. Vera District (Distrito *Verato*) (2695 km^2)
37c. Coria District (Distrito *Coriano*) (4321 km^2)
37d. Eastern Montes of Toledo District (Distrito *Montitoledano Oriental*) (3594 km^2)
37e. Villuercas Sierran District (Distrito *Serrano Villuerquino*) (1912 km^2)
37f. Cáceres Central District (Distrito *Cacereño Central*) (8804 km^2)
37g. São Mamede Sierran District (Distrito *Serrano São Mamedano*) (850 km^2)
37h. South Beira District (Distrito *Beirense Meridional*) (6687 km^2)
37i. Zêzere District (Distrito *Zezerense*) (1306 km^2)

38. Mariánica Range Sector (*Sector Cordillerano Mariánico*)

38a. East Mariánica District (Distrito *Mariánico Oriental*) (7508 km^2)
38b. Central Guadiana Sierran District (Distrito *Serrano Centroguadianés*) (5862 km^2)
38c. Pedroches and Alcudia District (Distrito *Pedrocheño-Alcudiense*) (11,352 km^2)
38d. Serena District (Distrito *Sereniano*) (8541 km^2)
38e. Llerena District (Distrito *Llerenense*) (5912 km^2)
38f. Tierra of Barros and Badajoz District (Distrito *Terrabarroseño-Pacense*) (4545 km^2)
38g. Aracena Sierran District (Distrito *Serrano Aracenense*) (3574 km^2)
38h. Andévalo District (Distrito *Andevalense*) (8021 km^2)
38i. Alentejo District (Distrito *Alentejano*) (19,765 km^2)

IId. Murcia and Almería Province (*Provincia Murciana-Almeriense*)

39. alicante and Murcia Sector (*Sector Alicantino-Murciano*)

39a. Alicante District (Distrito *Alicantino*) (1783 km^2)
39b. North Murcia District (Distrito *Murciano Septentrional*) (1636 km^2)
39c. South Murcia District (Distrito *Murciano Meridional*) (3874 km^2)

40. Almería Sector (*Sector Almeriense*)

40a. East Almería District (Distrito *Almeriense Oriental*) (2974 km^2)
40b. Gata Cape District (Distrito *Charidemo*) (276 km^2)
40c. West Almería District (Distrito *Almeriense Occidental*) (2289 km^2)
40d. Alhamilla Sierran District (Distrito *Serrano Alhamillense*) (384 km^2)

IIe. Bética Province (*Provincia Bética*)

41. Subbética Sector (*Sector Subbético*)

41a. Alcaraz Sierran District (Distrito *Serrano Alcaraceño*) (1348 km^2)
41b. Murcia Subbética District (Distrito *Subbético Murciano*) (2535 km^2)
41c. Cazorla Sierran District (Distrito *Serrano Cazorleño*) (2020 km^2)
41d. Segura Sierran District (Distrito *Serrano Segureño*) (1501 km^2)
41e. Mágina Sierran District (Distrito *Serrano Maginense*) (1900 km^2)
41f. Southwest Jaén Sierran District (Distrito *Serrano Giennense Suroccidental*) (1785 km^2)
41g. Córdoba Subbetica District (Distrito *Subbético Cordobés*) (1816 km^2)

42. Hoyas of Guadix and Baza Sector (*Sector Hoyano Accitano-Bastitano*)

42a. Hoya of Guadix District (Distrito *Hoyano Accitano*) (1447 km^2)
42b. Hoya of Baza District (Distrito *Hoyano Bastitano*) (1390 km^2)
42c. María Sierran District (Distrito *Serrano Mariense*) (1177 km^2)
42d. Estancias Sierran District (Distrito *Serrano Estanciano*) (1210 km^2)
42e. Filabres Sierran District (Distrito *Serrano Filábrico*) (1052 km^2)
42f. Baza Sierran District (Distrito *Serrano Bastitano*) (1118 km^2)

43. Nevada Sierran Sector (*Sector Serrano Nevadense*)

43a. East Nevada Sierran District (Distrito *Serrano Nevadense Oriental*) (665 km^2)
43b. High Nevada Sierran District (Distrito *Serrano Altonevadense*) (867 km^2)

44. Alpujarras and Gádor Sierran Sector (*Sector Alpujarreño-Serrano Gadorense*)

44a. Gádor Sierran District (Distrito *Serrano Gadorense*) (635 km^2)
44b. Alpujarras District (Distrito *Alpujarreño*) (930 km^2)

45. Granada and Almijara sierran Sector (*Sector Granadino-Serrano Almijarense*)

45*a*. **Vega** of Granada District (Distrito *Vegano Granadino*) (1817 km^2)
45b. Alfacar Sierran District (Distrito *Serrano Alfacarino*) (609 km^2)
45c. Trevenque Sierran District (Distrito *Serrano Trevenquino*) (290 km^2)
45d. Almijara Sierran District (Distrito *Serrano Almijarense*) (929 km^2)
45e. Tejeda Sierran District (Distrito *Serrano Tejedense*) (565 km^2)
45f. Axarquía District (Distrito *Axarquiense*) (1038 km^2)

46. Ronda Sector (*Sector Rondeño*)

46a. Antequera District (Distrito *Antequerano*) (2534 km^2)
46b. Ronda Sierran District (Distrito *Serrano Arundense*) (1724 km^2)
46c. Hoya of Málaga District (Distrito *Hoyano Malagueño*) (1685 km^2)
46d. Bermeja Sierran District (Distrito *Serrano Bermejense*) (372 km^2)

47. Campiña of guadalquivir Sector (*Sector Hispalense*)

47a. High Campiña District (Distrito *Altocampiñés*) (4988 km^2)
47b. Low Campiña District (Distrito *Bajocampiñés*) (10,653 km^2)
47c. Aljarafe District (Distrito *Aljarafeño*) (1363 km^2)
47d. Jerez and Medina Sidonia District (Distrito *Jerezano-Asidonense*) (4572 km^2)

IIf. Coastal Lusitania and west Andalusia Province (*Provincia Costera Lusitana-Andaluza occidental*)

IIfa. Divisorio Portuguese Subprovince (*Subprovincia Divisoria Portuguesa*)

48. Divisorio Portuguese Sector (*Sector Divisorio Portugués*)

48a. Littoral Beira District (Distrito *Beirense Litoral*) (3031 km^2)
48b. West Beira District (Distrito *Beirense Occidental*) (4661 km^2)
48c. Estremadura and Coimbra District (Distrito *Estremeño-Conimbricense*) (5115 km^2)
48d. Berlengas Islands District (Distrito *Insular Berlenguense*) (1 km^2)
48e. Lisboa District (Distrito *Ollisiponense*) (1375 km^2)
48f. Sintra Sierran District (Distrito *Serrano Sintrano*) (129 km^2)

IIfb. Cádiz and Sado Subprovince (*Subprovincia Gaditana-Sadense*)

49. Ribatejo and Sado Sector (*Sector Ribatagano-Sadense*)

49a. Ribatejo District (Distrito *Ribatagano*) (7494 km^2)
49b. Arrabida Sierran District (Distrito *Serrano Arrabidense*) (128 km^2)
49c. Sado District (Distrito *Sadense*) (4273 km^2)

50. Algarve and Monchique Sector (*Sector Algárvico-Monchiquense*)

50a. Algarve District (Distrito *Algárvico*) (1604 km^2)
50b. San Vicente Cape District (Distrito *Promontorio Vicentino*) (2 km^2)
50c. San Vicente Coastal District (Distrito *Costero Vicentino*) (450 km^2)
50d. Monchique Sierran District (Distrito *Serrano Monchiquense*) (5331 km^2)

51. Cádiz and Littoral Huelva Sector (*Sector Gaditano-Onubense Litoral*)

51a. Littoral Huelva District (Distrito *Onubense Litoral*) (2218 km^2)
51b. Marismas District (Distrito *Marismeño*) (1763 km^2)
51c. Littoral Cádiz District (Distrito *Gaditano Litoral*) (685 km^2)

52. algeciras and Aljibe Sector (*Sector Algecireño-Aljíbico*)

52a. Aljibe Sierran District (Distrito *Serrano Aljíbico*) (1389 km^2)
52b. Algeciras and Genal River District (Distrito *Algecireño-Genalense*) (956 km^2)
52c. Marbella District (Distrito *Marbellí*) (332 km^2)

5.4 Provincial Biogeographic Units and Their Realtionships with the Vegetation Series at Sector Level

A total of 8 provinces have been recognized in the Iberian Peninsula and Balearic Islands in the biogeographic typology given above. The sectors and biogeographic districts belonging to each province are shown.

Province by province, we show their sectors and list their potential natural vegetation units. For each unit of potential natural vegetation we indicate whether it is climatophilous, xerophilous, climato-temporihygrophilous, hygrophylous or whether it is a geopermaseries, as well as some specific or edaphic characteristic. (● presence, – absence).

The geographic positions of districts are clearly shown in the maps of provinces and subprovinces.

Biogeographic Typology of the Pyrenean Province at Sector Level (Figs. 5.4 and 5.5)

Ia. Pyrenean Province (*Provincia Pirenaica*)

Iaa. East Pyrenean Subprovince (*Subprovincia Pirenaica oriental*)

1. East Pyrenean Sector (*Sector Pirenaico Oriental*)

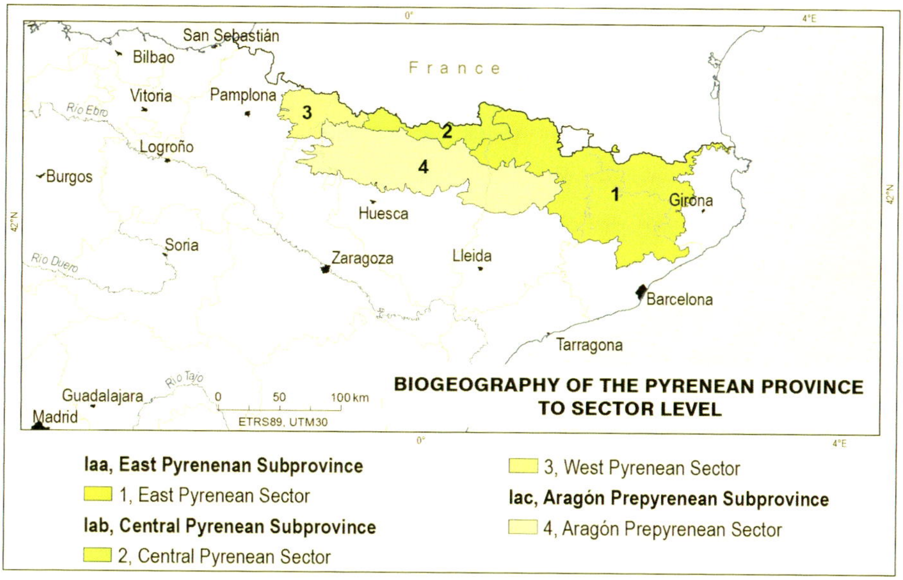

Fig. 5.4 Biogeographic map of the Pyrenean Province at sector level

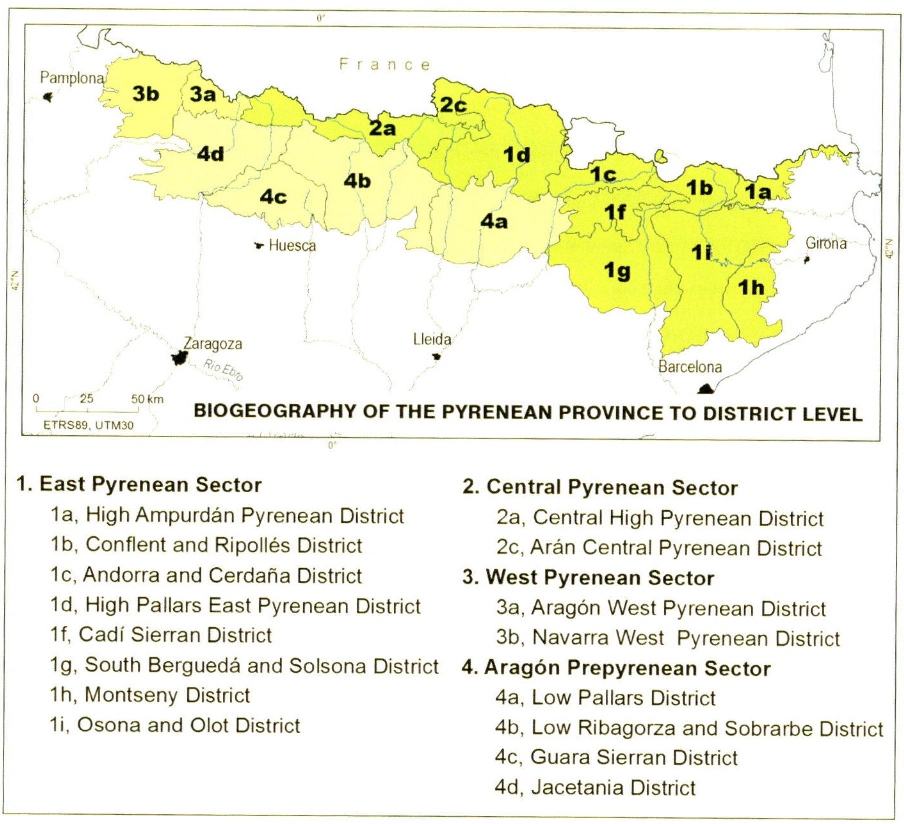

Fig. 5.5 Biogeographic map of the Pyrenean Province at district level

Iab. Central Pyrenean Subprovince (*Subprovincia Pirenaica central*)

2. Central Pyrenean Sector (*Sector Pirenaico Central*)
3. West Pyrenean Sector (*Sector Pirenaico occidental*)

Iac. Aragón Prepyrenean Subprovince (*Subprovincia Prepirenaica Aragonesa*)

4. aragón Prepyrenean Sector (*Sector Prepirenaico Aragonés*)

Sigmeta, geosigmeta and geopermasigmeta of the biogeographic sectors in the Pyrenean Province

Pyrenean province	1	2	3	4
Climatophilous sigmeta				
Veronico officinalis-Pino pyrenaicae S. (acidophilous)	●	–	–	–
Buxo sempervirentis-Abieti albae S. (neutro-acidophilous)	●	–	–	–
Geranio nodosi-Fago sylvaticae S. (acidophilous)	●	–	–	–
Lathyro linifolii-Querco petraeae S. (acidophilous)	●	–	–	–
Polygalo calcareae-Pino catalaunicae S. (basophilous)	●	–	–	○
Galio rotundifolii-Pino pyrenaicae S. (acidophilous)	–	●	–	–
Teucrio pyrenaici-Pino pyrenaicae S. (basophilous)	–	●	–	–
Pulmonario affinis-Abieti albae S. (sligth acidophilous)	–	●	–	–
Lysimachio nemorum-Fago sylvaticae S. (neutro-acidophilous)	–	●	–	–
Festuco altissimae-Abieti albae S. (neutro-acidophilous)	–	●	–	–
Emero majoris-Abieti albae S. (neutrophilous)	–	–	●	–
Goodyero repentis-Pino pyrenaicae S. (basophilous)	–	–	●	–
Pulmonario longifoliae-Querco fagineae S. (basophilous)	–	–	●	–
Roso arvensis-Querco pubescentis S. (basophilous)	–	–	●	–
Echinosparto horridi-Pino pyrenaicae S. (basophilous)	–	–	–	●
Pino uncinato-pyrenaicae S. (acidophilous)	–	–	–	●
Buxo sempervirentis-Querco subpyrenaicae S. (basophilous)	–	–	–	●
Pulsatillo fontqueri-Pino uncinatae S. (neutro-basophilous)	●	–	–	●
Prunello hastifoliae-Querco petraeae S. (acidophilous)	●	●	–	–
Luzulo nivae-Fago sylvaticae S. (acidophilous)	●	●	–	–
Rhododendro ferruginei-Abieti albae S. (acidophilous)	●	●	–	–
Arctostaphylo uvaeursi-Pino uncinatae S. (acidophilous)	●	●	–	–
Rhododendro ferruginei-Pino uncinatae S. (acidophilous)	●	●	–	–
Sorbo aucupariae-Fago sylvaticae S. (neutro-acidophilous)	●	–	●	–
Goodyero repentis-Abieti albae S.(acidophilous)	–	●	●	–
Pteridio aquilini-Querco pubescentis S. (neutro-acidophilous)	●	●	–	●
Helleboro occidentalis-Fago sylvaticae S. (neutro-acidophilous)	●	●	–	●
Scillo liliohyacinthi-Fago sylvaticae S. (neutro-acidophilous)	●	●	●	●
Buxo sempervirentis-Fago sylvaticae S. (neutrophilous)	●	●	●	●
Buxo sempervirentis-Querco pubescentis S. (basophilous)	●	●	●	●
Roso pendulinae-Fago sylvaticae S. (neutro-acidophilous)	●	●	●	●

(continued)

Pyrenean province	1	2	3	4
Climato-temporihygrophilous and xerophilous sigmeta				
Carici basilaris-Querco suberis S. (acidophilous)	●	–	–	–
Buxo sempervirentis-Querco rotundifoliae S. (basophilous)	–	–	–	●
Asplenio onopteridis-Querco ilicis S. (acidophilous)	●	●	–	●
Xerophilous sigmeta				
Astragalo salvatoris-Junipero macrocarpae S. (littoral)	●	–	–	–
Asplenio adiantinigri-Querco rotundifoliae S. (acidophilous)	●	–	–	–
Buxo sempervirentis-Junipero phoeniceae S. (basophilous)	–	–	–	●
Lonicero xylostei-Pino salzmannii S. (calco-dolomitic)	–	–	–	●
Daphno laureolae-Querco ilicis S. (basophilous)	●	●	–	●
Climato-temporihygrophilous sigmeta				
Doronico pardalianchis-Fraxino excelsioris S. (neutrofila)	●	–	–	–
Poo nemoralis-Tilio platyphylli S. (colluvio basophilous)	–	–	●	–
Brachypodio sylvatici-Fraxino excelsioris S. (neutrophilous)	–	–	●	–
Isopyro thalictroidis-Querco roboris S. (neutrophilous)	●	●	●	–
Climato-temporihygrophilous and hygrophilous sigmeta & geosigmeta				
Aceri opali-Querco petraeae S. (neutro-acidophilous)	●	–	–	–
Hyperico androsaemi-Ulmo glabrae S. (neutrophilous)	–	–	●	–
Violo mirabilis-Ulmo glabrae S. (neutrophilous)	–	–	–	●
Roso pendulinae-Acero pseudoplatani S. (acidophilous)	●	●	–	–
Hygrophilous geosigmeta				
Lamio flexuosi-Alno glutinosae Gs. (soft freshwater)	●	–	–	–
Carici pendulae-Salici atrocinereae Gs. (soft freshwater)	●	–	–	–
Carici pendulae-Alno glutinosae Gs. (soft freshwater)	●	–	–	–
Lithospermo purpureocaerulei-Ulmo minoris Gs. (hard freshwater)	●	–	–	–
Rusco aculeati-Fraxino angustifoliae Gs. (soft freshwater)	●	–	–	–
Salici atrocinereo-daphnoidis Gs. (hard freshwater)	●	–	–	–
Salici angustifolio-daphnoidis Gs. (hard freshwater)	–	–	–	●
*Agrostio stoloniferae-Myricario germanicae Gs. (*hard freshwater)	–	–	–	●
Equiseto hyemalis-Alno glutinosae Gs. (hard freshwater)	●	●	–	–
Lathraeo clandestinae-Populo nigrae Gs. (hard freshwater)	–	–	●	●
Veratro albi-Salici bicoloris Gs. (soft freshwater)	●	●	●	●
Salici lambertiano-angustifoliae Gs. (hard freshwater)	●	●	●	●
Betulo meridionalis-Salici albae Gs. (hard freshwater)	●	●	●	●
Permasigmeta and geopermasigmeta				
Armerio ruscinonensis Ps. (rock littoral)	●	–	–	–
Hieracio breviscapi-Festucoairoidis Gps. (acidophilous)	●	–	–	–
Oxytropido halleri-Kobresio myosuroidis Gps.(basophilous)	●	–	–	–
Saxifrago bryoidis-Minuartio sedoidis Gps. (acidophilous)	●	–	–	–
Oreochloo blankae-Carici curvulae Gps. (acidophilous)	–	●	–	–
Oxytropido foucaudii-Kobresio myosuroidis Gps. (basophilous)	–	●	–	–
Minuartio cerastiifoliae-Androsaco ciliatae Gps. (basophilous)	–	●	–	–
Minuartio sedoidis-Androsaco ciliatae Gps. (acidophilous)	–	●	–	–

1. East Pyrenean Sector, 2. Central Pyrenean Sector, 3. West Pyrenean Sector, 4. Prepyrenean Sector

Biogeographic Typology of the European Atlantic Province at Sector Level (Figs. 5.6 and 5.7)

Ib. European Atlantic Province (*Provincia Atlántica Europea*)

Iba. Cantabrian atlantic Subprovince (*Subprovincia Cantabroatlántica*)

5. Cantabrian and Basque Sector (*Sector Cántabro-Vascónico*)
6. North galicia and Asturias Sector (*Sector Galaico septentrional-Asturiano*)
7. Galicia and north Portugal Sector (*Sector Galaico-Portugués septentrional*)

Ibb. Orocantabrian Subprovince (*Subprovincia Orocantábrica*)

8. High Campoo and Carrión Sector (*Sector Altocampurriano-Carrionés*)
9. Picos of europa and Ubiña Sector (*Sector Picoeuropeano-Ubiñense*)
10. Laciana and Ancares Sector (*Sector Lacianiego-Ancarense*)

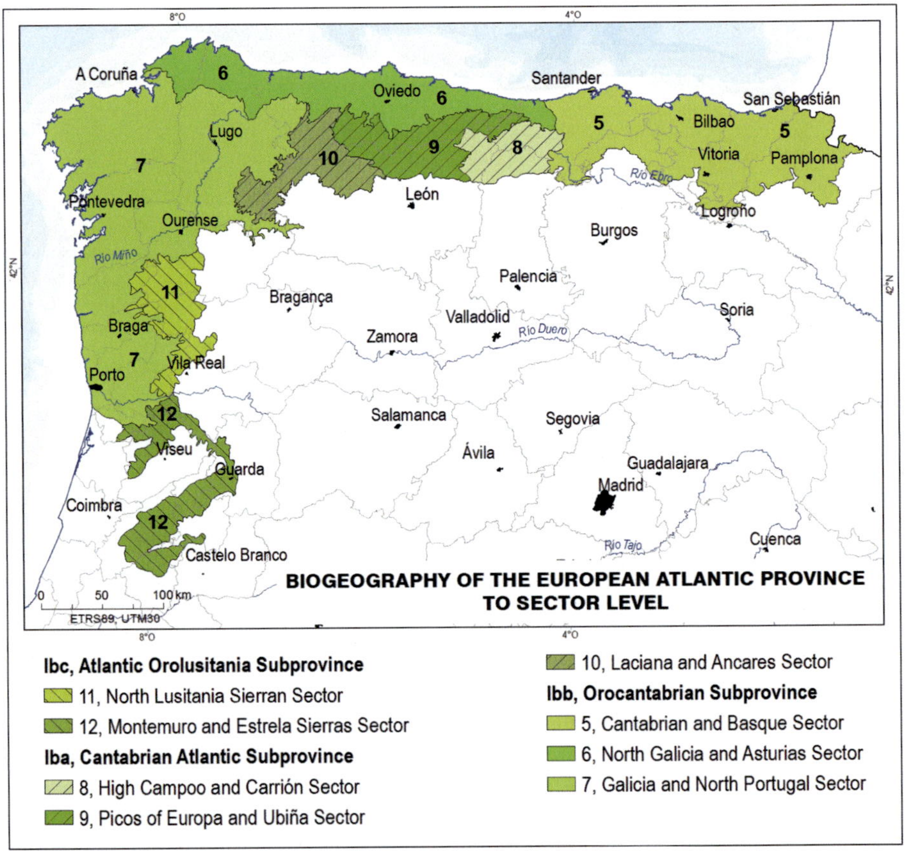

Fig. 5.6 Biogeographic map of the European Atlantic Province at sector level

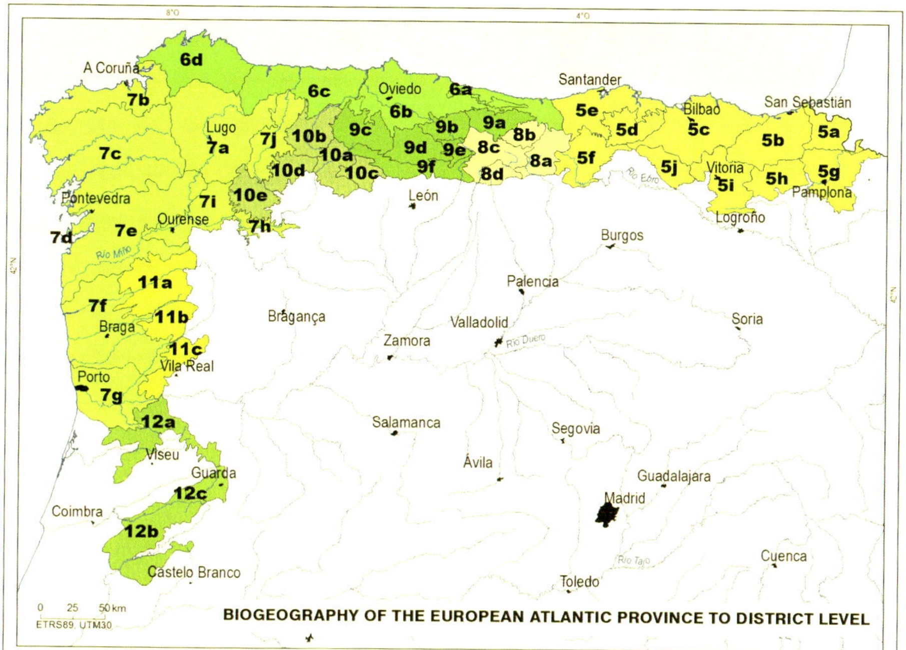

5. Cantabrian and Basque Sector
5a, Labourd and Baztán District
5b, Guipúzcoa District
5c, Biscay District
5d, Valnera and Trasmiera District
5e, North Cantabrian District
5f, South Cantabrian District
5g, Pamplona District
5h, Urbasa Sierran District
5i, Álava District
5j, Losa and Omecillo District

6. North Galicia and Asturias Sector
6a, Cuera and Sueve Sierras District
6b, Oviedo District
6c, Northwest Asturias District
6d, North Galicia District

7. Galicia and North Portugal Sector
7a, Lugo District
7b, Brigantium District
7c, Compostela District
7d, Cíes and Ons Islands District
7e, Low Miño and Pontevedra Sierran District
7f, Braga District
7g, Porto and Low Douro District
7h, Valdeorras District
7i, Orense and Lemos District
7j, Navia District

8. High Campoo and Carrión Sector
8a, High Campoo District
8b, Liébana District
8c, High Esla and Carrión District
8d, Espigüete Sierran District

9. Picos of Europa and Ubiña Sector
9a, Picos of Europa District
9b, Redes District
9c, Somiedo and Ubiña District
9d, Central Orocantabrian District
9e, Mampodre Sierran District
9f, Babia and Torío District

10. Laciana and Ancares Sector
10a, Laciana District
10b, High Narcea District
10c, Omaña District
10d, Ancares Sierran District
10e, Caurel Sierran District

11. North Lusitania Sierran Sector
11a, Peneda and Xurés Sierras District
11b, Barroso and Cabreira Sierras District
11c, Alvão and Marão Sierras District

12. Montemuro and Estrela Sierras Sector
12a, Montemuro and Caramulo Sierras District
12b, Estrela Sierran District
12c, Guarda District

Fig. 5.7 Biogeographic map of the European Atlantic Province at district level

Ibc. Atlantic Orolusitania Subprovince (*Subprovincia Orolusitana Atlántica*)

11. North lusitania Sierran Sector (*Sector Serrano Norlusitano*)
12. Montemuro and Estrela Sierras Sector (*Sector Serrano Montemuro-Estrelense*)

Sigmeta, geosigmeta and geopermasigmeta of the biogeographic sectors in the European Atlantic Province

European Atlantic province	5	6	7	8	9	10	11	12
Climatophilous sigmeta								
Saxifrago hirsutae-Fago sylvaticae S. (acidophilous)	●	–	–	–	–	–	–	–
Hyperico pulchri-Querco roboris S. (acidophilous)	●	–	–	–	–	–	–	–
Pulmonario longifoliae-Querco petraeae S. (acidophilous)	●	–	–	–	–	–	–	–
Pulmonario longifoliae-Querco fagineae S. (basophilous)	●	–	–	–	–	–	–	–
Roso arvensis-Querco pubescentis S. (basophilous)	●	–	–	–	–	–	–	–
Melampyro pratensis-Querco pyrenaicae S. (acidophilous)	●	–	–	–	–	–	–	–
Blechno spicant-Querco roboris S. (acidophilous)	–	●	–	–	–	–	–	–
Viburno tini-Querco roboris S. (acidophilous)	–	–	●	–	–	–	–	–
Rusco aculeati-Querco roboris S. (acidophilous)	–	–	●	–	–	–	–	–
Lonicero periclymeni-Querco pyrenaicae S. (acidophilous)	–	–	●	–	–	–	–	–
Lithodoro diffusae-Junipero sabinae S. (basophilous)	–	–	–	–	●	–	–	–
Junipero sabino-orocantabrica S. (basophilous)	–	–	–	–	●	–	–	–
Carici sylvaticae-Fago sylvaticae S. (neutrophilous)	–	–	–	–	●	–	–	–
Vaccinio myrtilli-Pino ibericae S. (acidophilous, relic)	–	–	–	–	●	–	–	–
Vaccinio myrtilli-Junipero alpinae S. (acidophilous)	–	–	–	–	–	●	–	–
Omphalodo nitidae-Fago sylvaticae S. (neutro-acidophilous)	–	–	–	–	–	●	–	–
Eryngio juresiani-Betulo celtibericae S. (acidophilous)	–	–	–	–	–	–	●	–
Vaccinio myrtilli-Querco roboris S. (acidophilous)	–	–	–	–	–	–	●	–
Lycoposdio clavati-Junipero alpinae S. (acidophilous)	–	–	–	–	–	–	–	●
Holco mollis-Querco pyrenaicae S. (acidophilous)	–	–	–	–	–	–	–	●
Saxifrago spathularis-Fago sylvaticae S. (acidophilous)	–	●	●	–	–	–	–	–
Carici caudatae-Fago sylvaticae S. (neutrophilous)	–	●	–	–	●	–	–	–
Daphno cantabricae-Arctostaphylo uvaeursi S. (basophilous)	–	–	–	●	●	–	–	–
Blechno spicant-Fago sylvaticae S. (acidophilous)	–	–	–	●	●	–	–	–
Vaccinio microphylli-Junipero alpinae S. (acidophilous)	–	–	–	●	●	●	–	–
Linario triornithophorae-Querco petraeae S. (acidophilous)	–	–	–	●	●	●	–	–
Luzulo henriquesii-Querco petraeae S. (acidophilous)	–	–	–	●	●	●	–	–
Linario triornithophorae-Querco pyrenaicae S. (acidophilous)	–	–	–	●	●	●	–	–
Avenello ibericae-Fago sylvaticae S. (acidophilous)	–	–	–	●	●	●	–	–
Saniculo europaeae-Ilici aquifolii S. (neutrophilous)	–	–	–	●	●	●	–	–
Avenello ibericae-Querco orocantabricae S. (acidophilous)	–	–	–	●	●	●	●	–
Luzulo henriquesii-Betulo celtibericae S. (acidophilous)	–	–	–	–	●	●	●	–
Epipactido helleborines-Fago sylvaticae S. (neutrophilous)	–	–	–	–	●	●	–	–

(continued)

European Atlantic province	5	6	7	8	9	10	11	12
Climato-temporihygrophilous and xerophilous sigmeta								
Spiraeo obovatae-Querco rotundifoliae S. (basophilous)	●	–	–	–	–	–	–	–
Physospermo cornubiensis-Querco suberis S. (acidophilous)	–	–	●	–	–	–	–	–
Cephalanthero longifoliae-Querco rotundifoliae S. (basophilous)	–	–	–	–	●	–	–	–
Berberido cantabricae-Querco fagineae S. (basophilous)	–	–	–	–	●	–	–	–
Genisto falcatae-Querco rotundifoliae S. (basophilous)	–	–	●	–	–	●	–	–
Xerophilous sigmeta								
Teucrio salviastri-Querco suberis S. (acidophilous)	–	–	●	–	–	–	–	–
Lithodoro diffusae-Querco rotundifoliae S. (basophilous)	–	–	–	–	●	–	–	–
Teucrio salviastri-Querco rotundifoliae S. (acidophilous)	–	–	●	–	–	–	–	●
Lauro nobilis-Querco ilicis S. (basophilous)	●	●	–	–	–	–	–	–
Climato-temporihygrophilous sigmeta								
Crataego laevigatae-Querco roboris S. (neutro-basophilous)	●	–	–	–	–	–	–	–
Fraxino angustifoliae-Ulmo glabrae S. (acidophilous)	–	●	–	–	–	–	–	–
Omphalodo nitidae-Fraxino angustifoliae S. (acidophilous)	–	–	●	–	–	–	–	–
Calluno vulgaris-Rhododendro pontici S. (acidophilous)	–	–	●	–	–	–	–	–
Hyperico androsaemi-Querco roboris S. (neutro-acidophilous)	–	–	●	–	–	–	–	–
Helleboro occidentalis-Tilio cordatae S. (neutrophilous)	–	–	–	–	●	–	–	–
Luzulo henriquesii-Pruno lusitanicae S.	–	–	–	–	–	–	●	–
Frangulo alni-Pruno lusitanicae S.	–	–	–	–	–	–	–	●
Polysticho setiferi-Fraxino excelsioris S. (neutrophilous)	●	●	–	–	–	–	–	–
Climato-temporihygrophilous and hygrophilous sigmeta & geosigmeta								
Hyperico androsaemi-Ulmo glabrae S. (neutrophilous)	●	–	–	–	–	–	–	–
Luzulo henriquesii-Acero pseudoplatani S. (neutro-acidophilous)	–	–	–	–	–	●	–	–
Hygrophilous geosigmeta								
Viburno lantanae-Ulmo minoris Gs. (hard freshwater)	●	–	–	–	–	–	–	–
Humulo lupuli-Alno glutinosae Gs. (hard freshwater)	●	–	–	–	–	–	–	–
Lonicero xylostei-Alno glutinosae Gs. (hard freshwater)	●	–	–	–	–	–	–	–
Hedero hibernicae-Fraxino angustifoliae Gs. (soft freshwater)	–	–	●	–	–	–	–	–
Senecioni bayonnensis-Alno glutinosae Gs. (soft freshwater)	–	–	●	–	–	–	–	–
Holoschoeno vulgaris-Salici arenariae Ps. (psammophilous)	–	–	●	–	–	–	–	–
Carici reuterianae-Betulo celtibericae Gs. (soft freshwater)	–	–	–	–	–	–	●	–
Salici lambertiano-salviifoliae Gs. (soft freshwater)	–	–	–	–	–	–	–	●

(continued)

European Atlantic province	5	6	7	8	9	10	11	12
Salici angustifolio-albae Gs. (soft freshwater)	●	●	–	–	–	–	–	–
Hyperico androsaemi-Alno glutinosae Gs. (hard freshwater)	●	●	–	–	–	–	–	–
Salici salviifoliae Gs. (soft freshwater)	–	●	●	–	–	–	–	–
Valeriano pyrenaicae-Alno glutinosae Gs. (soft freshwater)	–	●	–	–	–	●	–	–
Salici cantabrico-bicoloris Gs. (hard freshwater)	–	–	–	●	●	–	–	–
Salici cantabricae Gs. (hard freshwater)	–	–	–	●	●	–	–	–
Salici cantabrico-albae Gs. (hard freshwater)	–	–	–	●	●	–	–	–
Euphorbio hybernae-Fraxino excelsioris Gs. (hard freshwater)	–	–	–	●	●	–	–	–
Festuco giganteae-Fraxino excelsioris Gs. (hard freshwater)	–	–	–	●	●	●	–	–
Geopermasigmeta								
Spergulario rupicolae-Armerio depilatae Gps. (haloanemogenous rock littoral)	–	●	–	–	–	–	–	–
Junco trifidi-Oreochloo blankae Gps. (acidophilous)	–	–	–	●	–	–	–	–
Oxytropido neglectae-Kobresio myosuroidis Gps. (basophilous)	–	–	–	–	●	–	–	–
Leucanthemo crassifolii-Festuco pruinosae Gps. (haloanemogenous rock littoral)	●	●	–	–	–	–	–	–
Puccinellio maritimae-Sarcocornio perennis Gps. (halophilous mareal)	●	●	●	–	–	–	–	–
Otantho maritimi-Ammophilo australis Gps. (coastal dune)	●	●	●	–	–	–	–	–
Crithmo maritimi-Armerio pubigerae Gps. (haloanemogenous rock littoral)	–	●	●	–	–	–	–	–
Oxytropido neglecto-halleri Gps. (basophilous)	–	–	–	●	●	–	–	–
Teesdaliopsio confertae-Festuco eskiae Gps. (acidophilous)	–	–	–	●	●	●	–	–

5. Cantabrian and Basque Sector, 6. North Galicia and Asturias Sector, 7. Galicia and North Portugal Sector, 8. High Campoo and Carrión Sector, 9. Picos of Europa and Ubiña Sector, 10. Laciana and Ancares Sector, 11. North Lusitania Sierran Sector, 12. Montemuro and Estrela Sierran Sector

Biogeographic Typology of the Valencia-Provençe and Balearic Province at Sector Level

IIa. Valencia-Provençe and Balearic Province (*Provincia Valenciana-Provenzal-Balear*) (Figs. 5.8, 5.9 and 5.10)

IIaa. Catalonian and Provençe Subprovince (*Subprovincia Catalana-Provenzal*)

16. East Catalonian Sector (*Sector Catalán Oriental*)

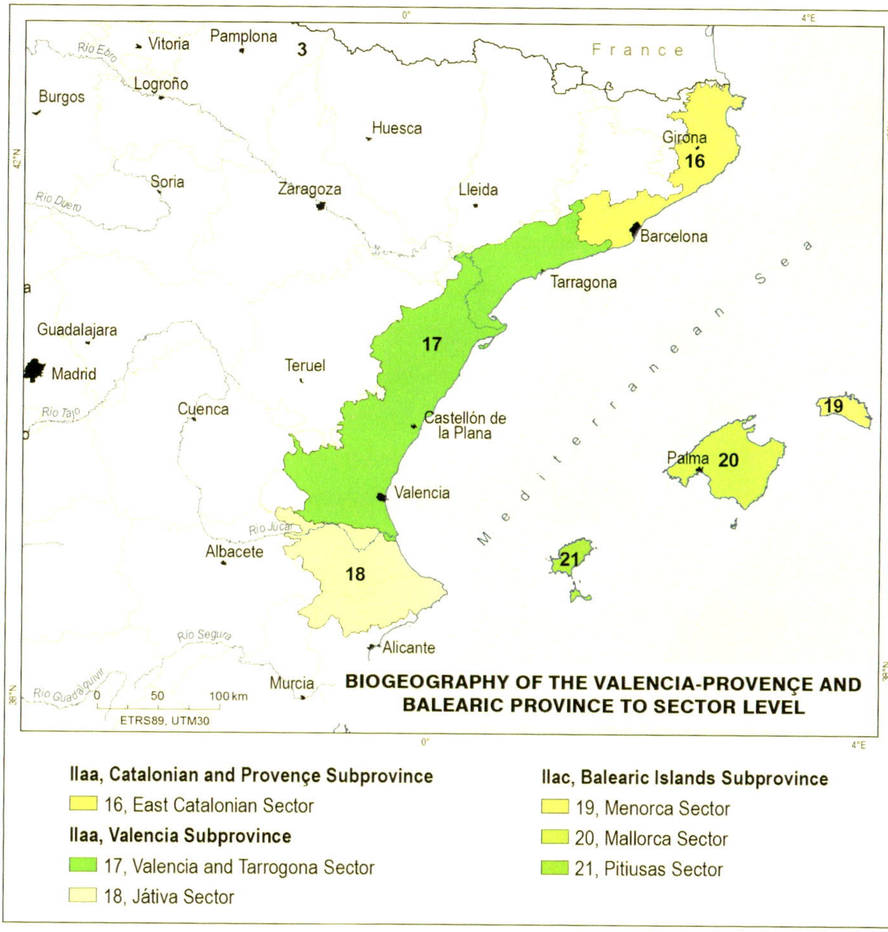

Fig. 5.8 Biogeographic map of the Valencia-Provençe and Balearic Province at sector level

IIab. VALENCIA Subprovince (*Subprovincia VALENCIANA*)

17. VALENCIA AND TARRAGONA Sector (*Sector VALENCIANO-TARRACONENSE*)
18. JÁTIVA Sector (*Sector SETABENSE*)

IIac. BALEARIC ISLANDS Subprovince (*Subprovincia BALEAR*)

19. MENORCA Sector (*Sector MENORQUÍN*)
20. MALLORCA Sector (*Sector MALLORQUÍN*)
21. PITIUSAS Sector (*Sector PITIÚSICO*)

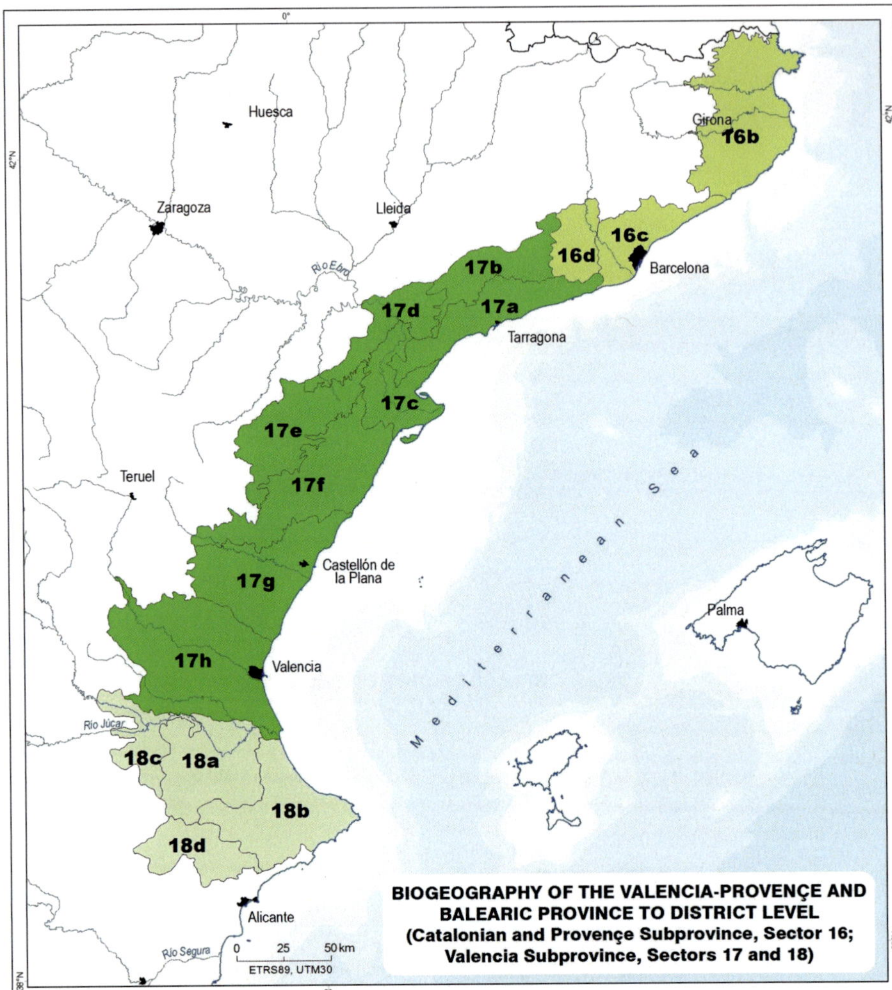

16. East Catalonian Sector
- 16b, Selva and Ampurdán District
- 16c, Vallés District
- 16d, High Penedés and Montserrat District

17. Valencia and Tarragona Sector
- 17a, Low Penedés and High Camp District
- 17b, Igualada and Prades District
- 17c, Low Ebro District
- 17d, Gandesa and Priorato District
- 17e, Beceite and Morella District
- 17f, Maestrat District
- 17g, Espadán Sierran and Castellón District
- 17h, Huerta of Valencia and Túria District

18. Játiva Sector
- 18a, Játiva District
- 18b, Alcoy and Denia District
- 18c, Allora and Cofrentes District
- 18d, Yecla and Villena District

Fig. 5.9 Biogeographic map of the Valencia-Provençe and Balearic Province at district level (Catalonian and Provençe subprovince and Valencia subprovince)

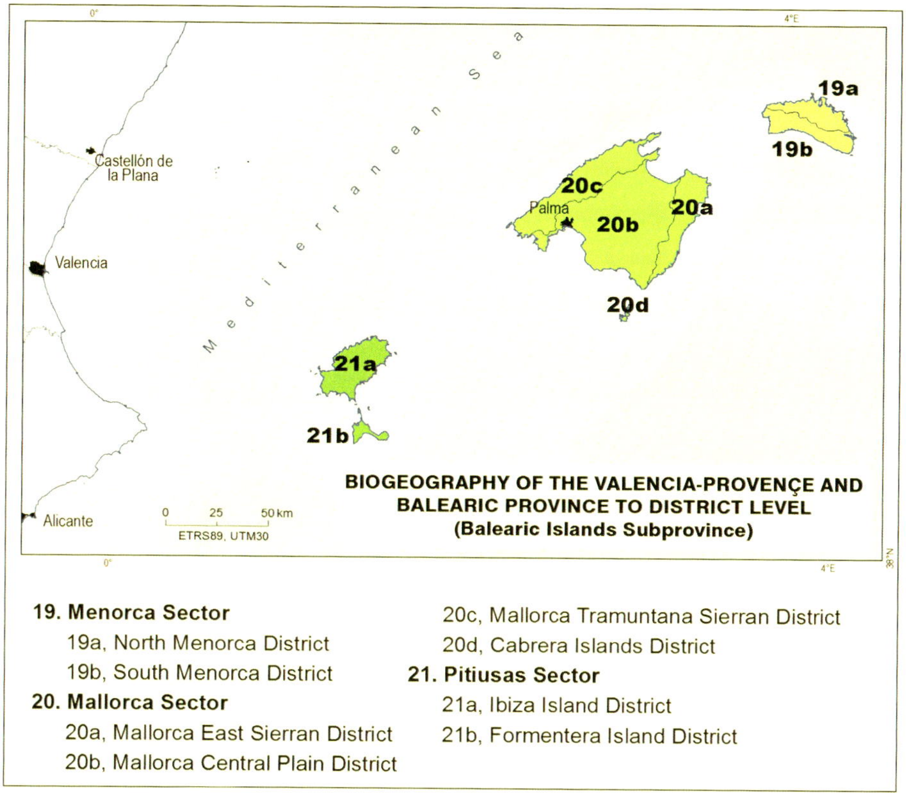

Fig. 5.10 Biogeographic map of the Valencia-Provençe and Balearic Province at district level (Balearic Isalnds subprovince)

Sigmeta, geosigmeta and geopermasigmeta of the biogeographic sectors in the Valencia-Provençe and Balearic province

VALENCIA-PROVENÇE and BALEARIC province	16	17	18	19	20	21
Climatophilous sigmeta						
Carici depauperatae-Querco pubescentis S. (acidophilous)	●	–	–	–	–	–
Primulo acaulis-Fago sylvaticae S. (neutrophilous)	–	●	–	–	–	–
Violo willkommii-Querco fagineae S. (basophilous)	–	●	–	–	–	–
Telino patentis-Querco fagineae S. (basophilous)	–	●	–	–	–	–
Cephalanthero rubrae-Querco pyrenaicae S. (acidophilous)	–	●	–	–	–	–
Fraxino orni-Querco fagineae S. (basophilous)	–	–	●	–	–	–
Viburno tini-Fraxino orni S. (basophilous)	–	–	●	–	–	–
Primulo balearicae-Acero granatensis S. (basophilous)	–	–	–	–	●	–
Cneoro tricocci-Ceratonio siliquae S. (basophilous)	–	–	–	–	●	–
Viburno tini-Querco ilicis S. (basophilous)	●	●	●	–	–	–
Asplenio onopteridis-Querco suberis S. (basophilous)	–	●	●	–	–	–

(continued)

Valencia-Provençe and Balearic province	16	17	18	19	20	21
Climatophilous and xerophilous sigmeta						
Arbuto unedonis-Pino halepensis S. (basophilous)	–	●	–	–	–	–
Asplenio onopteridis-Querco ilicis S. (acidophilous)	–	●	–	–	–	–
Prasio majoris-Oleo sylvestris S. (basophilous on leptosols)	–	–	–	●	–	–
Carici bracteosae-Querco ilicis S. (acidophilous, basophilous and dolomite)	–	–	–	●	–	–
Clematido cirrhosae-Querco rotundifoliae S. (basophilous and dolomite)	–	–	–	–	●	–
Cyclamini balearici-Querco ilicis S. (basophilous and dolomite)	–	–	–	–	●	–
Cneoro tricocci-Pistacio lentisci S. (basophilous)	–	–	–	–	–	●
Pistacio lentisci-Pino halepensis S. (basophilous)	–	●	●	–	–	–
Hedero helicis-Querco rotundifoliae S. (basophilous)	–	●	●	–	–	–
Rubio longifoliae-Querco rotundifoliae S. (basophilous)	–	●	●	–	–	–
Xerophilous sigmeta						
Daphno laureolae-Querco ilicis S. (basophilous)	●	–	–	–	–	–
Buxo sempervirentis-Pino catalaunicae S. (calco-dolomitic)	–	●	–	–	–	–
Telino patentis-Pino salzmannii S. (calco-dolomitic)	–	●	–	–	–	–
Phillyreo angustifoliae-Rhamno angustifolii S. (psammophilous)	–	●	–	–	–	–
Arctostaphylo crassifoliae-Pino catalaunicae S. (acidophilous)	–	●	–	–	–	–
Chamaeropo humilis-Junipero phoeniceae S. (basophilous)	–	–	●	–	–	–
Aro sagittifolii-Phillyreo rodriguezii S. (basophilous)	–	–	–	●	–	–
Rubio longifoliae-Junipero macrocarpae S. (psammophilous)	–	–	–	–	●	–
Genisto majoricae-Buxo balearicae S. (calco-dolomític)	–	–	–	–	●	–
Rhamno ludovicisalvatoris-Junipero turbinatae S. (dolomite)	–	–	–	–	●	–
Buxo sempervirentis-Junipero phoeniceae S. (basophilous)	●	●	–	–	–	–
Junipero turbinatae S. (coastal dune)	–	●	●	–	–	–
Rhamno infectorii-Junipero phoeniceae S. (basophilous)	–	●	●	–	–	–
Junipero turbinatae-Pino halepensis S. (basophilous & calco-dolomitic)	–	–	●	●	–	–
Clematido balearicae-Junipero turbinatae S. (coastal dune)	–	–	–	●	●	●
Climato-temporihygrophilous sigmeta						
Carici depressae-Querco canariensis S. (neutro-acidophilous)	●	–	–	–	–	–
Carici basilaris-Querco suberis S. (acidophilous)	●	–	–	–	–	–
Hygrophilous geosigmeta						
Lamio flexuosi-Alno glutinosae Gs. (soft freshwater)	●	–	–	–	–	–
Carici pendulae-Salici atrocinereae Gs. (soft freshwater)	●	–	–	–	–	–
Lithospermo purpureocaerulei-Ulmo minoris Gs. (hard freshwater)	●	–	–	–	–	–
Carici pendulae-Alno glutinosae Gs. (soft freshwater)	●	–	–	–	–	–
Erico erigenae-Salici pedicellatae Gs. (hard freshwater)	–	–	●	–	–	–
Vinco difformis-Fraxino angustifoliae Gs. (hard freshwater)	–	–	–	–	●	–
Vinco difformis-Populo albae Gs. (hard freshwater)	–	●	●	–	–	–
Coriario myrtifoliae-Salici angustifoliae Gs. (hard freshwater)	–	●	●	–	–	–
Saccharo ravennae-Tamarici canariensis Gs. (hard freshwater)	–	●	●	–	–	–
Populo albae Gs. (hard freshwater)	●	●	●	–	–	–

(continued)

Valencia-Provençe and Balearic province	16	17	18	19	20	21
Saponario officinalis-Salici lambertianae Gs. (hard freshwater)	●	●	●	–	–	–
Hedero helicis-Ulmo minoris Gs. (hard freshwater)	●	●	●	–	–	–
Geopermasigmeta						
Crithmo maritimi-Limonio dufourii Gps. (haloanemogenous rock littoral)	–	●	–	–	–	–
Crithmo maritimi-Limonio girardiani Gps. (haloanemogenous rock littoral)	–	●	–	–	–	–
Crithmo maritimi-Limonio rigualii Gps. (haloanemogenous rock littoral)	–	–	●	–	–	–
Limonio minuto-fontqueri Gps. (basophilous)	–	–	–	●	–	–
Limonio atruchio-minuti Gps. (haloanemogenous rock littoral)	–	–	–	●	–	–
Limonio caprariensis Gps (haloanemogenous rock littoral)	–	–	–	–	●	–
Crithmo maritimi-Limonio balearici Gps. (haloanemogenous rock littoral)	–	–	–	–	●	–
Limonio pseudodictyoclado-carregadorensis Gps. (haloanemogenous rock littoral)	–	–	–	–	●	–
Limonio pseudebusitani Gps. (haloanemogenous rock littoral)	–	–	–	–	–	●
Medicagini marinae-Ammophilo arundinaceae Gps. (coastal dune)	●	●	●	●	●	●
Limonio bellidifolii-Sarcocornio fruticosae Gps. (halophilous)	●	●	●	●	●	●

16. East Catalonian Sector, 17. Valencia and Tarragona Sector, 18. Játiva Sector, 19. Menorca Sector, 20. Mallorca Sector, 21. Pitiusas Sector

Biogeographic Typology of the Central Iberian Mediterranean Province at Sector Level (Figs. 5.11, 5.12, 5.13 and 5.14)

IIb. Central Iberian Mediterranean Province (*Provincia Mediterránea Ibérica Central*)

IIba. Low Aragón and high ebro Subprovince (*Subprovincia Bajoaragonesa-Altoebrense*)

22. Somontano Sector (*Sector Somontano*)
23. Bardenas and Monegros Sector (*Sector Bardenero-Monegrino*)
24. Rioja and Estella Sector (*Sector Riojano-Estellés*)
25. Cantabrian Castilian Sector (*Sector Castellano Cantábrico*)

IIbb. Oroiberian Subprovince (*Subprovincia Oroibérica*)

26. North Oroiberian Sierran Sector (*Sector Serrano Oroibérico Septentrional*)
27. South Oroiberian Sector (*Sector Oroibérico Meridional*)

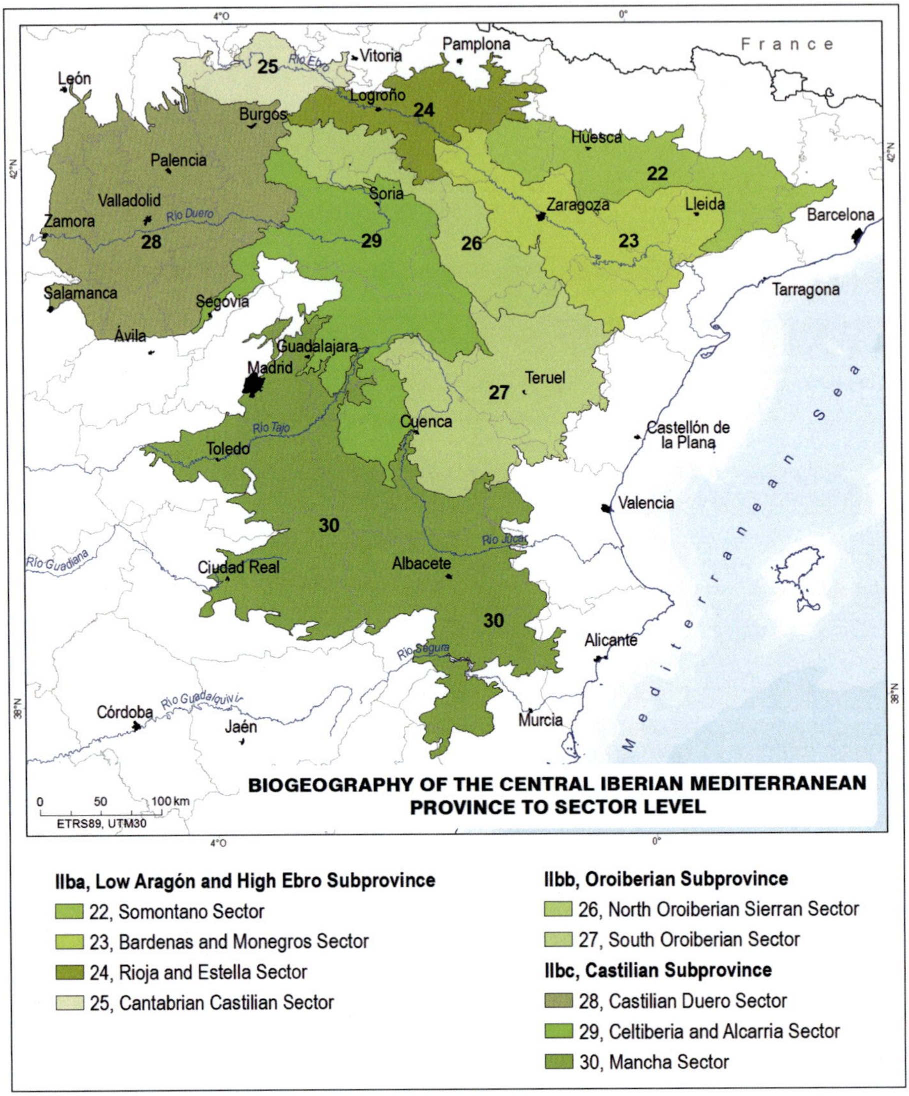

Fig. 5.11 Biogeographic map of the Central Iberian Mediterranean Province at sector level

IIbc. Castilian Subprovince (*Subprovincia Castellana*)

28. Castilian Duero Sector (*Sector Castellano Duriense*)
29. Celtiberia and Alcarria Sector (*Sector Celtibérico-Alcarreño*)
30. Mancha Sector (*Sector Manchego*)

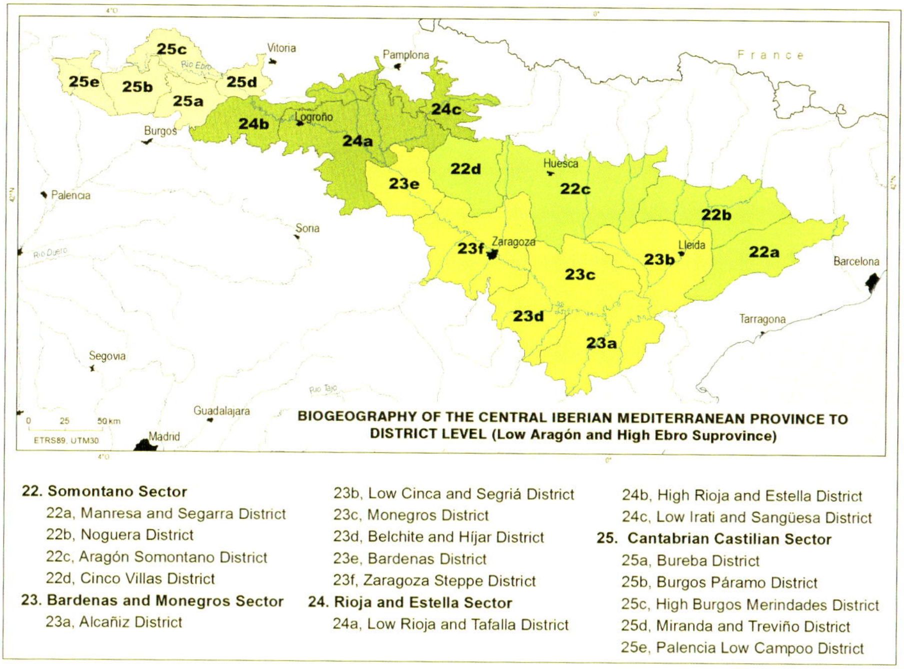

Fig. 5.12 Biogeographic map of the Central Iberian Mediterranean Province (Low Aragon and High Ebro) at district level

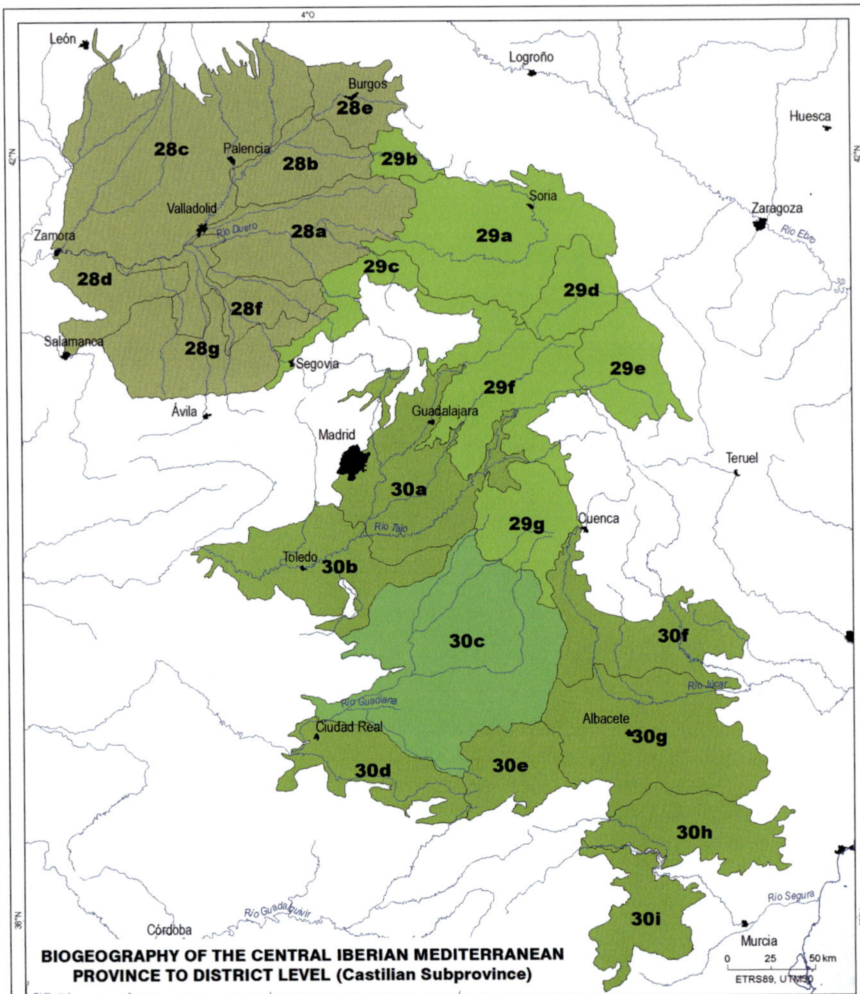

28. Castilian Duero Sector
- 28a, Castilian Ribaduero District
- 28b, Low Arlanza and Cerrato District
- 28c, Tierra of Campos District
- 28d, Tierras of Medina and Armuña District
- 28e, Arlanzón Basin District
- 28f, Tierra of Pinares District
- 28g, Tierra of Arévalo and Moraña District

29. Celtiberia and Alcarria Sector
- 29a, Soria District
- 29b, High Arlanza and Covarrubias District
- 29c, Segovia District
- 29d, High Jalón District
- 29e, Molina District
- 29f, High Alcarria District
- 29g, Obispalía District

30. Mancha Sector
- 30a, Low Madrid District
- 30b, Sagra District
- 30c, Sanjuán Mancha District
- 30d, Calatrava District
- 30e, Montiel District
- 30f, Cuenca Mancha District
- 30g, Júcar Mancha District
- 30h, Jumilla and Hellín District
- 30i, Espuña Sierran District

Fig. 5.13 Biogeographic map of the Central Iberian Mediterranean Province (Castilian subprovince) at district level

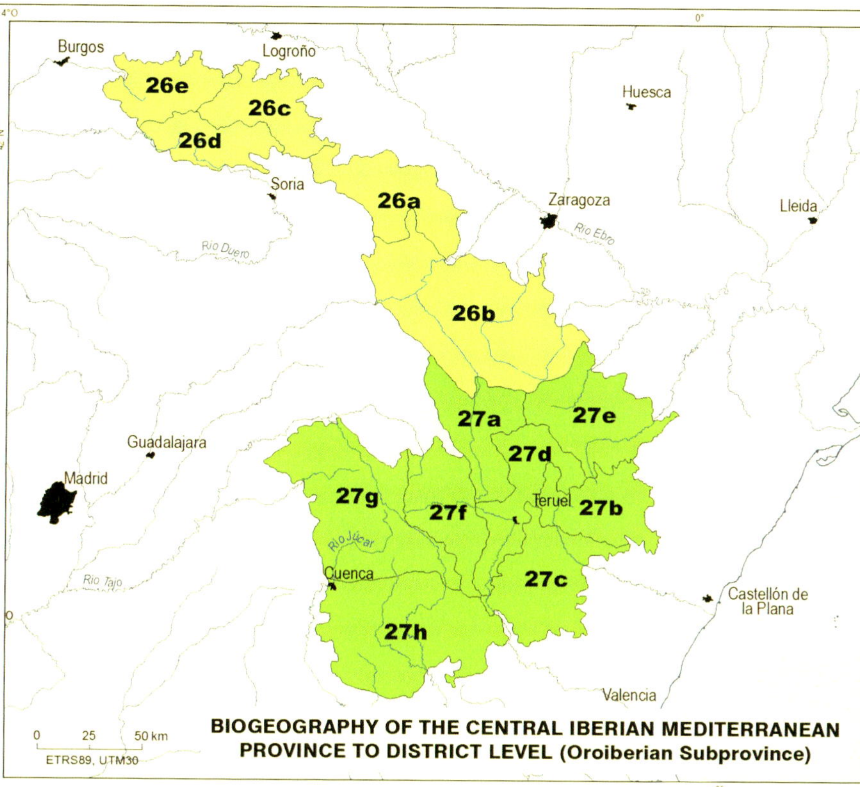

26. North Oroiberian Sierran Sector
- 26a, Moncayo Sierran District
- 26b, Bílbiles and Cucalón Sierran District
- 26c, Cameros Sierran District
- 26d, Urbión Sierran District
- 26e, Demanda Sierran District

27. South Oroiberian Sector
- 27a, High Jiloca District
- 27b, Gúdar Sierran District
- 27c, Javalambre Sierran District
- 27d, Ademuz and Teruel District
- 27e, Aliaga and Montalbán District
- 27f, Albarracín District
- 27g, North Cuenca Sierran District
- 27h, South Cuenca Sierran District

Fig. 5.14 Biogeographic map of the Central Iberian Mediterranean Province (Oroiberian subprovince) at district level

Sigmeta, geosigmeta and geopermasigmeta of the biogeographic sectors in the Central Iberian Mediterranean Province

CENTRAL IBERIAN MEDITERRANEAN province	22	23	24	25	26	27	28	29	30
Climatophilous sigmeta									
Violo willkommii-Querco fagineae S. (basophilous)	●	–	–	–	–	–	–	–	–
Vaccinio myrtilli-Pino ibericae S. (acidophilous, relic)	–	–	–	–	●	–	–	–	–
Carici sylvaticae-Fago sylvaticae S. (neutrophilous)	–	–	–	–	●	–	–	–	–
Avenello ibericae-Pino uncinatae S. (acidophilous)	–	–	–	–	●	–	–	–	–
Galio rotundifolii-Fago sylvaticae S. (acidophilous)	–	–	–	–	●	–	–	–	–
Vaccinio myrtilli-Junipero alpinae S. (acidophilous)	–	–	–	–	●	–	–	–	–
Saniculo europaeae-Ilici aquifolii S. (neutrophilous)	–	–	–	–	●	–	–	–	–
Junipero sabinae-Pino ibericae S. (basophilous)	–	–	–	–	–	●	–	–	–
Ononido aragonensis-Pino ibericae S. (basophilous)	–	–	–	–	–	●	–	–	–
Junipero sabinae-Pino uncinatae S. (basophilous)	–	–	–	–	–	●	–	–	–
Calluno vulgaris-Pino ibericae S. (basophilous)	–	–	–	–	–	●	–	–	–
Sileno melliferae-Querco faginae S. (basophilous)	–	–	–	–	–	●	–	–	–
Asparago acutifolii-Querco rotundifoliae S. (basophilous)	–	–	–	–	–	–	–	–	●
Galio odorati-Querco petraeae S. (acidophilous, relic)	–	–	–	●	●	●	–	–	–
Melico uniflorae-Betulo celtibericae S. (acidophilous)	–	–	–	●	●	●	–	–	–
Junipero hemisphaerico-thuriferae S. (basophilous)	–	–	–	●	●	●	–	–	–
Epipactido helleborines-Fago sylvaticae S. (neutrophilous)	–	–	–	●	●	–	–	–	–
Cephalanthero rubrae-Querco fagineae S. (basophilous)	–	–	–	–	–	–	●	●	●
Luzulo forsteri-Querco pyrenaicae S. (acidophilous)	–	–	–	●	●	●	–	●	–
Climatophilous and xerophilous sigmeta									
Junipero phoeniceo-thuriferae S. (calc-gypsophila)	–	●	–	–	–	–	–	–	–
Arbuto unedonis-Pino halepensis S. (basophilous)	–	●	–	–	–	–	–	–	–
Ephedro nebrodensis-Junipero sabinae S. (basophilous)	–	–	–	–	●	–	–	–	–

(continued)

Central Iberian Mediterranean province	22	23	24	25	26	27	28	29	30
Teucrio scorodoniae-Querco rotundifoliae S. (acidophilous)	–	–	–	–	●	–	–	–	–
Junipero thuriferae-Pino latisquamae S. (basophilous, calco-dolomitic)	–	–	–	–	–	●	–	–	–
Querco cocciferae-Pino halepensis S. (basophilous)	–	–	–	–	–	–	–	–	●
Rhamno lycioidis-Querco cocciferae S. (basophilous)	●	●	●	–	–	–	–	–	–
Querco rotundifoliae S.(basophilous)	●	●	●	–	●	–	–	–	–
Spiraeo obovatae-Querco fagineae S. (basophilous)	–	–	●	●	–	–	–	–	–
Junipero thuriferae-Querco rotundifoliae S. (basophilous)	–	–	–	–	–	●	●	●	●
Xerophilous sigmeta									
Rhamno lycioidis-Junipero phoeniceae S. (basophilous)	–	–	–	–	–	●	●	●	●
Climato-temporihygrophilous & hygrophilous sigmeta & geosigmeta									
Buxo sempervirentis-Querco rotundifoliae S. (basophilous)	●	–	–	–	–	–	–	–	–
Aceri campestris Fraxino excelsioris S. (neutro-acidophilous)	–	–	–	–	●	–	–	–	–
Hygrophilous geosigmeta									
Viburno lantanae-Ulmo minoris Gs. (hard freshwater)	–	–	–	●	–	–	–	–	–
Rubo lainzii-Salici atrocinereae Gs. (soft freshwater)	–	–	–	–	●	–	–	–	–
Astrantio majoris-Corylo avellanae Gs. (hard freshwater)	–	–	–	–	–	●	–	–	–
Erico erigenae-Salici pedicellatae Gs. (hard freshwater)	–	–	–	–	–	–	–	–	●
Rubio longifoliae-Nerio oleandri Gs. (very hard freshwater)	–	–	–	–	–	–	–	–	●
Humulo lupuli-Alno glutinosae Gs. (hard freshwater)	–	–	●	●	–	–	–	–	–
Populo nigrae-Salici neotrichae Gs. (soft freshwater)	–	–	–	–	–	–	●	●	–
Aro cylindracei-Ulmo minoris Gs. (soft freshwater)	–	–	–	–	–	–	●	●	–
Salici salviifoliae Gs. (soft freshwater)	–	–	–	●	●	●	–	–	–
Salici lambertiano-albae Gs. (hard freshwater)	–	–	–	●	●	●	–	–	–
Rubio tinctorum-Populo albae Gs. (hard freshwater)	●	●	●	–	–	–	–	●	●
Salici neotrichae Gs. (hard freshwater)	●	●	●	–	–	–	–	●	●
Opopanaco chironium-Ulmo minoris Gs. (hard freshwater)	●	●	●	●	●	●	●	●	●
Salici discoloro-angustifoliae Gs. (soft freshwater)	●	●	●	●	●	●	●	●	●

(continued)

Central Iberian Mediterranean province	22	23	24	25	26	27	28	29	30
Tamarici canariensis Gs. (soft freshwater)	●	●	●	●	●	●	●	●	●
Suaedo braunblanquetii-Tamarici boveanae Gs. (halophilous)	●	●	●	●	●	●	●	●	●
Suaedo braunblanquetii-Tamarici canariensis Gs. (halophilous)	●	●	●	●	●	●	●	●	●
Geopermasigmeta									
Antennario dioicae-Festuco curvifoliae Gps. (acidophilous)	–	–	–	–	●	–	–	–	–
Armerio microcephalae-Festuco aragonensis Gps. (acidophilous)	–	–	–	–	●	–	–	–	–
Suaedo braunblanquetii Gps. (halophilous)	●	●	●	–	–	–	–	–	–
Puccinellio caespitosae-Sauedo braunblanquetii Gps. (halophilous)	–	–	–	–	–	–	●	●	●

22. Somontano Sector, 23. Bardenas and Monegros Sector, 24. Rioja and Estella Sector, 25. Cantabrian Castilian Sector, 26. North Oroiberian Sierran Sector, 27. South Oroiberian Sector, 28. Castilian Duero Sector, 29. Celtiberia and Alcarria Sector, 30. Mancha Sector

Biogeographic Typology of the West Iberian Mediterranean Province at Sector Level (Figs. 5.15, 5.16 and 5.17)

IIc. West Iberian Mediterranean Province (*Provincia Mediterránea Ibérica Occidental*)

IIca. Carpetania and León Subprovince (*Subprovincia Carpetana-Leonesa*)

31. León Plain Sector (*Sector Planileonés*)
32. Bierzo and Sanabria Sector (*Sector Berciano-Sanabrés*)
33. Lusitanian Douro Sector (*Sector Lusitano Duriense*)
34. Salamanca Sector (*Sector Salmantino.*)
35. Guadarrama Sierran Sector (*Sector Serrano Guadarrámico*)
36. Bejar and Gredos Sierras Sector (*Sector Serrano Bejarano-Gredense*)

IIcb. Lusitania And Extremadura Subprovince (*Subprovincia Lusa-Extremadurense*)

37. Oretana Range and Tajo Sector (*Sector Cordillerano Oretano-Tagano*)
38. Mariánica Range Sector (*Sector Cordillerano Mariánico*)

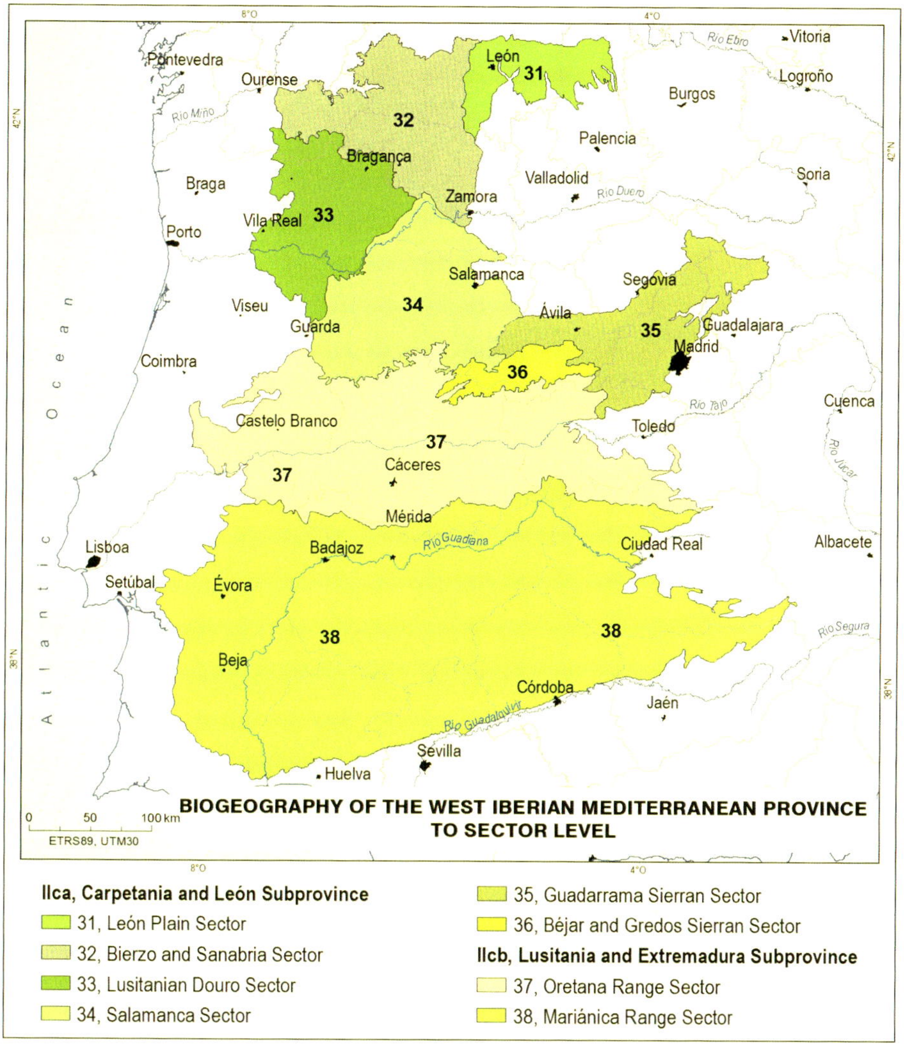

Fig. 5.15 Biogeographic map of the West Iberian Mediterranean Province at sector level

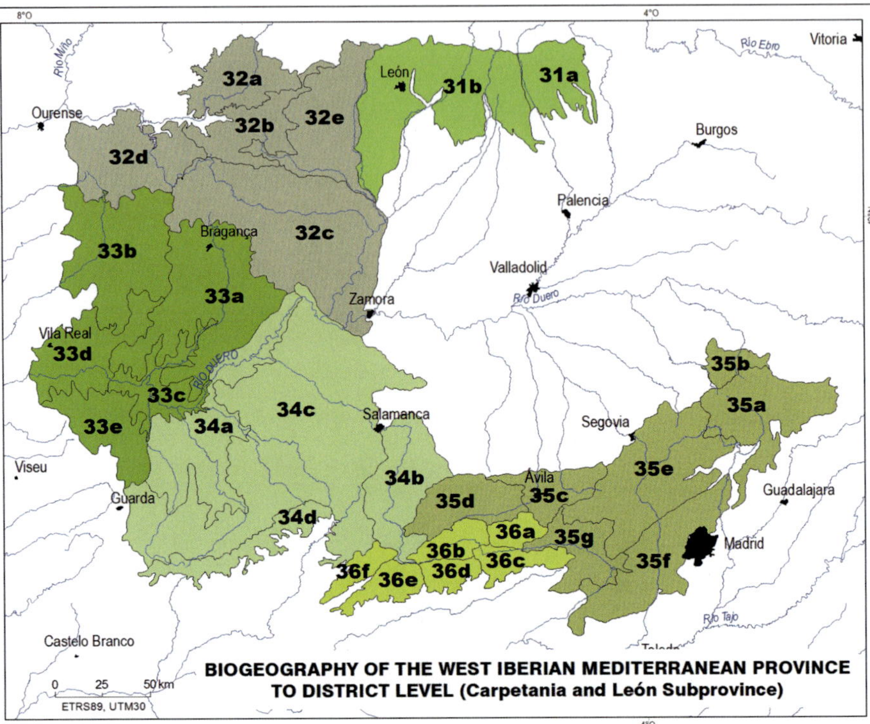

31. León Plain Sector
- 31a, Valdavia District
- 31b, León Páramo District

32. Bierzo and Sanabria Sector
- 32a, Bierzo District
- 32b, Cabrera and Montes of León District
- 32c, Zamora and Sanabria District
- 32d, Queixa Sierran District
- 32e, Maragatería District

33. Lusitanian Douro Sector
- 33a, Bragança District
- 33b, Chaves and Verín District
- 33c, Terraquente District
- 33d, Vila Real District
- 33e, North Beira District

34. Salamanca Sector
- 34a, Low Salamanca District
- 34b, Tormes District
- 34c, High Salamanca District
- 34d, Batuecas and Malcata Sierran District

35. Guadarrama Sierran Sector
- 35a, Ayllón Sierran District
- 35b, Riaza District
- 35c, Ávila District
- 35d, Corneja and Amblés District
- 35e, High Guadarrama Sierran District
- 35f, High Madrid District
- 35g, Central Alberche District

36. Béjar and Gredos Sierras Sector
- 36a, Ávila Paramera Sierran District
- 36b, Serrota Sierran District
- 36c, East Gredos Sierran District
- 36d, High Gredos Sierran District
- 36e, Tormantos Sierran District
- 36f, Béjar Sierran District

Fig. 5.16 Biogeographic map of the West Iberian Mediterranean Province (Carpetanian and León subprovince) at district level

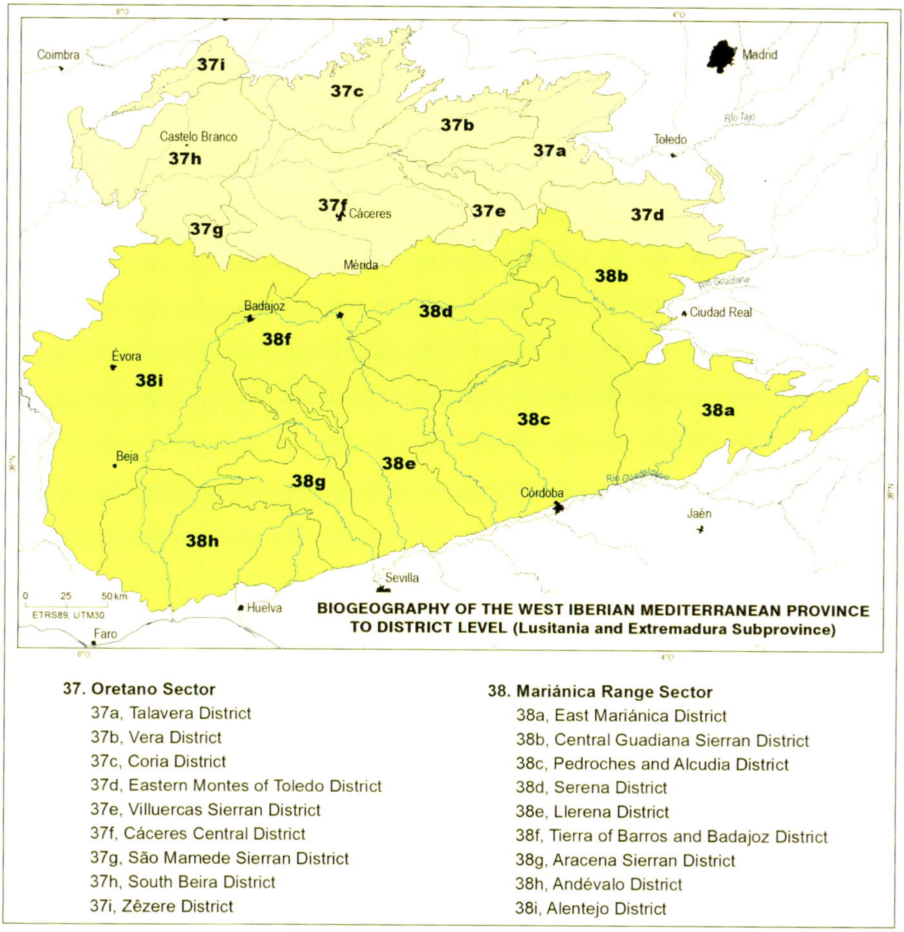

Fig. 5.17 Biogeographic map of the West Iberian Mediterranean Province (Lusitania and Extremadura subprovince) at district level

Sigmeta, geosigmeta and geopermasigmeta of the biogeographic sectors in the West Iberian Mediterranean province

West Iberian Mediterranean province	31	32	33	34	35	36	37	38
Climatophilous sigmeta								
Vaccinio myrtilli-Junipero alpinae S. (acidophilous)	–	●	–	–	–	–	–	–
Avenello ibericae-Querco orocantabricae S. (acidophilous)	–	●	–	–	–	–	–	–
Luzulo henriquesii-Betulo celtibericae S. (acidophilous)	–	●	–	–	–	–	–	–
Genisto sanabrensis-Junipero alpinae S. (acidophilous)	–	●	–	–	–	–	–	–
Genisto falcatae-Querco pyrenaicae S. (acidophilous)	–	●	–	–	–	–	–	–

(continued)

West Iberian Mediterranean province	31	32	33	34	35	36	37	38
Junipero lagunae-Querco suberis S. (acidophilous)	–	–	●	–	–	–	–	–
Hedero hibernicae-Querco fagineae S. (basophilous)	–	–	●	–	–	–	–	–
Galio odorati-Querco petraeae S. (acidophilous, relic)	–	–	–	–	●	–	–	–
Luzulo forsteri-Querco pyrenaicae S. (acidophilous)	–	–	–	–	●	–	–	–
Melico uniflorae-Betulo celtibericae S. (acidophilous)	–	–	–	–	●	–	–	–
Avenello ibericae-Pino ibericae S. (acidophilous)	–	–	–	–	●	–	–	–
Festuco merinoi-Querco pyrenaicae S. (acidophilous)	–	–	–	–	–	●	–	–
Smilaco asperae-Querco suberis S. (acidophilous)	–	–	–	–	–	–	●	–
Asparago aphylli-Querco suberis S. (acidophilous)	–	–	–	–	–	–	–	●
Lavandulo viridis-Querco suberis S. (acidophilous)	–	–	–	–	–	–	–	●
Doronico plantaginei-Querco canariensis S. (acidophilous)	–	–	–	–	–	–	–	●
Sanguisorbo hybridae-Querco broteroi S. (acidophilous)	–	–	–	–	–	–	–	●
Arisaro sinorrhini-Querco pyrenaicae S. (acidophilous)	–	–	–	–	–	–	●	●
Pulmonario longifoliae-Querco pyrenaicae S. (acidophilous)	●	–	–	–	●	–	–	–
Pyro bourgaeanae-Querco pyrenaicae S. (acidophilous)	–	●	–	●	–	–	–	–
Holco mollis-Querco pyrenaicae S. (acidophilous)	–	–	●	●	–	–	–	–
Pteridio aquilini-Pino ibericae S. (acidophilous)	–	–	–	–	●	●	–	–
Avenello ibericae-Junipero alpinae S. (acidophilous)	–	–	–	–	●	●	–	–
Asparago albi-Oleo sylvestris S. (acidophilous)	–	–	–	–	–	–	●	●
Sorbo torminalis-Querco pyrenaicae S. (acidophilous)	–	–	–	–	–	–	●	●
Arbuto unedonis-Querco pyrenaicae S. (acidophilous)	–	–	–	–	–	–	●	●
Sanguisorbo hybridae-Querco suberis S. (acidophilous)	–	–	–	–	–	–	●	●
Rhamno fontquerani-Querco rotundifoliae S. (basophilous)	–	–	–	–	–	–	●	●
Pistacio terebinthi-Querco broteroi S. (acidophilous)	–	–	–	–	–	–	●	●
Climatophilous and xerophilous sigmeta								
Junipero lagunae-Querco rotundifoliae S. (acidophilous)	●	–	–	–	●	–	–	–
Genisto hystricis-Querco rotundifoliae S. (acidophilous, relic)	–	●	●	●	–	–	–	–
Pyro bourgaeanae-Querco rotundifoliae S. (acidophilous)	–	–	–	–	–	–	●	●
Xerophilous sigmeta								
Rusco aculeati-Junipero lagunae S. (acidophilous, relic)	–	–	●	–	–	–	–	–
Festuco merinoi-Junipero lagunae S. (acidophilous)	–	–	–	–	–	●	–	–
Cytiso eriocarpi-Junipero lagunae S. (acidophilous)	–	–	–	–	–	–	●	–
Phlomido purpureae-Junipero turbinatae S. (acidophilous)	–	–	–	–	–	–	–	●

(continued)

West Iberian Mediterranean province	31	32	33	34	35	36	37	38
Phlomido purpureae-Pistacio lentisci S. (acidophilous)	–	–	–	–	–	–	–	●
Climato-temporihygrophilous sigmeta								
Clematido campaniflorae-Celtido australis S. (acidophilous)	–	–	●	–	–	–	–	–
Fraxino angustifoliae-Acero monspessulani S. (acidophilous)	–	–	●	–	–	–	–	–
Frangulo alni-Pruno lusitanicae S.	–	–	–	–	–	–	●	–
Campanulo primulifoliae-Rhododendro pontici S. (acidophilous)	–	–	–	–	–	–	–	●
Euphorbio monchiquensis-Querco canariensis S. (acidophilous)	–	–	–	–	–	–	–	●
Pyro bourgaeanae-Querco broteroi S. (acidophilous)	–	–	–	–	–	–	●	●
Oennatho crocatae-Querco pyrenaicae S. (acidophilous)	–	–	–	–	–	–	●	●
Climato-temporihygrophilous and hygrophilous sigmeta & geosigmeta								
Paeonio broteri-Abieti pinsapo S. (calco-dolomitic)	–	–	–	–	–	–	–	●
Hygrophilous geosigmeta								
Salici salviifoliae Gs. (soft freshwater)	●	–	–	–	–	–	–	–
Nerio oleandri-Salici pedicellatae Gs. (hard freshwater)	–	–	–	–	–	–	–	●
Campanulo primulifoliae-Alno glutinosae Gs. (soft freshwater)	–	–	–	–	–	–	–	●
Irido foetidissimae-Fraxino angustifoliae Gs. (soft freshwater)	–	–	–	–	–	–	–	●
Aro cylindracei-Ulmo minoris Gs. (soft freshwater)	●	●	–	●	–	–	–	–
Populo nigrae-Salici neotrichae Gs. (soft freshwater)	●	●	–	●	–	–	–	–
Rubo lainzii-Salici atrocinereae Gs. (soft freshwater)	–	●	–	●	●	●	–	–
Salici lambertiano-salviifoliae Gs. (soft freshwater)	–	–	–	●	●	●	●	–
Galio broteriani-Alno glutinosae Gs. (soft freshwater)	●	●	●	●	●	●	–	–
Querco pyrenaicae-Fraxino angustifoliae Gs. (soft freshwater)	●	●	●	●	●	●	–	–
Salici atrocinereae-Populo albae Gs. (hard freshwater)	–	–	–	–	–	–	●	●
Salici atrocinereo-australis Gs. (soft freshwater)	–	–	–	–	–	–	●	●
Pyro bourgaeanae-Flueggeo tinctoriae Gs. (soft freshwater)	–	–	–	–	–	–	●	●
Geopermasigmeta								
Teesdaliopsio confertae-Festuco summilusitanae Gps. (acidophilous)	–	●	–	–	–	–	–	–
Hieracio myriadeni-Festuco carpetanae Gps. (acidophilous)	–	–	–	–	●	–	–	–
Agrostio rupestris-Armerio bigerrensis Gps. (acidophilous)	–	–	–	–	–	●	–	–

31. León Plain Sector, 32. Bierzo and Sanabria Sector, 33. Lusitanian Douro Sector, 34. Salamanca Sector, 35. Guadarrama Sierran Sector, 36. Béjar and Gredos Sierran Sector, 37. Oretana Range and Tajo Sector, 38. Mariánica Range Sector

Biogeographic Typology of the Murcia and Almería Province at Sector Level (Figs. 5.18 and 5.19)

IId. Murcia and Almería Province (*Provincia Murciana-Almeriense*)

39. alicante and Murcia Sector (*Sector Alicantino-Murciano*)
40. Almería Sector (*Sector Almeriense*)

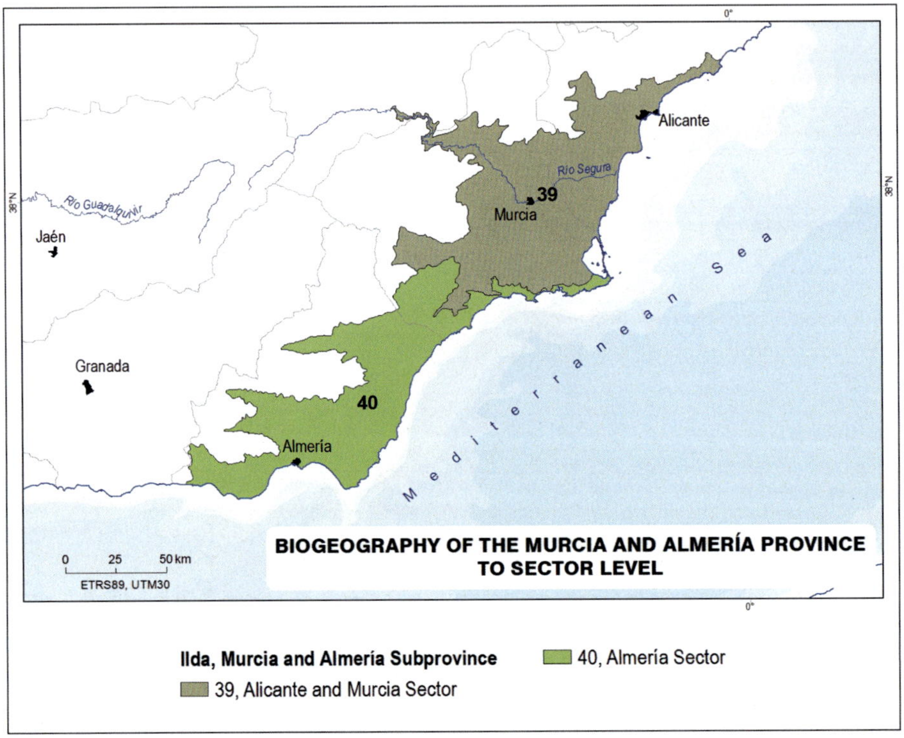

Fig. 5.18 Biogeographic map of the Murcia and Almería province at sector level

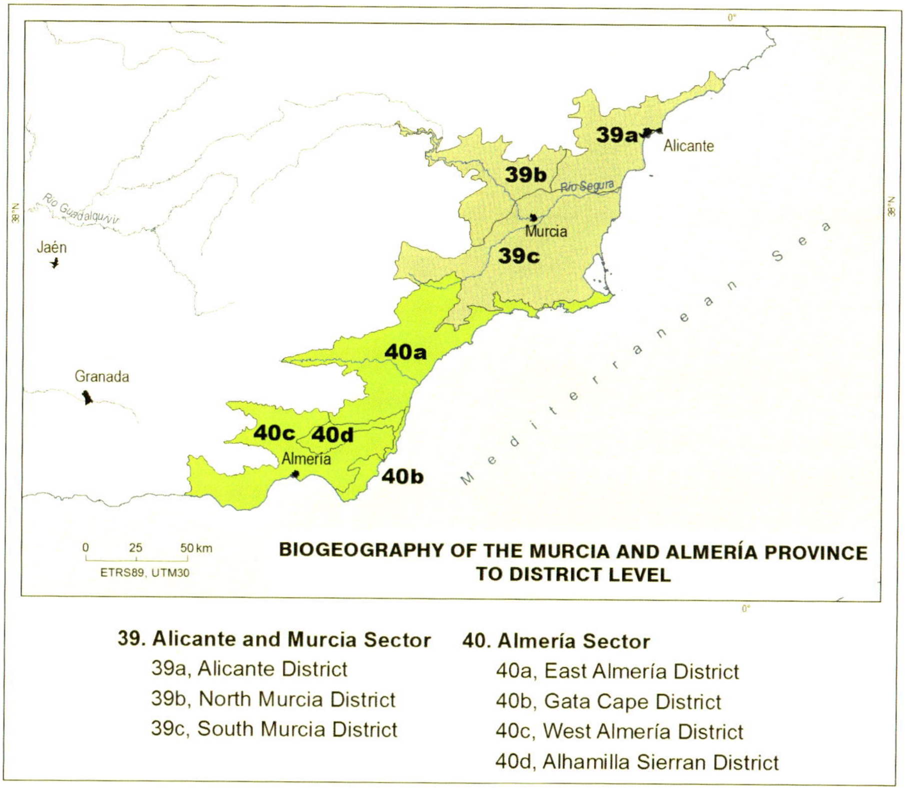

Fig. 5.19 Biogeographic map of the Murcia and Almería Province at district level

Sigmeta, geosigmeta and geopermasigmeta of the biogeographic sectors in the Murcia and Almería province

Murcia and Almería province	39	40
Climatophilous sigmeta		
Zizipho loti S. (psammophilous)	–	●
Zizipho loti-Mayteno europaei S. (basophilous)	–	●
Mayteno europaei-Periploco angustifoliae S. (basophilous)	–	●
Chamaeropo humilis-Junipero phoeniceae S. (basophilous)	●	●
Climatophilous and xerophilous sigmeta		
Rhamno capillaris-Periploco angustifoliae S. (basophilous)	●	–
Arisaro simorrhini-Tetraclinido articulatae S. (basophilous)	–	●
Chamaeropo humilis-Rhamno lycioidis S. (basophilous)	●	●
Querco cocciferae-Pino halepensis S. (basophilous)	●	●
Xerophilous sigmeta		
Corema to albi-Junipero macrocarpae (relict dune)	●	–
Rhamno angustifolii-Junipero turbinatae S. (dune)	●	●

(continued)

MURCIA AND ALMERÍA province	39	40
Climato-temporihygrophilous sigmeta		
Rubio longifoliae-Nerio oleandri S. (very hard freshwater)	●	●
Hygrophilous geosigmeta		
Zizipho loti-Nerio oleandri Gs. (very hard freshwater)	–	●
Lonicero biflorae-Populo albae Gs.	●	●
Geopermasigmeta		
Limonio cossoniani-Lycio intricati Gps. (haloanemogenous rock littoral)	●	●
Frankenio corymbosae-Arthrocnemo macrostachyi Gps. (halophilous littoral)	●	●
Cistancho phelypaeae-Sarcocornio fruticosae Gps. (halophilous)	●	●
Loto cretici-Ammophilo australis Gps. (coastal dune)	●	●

39. Alicante and Murcia Sector, 40. Almería Sector

Biogeographic Typology of the Bética Province at Sector Level (Figs. 5.20 and 5.21)

IIe. BÉTICA Province (*Provincia BÉTICA*)

41. SUBBÉTICA Sector (*Sector SUBBÉTICO*)
42. HOYAS OF GUADIX AND BAZA Sector (*Sector HOYANO ACCITANO-BASTITANO*)
43. NEVADA SIERRAN Sector (*Sector SERRANO NEVADENSE*)
44. ALPUJARRAS AND GÁDOR SIERRAN Sector (*Sector ALPUJARREÑO-SERRANO GADORENSE*)

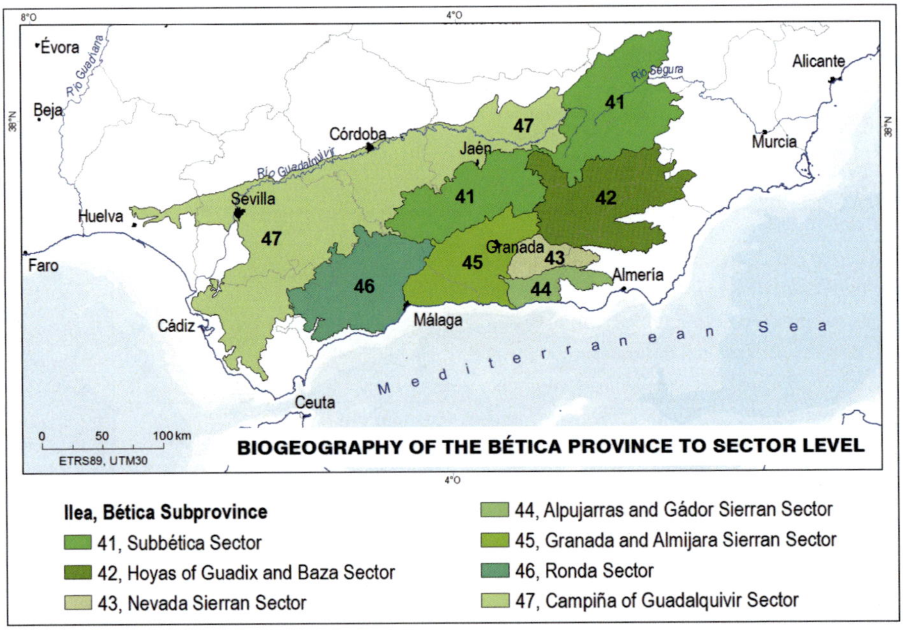

Fig. 5.20 Biogeographic map of the Bética Province at sector level

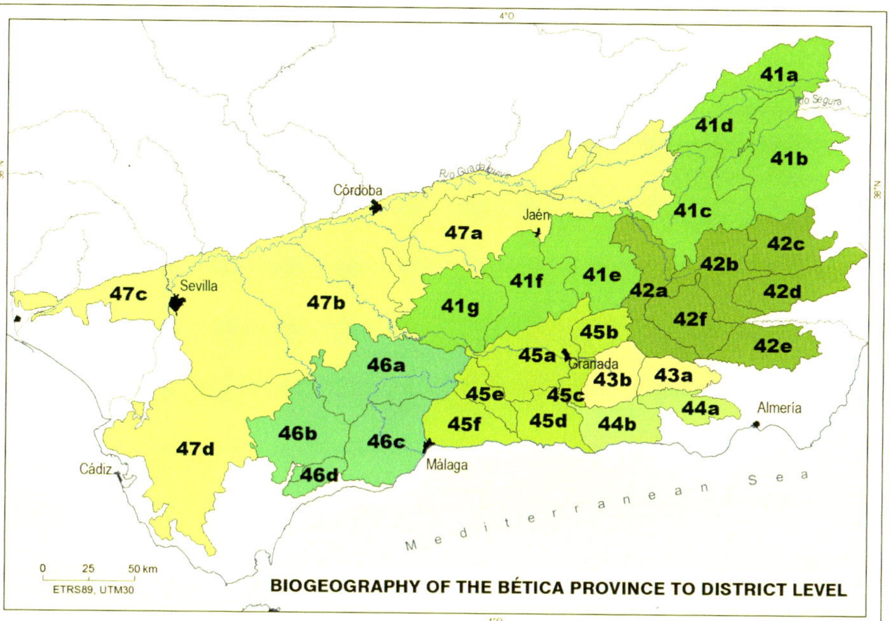

41. Subbética Sector
- 41a, Alcaraz Sierran District
- 41b, Murcia Subbética District
- 41c, Cazorla Sierran District
- 41d, Segura Sierran District
- 41e, Mágina Sierran District
- 41f, Southwest Jaén Sierran District
- 41g, Córdoba Subbética District

42. Hoyas of Guadix and Baza Sector
- 42a, Hoya of Guadix District
- 42b, Hoya of Baza District
- 42c, María Sierran District
- 42d, Estancias Sierran District
- 42e, Filabres Sierran District
- 42f, Baza Sierran District

43. Nevada Sierran Sector
- 43a, East Nevada Sierran District
- 43b, High Nevada Sierran District

44. Alpujarras and Gádor Sierran Sector
- 44a, Gádor Sierran District
- 44b, Alpujarras District

45. Granada and Almijara Sierran Sector
- 45a, Vega of Granada District
- 45b, Alfacar Sierran District
- 45c, Trevenque Sierran District
- 45d, Almijara Sierran District
- 45e, Tejeda Sierran District
- 45f, Axarquía District

46. Ronda Sector
- 46a, Antequera District
- 46b, Ronda Sierran District
- 46c, Hoya of Málaga District
- 46d, Bermeja Sierran District

47. Campiña of Guadalquivir Sector
- 47a, High Campiña District
- 47b, Low Campiña District
- 47c, Aljarafe District
- 47d, Jerez and Medina Sidonia District

Fig. 5.21 Biogeographic map of the Bética Province at district level

45. Granada and Almijara sierran Sector (*Sector Granadino-Serrano Almijarense*)
46. Ronda Sector (*Sector Rondeño*)
47. Campiña of guadalquivir Sector (*Sector Hispalense*)

Sigmeta, geosigmeta and geopermasigmeta of the biogeographic sectors in the Bética Province

Bética province	41	42	43	44	45	46	47
Climatophilous sigmeta							
Junipero sabinae-Pino latisquamae S. (basophilous)	●	–	–	–	–	–	–
Berberido hispanicae-Querco pyrenaicae S. (acidophilous)	●	–	–	–	–	–	–
Adeňocarpo decorticantis-Querco pyrenaicae S. (acidophilous)	–	–	●	–	●	–	–
Adenocarpo decorticantis-Querco suberis S. (acidophilous)	–	–	–	●	●	–	–
Oleo sylvestris-Querco suberis S. (acidophilous)	–	–	–	●	●	●	–
Adenocarpo decorticantis-Querco rotundifoliae S. (acidophilous)	–	●	●	●	●	●	–
Rhamno infectorii-Junipero sabinae S. (basophilous, calco-dolomitic)	●	●	●	●	●	●	●
Berberido hispanicae-Querco alpestris S. (basophilous, calco-dolomitic)	●	●	●	●	●	●	●
Aro neglecti-Oleo sylvestris S. (vertic soils)	●	●	●	●	●	●	●
Climatophilous and xerophilous sigmeta							
Junipero phoeniceae-Pino latisquamae S. (calco-dolomitic)	●	–	–	–	–	–	–
Berberido hispanicae-Junipero thuriferae S. (basophilous and calco-dolomitic)	●	–	–	–	–	–	–
Ephedro fragilis-Pino halepensis S. (basophilous)	–	●	–	–	–	–	–
Genisto versicoloris-Cytiso nevadensis S. (acidophilous)	–	●	–	–	–	–	–
Rhamno almeriensis-Pino halepensis S. (calco-dolomitic)	–	–	–	●	–	–	–
Mayteno europaei-Oleo sylvestris S. (basophilous)	–	–	–	–	●	–	–
Bunio macucae-Abieti pinsapo S. (ultramafic)	–	–	–	–	–	●	–
Querco cocciferae-Pino acutisquamae S. (ultramafic)	–	–	–	–	–	●	–
Daphno hispanicae-Pino nevadensis S. (calco-dolomitic)	–	●	–	–	●	–	–
Rhamno oleoidis-Querco rotundifoliae S. (basophilous, calco-dolomitic)	–	–	–	●	●	●	●
Pino acutisquamae S. (dolopsammophilous)	–	–	–	–	●	●	–
Paeonio coriaceae-Querco rotundifoliae S. (basophilous)	●	●	●	●	●	●	●
Berberido hispanicae-Querco rotundifoliae S. (basophilous, calco-dolomitic)	●	●	●	●	●	●	●
Xerophilous sigmeta							
Rhamno lycioidis-Pino halepensis S. (calco-dolomitic)	●	–	–	–	–	–	–
Berberido hispanicae-Junipero phoeniceae S. (basophilous & calco-dolomitic)	●	–	–	–	–	–	–
Chamaeropo humilis-Junipero phoeniceae S. (basophilous)	–	–	–	●	●	–	–
Cneoro tricocci-Buxo balearicae S. (calco-dolomitic)	–	–	–	–	●	–	–
Rhamno myrtifolii-Junipero phoeniceae S. (dolomite and dolopsammophilous)	–	–	–	–	●	●	–
Asparago horridi-Junipero turbinatae S. (calco-dolomitic)	–	–	–	–	–	●	–
Vinco difformis-Ceratonio siliquae S. (basophilous)	–	–	–	–	–	●	–
Climato-temporihygrophilous sigmeta							
Viburno tini-Querco alpestris S. (basophilous)	●	–	–	–	–	–	–

(continued)

Bética province	41	42	43	44	45	46	47
Oleo sylvestris-Querco alpestris S. (basophilous)	–	–	–	–	–	–	●
Daphno latifoliae-Acero granatensis S. (basophilous)	●	●	●	●	●	●	●
Hygrophilous geosigmeta							
Limonio delicatuli-Nerio oleandri Gs. (halophilous)	–	●	–	–	–	–	–
Aceri granatensis-Fraxino angustifoliae Gs. (soft freshwater)	–	–	●	–	–	–	–
Carici camposii-Salici atrocinereae Gs. (soft freshwater)	–	–	●	–	–	–	–
Erico terminalis-Salici angustifoliae Gs. (hard freshwater)	–	–	–	–	●	–	–
Crataego brevispinae-Populo albae Gs. (hard freshwater)	–	–	–	–	–	–	●
Crataego granatensis-Salici neotrichae Gs. (hard freshwater)	●	●	●	–	–	–	●
Galio viridiflori-Salici pedicellatae Gs. (hard freshwater)	–	–	–	●	●	–	–
Erico terminalis-Nerio oleandri Gs. (hard freshwater, serpentinícola)	–	–	–	●	●	–	–
Dorycnio recti-Salici pedicellatae Gs. (hard freshwater)	–	–	–	–	●	●	–
Salici pedicellatae-Populo albae Gs. (hard freshwater)	–	–	–	●	●	●	–
Nerio oleandri-Populo albae Gs. (hard freshwater)	●	●	●	●	●	●	●
Biaro carratracensis-Ulmo minoris Gs. (hard freshwater)	●	●	●	●	●	●	●
Suaedo braunblanquetii-Tamarici canariensis Gs. (lacustrine, halophilic)	●	●	●	●	●	●	●
Geopermasigmeta							
Erigeronto frigidi-Festuco clementei Gps. (acidophilous)	–	–	●	–	–	–	–
Crithmo maritimi-Limonio malacitani Gps. (haloanemogenous rock littoral)	–	–	–	●	●	●	–
Cistancho phelypaeae-Sarcocornio fruticosae Gps. (halophilous)	–	–	–	●	●	●	–
Loto cretici-Ammophilo australis Gps. (coastal dune)	–	–	–	●	●	●	–

41. Subbético Sector, 42. Hoyas of Guadix and Baza Sector, 43. Nevada Sierran Sector, 44. Alpujarras and Gádor Sierran Sector, 45. Granada and Almijara Sierran Sector, 46. Ronda Sector, 47. Campiña of Guadalquivir Sector

Biogeographic Typology of the Lusitania and West Andalusia Coastal Province at Sector Level (Figs. 5.22 and 5.23)

IIf. Andalusia and west Lusitania Coastal Province (*Provincia Costera Lusitana-Andaluza occidental*)

IIfa. Divisorio Portuguese Subprovince (*Subprovincia Divisoria Portuguesa*)

48. Divisorio Portuguese Sector (*Sector Divisorio Portugués*)

IIfb. Cádiz and Sado Subprovince (*Subprovincia Gaditana-Sadense*)

49. Ribatejo and Sado Sector (*Sector Ribatagano-Sadense*)
50. Algarve and Monchique Sector (*Sector Algárvico-Monchiquense*)
51. Cádiz and Littoral Huelva Sector (*Sector Gaditano-Onubense Litoral*)
52. Aljibe Sector (*Sector Aljíbico*) (Figs. 5.22 and 5.23)

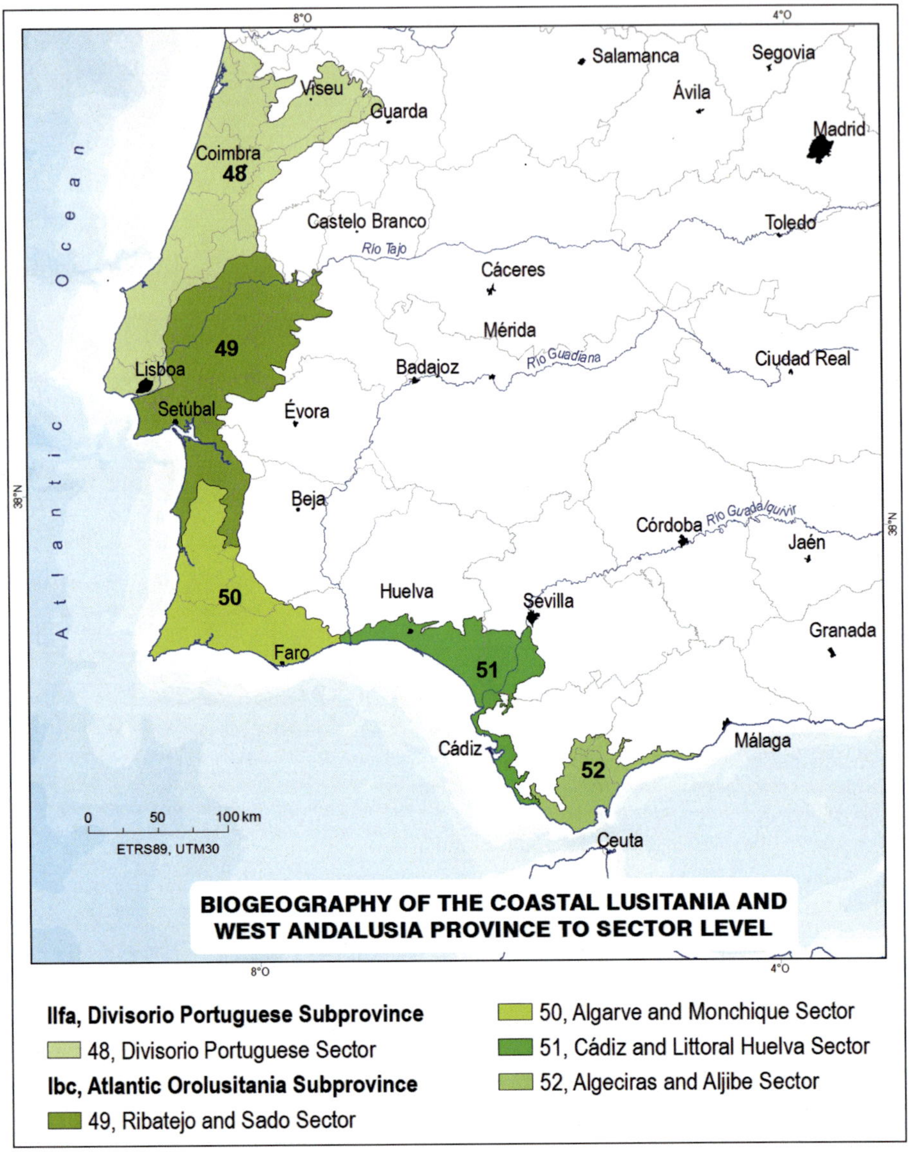

Fig. 5.22 Biogeographic map of the Lusitania and West Andalusia Coastal Province at sector level

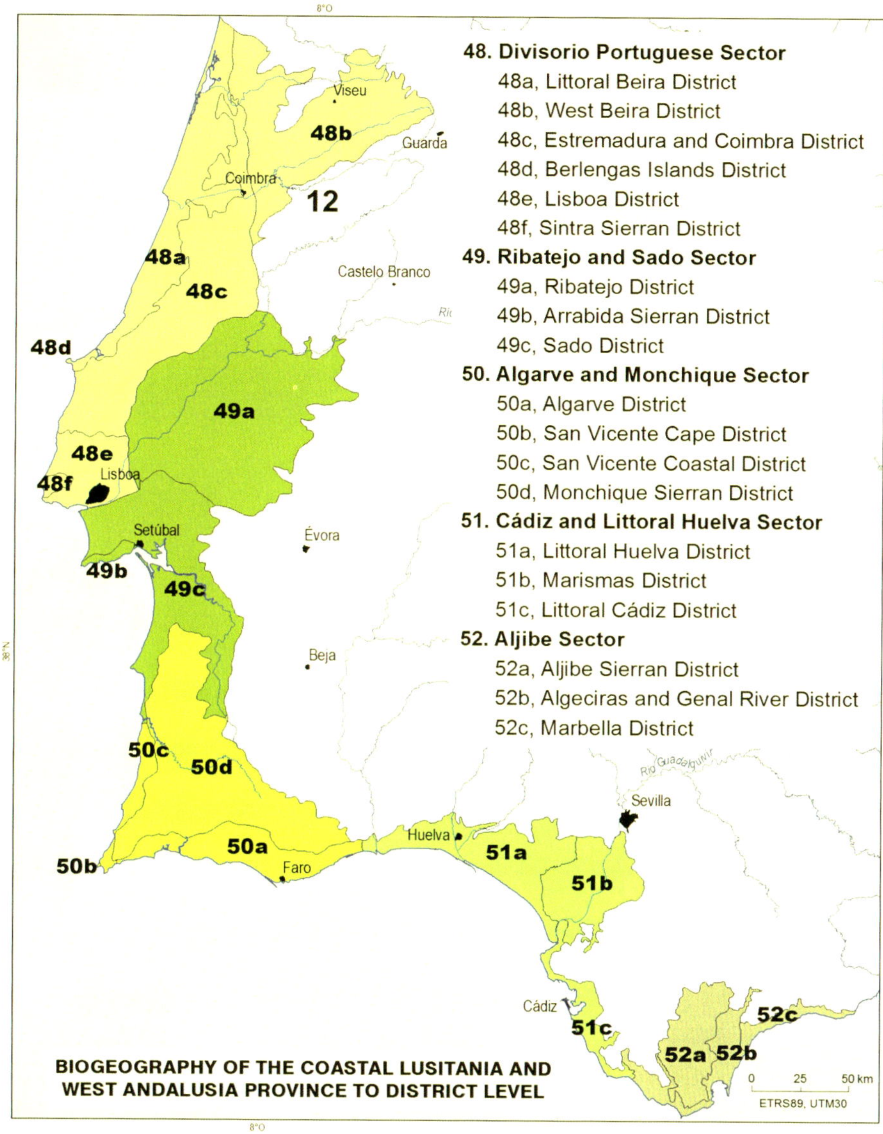

Fig. 5.23 Biogeographic map of the Lusitania and West Andalusia Coastal Province at district level

Sigmeta, geosigmeta and geopermasigmeta of the biogeographic sectors in the Lusitania and West Andalusia Coastal Province

Lusitania and West Andalusia Coastal province	48	49	50	51	52
Climatophilous sigmeta					
Sanguisorbo hybridae-Querco broteroir S. (acidophilous)	●	–	–	–	–
Asparago aphylli-Querco suberis S. (acidophilous)	–	●	–	–	–
Lavandulo viridis-Querco suberis S. (acidophilous)	–	–	●	–	–
Querco alpestris-broteroi S. (basophilous, relic)	–	–	●	–	–
Oleo sylvestris-Querco suberis S. (acidophilous)	–	–	–	●	–
Luzulo baeticae-Querco pyrenaicae S. (acidophilous)	–	–	–	–	●
Teucrio baetici-Querco suberis S. (acidophilous)	–	–	–	–	●
Viburno tini-Oleo sylvestris S. (basophilous)	●	●	–	–	–
Arisaro sinorrhini-Querco broteroi S. (basophilous)	●	●	–	–	–
Viburno tini-Querco rivas-martinezii S. (basophilous)	●	●	●	–	–
Arisaro sinorrhini-Querco pyrenaicae S. (acidophilous)	●	●	●	–	–
Climatophilous and xerophilous sigmeta					
Daphno gnidi-Junipero navicularis S. (acidophilous)	–	●	–	–	–
Aristolochio baeticae-Junipero turbinatae S. (basophilous)	–	–	●	–	–
Rhamno oleoidis-Querco rotundifoliae S. (basophilous, calco-dolomitic)	–	–	●	–	–
Xerophilous sigmeta					
Lonicero implexae-Querco rotundifoliae S. (basophilous)	●	–	–	–	–
Querco cocciferae-airensis S.	●	–	–	–	–
Vinco difformis-Lauro nobilis S. (basophilous)	●	–	–	–	–
Phlomido purpureae-Junipero turbinatae S. (basophilous)	–	–	●	–	–
Ulici argentei-Querco rotundifoliae S. (acidophilous)	–	–	●	–	–
Vinco difformis-Ceratonio siliquae S. (basophilous)	–	–	–	–	●
Osyrio quadripartitae-Junipero turbinatae S. (acidophilous)	●	●	–	–	–
Querco cocciferae-Junipero turbinatae S. (basophilous)	●	●	●	–	–
Climato-temporihygrophilous and xerophilous sigmeta					
Aro neglecti-Querco suberis S. (psammophilous)	●	●	●	●	●
Climato-temporihygrophilous sigmeta					
Oenantho crocatae-Querco pyrenaicae S.	●	–	–	–	–
Campanulo primulifoliae-Rhododendro pontici S. (acidophilous)	–	–	●	–	–
Euphorbio monchiquensis-Querco canariensis S. (acidophilous)	–	–	●	–	–
Rusco hypophylli-Querco canariensis S. (basophilous)	–	–	–	–	●
Vinco difformis-Ulmo minoris S. (acidophilous)	●	●	–	–	–
Ulici welwitschiani-Querco broteroi S. (basophilous)	●	–	●	–	–
Hygrophilous geosigmeta					
Holoschoeno vulgaris-Salici arenariae Gs. (psammophilous, relic)	●	–	–	–	–
Carici lusitanicae-Salici atrocinereae Gs. (soft freshwater)	–	●	–	–	–
Campanulo primulifoliae-Alno glutinosae Gs. (soft freshwater)	–	–	●	–	–
Viti sylvestris-Salici atrocinereae Gs. (dystrophic lentic freshwater)	–	–	–	●	–
Crataego brevispinae-Populo albae Gs. (hard freshwater)	–	–	–	–	●
Arisaro proboscidei-Alno glutinosae Gs. (soft freshwater)	–	–	–	–	●

(continued)

Lusitania and West Andalusia Coastal province	48	49	50	51	52
Rhododendro pontici-Alno glutinosae Gs. (soft freshwater)	–	–	–	–	●
Clematido campaniflorae-Salici neotrichae Gs. (soft freshwater)	●	●	–	–	–
Salici atrocinereo-australis Gs. (soft freshwater)	–	●	●	–	–
Irido foetidissimae-Fraxino angustifoliae Gs. (soft freshwater)	●	●	●	–	–
Rubio longifoliae-Coremo albi Gs. (psammophilous)	●	●	●	–	–
Equiseto telmateiae-Salici pedicellatae Gs. (hard freshwater)	–	–	–	●	●
Osyrio quadripartitae-Junipero turbinatae Gs. (coastal dune)	●	●	●	●	●
Geopermasigmeta					
Limonio emarginati Gps. (haloanemogenous rock littoral)	–	–	–	–	●
Puccinellio ibericae-Sarcocornio perennis Gps. (halophilous tidal)	●	●	●	●	●
Loto cretici-Ammophilo australis Gps. (coastal dune)	●	●	●	●	●

48. Divisorio Portuguese Sector, 49. Ribatejo and Sado Sector, 50. Algarve and Monchique Sector, 51. Cádiz and Littoral Huelva Sector, 52. Aljibe Sector

References

Alcaraz F (1984) Flora y vegetación del NE de Murcia. Publ. Univ. Murcia, Murcia, p 406

Alcaraz F (1996) Fitosociología integrada, paisaje y biogeografía. In: Loidi J (ed) Avances en Fitosociología. Universidad del País Vasco, Bilbao, pp 59–94

Alcaraz F, Díaz González TE, Rivas-Martínez S, Sánchez-Gómez P (1989) Datos sobre la vegetación del sureste de España: Provincia Biogeográfica Murciano-Almeriense. Itinera Geobotanica 2:1–133

Alcaraz F, Sánchez-Gómez P, de la Torre A (1991) Biogeografía de la provincia Murciano-Almeriense. Rivasgodaya 6:77–100

Asensi A, Díez Garretas B (1987) Andalucía occidental. In: Peinado M, Rivas-Martínez S (eds) La vegetación de España. Ser. Publ. Univ. Alcalá de Henares, Madrid, pp 197–230

Berastegi A, Darquistade A, García-Mijangos I (1997) Biogeografía de la España centro-septentrional. Itinera Geobotanica 10:149–182

Bolòs O 1967 Comunidades vegetales de las comarcas próximas al litoral situadas entre los ríos Llobregat y Segura. Mem R Acad Ci Artes Barcelona 38 (1). Barcelona.

Bolòs O (1987) Cataluña y la Depresión del Ebro. In: Peinado M, Rivas-Martínez S (eds) La vegetación de España. Ser. Publ. Univ. Alcalá de Henares, Madrid, pp 309–347

Braun-Blanquet J, Bolòs O (1958) Les groupements végétaux du bassin moyen de l'Ebre et leur dynamisme. Anales Est Exp Aula Dei 1–4:1–266

Cantó P (2007) Vegetation series as a tool for Biogeography: a case study of the central Iberian Peninsula. Phytocoenologia 37(3–4):417–442

Costa M (1987) El País Valenciano. In: Peinado M, Rivas-Martínez S (eds) La vegetación de España. Ser. Publ. Univ. Alcalá de Henares, Madrid, pp 381–307

Costa M (1997) Biogeografía. In: Izco J et al (eds) Botánica. Mc Graw Hill-Interamericana, Madrid, pp 683–742

Díaz González TE, Fernández Prieto JA (1988) Caracterización de las unidades fitogeográficas de Asturias. Monograf Inst Pirenaico de Ecología Jaca 4:517–528

Díaz González TE, Fernández Prieto JA (1994) El Paisaje Vegetal de Asturias. Itinera Geobot 8:5–242

Esteve F (1973) Vegetación y flora de las regiones central y meridional de la provincia de Murcia. Publ CEBAS, Murcia

Folch R (1986) La vegetació dels Països Catalans, 2ª edn. Ketres, Barcelona
Izco J (1987) Galicia. In: Peinado M, Rivas-Martínez S (eds) La vegetación de España. Ser. Publ. Univ. Alcalá de Henares, Madrid, pp 385–418
Ladero M (1987) La España Luso-Extremadurense. In: Peinado M, Rivas-Martínez S (eds) La Vegetación de España. Ser. Publ. Univ. Alcalá de Henares, Madrid, pp 455–486
Loidi J, Báscones JC (1995) Memoria del mapa de series de vegetación de Navarra 1:200.000. Publ. Gobierno de Navarra, Pamplona
Loidi J, Biurrun I, Herrera M (1997) La vegetación del centro-septentrional de España. Itinera Geobotanica 9:161–618
Martínez-Parras JM, Peinado M (1987) Andalucía Oriental. In: Peinado M, Rivas-Martínez S (eds) La vegetación de España. Ser. Publ. Univ. Alcalá de Henares, Madrid, pp 231–255
Molero Mesa J, Pérez-Raya F (1987) La Flora de Sierra Nevada. Avance sobre el catálogo florístico nevadense. Secr. Publ. Univ. Granada, Granada, p 397
Molina A, Loidi J, Fernández-González F (1993) Sobre las comunidades de matorral de la Depresión del Ebro España. Botanica Complutensis 18:11–50
Navarro Andrés F, Valle Gutiérrez C (1987) Castilla y León. In: Peinado M, Rivas-Martínez S (eds) La Vegetación de España. Serv. Ser. Publ. Univ. Alcalá de Henares, Madrid, pp 117–162
Navarro Sánchez G (1989) Contribución al conocimiento de la vegetación del Moncayo. Opuscula Bot Pharm Compl 5:5–64
Peinado M, Martínez Parras JM (1987) Castilla-La Mancha. In: Peinado M, Rivas-Martínez S (eds) La vegetación de España. Ser. Publ. Univ. Alcalá de Henares, Madrid, pp 163–196
Peinado M, Alcaraz F, Martínez-Parras JM (1992) Vegetation of Southeastern Spain. Flora et Vegetatio Mundi, X. J. Cramer, Berlin-Stuttgart, pp 487
Pérez Raya F, López Nieto JM, Molero Mesa J, Valle F (1990) Vegetación de Sierra Nevada Granada. Ayto. de Granada, Granada
Pons A, Quézel P (1985) The history of the flora and vegetation and past and present human disturbance in the Mediterranean region. In: Gómez Campo C (ed) Plant conservation in the Mediterranean area. Junk, Dordrecht, pp 25–43
Quézel P (1985) Definition of the Mediterranean region and origin of its flora. In: Gómez Campo C (ed) Plant conservation in the Mediterranean area. Junk, Dordrecht, pp 9–24
Rigual A (1972) Flora y vegetación de la provincia de Alicante. Inst. Est. Alicantinos, Alicante
Rivas Goday S (1964) Vegetación y flórula de la cuenca extremeña del Guadiana (Vegetación y flórula de la Provincia de Badajoz). Publ. Exma Dip. Pov. de Badajoz, Madrid, p 777
Rivas Goday S, Borja J (1961) Estudio de vegetación y flórula del macizo de Gúdar y Javalambre. Anales Inst. Bot. Cavanilles 19:3–550
Rivas Goday S, Rivas-Martínez S (1968) Matorrales y tomillares de la Península Ibérica comprendidos en la clase *Ononido-Rosmarinetea* Br.-Bl. 1947. Anales Inst Bot Cavanilles 25:5–201
Rivas-Martínez S (1963) Estudio de la vegetación y flora de las sierras de Guadarrama y Gredos. Anales Inst Bot Cavanilles 21(1):1–325
Rivas-Martínez S (1969) La vegetación de la alta montaña española. 1er. Simposio de Flora Europea 53–80. Sevilla
Rivas-Martínez S (1973) Avance sobre una síntesis corológica de la Península Ibérica. Baleares y Canarias Anales Inst Bot Cavanilles 30:68–87
Rivas-Martínez S (1981) Sobre la vegetación de la Serra da Estrêla (Portugal). Anales Real Acad Farmacia 47:435–480
Rivas-Martínez S (1985) Biogeografía y vegetación. Publ. Real Acad. Cienc. Exactas, Físicas y Nat, Madrid, p 103
Rivas-Martínez S (1987) Memoria y mapas de las series de vegetación de España (escala 1:400000). ICONA, Monografías, Madrid
Rivas-Martínez S (1988) Bioclimatología, Biogeografía y Series de Vegetación de Andalucía occidental. Lagascalia 15(extra):91–119
Rivas-Martínez S (2005) Avances en Geobotánica: discurso de Apertura del Curso Académico de la Real Academia Nacional del año 2005. Instituto de España. Real Academia Nacional de Farmacia, Madrid, p 142

Rivas-Martínez S, Loidi J (1999) Biogeography of the Iberian Peninsula. Itinera Geobotanica 13:49–68

Rivas-Martínez S, Pizarro J (1988) Datos sobre la vegetación y biogeografía de los Picos de Europa. Acta Bot Malacitana 13:201–208

Rivas-Martínez S, Díaz González TE, Fernández Prieto JA, Loidi J, Penas A (1984) La vegetación de la alta montaña cantábrica: Los Picos de Europa. Leonesas, León

Rivas-Martínez S, Fernández González F, Sánchez-Mata D (1987) El Sistema Central: de la Sierra de Ayllón a Serra da Estrela. In: Peinado M, Rivas-Martínez S (eds) La vegetación de España. Ser. Publ. Univ. Alcalá de Henares, Madrid, pp 419–451

Rivas-Martínez S, Fernández González F, Sánchez-Mata D (1990a) Endemic taxa of the Iberian Central System: distribution and ecology. In: Hernández Bermejo JE, Clemente M, Heywood VS (eds) Actas de la Conferencia Internacional Techniques for the conservation of threatened plant species in botanic gardens in the Mediterranean area (Córdoba, mayo 1987):179–184. Ed. Koeltz Scientific Books, Koenigstein, pp 205

Rivas-Martínez S, Lousã M, Díaz González TE, Fernández-González F, Costa JC (1990b) La vegetación del sur de Portugal (Sado, Alentejo y Algarve). Itinera Geobotanica 3:5–126

Rivas-Martínez S, Cantó P, Fernández-González F, Navarro C, Pizarro JM, Sánchez-Mata D (1990c) Biogeografía de la Península Ibérica, Islas Baleares y Canarias, vol 2. Publ. Dept. Biologia Vegetal Univ. Complutense de Madrid, Madrid, pp 1–5

Rivas-Martínez S, Asensi A, Molero Mesa J, Valle F (1991a) Endemismos vasculares de Andalucía. Rivasgodaya 6:5–76

Rivas-Martínez S, Báscones JC, Díaz González TE, Fernández-González F, Loidi J (1991b) Vegetación del Pirineo occidental y Navarra. Itinera Geobotanica 5:5–456

Rivas-Martínez S, Asensi A, Díez Garretas B, Molero Mesa J, Valle F (1997) Biogeographical synthesis of Andalusia southern Spain. J Biogeogr 24:915–928

Rivas-Martínez & coauthors Andalucía (Alfredo Asensi, Blanca Díez-Garretas, Joaquín Molero, Francisco Valle & Eusebio Cano) Aragón (Manuel Costa & María Luisa López), Asturias (Tomás E. Díaz & José Antonio F. Prieto), Baleares (Leonardo Llorens), Canarias (Marcelino del Arco, W. Wildpret, P.L. Pérez de Paz, O. Rodríguez, J.R. Acebes, A. García, V. E. Martín, J. A. Reyes Betancort, M. Salas, M.A. Díaz, J.A. Bermejo, R. González, M.V. Cabrera y S. García), Cantabria (José Antonio F. Prieto & Tomás E. Díaz), Castilla-La Mancha (Federico Fernández & Daniel Sánchez-Mata), Castilla y León (Ángel Penas), Cataluña (Ramón Masalles & Manuel Costa), Ceuta y Melilla (Alfredo Asensi & Blanca Díez-Garretas), Extremadura (Miguel Ladero & Ángel Amor), Galicia (Jesús Izco & Javier Amigo), La Rioja (Javier Loidi, José Antonio Molina & Gonzalo Navarro), Madrid (Paloma Cantó), Murcia (Francisco Alcaraz), Navarra (Javier Loidi & Juan Carlos Báscones), País Valenciano (Manuel Costa & Pilar Soriano), País Vasco (Javier Loidi) (2007) Mapa de series, geoseries y geopermaseries de vegetación de España (Memoria del mapa de vegetación potencial de España,). Parte I. Itinera Geobotanica 17: 5–436

Rivas-Martínez & coauthors Andalucía (Alfredo Asensi, Blanca Díez-Garretas, Joaquín Molero, Francisco Valle & Eusebio Cano) Aragón (Manuel Costa & Luis Villar), Asturias (Tomás E. Díaz & José Antonio F. Prieto), Baleares (Leonardo Llorens), Canarias (Marcelino del Arco, W. Wildpret, P.L. Pérez de Paz, O. Rodríguez, J.R. Acebes, A. García, V. E. Martín, J. A. Reyes Betancort, M. Salas, M.A. Díaz, J.A. Bermejo, R. González, M.V. Cabrera y S. García), Cantabria (José Antonio F. Prieto & Tomás E. Díaz), Castilla-La Mancha (Federico Fernández & Daniel Sánchez-Mata), Castilla y León (Ángel Penas, Luis Herrero & Sara del Río), Cataluña (Ramón Masalles & Manuel Costa), Ceuta y Melilla (Alfredo Asensi & Blanca Díez-Garretas), Extremadura (Miguel Ladero & Ángel Amor), Galicia (Jesús Izco & Javier Amigo), La Rioja (Javier Loidi & Gonzalo Navarro), Madrid (Paloma Cantó), Murcia (Francisco Alcaraz), Navarra (Javier Loidi & Juan Carlos Báscones), País Valenciano (Manuel Costa & Pilar Soriano), País Vasco (Javier Loidi) (2011a) Mapa de series, geoseries y geopermaseries de vegetación de España (Memoria del mapa de vegetación potencial de España). Parte II. Itinera Geobotanica 18 (1): 5–424

Rivas-Martínez S, Penas A, Rivas Sáenz S (2011b) Worldwide bioclimatic clasification systems. Glob Geobot. 1 (1):1–634 + 4 maps

Rivas-Martínez S, Navarro G, Penas A, Costa M (2011c) Biogeographic map of South America. A preliminary survey. Int J Geobot Res. 1 (1): 21–40+ Map

Rivas-Martínez S, Penas A, Díaz González TE, del Río S, Cantó P, Herrero L, Pinto Gomes C, Costa JC (2014) Biogeography of Spain and Portugal. Preliminary typological synopsis. Int J Geobot Res 4(1):1–64

Takhtajan A (1986) Floristic regions in the World. Univ. California Press

Villa L, Sesé JA, Ferrández JV (1997) Atlas de la flora del Pirineo Aragonés I. Cons. Protecc. Nat. Aragón, Inst. Est. Altoaragoneses, Huesca

Part II
Description of the Vegetation of the Territories

Chapter 6
The Lowlands and Midlands of Northwestern Atlantic Iberia

Javier Amigo, Manuel Antonio Rodríguez-Guitián, João Jose Pradinho Honrado, and Paulo Alves

Abstract The Iberian Cantabrian Atlantic biogeographical territory is a narrow strip of land, fallen L-shaped, which runs parallel to the coast from Pamplona (Spain) to near Aveiro (north Portugal), characterized by a wet and warm climate, with smooth winters and slight or absent drought in summer. Its relief is very varied and includes from coastal and inland plains to mountainous terrains with altitudes up to 1700 m. *Homo sapiens* has been present in this territory since the Upper Pleistocene but its influence on vegetation cover seems to have been very low until the Climate Optimum of the Holocene. Since this period human activities increased progressively and led to a wide deforestation of the territory, the expansion of non-arboreal seral communities and, during the last century, the introduction of a great variety of alien species for timber production and ornamental use. In spite of this great influence of man on the vegetal cover of the territory, some facts reveal its similarity to the rest of Atlantic Europe: (1) Supremacy of deciduous forests dominated by pedunculate oak and beech which are replaced by ash, maple, elm and linden trees in mixed forests, or alder, birch and willow in alluvial forests. (2) Replacing forests, seral scrub of thorny bushes or broom scrub (*Cytisus sp. pl.*) occur, with further degradation giving way to heathlands dominated by ericoid species but also with gorses (*Ulex sp. pl.*) and other thorny leguminosae (*Genista sp. pl.*). (3) Several types of meadows and other grasslands play a relevant role in the landscape and in traditional agricultural systems. (4) Different vegetation complexes typical of sandy deposits, rocky coasts and saltmarshes merge along its extensive shoreline. Nevertheless, there is a particular fact that differentiates these Atlantic territories from others located further north: the existence of many examples of evergreen vegetation, such as forests of holm oak (*Quercus ilex, Q. rotundifolia*), cork oak (*Quercus suber*) and laurel tree (*Laurus nobilis*), or scrub dominated by *Arbutus unedo* or *Phillyrea sp. pl.*, due to their proximity to the Mediterranean Region.

J. Amigo (✉) • M.A. Rodríguez-Guitián
Universidade de Santiago de Compostela, Santiago, Spain
e-mail: javier.amigo.vazquez@usc.es; manuelantonio.rodriguez@usc.es

J.J. Pradinho Honrado • P. Alves
Universidade de Porto, Porto, Portugal

J. Loidi (ed.), *The Vegetation of the Iberian Peninsula*, Plant and Vegetation 12,
DOI 10.1007/978-3-319-54784-8_6

6.1 Introduction

The territory described in this chapter is a land strip with a varying width, running parallel to the north and northwest coast of the Iberian Peninsula. Overall, it covers about 55,000 km^2, including the following administrative regions and provinces, from east to west (Fig. 6.1)

- Navarre, approximately its northwest corner until its capital Pamplona;
- Basque Country: the whole provinces of Guipúzcoa and Vizcaya, and roughly half the province of Álava;
- Cantabria and Asturias: the northern areas (60–70%) of these two regions, both holding only one province; the southern areas correspond to the mountains of the Cantabrian Range, which are included in a distinct biogeographic unit (see Chap. 7);
- Galicia, with roughly 26,000 km^2, is the region contributing the largest area, including two full provinces (A Coruña and Pontevedra) and a large part of the provinces of Lugo and Ourense; the highest mountains of these two provinces, at the borders with Asturias and Castile, are excluded and are also treated in Chap. 7;
- Portugal is represented by its northwest corner, including the whole provinces of Minho and Douro Litoral, a small neighbouring part of Trás-os-Montes, and part of Beira Litoral down to Serra do Buçaco, which is considered the southern limit of the European-Atlantic biogeographic province.

This land extension is shaped like a capital "L", rotated clockwise by 90°, with the shortest arm wider than the longest one. However, the linear distance to the Atlantic shoreline is hardly larger than 120 km.

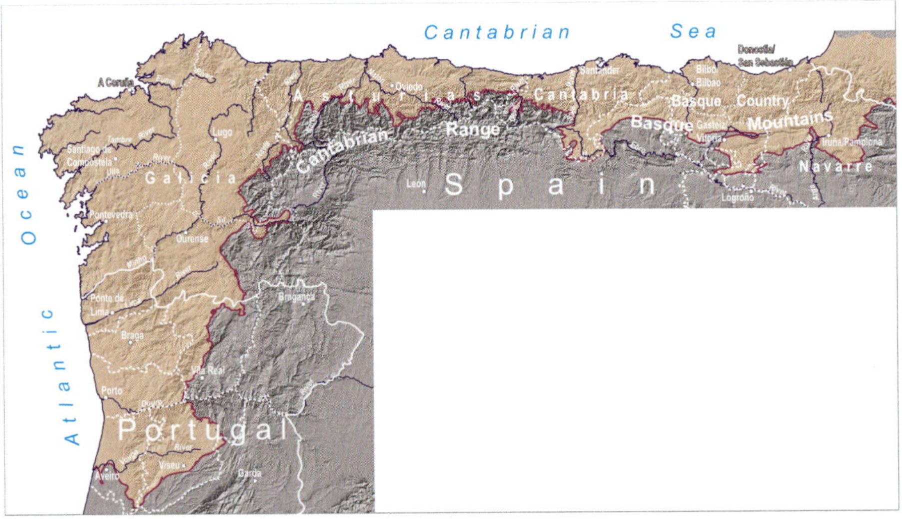

Fig. 6.1 Delimitation of administrative territories included in this chapter

6.1.1 Geomorphology and Soils

The geological substrates that characterize this territory include rocks of a rather variable age (>1100–2 million years). From a lithological perspective, this land mass is characterised by: the predominance of carbonate-rich rocks (limestones, dolomites, marls), alternating with siliceous metamorphic substrates (shales, slates, sandstones, quartzites) and some granite outcrops of various ages, in the centre-eastern half (centre-eastern Asturias, Cantabria, Basque Country and northern Navarre); the massive presence of palaeozoic metamorphic rocks (phyllites, slates, sandstones, quartzites, gneisses) in western Asturias and eastern Galicia; and the dominance of schists, gneisses and granitoids (leucogranites, diorites, syenites), along with several outcrops of alkaline (amphibolites, gabbros, eclogites) and ultramaphic (peridotites, granulites) rocks, in some cases serpentinized rocks, in western Galicia and northern Portugal (Vera 2004).

The topography is closely linked to this distribution of bedrock types, resulting from the collision of the Iberian and European tectonic plates during the Alpine orogeny. This process has deformed and raised palaeozoic substrates, already deformed by the Variscan orogeny and multiple erosive cycles. Later, several distensive phases have created numerous depressions within the mountainous ranges. Those depressions were filled with sediments during the second half of the Mesozoic and the Cenozoic. Examples of these are located south of the Basque-Cantabrian mountains (Llanada de Álava), the endorheic basin of Oviedo, and the depressions of inland Galicia (Terra Chá, Terra de Lemos, Val de Maceda). This territory has been affected by nival and periglacial phenomena, also in the coastal lowlands, during the cold periods of the Pleistocene. Glaciers were even formed in the main mountain ranges. During interglacial periods sea level rise allowed the formation of large marine deposits (*rasas*), currently located at a certain elevation above sea level.

Overall, the terrain morphology is more complex in the eastern half of the territory, due to the highest intensity of crustal deformations and to the largest variety of bedrock types with different resistance to erosive processes. In northern Navarre and eastern Basque Country there is the contact between the Pyrenees and the Basque-Cantabrian mountain range. The latter are oriented east-west and are considered the connection between the Pyrenees and the Cantabrian mountain range, with summits of over 1500 m (Ortzainzurieta 1570 m, Castro Valnera 1696 m). The Cantabrian Range dominates western Cantabria and Asturias, with many summits over 2000 meters. From the Cantabrian Range, several minor mountain ranges extend towards the north and into the Cantabrian Atlantic territories, often with summits above 1000 meters. The morphology of this sector includes sub-coastal hills and mountains, and coastal flatlands alternating with capes of resistant substrates (quartzites, marbles). When it reaches Galicia, the Cantabrian Range splits into two ridges. The northern one develops towards NNW, reaching the northern Galician mountains. The southern ridge, already outside Cantabrian Atlantic territories, surrounds the western part of the El Bierzo

Fig. 6.2 Lower areas close to sea have supported for centuries high population levels not compatible with a well conserved vegetation cover. North side of Ría de Arousa from A Curota (A Coruña, Galicia)

depression. The mountains of northern Galicia hold summits between 600–1000 m, close to the coast, and from their central sector a succession of north-south oriented mountains, known as the *Dorsal Gallega*, develops down to the valley of the river Miño. Southwards, at the border with Portugal, several mountains occur, from the Xurés/Gerês range (1538 m) to the Peneda, Alvão and Marão mountains, also reaching 1300–1400 m elevation. The Atlantic coast of Galicia is characterised by several "*rías*", separated by hill ranges (500–600 m) extending into the sea through a series of prominent capes and small archipelagos (Sisargas, Lobeiras, Sálvora, Ons, Cíes) (Fig. 6.2).

The boundaries of the main fluvial basins follow the divides of the Basque Mountains, the Cantabrian Mountains and the Galician-Portuguese Mountains. Between the Bidasoa River, ending at the border between Spain and France, and the Sor River, ending at the border between the provinces of A Coruña and Lugo, the rivers run to the Gulf of Biscay over short distances (the longest one, the Nalón River, is the only one over 100 km long). Towards the west, the main rivers draining into the Atlantic Ocean (Tambre, Ulla, Limia/Lima) are more than 100 km long, with the Miño/Minho reaching 350 km and the Duero/Douro 897 km. The Cantabrian Atlantic territories draining towards the Mediterranean comprise only small areas in the heads of the rivers Ebro and Zadorra.

The main processes of soil formation under the wet temperate climate that prevails in most of the territory are leaching of base cations, due to abundant rainfall, and organic enrichment on the upper layers resulting from the abundant biomass decaying from vegetation every year. For these reasons, the soils are usually acidic, especially when soils are formed from siliceous metamorphic materials, granitic rocks, or sediments derived from these. Often, soils formed on these materials in the highest mountains show a limited development and signs of podsolization, especially if developed on quartzites, sandstones, or leucogranites. In these areas, peatlands are frequent whenever topographic and moisture conditions are favourable. Nonetheless, in the less evolved soils (leptosols, umbrisols) on bedrocks providing abundant Ca or Mg, such as limestone, carbonated clays or serpentine rocks, pH values in the upper layers are kept close to neutral, even at locations with an annual rainfall above 1500–1700 mm. The soils of flat areas often present features derived from seasonal flooding (gleysols), especially when developed on fine textured sediments (vertisols).

6.1.2 Climatic Conditions

Due to the geographic rationale described above, climatic conditions throughout the area can be described based on two words: atlanticity and oceanity. This translates into climates with abundant rainfall, even in summer, due to the dominance of northern winds, and a narrow yearly temperature range. According to the bioclimatic classification of Rivas-Martínez, most of the Cantabrian Atlantic territory is included in the thermotemperate, mesotemperate and supratemperate climatic belts.

The mesotemperate is clearly the predominant belt, whereas the two other temperate belts are unequally distributed towards east and west. In the Cantabrian and Basque and North Galicia and Asturias sectors, the supratemperate is the second most frequent belt, with the thermotemperate limited to a very narrow strip along the coast and to isolated patches towards the east. Inversely, in the Galician and North Portugal sector, the second most widespread belt is the thermotemperate, whereas the supratemperate is limited to the scattered mountain ranges. The supratemperate belt again becomes remarkable in the North Lusitania Sierran sector.

In terms of ombroclimate, the most widespread regime is the humid belt, followed by the hyperhumid. Again, hyperhumid areas are more common towards the east. Nonetheless, in all four sectors there are records of mountain areas in the hyperhumid and even ultrahyperhumid belts. But there are also areas with lower rainfall values, in the subhumid belt, in the southern part of the Cantabrian and Basque sector and in inland stretches of wide fluvial valleys of the Galician and North Portugal sector.

The position of a bioclimatic border with the Mediterranean climate that characterizes most of this Atlantic territory, explains the fact that different parts of it hold some deficit in terms of summer rainfall. This shortage may last for two or even three consecutive months in summer, but in most cases it is compensated by spring rainfall and so the level of drought is not as high as the one recorded under the Mediterranean climates. This climatic peculiarity is called the Submediterranean variant of the Temperate climate, and it is quite widespread throughout the North Lusitania Sierran and Galician and North Portugal sectors, gradually disappearing towards the northern and eastern sectors. This variant also occurs in the southernmost subsector (Navarre and Álava área) of the Cantabrian and Basque sector. There are small portions of the territory that do hold a Mediterranean bioclimate (specifically in the mesomediterranean belt), in the Sil valley of inland Galicia and in the southern lowlands of the Galician and North Portugal sector. There are several publications with detailed maps of the thermo- and ombroclimatic belts for most of the territory (Loidi and Báscones 2006; Rodríguez-Guitián and Ramil-Rego 2007; Loidi et al. 2011; Monteiro-Henriques et al. 2015).

The oceanic character is another climatic peculiarity of all this territory, increasing towards areas that are located closer to the sea and at lower elevations. In fact, there are very few climatic stations along the coast between A Coruña and Guipúzcoa holding continentality indices above 10 units (measured as annual temperature ranges, according to the models of Rivas-Martínez), which are the lowest values across all of the Iberian Peninsula.

6.1.3 Biogeography

All the area is included in the Atlantic European province and, within it, in the Cantabrian Atlantic and Atlantic Orolusitania subprovinces. This is the southwest limit of the Eurosiberian region. This territory, stretching towards the north through French Aquitaine, is limited in the east, in its Spanish part, by another large biogeographic unit, the Cévennean-Pyrenean province, which begins in northeast Navarre. Towards the north and west, the limit is defined by the Atlantic Ocean throughout 2560 km of coastline. The southern limits are defined by its contact with the Mediterranean region in the eastern part (Navarre, Basque Country and Cantabria) and with the Orocantabric subprovince and the Galician-Leonese mountains (Chap. 7), from western Cantabria to eastern Galicia, from where the limit is again defined by the Mediterranean region, through northern Portugal down to the southwestern end of the province.

Together, the geomorphologic, climatic and biogeographic features define a favourable territory for hosting a flora with evident similarities to that of the remaining Atlantic and Central-European areas. The floristic peculiarities that discriminate these Iberian Cantabrian Atlantic lowlands from those territories north of the Pyrenees can be summarized as follows:

- It has a remarkable component of the Iberian endemic flora, especially in two environments: the coastal habitats with taxa that became specialized to

ecologically demanding biotopes, such as cliffs and dunes; and the supratemperate areas that provide habitats for the Pyrenean-Cantabrian flora.
- The proximity to the Mediterranean region has allowed the occurrence of numerous taxa with that origin among the Cantabrian Atlantic flora, taking advantage of favourable biotopes.

The biogeographic subdivision that will be used here is based on the proposals by Rivas-Martínez in their most recent version (Rivas-Martínez et al. 2014). Accordingly, we divide the territory into four sectors, from east to west: Cantabrian and Basque, North Galicia and Asturias, Galician and North Portugal and North Lusitania Sierran. In turn, for greater precision in the distribution of plant communities and their vegetation series, we divide these sectors into subunits, as already used by Rivas-Martínez in previous classifications (Rivas-Martínez 2007) with the rank of sub-sector; these subunits in the most recent proposal are classified in the category of "territories".

- The Cantabrian and Basque sector, which includes the eastern one-third of the total area and includes three subunits: the East Basque territory, including the western peaks of the Pyrenees down to the coast and most of Guipúzcoa from the coast up to the Basque Mountains; the Santanderian-Biscayan territory, corresponding to the lowland and coastal areas of Vizcaya and Santander; and the Navarran and Alavese territory, south of the two previous ones and less oceanic since it does not reach the Cantabrian coast.
- The North Galicia and Asturias sector, subdivided in: the Central-East Oviedo territory, covering the eastern half of Asturias (starting in the catchment of the river Narcea) and western Cantabria; and the North Galician-Asturian territory, which includes northern Galicia and the northern Atlantic catchments down to the river Eume.
- The Galician and North Portugal sector, including three subunits: the North Galician-Portuguese territory, covering most of Galicia draining to the Atlantic ocean, the South Galician-Portuguese territory, from the valley of the river Miño/Minho, at the border between Portugal and Spain, south to the vicinity of Aveiro, at the southwestern end of the Eurosiberian region; the Inner Galician territory, partially under mediterranean bioclimate, which includes two stretches of river valleys particularly interesting for their special thermal conditions: the Sil river which flows east-west until its junction with the Miño and supports mostly a Mediterranean-type bioclimate, and the Navia river representing a refuge for thermo-xericity despite emptying into the Cantabric Sea.
- The North Lusitania Sierran includes the mountains with the highest altitudes of northwest Portugal and some bordering with Galicia: Peneda, Gerês-Xurés, Larouco, Cabreira, Alvão and Marão. All these *serras*, previously considered within the Galician and North Portugal sector, form the Eurosiberian/Mediterranean chorologic border.

6.2 Human Influence

The European landscape has been conditioned since antiquity by the biological and cultural development of men. While during the Paleolithic those changes were mainly located along the Cantabrian Atlantic coast, with the use of fire and the predation on certain species. In the Holocene various agricultural and pastoral techniques were adopted. The footprint on the vegetation and landscape became increasingly clear: new species previously domesticated in Fertile Crescent were introduced and spread, while the destructive action on pristine ecosystems increased by the use of fire, land ploughing and extensive use of livestock. (Fig. 6.3)

Possibly these practices favoured the expansion of nitrophilous communities associated with roads in shady environments (alliances Aegopodion podagrariae, Galio aparines-Alliarion petiolatae, Balloto foetidae-Conion maculati, Bromo ramosi-Eupatorion cannabini from the class Galio aparines-Urticetea maioris) and cleared areas subject to regular trampling of human beings and livestock (alliances Arction lappae, Onopordion acanthii, Cirsion richteriano-chodati, class Artemisietea vulgaris). At the same time there was the expansion of nitrophilous shady communities associated with forest clearings caused by logging and wildfires

Fig. 6.3 Deforestation of more fertile lands is a consequence of human activities since Neolithic times. The "Alsasua corridor" from Puerto de Pagoeta (Álava). The arboreal vegetation in the foreground is a basophilous beech woodland (Carici sylvaticae-Fagetum sylvaticae). Down slope it borders on mesophytic Navarran-Alavese pedunculate oak woodlands (Crataego laevigatae-Quercetum roboris). In the background, extensive crop fields occupy lands covered by these latter oak woodlands many centuries ago

(alliances Atropion belladonae and Carici piluliferae-Epilobion angustifolii, class Epilobietea angustifolii).

Approximately 5000 years ago, semi-nomadic groups of humans began to organize in villages initiating the process of ruralizing the territory. There was an increase of land devoted to cultivation and livestock grazing at the expense of the area occupied by the original vegetation. There was a loss of woodlands, especially in territories near the coast, resulting in an open landscape where cultivated and managed areas were more common. This was the landscape that the Romans found upon arrival in these territories and that stayed during their rule.

In this period there was an expansion of plant communities adapted to certain recurring disturbances caused by human activities, such as those associated with rainfed cereal crops cultivation, road verges communities (alliances Scleranthion annui and Sisymbrion officinalis, class Stellarietea mediae) or nitrophilous communities growing in places of accumulation of organic litter (alliance Saginion procumbentis, class Polygono-Poetea annuae). The most characteristic species of this type of vegetation would be present in the different territories where human groups settled, but others quite possibly were introduced. It is very difficult to exactly establish now what the original source of this flora was.

The collapse of the Western Roman Empire (476 AD) marked the beginning of a period of strong political instability, in which these territories were overrun by various Germanic groups and then by the Muslims, and later by the Christian kingdoms. In some cases, those periods of instability led to the recovery of forest cover, but afterward, human actions increased in those territories and agricultural and pastoral activities had an important development at that time, generating cultural landscapes that persisted for centuries. At that historic moment we would have to place the establishment of numerous meadow plant communities (alliances Arrhenatherion elatioris and Cynosurion cristati of the class Molinio caeruleae-Arrhenatheretea elatioris) and the expansion of grasslands through extensive grazing of pastures in low (alliance Lolio perennis-Plantaginion majoris) and mountain areas (alliances Violion caninae and Campanulo herminii-Nardion strictae, class Nardetea strictae).

In the late fifteenth century, the Portuguese and Spanish began the exploration and colonization of the northern coasts of Africa and the East and West Indies. As a result of these actions, new species were progressively introduced in the Cantabrian Atlantic area as food, fiber, beverages or wood sources, as healing medicines or for ornamental purposes. The effects of this neophyte contingent are now evident either by replacing in many areas the floral elements coming from the Fertile Crescent, or affecting the many naturalized species of flora and fauna that have adapted successfully to the native natural and semi-natural habitats.

The course of action of human communities in nature was also modified with the development of the Industrial Revolution, by the mechanization of many activities previously done manually or with the help of animal power, with the emergence and use of agrochemicals such as fertilizers, and finally with the incorporation of

Fig. 6.4 Remains of xerophilous holm oak woodlands (Genisto hystricis-Quercetum rotundifoliae) and its mantle vegetation (Erico scopariae-Arbutetum unedonis) growing between vineyards on steep slope terraces in a mesomediterranean valley of Inner Galicia. Ribeira Sacra, Valle del Sil (Lugo)

genetically engineered organisms transformed by hybridization, artificial selection, transgenesis and cisgenesis.

As a result of the cultural changes mentioned, the current characteristics of the vegetation cover of the Cantabrian Atlantic territories can be summarized as:

- A significant reduction in the area covered by forests (Fig. 6.4) in the lowlands and less remote areas and a floristic and structural simplification of the woodland communities in the areas with a more intensive management. (Fig. 6.5)
- The regional extinction in some cases and a range reduction in others, of different tree species by anthropic causes.
- A remarkable expansion of the area occupied by shrub seral communities.
- The emergence and spread of segetal and ruderal communities related to agro-pastoral activities and road verges.
- The either conscious or casual introduction of a large set of plant invasive species. (Fig. 6.6)
- A substantial increase in the area occupied by forest plantations of exotic species of rapid growth and intensive management which causes serious ecological problems and modifications in ecosystems services (increased erosion, soil depletion, hydrological alterations, changes in composition and diversity in animal and plant species, etc.).

Fig. 6.5 Ancient management of woodlands to gather timber and fruits has led frequently to simplified (often monospecific) canopy woodlands, with a cleared up understory dominated by trivial species. Old chestnut (*Castanea sativa*) woodland at Torbeo (Lugo, Galicia)

6.3 Native Forests and Mantle Vegetation

In the following description of forests occurring in this biogeographic unit, we distinguish between climatophilous forests and edaphohygrophilous forests.

6.3.1 Climatophilous Forests

These forests grow on well drained soils and their floristic composition is mainly determined by climate (thermotypes, ombrotypes), topography (aspect and solar radiation) and soil nutrient content (nutrient-poor soils derived from granites or siliceous metamorphic substrates *vs*. more fertile soils derived from carbonated rocks or, less often, from basic and ultrabasic bedrocks).

The macroclimatic conditions prevailing in this biogeographical unit favour the predominant development of forests belonging to class Querco-Fagetea sylvaticae, with a canopy dominated by deciduous species of the genera *Quercus* and *Fagus*. In certain situations, other genera that include deciduous species (*Betula, Castanea, Acer, Fraxinus, Ulmus, Prunus, Tilia, Carpinus, Celtis*) or mesophytic evergreen

Fig. 6.6 Land uses in oceanic Cantabrian-Basque mountains. Gentle slopes near the villages are covered by alien conifer plantations (*Pinus radiata, Larix kaempferi*, etc.); far and less accessible terrains are still covered by deciduous woodlands, mainly dominated by beech (*Fagus sylvatica*). The tops of the mountains have grasslands grazed by sheep flocks in summer. North slope of the Sierra de Aralar (Guipúzcoa, Basque Country)

species (*Ilex aquifolium, Taxus baccata, Prunus lusitanica*) can reach considerable abundances in the canopy. Still, local factors such as a reduction in summer rainfall or the occurrence of soils with little development and water retention capacity will favour the presence of forests dominated by marcescent species (*Q. pyrenaica, Q. faginea, Q. pubescens*) or even by xerophytic evergreen species (*Q. ilex, Q. rotundifolia, Q. suber, Arbutus unedo*).

The elevation sequence of the main thermotypes (from the upper thermotemperate to the upper supratemperate) has a related diversity of forest types. The areas along the Cantabrian and Atlantic coasts (thermotemperate and lower mesotemperate thermotypes) host forests of a thermophilous nature, rich in evergreen species (*Arbutus unedo* and *Laurus nobilis* throughout the territory; *Quercus suber*, *Myrtus communis* or *Phillyrea angustifolia* toward its southwest end) as well as climbers (*Clematis vitalba, Hedera hibernica, H. helix, Lonicera periclymenum, Smilax aspera, Tamus communis*), that can penetrate through the valleys up to elevations of 400/500 m in coastal areas, or even 700/800 m in the inland flatlands and valleys. The native forest communities most frequent on siliceous soils of these bioclimatic belts are oak forests of *Quercus robur*, often with chestnut (*Castanea sativa*) and birch (*Betula pubescens* subsp. *celtiberica*) (alliance Quercion pyrenaicae). The associations with the widest occurrence in this territory are the Hyperico pulchri-Quercetum roboris in the Cantabrian and Basque

Fig. 6.7 Deciduous oak forests only cover wide surfaces in the least disturbed Cantabrian Atlantic landscapes. Acidophilous Cantabrian-Basque oak woodland (Hyperico pulchri-Quercetum roboris) at Iribas (Guipúzcoa, Basque Country)

sector, the Blechno spicant-Quercetum roboris in the North Galicia and Asturias sector, and the Rusco aculeati-Quercetum roboris in the thermo-mesotemperate, North Galician-Portuguese areas of Galicia and northern part of northwest Portugal (Izco et al. 1990a; Díaz-González and Fernández-Prieto 1994; Loidi et al. 1997a; Honrado 2003) (Fig. 6.7). In the South Galician-Portuguese territory, the Viburno tini-Quercetum roboris represents the transition between the Atlantic oak forests and the Mediterranean sclerophyllous forests (Rivas-Martínez 2011; Costa et al. 2012).

Within this same thermoclimatic belt, oak forests developed on fertile soils hold a higher canopy diversity, due to varying abundances of other meso-hygrophilous tree species, such as ash (*Fraxinus excelsior*, *Fraxinus angustifolia*), hazel (*Corylus avellana*), maple (*Acer pseudoplatanus*, *Acer campestre*), *Carpinus betulus* (exclusively at a few East Basquean locations), *Alnus glutinosa*, *Ulmus minor* or *Celtis australis*. In the eastern half of the territory these forests are included in two main associations: the Polysticho setiferi-Fraxinetum excelsioris in the Central-East Oviedo surroundings and in the Cantabrian-Basque areas draining to the Cantabrian Sea, and the Crataego laevigatae-Quercetum roboris in the Navarran-Alavese areas (Díaz-González and Fernández-Prieto 1994; Loidi et al. 1997a, 2011). In the western areas, a single association has been described so far, distributed across the hilly and mountainous areas of the South Galician-Portuguese territory (Hyperico androsaemi-Quercetum roboris, Honrado et al. 2002); however, forests with this type of ecology occur in several areas of northern and central Galicia

(Rodríguez-Guitián 2005). All these community types are included in the alliance Pulmonario longifoliae-Quercion roboris, vicarious of the centre-european Carpinion betuli.

Spread among these predominant forest types, others can be observed at locations with specific climatic or edapho-topographic conditions: holm oak forests of *Quercus ilex* on carbonated lithosols close to the coast and of *Q. rotundifolia* in the inland sheltered valleys (alliance Quercion ilicis, class Quercetea ilicis); beech forests on steep slopes facing north, at locations with a high cloudiness and summer rainfall, especially in the eastern half of the territory (Ilici aquifolii-Fagion sylvaticae); and oak forests of *Q. pyrenaica* on siliceous sandy soils (Melampyro pratensis-Quercetum pyrenaicae, Lonicero periclymeni-Quercetum pyrenaicae; alliance Quercion pyrenaicae), or when on calcareous soils, dominated by *Q. faginea* (Pulmonario longifoliae-Quercetum fagineae) or *Q. pubescens* (Roso arvensis-Quercetum pubescentis), included in the alliance Quercion pubescenti-petraeae, in areas with some summer drought at some distance from the coast (Díaz-González and Fernández-Prieto 1994; Loidi et al. 1997a, 2011; Díaz-González 2010a).

The Cantabrian Atlantic areas with the lowest rainfall values are located in the mid and lower segments of the river Sil, the main tributary of the river Miño, where total annual rainfall is not higher than 800 mm. Here, the steepest and sunniest slopes host holm oak woods (Genisto hystricis-Quercetum rotundifoliae) and cork oak forests (Physospermo cornubiensis-Quercetum suberis), which become more widespread towards the east, in the municipalities of Valdeorras and El Bierzo, with a pronounced rain shadow effect (Rivas-Martínez et al. 1984; Rivas-Martínez 1987; Romero-Buján 1993; Izco et al. 1990b). Cork oak forests are also present, although less frequently, in the valley of the river Navia (Asturias/Galicia) and in several localities of inland Cantabria and coastal Basque Country (Aedo et al. 1991; Aseginolaza et al. 1989; Díaz-González 2010a). Nonetheless, all these evergreen forests are under strong oceanic (atlantic) influence, as revealed by the presence of species such as *Quercus robur, Glandora prostrata, Daboecia cantabrica, Holcus mollis, Stellaria holostea, Teucrium scorodonia*, etc. (Figs. 6.8 and 6.9)

Throughout this territory with low elevations and close to the coast it is possible to find fragments of other types of forests, often in mosaic with the previous ones. Those forests develop at sites with soil conditions that are not adequate for deciduous oaks (*Quercus robur*, *Q. petraea*) or for beech (*Fagus sylvatica*): sheltered, shady slopes with rocky soils or abundant rock outcrops. These biotopes are occupied by a wide set of tree species that are infrequent in the previous forest types, such as limes (*Tilia platyphyllos, T. cordata*), wych elm (*Ulmus glabra*), maples (*Acer pseudoplatanus, A. platanoides*), wild cherry (*Prunus avium*) and ashes (*Fraxinus excelsior, F. angustifolia*). In recent years several types of these mixed forests occurring on slopes or cliffs in the northern Iberian Peninsula have been described (Rivas-Martínez 2002a; Costa et al. 2012; Biurrun et al. 2011), and included within the alliance Tilio platyphylli-Acerion pseudoplatani, revealing their highly variable floristic composition throughout the territory. The Hyperico androsaemi-Ulmetum glabrae occurs in Cantabrian-Basque siliceous areas (Biurrun

Fig. 6.8 Even though the Cantabrian Atlantic territories of Iberia are affected by a pre-eminent oceanic climate, many "islands" of woodlands dominated by different typical Mediterranean Fagaceae can be found. Marcescent basophilous *Quercus faginea* woodland (Pulmonario longifoliae-Quercetum fagineae) on limestone; south facing slope at the Sierra de Aralar (Navarra)

Fig. 6.9 Another example of Mediterranean woodland vegetation, the mesophytic cork-oak woodland (Physospermo cornubiensis-Quercetum suberis) at the river Sil valley in Valdeorras (Ourense, Galicia)

et al. 2011), whereas the Fraxino angustifoliae-Ulmetum glabrae is found in inland mountainous areas of northern Portugal included in the North Lusitania Sierran sector (Monteiro-Henriques et al. 2010). Other authors have published data on similar communities from several locations in Galicia (Amigo et al. 1994; Rodríguez-Guitián et al. 2000; Rodríguez-Guitián 2005).

The areas located in the upper mesotemperate and supratemperate thermotypes host a lower diversity of forest types due to the colder temperature regimes. Beech forests are remarkable for their ecological amplitude and vitality, occurring on siliceous (alliance Ilici aquifolii-Fagion sylvaticae) as well as limestone soils (alliance Fagion sylvaticae), and being dominant in mountain areas above 800/900 m in the eastern part of the territory. Toward the west, beech forests become increasingly less frequent in the Cantabrian Atlantic territories, although along the Orocantabrian mountain range they maintain their continued presence until its western end. The current western limit of these forests in the Atlantic fringe of Europe is represented by the beech woods in the upper catchment of the river Eo (Lugo, Galicia) (Rodríguez-Guitián et al. 2001). The diversity of beech forests has been described as follows:

- association Saxifrago hirsutae-Fagetum sylvaticae: Cantabrian and Basque sector on siliceous substrates (Loidi 1983; Loidi et al. 1997a)
- association Saxifrago spathularis-Fagetum sylvaticae: Central-East Oviedo and North Galician-Asturian territories on siliceous substrates (Rodríguez-Guitián et al. 2003; Rodríguez-Guitián 2005, 2006)
- association Carici sylvaticae-Fagetum sylvaticae: Cantabrian and Basque sector on limestone/carbonated substrates (Loidi et al. 1997a, 2011)
- association Carici caudatae-Fagetum sylvaticae: Central-East Oviedo territory on limestone/carbonated substrates (Rivas-Martínez 2011)
- association Epipactido helleborines-Fagetum sylvaticae: Navarran-Alavese and East Basque territories on limestone/carbonated substrates (Berastegi et al. 1997a; Loidi et al. 1997a, 2011). (Figs. 6.10 and 6.11)

As the importance of beech forests decreases in the landscapes of meso-supratemperate areas, that of oak forests increases, especially in the case of *Quercus robur*. Typical sub-associations of both the Blechno spicant-Quercetum roboris and Rusco aculeati-Quercetum roboris, and the association Vaccinio myrtilli-Quercetum roboris occupy these belts in siliceous areas of the North Galician-Asturian, North Galician-Portuguese and North Lusitania Sierran sectors, respectively (Izco et al. 1990a; Costa et al. 1998a). The mountains of inner Cantabria and Álava host forests of *Quercus petraea* on siliceous soils (association Pulmonario longifoliae-Quercetum petraeae, alliance Ilici aquifolii-Fagion sylvaticae) (Loidi et al. 1997a).

With a much wider distribution, short-sized forests of *Ilex aquifolium* (alliance Ilici aquifolii-Fagion sylvaticae) occur at the edges of beech and oak forests, in areas that are grazed by cattle and sheep (Aseginolaza et al. 1989; Aedo et al. 1991; Rodríguez-Guitián 2005). Also scattered throughout the mountains in the territory are the small woodlands of *Taxus baccata*, which hold a remarkable biogeographic

Fig. 6.10 Common beech woodlands are more frequent and extensive in the central-eastern part of the territory but still are present as small woods in NE Galicia. Calcicolous beech woodlands (Carici sylvaticae-Fagetum sylvaticae) on limestone (foreground) and silicicolous beech woodlands (Saxifrago spathularis-Fagetum sylvaticae) on slates and quartzites (background); Saja Valley (Cantabria)

significance (Fernández-Prieto and Díaz-González 2003; Díaz-González 2010b; Ihobe 2011; Rodríguez-Guitián et al. 2010).

With a very localized and fragmented distribution, in the highest mountains with edaphic conditions unsuitable for beech or oaks, forests can be dominated by birch (*Betula pubescens* subsp. *celtiberica*) and rowan (*Sorbus aucuparia*), related to frequent snowfall (Herrera 1995; Rodríguez-Guitián 2005). These forests resemble those that are common in the upper supratemperate areas of the Orocantabrian siliceous mountains, which are included in the alliance Ilici aquifolii-Fagion sylvaticae.

When the forests described above suffer regeneration processes, opening of clearings due to natural phenomena (storms, avalanches, landslides, etc.) or to human logging, woody formations dominated by fast-growing tree species can develop (pre-climax forests and edge woodlands). These formations maintain the nemoral conditions required by the flora typical of mature forests while such mature forests do not regenerate at those spots. Under certain conditions, similar formations can occur spontaneously, as permanent communities, at the edges of forest patches when these reach areas that are not suitable for their characteristic dominant

Fig. 6.11 Fern-rich understory in a mesotemperate acidophilous North Galician-Asturian beech woodland (Saxifrago spathularis-Fagetum sylvaticae); Belmonte (Asturias)

trees, such as soils with abundant rocks or low water retention capacity. They can also develop on former farmland that has suffered abandonment.

Similar to the mature forests they replace, these pre-climax forests and edge woodlands hold physiognomic and floristic features that differ depending on whether they occur on climatic or topo-edaphic conditions. In general, pre-climax forests on siliceous soils are dominated by heliophilous species with low nutrient demands (*Salix atrocinerea*, *Betula pubescens* subsp. *celtiberica*, *Sorbus aucuparia*) or by species with a high capacity to resprout (*Corylus avellana*, *Ilex aquifolium*). These communities are included in the Betulion fontqueri-celtibericae alliance, such as the Cantabrian-Basque association Rhamno frangulae-Betuletum celtibericae or the North Galician-Asturian and North Galician-Portuguese Holco mollis-Betuletum celtibericae (Amigo and Romero-Buján 1998; Berastegi et al. 1997b). Often the regeneration of Cantabrian Atlantic silicicolous forests over heathland and broom scrub includes a community dominated by *Frangula alnus* and *Pyrus cordata* (Frangulo alni-Pyretum cordatae association), included in the alliance Frangulo-Pyrion cordatae (class Rhamno cathartici-Prunetea spinosae) (Herrera et al. 1991). On limestone soils, the elimination of the dominant trees (ash, oak, maple, beech, lime) favours the formation of small, closed woodlands dominated by *Corylus avellana* and *Crataegus monogyna* (Betulion fontqueri-celtibericae alliance) (Amigo et al. 1994). These pre-climax forests can be replaced by thorn scrub which most often include *Crataegus monogyna* and *Prunus spinosa*, as well as several species of the genera *Rubus* and *Rosa* (alliance Pruno-Rubion ulmifolii, class Rhamno-Prunetea) (Arnáiz and Loidi 1983; Giménez de Azcárate et al. 1996; Costa et al. 2012).

Finally, one should mention the existence of pre-climax forest communities dominated by small evergreen trees (laurel, strawberry tree, holly). The dominant species include *Laurus nobilis, Rhamnus alaternus, Arbutus unedo, Olea europea* var. *sylvestris, Phillyrea latifolia, Ph. angustifolia, Prunus lusitanica*, etc., which occur scattered along those bioclimatic belts that are dominated by thermophilous deciduous forests (thermotemperate oak forests) and by xeric-evergreen woodlands (of cork oak or holm oak). Roughly ten associations have been described to include this type of formations, on siliceous as well as limestone soils, within the alliance Arbuto unedonis-Laurion nobilis (class Quercetea ilicis) (Bueno and Fernández-Prieto 1991; Aguiar and Capelo 1995; Loidi et al. 1997a; Costa et al. 2001; Rodríguez-Guitián et al. 2007; Pulgar and Manso 2010).

6.3.2 Edaphohygrophilous Forests

These are forests developed on soils that are saturated with moisture through most of the year. Their physiognomic and floristic features depend upon a large variety of environmental factors, such as air content in the soil, distance to river courses, or river turbulence regimes. Despite the fact that most of the catchments in this territory develop over small distances, as described above, they usually exhibit large variations in terms of flow, with maxima following the autumn rainfall and the snowmelt in spring, and relatively sharp low flows in summer, especially towards the west. This explains, together with the variety of thermotypes and bedrock types, the diversity of these forest types in the territory.

The distribution of these communities does not follow a simple pattern throughout the territory. In the eastern part of the Cantabrian Atlantic subprovince, the upstream segments (upper mesotemperate and supratemperate belts) of rivers draining into the Cantabrian Sea usually flow under a canopy of beech forests, which produce dense over-shading and prevent the formation of real riparian forests. Where the rivers leave the beech forests and enter the domain of the oak forests, their edges carry forests dominated by ash or alder (Hyperico androsaemi-Alnetum glutinosae). These forests occur almost continuously downstream until the river mouth (Amigo et al. 1987). The abundance of carbonated substrates favours the occurrence of many species of the order Fagetalia sylvaticae inside these forests, which are therefore included in the alliance Alnion incanae (class Salici purpureae-Populetea nigrae). Riparian communities dominated by large willows (*Salix alba, Salix fragilis*) (alliance Salicion albae, class Salici-Populetea) are only found on flood banks of the downstream segments of the largest rivers in this sector (Asón, Pas, Miera, Sella, Deva, Nalón, Narcea) (Díaz-González and Fernández-Prieto 1994; Loidi et al. 1997a).

In northwest Asturias and northern Galicia, the scarcity of beech forests allows a higher diversity of these riparian forests along stream banks. The upstream segments are occupied by woodlands dominated by hazel (*Corylus avellana*) and grey willow (*Salix atrocinerea*) (association Hyperico androsaemi-Coryletum avellanae). These are replaced downstream by ash forests with *Fraxinus excelsior* and *Acer pseudoplatanus* (association Valeriano pyrenaicae-Fraxinetum excelsioris) or by birch forests of *Betula pubescens* subsp. *celtiberica* with *Corylus avellana* and *Frangula alnus* (association Violo palustris-Betuletum pubescentis), and further downstream by silicicolous alder forests with ash (*Fraxinus excelsior*) of the association Valeriano pyrenaicae-Alnetum glutinosae (Amigo et al. 1987; Rodríguez-Guitián 2010; Díaz-González 2010a). All these communities are included in the alliance Alnion incanae.

The rivers of Galicia and Northern Portugal that drain into the Atlantic Ocean run through hilly landscapes that favour the development of silicicolous alder forests on river banks as from the upstream areas of their catchments (Senecioni bayonnensis-Alnetum glutinosae) (Amigo et al. 1987). In the mid and lower segments of the largest rivers (Minho, Lima, Douro) there are communities dominated by shrubby willows (*Salix alba, Salix eleagnos* subsp. *angustifolia, Salix salviifolia, Salix neotricha*) that develop along the alder forests and towards the riverbed on the pebble accumulations that are above the water surface during summer (alliance Salicion salviifoliae, class Salici-Populetea). In the mountains of Peneda and Gerês, stream banks are colonized by birch forests (Carici reuterianae-Betuletum celtibericae) (Honrado et al. 2003). This wide set of riparian forests hosts a number of endemic plant species (*Omphalodes nitida, Narcissus cyclamineus, Senecio nemorensis* subsp. *bayonnensis*) that are characteristic of the alliance Osmundo regalis-Alnion glutinosae, just as *Fraxinus angustifolia* among the tree species.

Finally, the segment of the river Ebro between Reinosa and Valle de Sedano (at the southern end of Cantabria), as well as its tributaries running through the Navarran-Alavese territory, host a peculiar set of woody riparian vegetation which includes alder forests of the association Humulo lupuli-Alnetum glutinosae (alliance Populion albae) on the river banks and willow scrub dominated by *Salix neotricha* (Salicetum neotrichae) (alliance Salicion discoloro-neotrichae) in the flood channel and on pebble islets that stay above the water level during summer (Biurrun 1999).

Altogether, riparian forests of the Cantabrian Atlantic subprovince are richest in species of vascular plants, especially in nemoral specialists. In addition to this remarkable floristic value most of the known Cantabrian Atlantic populations of several macaronesian ferns (*Culcita macrocarpa, Dryopteris guanchica, Vandenboschia speciosa, Woodwardia radicans*) and of the tertiary relict *Prunus lusitanica* are found in these forests, especially in their sheltered and hardly accessible segments.

Due to the abrupt topography of the majority of catchments in this territory, large alluvial flatlands submitted to seasonal flooding are scarce. Moreover, due to their high soil fertility and accessibility, these flatlands preserve only a very small fraction of their primeval forest cover, which has been replaced for centuries by croplands, and more recently by urban and industrial areas. The scarce data available from these forest types describe woodlands dominated by mixtures of oaks (*Quercus robur*, *Q. pyrenaica*), ashes (*Fraxinus excelsior*, *F. angustifolia*), alder (*Alnus glutinosa*), grey willow (*Salix atrocinerea*) and smooth-leaved elm (*Ulmus minor*), with abundant climbers and hygrophilous plants (Loidi et al. 1997a; Biurrun 1999; Amigo et al. 2009; Costa et al. 2012). The best representations of this group of forests are found in the Navarran-Alavese territory (associations Carici pendulae-Fraxinetum excelsioris and Viburno lantanae-Ulmetum minoris), in the Galician catchments of the rivers Miño and Sil (association Hedero hibernicae-Fraxinetum angustifoliae) and in northern Portugal (association Omphalodo nitidae-Fraxinetum angustifoliae). The first of these communities is included in the alliance Alnion incanae, whereas all the others belong to the alliance Populion albae (class Salici-Populetea).

The flatlands that are more prone to flooding promote the formation of marsh and mire environments, in the margins of which one can find forests that are developed on more or less acidic, highly organic substrates (alliance Alnion glutinosae, class Alnetea glutinosae). The situations where these forests are more frequent are located in the landward parts of estuaries, floodplains, and ancient meanders of the main rivers and lagoon systems on the coast and also inland on sedimentary environments. These forests are dominated by alder (*Alnus glutinosa*) with variable abundance of willow (*Salix atrocinerea*), thermophilous plants (*Calystegia sepium*, *Laurus nobilis*), large sedges (*Carex paniculata* subsp. *lusitanica*, *Carex pendula*) and ferns typical of these environments (*Athyrium filix-femina*, *Dryopteris carthusiana*, *Osmunda regalis*, *Thelypteris palustris*). The association Carici lusitanicae-Alnetum glutinosae is found from the Basque Country down to southern Galicia and northern Portugal, where it is gradually replaced toward the south by the association Carici lusitanicae-Salicetum atrocinereae. In some areas with a colder climate (e.g. in the mountains of northern Galicia), alder and many of the above-mentioned vascular plants disappear from these marsh forests, being replaced by birch (*Betula pubescens* subsp. *celtiberica*) and *Myrica gale*, with an herbaceous undergrowth of *Deschampsia cespitosa* subsp. *subtriflora*, *Carex laevigata* or *Molinia caerulea* and bryophytes of the genus *Sphagnum* at ground level (Izco et al. 2001b).

Edge communities of edaphohygrophilous forests differ depending on climate and topo-edaphic conditions. In the upstream areas, communities dominated by *Erica arborea* and *Frangula alnus* are common throughout all the Cantabrian Atlantic territories. Where rivers run through flatlands, thorn scrub usually develops, recognisable by the abundance of *Prunus spinosa*, *Crataegus monogyna* and *Frangula alnus* (class Rhamno-Prunetea). The association Rubo ulmifolii-Rosetum corymbiferae occurs at forest edges of floodplains in the Inner Galician territory (Giménez de Azcárate et al. 1996; Amigo et al. 2009). In the case of marsh and mire forest dominated by birch, their degradation is usually followed by the development of tall scrub with *Myrica gale* and heather species typical of damp soils (*Erica mackaiana*, *Erica tetralix*, *Erica ciliaris*) (Izco et al. 2001b).

6.4 Seral Scrub

The succession stages of different climax woodlands depend on the degree of disturbance that affects those formations. The regressive succession of the different broad-leaved woodlands can be synthesized considering two levels of disturbance in these forests: a moderate level at which most of the soil conditions are maintained and a severely degraded level wherein there is a loss of organic material from the top horizons, and nutrient depletion caused by erosion.

6.4.1 Shrubby Formations After Moderate Disturbance

Where there has been a total or partial deforestation in a woodland area but the soil characteristics have not been excessively deteriorated, the open spaces allow the development of thickets. These formations of small phanerophytes quickly become dense and will rapidly succeed in restoring the forest if the disturbance ceases to exist. These thickets of early-stage seral shrubs can be classified into two major types: prickly thickets on meso- to eutrophic soils that correspond to the Rhamno catharticae-Prunetea spinosae class and heliophytic thickets formed by retamoid shrubs of the Cytisetea scopario-striati class. Often the species that form these communities occur as isolated individuals within forests, as a result of clearings caused by falling trees or other occasional disturbances. They are more often displayed as woody fringes, creating a buffer zone between the forest and the other vegetation. They also appear as linear features outside the forest ecosystems, used by farmers to create boundaries between farmland and grassland where the livestock grazes.

Meso-eutrophic woodlands developed on base-rich soils originating from limestone have Rhamno catharticae-Prunetea spinosae thickets. In the Navarran-Alavese, Santanderian-Biscayan and Central-East Oviedo territories, those communities are dominated by thorny deciduous shrubs with seeds dispersed mainly by birds (*Prunus spinosa*, *Cornus sanguinea*, *Ligustrum vulgare*, *Crataegus monogyna*, *Euonymus europaeus*, *Rosa sp. pl.*, etc.). The most common association is the Tamo communis-Rubetum ulmifolii which also reaches the North Galicia and Asturias and even the Galician and North Portugal sectors (Fig. 6.12). In the Navarran-Alavese area there is another association exclusively related to mesophytic oak woods, called Rhamno catharticae-Crataegetum laevigatae, with shrubs such as *Rhamnus cathartica*, *Viburnum lantana* or *Crataegus laevigata*. Both associations belong to the alliance Pruno spinosae-Rubion ulmifolii, occurring in thermotemperate and mesotemperate, but reaching even mesomediterranean thermotypes (Berastegi et al. 1997a; Loidi et al. 1997a). These formations, dominated by thorny shrubs, can sometimes reach the supratemperate belt, where they might include some characteristic species of the Rhamno alpini-Berberidion vulgaris alliance, such as *Berberis vulgaris*, *Rhamnus cathartica* or *Rosa villosa*,

Fig. 6.12 Thorny scrub (Tamo communis-Rubetum ulmifolii) on deep limestone soils used as boundary between pastures and arable lands. In the background, holm oak (*Quercus rotundifolia*) woodland on stony limestone soil. Becerreá Valley (Lugo, Galicia)

as has been reported from the Santanderian-Biscayan territory (Onaindía 1986). They can also form less dense thickets belonging to the Rhamno catharticae-Prunetea spinosae but with a low proportion of spiny shrubs, being part of the seral stages of oligotrophic temperate woodlands. Such formations are known from the Cantabrian and Basque to the Galician and North Portugal sectors and form the association Frangulo alni-Pyretum cordatae (Herrera et al. 1991). It is worth mentioning that often in the Galician and North Portugal sector, the Frangulo alni-Pyretum cordatae is replaced by similar communities dominated by brambles (*Rubus* spp.), and in the North Lusitania Sierran sector those communities are enriched by several endemic species of brambles in the mountain areas, such as *Rubus vagabundus*, *Rubus lainzii*, *Rubus brigantinus* and *Rubus henriquesii*.

As part of the seral stages of acidophilous *Quercus robur*, *Quercus pyrenaica* or *Fagus sylvatica* woodlands, there are other types of thickets from Navarre to Portugal formed by retamoid species (non-leafy evergreen shrubs with flexible photosynthetic branches) such as *Cytisus commutatus*, *Cytisus scoparius*, *Cytisus cantabricus*, *Cytisus grandiflorus*, *Adenocarpus lainzii*, *Genista florida* subsp. *polygaliphylla* in communities that belong to the Cytisetea scopario-striati class; most of the broom species are non-leafy evergreen, an adaptation to exposed environments which allows controlling water loss by evapotranspiration. In the

Cantabrian and Basque sector these broom formations belong to the Ulici europaei-Cytisetum commutati association, and in the Santanderian-Biscayan territory to the Cytiso scoparii-Genistetum polygaliphyllae. In the western areas of the North Galician-Asturian territory they are the most conspicuous and common type of heliophilous seral formations. There are several examples of communities dominated by different species of *Cytisus:* the Cytisetum striati, which is most widespread from the Central-East Oviedo area to the South Galician-Portuguese territories; the Ulici europaei-Cytisetum ingramii in the North Galician-Asturian territory; and even the Lavandulo sampaioanae-Cytisetum multiflori in the Inner Galicia territory with a mesomediterranean climate. In areas with less oceanic influence (again in the same Inner Galicia or North Lusitania Sierran sector) there are broom communities of the Cytiso striati-Genistetum polygaliphyllae association and in the lowland areas of the South Galician-Portuguese territory the Cytisetea communities can incorporate thermophilic *Cytisus* species such as *Cytisus grandiflorus*. In this Galician and North Portugal sector, on heavily mobilized anthrosols derived from granites, the *Cytisus* communities can be replaced by *Adenocarpus lainzii*, which appears to be more competitive there. The class Cytisetea scopario-striati also encompasses other communities which are not dominated by retamoid species but by white heather (*Erica arborea*), such as the Avenello flexuosae-Ericetum arboreae in hyperhumid areas, or the Genisto falcatae-Ericetum arboreae under temperate sub-Mediterranean climate. Those original tall heather formations are also present in the Cantabrian and Basque sector where the Pteridio aquilini-Ericetum arboreae is one of the successional stages of acidophilous *Fagus*-woodlands. By contrast, in the westernmost sectors the occurrence of thorny formations of the Pruno-Rubion ulmifolii is scarce and fragmented, which is consistent with the shortage of basic or ultrabasic substrates (Díaz-González and Fernández-Prieto 1994; Loidi et al. 1997a; Izco et al. 1999; Costa et al. 2000; Rodríguez-Guitián et al. 2003; Costa et al. 2004).

6.4.2 *Shrubby Formations After Prolonged Disturbance*

Anthropogenic pressure caused by the removal of large areas of forest and the prevention of progressive succession by fire and grazing leads to the formation of woody seral stages dominated by low shrubs, denoting a substantial change in soil quality: the upper organic horizons are lost by erosion and high rainfall causes the migration of nutrients to the lower layers, so that the soil will lose fertility and become more acid. This process has resulted in the proliferation of a peculiar type of scrublands throughout Atlantic Europe: heathlands. They are the result of man-vegetation interaction over the last 4000 years, which resulted not only in a particular and very recognizable type of plant communities, but also in similar forms of exploitation and management of the vegetation by repeated logging or burning (Diemont et al. 1996; Webb 1998).

These heathlands are named as such because of the abundance of different species of the heather family (*Erica, Calluna, Daboecia*). In the Cantabrian Atlantic territories these species co-dominate with species of thorny woody shrubs of the genus *Ulex*; depending on their respective dominance they are commonly called heath-gorse formations or just gorse formations (Fig. 6.13). These communities belong to the class Calluno vulgaris-Ulicetea minoris and achieve in this territory the highest richness of associations throughout Europe (Díaz-González 1998; Izco et al. 2007, Loidi et al. 2007, 2010). The diversification of the Calluno-Ulicetea class in our study area assumes the presence of two major alliances: heath-gorse formations under humid and hyperhumid ombrotypes, and dynamically related to oligotrophic forests, which is present throughout the territory, and belongs to the alliance Daboecion cantabricae; and another subhumid to humid acidophilous scrub, that is well adapted to the sub-Mediterranean climate of most of the Galician and North Portugal and North Lusitania Sierran sectors, and is classified as the Ericion umbellatae alliance. The dominant species of the Daboecion cantabricae are *Daboecia cantabrica*, *Erica vagans*, *Ulex gallii* subsp. *breoganii*, *Erica mackaiana*, *Erica ciliaris* and *Thymelaea coridifolia*. This alliance includes up to 18 different associations in the Cantabrian Atlantic area that result from different combinations

Fig. 6.13 Heath (*Erica cinerea*) and gorse (*Ulex latebracteatus* subsp. *izcoi*) as main floristic components of an acidophilous Cantabrian Atlantic seral scrub. Galician and North Portugal heathland (Ulici izcoi-Ericetum cinereae) at Bueu (Pontevedra, Galicia). (Photo by M.I. Romero-Buján)

of biogeographic units and a variation of ombrotypes, substrates, etc. In the Cantabrian and Basque and Central-East Oviedo territories, the Erico vagantis-Ulicetum europaei association is very common on decarbonated basic substrates. Also the Daboecio cantabricae-Ulicetum cantabricae is common, thriving best under colder conditions and reaching higher altitudes. In the North Galician-Asturian territory the endemic association Ulici breoganii-Ericetum mackaianae occurs, and in the Galician and North Portugal sector the association Ulici izcoi-Ericetum cinereae is very conspicuous in the landscape. In both territories the gorse may disappear with the decrease of oceanity, and is being replaced by communities dominated by heather and forming the Pterosparto cantabricae-Ericetum aragonensis association (a very abundant association in supratemperate territories of the Orocantabric subprovince), or the Carici asturicae-Ericetum aragonensis, typical of the Serra do Gerês in the North Lusitania Sierran sector (Díaz-González and Fernández-Prieto 1994; Herrera 1995; Loidi et al. 1997b; Izco et al. 1999; Honrado et al. 2004). (Figs. 6.14 and 6.15)

On the other hand, the alliance Ericion umbellatae comprises mainly associations present in the Mediterranean area of West Iberia with a humid ombrotype and penetrating into some temperate areas (sub-Mediterranean variant) of the northwest quadrant, particularly in the Galician and North Portugal and North Lusitania Sierran sectors. Their communities are characterized by the presence of *Ulex*

Fig. 6.14 Hyperhumid heath-gorse community (Gentiano pneumonanthes-Ericetum mackaianae) on peaty soils in a supratemperate area of the northern Galician-Asturian mountains. Xistral Range (Lugo, Galicia)

Fig. 6.15 Flowering heathland colours: red (*Erica australis* subsp. *aragonensis*) and yellow (*Pterospartum tridentatum* subsp. *cantabricum*) at the Leboreiro Range (Galicia/Portugal). Heath communities like this occur widespread on granitic Inner Galician and North Lusitania Sierran territories

minor, *Ulex micranthus*, *Ulex latebracteatus*, *Pterospartum tridentatum* subsp. *lasianthum*, *Genista triacanthos*, *Cistus psilosepalus* or *Thymelaea broteriana*. The Pterosparto lasianthi-Ericetum cinereae and Ulici micranthi-Pterospartetum lasianthi can be considered the most widespread associations within this biogeographic unit (Pulgar 1999; Honrado et al. 2004; Costa et al. 2008); in the South Galician-Portuguese territory the latter occurs with the endemic gorse *Ulex micranthus* and prickled broom (*Pterospartum tridentatum s.l.*) on schist substrates with an Atlantic influence, but on granites the Ulicetum latebracteato-minoris association dominates areas with an oceanic influence and is also dominated by an endemic gorse (*Ulex latebracteatus*).

Among the diversity of the Cantabrian Atlantic heath-gorse communities we must consider other associations specialized in highly selective ecological environments. This is the case with the weakly halophytic (splashed by marine salt spray) heath-gorse of the coastal cliffs that belongs to the alliance Dactylo maritimae-Ulicion maritimi; or the hydrophilic heathlands included in the Genistion micrantho-anglicae alliance. Both cases must be interpreted as permanent communities (permaseries) and not as seral heaths, and shall be addressed in Sect. 6.6.

In the eastern territories beyond Central-East Oviedo the dominance of calcareous substrates favours the presence of other shrubs that have a similar ecologic role as heathlands of the neutral-basophilic series. There is no strong acidification of the soil and loss of cations as in the case of the heathlands, but the loss of soil in the top layers by erosion can be similar. The presence of cushion chamaephytes gives a characteristic physiognomy to these scrub formations classified in the Festuco hystricis-Ononidetea striatae class. Two associations are recognized within this class, both dominated by a shrubby, cushion-shaped legume, *Genista hispanica* subsp. *occidentalis*: in the East Basque and Navarran-Alavese territories these spiny formations, belonging to the Helictotricho cantabrici-Genistetum occidentalis association, can be considered a successional stage towards mesotrophic *Fagus sylvatica* and *Quercus robur* woodlands in both sectors, or of the holm oak woodlands (dominated by the evergreen *Quercus ilex*) in the Cantabrian and Basque sector. In the Central-East Oviedo area and Santanderian-Biscayan territories these spiny formations have gorse (*Ulex europaeus*) in their composition, forming the Ulici europaei-Genistetum occidentalis association. Both associations are belong to the alliance Genistion occidentalis (Díaz-González and Fernández-Prieto 1994; Loidi et al. 1997a; Loidi and Báscones 2006).

Besides those communities widely distributed in the temperate Cantabrian Atlantic territories (thorny scrub, broom formations and heathlands), we should mention the singular and sporadic presence of acidophilous shrubs dominated by rock-roses (*Cistus sp. pl.*); they are the ecological vicariant of heathlands in the Mediterranean territory, with subhumid to semi-arid ombrotypes. These communities are concentrated in the Galician-Portuguese mesomediterranean part and the subhumid part of the Sil river valley, with evergreen forests as potential vegetation (*Quercus rotundifolia*, *Quercus suber*), although its extension in terms of area does not exceed 300 km^2. They are dominated by *Cistus ladanifer* or *Cistus populifolius* and there are at least two associations within the alliance Ulici argentei-Cistion ladaniferi of the typical Mediterranean class Cisto-Lavanduletea stoechadis (Izco and Ortiz 1985).

There are other examples of shrub communities dominated by species of undeniable Mediterranean character that can be found from the Basque Country to Portugal, as seral stages of xero-thermophilic forest formations (acidophilous *Quercus pyrenaica*, and *Quercus suber*, and calcareous *Quercus ilex* woodlands) or as permanent communities on rocky outcrops of various kinds. Those are recognized by the dominance of *Arbutus unedo*, or sometimes by *Phillyrea latifolia*, and form a group of Cantabrian Atlantic associations that fall in the Arbuto unedonis-Laurion nobilis alliance of the Quercetea ilicis class (Díaz-González and Fernández-Prieto 1994; Loidi et al. 1994).

6.5 Herbaceous Vegetation

The abundance of precipitation and generally mild temperatures that characterize the Cantabrian Atlantic territory are especially favourable for the development of perennial herbaceous plant communities. However, on the long term vegetation dominated by herbaceous plants is less competitive than woody vegetation, so the presence of herbaceous communities at a particular location corresponds to one of the following circumstances: it can be a seral stage resulting from the removal of forest or shrub cover and will eventually be replaced by those types of vegetation, except in the case of secondary succession that is blocked by anthropic pressure, or it is a permanent community that represents the most stable situation in a specific biotope subject to very limiting ecological factors (pioneer communities on soils in their early stages of formation, mountain tops affected by wind and cryoturbation, permanently flooded areas, cracks in rock walls with little weathering, etc.). Some of this vegetation is associated with stressful environments such as sand dunes, coastal salty habitats, or peatlands. This will be discussed in a later Sect. 6.6. In this section we will address grasslands communities that are very noticeable in the landscape because their maintenance is associated with agricultural and pastoral practices. We shall also describe some of the rupicolous communities in the end.

6.5.1 Semi-natural Managed Grasslands

Perennial grasslands periodically mown to provide hay for livestock are one of the most characteristic elements of the Cantabrian Atlantic landscape. The diversity of grasslands results from the different lithological conditions from which the soil is derived. Topographic variability (in valleys but also on steep slopes) and the water regime of the soils (natural or managed) also contribute to shape this variety. These factors influence the type of existing herbaceous vegetation more profoundly than biogeographic and bioclimatic conditions affecting the flora of such habitats. Additionally, we should consider the intensity of man's management practices such as mowing, planting, adding depleted nutrients, selecting the type of livestock that grazes those grasslands, etc.; all of those contribute to the variety in semi-natural grassland communities comprised within the Molinio caeruleae-Arrhenatheretea elatioris class.

Throughout the entire Cantabrian Atlantic area hay meadows have been described from non-hydromorphic soils, with periodical hay cutting and in some cases with subsequent manuring, which are part of farming practices undertaken for centuries (Tüxen and Oberdorfer 1958; Rivas Goday and Rivas-Martínez 1963; Braun-Blanquet 1967; Bellot 1968; Teles 1970; Díaz-González and Fernández-Prieto 1994; Berastegi 2013). Examples are the meadows belonging to the Arrhenatherion elatioris alliance: those of the Malvo moschatae-Arrhenatheretum elatioris association, existing all over the Cantabrian and Basque and North Galicia and Asturias sectors, can be considered as meadows with a relatively great diversity of herbaceous species per unit area (Izco and Guitián 1984). The North Galician-Asturian territory, and especially the Galician and North Portugal and North Lusitania Sierran sectors, harbours another association, the Agrostio castellanae-Arrhenatheretum elatioris, which is somewhat less rich in species but it behaves like the vicarious version of the previous association in the occidental Cantabrian

Atlantic (Teles 1970); for their persistence in the North Lusitania Sierran sector (where the sub-Mediterranean climatic influence is more pronounced) these managed grasslands are irrigated by very small canals during part of the year and are never grazed during winter.

Other semi-hygrophilous grasslands, resulting from a combined management of mowing and grazing by cattle, are found throughout the territory, and are included within the alliance Cynosurion cristati. They are usually less diverse in species richness than the previous alliance and show a greater dominance of grass species. This alliance contains at least four associations: the Lino biennis-Cynosuretum cristati on deep soils in the thermotemperate and mesotemperate thermotypes from Navarre to Galicia; the Merendero pyrenaicae-Cynosuretum cristati in the Cantabrian and Basque and North Galicia and Asturias sectors but in the supratemperate thermotype; the Caro verticillati-Cynosuretum cristati also in the thermo/mesotemperate thermotype but on oligotrophic soils of the North Galician-Asturian territory and Galician and North Portugal sector; and the Anthemido nobilis-Cynosuretum cristati in meso/supratemperate thermotypes in the North Lusitania Sierran sector and South Galician-Portuguese territory (Bellot and Casaseca 1956; Tüxen and Oberdorfer 1958; Teles 1970). (Figs. 6.16 and 6.17)

Fig. 6.16 Hay meadows (Molinio-Arrhenatheretea) were a common element of the Cantabrian Atlantic landscape since the Middle Ages, but nowadays their extension and use have been reduced considerably. Cynosurion cristati meadow at late spring time in Friol (Lugo, Galicia) (Photo by M.I. Romero-Buján)

Fig. 6.17 Mown meadows (Lino biennis-Cynosuretum cristati) at different stages of hay collection at Abadín (Lugo, Galicia)

Since the middle of last century it has been found that the abandonment of traditional forms of management in the grasslands of the Arrhenatherion elatioris has led to their transformation into Cynosurion cristati pastures. This transformation has affected more strongly the territories with basic-neutral soils, where Arrhenatherion elatioris meadows were more common, such as in the Cantabrian and Basque sector and the Central-East Oviedo territory. In the South Galician-Portuguese territory and in the North Lusitania Sierran sector there were recent changes in landscape management and most of the Cynosurion cristati meadows have disappeared due to the abandonment of the practice of closing off the pastures before spring begins, following the cattle grazing season in winter.

6.5.2 *Poorly Managed Wet Meadows*

The presence of hygrophilous meadows, in many cases exploited for hay, is common throughout the territory. These grasslands are located in areas of high rainfall, little or no summer drought and gentle topography, conditions involving a prolonged waterlogging of the soils. The floristic composition of these meadows shows a relatively high frequency and even dominance of species of the genera

Juncus and *Carex*, grasses (*Deschampsia*, *Molinia*) and other hygrophilous and meso-hygrophilous forbs. Up to six different thermo/meso/supratemperate associations of oligotrophic grasslands belonging to the alliance Juncion acutiflori have been described. But in the Cantabrian and Basque sector and Central-East Oviedo territory there are also examples of hygrophilous grasslands on base-rich neutral soils, belonging to the alliance Molinion caeruleae, and even Calthion palustris, which are indicators of a lesser degree of oceanity.

Among the wide variety of hygrophilous meadows in our territory there are dense grasslands of tall sedges and rushes (*Schoenus nigricans*, *Scirpoides holoschoenus* or *Juncus acutus*) developed on soils with marked temporal hydromorphism, typical of the alliance Molinio caeruleae-Holoschoenion vulgaris. Although this alliance mainly occurs in the Mediterranean region, it may be present in several sectors along the Cantabrian Atlantic subprovince at particular sites with heavy clay soils, and in coastal dune slacks.

6.5.3 Other Grasslands

The pastures addressed so far, despite their response to certain habitat conditions, have been maintened over time due to human activities which mainly focused on the harvest and use of biomass to sustain livestock combined with possible direct grazing. But throughout the territory we find other communities, belonging to the same class Molinio-Arrhenatheretea, that respond to conditions of increased supply of nutrients such as nitrogen or phosphorus, combined with trampling as a result of greater grazing pressure (Loidi et al. 1997a; Biurrun 1999; Honrado 2003; Berastegi 2013). The floristic differences produced in grasslands under such conditions have given rise to other community types:

- From the entire Cantabrian Atlantic subprovince some grasslands of the Lolio perennis-Plantaginion majoris have been described. They formed under strong trampling conditions that led to soil compaction and nitrification by cattle, and that resulted in an increase of rosette hemicryptophytes (*Plantago sp. pl.*, *Hypochaeris radicata*, *Bellis perennis*, *Chamaemelum nobile*, etc.).
- Potentillion anserinae communities are known from the East Basque to the South Galician-Portuguese territories. They occur as nitrified hygrophilous meadows, and are often rich in rushes that colonize river channels that are subject to regular flooding which results in sediment deposition.
- There are also perennial grasslands on loamy nitrified soils, belonging to the alliance Paspalo distichi-Polypogonion viridis. Although these communities are more common in the Mediterranean region, in the Cantabrian Atlantic territory they have been reported from estuaries with regular tidal debris deposition.
- Occurring in small areas in the territory, but with a very distinct physiognomy, are Trifolio fragiferi-Cynodontion dactyli pastures, grassland communities dominated by *Cynodon dactylon*. Although the optimum bioclimate of these

communities is Mediterranean, they appear rather frequently on compacted sandy soils with temporal hydromorphy, such as stabilized dune systems and areas close to the coast, predominantly in the thermotemperate belt.

- Even though dry grasslands are very common in Mediterranean territories, several associations can be found in sub-Mediterranean climates. Most of these communities belong to the class Stipo giganteae-Agrostietea castellanae, and incorporate associations typical of deep soils, while others can occur on shallow soils or outcrops. Several grassland communities occurring on deep soils are dominated by *Agrostis castellana*, and belong phytosociologically to the Agrostion castellanae alliance. They can be found in the South Galician-Portuguese territory.

6.5.4 Calcareous Meso/Xerophytic Grasslands

The grassland communities referred to above generally thrive on moist, deep soils, sometimes even temporary waterlogged. But there are also many natural mesophytic or xerophytic grasslands with dense swards, occurring on soils without temporal hydromorphy. They belong to the Festuco valesiacae-Brometea erecti class and are abundant in subsectors dominated by calcareous substrates (Central-East Oviedo territory and Cantabrian and Basque sector), occurring sporadically in western subsectors. Although such grasslands belonging to various associations are known from the neighboring Pyrenean and Orocantabrian subprovinces, in the Cantabrian Atlantic territory two main associations occur: the most widely represented is the Seselio cantabrici-Brachypodietum rupestris, usually characterized by the dominance of *Brachypodium pinnatum* subsp. *rupestre*; and on loamy soils in the Navarran-Alavese territory the Calamintho acini-Seselietum montani association is found (Loidi et al. 1997a; Berastegi 2013). Both associations belong to the alliance Potentillo montanae-Brachypodion rupestris. (Fig. 6.18)

In the Cantabrian and Basque territory dry, calcareous grassland communities also occur. Their typical floristic composition consists of a mixture of caespitose hemicryptophytes and dwarf chamaephytes that are adapted to environments subjected to cryoturbation on high mountain tops. These formations belong to the Festuco hystricis-Ononidetea striatae class, often characterized by the abundance of endemic species. In the Cantabrian and Basque sector they occur in at least two associations in the thermo and mesotemperate belts. Another association within this class, the Helictotricho cantabricae-Seslerietum hispanicae, is not linked to cryoturbation but occurs on slopes and rocky surfaces with large crevices and shallow soils. And in the supratemperate belt the association Carici ornithopodae-Teucrietum pyrenaici, adapted to mountain ridges with heavy snowfall, occurs (Loidi 1983; Loidi et al. 1997a).

Fig. 6.18 Calcareous grassland (Calamintho-Seselietum montani) in the foreground; along with hay meadows (Cynosurion cristati, in the background) they have been the traditional way of human exploitation of the potential area of mesophilous and calciphilous Navarran-Alavese oak woodlands (Roso-Quercetum pubescentis and Crataego-Quercetum roboris). Arbizu, Urbasa-Andía Range (Navarra) (Photo by Monika Janišová)

6.5.5 Hygrophilous Caespitose Acidophilous Grasslands

We now address other communities of dense but low-growing grasslands, typical of acid soils in areas that remain cool in summer. They are included in the Nardetea strictae class and grow on both siliceous and calcareous substrates, but in the latter case only if rainfall is so intense that the cations are leached to deeper soil levels. Such grasslands have a high regional diversity dependent on several factors, such as altitude, radiation, and soil composition. Although the greatest diversity of grasslands of the Nardetea strictae class in the Iberian Peninsula is found in the orotemperate and oromediterranean areas of all mountain ranges, in the Cantabrian Atlantic areas they are quite frequent at meso- and supratemperate levels, which have high rainfall values. Up to six associations have been described from the four sectors of the territory concerned here (three Cantabrian Atlantic sectors plus North Lusitania Sierran), all belonging to the alliance Violion caninae (Darquistade et al. 2004; Izco et al. 2009). They are characterized by an abundance of grass or graminoid species, including *Nardus stricta*, *Danthonia decumbens*, *Agrostis curtisii*, *Agrostis hesperica*, *Carex binervis*, *Carex pilulifera* and *Juncus squarrosus*, and are accompanied by small forbs such as *Potentilla sterilis*, *Pedicularis sylvatica*, *Galium saxatile*, *Carum verticillatum*, *Serratula tinctoria* subsp. *seoanei* and *Gentiana pneumonanthe*.

6.5.6 Rupicolous Vegetation

This type of vegetation does not strictly consist of herbaceous plants, but also contains dwarf chamaephytes, some of which especially grow in small cracks and crevices. The endemic component of the flora typical of these habitats is very significant. Much of the rupicolous vegetation belongs phytosociologically to the class Asplenietea trichomanis. Within the Cantabrian Atlantic territories calcareous rocks have bigger dimensions in both length and height, than those of the siliceous type, and thus are more favourable for the development of communities that colonize crevices. Therefore, the Central-East Oviedo territory and the Cantabrian and Basque sector are the main areas with a great variety of communities in this category (Rivas-Martínez et al. 1984; Díaz-González and Fernández-Prieto 1994; Loidi et al. 1997a).

In the granitic outcrops of the Serra do Gerês (North Lusitania Sierran sector) the association Phalacrocarpo oppositifoliae-Silenetum acutifoliae appears in the supratemperate belt under sub-Mediterranean climate; within the humid ombrotype and extending from thermotemperate to mesomediterranean the association Linario glabrescentis-Cheilanthetum tinaei occurs. These silicicolous associations are classified in the Saxifragion willkommianae and Cheilanthion hispanicae alliances, respectively. These two alliances have a distinct Iberian Mediterranean distribution, but are represented in sub-Mediterranean territory in specific areas of the Galician and North Portugal and North Lusitania Sierran sectors, in xeric biotopes (Fernández-Areces et al. 1987; Honrado et al. 2012).

The calcicolous communities are much more adapted to humid and hyperhumid ombrotypes. In the Navarran-Alavese territory an association, the Violo biflorae-Saxifragetum paucicrenatae occurs. It belongs to an alliance (Violo biflorae-Cystopteridion alpinae) typical of cracks in the humid orotemperate ombrotypes with prolonged snow cover in Pyrenean and Orocantabrian territories. But the most diversified alliance is the Saxifragion trifurcato-canaliculatae, endemic to the South European Atlantic territory (*sensu* Rivas-Martínez et al. 2014) and represented by up to five associations: the Centrantho lecoqii-Phagnaletum sordidae is located in the mesotemperate part of the Santanderian-Biscayan territory, while the Dethawio tenuifoliae-Potentilletum alchimilloidis and Drabo dedeanae-Saxifragetum trifurcatae have supratemperate distributions throughout the Cantabrian and Basque sector (Loidi et al. 1997a). Finally, there are two associations that are best represented in Orocantabrian subprovince, but occur here and there also in the Central-East Oviedo territory: the Crepido asturicae-Campanuletum legionensis and the

Saxifragetum paniculato-trifurcatae; both include specialists plants of vertical limestone walls, namely *Saxifraga trifurcata*, *Crepis albida* subsp. *asturica*, *Campanula arvatica*, *Agrostis schleicheri* and *Antirrhinum braun-blanqueti*.

The Petrocoptidetum glaucifoliae association can be found in the Central-East Oviedo territory as well, although its optimum is Orocantabrian, and it encompasses calcicolous rupicolous communities growing on calcareous vertical walls many times exceeding 90 degrees; this habitat of overhang limestone crags (*balmes* in Spanish) is more restrictive than the previous, and provides a refuge for highly specialized communities included in a different class: the Petrocoptido pyrenaicae-Sarcocapnetea enneaphyllae (Rivas-Martínez et al. 2002).

6.6 Littoral Zone and Peatlands

Considering the geomorphological and climatic characteristics of this geographical territory, there are two major types of azonal vegetation that are particularly remarkable in terms of presence in the various coastal landscape types, each of them with its own characteristic floristic and phytocoenotical diversity: coastal vegetation in all its forms and peatland environments.

6.6.1 Coastline: Three Kinds of Habitats Present

The length of the Cantabrian Atlantic coast is over 2500 km long and consists of a succession of three major ecological environments with different balances of erosion/sedimentation: the cliffs, sand beaches and saltmarshes. Each of these three environments offers ideal conditions for highly specialized types of vegetation: the halochasmophytic vegetation on the cliffs and rocky shores, the psammophilous vegetation on sandy coasts, and halophytic vegetation in river estuaries and other areas with brackish water. On all these environments there are quite a few studies about their vegetation types (Loriente 1974; Fernández-Prieto and Loidi 1984; Izco et al. 1993a, 2001a; Díaz-González and Fernández-Prieto 1994; Izco and Sánchez 1996; Bueno 1997; Loidi et al. 1997a; Costa et al. 1998b; Neto et al. 2007; Díaz-González 2009).

6.6.1.1 Rocky Shores

On rocky shores the processes of marine abrasion or deposition of large boulders are the dominant ones. Although the Cantabrian Atlantic rocky shores are formed by various lithological materials whose resistance to erosion is variable, the appearance of the vegetation that colonizes the cliffs is mainly due the dynamics of waves, usually stronger during the winter, and the salty spray splash. Both factors combined make a series of vegetation zones that indicate different levels of exposure to the mechanical action of waves and wind force charged with salts. There is a first line of discontinuous vegetation consisting mainly of chasmophytic hemicryptophytes and some succulent or suffruticose plants; various species of the Plumbaginaceae are part of those communities (genera *Armeria* and *Limonium*) some of them endemic to the territory. At least five different associations were recognized from northern Portugal to the Basque coast, all belonging to the alliance Crithmo maritimi-Armerion maritimae (Crithmo maritimi-Limonietea class). A second strip on the cliffs, that is safe from the mechanical action of waves, allows the establishment of a soil layer that can accommodate a dense graminoid cover, despite the salt spray. These grasslands are often dominated by the grass *Festuca pruinosa*, and are described in 4 different associations that are classified in the same phytosociological alliance referred to earlier (Fig. 6.19). Finally, at the top of the cliffs, where the rock weathering is more advanced, a type of shrubland, formed by cushion chamaephytes and dominated by gorse (*Ulex sp. pl.*) and heather (*Erica sp. pl.*) is present, being included in the Calluno vulgaris-Ulicetea minoris class (see Sect. 6.4). The presence in those communities of a unique set of genotypes and endemic species, adapted to aerohaline environments (*Rumex acetosa* subsp. *biformis*, *Angelica pachycarpa*, *Daucus carota* subsp. *gummifer*, *Leucanthemum*

Fig. 6.19 Halophytic plant communities of Cantabrian cliffs (Crithmo maritimi-Armerion maritimae) include small cushion-shaped chamaephytes and dense grassland dominated by *Festuca pruinosa*. Both communities at Cabo Peñas (Asturias)

Fig. 6.20 Heather-gorse scrub, hemisphaerically shaped by marine wind (Dactylo maritimae-Ulicion maritimi). Cisto salviifolii-Ulicetum humilis at Cabo Vilán (A Coruña, Galicia) (Photo by M.I. Romero-Buján)

crassifolium), reveals that they are not part of seral shrub formations but are permanent communities (permaseries). This set of peculiar plants justifies their inclusion in a distinctive alliance, which also comprises five different associations ranging from the Cisto salviifolii-Ulicetum humilis in the Galician and North Portugal sector to the Ulici humilis-Ericetum vagantis in the Cantabrian and Basque one. (Fig. 6.20)

6.6.1.2 Sandy Shores

In the coastal sections where the deposition of coarse to medium-sized materials is predominant (pebbles or sand), sandy beaches are built. Here, the wind often carries sand particles inland, which form a dune system when they encounter an obstacle. These environments are well known for their difficulty in terms of colonization by plants, since to the selective factors already mentioned (waves, salt spray and sand burial caused by the wind) we must add an unstable and very low water holding capacity of the substrate. A distinct group of habitats are recognized universally in these dune systems dividing them into vegetation belts parallel to the tide line depending on wind action and the distance to the sea: embryonic dune, moving dune (or front dune) and fixed dune (or grey dune), with the possibility of having a fossilized dune further inland (brown dune). With the exception of fossil dunes, which can support a higher biomass and woody vegetation, the various communities that colonize the dune strips are highly specialized and integrated into the Euphorbio paraliae-Ammophiletea australis class. The embryonic dune corresponds to a single graminoid community with discontinuous coverage along the entire coast: the Euphorbio paraliae-Elytrigietum boreo-atlanticae. Behind this community, a distinct frontal dune stands out conspicuously, formed by the perennial grass *Ammophila arenaria* subsp. *australis*. This plant has a notable root system that possesses vertical rhizomes and can resist sand mobility and consequent burial. Also in this case the association that is recognized is the same for all Iberian Cantabrian Atlantic beaches: the Otantho maritimi-Ammophiletum australis. Finally, in the space behind the front dune, in which the decrease of the selective stress caused by wind allows the establishment of an environment with a greater plant diversity (grey dune), the communities are dominated by hemicryptophytes and chamaephytes. In the grey dunes of the Galician and North Portugal sector grow communities of the Helichrysion picardii alliance with suffruticose Iberian endemics of western Mediterranean origin (*Iberis procumbens*, *Helichrysum picardii*, *Artemisia crithmifolia*, *Scrophularia frutescens*, *Alyssum gallaecicum*); however, from the North Galician-Asturian territory to the Cantabrian and Basque sector, the vegetation of these grey dunes is somewhat poorer, and is placed in the alliance Euphorbio portlandicae-Helichrysion maritimi.

6.6.1.3 Salt Marshes

These are low muddy areas with a predominance of sedimentation of fine particles (silt and clay) carried by slow-flowing river water. They are much less exposed to wind and salt spray in comparison with the two previous cases, and are daily flooded by brackish water, mostly associated with rivers estuaries, with acid sulfate soils. Where the tide rises and falls they are usually traversed by numerous channels. The continuous mixing in different proportions of freshwater and saltwater, with daily and seasonal variations and various degrees of oxygenation of the soil, are the key factors that influence the distribution of plant communities in these environments.

There are several detailed studies on the diversity of plant communities in a great number of Cantabrian Atlantic saltmarshes (Izco et al. 1993b; Herrera 1995; Izco and Sánchez 1996; Bueno 1997). A simplified model should distinguish at least two different ecological positions:

- An external halophytic saltmarsh subjected to a greater influx of salt water that floods the vegetation twice a day. Following a high to low tidal flooding gradient, we can distinguish this sequence of communities: the Zosteretum noltii submerged beds, grass formations of the Spartinetum maritimae, succulent chamaephytes communities of the Puccinelio maritimae-Sarcocornietum perennis and, in less flooded zones, the Halimionetum portulacoidis shrubland. All these associations are present from the Basque Coast to Portugal. Where the

Fig. 6.21 Saltmarshes at Corrubedo Natural Park (A Coruña, Galicia) showing dominant chamaephytes such as *Halimione portulacoides* and *Sarcocornia perennis*

influence of flooding becomes smaller, the phytocoenotical biodiversity can increase because of the presence of succulent therophytic communities that normally occupy small areas, such as the Salicornietum fragilis or Salicornietum dolichostachyae. (Fig. 6.21)

– An internal sub-halophytic saltmarsh, where the river inputs are predominating compared with seawater entry, and thus less salty water (brackish). These environments are common in all estuaries, often dominated by rush beds such as the Junco maritimi-Caricetum extensae or Agrostio stoloniferae-Juncetum maritimi or reeds/tall sedges communities of the Armerion maritimae alliance. Communities with much lower biomass can also occur in those mosaics, such as the Enteromorpho intestinalidis-Ruppietum maritimae present in channels and small ponds with slow water evaporation, or the four associations of the alliance Limonio ovalifolii-Frankenion laevis, which occur on sandy soils that drain easily at low tide.

6.6.2 Peatlands

The warm and humid climate that characterized the Cantabrian Atlantic territories throughout the Holocene has favoured the formation of a large number of peat wetlands. The variety of geomorphological environments that currently characterizes these ecosystems can be summarized as: blanket bogs, slope raised bogs, large depressions and valley peatlands and corrie bogs.

The variety of climatic and pedo-topographical conditions where these wetlands appear influences their extent, complexity and number of ecological environments represented (bogs, swamp forests, wet heaths and grasslands, communities of clear running waters, ponds, lagoons, etc.). Presently, the Cantabrian Atlantic territory of Galicia and the west part of Asturias are the areas with the greatest number and diversity of peaty areas in the Iberian Peninsula (Ramil-Rego et al. 1996; Izco et al. 2001b).

Traditionally the vegetation of the peat areas has been divided into two types of vegetation classes based on floristic criteria and degree of mineralization of the water that goes through these systems: the Oxycocco palustris-Sphagnetea magellanici (very acidic bog vegetation, with or without tall hummocks) and the Scheuchzerio palustris-Caricetea nigrae (peatland vegetation from acidic to basic substrates, with generally a flat topography). However, there are intermediate situations in which the inclusion in one of the two classes mentioned is difficult from a phytosociological point of view. When peat systems cover large areas they often form complex mosaics with other vegetation types (classes Alnetea glutinosae, Molinio-Arrhenatheretea, Littorelletea uniflorae, Calluno-Ulicetea, Montio fontanae-Cardaminetea amarae) whose diversification and local distribution is strongly influenced by various environmental factors (Moore 1968; Ellenberg 1988; Rivas-Martínez 2011; Costa et al. 2012).

6.6.2.1 Blanket Bogs

These bogs develop under very particular conditions: mountain tops of flat topography or with gentle slopes, on nutrient-poor rocks (quartzites, sandstones, granitic rocks) and subjected to a strong oceanic climate (hyperoceanic). Although in the early Holocene this type of wetland covered large areas of the Cantabrian Atlantic mountains, subsequent vegetation dynamics and human disturbance caused their disappearance or colonization by other vegetation (forests, wet heaths and grasslands, forest plantations). Today the Iberian active blanket bogs cover a few mountain ranges of northern Galicia (Xistral and Capelada mountain range, Montes do Buio) where they occupy a restricted area of about 2700 hectares (Izco et al. 2001b; Rodríguez-Guitián et al. 2009), and small pockets in some central-eastern mountain ranges such as the Monte Zalama from the Santanderian-Biscayan territory (Heras and Infante 2004). These active blanket bogs are the most southwestern ones on the European mainland, sustaining endemic vegetation types and containing valuable paleobotanic information concerning the environmental changes that occurred in the territory during the past 10,000 years (Ramil-Rego 1992; Muñoz Sobrino et al. 2005).

The Galician blanket bog vegetation develops over peat deposits that are generally 0.5 to 5.0 m thick (Izco et al. 2001b); the vegetation is mainly composed of

Fig. 6.22 Summertime appearance of NW Iberian active blanket-bog communities (Carici durieui-Eriophoretum angustifolii, Molinio caeruleae-Eriophoretum angustifolii) at the top of the Xistral Range (Lugo, Galicia)

Eriophorum angustifolium, *Carex durieui*, *Molinia caerulea*, *Avenella flexuosa*, *Trichophorum cespitosum*, etc., evidencing the limited role of bryophytes and heather (*Erica mackaiana*, *Calluna vulgaris*) unlike other types of blanket bog in western Europe (Fig. 6.22). The communities described in Galicia are assigned to the endemic Iberian alliance Erico mackaianae-Sphagnion papillosi (Rodríguez-Guitián et al. 2009). At Mount Zalama the active part of the blanket bog is covered by a floristically singular herbaceous-woody vegetation dominated by *Calluna vulgaris*, *Erica tetralix*, *Eriophorum vaginatum*, *Molinia caerulea*, *Daboecia cantabrica*, *Hypnum cupressiforme* and *Dicranum scoparium* (Heras and Infante 2004).

6.6.2.2 Slope Peatlands

These are wetlands of small size but relatively frequent in the territory, particularly in the central-western Cantabrian Atlantic territories. Usually they form on the lower and middle slopes as a result of resurgences of subsurface waters, which favour peat formation processes, whose activity is maintained as long as there is a supply of water from the springs. Where the water is low in nutrients these wetlands usually consist of small ponds in their upper parts, feeding down-slope *Sphagnum*-rich raised bogs belonging to the alliances Ericion tetralicis and Erico mackaianae-

Sphagnion papillosi (Figs. 6.23 and 6.24). There are variations of these acidic peaty systems that develop into forested peatlands dominated by *Betula pubescens* subsp. *celtiberica*, accompanied by *Salix atrocinerea*, *Frangula alnus* and *Myrica gale* in the North Galician-Asturian territory (Izco et al. 2001b) and by *Osmunda regalis* in the East Basque territory (Peralta et al. 2013).

Herbaceous formations dominated by bryophytes appear in areas with gentle slopes with an abundant supply of mineral-rich water (*Eucladium* spp., *Cratoneurion* spp.), belonging to the alliance Cratoneurion commutati (class Montio-Cardaminetea). In permanent waterlogged areas with high ions concentrations, alkaline fens occur with *Schoenus nigricans*, *Carex davalliana*, etc. (alliance Caricion davallianae, class Scheuchzerio-Caricetea nigrae). When the water resurges on slopes of strong inclination or rocky walls, communities composed by brown mosses and ferns occur, being assigned to the alliance Adiantion capilli-veneris (class Adiantetea capilli-veneris) and characterized by the formation of carbonate precipitates (tufa or travertine).

6.6.2.3 Sedimentary Depressions and Valley Bogs

These wetlands are mainly associated with areas affected by regular fluctuations in their water tables as a result of changes in the flow of adjacent rivers. They typically

Fig. 6.23 Different types of peatlands favour different plant communities. *Sphagnum*-rich cushion-shaped peatlands (Erico mackaianae-Sphagnion papillosi) at the Xistral Range (Lugo, Galicia) (Photo by M.I. Romero-Buján)

Fig. 6.24 Acidic valley bottom complex mire dominated by cotton sedge (*Eriophorum angustifolium*). Abadín (Lugo, Galicia) (Photo by M.I. Romero-Buján)

cover large areas, although many of them have been transformed for centuries in areas of agricultural use or for livestock grazing by opening drainage channels. In the best preserved areas the majority of communities in these peat systems are similar to those described for mountain peatlands with the prominence of certain vegetation types linked to drainage channels (Eleocharition multicaulis), ponds and small lakes (Littorellion uniflorae, Utricularion vulgaris), and megaforbic communities (Filipendulion ulmariae, Magnocaricion elatae) together with bog woodlands (class Alnetea glutinosae).

6.6.2.4 Corrie Bogs

Corries (or tarns) are small pools placed in depressions originating from glacial plucking. This type of environment is relatively scarce in the Cantabrian Atlantic mountain ranges, especially when compared with the neighboring Orocantabrian subprovince. The best examples are found in the mountains of the North Lusitania Sierran sector as well as in the mountains along the boundary between Cantabria, Burgos and the Basque Country, although many of them have been much altered since antiquity, because of summer grazing by cattle. The corries are usually found in different stages of sedimentation; some of them may become dry during the summer, but others remain with open water throughout the year, with oligotrophic, floating communities (alliance Caricion nigrae) or transitional acid bogs with *Sphagnum* hummocks (alliance Ericion tetralicis) at their edges.

6.7 Vegetation Series

The idea of vegetation series originated from the notions proposed by Clements (1916), when exposing all the explanatory conceptual contributions regarding the replacement of plant communities in the same physical space in what is known as vegetation dynamism. More recently, as a result of the great development of Phytosociology as a method of analysis of vegetation, a precise definition was established: "A Vegetation Series or Sigmetum is a geobotanic notion that tries to express all the plant communities, or collection of stages, that can be found in similar tessellar places as a result of the succession processes" (Rivas-Martínez 2002b). This definition is based on the concept of "tessella", a geographical territory of greater or lesser extent, environmentally homogenous, with one association as potential climax vegetation and therefore a certain sequence of communities of substitution (Rivas-Martínez 2005). The vegetation series (or sigmetum), besides being the basic unit of Dynamic Phytosociology, is an especially useful tool for mapping the natural environment and for interpreting landscape dynamics in areas, such as described here, where the interaction of human populations with their environment has been intense since prehistoric times.

In general the mature stage in ecological succession, also known as climax, is represented by a woodland association. Those are the vegetation types with more biomass and greater structural complexity towards which, over time, all seral communities of each tesela in a progressive succession evolve, in the absence of disturbances. Most of the vegetation series are named after the mature stage woodland association, which represents the Potential Natural Vegetation (PNV) for each particular series, commonly referred to as *Cabeza de Serie* in Spanish geobotanical literature.

Based on the response of the vegetation series to the environmental conditions provided by the territory's macroclimate, or if the territory is submitted or conditioned by excess or shortage in terms of water availability (due to topography, a special geological substrate, proximity to watercourses, etc.), we can distinguish four different types of series (Loidi et al. 2011):

- Climatophilous: series on mature soils with water intake only from rain.
- Temporihygrophilous: edaphophilous series on wet soils but only for part of the year, well drained during the dry season.
- Edaphohygrophilous: edaphophilous series on soils that are unusually wet or flooded most of the year, such as riverbanks, marsh areas, etc.
- Edaphoxerophilous: edaphophilous series on especially dry or xeric soils.

Table 6.1 lists all series recognized in the Cantabrian Atlantic territory with climax woodlands as PNV, divided according to their presence in the different biogeographic subsectors.

Table 6.1 Biogeographical distribution (territories) of vegetation series with climax woodlands as PNV in the Cantabrian Atlantic territory

		NLS	S G-P	N G-P	*I G*	N G-A	CEOv	Sa-Bi	E B	*Na-Al*
	Climatophilous & temporihygrophilous ♦ series									
	Common oak (*Quercus robur*) series									
1	*Hyperico androsaemi-Querco roboris-S.*		+	+						
2	*Rusco aculeati-Querco roboris-S.*	+	+	+	+					
3	*Viburno tini-Querco roboris-S.*		+							
4	*Vaccinio myrtilli-Querco roboris-S.*	+								
5	*Blechno spicantis-Querco roboris-S.*					+	+			
6	*Hyperico pulchri-Querco roboris-S.*							+	+	+
7	♦ *Crataego laevigatae-Querco roboris-S.*									+
	Sessile oak (*Q. petraea*) series									
8	*Pulmonario longifoliae-Querco petraeae-S.*								+	
	Pyrenean oak (*Q. pyrenaica*) series									
9	*Lonicero periclymeni-Querco pyrenaicae-S.*				+	+				
10	*Melampyro pratensis-Querco pyrenaicae-S.*							+	+	+
	Downy oak (*Q. pubescens*) series									
11	*Roso arvensis-Querco pubescentis-S.*									+
	Portuguese oak (*Q. faginea*) series									
12	*Pulmonario longifoliae-Querco fagineae-S.*							+		+
	Common beech (*Fagus sylvatica*) series									
13	*Saxifrago spathularis-Fago sylvaticae-S.*				+	+				
14	*Carici caudatae-Fago sylvaticae-S.*						+			
15	*Saxifrago hirsutae-Fago sylvaticae-S.*							+	+	+
16	*Carici sylvaticae-Fago sylvaticae-S.*							+	+	+
17	*Epipactido helleborines-Fago sylvaticae-S.*								+	+
	Mixed ash/elm/oak (*Fraxinus*/*Ulmus*) series									
18	♦ *Hedero hibernicae-Fraxino angustifolii-S.*		+	+	+					
19	♦ *Omphalodo nitidae-Fraxino angustifolii-S.*		+							
20	♦ *Fraxino angustifolii-Ulmo glabri-S.*		+							
21	*Polysticho setiferi-Fraxino excelsioris-S.*						+	+	+	
22	♦ *Hyperico androsaemi-Ulmo glabri-S.*								+	
23	♦ *Viburno lantanae-Ulmo minoris-S.*									+
	Birch (*Betula celtiberica*) series									
24	*Eryngio juresiani-Betulo celtibericae-S.*	+								
	Cork oak (*Q. suber*) series									
25	*Physospermo cornubiensis-Querco suberis-S.*				+					
	Holm oak (*Q. rotundifolia*) series									
26	*Genisto hystricis-Querco rotundifoliae-S.*				+					
	Edaphoxerophilous series									
27	*Arenario montanae-Querco suberis-S.*				+					
28	*Genisto falcatae-Querco rotundifoliae-S.*				+					
29	*Lauro nobilis-Querco ilicis-S.*						+	+	+	
	Edaphohygrophilous series									
	Black alder (*Alnus glutinosa*) series									
30	*Senecioni bayonnensis-Alno glutinosae-S.*		+	+	+					
31	*Carici lusitanicae-Alno glutinosae-S.*		+	+	+	+	+	+		
32	*Galio broteriani-Alno glutinosae-S.*	+								
33	*Valeriano pyrenaicae-Alno glutinosae-S.*				+	+				
34	*Hyperico androsaemi-Alno glutinosae-S.*						+	+	+	
35	*Lonicero xylostei-Alno glutinosae-S.*									+

(continued)

Table 6.1 (continued)

	Ash (*Fraxinus excelsior*) series									
36	*Valeriano pyrenaicae-Fraxino excelsioris-S.*					+				
37	*Carici pendulae-Fraxino excelsioris-S.*									+
	Birch (*Betula pubescens s.l.*) series									
38	*Violo palustris-Betulo pubescentis-S.*			+		+				
39	*Carici reuterianae-Betulo celtibericae-S.*	+								
	Willows (*Salix sp. pl.*) series									
40	*Salico salviifoliae-S.*		+	+	+					
41	*Salico angustifolio-albae-S.*						+	+		
42	*Salico discoloro-angustifoliae-S.*									+
	Woodland Permaseries									
	Laurel tree (*Laurus nobilis*) series									
43	*Omphalodo nitidae-Lauretum nobilis*		+	+						
44	*Calluno vulgaris-Lauretum nobilis*			+		+	+			
45	*Holco mollis-Lauretum nobilis*					+				
46	*Tamo communis-Lauretum nobilis*					+	+			
47	*Hedero helicis-Lauretum nobilis*						+	+		
	Olive (*Olea europaea*) serie									
48	*Lithodoro diffusae-Oleetum europaeae*						+	+		

The abbreviation "-S." at the end of every woodland community name, means *Sigmetum*. Columns correspond to: North Lusitania Sierra (NLS), South Galician-Portuguese (SG-P), North Galician-Portuguese (NG-P), Inner Galician (I G), North Galician-Asturian (NG-A), Central-East Oviedo (CEOv), Santanderian-Biscayan (Sa-Bi), East Basque (E B) and Navarran-Alavese (Na-Al). (Abbreviated names in italics: non-coastal territories)

Throughout the Cantabrian Atlantic subprovince it is very common that the woodlands which represent the PNV are much altered or completely gone from the territory and replaced most of the times by shrub stages (see Sect. 6.4). Although the diversity of seral shrub communities replacing mature woodlands in a territory is quite large, good correlations between them can be found when addressing this diversity at the level of phytosociological alliances, as shown in Table 6.2.

Regarding the climatophilous series, two categories can be recognized associated with the richness and availability of nutrients in the soil where they thrive: the baso-neutrophilous series on base-rich substrates, and the acidophilous series on soils poor in cations. As a rule, the former ones have seral stages with spiny/thorny shrub communities (class Rhamno-Prunetea spinosae) belonging to the alliance Pruno spinosae-Rubion ulmifolii, or even the Rhamno alpini-Berberidion vulgaris in the supratemperate belt. By contrast, the acidophilic series usually have as a first seral stage a broom formation (class Cytisetea scopario-striati, alliance Ulici europaei-Cytision striati or Cytision multiflori); but they can also be replaced by more oligotrophic shrub formations of the Rhamno-Prunetea class, such as the thickets of the alliance Frangulo alni-Pyrion cordatae. Naturally, most of the baso-neutrophilous series occur in the central-eastern territories where calcareous substrates are abundant: the Navarran-Alavese, East Basque, Santanderian-Biscayan and Central-East Oviedo territories. In both types of climatophilous series, either acidophilous or baso-neutrophilous, a pre-climax community may develop as a response to forest clearings resulting from natural or artificial causes,

Table 6.2 Main seral corresponding stages of vegetation series present in the Cantabro Atlantic territory. Table A: Climatophilous series, temporihygrophilous & edaphoxerophilous series, and Table B: Edaphohygrophilous series & Woodland permaseries

A

Serie number	Mature woodland	Secondary woodland	Tall heath/ scrub	Low heath/ scrub	Herbaceous vegetation
1	*Hyperico androsaemi-Quercetum roboris*	*Corylus avellana* community	*Ulici-Cytision striati*	*Ericion umbellatae* +	*Cynosurion cristati*
			Frangulo-Pyrion cordatae	*Daboecion cantabricae*	
2	*Rusco aculeati-Quercetum roboris*	*Betulion fontqueri-celtibericae*	*Ulici-Cytision striati* + *Arbuto-Laurion*	*Daboecion cantabricae*	*Cynosurion cristati*
3	*Viburno tini-Quercetum roboris*	*Cytiso grandiflori-Arbutetum unedonis*	*Ulici-Cytision striati*	*Ericion umbellatae*	*Cynosurion cristati*
4	*Vaccinio myrtilli-Quercetum roboris*	*Betulion fontqueri-celtibericae*	*Ulici-Cytision striati*	*Daboecion cantabricae*	*Cynosurion cristati*
5	*Blechno spicantis-Quercetum roboris*	*Betulion fontqueri-celtibericae*	*Frangulo-Pyrion cordatae*	*Daboecion cantabricae*	*Cynosurion cristati*
			Ulici-Cytision striati		
6	*Hyperico pulchri-Quercetum roboris*	*Betulion fontqueri-celtibericae*	*Frangulo-Pyrion cordatae*	*Daboecion cantabricae*	*Cynosurion cristati*
			Arbuto-Laurion nobilis		
7	*Crataego laevigatae-Quercetum roboris*	Unknown	*Pruno-Rubion ulmifolii*	Unknown	*Potentillo-Brachypodion Cynosurion cristati*
8	*Pulmonario longifoliae-Quercetum petraeae*	*Betulion fontqueri-celtibericae*	*Cytision multiflori*	*Daboecion cantabricae*	*Cynosurion cristati*
9	*Lonicero periclymeni-Quercetum pyrenaicae*	*Betulion fontqueri-celtibericae*	*Ulici-Cytision striati* + *Arbuto-Laurion*	*Daboecion cantabricae*	*Cynosurion cristati*

(continued)

Table 6.2 (continued)

A					
Serie number	Mature woodland	Secondary woodland	Tall heath/ scrub	Low heath/ scrub	Herbaceous vegetation
10	*Melampyro pratensis-Quercetum pyrenaicae*	*Betulion fontqueri-celtibericae*	*Frangulo-Pyrion cordatae*	*Daboecion cantabricae*	*Cynosurion cristati*
11	*Roso arvensis-Quercetum pubescentis*	*Pinus sylvestris* community	*Pruno-Rubion ulmifolii*	*Genistion occidentalis*	*Potentillo-Brachypodion* *Arrhenatherion elatioris*
12	*Pulmonario longifoliae-Quercetum fagineae*	Unknown	*Pruno-Rubion ulmifolii*	*Genistion occidentalis*	*Potentillo-Brachypodion*
13	*Saxifrago spathularis-Fagetum sylvaticae*	*Betulion fontqueri-celtibericae*	*Ulici-Cytision striati*	*Daboecion cantabricae*	*Cynosurion cristati*
14	*Carici caudatae-Fagetum sylvaticae*	*Betulion fontqueri-celtibericae*	*Pruno-Rubion ulmifolii* *Rhamno-Berberidion*	*Genistion occidentalis*	*Potentillo-Brachypodion*
15	*Saxifrago hirsutae-Fagetum sylvaticae*	*Betulion fontqueri-celtibericae*	*Cytision multiflori*	*Daboecion cantabricae*	*Violion caninae*
16	*Carici sylvaticae-Fagetum sylvaticae*	*Betulion fontqueri-celtibericae*	*Pruno-Rubion ulmifolii* *Rhamno-Berberidion*	*Genistion occidentalis*	*Potentillo-Brachypodion Arrhenatherion elatioris*
17	*Epipactido helleborines-Fagetum sylvaticae*	*Betulion fontqueri-celtibericae*	*Pruno-Rubion ulmifolii*	*Genistion occidentalis*	*Potentillo-Brachypodion*
18	*Hedero hibernicae-Fraxinetum angustifolii*	Unknown	*Pruno-Rubion ulmifolii*	Unknown	*Cynosurion cristati*
19	*Omphalodo nitidae-Fraxinetum angustifolii*	Unknown	*Pruno-Rubion ulmifolii*	Unknown	*Cynosurion cristati*
20	*Fraxino angustifolii-Ulmetum glabrae*	Unknown	*Pruno-Rubion ulmifolii*	Unknown	*Cynosurion cristati*

(continued)

Table 6.2 (continued)

A

Serie number	Mature woodland	Secondary woodland	Tall heath/ scrub	Low heath/ scrub	Herbaceous vegetation
21	*Polysticho setiferi-Fraxinetum excelsioris*	*Betulion fontqueri-celtibericae*	*Pruno-Rubion ulmifolii*	*Genistion occidentalis*	*Arrhentherion elatioris* *Cynosurion cristati*
22	*Hyperico androsaemi-Ulmetum glabrae*	*Betulion fontqueri-celtibericae*	*Pruno-Rubion ulmifolii*	Unknown	Unknown
23	*Viburno lantanae-Ulmetum minoris*	Unknown	*Pruno-Rubion ulmifolii*	Unknown	*Brachypodion phoenicoidis*
24	*Eryngio juresiani-Betuletum celtibericae*	Unknown	*Cytision multiflori*	*Daboecion cantabricae*	*Violion caninae*
25	*Physospermo cornubiensis-Quercetum suberis*	*Ericion arboreae*	*Pruno-Rubion ulmifolii* *Ulici-Cytision striati*	*Ericion umbellatae*	*Agrostion castellanae*
26	*Genisto hystricis-Quercetum rotundifoliae*	Non-existent	*Ericion arboreae*	*Ulici-Cistion ladaniferi*	*Agrostion castellanae*
27	*Arenario montanae-Quercetum suberis*	Non-existent	*Ulici-Cytision striati*	*Daboecion cantabricae*	*Sedion anglici*
28	*Genisto falcatae-Quercetum rotundifoliae*	Non-existent	*Pruno-Rubion ulmifolii*	Non-existent	*Potentillo-Brachypodion*
29	*Lauro nobilis-Quercetum ilicis*	*Arbuto-Laurion*	*Pruno-Rubion ulmifolii*	*Genistion occidentalis*	*Potentillo-Brachypodion*

B

Serie number	Mature woodland	Tall scrub	Herbaceous vegetation
30	*Senecioni bayonnensis-Alnetum glutinosae*	*Pruno-Rubion ulmifolii*	*Cynosurion cristati, Juncion acutiflori*
31	*Carici lusitanicae-Alnetum glutinosae*	*Salix atrocinerea community*	*Magnocaricion elatae*

(continued)

Table 6.2 (continued)

B			
Serie number	Mature woodland	Tall scrub	Herbaceous vegetation
32	*Galio broteriani-Alnetum glutinosae*	*Pruno-Rubion ulmifolii*	*Cynosurion cristati, Juncion acutiflori*
33	*Valeriano pyrenaicae-Alnetum glutinosae*	*Pruno-Rubion ulmifolii*	*Filipendulion ulmariae, Cynosurion cristati*
34	*Hyperico androsaemi-Alnetum glutinosae*	*Pruno-Rubion ulmifolii*	*Filipendulion ulmariae, Calthion palustris*
35	*Lonicero xylostei-Alnetum glutinosae*	*Pruno-Rubion ulmifolii*	*Filipendulion ulmariae, Potentillion anserinae*
36	*Valeriano pyrenaicae-Fraxinetum excelsioris*	*Pruno-Rubion ulmifolii*	*Cynosurion cristati, Juncion acutiflori*
37	*Carici pendulae-Fraxinetum excelsioris*	*Pruno-Rubion ulmifolii*	*Filipendulion ulmariae, Potentillion anserinae*
38	*Violo palustris-Betuletum pubescentis*	*Erica arborea-Frangula alnus* community?	*Cynosurion cristati, Juncion acutiflori*
39	*Carici reuterianae-Betuletum celtibericae*	*Frangulo-Pyrion cordatae*	*Cynosurion cristati, Juncion acutiflori*
40	*Salicetum salviifoliae*	Non-existent	*Glycerio-Sparganion neglecti*
41	*Salicetum angustifolio-albae*	Non-existent	*Potentillion anserinae,*
42	*Salicetum discoloro-angustifoliae*	Non-existent	*Senecionion fluviatilis*
43	*Omphalodo nitidae-Lauretum nobilis*	Non-existent	Unknown
44	*Calluno vulgaris-Lauretum nobilis*	Non-existent	Unknown
45	*Holco mollis-Lauretum nobilis*	Non-existent	Unknown
46	*Tamo communis-Lauretum nobilis*	Non-existent	*Potentillo-Brachypodion*
47	*Hedero helicis-Lauretum nobilis*	Non-existent	*Potentillo-Brachypodion*
48	*Lithodoro diffusae-Oleetum europaeae*	Non-existent	*Potentillo-Brachypodion*

Series numbers are the same as used in Table 6.1

and in which there is no alteration in terms of soil conditions; such pre-climax communities are often dominated by *Betula pubescens* subsp. *celtiberica* and *Salix sp. pl.* in oligotrophic series and *Corylus avellana* in the meso-eutrophic ones, but in either case the communities described until recently were placed in the same alliance Betulion fontqueri-celtibericae. (Figs. 6.25 and 6.26)

The Mediterranean evergreen and sclerophyllous forests communities can be considered as biogeographic singularities, occurring as edaphoxerophilous series

Fig. 6.25 Different beech woodland series are derived from different geological substrates. Calcicolous beech woodland series (Carici sylvaticae-Fago sylvaticae sigmetum) with thorny scrub (Pruno-Rubion ulmifolii), grazed pastures (Potentillo-Brachypodion rupestris) and with visible hints of limestone at Llano de Urbasa (Navarra)

Fig. 6.26 Silicicolous beech woodland series (Saxifrago spathularis-Fago sylvaticae sigmetum) with the upper border of the forest mantled by birch prewoodland (Betulion fontqueri-celtibericae), shrubland of *Erica arborea* (Ulici-Cytision striati) and a mosaic of *Erica mackaiana* heathland (Daboecion cantabricae) and hygrophilous grassland (Violion caninae) where a cow is grazing in the foreground; Carondio Range (Allande, Asturias)

Fig. 6.27 Edaphoxerophilous cork oak series (Arenario montanae-Querco suberis sigmetum) developed at rocky and southerly exposed slopes at the Navia river valley; Tamagordas (Asturias)

(Lauro nobilis-Querco ilicis-S., Genisto falcatae-Querco rotundifoliae-S.) but also as climatophilous series associated with the exceptional mesomediterranean bioclimate (Genisto hystricis-Querco rotundifoliae-S., Physospermo cornubiensis-Querco suberis-S.) present only in the Inner Galician territory (Fig. 6.27). The seral shrub stages can range from thorny communities of the Pruno spinosae-Rubion ulmifolii on limestone substrates, to retamoid scrub of the Cytision multiflori and even shade-tolerant thickets of the Ericion arboreae under a mesomediterranean climate. In the holm oak calcareous woodlands of the Lauro-Quercetum ilicis we can find pre-climax arboreal formations of the Arbuto unedonis-Laurion nobilis alliance, from the Quercetea ilicis class.

For identification and mapping of the different vegetation series and in the absence of the mature stage (the woodland community heading this series) or its pre-climax shrub mantles, the shrub communities replacing those referred to above in a situation of more intense anthropogenic degradation can be used as a diagnostic element. The baso-neutrophilous series have scrublands of the alliance Genistion occidentalis (class Festuco hystricis-Ononidetea striatae), which are completely absent in acidophilic series, where they are replaced by heathlands of the Daboecion cantabricae or, in areas with a temperate sub-Mediterranean climate, by heathlands of the Ericion umbellatae (class Calluno-Ulicetea). The original mesomediterranean series of *Quercus rotundifolia* woodlands are also an exception in this case, being replaced by scrub of a more sclerophyllous nature of the alliance Ulici argentei-Cistion ladaniferi (class Cisto-Lavanduletea stoechadis).

In many cases edaphohygrophilous woodlands have been replaced by meadows or other herbaceous communities, for agricultural and pastoral purposes; therefore, the identifiable replacement communities of those woodlands are woody shrub formations normally occurring in edges of the woodland or a seral stage of perennial grasslands (Molinio-Arrhenatheretea or Festuco-Brometea classes, see Sect. 6.5).

Riparian series with willows (*Salix sp. pl.*) woodlands as PNV, colonizing streambanks, often lack the shrub stage replacement, having only a perennial herbaceous community as substitute.

Just as in the edaphohygrophilous series, the climax woodlands of the temporihygrophilous series have been seriously affected by deforestation for the sake of croplands or permanent grasslands. In many cases this prevents the identification of the progressive seral stages and the dynamics of the whole series. In others, the PNV is only identifiable from scarce individuals of the association persisting in hedges inserted between extensive crop fields and grasslands for livestock grazing. This is a process which particularly affects the series dominated by *Ulmus minor* and *Fraxinus angustifolia* (Table 6.2-A, series 18, 19, 20 & 23).

6.7.1 Woody Permaseries

Throughout the entire Cantabrian Atlantic territory, especially near the shoreline and other thermoxerophilous environments, some communities have been located and studied whose origin and floristic affinities may relate them to the typical Mediterranean evergreen vegetation (class Quercetea ilicis) (Bueno & Fernández-Prieto 1991, Loidi et al. 1994, Honrado et al. 2003, Álvarez-Arbesú 2005, Rodríguez-Guitián et al. 2007). These are communities dominated by evergreen trees, mainly *Laurus nobilis* and exceptionally *Olea europea* var. *sylvestris*, in small patches of woodlands in particularly dry or steep and rocky biotopes. They behave as permanent communities, thus without any other woody seral stage than the present community. Although they are confined to small sporadic spots, their importance is enormous as they are relict vegetation reflecting paleoclimatic conditions dating back to the Pleistocene (Rodríguez-Guitián et al. 2007). All these associations belong to the alliance Arbuto unedonis-Laurion nobilis included in Table 6.2-B as permaseries (Permasigmetum). It should be noted that such relict formations occur in almost all coastal biogeographical units referred to as "territories" (Fig. 6.28). Several of these permaseries have been recognized and described in relatively recent times, so, understandably, only the composition of their mature formations has been studied and much is unknown about the seral herbaceous communities, as these biotopes are small and have very fragmented geographic distributions.

The set of woody Permaseries occurring in the form of very small patches of woodland can be completed with some, also evergreen, formations dominated by *Arbutus unedo*. They were described from the Basque Country to Portugal as more

Fig. 6.28 Permaseries of laurel-tree woodlands (Calluno vulgaris-Lauro nobilis sigmetum) on a low siliceous littoral cliff at O Vicedo (Lugo, Galicia)

or less dense and frequent formations, constituting various associations integrated mostly in the alliance Arbuto unedonis-Laurion nobilis (Díaz-González and Fernández-Prieto 1994; Loidi et al. 1994; Aguiar and Capelo 1995; Rodríguez-Guitián et al. 2007). But in most cases such formations have been described as seral stages of mature forests and only exceptionally as permanent communities found on rocky outcrops and other geomorphologically particular positions.

References

Aedo C, Diego C, García Codrón JC, Moreno G (1991) El bosque en Cantabria. Universidad de Cantabria. Asamblea Regional de Cantabria, Santander

Aguiar C, Capelo J (1995) Anotação sobre a posição fitossociológica dos medronhais do alto Minho. Silva Lusitana 3:123–125

Álvarez-Arbesú R (2005) La cubierta vegetal de los acantilados asturianos. Publ. Dpto. Biol. Org. Sists. Universidad de Oviedo, p 199

Amigo J, Romero-Buján MI (1998) Abedulares de origen antrópico en Galicia: caracterización fitosociológica. Studia Botanica 17:37–51

Amigo J, Guitián J, Fernández-Prieto JA (1987) Datos sobre los bosques ribereños de aliso (Alnus glutinosa) cántabro-atlánticos ibéricos. Publ. Univ. La Laguna. Ser. Informes 22:159–176

Amigo J, Giménez de Azcárate J, Romero-Buján MI (1994) Omphalodo nitidae-Coryletum avellanae, a new mesophytic woodland community of the northwest Iberian Peninsula. Botanica Helvetica 104:103–122
Amigo J, Pulgar I, Izco J (2009) Evidence of riverside ash tree forests in southern Galicia (northwestern Spain). Lazaroa 30:181–189
Arnáiz C, Loidi J (1983) Sintaxonomía de la Pruno-Rubion ulmifolii (Prunetalia) en España. Lazaroa 4:17–22
Aseginolaza C, Gómez D, Lizaur X, Montserrat G, Morante G, Salaverria M, Uribe-Echebarria P (1989) Vegetación de la Comunidad Autónoma del País Vasco. Gobierno Vasco, Vitoria-Gasteiz, p 361
Bellot F (1968) La vegetación de Galicia. Anal. Inst. Bot. Cavanilles 24:3–306
Bellot F, Casaseca B (1956) Contribución al estudio fitosociológico de los prados gallegos. Trab. Jard. Bot. Santiago de Compostela 8:1–40
Berastegi A (2013) Prados y pastizales en Navarra: descripción, tipificación y ecología. Guineana 19:1–505
Berastegi A, Darquistade A, García-Mijangos I (1997a) Biogeografía de la España centro-septentrional. Itinera Geobotanica 10:149–182
Berastegi A, García-Mijangos I, Darquistade A, Loidi J (1997b) Nuevos datos sobre los bosques secundarios (prebosques) del sector cántabro-euskaldún. Lazaroa 18:165–172
Biurrun I (1999) Flora y vegetación de los ríos y humedales de Navarra. Guineana 5:1–338. Universidad del País Vasco
Biurrun I, Campos JA, García-Mijangos I, Herrera M, Loidi J (2011) Nuevos datos sobre los bosques de barrancos y pies de cantil (Tilio-Acerion) del País Vasco y regiones limítrofes. Actes del IX Col-loqui Internacional de Botànica Pirenaico-Cantàbrica a Ordino, Andorra, pp 64–74
Braun-Blanquet J (1967) Vegetationsskizzen aus dem Baskenland mit Ausblicken auf das weitere Ibero-Atlantikum. II Teil. Vegetatio 14(1–4):1–126
Bueno A (1997) Flora y vegetación de los estuarios asturianos. Cuadernos de Medio Ambiente, Naturaleza. Oviedo 3:1–334
Bueno A, Fernández-Prieto JA (1991) Acebuchales y lauredales de la costa cantábrica. Lazaroa 12:273–302
Clements FE (1916) Plant succession: An analysis of the development of the vegetation. Carnegie Institute, Publication n° 242. Washington DC, p 515
Costa J C, Aguiar C, Capelo J, Lousã M, Neto C (1998a) Biogeografia de Portugal Continental. Quercetea 0:5–56
Costa JC, Capelo J, Lousã M, Espírito Santo MD (1998b) Sintaxonomia da vegetação halocasmofítica das falesias marítimas portuguesas (Crithmo-Staticetea Br.-Bl. 1947). Itinera Geobotanica 11:227–247
Costa JC, Capelo J, Lousã M, Antunes JH, Castro Aguiar C, Izco J, Ladero M (2000) Nota acerca os giestais da Ulici europaei-Cytision striati Rivas-Martínez, Díaz, Fernández-González & Loidi em Portugal Continental. Silva Lusitana 8(1):120–128
Costa JC, López MC, Capelo J, Lousã M (2001) Sintaxonomia das comunidades de Prunus lusitanica L. subsp. lusitanica no ocidente da Península Ibérica. Silva Lusitana 8(2):253–263
Costa JC, Aguiar C, Capelo J, Lousã M, Antunes J, Castro H, Honrado J, Izco J, Ladero M (2004) A clase Cytisetea scopario-striati em Portugal Continental. Quercetea 4:45–70
Costa JC, Honrado J, Monteiro-Henriques T, Neto C, Aguiar C (2008) As comunidades de Pterospartum tridentatum sensu lato em Portugal Continental. Silva Lusitana 16(1):123–127
Costa JC, Neto C, Aguiar C, Capelo J, Espírito Santo MD, Honrado J, Pinto-Gomes C, Monteiro-Henriques T, Sequeira M, Lousã M (2012) Vascular plant communities in Portugal (continental, the Azores and Madeira). Global Geobotany 2:1–180
Darquistade A, Berastegi A, Campos JA, Loidi J (2004) Pastizales supratemplados cántabro-euskaldunes de Agrostis curtisii: caracterización y encuadre fitosociológico. Silva Lusitana 12 (2):135–149
Díaz-González TE (1998) Síntesis de la vegetación arbustiva de Europa Occidental. I: Brezales (Calluno-Ulicetea). Itinera Geobotanica 11:7–31

Díaz-González TE (2009) Caracterización de los Hábitats de Interés Comunitario (Red Natura 2000) existentes en el Principado de Asturias. I: Hábitats litorales halófilos (dunas, acantilados y marismas). Bol. Cien. Nat. R.I.D.E.A. 50:223–280

Díaz-González TE (2010a) Caracterización de los hábitats de interés comunitario (Red Natura 2000) existentes en Asturias. II: Bosques y arbustedas arborescentes. Bol. Cien. Nat. R.I.D.E. A. 51:213–276

Díaz-González TE (2010b) Tejos y tejedas: un patrimonio natural y cultural que debemos conservar y proteger. Revista Peña Santa 6:72–82

Díaz-González TE, Fernández-Prieto JA (1994) La vegetación de Asturias. Itinera Geobotanica 8:243–528

Diemont WH, Webb NR, Degn HJ (1996) A pan-European view on heathland conservation. Proceedings of the National Heathland Conference 1996. English Nature, Peterborough

Ellenberg HH (1988) Vegetation ecology of Central Europe, 4th edn. Cambridge University Press, Cambridge, p 753

Fernández-Areces MP, Pérez-Carro FJ, Díaz-González TE (1987) Estudio del Cheilanthion hispanicae Rivas Goday 1955 em. Sáenz de Rivas & Rivas-Martínez 1979 y comunidades afines, en el sector Orensano-Sanabriense (Provincia Carpetano-Ibérico-Leonesa). Lazaroa 7:207–220

Fernández-Prieto JA, Díaz-González TE (2003) Las clasificaciones de los hábitats naturales de la Unión Europea y la Directiva Hábitats. Las formaciones leñosas altas atlánticas ibéricas. Naturalia Cantabricae 2:25–32

Fernández-Prieto JA, Loidi J (1984) Estudio de las comunidades vegetales de los acantilados costeros de la cornisa cantábrica. Documents Phytosociologiques 8:185–218

Giménez de Azcárate J, Romero-Buján MI, Amigo J (1996) Los espinales de la Pruno-Rubion ulmifolii en Galicia. Lazaroa 16:89–104

Heras FT, Infante M (2004) La turbera cobertor del Zalama (Burgos-Vizcaya): un enclave único en riesgo de desaparición. Est. Mus.Cienc. Nat. de Álava 18-19:49–57

Herrera M (1995) Estudio de la vegetación vascular de la cuenca del Río Asón. Guineana 1. Universidad del Pais Vasco, p 435

Herrera M, Fernández-Prieto JA, Loidi J (1991) Orlas arbustivas oligotróficas cantábricas: Frangulo alni-Pyretum cordatae. Stvdia Botanica 9:17–23

Honrado J (2003) Flora e Vegetação do Parque Nacional da Peneda-Gerês. Ph. Thesis. Faculdade de Ciências, Universidade do Porto, Porto, p 540

Honrado J, Alves P, Nepomuceno-Alves H, Barreto-Caldas F (2002) Ten new syntaxa from the Miniensean biogeographic subsector (Northwestern Portugal). In: Notas do Herbário da Estação Florestal Nacional (LISFA). Silva Lusitana 10(2):247–259

Honrado J, Alves P, Aguiar C, Ortiz S (2003) Juresian riparian birch woodlands: Carici reuterianae-Betuletum celtibericae ass. nova. In: Notas do Herbário da Estação Florestal Nacional (LISFA). Silva Lusitana 11:237–241

Honrado J, Alves P, Nepomuceno-Alves H, Barreto-Caldas F (2004) A vegetação do alto Minho. Esboço fitossociológico da vegetação natural do extremo Noroeste de Portugal (sectores Galaico-Português e Geresiano). Quercetea 5:3–102

Honrado J, Pulgar I, Alves P, Ortiz S (2012) Phalacrocarpo oppositifolii-Silenetum acutifoliae ass. nova hoc loco. In Costa JC Neto C Aguiar C, Capelo J Espírito Santo MD Honrado J, Pinto-Gomes C, Monteiro-Henriques T, Sequeira M, Lousã M (eds) Vascular plant communities in Portugal (continental, the Azores and Madeira). Global Geobotany 2:97–98

Ihobe (Sociedad Pública del Departamento de Medio Ambiente, Planificación Territorial, Agricultura y Pesca del Gobierno Vasco) (2011) Primera evaluación del estado de conservación de los hábitats de bosque de interés comunitario en el País Vasco. Bilbao

Izco J, Guitián J (1984) Los prados de siega con Malva moschata (Arrhenatherion elatioris) en Galicia. Pastos 12(2):255–264

Izco J, Ortiz S (1985) El mosaico pastizal-esteval (jaral de Cistus ladanifer L.) en Galicia. Bol. Soc. Brot. 58:115–138

Izco J, Sánchez JM (1996) Los medios halófilos de la ría de Ortigueira (A Coruña, España). Vegetación de dunas y marismas. Thalassas 12:63–100

Izco J, Amigo J, Guitián J (1990a) Los robledales Galaico-Septentrionales. Acta Bot. Malacitana 15:267–276

Izco J, Amigo J, Guitián J (1990b) Composición, relaciones y sistematización de los bosques esclerofilos del noroeste ibérico. Not. Fitosoc. 22:83–114

Izco J, Guitián P, Sánchez JM (1993a) Análisis y clasificación de las comunidades vegetales vivaces de las dunas vivas gallegas. Rev. Acad Galega Ciencias 12:79–104

Izco J, Guitián P, Sánchez JM (1993b) La marisma superior cántabro-atlántica meridional: estudio de las comunidades de Juncus maritimus y de Elymus pycnanthus. Lazaroa 13:149–169

Izco J, Amigo J, García-San León D (1999) Análisis y clasificación de la vegetación leñosa de Galicia (España). Lazaroa 20:29–47

Izco J, Amigo J, García San-León D (2001a) Análisis y clasificación de la vegetación de Galicia (España) II. La vegetación herbácea. Lazaroa 21:25–50

Izco J, Díaz-Varela R, Martínez-Sánchez S, Rodríguez-Guitián MA, Ramil-Rego P, Pardo-Gamundi I (2001b) Análisis y valoración de la Sierra de O Xistral: un modelo de aplicación de la Directiva Hábitat en Galicia. Consellería de Medio Ambiente. Xunta de Galicia. Santiago de Compostela, p 162

Izco J, Amigo J, Ramil-Rego P, Díaz-Varela R (2007) Brezales: biodiversidad, usos y conservación. Recursos Rurais 2:1–19

Izco J, Amigo J, Pulgar I (2009) Violion caninae grasslands (Nardetea strictae) in the North and North-West of Spain. Acta Bot. Gallica 156(3):437–454

Loidi J (1983) Datos sobre la vegetación de Guipúzcoa (País Vasco). Lazaroa 4:63–90

Loidi J, Báscones JC (2006) Memoria del Mapa de Series de Vegetación de Navarra. E 1:200.000. Departamento de Medio Ambiente, Ordenación del Territorio y Vivienda. Gobierno de Navarra.

Loidi J, Herrera M, Olano JM, Silván F (1994) Maquis vegetation in the eastern Cantabrian coastal fringe. J. Veg. Sci. 5(4):533–540

Loidi J, Biurrun I, Herrera M (1997a) La vegetación del centro-septentrional de España. Itinera Geobotanica 9:161–618

Loidi J, García-Mijangos I, Herrera M, Berastegi A, Darquistade A (1997b) Heathland vegetation of the northern-central part of the Iberian Peninsula. Folia Geobot. Phytotax. 32:259–281

Loidi J, Biurrun I, Campos JA, García-Mijangos I, Herrera M (2007) A survey of heath vegetation of the Iberian Peninsula and Northern Morocco: a biogeographic and bioclimatic approach. Phytocoenologia 37(3–4):341–370

Loidi J, Biurrun I, Campos JA, García-Mijangos I, Herrera M (2010) A biogeographical analysis of the European Atlantic lowland heathlands. J. Veg. Sci. 21:832–842

Loidi J, Biurrun I, Campos J A, García-Mijangos I, Herrera M (2011) La vegetación de la Comunidad Autónoma del País Vasco. Leyenda del mapa de series de vegetación a escala 1:50.000. Ed. Universidad del País Vasco (electronic version).

Loriente E (1974) Vegetación y flora de las playas y dunas de la provincia de Santander. Inst. Cul. Cantabria, Dip. Prov. Santander, p 287

Monteiro-Henriques T, Costa JC, Bellu A, Aguiar C (2010) Fraxino angustifoliae-Ulmetum glabrae: an original endemic and extremely localized forest from mainland Portugal. Braun-Blanquetia 46:323–327

Monteiro-Henriques T, Martins MJ, Cerdeira JO, Silva P, Arsénio P, Silva Á, Bellu A, Costa JC (2015) Bioclimatological mapping tackling uncertainty propagation: application to mainland Portugal. Int. J. Climatol. doi:10.1002/joc.4357

Moore JJ (1968) A classification of the bogs and wet heaths of northern Europe (Oxycocco-Sphagnetea Br.-Bl. & Tx. 1943). Ver. Int. Symp. Pflanzensoz. Stolzenau/Weser 1964:306–320

Muñoz Sobrino C, Ramil-Rego P, Gómez-Orellana L, Díaz-Varela RA (2005) Palynological data on major Holocene climatic events in NW Iberia. Boreas 34(3):381–400

Neto C, Costa JC, Honrado J, Capelo J (2007) Phytosociologic associations and Natura 2000 habitats of Portuguese coastal sand dunes. Fitosociologia 44 (2) suppl. 1:29–35

Onaindía M (1986) Ecología vegetal de las Encartaciones y Macizo del Gorbea (Vizcaya). Tesis Doctoral, Univ. País Vasco, p 271

Peralta J, Biurrun I, García-Mijangos I, Remón JL, Olano JM, Lorda M, Loidi J, Campos JA 2013 Manual de Hábitats de Navarra. Departamento de Desarrollo Rural, Medio Ambiente y Administración Local. Gobierno de Navarra, Pamplona, p 576

Pulgar I (1999) La vegetación de la Baixa Limia y Sierras del entorno. Tesis Doctoral. Fac. Farmacia, Univ. Santiago de Compostela, p 369

Pulgar I, Manso D (2010) Datos sobre Prunus lusitanica L. (Rosaceae) en Galicia. Nova Acta Cient. Comp. (Biol.): 33–47

Ramil-Rego P (1992) La vegetación cuaternaria de las Sierras Septentrionales de Lugo a través del análisis polínico. PhD. Facultad de Biología. Universidade de Santiago de Compostela, p 356

Ramil-Rego P, Rodríguez-Guitián MA, Rodríguez-Oubiña J (1996) Valoración de los humedales continentales del NW Ibérico: caracterización hidrológica, geomorfológica y vegetacional de las turberas de las Sierras Septentrionales de Galicia. In: Pérez-Alberti A & Martínez-Cortizas A (eds) Avances en la reconstrucción paleoambiental de las áreas de montaña lucenses. Monografías G.E.P. n° 1:166–187. Diputación Provincial de Lugo

Rivas Goday S, Rivas-Martínez S (1963) Estudio y clasificación de los pastizales españoles. Publ. Ministerio de Agricultura, Madrid, p 269

Rivas-Martínez S (1987) Memoria del Mapa de Series de Vegetación de España. Serie Técnica n° 1:9–208. ICONA. Madrid.

Rivas-Martínez S (2002a) High syntaxa of Spain and Portugal and their characteristic species. In: Rivas-Martínez S, Díaz-González TE, Fernández-González F, Izco J, Loidi J, Lousã M, Penas A (eds) Vascular plant communities of Spain and Portugal, Addenda to the syntaxonomical checklist of 2001. ItineraGeobotanica 15(2):434–922

Rivas-Martínez S (2002b) Review of definitions. In: Rivas-Martínez (on-line) Worldwide Bioclimatic Classification System. http://www.globalbioclimatics.org/book/review.htm. Accesed on 14 Feb 2015

Rivas-Martínez S (2005) Avances en Geobotánica. In: Rivas-Martínez (on-line) Worldwide Bioclimatic Classification System. http://www.globalbioclimatics.org/book/ranf2005.pdf. Accesed on 14 Feb 2015

Rivas-Martínez S (2007) Mapa de series, geoseries y geopermaseries de vegetación de España. Memoria del mapa de vegetación potencial de España. Parte I. Itinera Geobotanica 17:5–435

Rivas-Martínez S (2011) Mapa de series, geoseries y geopermaseries de vegetación de España. [Memoria del mapa de vegetación potencial de España] Parte II. Itinera Geobotanica 18 (1,2):1–800

Rivas-Martínez S, Díaz-González TE, Fernández-Prieto JA, Loidi J, Penas A (1984) La vegetación de la Alta Montaña Cantábrica. Los Picos de Europa. Ediciones Leonesas, León, p 300

Rivas-Martínez S, Cantó P, Izco J (2002) Petrocoptido pyrenaicae-Sarcocapnetea enneaphyllae classis nova hoc loco. In: Rivas-Martínez S, Díaz-González TE, Fernández-González F, Izco J, Loidi J, Lousã M, Penas A (eds) Vascular plant communities of Spain and Portugal, Addenda to the syntaxonomical Checklist of 2001. Itinera Geobotanica 15(1):156–166

Rivas-Martínez S, Penas A, Díaz-González TE, Del Rio S, Cantó P, Herrero L, Pinto-Gomes C, Costa JC (2014) Biogeography of Spain and Portugal. Preliminary typological synopsis. International Journal of Geobotanical Research 4:1–64

Rodríguez-Guitián MA (2005) Avaliación da diversidade sílvica do subsector galaico-asturiano septentrional: tipos de bosques, valor para a conservación e principais ameazas. Recursos Rurais. Serie cursos e monografias do IBADER 2:23–44

Rodríguez-Guitián MA (2006) Acerca de la identidad fitosociológica de los hayedos silicícolas sublitorales del centro de la cornisa cantábrica. Lazaroa 27:59–78

Rodríguez-Guitián MA (2010) Temperate riverside forests without alder trees at the NW of the Iberian peninsula: ecology, phytosociological profile and interest for preservation policies. Lazaroa 31:9–37

Rodríguez-Guitián MA, Ramil-Rego P (2007) Revisión de las clasificaciones climáticas aplicadas al territorio gallego desde una perspectiva biogegráfica. Recursos Rurais 1(3):31–53

Rodríguez-Guitián MA, Amigo J, Romero-Franco R (2000) Aportaciones sobre la interpretación, ecología y distribución de los bosques supratemplados naviano-ancarenses. Lazaroa 21:51–71
Rodríguez-Guitián MA, Ferreiro J, Negral MA, Merino A (2001) Distribución y ecología del haya (Fagus sylvatica L.) en el Subsector Galaico-Asturiano Septentrional (NW Ibérico). Actas del III Congreso Forestal Español. Mesas 1 y 2:201–207. Granada
Rodríguez-Guitián MA, Real C, Amigo J, Romero-Franco R (2003) The Galician-Asturian beechwoods (Saxifrago spathularidis-Fagetum sylvaticae): description, ecology and differentiation from other Cantabrian woodland types. Acta Bot Gallica 200:15–36
Rodríguez-Guitián MA, Ramil-Rego P, Romero-Franco R (2007) Caracterización ecológica y florística de las comunidades lauroides del occidente de la Cornisa Cantábrica. Lazaroa 28:35–65
Rodríguez-Guitián MA, Ramil-Rego P, Real C, Díaz-Varela R, Ferreiro da Costa J, Cillero C (2009) Caracterización vegetacional de los complejos de turberas de cobertor activas del SW europeo. In: Llamas F, Acedo C (eds) Botánica Pirenaico-Cantábrica en el siglo XXI. Área de Publicaciones. Universidad de León, León, pp 633–653
Rodríguez-Guitián MA, Ramil-Rego P, Díaz-Varela R, Pereira-Espinel J, Real C (2010) Los bosques dominados por Taxus baccata L. del extremo occidental de la Cordillera Cantábrica: caracterización, valor de conservación y amenazas. Botànica Pirenaico-cantàbrica. Actes del IX Col·loqui Internacional de Botànica Pirenaico-Cantàbrica: 367–378. Ordino. Andorra.
Romero-Buján MI (1993) La vegetación del valle del río Cabe (Terra de Lemos, Lugo). PhD, Facultad Biología, Univ. Santiago de Compostela, p 279
Teles AN (1970) Os lameiros de montanha do norte de Portugal. Subsídios para a sua caracterização fitossociológica e química. Agronomia Lusitana 31:5–132
Tüxen R, Oberdorfer E (1958) Die Pflanzenwelt Spaniens. II Teil. Eurosibirische Phanerogamen-Gesellschaften Spaniens. Veröff Geobot Inst Rübel Zürich 32:1–298
Vera JA (ed) (2004) Geología de España. Sociedad Geológica de España-Instituto Geológico y Minero de España, Madrid, p 890
Webb NR (1998) The traditional management of European heathlands. J Appl Ecol 35:987–990

Chapter 7
The High Mountain Area of Northwestern Spain: The Cantabrian Range, the Galician-Leonese Mountains and the Bierzo Trench

Tomás Emilio Díaz González and Ángel Penas

Abstract The Cantabrian Range, the Galician-Leonese mountains and the Bierzo trench comprise a territory in the north of the Iberian Peninsula on a wide variety of rocks and substrates and with significant climatic biodiversity. The area is covered by the two macrobioclimates existing in the whole of the Iberian Peninsula and the Balearic Islands: Mediterranean and Temperate, and is part of the Eurosiberian and Mediterranean biogeographical regions This variability in geology, soil and bioclimate has given rise to a highly diverse vegetation that we synthesise as follows. In the Cantabrian Range: (1) Cryorotemperate grasslands. (2) -Oro-cryorotemperate vegetation on rock fields and screes. (3) Supra-orotemperate grasslands and high-mountain pastures. (4) Cantabrian high-mountain scrubland. (5) Peat-bog vegetation. (6) Forests and woodlands dominated by *Fagus sylvatica, Quercus petraea, Betula celtiberica, Quercus orocantabrica, Quercus rotundifolia, Quercus pyrenaica, Quercus faginea, Juniperus thurifera, Ilex aquifolium, Tilia platyphyllos* and *Tilia cordata*, oro-Cantabrian acer woods with ash, and riparian forests and woodlands. (7) Meso-supratemperate scrublands (broom communities, spiny scrublands, heathlands, broom scrublands). (8) Meso-supratemperate meadows and grasslands. (9) Other vegetation types (aquatic vegetation, herbaceous ruderal vegetation, and others).

In the Galician-Leonese mountains and Bierzo trench: (1) High mountain vegetation (vegetation on rock fields and scree, supra-oromediteranean and temperate submediterranean scrublands and grasslands and grasslands of *Nardus stricta*. (2) Forest and woodland dominated by *Betula celtiberica, Quercus orocantabrica, Quercus rotundifolia, Quercus pyrenaica*, and riparian forests and woodlands. (3) Meso-supramediterranean and supratemperate submediterranean scrublands

T.E. Díaz González (✉)
Department of Organisms and Systems, Biology Faculty, University of Oviedo, E-33071 Oviedo, Spain
e-mail: tediaz@uniovi.es

Á. Penas
Department of Biodiversity and Environmental Management (Area of Botany), Mountain Livestock Farming Institute (Joint Centre CSIC-ULE), Faculty of Biological and Environmental Sciences, University of León, Campus of Vegazana, s/n, León E-24071, Spain

J. Loidi (ed.), *The Vegetation of the Iberian Peninsula*, Plant and Vegetation 12,
DOI 10.1007/978-3-319-54784-8_7

(broom communities, spiny nanoshrublands, heathlands and other types of scrubland). (4) Meso-supramediterranean meadows and grasslands. (5) Other vegetation types (aquatic vegetation, meso-supramediterranean rock field vegetation, annual grasslands and herbaceous ruderal vegetation and others).

7.1 Introduction: The Cantabrian Range

Following the proposals of Alonso et al. (2007), Aramburu and Bastida (1995) and Vera (2004) and the bibliography cited by these authors, the Cantabrian Range constitutes a coastal mountain range stretching over 400 km and following the northern coastline of the Iberian Peninsula between Galicia and the Pyrenees. Geographically, it is the westward prolongation of the Pyrenees through the Basque-Cantabrian mountains until the highest elevations in the central sector (provinces of Asturias, León, Cantabria and Palencia), where it reaches altitudes of over 2500 m in the Picos de Europa. Geologically, the mountain range was raised during the alpine orogenesis, and is flanked to the south by two synorogenic foreland basins, the Duero and Ebro river basins, located before the tectonic front of the mountain chain and whose infill may have a depth of as much as 3500 m of Tertiary continental sediments. The northernmost limit of the mountain range is delimited by the Cantabrian coast, although the geological features defined on land may extend along the continental platform and shelf until the abyssal plain of the Bay of Biscay.

From a geo-structural point of view, three sectors of domains can be distinguished in the Cantabrian Range. To the east is the Basque-Cantabrian domain in connection with the Pyrenees, with imposing Mesozoic series and more moderate reliefs that rarely exceed 1500 m. In the centre stands the feature known as the Asturian massif, which has the highest reliefs and where a much thinner Mesozoic covering has been mostly eroded to reveal a Palaeozoic basement raised during the alpine deformation. The western sector, located on the border between Asturias and Galicia, again has lower elevations; here there are no Mesozoic sediments and the alpine deformation is evident in some Tertiary basins delimited by faults.

7.1.1 The Relief in the Asturian Massif

The most notable topographic feature of this massif is a very extensive area with summit levels over 1800 m. The most important alpine fault producing the reliefs in the central sector of the Cantabrian Range is the Falla de León, or Leonese fault. To the north of this structure the elevation of the Palaeozoic basement (summit level over 2100 m) is greater than to the south (where the summit level tends to be around 1800 m). On the northern lip the highest peaks are Peña Ubiña (2417 m), Peña Prieta (2536 m) and the peaks of Espigüete (2450 m) and Curavacas (2520 m). The main peaks in the Asturian massif are generally formed by extensive outcrops of carboniferous limestone (Caliza de Montaña), which usually produce the highest reliefs due to differential erosion. Of the calcareous ranges that stand out above their adjacent areas due to their lesser differential erosion, the most emblematic in the Cantabrian Range is without question the Picos de Europa massifs, in which are located the highest mountain summit in the whole Cantabrian Range (Torrecerredo, 2648 m). Other mountain ranges in the same formation are Aramo and La Sobia.

At the south of the Cantabrian Range lies the Duero basin, and towards the west the El Bierzo trench, currently occupied by the basin of the Sil River, and which lies on a Tertiary basin flanked by alpine thrust faults. Southwards of the latter there are the Montes de León (Leonese Mountains), with heights of up to 2185 m at the Teleno summit.

7.1.2 The Mountains in the Basque-Cantabrian Zone

The mountains in the Basque-Cantabrian zone, which are mostly out of the vegetation description of this chapter, have a lower altitude than in the Asturian massif and the main hill ranges are arranged in an approximately east-west or northwest-southeast direction, following the outlines of the tectonic structures.

Towards the east, in the provinces of Burgos and Cantabria, the most notable structural feature is the alternation of large raised table mountains (Páramo de Masa, Las Loras and others, also known as the Burgos platform). To the south of the Burgos platform, the structure at the forefront of the mountain range changes laterally into a series of folds in a northwest-southeast direction. In this sector of the mountain range to the north of the province of Burgos, altitudes rarely exceed 1200 m (the maximum height is Peña Amaya, at 1362 m, on the southern border of the mountain range), in contrast with the numerous summits of over 2000 m immediately to the northwest in the province of Palencia, in the transition to the Asturian massif (Peña Labra, Pico Tres Mares-Alto Campoo, Sierra de Híjar).

The easternmost area of the Basque-Cantabrian zone represents the transition to the Pyrenees, where the typical Pyrenean structural domains can be differentiated. The main mountain ranges in the southern part (Cantabria, Obarenes, Tesla) of this eastern sector of the Basque-Cantabrian zone rarely exceed altitudes of 1300 m. The trenches located to the north of these ranges (Synclinals of Villarcayo and Miranda-Treviño) are basins with deposits of Tertiary sediments caused by their own erosion and the erosion of the mountains located to the north, which constitute the watershed for the mountain range in this sector and the site of the highest summits in the Basque-Cantabrian zone (Castro Valnera at 1717 m, Gorbeia, Aizkorri and others).

7.1.3 The Galician Mountains

In western Asturias and inner eastern Galicia the structural model described for the central sector of the mountain range ceases to be applicable. In this western sector, the alpine compression has been resolved along faults in a northwest-northeast direction. Thus in the northwest of Galicia, the Tertiary basins are linked to the aforementioned faults in a north-west direction, producing vertical shears of up to 500 m and horizontal shears of 1 km. Some of the faults in the north-east direction are currently active and also have Tertiary basins and mountain alignments following the same direction, such as the ranges of Ancares, El Caurel, Faro and others. Also evident on the topographical map in this western sector of the Cantabrian Range are other mountain ranges oriented in a direction that changes progressively from north-south near the coastline, to northwest-southeast towards the south, and continuing even towards the Montes de León. These ranges trace the shape of what is known as the Asturian Arc.

Although tectonics are the main cause of the most significant features in the relief, it is also shaped by other processes that have left a clear imprint on the mountains and valleys we see today. Erosion, fundamentally fluvial and glacial, plays a key role in the evolution of mountain ranges in general, and of the Cantabrian range in particular. We have already noted the importance of fluvial erosion in modelling the current relief and shifting the watershed in the mountain range towards the south. The rivers on the north face have a greater erosive capacity due to the proximity of their base level in the Cantabrian Sea, and this is evidenced by the multitude of deep channels that can be seen on this face of the mountain range. The most spectacular example is most certainly the Cares River gorge in the calcareous massif of the Picos de Europa, whose walls have a gap of more than 2000 m of elevation from the river bed to the summits of Torrecerredo (2648 m) and Peña Santa de Castilla (2596 m) (Figs. 7.1 and 7.13). Other significant gorges on the north face of the range are Los Beyos and La Hermida, formed by the Sella and Deva Rivers on the western and eastern edges of the Picos de Europa respectively. On the south face, the only significant river channel is that of the Sil River, whose

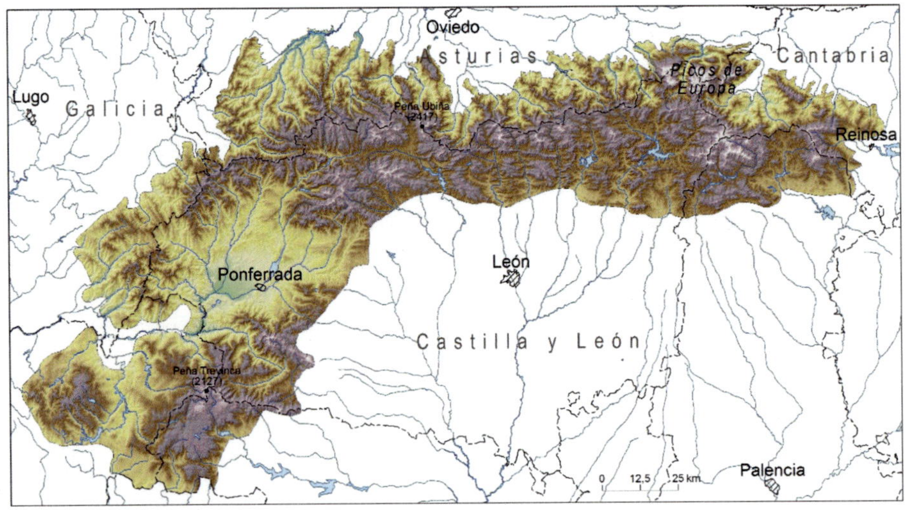

Fig. 7.1 Physical map of the territory

channel –only a few kilometres from its source in the region of Babia in León– already has a depth of over 80 m, becoming much more pronounced downriver in the deep broad canyons of the Ribeira Sacra in the province of Orense.

During the Pleistocene, a succession of glacial periods also made a significant mark on the relief. During the last glacial maximum, most of the summits were covered by ice, and its presence has been documented on the northern face starting at altitudes of between 950 and 400 m. Due to the greater insolation the ice level was higher on the south face; in the Babia region it has been calculated from an altitude of approximately 1500 m, with the ice sheet attaining a thickness of up to 400 m. These glaciers have given rise to a multitude of characteristic erosive forms such as the typical U-shaped valleys, glacial cirques, horns (Pico Cornón, etc.), and to large quantities of till deposits on valley floors, or forming the characteristic moraine arcs. Most of the emblematic lakes in the Cantabrian Range (Somiedo Lakes, Covadonga Lakes, and others) are of glacial origin.

Lithology The age of the rocks in the Cantabrian Range span from the Precambrian (>541 Ma) to the Quaternary eras, and can be divided into three groups, each of which delivers one part of the geological history of this region:

Pre-Cambrian rocks: these appear throughout an arc-shaped band with a width of no more than 20 km (Narcea Antiform), running through the locations of Cudillero, Tineo, Cangas del Narcea and Puerto de Leitariegos. From the southern limit of Asturias, the Narcea Antiform penetrates into the province of León, and adopts a WNW-ESE direction between Villablino and Barrios de Luna. The pre-Cambrian rocks consist of lutite and sandstone (Narcean slate), interspersed with an abundance of volcanic and plutonic rocks.

Palaeozoic rocks: these are found on both sides of the Narcea Antiform and discordantly on pre-Cambrian rocks. The rocks in the eastern part form what is known as the Cantabrian Zone, and can be dated to between the Cambrian and Carboniferous (Pennsylvanian) eras. Until the upper Devonian, the Palaeozoic rocks in this area consist of units of limestone, dolomite, sandstone, quartzite and lutite, deposited on the marine continental platform.

The Palaeozoic rocks located to the west of the Narcea Antiform form part of what is known as the Leonese-Western Asturian Zone. Almost all these rocks date from between the Cambrian and the Silurian era, and consist mainly of slate and quartzite deposited on a deeper marine medium. The lower part of the series contains some calcareous units (Vegadeo limestone). Overlying this whole lithological series in the western sector of Asturias is the discordant presence of sandstone and lutite from the Stephanian era containing coal seams that are still mined in the areas of Cangas del Narcea, Degaña and Villablino.

Permian, Mesozoic and Cenozoic rocks: These units represent the Alpine Cycle. Permian and Mesozoic rocks can be seen almost continuously along the central and northern parts of Asturias, and form the western part of what is known as the Basque-Cantabrian basin. There are other more limited outcrops at several points on the Asturian and Leonese faces of the Cantabrian Range, comprising limestone, marl, sandstone, conglomerate and lutite, mainly deposited on the southernmost edge of the basin, which was originally continental and then evolved into a shallow marine basin. The environment in which the sedimentation of these materials occurred was also conducive to life, as evidenced by the abundant fossil remains of brachiopods, lamellibranchia, cephalopods and dinosaurs. Overlying the Mesozoic materials along a band running E-W from Grado to Cangas de Onís and to the south of the mountain range are Cenozoic rocks.

7.1.4 The Galician-Leonese Mountains

Located in the north-western corner of the Iberian Peninsula, they form part of the Hesperian Shield. Their uplift corresponds to the reactivation of late Hercynian faults during the Alpine orogeny, which caused a tectonic of raised and sunken blocks, originating fosses and massifs respectively. The mountain alignments have a general E-W direction and can be divided into two groups, separated by the long trench of the Sil river and comprising the areas of El Bierzo and Valdeorras. Noteworthy in the northern group are the massifs of El Caurel (Pájaro 1616 m) and Ancares (Peñarrubia 1826 m, Miravalles 1969 m), with a SW-NE orientation and connecting with the great Cantabrian Range of which they to some degree constitute its western extreme. South of the Sil trench there is an assembly of mountain alignments which form part of the interfluve between the Sil-Miño basin and the Duero valley. One of the most notable reliefs is the Sierra de Queixa (Cabeza de Manzaneda 1778 m), which is continued eastwards in the huge massifs of the Montes de León with the alignments of Sierra Segundera (Peña Trevinca 2096 m) –La Cabrera (Vizcondillo 2144 m) and Montes Aquilianos (Cabeza de la Yegua 2135 m)– and Sierra del Teleno (2188 m); this whole group is delimited on the southern side by the Sierra de la Culebra, with lower elevations (Peña Mira 1243 m).

The lithology of these mountains is predominantly siliceous, with quartzite, slate and some granitic outcrops. In the Ancares area there are some bars of Palaeozoic limestone which give rise to narrow canyons when traversed by the Sil river.

7.2 Biogeography

The biogeographical territories, according to units and codes proposed by Rivas-Martínez et al. (2014), are the following:

IB. Atlantic-Central European Subregion (*Subregión Atlántica-Centroeuropea*)

Ib. European Atlantic Province (*Provincia Atlántica Europea*)

Ibb. Oro-Cantabrian Subprovince (*Subprovincia Orocantábrica*)

7. Alto Campoo and Carrión Sector (*Sector Altocampurriano-Carrionés*)
   - 7a. Alto Campoo District (*Distrito Altocampurriano*)
   - 7b. Liébana District (*Distrito Lebaniego*)
   - 7c. Alto Esla and Carrión District (*Distrito Altoeslano-Altocarrionés*)
   - 7d. Sierra de Espigüete District (*Distrito Serrano Espigüeteño*)

8. Picos de Europa and Ubiña Sector (*Sector Picoeuropeano-Ubiñense*)
   - 8a. Picos de Europa District (*Distrito Picoeuropeano*)
   - 8b. Redes District (*Distrito Redesano*)
   - 8c. Somiedo and Ubiña District (*Distrito Somedano-Ubiñense*)
   - 8d. Central Oro-Cantabrian District (*Distrito Orocantábrico Central*)
   - 8e. Sierra de Mampodre District (*Distrito Serrano Mampodrense*)
   - 8f. Babia and Torío District (*Distrito Babia-Torío*)

9. Laciana and Ancares Sector (*Sector Lacianiego-Ancarense*)
   - 9a. Laciana District (*Distrito Lacianiego*)
   - 9b. Alto Narcea District (*Distrito Altonarceense*)
   - 9c. Omaña District (*Distrito Omañés*)
   - 9d. Sierra de Ancares District (*Distrito Serrano Ancarense*)
   - 9e. Sierra del Caurel District (*Distrito Serrano Caureliano*)

IIc. West Iberian Mediterranean Province (*Provincia Mediterránea Ibérica Occidental*)

IIca. Carpetan-Leonese Subprovince (*Subprovincia Carpetana-Leonesa*)

33. Bierzo and Sanabrian Sector (*Sector Berciano-Sanabrés*)
   - 33a. Bierzo District (*Distrito Berciano*)
   - 33b. Cabrera and Montes de León District (*Distrito Cabreireño-Monteleonés*)
   - 33c. Zamoran-Sanabrian District (*Distrito Zamorano-Sanabrés*)
   - 33d. Sierra de Queixa District (*Distrito Serrano Queixense*).

Each one of these sectors is characterised below with the various climatophilous, edaphoxerophilous, climato-temporihygrophilous and edaphohygrophilous vegetation series and geopermaseries, following the proposals established by Rivas-Martínez et al. (2007, 2011a).

European Atlantic & West Iberian Mediterranean Province	7	8	9	33
Climatophilous Series				
Carici caudatae-Fago sylvaticae S. (neutrophilous)	–	•	–	–
Lithodoro diffusae-Junipero sabinae S. (basophilous)	–	•	–	–
Junipero sabino-orocantabrica S. (basophilous)	–	•	–	–
Carici sylvaticae-Fago sylvaticae S. (neutrophilous)	–	•	–	–
Vaccinio myrtilli-Pino ibericae S. (acidophilous, relict)	–	•	–	–
Omphalodo nitidae-Fago sylvaticae S. (neutro-acidophilous)	–	–	•	-
Genisto sanabrensis-Junipero alpinae S. (acidophilous)	–	–	–	•
Genisto falcatae-Querco pyrenaicae S. (acidophilous)	–	–	–	•

(continued)

European Atlantic & West Iberian Mediterranean Province	7	8	9	33
Daphno cantabricae-Arctostaphylo uvaeursi S. (basophilous)	•	•	–	–
Blechno spicant-Fago sylvaticae S. (acidophilous)	•	•	–	–
Vaccinio microphylli-Junipero alpinae S. (acidophilous)	•	•	•	
Linario triornithophorae-Querco petraeae S. (acidophilous)	•	•	•	–
Luzulo henriquesii-Querco petraeae S. (acidophilous)	•	•	•	–
Avenello ibericae-Fago sylvaticae S. (acidophilous)	•	•	•	–
Saniculo europaeae-Ilico aquifolii S. (neutrophilous)	•	•	•	–
Epipactido helleborines-Fago sylvaticae S. (neutrophilous)	–	•	•	–
Vaccinio myrtilli-Junipero alpinae S. (acidophilous)	–	–	•	•
Luzulo henriquesii-Betulo celtibericae S. (acidophilous)	–	•	•	•
Linario triornithophorae-Querco pyrenaicae S. (acidophilous)	•	•	•	•
Avenello ibericae-Querco orocantabricae S. (acidophilous)	•	•	•	•
Climatophilous and edaphoxerophilous series				
Cephalanthero longifoliae-Querco rotundifoliae S. (basophilous)	–	•	–	–
Berberido cantabricae-Querco fagineae S. (basophilous)	–	•	–	–
Genisto falcatae-Querco rotundifoliae S. (basophilous)	–	–	•	–
Genisto hystricis-Querco rotundifoliae S. (acidophilous)	–	–	–	•
Edaphoxerophilous series				
Lithodoro diffusae-Querco rotundifoliae S. (basophilous)	–	•	–	–
Climato-temporihygrophilous series				
Helleboro occidentalis-Tilio cordatae S. (neutrophilous)	–	•	–	–
Querco pyrenaicae-Fraxino angustifoliae S. (soft freshwater)	–	–	–	•
Climato-temporihygrophilous and edapho-hygrophilous series				
Luzulo henriquesii-Acero pseudoplatani S. (neutro-acidophilous)	–	–	•	–
Edapho-hygrophilous series				
Hyperico androsaemi-Alno glutinosae S. (soft freshwater)	–	•	–	–
Valeriano pyrenaicae-Alno glutinosae S. (soft freshwater)	–	–	•	–
Salici cantabrico-bicoloris S. (hard freshwater)	•	•	–	–
Salici cantabricae S. (hard freshwater)	•	•	–	–
Salici cantabrico-albae S. (hard freshwater)	•	•	–	–
Euphorbio hybernae-Fraxino excelsioris S. (hard freshwater)	•	•	–	–
Festuco giganteae-Fraxino excelsioris S. (hard freshwater)	•	•	•	–
Aro cylindracei-Ulmo minoris S. (soft freshwater)	–	–	–	•
Populo nigrae-Salici neotrichae S. (soft freshwater)	–	–	–	•
Galio broteriani-Alno glutinosae S. (soft freshwater)	–	–	–	•
Geopermaseries				
Junco trifidi-Oreochloo blankae GPS. (acidophilous)	•	–	–	–
Oxytropido neglectae-Kobresio myosuroidis GPS. (basophilous)	–	•	–	–
Oxytropido neglecto-halleri GPS. (basophilous)	•	•	–	–
Teesdaliopsio confertae-Festuco eskiae GPS. (acidophilous)	•	•	•	–
Teesdaliopsio confertae-Festuco summilusitanae GPS. (acidophilous)	–	–	–	•

7.3 Bioclimatology

The bioclimatology of the territory is shown in maps of Figs. 7.2 and 7.3 with the distribution of the thermotypes and ombrotypes respectively being consistent with Rivas-Martínez's typology (Rivas-Martínez et al. 2011b). For some thermopluviometric stations of the territory, the following tables indicate the values of the climate parameters and bioclimatic indices that allow us to establish the corresponding bioclimatic diagnoses of the territory. We also define the relationship between the different types of potential natural vegetation and their bioclimatology.

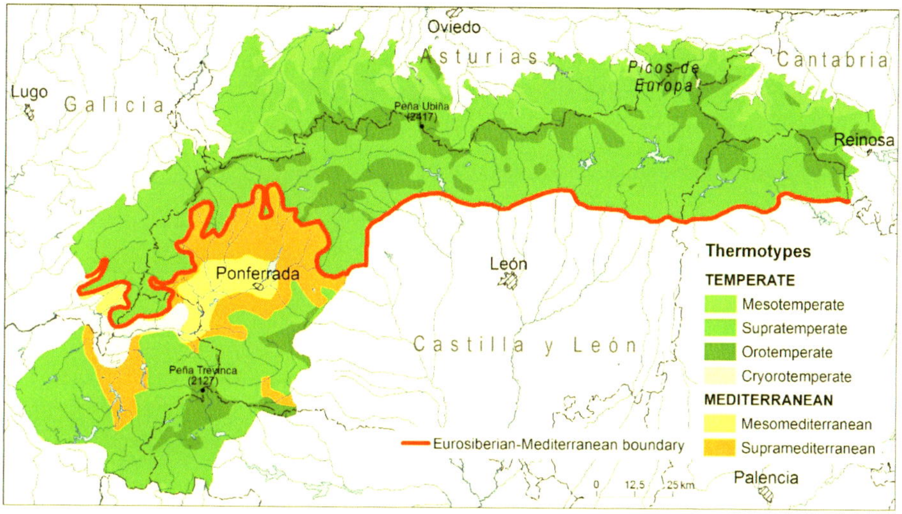

Fig. 7.2 Map of thermotypes of the Cantabrian range, Bierzo Trench and Leonese Mountains

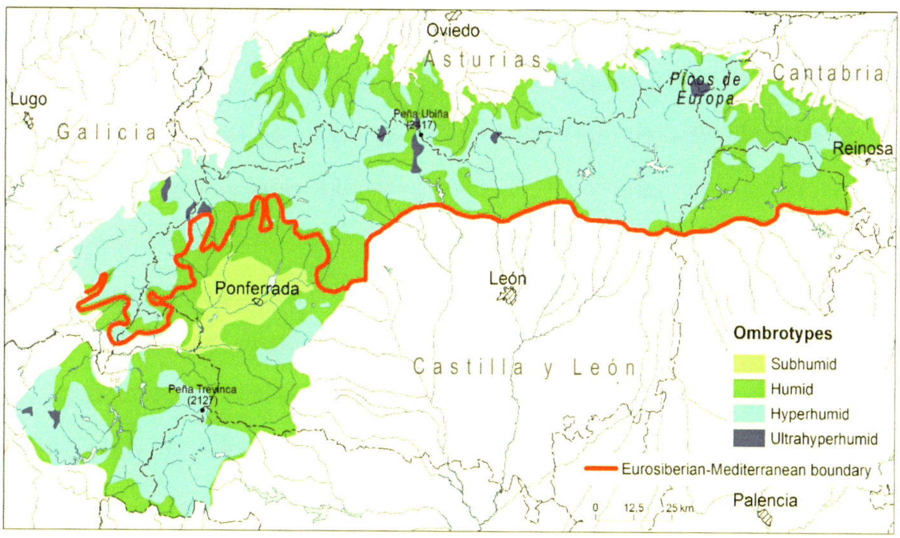

Fig. 7.3 Map of ombrotypes of the Cantabrian range, Bierzo Trench and Leonese Mountains

Bioclimatic stages	Thermotypes						Ombrotypes			
Vegetation series and geopermaseries	Mm	Sm	Mt	St	Ot	Ct	dr	sh	h	hh
Climatophilous series										
Genisto falcatae-Querco pyrenaicae S. (acidophilous)	–	•	–	•	–	–	–	•	•	–
Linario triornithophorae-Querco pyrenaicae S. (acidophilous)	–	–	•	•	–	–	–	–	•	–
Saniculo europaeae-Ilico aquifolii S. (neutrophilous)	–	–	•	•	–	–	–	–	•	–
Epipactido helleborines-Fago sylvaticae S. (neutrophilous)	–	–	•	•	–	–	–	–	•	–
Carici sylvaticae-Fago sylvaticae S. (neutrophilous)	–	–	•	•	–	–	–	–	•	•
Blechno spicant-Fago sylvaticae S. (acidophilous)	–	–	•	•	–	–	–	–	•	•
Junipero sabino-orocantabrica S. (basophilous)	–	–	–	•	–	–	–	•	•	–
Omphalodo nitidae-Fago sylvaticae S. (neutro-acidophilous)	–	–	–	•	–	–	–	–	•	–
Linario triornithophorae-Querco petraeae S. (acidophilous)	–	–	–	•	–	–	–	–	•	–
Luzulo henriquesii-Querco petraeae S. (acidophilous)	–	–	–	•	–	–	–	–	•	–
Carici caudatae-Fago sylvaticae S. (neutrophilous)	–	–	–	•	•	–	–	–	•	•
Lithodoro diffusae-Junipero sabinae S. (basophilous)	–	–	–	•	•	–	–	–	•	•
Luzulo henriquesii-Betulo celtibericae S. (acidophilous)	–	–	–	•	•	–	–	–	–	•
Avenello ibericae-Querco orocantabricae S. (acidophilous)	–	–	–	•	•	–	–	–	–	•
Genisto sanabrensis-Junipero alpinae S. (acidophilous)	–	–	–	–	•	–	–	–	•	•
Vaccinio myrtilli-Junipero alpinae S. (acidophilous)	–	–	–	–	•	–	–	–	•	•
Vaccinio myrtilli-Pino ibericae S. (acidophilous, relict)	–	–	–	–	•	–	–	–		•
Daphno cantabricae-Arctostaphylo uvaeursi S. (basophilous)	–	–	–	–	•	–	–	–	–	•
Vaccinio microphylli-Junipero alpinae S. (acidophilous)	–	–	–	–	•	–	–	–	–	•
Avenello ibericae-Fago sylvaticae S. (acidophilous)	–	–	–	–	•	–	–	–	–	•
Climatophilous and edaphoxerophilous series										
Cephalanthero longifoliae-Querco rotundifoliae S. (basophilous)	•	–	•	•	–	–	–	•	•	–
Genisto hystricis-Querco rotundifoliae S. (acidophilous)	•	•	–	–	–	–	•	•	•	–
Genisto falcatae-Querco rotundifoliae S. (basophilous)	–	–	•	•	–	–	–	•	•	–

(continued)

Bioclimatic stages	Thermotypes						Ombrotypes			
Vegetation series and geopermaseries	Mm	Sm	Mt	St	Ot	Ct	dr	sh	h	hh
Berberido cantabricae-Querco fagineae S. (basophilous)	–	–	–	•	–	–	–	–	•	–
Xerophilous series										
Lithodoro diffusae-Querco rotundifoliae S. (basophilous)	–	–	–	•	–	–	–	•	•	–
Climato-temporihygrophilous series										
Querco pyrenaicae-Fraxino angustifoliae S. (soft freshwater)	–	•	–	–	–	–	•	•	•	–
Helleboro occidentalis-Tilio cordatae S. (neutrophilous)	–	–	•	•	–	–	–	–	•	–
Climato-temporihygrophilous and edapho-hygrophilous series										
Luzulo henriquesii-Acero pseudoplatani S. (neutro-acidophilous)	–	–	–	•	–	–	–	–	•	–
Hygrophilous series										
Aro cylindracei-Ulmo minoris Gs. (soft freshwater)	–	•	–	–	–	–	•	•	–	–
Populo nigrae-Salici neotrichae Gs. (soft freshwater)	–	•	–	–	–	–	•	•	–	–
Galio broteriani-Alno glutinosae Gs. (soft freshwater)	–	•	–	–	–	–	•	•	•	–
Hyperico androsaemi-Alno glutinosae S. (soft freshwater)	–	–	•	–	–	–	–	•	•	–
Valeriano pyrenaicae-Alno glutinosae Gs. (soft freshwater)	–	–	•	•	–	–	–	•	•	–
Festuco giganteae-Fraxino excelsioris Gs. (hard freshwater)	–	–	•	•	–	–	–	•	•	–
Salici cantabricae Gs. (hard freshwater)	–	–	•	•	–	–	–	•	•	–
Salici cantabrico-albae Gs. (hard freshwater)	–	–	–	•	–	–	–	–	•	–
Euphorbio hybernae-Fraxino excelsioris Gs. (hard freshwater)	–	–	–	•	–	–	–		•	•
Salici cantabrico-bicoloris Gs. (hard freshwater)	–	–	–	–	•	–	–	–		•
Geopermaseries										
Oxytropido neglecto-halleri Gps. (basophilous)	–	–	–	–	•	•	–	–	•	•
Teesdaliopsio confertae-Festuco summilusitanae Gps. (acidophilous)	–	–	–	–	•	–	–	–	•	•
Teesdaliopsio confertae-Festuco eskiae Gps. (acidophilous)	–	–	–	–	•	–	–	–	–	•
Junco trifidi-Oreochloo blankae Gps. (acidophilous)	–	–	–	–	–	•	–	–	–	•
Oxytropido neglectae-Kobresio myosuroidis Gps. (basophilous)	–	–	–	–	–	•	–	–	–	•

Mm mesomediterranean, *Sm* supramediterranean, *Mt* mesotemperate, *St* supratemperate, *Ot* orotemperate, *Ct* cryorotemperate, *dr* dry, *sh* subhumid, *h* humid, *hh* hyperhumid

Locality	Prov	Alt. (m)	Lat.	Long.	POT	POP	T	M	m	P	Itc	Tp	Ic	Io	Ios_2	Ios_3	Ios_4
Leitariegos	O	1525	43° 8′ N	6° 25′ W	25	25	5.0	4.4	−6.6	1739	28	618	13.4	22.8	5.04	6.72	9.61
La Foz de Morcín	O	260	43° 15′ N	5° 52′ W	31	36	13.2	12.0	2.2	1250	274	1582	12.7	7.9	3.71	4.11	5.03
Amieva	O	700	43° 13′ N	5° 1′ W	39	31	10.4	8.4	3.8	1866	226	1251	13.4	14.9	5.18	5.63	7.46
Cangas de Narcea	O	670	43° 2′ N	6° 40′ W	32	26	10.3	8.5	−0.2	2042	186	1233	13.8	16.6	2.43	4.08	5.59
Potes	S	280	43° 9′ N	4° 37′ W	24	22	13.2	11.4	1.8	637	264	1582	13.4	4.03	1.08	1.29	1.74
Triollo	P	1299	42° 59′ N	4° 41′ W	26	26	8.3	4.3	−4.5	1166	80	996	16.6	10.2	2.07	2.88	4.11
Cervera de Pisuerga	P	1013	42° 51′ N	4° 29′ W	37	37	8.7	6.3	−3.5	981	115	1049	16.0	9.35	1.95	2.52	3.66
Isoba	Le	1540	43° 3′ N	5° 21′ W	27	27	5.5	2.8	−3.2	1347	49	667	13.5	18.3	4.38	4.94	7.51
Maraña	Le	1253	43° 3′ N	5° 10′ W	43	39	8.1	5.6	−3.9	1414	58	972	15.3	14.5	2.61	3.29	5.22
Ponferrada	Le	555	42° 38′ N	6° 34′ W	49	49	12.7	8.2	1.2	638	221	1522	16.8	4.2	0.99	1.30	178
La Baña	Le	1040	42° 16′ N	6° 40′ W	32	38	10.5	7.9	−1.0	1071	174	1260	15.6	8.5	1.83	2.20	2.83
Puebla de Sanabria	Za	960	42° 3′ N	6° 38′ W	38	38	9.6	7.2	−2.4	998	143	1151	15.5	8.67	1.10	150	2.41
Presa Cardena	Za	1600	42° 8′ N	6° 47′ W	37	37	5.8	2.8	−5.1	1578	35	718	15.7	14.0	2.14	3.66	5.98
Porto	Za	1531	42° 10′ N	6° 54′ W	67	74	6.7	3.5	−3.7	1588	65	802	15.2	17.5	2.57	3.81	5.8
Parafita de Queija	Or	1194	42° 15′ N	7° 21′ W	48	42	9.1	6.9	−0.7	1271	153	1090	13.1	11.7	1.92	2.63	4.29

Prov Province, *Alt. (m)* Altitude (m), *Lat.* Latitude, *Long.* Longitude, *POT* Thermal observation period (year), *POP* Rainfall observation period (year), *T* Average temperature, *M* Average temperature of the maximums of the coldest month, *m* Average temperature of the minimums of the coldest month, *P* Average precipitation, *Itc* Compensated thermicity index, *Tp* Positive temperature, *Ic* Continentality index, *Io* annual ombrothermic index, Ios_2 bimonthly ombrothermic index, Ios_3 quarterly ombrothermic index, Ios_4 four-monthly ombrothermic index (definitions of these parameters are given in Chap. 2)

Locality	Prov.	Alt. (m)	Lat.	Long.	Bioclimatic diagnosis
Leitariegos	O	1525	43° 8′ N	6° 25′ W	Temperate oceanic low orotemperate upper hyperhumid weak semihyperoceanic
La Foz de Morcín	O	260	43° 15′ N	5° 52′ W	Temperate oceanic low mesotemperate low humid weak semihyperoceanic
Amieva	O	700	43° 13′ N	5° 1′ W	Temperate oceanic upper mesotemperate low hyperhumid weak semihyperoceanic
Cangas de Narcea	O	670	43° 2′ N	6° 40′ W	Temperate oceanic submediterranean low supratemperate low hyperhumid weak semihyperoceanic
Potes	S	280	43° 9′ N	4° 37′ W	Mediterranean pluviseasonal oceanic upper mesomediterranean low subhumid weak semihyperoceanic
Triollo	P	1299	42° 59′ N	4° 41′ W	Temperate oceanic submediterranean upper supratemperate upper humid weak euoceanic
Cervera de Pisuerga	P	1013	42° 51′ N	4° 29′ W	Temperate oceanic submediterranean upper supratemperate upper humid weak euoceanic
Isoba	Le	1540	43° 3′ N	5° 21′ W	Temperate oceanic low orotemperate upper hyperhumid semihyperoceanic
Maraña	Le	1253	43° 3′ N	5° 10′ W	Temperate oceanic low orotemperate upper hyperhumid weak semihyperoceanic
Ponferrada	Le	555	42° 38′ N	6° 34′ W	Mediterranean pluviseasonal oceanic upper mesomediterranean low subhumid weak euoceanic
La Baña	Le	1040	42° 16′ N	6° 40′ W	Temperate oceanic submediterranean low supratemperate low humid weak euoceanic
Puebla de Sanabria	Za	960	42° 3′ N	6° 38′ W	Mediterranean pluviseasonal oceanic upper supramediterranean low humid weak euoceanic
Presa Cardena	Za	1600	42° 8′ N	6° 47′ W	Temperate oceanic submediterranean low orotemperate low hyperhumid weak euoceanic
Porto	Za	1531	42° 10′ N	6° 54′ W	Temperate oceanic submediterranean upper supratemperate low hyperhumid weak euoceanic
Parafita de Queija	Or	1194	42° 15′ N	7° 21′ W	Temperate oceanic submediterranean low supratemperate upper humid semihyperoceanic

7.4 The Plant Landscape of the Cantabrian Range (Oro-Cantabrian Subprovince)

The vegetation typology used for the description of the vegetation in this territory is that established by Rivas-Martínez et al. (1999, 2001, 2002a, b)

7.4.1 High-Mountain Vegetation (Cryoro and Orotemperate)

7.4.1.1 Cryorotemperate Grasslands

The **cryorotemperate (alpine) grasslands of *Kobresia myosuroides* (*Elyna myosuroides*) in the Picos de Europa sector** (Oxytropido neglectae-Kobresietum myosuroidis) represent the mature stage of the climactic (chionophobous) vegetation series in the hyperhumid cryorotemperate belt of the Picos de Europa, and occupy deep calcified soils of a moderately alpine type. They are characterised by the presence of *Kobresia myosuroides* (*Elyna myosuroides*), *Oxytropis neglecta (Oxytropis pyrenaica)*, *Carex capillaris*, *Carex sempervirens, Silene acaulis* and *Erigeron uniflorus* subsp. *picoeuropaeanus*, among others. Peaks and ridges in high calcareous mountains over 2000 m with no permanent snow cover or intense temporary hydromorphia tend to suffer an alpinising effect due to wind and geliturbation, which causes the climatophilous creeping juniper woods of the Juniperion alpinae –corresponding to the upper orotemperate hyperhumid horizon– to be replaced by upper orotemperate oro-Cantabrian grasslands (upper subalpine) of *Oxytropis halleri* (Oxytropidetum neglecto-halleri). These permanent climatophilous and anemochionophobous meadows, rich in pulvinate rosulate nanochamaephytes of the Armerion cantabricae, present a series of territorial endemics and a large group of alpine and Pyrenean species. Frequent in these environments are *Androsace lactea, Arenaria purpurascens, Astragalus australis, Callianthemum coriandrifolium, Carex capillaris, Carex rupestris, Dethawia tenuifolia, Gentiana nivalis, Gentiana occidentalis, Geum pyrenaicum, Horminium pyrenaicum, Lotus alpinus, Minuartia verna, Oxytropis foucaudii, Oxytropis halleri, Oxytropis neglecta, Pedicularis foliosa, Polygonum viviparum, Ranunculus gouanii, Ranunculus thora* and *Silene acaulis*, among others.

7.4.1.2 Oro-Cryorotemperate Vegetation on Rock Cliffs and Screes

In supra-orotemperate territories in the Picos de Europa and Ubiña sector in a hyperhumid ombroclimate, the fissures in calcareous rocks are colonised by various chasmophytic communities with sparse coverage (Anemono pavonianae-Saxifragetum canaliculatae) that are floristically characterised by the presence of *Saxifraga canaliculata, Anemone pavoniana*, *Valeriana apula, Hieracium lainzii, Saxifraga babiana* var. *septentrionalis*, *Crepis albida* subsp. *asturica, Campanula arvatica* subsp. *arvatica* and *Saxifraga paniculata*, among others. Fissures in calcareous limestone pavements on oro-cryorotemperate summits in the Picos de Europa and Ubiña sector are colonised by various chasmophytic communities (Potentillo asturicae-Valerianetum apulae) whose floristic composition includes *Potentilla nivalis* subsp. *asturica*, *Valeriana apula, Saxifraga* x *rivas-martinezii, Saxifraga canaliculata, Silene ciliata, Hieracium bombycinum* and *Globularia repens*, in some cases along with glericolous plants such as *Galium pyrenaicum*, or with high-mountain tufts such as *Silene acaulis* or *Festuca burnatii*. The peculiar conditions of the territory in the Picos de Europa sector give rise to the colonisation of the fissures in oro-cryorotemperate calcareous rock fields with a hyperhumid character by chasmophytic vegetation (Saxifrago felineri-Dethawietum tenuifoliae) with sparse coverage, with an abundance of plants such as *Dethawia tenuifolia*, *Saxifraga aretioides* subsp. *felineri*, *Asperula hirta*, *Reseda glauca*, *Potentilla nivalis* subsp. *asturica*, *Silene ciliata*, *Saxifraga canaliculata*, *Campanula arvatica* subsp. *arvatica* and *Globularia repens*, among others.

In the fissures of shady, generally oozing and chionophilous limestone rocks, in the upper supratemperate, oro- and cryorotemperate territories of the Picos de Europa and Ubiña sector the vegetation (Campanulo arbaticae-Saxifragetum paucicrenatae) is characterised by the presence of *Saxifraga hirsuta* subsp. *paucicrenata*, *Campanula arvatica*, *Viola biflora*, *Asplenium viride*, *Cystopteris fragilis* subsp. *pseudoregia* and *Cystopteris alpina*.

In siliceous rocks in more continentalised southern oro-Cantabrian humid-hyperhumid supra-orotemperate and frequently submediterrranean territories, the fissures are colonised by chasmophytic communities (Murbeckiello boryi-Sperguletum rimarum) characterised by the presence of *Murbeckiella boryi* subsp. *boryi* and *Spergula viscosa* subsp. *pourretii* (*Spergula rimarum*), among others; this vegetation occurs spread throughout Carpetan, Sierra Iberian and southern oro-Cantabrian territories (Fig. 7.4).

Calcareous fissures in cave ceilings and walls, cavities and overhangs protected by ledges in humid to hyperhumid meso-cryorotemperate territories in the Picos de Europa and Ubiña sector are occupied by species-poor chasmophytic rupicolous and spelucicolous communities (Petrocoptidetum glaucifoliae) that are floristically characterised by a predominance of *Petrocoptis pyrenaica* subsp. *glaucifolia*, and to a lesser degree by *Asplenium csikii (Asplenium trichomanes* subsp. *pachyrachis*) in the easternmost territories.

Calcareous limestone pavements with surface gelifractures covered with small or medium-sized pebbles and with a short period of snow cover (chionophobous) in cryorotemperate territories in the Picos de Europa sector are colonised by various glericolous communities with sparse coverage (Galio pyrenaici-Salicetum fontqueri). This scree vegetation is floristically characterised by the presence of *Salix breviserrata* subsp. *fontqueri*, *Galium pyrenaicum*, *Iberis carnosa* subsp. *lereschiana*, *Festuca glacialis*, *Jasione cavanillesii* and *Saxifraga oppositifolia*. In oro-cryorotemperate (and even upper supratemperate) territories in the Picos de Europa and Ubiña sector, the deposits of coarse stones and semi-fixed calcareous blocks with a long period of snow cover, and whose soils may remain moist for a large part of the year (chionophilous and hygrophilous), are colonised by glericolous communities with sparse coverage (Epilobio anagallidifolii-Doronicetum braun-blanquetii). They are floristically characterised by the presence of *Doronicum viscosum*, *Campanula arvatica* subsp. *arvatica*, *Epilobium anagallidifolium*, *Cystopteris fragilis* subsp. *pseudoregia*, *Arabis alpina* var. *cantabrica* and *Linaria filicaulis* subsp. *filicaulis*. Chionophobous medium-sized mobile limestone distributed throughout oro-Cantabrian upper supratemperate and

Fig. 7.4 Oro- and Cryorotemperate belt on siliceous substrates: Screes and rocky summits around Cubil del Can, Palencia, in the Lechada valley, León (Photo by Sara del Río)

orotemperate territories (Picos de Europa and Ubiña and Alto Campoo and Carrión sectors) are colonised by various glericolous communities with sparse coverage (Linario filicaulis-Crepidetum pygmaeae). The most frequent plants in these environments are *Linaria filicaulis* subsp. *filicaulis*, *Crepis pygmaea*, *Iberis carnosa* subsp. *lereschiana*, *Silene vulgaris* subsp. *prostrata*, *Rumex scutatus*, *Arabis alpina* var. *cantabrica*, *Galium pyrenaicum* and *Ranunculus parnassifolius* subsp. *favargeri*.

Small-sized rock deposits with varying degrees of mobility with persistent snow coverage –ensuring soils remain constantly moist almost all year round (chionophilous and hygrophilous)– are colonised by the typical glericolous vegetation (Ranunculo leroyi-Saxifragetum praetermissae) of the high-mountain Picos de Europa and Ubiña sector (oro-cryorotemperate belts with a humid-hyperhumid ombroclimate), floristically characterised by the presence of *Saxifraga praetermissa*, *Ranunculus alpestris* subsp. *leroyi*, *Epilobium anagallidifolium*, *Cystopteris fragilis* subsp. *pseudoregia* and *Arabis alpina* subsp. *cantabrica*.

Chionophobous and thermophilous semi-fixed scree beds with a siliceous nature formed by small or medium-sized rocks are colonised by glericolous vegetation with sparse coverage (Cryptogrammo crispae-Silenetum gayanae) distributed throughout the orotemperate (occasionally supratemperate) thermotype in territories in the Laciana and Ancares and Ubiña sectors, reaching as far as the oromediterranean belt in the Bierzo-Sanabrian sector. It is floristically characterised by the presence of *Silene foetida* subsp. *gayana*, *Cryptogramma crispa*, *Valeriana montana*, *Rumex suffruticosus* and *Solidago virgaurea* subsp. *alpestris*. Scree beds with coarse siliceous stones and an abundant soil matrix are found throughout the oro-Cantabrian upper supratemperate and upper supramediterranean Bierzo-Sanabrian territories and are colonised by glericolous communities with sparse coverage (Linario glabrescentis-Rumicetum suffruticosi), and characterised by the presence of *Rumex suffruticosus*, *Linaria saxatilis* subsp. *glabrescens*, *Phalacrocarpum oppositifolium*, *Dryopteris oreades*, *Cryptogramma crispa* and *Eryngium duriaei*. Glericolous vegetation with sparse coverage (Sesamoido pygmaeae-Silenetum gayanae) grows on small chionophobous earth-covered slate chips, and includes numerous elements with a creeping biotype distributed throughout oromediterranean territories in the Bierzo-Sanabrian sector and upper supratemperate and orotemperate territories in the Laciana-Ancares and Ubiña subsector (Picos de Europa and Ubiña sector). It is floristically characterised by the presence of *Silene foetida* subsp. *gayana*, *Eryngium durieui*, *Sesamoides minor*, *Rumex suffruticosus* and *Phalacrocarpum oppositifolium*.

The interstices in the deposits of large calcareous blocks produced by periglacial gelifraction and the erratic movements of high-mountain glaciers, and in fissures in limestone pavements found throughout the oro-Cantabrian upper supratemperate and orotemperate territories are colonised by glericolous pteridophytic vegetation (Cystopterido pseudoregiae-Dryopteridetum submontanae) floristically characterised by the presence of *Dryopteris submontana*, *Polystichum lonchitis*, *Polystichum aculeatum*, *Cystopteris fragilis* subsp. *pseudoregia* and *Polystichum x illyricum*. In contrast, the interstices in the deposits of large siliceous blocks (micaschist and quartzite) are colonised by glericolous pteridophytic communities (Cryptogrammo crispae-Dryopteridetum oreadis) typical of the supra-orotemperate and supra-oromediterranean territories in the Iberian mountains (Pyrenean, oro-Cantabrian, oro-Iberian and even in the Galician-Leonese mountains), floristically characterised by the presence of *Dryopteris oreades*, *Cryptogramma crispa* and *Dryopteris expansa*.

In areas of heavy snow accumulations on limestone substrates in oro-cryorotemperate territories in the Picos de Europa, on coarse pergelic, basic or neutral soils in which the snow cover persists from late September to July or August, there are several humid-hyperhumid chionophilous communities (Ranunculo leroyi-Gnaphalietum hoppeani) that are floristically characterised by the presence of *Ranunculus alpestris* subsp. *leroyi*, *Omalotheca hoppeana*, *Carex parviflora*, *Veronica aphylla*, *Veronica alpina*, *Sagina nevadensis*, *Epilobium anagallidifolium* and *Potentilla brauniana*. Under the same bioclimatic and topographic conditions, but on siliceous substrates, the communities correspond to the Sedion candollei, which includes chionophilous hemicryptophyte and therophyte succulent altioreina communities, growing on coarse soils with persistent snow cover for at least 9 months of the year, and characterised by the presence of *Sedum candollei*, and which are also present on the Galician-Leonese mountains in the submediterranean orotemperate belt.

7.4.1.3 Supra-Orotemperate Grasslands and High-Mountain Pastures

In the territories of Somiedo and Babia–Torío, the basophilous and chionophilous grasslands that occupy biotopes with a long period of snow cover in the upper supratemperate and orotemperate belts correspond to the association Pediculari comosae-Caricetum sempervirentis, which occupies deep soils that are decarbonated on the surface, and is floristically characterised by the presence of *Pedicularis comosa* subsp. *comosa*, *Carex sempervirens*, *Sesleria albicans*, *Anemone pavoniana*, *Armeria cantabrica*, *Polygonum viviparum*, *Pedicularis pyrenaica* subsp. *pyrenaica*, *Luzula nutans* and *Arenaria purpurascens*, among others. This grassland is a geovicariant of the Pediculari fallacis-Armerietum cantabricae, a basophilous and chionophilous grassland with a central-eastern oro-Cantabrian orotemperate distribution, although it may sporadically appear in favourable seasons in both the cryorotemperate and the supratemperate thermotypes, and whose floristic composition includes *Armeria cantabrica*, *Pedicularis pyrenaica* subsp. *fallax*, *Jasione cavanillesii*, *Anemone pavoniana*, *Carex sempervirens*, *Poa alpina*, *Arenaria purpurascens*, *Alchemilla plicatula*, *Sesleria albicans*, *Silene acaulis*, *Polygonum viviparum*, *Aquilegia pyrenaica* subsp. *discolor* and *Helictotrichon sedenense*, among others.

The eastern oro-Cantabrian grasslands of *Brachypodium rupestre* (Bromo erecti-Caricetum brevicollis) extend throughout oro-Cantabrian lower supratemperate and orotemperate humid-hyperhumid basophilous territories; these mesophytic grasslands of *Brachypodium rupestre* are characterised by the presence of *Carex brevicollis*, *Bromus erectus*, *Eryngium bourgatii*, *Potentilla tabernaemontani*, *Thymus praecox* subsp. *britannicus*, *Plantago media*, *Helianthemum nummularium*, *Phyteuma orbiculare* susp. *ibericum*, *Carex humilis*, and others.

The psychroxerophilous basophilous grasslands of *Festuca burnatii* (Festucetum burnatii) in the Picos de Europa are distributed throughout humid-ultrahyperhumid supra-oro-cryorotemperate areas that prosper in the spurs and calcareous lithosols of the eastern Picos de Europa and Ubiña sector. They are characterised by the presence of *Oreochloa confusa*, *Festuca burnatii*, *Saxifraga conifera*, *Arenaria aggregata* subsp. *cantabrica*, *Koeleria vallesiana* s.l., *Festuca hystrix*, *Draba cantabriae* subsp. *cantabriae*, *Helianthemum urrielense*, *Draba dedeana* and *Carex humilis*, among others, and mainly constitute the substitution stage of the orotemperate creeping juniper woods of the Daphno-Arctostaphyletum uva-ursi on limestone. The Saxifrago coniferae-Festucetum burnatii is found to the west of these communities and constitute the humid-hyperhumid supra-orotemperate psychroxerophilous grasslands of *Festuca burnatii* that grow in spurs and calcareous lithosols in the western Ubiña sector, with a disjunction in the Montes Aquilianos (supra-orotemperate submediterranean Bierzo-Sanabrian sector). They are characterised by *Saxifraga conifera*, *Armeria bigerrensis* subsp. *legionensis*, *Centaurea janeri* subsp. *babiana*, *Draba cantabriae* subsp. *izcoi*, *Festuca hystrix*, *Helianthemum croceum* subsp. *cantabricum*, *Helianthemum canum* subsp. *cantabricum*, *Koeleria vallesiana* s.l. and others, and are the substitution stage of the orotemperate creeping juniper woods of the Daphno-Arctostaphyletum uva-ursi.

On siliceous substrates in oro-Cantabrian orotemperate territories there is a presence of hard-leaved psychroxerophilous grasslands of the Teesdaliopsio confertae-Festucetum eskiae, dominated by *Festuca eskia*, *Luzula caespitosa*, *Agrostis tileni* and others; they constitute the mature stage of the oro-Cantabrian upper orotemperate silicicolous hyperhumid geopermaseries in the altioreina area of the chionophobous meadows of *Festuca eskia* with *Teesdaliopsis conferta*, although they may sometimes also represent the substitution stage of the creeping juniper woods of the Junipero alpinae-Vaccinietum microphylli. In oro-cryoromediterranean silicicolous Bierzo-Sanabrian, western-Carpetan and orotemperate Laciana and Ancares territories these hard grasslands are replaced by those in the association Teesdaliopsio confertae-Festucetum summilusitanae, characterised by the presence of *Festuca summilusitana* and the absence of *Festuca eskia*. They constitute the mature stage of the orophilous Bierzo-Sanabrian orotemperate upper submediterranean humid-hyperhumid silicicolous geopermaseries of the chionophobous grasslands of *Festuca summilusitana* with *Teesdaliopsis conferta*. In siliceous territories in the supra-orotemperate thermotype in the Laciana and Ancares sector there are pioneer grasslands in the association Agrostio duriaei-Festucetum rivas-martinezii; these perennial caespitose grasslands are characterised by the presence of *Agrostis durieui* and *Festuca rivas-martinezii*.

Pastures dominated by succulent plants (crassifolious) occur throughout the oro-Cantabrian territories, with a particular predominance of the association Agrostio durieui-Sedetum pyrenaici, floristically characterised by the presence of *Sedum anglicum* subsp. *pyrenaicum*, *Agrostis durieui* and *Sedum brevifolium*, which colonise incipient lithosols collected on ledges and siliceous rocky outcrops in the subhumid-hyperhumid supra-orotemperate belts in the oro-Cantabrian province, and in the Bierzo- Sanabrian sector. In areas in the Picos de Europa and Ubiña sector there are pioneer crassifolious communities characterised by *Sedum anglicum* subsp. *pyrenaicum* and *Sedum album* subsp. *micranthum* (Sedetum micrantho-pyrenaici) which colonise incipient lithosols collected on ledges and siliceous rocky outcrops (conglomerate and sandstone) in the subhumid-hyperhumid supra- and orotemperate belts. In Laciana-Ancares and Babia-Torío territories there are pioneer crassifolious communities characterised by the endemic *Spergula viscosa* subsp. *pourretii* (*Spergula rimarum*) (Sedo brevifolii-Sperguletum rimarum) which colonise incipient lithosols collected on ledges and siliceous rocky outcrops (conglomerate and sandstone) in the subhumid-hyperhumid supra and orotemperate belts.

There is a wide diversity of grasslands with *Nardus stricta* or high-mountain pastures in oro-Cantabrian territories. The high-mountain pastures of the Polygalo edmundii-Nardetum are fundamentally orotemperate in the Picos de Europa and Ubiña and Alto Campoo and Carrión sector, and grow on flat protected areas or on the bottom of sinkholes, where there is a substantial accumulation of snow which often persists until early summer (Fig. 7.5). They are associated with deep

Fig. 7.5 View of *Nardus stricta* grasslands of the Polygalo edmundii-Nardetum strictae at Horcadina de Covarrobres, Picos de Europa Central Massif, Cantabria (Photo by Sara del Río)

decarbonated soils that are rich in lime and with a high degree of moisture, and are floristically characterised by the presence of *Nardus stricta*, *Festuca nigrescens* subsp. *microphylla, Trifolium thalii*, *Polygala edmundii, Phleum alpinum*, *Plantago alpina*, *Carex macrostyla* and *Jasione laevis*, among others. The high-mountain pastures of the Geranio subargenti-Nardetum strictae are distributed throughout hyperhumid orotemperate and upper supratemperate territories in the Ubiña sector, and thrive on base-poor soils containing calcareous materials, formed from well-established scree which confers a degree of stoniness and stronger drainage; these soils are therefore more filtering than the usual *Nardion* soils and thus less hydromorphic; the floristic combination in these high-mountain pastures is *Geranium subargenteum*, *Nardus stricta*, *Festuca nigrescens* subsp. *microphylla*, *Meum athamanticum* and *Plantago serpentina*. In narrow bands on peaty soils beside springs and streams there are extensions of dense pastures dominated by *Festuca rivularis* (Campanulo herminii-Festucetum rivularis); the water flow – permanent or temporary– maintains the soil profile waterlogged or at least saturated with water during the growth period, and the moving character of the water guarantees its oxygenation. It also withstands prolonged snow cover thanks to its topographic situation (river beds and trenches), which contributes to the temporary flooding it undergoes every thawing season; these reophilous and chionophilous high-mountain pastures are known from the Iberian siliceous high mountains in the (supra) oro- and cryorotemperate belts. In addition to *Festuca rivularis* these environments frequently include mainly *Nardus stricta* and *Campanula herminii*;

it is distributed throughout mountain areas in the Cordillera Central (Estrellan, Bejar-Gredos and Guadarramean sectors), the Bierzo-Sanabrian sector, and the siliceous massifs of the oro-Cantabrian subprovince. In the territory of Los Ancares there is a pasture growing on deep soils with temporary hydromorphia with its optimum in orotemperate Ancares territories (Campanulo herminii-Trifolietum alpini), characterised by the presence of *Campanula herminii, Trifolium alpinum, Nardus stricta, Jasione laevis* subs*p. carpetana, Meum athamanticum, Poa alina* subsp. *legionensis, Plantago alpina* and *Festuca nigrescens* subs*p. microphyll*a. The hydrophilous high-mountain or hygro-peaty silicicolous pastures in the upper supra-, oro- and cryorotemperate belts in oro-Cantabrian, oro-Iberian-Sorian and Carpetan-Leonese territories (Luzulo carpetanae-Pedicularietum sylvaticae) are located in trenches liable to waterlogging by runoff water and in contact with peat bogs and lakes, and characterised by the presence of *Juncus squarrosus, Nardus stricta, Pedicularis sylvatica* and *Luzula campestris* subsp. *carpetana*, among others. In contrast, in drier siliceous soils in upper supratemperate oro-Cantabrian territories there are high-mountain pastures of the Thymelaeo dendrobryii-Nardetum strictae that are floristically characterised by the presence of *Thymelaea dendrobryum, Carex asturica, Nardus stricta, Avenula lodunensis, Luzula nutans, Meum athamanticum, Dianthus langeanus* and occasionally *Crocus carpetanus* and *Luzula caespitosa*.

On rain-drenched paths in alpine-Pyrenean-Cantabrian high mountains (upper supratemperate oro- and cryorotemperate territories) there are sparse grasslands (Plantagini majoris-Poetum supinae) characterised by the presence of *Poa supina*, and extending throughout eastern oro-Cantabrian areas in the Cantabrian Range. The Central-western oro-Cantabrian and Bierzo-Sanabrian territories are occupied by vicariant grasslands of the Spergulario capillaceae-Poetum supinae which extend throughout upper supratemperate, orotemperate and oromediterranean areas, and are floristically characterised by the presence of *Spergularia capillacea* and *Poa supina*.

7.4.1.4 Oro-Cantabrian High-Mountain Scrubland (Subalpine Juniper Woods, Broom Scrubland and Heathlands)

On calcareous substrates in the orotemperate belt in the southern high-mountain Picos de Europa and Ubiña sector there are chionophobous scrublands with a strong presence of creeping juniper (*Juniperus alpina*), along with bearberry (*Arctostaphylos uva-ursi*) and spurge-laurel (*Daphne laureola* subsp. *cantabrica*), to which are associated *Rosa pendulina* and *Cotoneaster integerrimus* var. *pyrenaica*. These oro-Cantabrian subalpine creeping juniper woods with bearberry and spurge-laurel (Daphno cantabricae-Arctostaphyletum uvae-ursi) represent the mature stage of the hyperhumid orotemperate (subalpine) climatophilous series in all the oro-Cantabrian calcareous mountains, and occupy crests, ledges and generally biotopes with snow cover for short periods only.

In similar substrates but on southern slopes in territories in the Picos de Europa-Ubiña sector there is a scrubland also with a conspicuous presence of *Juniperus alpina*, but with a high participation of *Juniperus sabina*, the fleshy-leaved bearberry (*Arctostaphylos crassifolia*), Cantabrian barberry (*Berberis vulgaris* subsp. *cantabrica*) and *Glandora diffusa* (*Lithodora diffusa*), among others. These oro-Cantabrian subalpine creeping juniper woods (Arctostaphylo crassifoliae-Juniperetum sabinae) are distributed throughout calcicolous and chionophobous, humid-hyperhumid upper supratemperate, orotemperate and often submediterranean territories in the southern Picos de Europa and Ubiña sector. This creeping juniper woodland has a dispersed and relict distribution, and can also be found on the northern oro-Cantabrian face, on calcareous spurs, edges of sinkholes and karstic areas; and in the Picos de Europa, the north of the Peñas de Orníz-Calabazosa hill range, and in the Picos Albos in Saliencia.

On orotemperate base-poor substrates in the Cantabrian Range the dominant shrubby vegetation comprises oro-Cantabrian subalpine chionophobous silicicolous juniper woods with black-blueberry (Vaccinio microphylli-Juniperetum alpinae). These scrublands are dominated by creeping juniper (*Juniperus alpina*) and the blueberries *Vaccinium uliginosum* subsp. *microphyllum* and *Vaccinium myrtillus*, along with heath (*Calluna vulgaris*) and represent the mature stage of the hyperhumid oro-Cantabrian orotemperate silicicolous climactic series, preferentially occupying well-drained soils on crests and stony slopes in contact with orotemperate silicicolous psychroxerophilous grasslands.

In the same orotemperate belt in the Cantabrian Range, but in trenches and broad basins on slate, sandstone or quartzite with deep soils with a substantial humiferous horizon there is a presence of oro-Cantabrian subalpine silicicolous chionophilous juniper woods with heath and blueberries (Erico tetralicis-Vaccinietum microphylli). In these dense and low-growing scrublands there is a predominance of the heaths *Calluna vulgaris* and *Erica tetralix*, blueberries *(Vaccinium myrtillus, Vaccinium uliginosum* subsp. *microphyllum)* and creeping juniper *(Juniperus alpina)*. This chionophilous community (snow-covered for several months of the year) has its optimum development in the orotemperate (subalpine) belt in oro-Cantabrian territories, generally in contact with high-mountain pastures in the alliance Campanulo herminii-Nardion. In escarpments and on windblown sites, these juniper woods are substituted by those of the Vaccinio microphylli-Juniperetum alpinae.

A very singular type of oro-Cantabrian subalpine creeping scrubland is represented by the communities dominated by *Empetrum nigrum* subsp. *hermaphroditum* (a tetraploid ericoid species distributed throughout the circumarctoboreal and lower circumtemperate and altioreine areas) which are very scarce and relict in the Cantabrian Range, and located topographically in shady areas in the cryorotemperate bioclimatic belt (lower alpine) in the Redes district (Picos de Europa and Ubiña sector). These creeping acidophilous scrublands with *Empetrum hermaphroditum* and *Vaccinium microphyllum* (Avenello ibericae-Empetretum hermaphroditi) represent a shady, slightly chionophilous climatophilous permaseries growing on cryptopodsolic dystrophic rankeriform cryoturbic and shaded leptosols. They are floristically characterised by the presence of *Empetrum nigrum* subsp. *hermaphroditum* and *Vaccinium uliginosum* subsp. *microphyllum*, characteristic species of the class Loiseleurio-Vaccinietea, which coexist with western Iberian oreine (high mountain) and oro-Iberian endemic silicicolous elements such as *Avenella iberica, Agrostis tileni, Luzula caespitosa* and *Leontodon cantabricum* that are absent from the Pyrenees. *Calluna vulgaris, Vaccinium myrtillus* and *Juncus trifidus* are not infrequent in these scrublands.

The broom scrubland communities generally have high coverage dominated by retamoid aphyllous nano- or microphanerophytic legumes growing on deep meso-oligotrophic siliceous soils, with mull humus and without temporary hydromorphia; they constitute the edge or first substitution stage of several types of climatophilous woodlands, and although their optimum is attained in Iberian siliceous territories, they are distributed throughout Mediterranean-Ibero-Atlantic, oro-Cantabrian-Atlantic and western Pyrenean areas, in the meso-supramediterranean and thermo-, meso- and supratemperate belts. In the oro-Cantabrian Subprovince, the broom scrubland in upper supratemperate and orotemperate territories corresponds to the association Carici asturicae-Genistetum obtusirameae, which has its optimum in oro-Cantabrian centre-western siliceous areas, particularly in the more oceanic and rainy areas on the northern slope of the Cantabrian Range; these broom scrublands are floristically characterised by the presence of *Genista obtusiramea, Carex asturica* and *Juniperus alpina.* In more continental and less rainy oro-Cantabrian areas, especially the upper supratemperate and orotemperate Carrión and Ubiña territories, these broom scrublands are substituted by those of the Cytiso oromediterranei-Genistetum obtusirameae, characterised floristically by the presence of *Avenella iberica, Cytisus oromediterraneus, Genista obtusiramea* and *Juniperus alpina.*

Below these thermoclimatic levels, broom scrublands constitute the shrubby edge of several forest formations, and in the upper supratemperate (occasionally extending to the lower orotemperate) belt in the Ubiña and Alto Campoo and Carrión sectors the predominant association is the Cytiso cantabrici-Genistetum obtusirameae (Fig. 7.6), typical of hyperhumid and ultrahyperhumid areas and characterised by the presence of *Cytisus cantabricus, Genista obtusiramea, Genista florida* subsp. *polygaliphylla, Erica arborea, Gentiana lutea* subsp. *lutea, Avenula sulcata, Vaccinium myrtillus* and *Avenella iberica,* among others. In the territories in the western oro-Cantabrian area (Laciana-Ancares), these broom scrublands are substituted by those of the Genistetum obtusirameo-polygaliphyllae, characterised floristically by the presence of *Genista obtusiramea, Genista florida* subsp. *polygaliphylla, Erica arborea, Cytisus scoparius* subsp. *scoparius* and *Gentiana lutea* var. *aurantiaca.*

The heathlands in the siliceous oro-Cantabrian high mountains correspond to the association Carici asturicae-Callunetum vulgaris. These scrublands with *Vaccinium myrtillus* and *Genista pilosa* occupy dry soils with scarce snow cover originating on quartzite and sandstone and distributed throughout the upper supratemperate and

Fig. 7.6 Broom scrub of the Cytiso cantabrici-Genistetum obtusirameae and some remnants of the orocantabrian oak forests of the Avenello ibericae-Quercetum orocantabricae near the summit of the San Glorio pass, Puerto de San Glorio, León (Photo by Sara del Río)

orotemperate territories; their floristic composition includes, in addition to the species indicated, *Carex asturica, Thymelaea dendrobryum*, *Carex pilulifera* and *Hypericum burseri*, and they contact with the high-mountain pastures of the *Campanulo herminii-Nardion strictae* and the broom scrubland of *Genista obtusiramea*.

7.4.1.5 Peat-Bog Vegetation

The supra-orotemperate oro-Cantabrian high bulging peat bogs with heaths and sphagnum (Calluno vulgaris-Sphagnetum capillifolii) have a series of generally clearly defined bulges which often stand out by as much as 1 m above the water level in the pools, and which are in the terminal phase of the colonisation process. In addition to their distinct bulge, these acidophilous communities are physiognomically characterised by the reddish colour of the sphagnum of which they are formed (mainly *Sphagnum capillifolium* and to a lesser degree *Sphagnum magellanicum*, *Sphagnum russowii*, *Sphagnum subnitens* and *Sphagnum papillosum*), along with a substantial participation of heaths (*Erica tetralix* and *Calluna vulgaris*) and blueberries (*Vaccinium myrtillus*). The graminoids growing in these bulging peat bogs include particularly *Trichophorum cespitosus* subsp. *germanicus*, *Carex nigra* and

Juncus squarrosus, among others such as *Drosera rotundifolia*. The oro-Cantabrian supra-orotemperate high flat peat bogs with sundew and bog asphodel (Narthecio ossifragi-Sphagnetum tenelli) form a fairly deep and continuous peaty carpet – without going so far as to form bulges–, where there is a degree of water runoff. It is characterised by the dominance of *Sphagnum tenellum*, *Sphagnum subsecundum* var. *rufescens*, and to a lesser degree by *Sphagnum papillosum*, along with *Erica tetralix*, *Trichophorum cespitosus* subsp. *germanicus*, *Narthecium ossifragum*, *Drosera rotundifolia*, *Juncus squarrosus*, *Carex nigra* and *Carex echinata*, among others. When the peat bogs have some degree of mineralisation, the biotope is occupied by oro-Cantabrian supra-orotemperate high peat bogs with sundew and marsh asphodel (Drosero anglicae-Narthecietum ossifragi), characterised by the presence of *Drosera anglica (Drosera longifolia)*, *Eriophorum angustifolium*, *Utricularia minor*, *Narthecium ossifragum*, *Trichophorum caespitosus* subsp. *germanicus*, *Sphagnum subsecundum* var. *rufescens*, *Riccardia incurvata*, *Drosera rotundifolia* and *Sphagnum cuspidatum*, among others.

The barely turfophilous biotopes in the Cantabrian high mountains are occupied by oro-Cantabrian supra-orotemperate peat bogs with heaths (Erico tetralicis-Trichophoretum germanici). These are somewhat drier than the biotopes colonised by the flat peat bogs and sometimes grow on fossil peat bogs. A characteristic of this vegetation type is the fact that every year it undergoes intense and prolonged water runoff and desiccation with varying degrees of severity during the summer. It is characterised by the presence of *Trichophorum caespitosus* subsp. *germanicus*, *Narthecium ossifragum*, *Carex echinata*, *Parnassia palustris*, *Erica tetralix*, *Drosera rotundifolia* and *Juncus squarrosus*, among others. The Carpetan-Iberian-Leonese and oro-Cantabrian non-mossy peat bogs (Caricetum echinati-nigrae) grow in peaty areas with flowing water in oro-Cantabrian upper supratemperate and orotemperate territories where there is no development of sphagnum peat. These oligotrophic water communities occur throughout the Carpetan, Bierzo-Sanabrian and oro-Cantabrian territories and are dominated by *Carex echinata*, *Carex nigra*, *Drosera rotundifolia* and *Viola palustris* subsp. *palustris*, among others.

Oro-Cantabrian supra and orotemperate silted peat bogs (Carici nigrae-Sphagnetum recurvi) constituted by *Sphagnum flexuous (S.* complex *recurvum)*, *Carex nigra*, *Carex echinata*, *Polytrichum commune*, *Drepanocladus fluitans* and others grow on the edges of silted pools. These Cantabrian high-mountain pools in areas of sandstone and quartzite, generally without muddy soils on their floors, are gradually being invaded by populations –generally pure or almost pure– of *Sphagnum flexuosum*. As the process advances and the typical floating peat forms, other elements –including particularly carices– begin to move in and finally form the association Carici nigrae-Sphagnetum recurvi. The colonisation of pools and lakes with muddy soils may also occur through communities dominated by *Carex rostrata*, whose rhizomes, along with those of *Potentilla palustris* and *Caltha palustris*, form a solid framework for the edification of a soil capable of supporting oro-Cantabrian supra-orotemperate lacustrine peat pastures (Potentillo palustris-Caricetum nigrae), in which there is a constant presence of *Potentilla palustris*, *Carex nigra*, *Carex echinata*, *Carex rostrata*, *Sphagnum flexuosum* (*Sphagnum recurvum* var. *amblyphyllum*), *Drepanocladus fluitans*, etc.

The most frequent eutrophic peaty communities in Cantabrian high-mountain areas correspond to the supra-orotemperate eutrophic peat bogs in the Picos de Europa-Ubiña sector (Pinguiculo grandiflorae-Caricetum lepidocarpae). These are low turfophilous eutrophic communities (rich in calcium ions) characterised by the presence of *Carex lepidocarpa*, *Pinguicula grandiflora*, *Carex davalliana*, *Juncus alpestris*, *Carex nigra*, *Eleocharis quinqueflora* and *Parnassia palustris*, among others (Fig. 7.7). Less frequent are the supra-orotemperate eutrophic peat bogs with Somiedo centaury (Centaurio somedani-Caricetum davallianae) typical of cold emerging waters, on slightly sloping terrain above 1400 m in Somiedo and Babia-Torío territories, with *Centaurium somedanum* (a Somiedo and Babia endemism), *Carex davalliana*, *Selaginella selaginoides*, *Pinguicula grandiflora*, *Carex panicea*, *Epilobium alsinifolium* and *Carex lepidocarpa*, among others. The initial or floating oro-cryorotemperate turfophilous communities typical of low peat bogs and hygro-peaty mineral, neutrophilous or slightly acidophilous (acidoclines) soils (Equiseto variegati-Salicetum hastatellae) can also be considered uncommon in the Cantabrian Range. They have a Pyrenean distribution with disjunctions in the Picos de Europa sector and are characterised by the presence of *Salix hastatella* subsp. *picoeuropeana, Equisetum variegatum* and *Carex lepidocarpa*, among others.

Fig. 7.7 Calcareous peat bog and pool with *Sparganium angustifolium* in Vega de Liordes, Picos de Europa, Cantabria (Photo by Sara del Río)

7.4.2 *The Woodlands of the Cantabrian Range*

7.4.2.1 *Fagus sylvatica* Woodland (Beechwoods)

Possibly the most representative woodlands in the mountain areas of the Cantabrian Range between altitudes of 700 and 1900 m are beechwoods (Fig. 7.8). These woodlands, in which beech (*Fagus sylvatica*) is the dominant species, occupy areas with high precipitation (above 1400 mm of rainwater annually), which explains their preference for shady slopes with less insolation, and sites that are prone to summer mists that can compensate for the lack of direct precipitation in the summer months.

In calcareous areas in the centre of the Cantabrian Range and its northern spurs (essentially from Somiedo to the eastern end of the Redes Nature Reserve), the beechwood is known as beechwood with mercurial (Carici sylvaticae-Fagetum sylvaticae), and also extends throughout meso-supratemperate oceanic territories on the Cantabrian coast as far as the mountain areas of the Basque Country and the Iberian System (Fig. 7.9). These woodlands grow on rich eutrophic soils (humic-calcic cambisols or calcic luvisol) formed on limestone-type rocks. Although the dominant and almost exclusive species in the tree layer is beech, there may be a sporadic presence of sessile oak (*Quercus petraea*), ash (*Fraxinus excelsior*), acer (*Acer pseudoplatanus*) and yew (*Taxus baccata*). The understorey contains small trees and shrubs such as *Sorbus aria*, *Ilex aquifolium*, *Corylus avellana* and

Fig. 7.8 Extensive beech forests seen from the Piedrasluengas pass (Palencia) with the eastern massif of the Picos de Europa at the horizon (Photo by Sara del Río)

Fig. 7.9 Calcareous beech forest in autumn, Puerto de Ventana, at the foot of Peña Ubiña, León (Photo by Sara del Río)

Crataegus monogyna. Shrubs and sub-shrubs are also scarce –which facilitates movement through the understorey, in contrast to other deciduous woodlands such as pedunculate or oakwoods–, and there is a frequent presence of *Daphne laureola* and occasionally *Genista occidentalis* and *Erica vagans*. Its dense herbafruticose stratum is composed of numerous more or less demanding plants on rich soils, such as the grasses *Melica uniflora*, *Hordelymus europaeus* and *Brachypodium sylvaticum*; the ferns *Polystichum aculeatum* and *Dryopteris dilatata*; the cyperaceae *Carex sylvatica*, and further *Galium odoratum*, *Mercurialis perennis*, *Helleborus viridis* subsp. *occidentalis*, *Lilium martagon*, *Corydalis bulbosa*, *Scilla lilio-hyacinthus* and *Veronica montana*, among others; and typically nemoral plants such as *Oxalis acetosella*, *Poa nemoralis*, *Crepis lampsanoides*, *Euphorbia dulcis*, *Euphorbia amygdaloides*, *Euphorbia hyberna*, *Anemone nemorosa*, *Saxifraga hirsuta*, *Ranunculus tuberosus*, etc.

The beechwoods of the Picos de Europa and Sierra de Cuera (Carici caudatae-Fagetum sylvaticae) grow on karstified limestone in the Picos de Europa and Sierra del Cuera (now in Cantabrian-Atlantic territory). They are chionophilous, shallower and drier, and found on limestone massifs often in steeply sloping areas. These are lower meso-orotemperate humid-hyperhumid beechwoods with abundant rocky outcrops and a significant accumulation of undecomposed litterfall which extend as far as the Redes and Mampodre districts in the west. The tree layer of this woodland formation consists of beech trees with a smaller stature and more misshapen stems; the shrubby layer has a predominance of *Rhamnus alpina* and

Fig. 7.10 South-facing basophilous and xerophilous beech forest of the Epipactido helelborines-Fagetum sylvaticae and shrubland of the Cytiso scoparii-Genistetum polygaliphyllae, Alto de Aralla, León (Photo by Sara del Río)

even *Ribes alpinum*, *Berberis vulgaris* subsp. *cantabrica* and *Genista legionensis*. The herb layer is less dense and diverse then in the typical facies, with a characteristic presence of the grasses *Sesleria albicans* and *Helictotrichon cantabricum*, along with the cyperaceae *Carex caudata* and *Carex brevicollis*, the umbellifera *Pimpinella siifolia*, and numerous orchids in the genera *Cephalanthera* and *Epipactis*.

In the southern calcareous territories of the Cantabrian Range there are xerophilous and basophilous beechwoods with orchids (Epipactido helleborines-Fagetum sylvaticae), whose general distribution is southern oro-Cantabrian-Atlantic and in the Sierra Ibérica (Fig. 7.10). They extend throughout the meso-supratemperate belts in humid-hyperhumid ombroclimates, on generally fairly deep base-rich, preferably south-facing soils on steep slopes and in highly karstified areas. These woodlands are richer in spiny shrubs in their outer edges and understorey than the ombrophilous woodlands on deeper soils, and they have a fairly dense herb layer, particularly due to the presence of certain herbs such as *Brachypodium rupestre*, *Laserpitium eliasii*, *Hepatica nobilis* and various orchids such as *Epipactis helleborine*, *Cephalanthera damasonium* and *Neottia nidus-avis*.

While the base-rich substrates (limestone) in oro-Cantabrian areas with high precipitation are the most suitable biotopes for the development of the eutrophic beechwoods mentioned previously, the beechwoods under an identical ombroclimate but on base-poor substrates (siliceous soils, mainly sandstone and slate) are the oro-Cantabrian beechwoods with birch, corresponding to the association Blechno spicant-Fagetum sylvaticae. These beechwoods on poor soils (oligotrophic and acidophilous) are exclusive to the centre and east of the Cantabrian Range at altitudes of between 900 and 1900 m (supratemperate bioclimatic belt), occupying areas with very high precipitation where there is no water deficit in summer. This alternation between oligotrophic and eutrophic beechwoods is repeated throughout the length and breadth of the central and eastern areas of the Cantabrian Range and is generally manifested not only in the floristic composition of the understorey but also in its substitution stages. The central-eastern Cantabrian acidophilous beechwood grows in meso-supratemperate belts with a humid-hyperhumid ombroclimate, on poor soils of the dystic cambisol or humic cambisol types, where the Celtiberian birch (*Betula celtiberica*) is an abundant element in both its young and adult phases; these types of forest are therefore generally designated as oro-Cantabrian beechwoods with birch (Blechno spicant-Fagetum sylvaticae). The shrub layer is not usually dense, and includes *Ilex aquifolium* and *Sorbus aucuparia*. Shrubs and sub-shrubs are also scarce, although there tends to be a presence of *Vaccinium myrtillus* and *Erica arborea*. The herb layer is much denser and has a predominance of acidophilous plants such as *Luzula sylvatica* subsp. *henriquesii*, *Saxifraga spathularis*, *Saxifraga* x *polita*, *Avenella flexuosa*, *Blechnum spicant* and *Galium rotundifolium*, among others; a floristic combination which is very similar to that of the more sciophilous sessile oakwoods with birch (Fig. 7.11). These beechwoods with birch are dynamically related with young woodlands or pre-forests of birch, rowan and willow (*Salix caprea*, *Salix*

Fig. 7.11 Beech forest of the Blechno spicant-Fagetum sylvaticae in Gabanceda Valley, Picos de Europa Western Massif, León (Photo by Sara del Río)

atrocinerea). Elsewhere, if the mature beech stands are clearcut and there is no significant soil erosion, the sessile oak may gain ground at the expense of the beech, and thus the management system applied in the woodland is occasionally the determining factor in favouring the predominance of either beech or sessile oak.

One very singular type of beechwood in the Cantabrian Range grows at its western end in the supratemperate belt, mainly in the Sierra del Caurel (Sierra del Caurel District) and throughout the Sierra de Ancares (Ancares District). As with other beechwoods, these western oro-Cantabrian beechwoods with spurge-laurel (Omphalodo nitidae-Fagetum sylvaticae) occupy shady areas with high precipitation (hyperhumid) and low insolation, and are therefore relatively infrequent, probably due to the strong Mediterranean character of the area. They grow on rich slate or marbleised limestone overlaid with mature calcium-rich soils with a loamy texture (neutro-acidophilous mesotrophic soils), as is evidenced by the peculiar floristic composition of the understorey. The tree layer is dominated by beech, although there may be an occasional and sporadic presence of Cantabrian oak (*Quercus orocantabrica*) and Celtiberian birch (*Betula celtiberica*). Although not dense, the shrub layer is diverse, with hazel, holly, rowan and yew growing simultaneously. In the sub-shrub layer there is a very considerable mixture of shrubs typical of poor soils (oligotrophic) such as *Vaccinium myrtillus*, along with others with a marked eutrophic character (on soils that are richer in nutrients) such as *Daphne laureola*, very common in this type of beechwoods. The herb layer of these beechwoods is dense and diverse, and includes the concurrent presence of plants with a north-western Iberian distribution typical of poor soils (*Omphalodes nitida*, *Saxifraga spathularis*, *Luzula sylvatica* subsp. *henriquesii*) and other more general oligotrophic plants (*Euphorbia hyberna*, *Ceratocapnos claviculata*, etc.), along with plants on rich soils such as *Melica uniflora*, *Mercurialis perennis*, *Sanicula europaea*, *Lilium martagon* and *Galium odoratum*. Ferns are not infrequent in this type of forest, particularly *Blechnum spicant*, *Athyrium filix-femina*, *Polystichum setiferum*, *Phyllitis scolopendrium* and *Dryopteris affinis*.

In oro-Cantabrian lower orotemperate territories with acidophilous soils and a hyperhumid ombroclimate there are climatophilous and chionophilous beechwoods known as oro-Cantabrian orotemperate oligotrophic beechwoods with *Avenella ibericae* (Avenello ibericae-Fagetum sylvaticae), which may topographically descend as far as the upper supratemperate hyperhumid horizon. These beechwoods preferentially occupy north-facing slopes and riverbeds, particularly in areas where the summer mists persist or are frequent for several hours a day during the hottest months. Its summer continentality is therefore lower than in the supra-orotemperate series of oro-Cantabrian oakwoods (Avenello ibericae-Querco orocantabricae sigmetum) with which it shares the subalpine silicicolous forest territories on both slopes of the Cantabrian Range. Among the differential species of the Avenello ibericae-Fagetum sylvaticae, as opposed to the oro-Cantabrian supratemperate silicicolous beechwoods of the Blechno spicant-Fagetum sylvaticae, it is worth noting *Avenella iberica*, *Daphne laureola* var. *cantabrica*, *Doronicum pubescens*, *Dryopteris expansa*, *Gymnocarpium dryopteris*, *Luzula* x *somedana* (*L. nutans* x *L. henriquesii*), *Quercus orocantabrica*, *Quercus* x *puentei*

(*Q. orocantabrica* x *Q. petraea*), *Rosa pendulina* and *Saxifraga* x *polita* (*S. spathularis* x *S. hirsuta*), among others.

7.4.2.2 *Quercus petraea* Woodland (Sessile Oakwoods)

One of the fundamental traits of the mountain territories in the Cantabrian Range (oro-Cantabrian) is the presence of sessile oakwoods (*Quercus petraea*) on both richer and poorer soils. There are two types of woodland dominated by this oak and those are exclusive to the Cantabrian Range: one associated with ombrophilous environments and the other with sunny locations; both types of woodland are widely represented in the oro-Cantabrian territory, although always in areas with high precipitation, with more than 1400 mm of rainwater annually.

The first have a marked abundance of birch (*Betula celtiberica*) and are designated silicicolous sessile oakwoods on shady sites with birch (Luzulo henriquesii-Quercetum petraeae). They are found mainly in mountain areas above 700 m and generally not over 1900 m (meso- and supratemperate bioclimatic belts), preferentially occupying areas of moist shade with little insolation; that is in topographic situations where they are liable to receive abundant rainfall (hence their classification as ombrophilous) and with frequent summer mists. They grow on siliceous substrates (sandstone, slate, etc.) overlaid with nutrient-poor acid cambisol-type soils. The tree layer in the oro-Cantabrian (humid-hyperhumid) silicicolous ombrophilous sessile oakwoods is dominated in the mature stages by sessile oak and Celtiberian birch (*Betula celtiberica*), and shrubs and small trees such as *Sorbus aucuparia*, *Corylus avellana* and *Ilex aquifolium*. In moister biotopes there may be an abundance of beech (*Fagus sylvatica*), representing a transition in both floristic and ecological terms towards oro-Cantabrian oligotrophic beechwoods. The understorey has a presence of *Erica arborea*, *Vaccinium myrtillus* and, among the herbs (whose structure is compact and closed) there is predominance of *Luzula sylvatica* subsp. *henriquesii*, which occasionally forms dense and continuous populations along with plants with a north-western Iberian distribution and a marked acidophilous character, such as the rushes mentioned earlier and *Saxifraga spathularis*. These grow alongside nemoral herbs such as *Oxalis acetosella*, *Valeriana montana* and *Avenella flexuosa*, and ferns such as *Dryopteris dilatata*, *Dryopteris filix-mas*, *Blechnum spicant*, etc. These sessile oakwoods are dynamically related to young woodlands dominated by birch (*Betula celtiberica*), in which there is a frequent presence of other pioneer trees such as rowan (*Sorbus aucuparia*) and willow (*Salix caprea*, *Salix atrocinerea*). Floristically these pre-forests are scarcely differentiated from the birchwoods in Cantabrian high-mountain areas. The constant management practice of thinning most likely favours beech over sessile oak, and contributes to enhancing the similarities of these woods with oro-Cantabrian oligotrophic beechwoods. The best examples of sessile oakwoods with birch can be found in the Muniellos National Biological Reserve and its surroundings in the valleys of Monasterio de Hermo and de Monasterio del Coto (Asturias).

The structural characteristics and floristic composition of the oro-Cantabrian silicicolous sessile oakwoods on sunny sites (Linario triornithophorae-Quercetum petraeae) are clearly differentiated from sciophilous sessile oakwoods. These woodlands grow at altitudes of between 700 and 1900 m in sunny sites with an annual precipitation of between 700 and 1400 mm of rainwater, on siliceous substrates with dry oligotrophic leptosol-type soils. These xerophilous woodlands contain the same small trees and shrubs as the sciophilous sessile oakwoods. The mature stages of these forest structures consist of a tree layer formed by sessile oak (*Quercus petraea*) which occasionally exists alongside Cantabrian oak (*Quercus orocantabrica*) and birches (*Betula celtiberica*) or Pyrenean oaks (*Quercus pyrenaica*), in lower and varying proportions. In the shrubby and sub-shrubby layer there may be a presence of *Corylus avellana*, *Ilex aquifolium*, *Frangula alnus*, *Erica arborea*, *Erica australis* subsp. *aragonensis*, *Cytisus scoparius* and *Vaccinium myrtillus*, with a notable presence of *Sorbus aria*. The shrubs abundant in more open woods include several heaths (various species of the genus *Erica*, *Daboecia cantabrica*, *Calluna vulgaris*) and dwarf furze (*Ulex gallii*). *Lonicera periclymenum* is frequent among the creepers. The herb stratum is dense and abundant, as this is a forest type that is generally fairly open, and it is here where the differences between both types of silicicolous sessile oakwoods can be seen most clearly: in the xerophilous oakwoods described here there is a total lack or scarcity of plants such as *Luzula sylvatica* subsp. *henriquesii* and *Saxifraga spathularis*, and an abundance of others such as *Luzula lactea*, *Pseudarrhenatherum longifolium*, *Arenaria montana*, etc. The most common species include the grasses *Avenella flexuosa*, *Festuca paniculata* subsp. *multispiculata*, *Pseudarrhenatherum longifolium* and *Holcus mollis*, *Teucrium scorodonia*, *Omphalodes nitida* and *Linaria triornithophora*; these last two have a north-western Iberian distribution. The common fern (*Pteridium aquilinum*) is practically the only frequent species in the understorey. Sessile oakwoods on sunny sites proliferate on crests and on sunny slopes in rainy areas of the Cantabrian Range (a territory to which they are exclusive), and are replaced by Pyrenean oakwoods in less rainy areas.

7.4.2.3 *Betula celtiberica* Woodland (Birchwoods)

Woodland formations dominated by Celtiberian birch (*Betula celtiberica*) are fairly frequent in rainier Cantabrian siliceous territories and constitute the young phases of most of the potential forests in these territories. Birchwoods are also the potential vegetation in oro-Cantabrian siliceous mountains on shady slopes located at altitudes of between 1700 and 2100 m (upper supratemperate and lower orotemperate bioclimatic belts), with an annual precipitation of significantly above 1400 mm (in many cases over 2000 mm) of rainwater. In these environments where snow persists over long periods of time, birchwoods (Luzulo henriquesii-Betuletum celtibericae) –which are exclusive to the Cantabrian Range and the Montes de León– constitute the upper limit of the forest. These birchwoods rarely reach tall heights (generally not exceeding 15 m) and their tree canopy is relatively open, thus enabling light to penetrate. They grow on nutrient-poor brown-earth type oligotrophic and slightly peaty soils with substantial amounts of undecomposed organic matter. In the tree layer there is a predominance of the Celtiberian birch (*Betula celtiberica*), with the occasional presence of sessile oak (*Quercus petraea*), beech (*Fagus sylvatica*), yew (*Taxus baccata*), rowan (*Sorbus aucuparia*), whitebeam (*Sorbus aria*) and holly (*Ilex aquifolium*), among other large and small trees. The penetration of light through the tree canopy in these woodlands favours the presence of a dense understorey of *Erica arborea*, *Vaccinium myrtillus* and *Calluna vulgaris*, along with numerous ferns (*Dryopteris dilatata*, *Dryopteris filix-mas*, *Dryopteris expansa*, *Pteridium aquilinum*, *Lastrea limbosperma* and *Blechnum spicant*) and acidophilous herbs such as *Luzula sylvatica* subsp. *henriquesii*, *Avenella flexuosa*, *Saxifraga spathularis*, *Melampyrum pratense*, *Poa chaixii* and *Polygonatum verticillatum*. The nemoral species present in most of the woodlands with these characteristics include particularly *Stellaria holostea*, *Poa nemoralis*, *Oxalis acetosella*, *Euphorbia hyberna*, *Euphorbia amygdaloides*, *Euphorbia dulcis*, *Crepis lampsanoides*, *Anemone nemorosa*, etc. The young stages of these birchwoods (pre-forests) are constituted by *Sorbus aucuparia*, *Erica arborea* and broom. In higher areas –now in the orotemperate belt– the forest is very open and rich in *Erica arborea* and *Calluna vulgaris*, with an admixture of *Vaccinium uliginosum* subsp. *microphyllum*. One very notable aspect of birchwoods in higher areas is the appearance of twisted or fallen trees, in response to the extreme climatic conditions in which they grow; this leads to their gradual replacement by the climactic scrublands typical of Cantabrian high-mountain areas –that is, in the orotemperate bioclimatic belt–, thus leading to the eventual disappearance of the forest. As occurs in particular types of beechwoods, birchwoods on steeply sloping mountainsides are generally criss-crossed by extremely moist river beds in which snow accumulates during winter. These conditions imply that the soils are richer in forest mull and the woods are enriched with plants with a megaforbic character (vigorous tall perennial grasses with abundant, large leaves capable of receiving water drops from the splashing of waterfalls and streams and rivers, in whose environments they grow profusely) such as *Cicerbita plumieri*, *Veratrum album*, *Angelica major*, *Aconitum vulparia* subsp. *neapolitanum* and *Adenostyles alliariae* subsp. *pyrenaica*, in addition to certain more demanding ferns on nutrient-rich soils such as *Lastrea limbosperma* and *Athyrium filix-femina*.

7.4.2.4 *Quercus orocantabrica* Woodland (Cantabrian Oakwoods)

The oakwoods of *Quercus orocantabrica* (Avenello ibericae-Quercetum orocantabricae) grow in and are exclusive to the mountains in the arc formed by the Cantabrian Range, the Sierra de Queixa and the mountain massifs of Cabrera and Sanabria, always above 1000 m –supratemperate (montane) and lower orotemperate (lower subalpine) bioclimatic belts– with precipitation of over 1400 mm of rainwater annually (hyperhumid); they have their optimum on soils of the dystic cambisol or cambic podsol types, originating on siliceous substrates (slate, sandstone, etc.). In the mature stages the tree layer is formed by Cantabrian oak (*Quercus orocantabrica*) and in much lower and varying proportions by Celtiberian birches (*Betula celtiberica*), Pyrenean oak (*Quercus pyrenaica*), sessile oak (*Quercus petraea*) and hybrids of *Quercus orocantabrica* with these two last species. In the shrub and sub-shrub layer of these woodland formations there is an abundance of *Erica arborea* and *Vaccinium myrtillus*, and the possible presence of *Corylus avellana*, *Ilex aquifolium*, *Frangula alnus*, *Sorbus aucuparia*, *Erica australis* subsp. *aragonensis* and *Cytisus scoparius*. The predominant shrubs in the more open Cantabrian oakwoods include a range of heaths (several species in the genus *Erica*, *Daboecia cantabrica*, *Calluna vulgaris*) and dwarf furze (*Ulex gallii*). The herb layer is dense and abundant, as this type of wood tends to be fairly open. The common fern (*Pteridium aquilinum*) is practically the only pteridophyte frequent in the understorey. The most prevalent species include herbs with an acidophilous preference and the grasses *Avenella flexuosa*, *Festuca paniculata* subsp. *multispiculata*, *Pseudarrhenatherum longifolium* and *Holcus mollis*, along with *Omphalodes nitida*, *Melampyrum pratense* and *Luzula lactea*, and other nemoral species with a broader distribution such as *Stellaria holostea*, *Arenaria montana*, *Teucrium scorodonia*, and others.

7.4.2.5 *Ilex aquifolium* Woodland (Oro-Cantabrian and Iberian Holly Woods)

The oro-Cantabrian and Iberian holly woods (Saniculo europaeae-Ilicetum aquifolii) are woodland formations with varying degrees of openness comprising holly (*Ilex aquifolium*) with frequent beechwood plants in their understorey (*Melica uniflora*, *Mercurialis perennis*, *Hedera helix* and European *Sanicula*, among others). These forest formations extend throughout the territories in the Iberian System and the Cantabrian Range, essentially in the supratemperate belt with a humid to hyperhumid ombroclimate on soils with a clayey texture. The presence of holly woods has been favoured by high-mountain livestock activities, as these structures were used by humans as a shelter for their animals, which explains their frequent presence in mountain passes. The most notable examples in the Cantabrian Range include particularly the holly woods in the area around the Ubiña massif, in Puertos de Agüeria in Quirós (Asturias).

7.4.2.6 *Tilia platyphyllos* and *Tilia cordata* Communities (Oro-Cantabrian Lime Woods with *Ulmus glabra* and *Fraxinus excelsior*)

In the territories in the Cantabrian Range below altitudes of 800–900 m (that is, in the meso- and supratemperate bioclimatic belts) in a humid-hyperhumid ombroclimate on calcareous substrates and mainly on colluvia on slopes with rich soils, the potential vegetation corresponds to mixed woodland with limes, sessile oak, ash and acer –among other tree elements–, distributed fundamentally through the northern valleys in the centre and east of the Cantabrian Range with a presence in some calcareous outcrops in the west (territories in the Laciana and Ancares sector in the Narcea river basin). These lime, sessile oak and ash woods (Helleboro occidentalis-Tilietum platyphylli (=Mercurialidi perennis-Fraxinetum excelsioris) grow best in deep cool mature nutrient-rich soils on colluvia, and have their optimum on eutrophic brown soils; certain facies may grow on drier soils with fewer calcareous outcrops. Their tree layer is very diverse, and *Tilia platyphyllos* and *Tilia cordata* feature strongly along with other trees such as sessile oak (*Quercus petraea*), common ash (*Fraxinus excelsior*), acer (*Acer pseudoplatanus*) and mountain elm (*Ulmus glabra*). Beech may be present in rainier areas (*Fagus sylvatica*). The shrubby and sub-shrubby strata are generally rich and diverse, with *Corylus avellana*, *Crataegus monogyna*, *Prunus spinosa*, *Ligustrum vulgare*, *Ilex aquifolium*, *Cornus sanguinea*, roses (several species in the genus *Rosa*) and brambles (several species in the genus *Rubus*). The most frequent creepers are *Lonicera periclymenum*, *Hedera helix*, *Tamus communis* and *Clematis vitalba* and, in warmer areas, *Rubia peregrina* and *Smilax aspera*. The scrub layer tends to be sparse, although there may be a presence of *Ruscus aculeatus*, *Erica vagans* or *Daphne laureola*. In the herb stratum, also very diverse, the soil richness is evidenced through demanding plants such as the fern *Polystichum setiferum* and other herbaceous nemorals characteristic of forests with rich soils (eutrophic): *Carex sylvatica*, *Arum italicum*, *Helleborus viridis* subsp. *occidentalis*, *Mercurialis perennis*, *Primula vulgaris*, *Ornithogalum pyrenaicum*, *Prunella vulgaris*, *Sanicula europaea*, *Saxifraga hirsuta*, *Brachypodium sylvaticum*, *Melica uniflora*, *Hypericum androsaemum*, *Galium odoratum*, *Lilium martagon*, etc., in addition to other ferns such as *Phyllitis scolopendrium* and *Dryopteris affinis*. Several facies can be recognised in the heart of these mixed woodlands, in addition to the characteristic facies (with *Tilia cordata*) of karst soils. In less rainy areas (below 1150 mm of rainwater a year), the abundance of *Quercus pyrenaica* is indicative of the facies with Pyrenean oaks. In contrast, in rainier areas (precipitation over 1400 mm of rainwater a year), as indicated earlier, the forest is enriched with *Fagus sylvatica* (facies with beech). In the Somiedo and Valdeón (León) districts where the colluvia consists of finer elements, *Quercus faginea* forms part of this type of mixed woodland, and constitutes a particular facies. It is fairly frequent to find young stages of the forest that are rich in ash and acer and without any significant presence of sessile oaks. An abundance of chestnuts (*Castanea sativa*) is sometimes also found in the tree layer in these woodlands.

7.4.2.7 Riparian Woodlands (Alder, Ash and Willow Woods)

The banks of rivers and streams flowing through valleys in territories with an oceanic climate along the Cantabrian coast (northern Galician-Asturian territories and more northern areas in the Laciana and Ancares sector) are colonised by riparian woodlands with alders. The limiting factor for their development is the presence of water in the deeper levels of the soil, and these riparian alders and their edges and substitution stages grow on moister, temporarily waterlogged riverbed soils. In these riparian woodlands the alder (*Alnus glutinosa*) is the dominant and characteristic tree, and its survival is conditioned by the fact that its roots are almost perpetually steeped in water. To the west of the Narcea river these riparian woodlands correspond to the so-called western riparian alder woods (Valeriano pyrenaicae-Alnetum glutinosae), which are exclusive to the territories that extend from the north of Lugo to the Narcea basin (western zone of the Galician-Asturian sector), penetrating into the northern inland valleys of the western territories of the Cantabrian Range (Laciana-Ancares sector), from sea level to an altitude of 700–800 m. They grow on anmoriform-type soils, characterised by the presence of a blueish-grey or greenish gley horizon in their profile, and continuously steeped in water, and whose colouring is due to iron in a ferrous state owing to the reduction caused by water that is rich in organic matter. They have a high floristic diversity, as the alder is accompanied by ash (*Fraxinus excelsior*), grey willow (*Salix atrocinerea*) and trees that are common in other types of woodland such as birch (*Betula celtiberica*), mountain elm, wych elm (*Ulmus glabra*) and white willow (*Salix alba*). The herb stratum has a frequent presence of plants adapted to conditions of high soil moisture such as the sedge *Carex acuta* subsp. *reuteriana*, royal fern (*Osmunda regalis*), Pyrenean valerian (*Valeriana pyrenaica*), *Athyrium filix-femina* and bittersweet nightshade (*Solanum dulcamara*). It is uncommon to find well-formed mature alders. They generally appear as narrow rows of trees forming linear woodlands which have usually been cut to obtain land suitable for farming. It is not infrequent to find allochthonous tree species such as poplar or black locust among the alders. Occasionally, due to the narrowness of the river or streambed, the canopies of the alders on both banks make contact and form what are known as gallery woodlands, creating a highly shady and moist environment in the heart of the river.

The riparian woodlands with alders that extend to the east of the Narcea basin along the whole Cantabrian coastline through to Aquitaine and Les Landes in France –that is, through the centre-eastern Cantabrian-Atlantic areas, with slight penetrations in the territories in the Picos de Europa-Ubiña sectors–have a notably different floristic composition from the aforementioned western riparian woodlands. They therefore correspond to another type of riparian woodlands, namely eastern riparian alder woods (Hyperico androsaemi-Alnetum glutinosae), which occupy the meso- and supratemperate bioclimatic belts. The soils on the valley floors where they grow are conditioned by the type of rock beneath, and the frequent predominance of calcareous rocks means the soils are rich in bases and the waters therefore range from slightly hard to hard. In addition to the alder, there is a common presence of other flat-leaved deciduous trees such as *Fraxinus excelsior*, *Quercus robur*, *Ulmus glabra* and several species of *Salix*, with beech in environments with more or less permanent summer mists. Frequent and abundant among the shrubs are brambles in the genus *Rubus*, *Corylus avellana*, *Sambucus nigra*, *Laurus nobilis*, *Cornus sanguinea* and *Frangula alnus*. The understorey is more diverse, and features numerous plants that are demanding in terms of nutrition and soil moisture such as *Hypericum androsaemum, Carex pendula, Carex remota, Festuca gigantea, Lysimachia nemorum, Solanum dulcamara, Circaea lutetiana, Saxifraga hirsuta, Lamiastrum galeobdolon, Lathraea clandestina* and the ferns *Athyrium filix-femina* and *Dryopteris dilatata*, among others.

On the banks of rivers and streams in the valleys and mountains in the interior of the Cantabrian Range, with a more continental character than those of the Cantabrian coastline (northern oro-Cantabrian territories), the riparian woodlands are lacking in alders, and are dominated by the common ash (*Fraxinus excelsior*). These oro-Cantabrian northern riparian ash woods (Festuco giganteae-Fraxinetum excelsioris) without alders are widespread in the rainier oro-Cantabrian territories whose climactic vegetation corresponds to beechwoods and sessile oakwoods with birch. The disappearance of alder from the riparian woodlands in these environments is particularly notable in the rivers in the Narcea basin –such as the Naviego, Muniellos or Coto–, or in the Navia basin, such as the Ibias. There is a curious correspondence between the disappearance of *Quercus pyrenaica* and *Alnus glutinosa* at higher altitudes. These riparian woodlands are exclusive to riverbeds and riverbanks in the northern territories of the Cantabrian Range, from the valley bottoms to an altitude of approximately 1700 m (meso and supratemperate bioclimatic belts). The soils on which they grow are rich and moist most of the year and the tree layer in these riparian woodlands without alder is dominated by *Fraxinus excelsior*, which grows alongside the occasional presence of *Acer pseudoplatanus, Ulmus glabra, Betula celtiberica, Fagus sylvatica, Tilia platyphyllos* and *Quercus petraea*. In the shrub and sub-shrub layer there is an abundance of *Corylus avellana, Salix atrocinerea, Salix caprea* and *Rubus*. The herb stratum otherwise has a very similar floristic composition to that of alder woods, with a frequent presence of *Carex remota, Carex laevigata, Athyrium filix-femina, Osmunda regalis, Festuca gigantea, Valeriana pyrenaica, Circaea lutetiana, Chaerophyllum hirsutum, Silene dioica, Luzula sylvatica* subsp. *henriquesii, Omphalodes nitida, Saxifraga spathularis, Holcus mollis* and *Teucrium scorodonia*.

Southern riparian oro-Cantabrian ash woods (Euphorbio hybernae-Fraxinetum excelsioris) constitute the riparian woodlands in the upper semicontinental supratemperate territories on the southern slope of the Cantabrian Range, occupying the beds of rivers with slightly hard waters on deep soils. The tree layer features a predominance of *Fraxinus excelsior* along with *Prunus padus, Fagus sylvatica, Ulmus minor* and *Populus nigra*, whereas in the shrubby layer there is a relatively high frequency of *Viburnum lantana, Crataegus monogyna, Corylus avellana, Ribes petraeum, Sorbus aria* and *Rosa corymbifera*, among others. Frequent in the herb stratum are *Euphorbia hyberna, Helleborus viridis* subsp. *occidentalis, Allium ursinum, Symphytum tuberosum, Lathraea clandestina, Anemone ranunculoides, Lilium martagon, Carum carvi* and *Aquilegia vulgaris*, among others. This type of woodland is today very strongly reduced in area due to intense farming and livestock activities (Herrero et al. 2002).

In southern oro-Cantabrian supratemperate territories, tree-willow woods with white willow (Salicetum cantabrico-albae) can be found growing along river courses with hard or slightly hard waters. They have a predominance of both tree (*Salix alba*, *Salix euxina* (*S. fragilis*) and *Salix* x *rubens* (*S. alba* x *S. euxina*), and shrub willows *Salix cantabrica*, *Salix atrocinerea*, *Salix* x *expectata* (*S. atrocinerea* x *S. cantabrica*), *Salix purpurea* subsp. *lambertiana*, *Salix triandra* subsp. *discolor*, *Salix caprea* and *Salix elaeagnos* subsp. *angustifolia*, among others. In the herb stratum of these riparian formations there is a frequent presence of *Brachypodium sylvaticum*, *Rubus caesius*, *Mentha longifolia* and *Juncus effusus*. On gleyic fluvisols in troughs in broad continental valleys, common ash and cherry, alder (Euphorbio hybernae-Fraxinetum excelsioris) grow adjacent to these white willows with Cantabrian willows, along with the best surviving hay meadows in the mountain range (Malvo moschatae-Arrhenatheretum bulbosi and Bromo commutati-Polygonetum bistortae).

Cantabrian tree-willow woods (Salicetum cantabricae) colonise the banks of oro-Cantabrian meso-supratemperate rivers and watercourses that are subject to major spring flooding. They grow on both stony, and silty substrates, generally rich in calcium carbonate. The shrub layer is characterised by the presence of *Salix cantabrica*, *Salix elaeagnos* subsp. *angustifolia*, *Salix triandra* subsp. *discolor*, *Salix purpurea* subsp. *lambertiana*, *Salix atrocinerea*, *Salix caprea*, *Salix* x *expectata* (*S. atrocinerea* x *S. cantabrica*) and *Salix* x *multidentata* (*S. atrocinerea* x *S. triandra* subsp. *discolor*), among others. In the herb stratum there is an abundance of *Mentha longifolia*, *Senecio aquaticus*, *Equisetum palustre*, *Calamagrostis pseudophragmites*, etc.

Cantabrian tree-willow woods with *Salix bicolor* (Salicetum cantabrico-bicoloris) are hygrophilous, calcicolous and chionophilous tree stands typically found on the banks of streams with slightly hard water that extend throughout orotemperate (subalpine) territories with a hyperhumid ombroclimate in the oro-Cantabrian sub-province, particularly in the territories of Somiedo and Carrión. They are floristically characterised by the dominance of *Salix bicolor*, *Salix cantabrica* and the hybrid of the two, *Salix* x *vazquezii* (*S. bicolor* x *S. cantabrica*), and are in contact with megaforb communities with a presence of *Adenostyles alliariae* subsp. *pyrenaica*, *Ranunculus aconitifolius*, *Veratrum album* and *Aconitum vulparia* subsp. *neapolitanum*, among others. This community forms part of the high shrublands of chionophilous orotemperate riparian willow and green alder with an alpine-Pyrenean-Balcanic and oro-Cantabrian relict distribution.

There is an occasional presence in eastern oro-Cantabrian territories of tree-willow woods with *Salix salviifolia* and *Salix cantabrica* (Salicetum salviifolio-cantabricae), typical of riparian watercourses that do not dry out during the summer months, and with slightly hard to soft waters. Willows with a widespread distribution extend throughout the territories in León (plani-Leonese, Maragatan and Sanabrian areas) and can be seen to reach –along the Ebro river basin– as far as territories in the Sierra de Campoo and in the southern Cantabrian or Castilian-Cantabrian sector, and possibly areas near the Sierra de la Demanda in the supratemperate supramediterranean or submediterranean belts. Its tree stratum includes *Salix cantabrica*, *Salix salviifolia*, *Salix* x *legionensis* (*S.* x *coenocarpetana S. salviifolia* x *S. cantabrica*), *Salix elaeagnos* subsp. *angustifolia*,

Salix x *pseudoelaeagnos* (*S. salviifolia* x *S. elaeagnos* subsp. *angustifolia*), *Salix* x *expectata* (*S. cantabrica* x *S. atrocinerea*) and *Salix* x *secalliana* (*S. salviifolia* x *S. atrocinerea*), among others.

7.4.2.8 Oro-Cantabrian Acer Woods with Ash

Mixed woodlands dominated by acer (*Acer pseudoplatanus*) are distributed throughout areas in the siliceous mid-mountains in the Cantabrian Range (supratemperate oro-Cantabrian territories with a humid-hyperhumid ombroclimate), although they are more abundant in the west of this area (Laciana-Ancares sector). These oro-Cantabrian acer woods with ash occupy gorges on cool, well-drained soils with mull humus and that are relatively rich in nutrients, mainly on colluvia in the lower areas of the slopes, or on former terraces or riverbanks with large blocks. They are associated with mountainside deposits containing a mixture of coarse and fine materials and are therefore well aerated and drained. Precisely the fact that these are more fertile soils has led to their intensive use for cultivation and grazing, and it is therefore very uncommon to find mature woodlands there, and when they do exist, they tend to have a limited extension and represent young stages.

Oro-Cantabrian acer woodlands with ash (Luzulo henriquesii-Aceretum pseudoplatani) have a fairly diverse tree layer with a codominance of different tree species, mainly *Acer pseudoplatanus*, *Prunus avium*, *Fraxinus excelsior*, *Corylus avellana* and *Betula celtiberica*. Other trees that may be present include *Sorbus aucuparia* and several species of *Quercus*, generally *Q. petraea*, and to a lesser degree *Quercus pyrenaica*, *Ulmus glabra* and *Tilia platyphyllos*. The shrub layer may be profuse, mainly in the younger stages of the woodland, with *Corylus avellana*, *Rubus* sp. pl. and *Salix atrocinerea* as the most abundant plants. The understorey frequently contains plants with strict nutrient requirements such as *Polystichum setiferum, Mercurialis perennis, Milium effusum, Brachypodium sylvaticum*, and other more general nemoral and acidophilous plants including *Omphalodes nitida, Physospermum cornubiense, Linaria triornithophora, Luzula sylvatica* subsp. *henriquesii, Saxifraga spathularis, Hypericum pulchrum, Vaccinium myrtillus* and *Holcus mollis*, among others. The areas with potential vegetation corresponding to these woodlands are intensively used in the valleys in the west of the Cantabrian Range; villages and towns have tended to become established on or nearby these sites, and their soils are used mainly for market gardening and intensively farmed meadows.

7.4.2.9 *Quercus rotundifolia* (Holm Oak) Woodlands (Oro-Cantabrian Holm-Oakwoods)

The relict forests of *Quercus rotundifolia* and *Quercus ilex* subsp. *gracilis*, essentially northern oro-Cantabrian and mesotemperate (coline), correspond to what are known as oro-Cantabrian holm-oakwoods (Cephalanthero longifoliae-Quercetum rotundifoliae). The floristically richest of these are found in south-eastern oro-Cantabrian continental areas (Deva river basin, Cantabria), and they become impoverished in the northern and western oro-Cantabrian territories (basins of the Pajares, Quirós, Teberga and Pigüeña rivers). They grow on sunny and calcareous slopes in territories located below altitudes of 1700–1900 m (meso- and supratemperate belts) on the northern face of the Cantabrian Range, from the Narcea river basin until the Picos de Europa. Although they do not cover large extensions, they have an important scenic and paleoclimatic value as their presence evokes the climate events undergone by this territory throughout its existence, and particularly the migrations of flora and Mediterranean vegetation towards the northern areas of the Iberian Peninsula. They grow on dry, generally pinkish or light brown eutrophic soils that form on steep calcareous terrain in gorges, karst zones or on well-drained calcareous colluvia. Occasionally in the tree layer there is a frequent presence of *Fraxinus excelsior* and *Prunus mahaleb*. Among the shrubs that form the understorey there is a low abundance of evergreen elements such as *Rhamnus alaternus*, and a frequent presence of *Corylus avellana*, *Ligustrum vulgare*, *Phillyrea latifolia*, *Pistacia terebinthus* and *Arbutus unedo*.

In the territories located to the south of the Cantabrian Range, the woodlands of *Quercus rotundifolia* correspond to southern oro-Cantabrian holm-oakwoods (Lithodoro diffusae-Quercetum rotundifoliae). These relict and calcicolous forest formations grow in supratemperate territories with a submediterranean character and a subhumid-humid ombroclimate, and are characterised by the presence in the tree layer of *Quercus rotundifolia* and sometimes *Juniperus thurifera* subsp. *orocantabrica*. Frequent in the shrub layer are *Juniperus sabina*, *Glandora diffusa* (*Lithodora diffusa*), *Teucrium pinnatifidum* and *Crataegus monogyna*, among others. The herb stratum commonly includes *Saponaria ocymoides*, *Rubia peregrina*, *Carex halleriana*, *Brachypodium rupestre*, *Bromus erectus*, *Helleborus viridis* subsp. *occidentalis*, etc. Also frequent in the understorey and on the edges of these forest communities are pulvinate caespitose scrublands in the order Festuco hystricis-Poetalia ligulatae.

7.4.2.10 *Quercus faginea* Woodlands (Oro-Cantabrian Gall-Oakwoods)

Iberian Portuguese oakwoods (*Quercus faginea*) with Cantabrian barberry (*Berberis vulgaris* subsp. *cantabrica*) (Berberido cantabricae-Quercetum fagineae) are found in relict form at low altitudes on sunny slopes in Valdeón, and dispersed in other inland limestone valleys in the northern Picos de Europa-Ubiña sector (occasionally extending as far as Somiedo). These oro-Cantabrian Portuguese oakwoods occupy carbonate-rich substrates in the mesomontane (coline) belt, sometimes reaching the lower supratemperate (montane) belt, in humid ombroclimates with a strong submediterranean character. In addition to Portuguese oak, the tree layer contains *Fraxinus excelsior*, *Corylus avellana*, *Prunus mahaleb*, *Sorbus intermedia*, *Tilia platyphyllos* and *Crataegus monogyna*. The understorey has an abundance of Cantabrian barberry (*Berberis vulgaris* subsp. *cantabrica*), *Genista legionensis*, *Laserpitium nestleri* subsp. *flabellatum*, *Glandora diffusa* (*Lithodora diffusa*), *Pimpinella siifolia*, *Ribes alpinum*, *Primula vulgaris* subsp. *columnae*, *Hepatica nobilis*, *Helleborus viridis* subsp. *occidentalis*, *Ilex aquifolium*, *Mycelis muralis* and *Mercurialis perennis*, among others. In the headwaters of the Cares river where the type association is found, they are represented by a group of microforests in recovery, with trees of about 10 m tall and trunks with a diameter of 30 cm –the result of old cutting operations–, which have as their outer edge a shrubland of barberry and spiny shrubs (Pruno spinosae-Berberidetum cantabricae) and a heathland of purple gromwell and rambling heaths (Lithodoro diffusae-Genistetum legionensis) on the forest floor.

7.4.2.11 *Quercus pyrenaica* Woodlands (Oro-Cantabrian Pyrenean Oakwoods)

Pyrenean oakwoods (forests of *Quercus pyrenaica*) in the meso-supratemperate oro-Cantabrian territories (coline and montane) with a subhumid ombroclimate correspond to the association Linario triornithophorae-Quercetum pyrenaicae. The most notable phenological aspect of these woodlands is that, in autumn and most of winter, the dry leaves of the Pyrenean oak remain on the branches (marcescent leaves), giving these woodland formations a very distinctive appearance. These Pyrenean oak forests grow on dry, non-podsolized oligotrophic soils (central-European Pyrenean oakwood brown earth or brown ranker) with an A-horizon that is humus-poor and rendered brown by the high iron hydroxide content, and a poorly developed B- horizon, tough that is often absent. The tree layer in the Pyrenean oakwoods is dominated by *Quercus pyrenaica*, although in some cases there is a presence of sessile oak (*Quercus petraea*) and its hybrids. In the shrub stratum there is a frequent presence of *Frangula alnus* and *Pyrus cordata*. Pyrenean oakwoods are rarely dense as this is a tree that requires much light in its juvenile stage and provides scarce coverage in the adult stage. This aids the diversification of the herb stratum, as other plants of a heliophilous nature thrive alongside the typical species of these woodlands (*Holcus mollis*, *Stellaria holostea*, *Physospermum cornubiensis*, *Avenella flexuosa*, *Teucrium scorodonia*, *Melampyrum pratense*, etc.).

7.4.2.12 *Juniperus thurifera* Woodland (Oro-Cantabrian Spanish Juniper Woods)

Sunny calcareous slopes on the southern face of the Cantabrian Range with a marked submediterranean continental character and a subhumid-humid ombroclimate serve as a refuge for oro-Cantabrian Spanish juniper woods (Juniperetum sabino-orocantabricae). These relict formations of Spanish juniper woodland (Fig. 7.12) are the westernmost of their kind on the Iberian Peninsula. They are located at altitudes of around 1150–1300 m (supratemperate territories) and occupy sunny

Fig. 7.12 Juniper woodland of the Juniperetum sabino-orocantabricae on a rocky south-facing calcareous slope. The opposite slope is covered by beech forest of the Epipactido helelborines-Fagetum sylvaticae, Crémenes, Esla river valley close to Cistierna, León (Photo by Sara del Río)

mountain sides, areas with slopes of 30%, and relatively undeveloped soil. The most important forest stands are located in León (Crémenes and the upper basin of the Luna river) and Palencia (Velilla del Río Carrión and the upper basin of the Pisuerga river) (Torío-Babia and Alto Esla-Carrión districts). The tree layer in these open woodlands is dominated by *Juniperus thurifera* subsp. *orocantabrica*, whereas the understorey contains an abundance of *Juniperus sabina*, *Juniperus alpina*, *Berberis vulgaris* subsp. *cantabrica*, *Glandora diffusa* (*Lithodora diffusa*), *Thymus mastichina*, *Prunus spinosa* and *Santolina semidentata*, among other plants. The considerable summer thermoxericity of the territory favours the presence of spiny scrublands with barberry (Pruno spinosae-Berberidetum cantabricae) in the broom scrublands of the Lithodoro diffusae-Genistetum scorpii and the oro-Cantabrian psychroxerophilous grasslands of the Arenario cantabricae-Festucetum hystricis that appear in the clearings of Spanish juniper woods.

7.4.3 *Meso- and Supratemperate Scrublands*

7.4.3.1 Broom Scrublands

In the meso-supratemperate territories in the subhumid to hyperhumid Ubiña and Alto-Campoo-Carrión sectors the broom scrublands correspond to the association

Cytiso cantabrici-Genistetum polygaliphyllae and are characterised floristically by the presence of *Cytisus cantabricus*, *Genista florida* subsp. *polygaliphylla*, *Erica arborea*, *Pteridium aquilinum* and *Orobanche rapum-genistae*, among others. In contrast, the western part of the oro-Cantabrian province and the Lugan, Iberian-Sorian, Leonese, Orensian-Sanabrian, Alto-Salamancan and Castilian-Cantabrian territories are characterised by the common occurrence of the Cytiso scoparii-Genistetum polygaliphyllae, a large supratemperate and supramediterranean association that constitutes the shrubby edge of beechwoods and Pyrenean oakwoods over large areas; it is floristically characterised by the dominance of *Genista florida* subsp. *polygaliphylla, Cytisus scoparius, Erica arborea, Pteridium aquilinum* and *Adenocarpus complicatus* subsp. *complicatus.*

7.4.3.2 Spiny Scrublands

The spiny shrublands dominated by *Berberis vulgaris* subsp. *cantabrica* and *Rhamnus alpina* in the supratemperate belt of the Picos de Europa-Ubiña sector correspond to the association Pruno spinosae-Berberidetum cantabricae. They grow on calcareous substrates (rocky or rock-wall biotopes) and have a frequent presence of *Amelanchier ovalis*, *Prunus spinosa*, *Ribes alpinum*, *Rosa villosa* and *Rhamnus catharticus*, among others. They constitute the spiny edges of beechwoods, Spanish juniper woods and sessile oakwoods. The mesotemperate subhumid-humid oro-Cantabrian spiny scrublands can be referred to the association Rosetum corymbifero-vosagiacae, characterised by the predominance of *Rosa vosagiaca, Berberis vulgaris* subsp. *cantabrica, Prunus mahaleb* and *Ribes alpinum*; this spiny scrubland grows on siliceous substrates and constitutes the edges of Pyrenean oakwoods and birchwoods.

7.4.3.3 Heathlands

The silicicolous serial communities formed by meso-, micro- and nanofruticose plants (heathlands, heathland-cistus thickets, heathland-gorse thickets, etc.) are found on hygro-peaty soils, with an abundance of species that form a highly acidic organic matter that tends to podsolize the soils. The soils are generally eroded dystic cambisols or ferric podsols with highly acidic coarse humus, and may occasionally be gleyic or have the properties of stagnosols. These fruticose communities are typical of thermo-orotemperate and thermo-supramediterranean subhumid-ultrahyperhumid, oceanic and hyperoceanic bioclimates with a European-Atlantic sub-Atlantic, Pyrenean-Cevennian, Mediterranean Ibero-Atlantic, oro-Iberian and Tingitanian distribution. The notable extension acquired by these heathlands is due to the destruction by felling, fire and overgrazing of the original forests, the high shrublands forming tangel or mulliform humus, and the productive pastures. In the meso-supratemperate territories in the Laciana and Ancares sector, the most representative heathland-gorse thicket corresponds to the association Halimio alyssoidis-Ulicetum breoganii, which occasionally reaches northern Galician-Asturian territories; it is floristically characterised by the presence of *Ulex gallii* subsp. *breoganii*, *Halimium alyssoides*, *Daboecia cantabrica*, *Erica cinerea*, *Calluna vulgaris*, *Agrostis curtisii* and *Pseudarrhenatherum longifolium*. In centre-eastern oro-Cantabrian humid-hyperhumid meso-supratemperate territories, this heathland-gorse thicket is replaced by communities of the Vaccinio myrtilli-Ulicetum gallii which occasionally reach territories in the Oviedo area and extend throughout the Cantabrian-Basque territories; these heathlands-gorse thickets range from acidophilous to moderately neutrophilous, and grow on soils of varying depths with mor humus and more rarely peaty mor; they are dominated by *Ulex gallii* subsp. *gallii*, *Erica vagans*, *Daboecia cantabrica*, *Calluna vulgaris*, *Erica cinerea*, *Pseudarrhenatherum longifolium*, *Cirsium filipendulum*, *Agrostis curtisii* and *Vaccinium myrtillus*, among others. The most widespread heathland throughout the oro-Cantabrian area corresponds to the association Pterosparto cantabricae-Ericetum aragonensis: these are heathlands dominated by red heath (*Erica australis* subsp. *aragonensis*) which grows throughout the Cantabrian Range from Los Ancares through to the massif of Peña Labra in Campoo (fundamentally oro-Cantabrian), in the mid and upper levels of the supratemperate belt, and in the westernmost mesotemperate territories on siliceous substrates, where they are able to withstand the summer drought. This association constitutes a transitional unit between the *Daboecion cantabricae* and *Ericion umbellatae*, as its floristic composition includes elements such as *Daboecia cantabrica*, *Halimium umbellatum*, *Halimium alyssoides*, *Erica umbellata*, *Erica cinerea*, *Pterospartum tridentatum* subsp. *cantabricum* and *Polygala microphylla*, except in the ecotones with the hygrophilous communities of the Genistion micrantho-anglicae, on the southern oro-Cantabrian limits in contact with the Mediterranean region. In hydromorphic biotopes with highly oligotrophic podsolized soils there is a presence of heathlands of the Genisto anglicae-Ericetum tetralicis typical of the supramediterranean Bierzo-Sanabrian, Leonese and supratemperate oro-Cantabrian territories; they are floristically characterised by the presence of *Erica tetralix*, *Genista anglica*, *Genista micrantha*, *Calluna vulgaris*, *Halimium alyssoides* and *Agrostis curtisii*.

7.4.3.4 Pulvinate Scrublands

Those scrublands comprise communities characterised by *Genista occidentalis*, *Genista legionensis* and a broad range of oro-Cantabrian-Atlantic and frequently submediterranean or supramediterranean thermo-orotemperate subhumid-hyperhumid calcicolous endemisms with an oro-Cantabrian, Oviedan, Cantabrian-Basque and Castilian-Cantabrian distribution and with disjunctions in the Sierra Ibérica and the western Pyrenees. They belong to the Genistion occidentalis alliance and grow on rendziniform leptosols, leptic luvisols and eutric colluvials, in addition to representing permanent communities on rocky habitats, particularly orotemperate. These scrublands have a broad serial significance in the deciduous and evergreen woodlands corresponding to the calcicolous and calco-dolomitic territorial macroseries and in different alliances. The pulvinate or hedhehog scrublands of the Picos de Europa and eastern Ubiña sector (which occasionally reach the Cantabrian-Atlantic territory of the Sierra de Cuera and Sueve District) correspond to the association Lithodoro diffusae-Genistetum legionensis; these basophilous hedgehog gorse scrublands (*Genista legionensis* and *Genista occidentalis*) are distributed throughout subhumid to ultrahyperhumid meso- to orotemperate territories and are floristically characterised –in addition to broom– by the presence of *Helianthemum urrielense*, *Glandora diffusa*, *Teucrium pyrenaicum*, *Helianthemum canum* subsp. *cantabricum*, *Globularia nudicaulis*, *Thymelaea ruizii* and *Euphorbia flavicoma* subsp. *occidentalis*, among others. The centre-eastern oro-Cantabrian broom scrublands (Ubiña and Alto Campoo-Carrión territories) belong to the association Lithodoro diffusae-Genistetum occidentalis; these mesotemperate to supratemperate humid to hyperhumid basophilous broom scrublands are characterised by the presence of *Genista occidentalis*, *Teucrium pyrenaicum*, *Helianthemum canum* subsp. *cantabricum*, *Koeleria vallesiana*, *Helianthemum nummularium*, *Helianthemum croceum* subsp. *cantabricum* and *Teucrium chamaedrys*, among other species. In the warmer and more continental southern areas of the Cantabrian Range (Babia-Torío District), there are scrublands of the Lithodoro diffusae-Genistetum scorpii, which form part of the Phoenician juniper woodland series that are floristically characterised by the presence of *Genista scorpius*, *Genista legionensis*, *Glandora diffusa*, *Thymus mastichyna*, along with other elements with a Mediterranean character.

7.4.4 Meso- and Supratemperate Meadows and Grasslands

The supratemperate oro-Cantabrian and northern Galician-Asturian early-colonising grasslands of the Diantho langeani-Festucetum rivas-martinezii constitute isolated herbaceous communities subject to frequent cryoturbation processes and dominated by pulvinate chamaephytes and hemicryptophytes colonising lithosols on slate, sandstone and quartzite. This association is characterised by the endemisms *Dianthus langeanus* and *Festuca rivas-martinezii* and has its optimum in the humid supratemperate fringe of the Laciana and Ancares sector, where the potential vegetation is the *Quercus pyrenaica* woodland (Pyrenean oakwoods) of the Linario triornithophorae-Querco pyrenaicae sigmetum. This habitat is frequently in contact with the crassifolious grasslands of the association Agrostio durieui-Sedetum pyrenaici. These grasslands are replaced in the Bierzo-Sanabrian sector and in more continental oro-Cantabrian areas by the siliceous grasslands of the Scleranthо perennis-Plantaginetum radicatae, characterised by the presence of *Plantago radicata*, *Koeleria caudata* subsp. *crassipes* and *Scleranthus perennis*, among others.

The annual ephemeral grasslands that colonise calcareous soils in the oro-Cantabrian and Cantabrian-Atlantic sub-provinces in the subhumid-humid thermo-supratemperate belts, with Bierzo-Sanabrian mesomediterranean humid ombroclimate disjunctions corresponding to the Minuartio hybridae-Saxifragetum tridactylites –albeit with floristically impoverished communities–, and are characterised by the presence of *Saxifraga tridactylites*, *Minuartia hybrida*, *Asterolinon linum-stellatum*, *Arenaria serpyllifolia*, *Brachypodium dystachyon*, *Campanula erinus*, *Hornungia petraea* and *Linum strictum*, among other species.

The Cantabrian-Atlantic grasslands of *Brachypodium rupestre* (Seselio cantabrici-Brachypodietum rupestris) are basophilous mesophytic grasslands dominated by *Brachypodium rupestre* with an oceanic, thermo-mesotemperate subhumid-hyperhumid influence, commonly found in the Cantabrian-Basque sector, the eastern territories of the northern Galician-Asturian sector, and the Picos de Europa district. There are very widespread on both limestone substrates and flysch, and fill practically all the remaining spaces between the other communities that form the landscape mosaic in the coline and montane environments in these sectors. Floristically, in addition to *Brachypodium pinnatum* subsp. *rupestre*, they are characterised by the presence of *Bromus erectus*, *Seseli cantabricum*, *Scabiosa columbaria*, *Carlina vulgaris*, *Asperula aristata*, *Carduus argemone*, *Hieracium pilosella*, *Picris hieracioides*, among others. The abandonment of the hay meadows has given rise to a proliferation of this type of grasslands, which precede the entry of the forest mantle or edge stages. In eastern oro-Cantabrian territories these grasslands correspond to the association Bromo erecti-Caricetum brevicollis, typical of supratemperate and lower orotemperate humid-hyperhumid territories, and characterised by *Carex brevicollis*, *Bromus erectus*, *Eryngium bourgatii*, *Potentilla tabernaemontani*, *Thymus praecox* subsp. *britannicus*, *Plantago media*, *Helianthemum nummularium*, *Phyteuma orbiculare* susp. *ibericum*, *Carex humilis*, etc. In western oro-Cantabrian areas these grasslands are replaced by those in the association Helianthemo cantabrici-Brometum erecti, which spread throughout the subhumid-hyperhumid meso-supratemperate areas and are floristically characterised by the presence of *Bromus erectus*, *Helianthemum croceum* subsp. *cantabricum*, *Acinos alpinus* subsp. *pyrenaicus*, *Plantago media*, *Seseli montanum*, *Helianthemum nummularium*, *Koeleria vallesiana*, *Teucrium pyrenaicum*, *Thesium pyrenaicum*, *Pimpinella tragium* subsp. *lithophila*, *Helianthemum canum* subsp. *cantabricum*, *Arenaria grandiflora* s.l., *Eryngium bourgatii*, *Potentilla montana*, etc. The calcicolous grasslands in the Laciana and Ancares sector with the rosette chamaephyte *Erodium glandulosum* (Koelerio vallesianae-Erodietum glandulosi) are very singular, and thrive in the supratemperate humid-hyperhumid lithocalcareous soils in this territory.

The Cantabrian supratemperate hay meadows on soils with temporary hydromorphia, with systematic agricultural management consisting of mowing and fertilising, correspond to the association Bromo commutati-Polygonetum bistortae and grow throughout the oro-Cantabrian and Carpetan territories. They are characterised by the presence of *Polygonum bistorta*, *Bromus commutatus*, *Stellaria graminea*, *Prunella grandiflora* subsp. *pyrenaica*, *Cirsium palustre*, *Lychnis flos-cuculi*, *Lotus pedunculatus*, *Carum verticillatum*, *Trisetum flavescens*,

Alopecurus pratensis, *Cynosurus cristatus*, *Carum carvi* and *Sanguisorba officinalis*, among others.

The north-western oligotrophic hay meadows (Agrostio castellanae-Arrhenatheretum bulbosi) are subjected to little management and extend throughout north-western Iberian meso-supratemperate territories (northern Galician-Asturian, Galician-northern Portuguese and the Laciana and Ancares sector). They are distinguished by their finicolous character within the alliance in both soils and biogeographical aspects, which implies floristic poverty. They have a frequent presence of *Malva colmeiroi*, *Holcus mollis*, *Arrhenatherum elatius* subsp. *bulbosum*, *Dactylis glomerata*, *Holcus lanatus*, *Rhinanthus minor*, *Ornithopus* sp. pl., *Prunella grandiflora* subsp. *pyrenaica*, *etc*. Cantabrian mesophilous hay meadows (Malvo moschatae-Arrhenatheretum bulbosi) are relatively underexploited and occur throughout the supratemperate territories in the Cantabrian-Basque, Castilian-Cantabrian, Oviedo, Picos de Europa-Ubiña and Laciana and Ancares sectors. In mesotemperate territories these meadows are becoming increasingly less frequent due to the type of intensive production methods to which they are subjected, and as a result they are transformed into meadows of the Cynosurion cristati. The most frequent species are *Arrhenatherum elatius* subsp. *bulbosum*, *Malva moschata* var. *geraniifolia*, *Avena pubescens*, *Anthoxanthum odoratum*, *Briza media*, *Lathyrus pratensis*, *Dactylis glomerata*, *Trisetum flavescens*, *Crepis capillaris*, *Trifolium pratense*, *Holcus lanatus*, etc.

Cantabrian supratemperate mesophilous meadows (Merendero pyrenaicae-Cynosuretum cristati) are intensively grazed and are floristically characterised by the presence of *Cynosurus cristatus*, *Lolium perenne*, *Trifolium repens*, *Plantago media*, *Merendera pyrenaica*, *Phleum bertolonii*, etc., along with other elements of the Nardetalia (pastures) such as *Nardus stricta*, *Danthonia decumbens*, etc. These supratemperate meadows are spread throughout the Cantabrian-Atlantic, oro-Cantabrian and western Pyrenean territories and are frequently found in mountain passes that are subjected to intensive grazing by cattle, horses and sheep.

Communities with a meadow-like appearance that thrive on many sites that are frequented by humans and animals constitute grazing meadows with a roadside character (Lolietum perennis) that are relatively productive and can be used for grazing. They are distributed throughout the thermo-, meso- and supratemperate parts of the Eurosiberian and particularly the moist sites of the Mediterranean region. They are characterised by the presence of *Lolium perenne*, *Plantago major*, *Trifolium repens*, *Plantago lanceolata*, *Taraxacum ekmanii*, *Bellis perennis* and *Festuca arundinacea*, among others.

On shady forest or pre-forest paths subjected to light trampling, on soils with varying degrees of sandiness, there are some sciophilous roadside meadows with a woodland-type character (Juncetum tenuis). These more or less dense herbaceous communities are floristically characterised by the dominance of the North American neophyte *Juncus tenuis*, and are found throughout the Eurosiberian region from central Europe to Iberia.

Hygrophilous meadows-reed beds, growing on rich soils with a high water table almost all year round –although not waterlogged– and whose management basically consists of mowing and fertilising, form part of the association Loto pedunculati-Juncetum conglomerati. They grow throughout Cantabrian-Atlantic and oro-Cantabrian mesotemperate territories and, in addition to *Juncus conglomeratus* and *Juncus effusus*, their composition includes numerous grasses such as *Holcus lanatus*, *Gaudinia fragilis*, *Anthoxanthum odoratum*, *Poa trivialis*, and others, along with other mesohygrophilous meadow species. In southern oro-Cantabrian territories the oligotrophic meadows-reed beds belong to the association Deschampsio hispanicae-Juncetum effusi. These communities are used for grazing, and have a predominance of reeds (essentially *Juncus effusus*) along with other hygrophilous plants of the order *Molinietalia*, such as *Juncus acutiflorus*, *Carum verticillatum*, *Lotus pedunculatus*, etc. Although the association has its optimum in southern oro-Cantabrian territories, it penetrates into some Mediterranean enclaves in the plani-Leonese and Bierzo-Sanabrian sectors. Atlantic oligotrophic meadows-reed beds (Senecioni aquatici-Juncetum acutiflori) are not very intensively managed and maintained by mowing and grazing, causing them to be invaded by reeds and other oligotrophic and hygrophilous plants, with an absence or scarcity of higher-quality meadow species, and the frequent participation of the characteristic plants of turfophilous communities. They grow from at least Ireland as far as the northern Iberian Peninsula throughout the thermo-supratemperate Cantabrian-Atlantic and Laciana and Ancares sector, and the main species are *Juncus acutiflorus*, *Juncus effusus*, *Senecio aquaticus*, *Carex echinata*, *Molinia coerulea*, *Carum verticillatum*, *Carex panicea*, *Holcus lanatus*, *Lotus pedunculatus*, *Poa trivialis*, *Potentilla erecta*, *Carex binervis*, etc. Oro-Cantabrian supratemperate hygronitrophilous reed beds (Mentho longifoliae-Juncetum inflexi) are dense grasslands with a hygrophilous nature growing on nitrified marshy soils, generally compacted due to trampling by livestock and with soil moisture during most of the year. These communities are widely distributed throughout Eurosiberian areas and occasionally penetrate into Mediterranean areas. They grow preferentially on the edges of small watercourses and other moist sites, in contact with more or less moist meadows and grasslands, or else with grasslands in the class Phragmitetea. Floristically they are characterised by the presence of *Juncus inflexus*, *Mentha longifolia*, *Rumex crispus*, *Carex hirta*, *Ranunculus repens*, *Agrostis stolonifera* and *Juncus articulatus*, along with various meadow species such as *Prunella vulgaris*, *Plantago lanceolata*, *Dactylis glomerata*, *Trifolium pratense*, etc.

The tuft-like plant formations dominated by *Molinia caerulea* subsp. *arundinacea* and *Schoenus nigricans* constitute the Cantabrian hygrophilous grasslands (Molinio arundinaceae-Schoenetum nigricantis) typical of springs, waterfalls and gorges with highly carbonated water all year round. They appear occasionally in thermo- and mesotemperate areas along the Cantabrian coastline from the Basque Country towards Asturias and represent finicolous aspects of the alliance Molinio-Holoschoenion.

Hygronitrophilous pastures (Potentillo anserinae-Agrostietum stoloniferae) are found growing in southern oro-Cantabrian territories and occupy sandy or stony soils in riverbeds prone to flooding, under the influence of anthropozoogenic actions. They are characterised by the presence of *Agrostis stolonifera*, *Potentilla anserina* and *Senecio aquaticus*, among others. Their optimum development is reached in the humid-hyperhumid supratemperate belt in territories in the Ubiña sector.

7.4.5 Other Vegetation Types

7.4.5.1 Aquatic Communities (More Unusual)

The aquatic communities of *Callitriche brutia* and *Ranunculus peltatus* (Callitricho brutiae-Ranunculetum peltati) grow in deep cold stagnant oligotrophic neutro-acidophilous waters; they are found in cold and neutral or acid pools, springs and shallow lakes (which eventually dry up) in the mountains in the central and northern Iberian Peninsula, occasionally reaching as far as the central Picos de Europa-Ubiña, Laciana and Ancares, and north-western Asturian sectors. On the edges of upper supratemperate and orotemperate oro-Cantabrian lakes there are dwarf waterlogged pastures of water-starwort (Spargano angustifolii-Callitrichetum platycarpae) floristically defined by the presence of *Sparganium angustifolium*, *Isoetes velatum* subsp. *asturicense*, *Callitriche* sp. pl. and others (Fig. 7.13).

Sciophilous cold-water meadows dominated by *Cardamine raphanifolia* and *Chrysosplenium oppositifolium* (Cardamino flexuosae-Chrysosplenietum oppositifolii) grow in the oro-Cantabrian, Galician-Asturian and Cantabrian-

Fig. 7.13 Glaciar lake in La Baña, Sierra de La Cabrera, León (Photo by Sara del Río)

Basque meso-supratemperate belts and are floristically characterised by the presence of *Cardamine raphanifolia*, *Chrysosplenium oppositifolium*, *Stellaria alsine* and *Cardamine flexuosa*, among other plants. Sciophilous cold-water meadows dominated by *Saxifraga lepismigena* (Saxifragetum lepismigenae) colonise permanently oozing siliceous rocks and waterfalls with a western oro-Cantabrian and Galician-Asturian distribution in both mesotemperate and supratemperate territories; they are floristically characterised by the presence of *Saxifraga lepismigena*, *Chrysosplenium oppositifolium*, *Cardamine raphanifolia* and *Stellaria alsine*, among others. Water meadows of *Stellaria uliginosa* and *Montia fontana* subsp. *chondrosperma* (Stellario uliginosae-Montietum variabilis) grow on the banks of streams and emerging watercourses in oro-Cantabrian montane territories; these heliophilous communities floristically are characterised by the presence of *Montia fontana* subsp. *chondrosperma*, *Stellaria alsine* and *Epilobium alsinifolium*, among other species. In oromediterranean and orotemperate Carpetan-Leonese and oro-Cantabrian flowing streams with highly oligotrophic waters there are oligotrophic water meadows with *Myosotis stolonifera* and *Veronica langei* (Myosotidetum stoloniferae), floristically characterised by the presence of *Myosotis stolonifer*a, *Festuca rivularis* and *Veronica langei*, among other species. The water meadows dominated by *Saxifraga stellaris* subsp. *alpigena* and *Myosotis stolonifera* (Stellario alsines-Saxifragetum alpigenae) constitute communities that are typical of very cold springs and their nearby streams and are practically undisturbed by animals, with a Carpetan-Leonese orotemperate oro-Cantabrian and oro-cryoromediterranean distribution; they are floristically characterised by the presence of *Saxifraga stellaris* subsp. *robusta* (*Saxifraga stellaris* subsp. *alpigena*), *Epilobium alsinifolium*, *Stellaria alsin*e, *Festuca rivularis* and *Myosotis stolonifera*, among others, in addition to an abundance of briophytic companion species with a frequent presence of *Aulacomnium palustre*, *Philonotis fontana*, *Bryum pseudotriquetum*, etc.

Dwarf watercress formations (Glycerio declinatae-Apietum repentis) are communities dominated by *Apium repens* growing on emerging springs and the banks of rivers and streams, occasionally on base-poor substrates grazed by livestock; on the Iberian Peninsula they are restricted to the supramediterranean and supratemperate belts in the northern half, occasionally attaining the western oro-Cantabrian supratemperate territories. The formations of catabrosa and small sweet grass (Glycerio declinatae-Catabrosetum aquaticae) are Iberian communities typical of shallow calcareous and nitrified waters, and characterised by the presence of *Glyceria declinata*, *Glyceria fluitan*s, *Catabrosa aquatica*, *Veronica beccabunga*, etc., which extend as far as oro-Cantabrian thermo-supratemperate territories.

Flooded bladderwort grasslands (Sphagno-Utricularietum minoris) are distrophic and oligotrophic floating communities growing in small marshy ponds, and occasionally appearing in the middle of peat bogs with sphagnum in the association Drosero anglicae-Narthecietum ossifragi in oro-Cantabrian areas.

Oro-Cantabrian megaforbic grasslands (Senecioni laderoi-Filipenduletum ulmariae) are pastures formed by tall hygrophilous hemicryptophytes with a megaforb appearance and an oro-Cantabrian distribution, characterised by the presence of *Filipendula ulmaria*, *Epilobium hirsutum*, *Lysimachia vulgaris*, *Senecio laderoi*, *Equisetum palustris*, *Poa trivialis*, *Polygonum bistorta*, *Trisetum flavescens*, *Lychnis flos-cuculi* and *Caltha palustris*, among others.

7.4.5.2 Nitrophilous Communities

The perennial nitrophilous grasslands in the association Chenopodio bonihenrici-Senecietum nebrodensis grow in resting areas for livestock in the oro-Cantabrian and Carpetan mountains. These are communities located on deep cool soils in both cattle shelters and on dumps and places near human communities distributed throughout upper supratemperate and orotemperate oro-Cantabrian and oromediterranean Carpetan territories, and are characterised by the presence of *Chenopodium bonus-henricus*, *Senecio nebrodensis* (*Senecio duriaei*), *Urtica dioica*, *Geranium pyrenaicum*, etc. In meso-supratemperate Cantabrian-Atlantic and oro-Cantabrian areas the perennial nitrophilous grasslands in livestock shelters and on dumps and places near human communities correspond to the Malvo mauritianae-Rumicetum obtusifolii, and are characterised by the presence of *Malva sylvestris* (=*Malva sylvestris* subsp. *mauritiana*), *Arctium minus*, *Urtica dioica*, *Lamium maculatum*, *Marrubium vulgare*, *Rumex obtusifolius*, *Conium maculatum*, *Echium vulgare*, *Artemisia vulgaris*, *Ballota nigra* subsp. *foetida* and *Galium aparine*, among others.

The upper supratemperate calcareous thistle scrublands in the Ubiña sector (Carduo nutantis-Cirsietum chodati) are nitrophilous communities with large thistles that grow on disturbed soils, road verges and in livestock shelters, and are characterised by the presence of *Carduus nutans* var. *phyllolepis* and *Cirsium eriophorum* subsp. *chodati*. The upper supratemperate calcareous thistle scrublands in the Picos de Europa (Cirsio chodati-Carduetum cantabrici) have their optimum on well-drained disturbed calcareous soils on road verges and in livestock shelters; this association is characterised by the presence of *Carduus cantabricus*, *Cirsium eriophorum* subsp. *chodati*, *Carduus nutans* var. *phyllolepis*, *Verbascum thapsus* and *Urtica dioica*, among others. In upper supratemperate and orotemperate oro-Cantabrian siliceous territories (and in hard limestone territories in El Caurel), the nitrophilous communities of large thistles correspond to the association Cirsio chodati-Carduetum carpetani, and grow preferentially on dumps, verges and in cattle shelters and are characterised by the presence of *Carduus carpetanus*, *Cirsium eriophorum* subsp. *chodati*, *Carduus nutans* var. *phyllolepis* and *Cirsium vulgare*.

Nitrophilous and pyrophytic grasslands with asphodels (Asphodelo arrondeaui-Epilobietum angustifolii) are herbaceous, heliophilous and nitrophilous communities that grow in forests clearings and scrublands that have recently been cleared or burned, and where there is therefore a significant organic matter input. This association is distributed throughout Bierzo-Sanabrian supratemperate territories in the Laciana and Ancares and centre-western Picos de Europa-Ubiña sector. It is characterised by the presence of *Epilobium angustifolium*, *Luzula lactea*, *Avenula flexuosa* s.l., *Eryngium duriaei*, *Asphodelus macrocarpus* subsp. *arrondeaui*, *Digitalis purpurea*, etc.

On earthy trampled soils there is a presence of summer flowering meadows in which *Matricaria discoidea* (*Matricaria matricarioides*) tends to be the dominant species, along with *Poa annua*, *Polygonum arenastrum* and *Coronopus didymus*. These communities (Matricario-Polygonetum arenastri) are distributed throughout the thermo-, meso- and supratemperate Eurosiberian and supra- and oromediterranean territories. In contrast, on trampled calcareous clay soils there are meadows (Poo annuae-Coronopodetum squamati) with a dominance of *Coronopus squamatus* along with *Poa annua* and *Polygonum arenastrum*, among other plants, which occasionally appear in the Picos de Europa-Ubiña and Oviedan sectors.

In the relatively infrequent cereal crops in the supratemperate territories in this area there are weeds belonging to the association Linario elegantis-Anthoxanthetum aristati. These therophytic grasslands with a summer phenology are typical of cereal crops (rye, spelt, etc.) growing on soils with coarse elements and a sandy-silty texture in the supramediterranean belt in the Bierzo-Sanabrian sector, and in the supratemperate belt in the Laciana and Ancares sector; they reach as far as northern Galician-Asturian territories. The association is characterised by the presence of *Linaria elegans*, *Sedum arenarium*, *Aphanes australis*, *Arnoseris minima* and *Anthoxanthum aristatum*, among others.

On the boundaries and forest edges in this territory there are forb communities (Galio aparines-Anthriscetum sylvestris) with a mesic character comprising large perennial herbs including an abundance of umbellifers (such as *Anthriscus sylvestris*), on cool to moist soils generally affected by hydromorphia and which extend from the mesotemperate to the supratemperate belt. On highly nitrified mesotemperate and meso-supramediterranean sites on cool deep soils there are nettle beds with *Sambucus ebulus* belonging to the association Urtico dioicae-Sambucetum ebuli.

On the edges of mesophilous beechwoods in the Picos de Europa-Ubiña sectors there are herb communities (Centaureo nemoralis-Origanetum vulgaris) characterised by the presence of *Trifolium medium*, *Centaurea nemoralis*, *Vicia orobus*, *Lathyrus latifolius* and *Geranium sanguineum* that are distributed throughout the meso- and supratemperate belts. The edges of siliceous forests in the oro-Cantabrian and northern Galician-Asturian territories (essentially oligotrophic beechwoods, Pyrenean oakwoods and oligotrophic pedunculated oakwoods) are occupied by the herb communities of the Omphalodo nitidae-Linarietum triornithophorae and characterised by the presence of *Linaria triornithophora*, *Omphalodes nitida*, *Hypericum perforatum*, *Teucrium scorodonia*, *Clinopodium vulgare*, *Lathyrus niger* and *Campanula rapunculus*, among others.

The nitrophilous megaforbic vegetation growing in snow pockets formed at the foot of north-facing calcareous shelves after the thaw correspond to the association Aconito neapolitani-Myrrhidetum odoratae. These communities are found in the upper supratemperate and orotemperate territories in the Picos de Europa-Ubiña sectors, and are floristically characterised by the presence of *Myrrhis odorata*, along with *Aconitum vulparia* subsp. *neapolitanum*, *Adenostylis alliariae* subsp. *pyrenaica*, *Scrophularia alpestris* and *Veratrum album*, among others. They are also in contact with the communities of the Violo biflorae-Cystopteridion alpinae on fissures in shady, oozing chionophilous rocks. In siliceous substrates in these territories these communities are substituted by the megaforbic nitrophilous communities of the Allio victorialis-Adenostyletum pyrenaicae, which colonise shady cracks and the base of ledges, particularly on siliceous substrates and always on soils that are cool in summer and with longer snow cover. They are characterised by the presence of *Adenostylis alliariae* subsp. *pyrenaica*, *Allium victorialis*, *Veratrum album*, *Aconitum vulparia* subsp. *neapolitanum*, *Cicerbita plumieri* and *Athyrium distentifolium*. In supratemperate oro-Cantabrian areas the megaforbic communities correspond to the association Myrrhido odoratae-Valerianetum pyrenaicae, and are floristically characterised by the presence of *Valeriana pyrenaica*, *Adenostylis alliariae* subsp. *pyrenaica*, *Chaerophyllum hirsutum*, *Tozzia alpina*, *Cicerbita plumieri* and *Hugueninia tanacetifolia* subsp. *suffruticosa*. The megaforbic communities of the Aconitetum neapolitano – vulgaris grow on the banks of streams and rocky torrents in open calcareous areas in territories in the upper supratemperate Picos de Europa-Ubiña sectors in a hyperhumid ombroclimate, and are floristically characterised by the presence of *Aconitum napellus* subsp. *vulgare*, *Aconitum vulparia* subsp. *neapolitanum*, *Chaerophyllum hirsutum* and *Senecio aquaticus*, along with the presence of plants such as *Mentha longifolia*, which highlight its somewhat nitrophilous character due to grazing in the area.

The vegetation of the Cantabrian Range is documented by a large set of publications among which we can highlight those by Díaz González (2009, 2014a, b), Díaz González and Fernández Prieto (1988, 1994a, 1999, 2006), Díaz González et al. (2005, 2014), Navarro Andrés and Díaz González (1977) and Rivas-Martínez et al. (1971, 1984)

7.5 The Plant Landscape of the Galician-Leonese Mountains and the Bierzo Trench (Bierzo-Sanabrian Sector)

7.5.1 Oromediterranean and Temperate Submediterranean High-Mountain Vegetation

7.5.1.1 Rock Field and Scree Vegetation

In Bierzo-Sanabrian territories, and more specifically in the orotemperate submediterranean thermotype, always on base-poor substrates and in the fissures of siliceous rocks, there are plant communities with sparse coverage in which there is an almost constant presence of *Spergula rimarum*, along with *Murbeckiella boryi*, *Sedum hirsutum* and some other plants typical of nearby scree and stony areas (Murbeckiello boryi-Sperguletum rimari).

The edaphogenetic processes that fracture of the rocks produce a configuration of scree consisting of medium-sized or small clasts which are occupied by different plant communities. Thus on orotemperate chionophilous slate flakes in the Bierzo-Sanabrian and Laciana and Ancares sectors we find the presence of *Silene foetida* subsp. *gayana*, *Ranunculus cabrerensis* subsp. *cabrerensis*, *Linaria alpina* subsp. *alpina* and *Cryptogramma crispa* (Cryptogrammo crispae-Ranunculetum cabrerensis). The plant community formed by *Silene foetida* subsp. *gayana*, *Rumex suffruticosus*, *Senecio pyrenaicus* and *Cryptogramma crispa* (Cryptogrammo crispae-Silenetum gayanae), with a chionophobous character and growing on medium or small slate and quartzite scree with a more abundant earthy matrix than in the previous case, has an upper supratemperate and orotemperate distribution in the Bierzo-Sanabrian, Laciana and Ancares and Picos de Europa-Ubiña sectors. On shingle with coarse stones with an abundant earthy matrix there is predominance of oro-Cantabrian and Bierzo-Sanabrian communities with a supratemperate upper orotemperate hyperhumid optimum (Linario glabrescentis-Rumicetum suffruticosi) which include *Rumex suffruticosus*, *Phalacrocarpum oppositifolium*, *Linaria saxatilis* subsp. *glabrescens* and *Eryngium duriaei*.

The territories in the Bierzo and Ancares sector are also home to communities that establish on small earth-covered slate beds with a chionophobous character (Sesamoido pygmaiae-Silenetum gayanae), also distributed throughout the orotemperate territories in the Laciana and Ancares and southern Picos de Europa-Ubiña sectors, and characterised by the absence of pteridophytes and the presence of *Sesamoides minor*, *Silene foetida* subsp. *gayana* and *Paronychia polygonifolia*.

In the interstices in deposits of large semipermanent siliceous blocks in the western European high mountains there is a glericolous community with low coverage that is widely distributed throughout the high Eurosiberian and Iberian Mediterranean mountains, and characterised by the presence of two ferns: *Cryptogramma crispa* and *Dryopteris oreades* (Cryptogrammo crispae-Dryopteridetum oreadis). This community –depending on the territory under study– presents a wide variability and is enriched with different species; thus in territories in the Bierzo and Ancares sector it includes the presence of *Agrostis tileni*, and *Doronicum diazii* in the Laciana and Ancares sector.

7.5.1.2 Oromediterranean and Temperate Submediterranean High-Mountain Scrublands

Above altitudes of 1800 m on escarpments and rocky quartzite and slate soils preferentially on crests and ledges in the mountains of La Cabrera and the Montes Aquilianos, there is a presence of scrublands dominated by *Juniperus alpina*, along with *Genista sanabrensis*, *Avenella iberica* and *Cytisus oromediterraneus*. These creeping juniper woods of Sanabrian broom (Genisto sanabrensis-Juniperetum alpinae) represent the mature stage of the orotemperate submediterranean hyperhumid climatophilous series in all the Bierzo-Sanabrian mountains.

On similar substrates but on better-formed soils, there is another type of juniper woods, also present in the more continental and southerly territories in the oro-Cantabrian subprovince. They are dominated by *Juniperus alpina*, accompanied by *Avenella iberica, Cytisus oromediterraneus* and *Vaccinium myrtillus*, which – together with taxa with a marked western Iberian character such as *Phalacrocarpum oppositifolium* var. *anomalum*, *Gentiana lutea* var. *aurantiaca*– define the community Vaccinio myrtilli-Juniperetum alpinae, whose preferential distribution is in the Sierra Ibérica. These Bierzo-Sanabrian creeping juniper woods (Vaccinio myrtilli-Juniperetum alpinae) have a southern oro-Cantabrian chionophobous, orotemperate submediterranean and humid-hyperhumid distribution.

In both cases, and on orotemperate submediterranean Bierzo-Sanabrian base-poor substrates, the aforementioned dominant shrubby vegetation occasionally presents a modified plant configuration based on the persistence of the snow cover. It is colonised by plants such as *Vaccinium uliginosum* subsp. *microphyllum*, which indicate an ecotone with chionophilous high shrublands occupying trenches and broad basins on slate, sandstone and quartzite with deep soils with a significant humiferous horizon, where there is a presence of orotemperate submediterranean Bierzo-Sanabrian and oro-Cantabrian silicicolous chionophilous creeping juniper woods with heath and blueberries (Erico tetralicis-Vaccinietum microphylli). In these dense low-growing scrublands there is a predominance of *Calluna vulgaris*, *Erica tetralix, Vaccinium myrtillus*, *Vaccinium uliginosum* subsp. *microphyllum* and *Juniperus alpina*.

A very particular type of orotemperate submediterranean scrubland present in these territories is formed by *Echinospartum ibericum*, along with *Armeria ciliata*, *Festuca summilusitana* and *Cytisus oromediterraneus*, and growing on hard slate and quartzite ledges in humid-hyperhumid upper supratemperate and orotemperate submediterranean belts. These acidophilous scrublands with *Echinospartum ibericum* and *Armeria ciliata* (Armerio ciliatae-Echinospartetum iberici) represent a Bierzo-Sanabrian edaphoxerophilous and chionophobous permaseries.

7.5.1.3 Supra-Oromediterranean and Temperate Submediterranean Grasslands

As a substitution stage of the scrublands of the Genisto sanabrensis-Juniperetum alpinae, or as a permaseries, these orotemperate humid-hyperhumid submediterranean Bierzo-Sanabrian territories are home to psychroxerophilous silicicolous grasslands characterised by the presence of *Festuca summilusitana*, *Luzula caespitosa*, *Teesdaliopsis conferta*, often accompanied by *Agrostis tileni*, *Dianthus langeanus* and *Jurinea humilis*, among others (Teesdaliopsio confertae-Festucetum summilusitanae). These communities with a climatophilous character are sometimes in direct contact with high-mountain pastures of *Nardus stricta*, where there is a frequent presence of plants such as *Trifolium alpinum* and *Plantago alpina*.

The Rozadais geological formation –essentially constituted by slate but with a significant presence of limestone boulders– is present in a small area with a submediterranean orotemperate thermotype in Bierzo-Sanabrian territories where its soils support plants with a basophilous nature such as *Sideritis lurida* (endemism), *Arenaria grandiflora* subsp. *grandiflora*, *Erodium glandulosum* etc., and other acidophilous plants such as *Avenella iberica*, *Festuca summilusitana* etc., forming the Sideritido luridae-Arenarietum grandiflorae grasslands (González de Paz 2012).

On shallow slate and quartzite soils in supramediterranean and submediterranean supratemperate thermotypes it is common to find open grasslands formed by low-growing pulvinate and caespitose chamaephytes such as *Dianthus laricifolius* subsp. *merinoi*, *Plantago radicata*, *Pilosella castellana*, *Jasione sessiliflora* and others (Diantho merinoi-Plantaginetum radicatae).

In contrast, on lithocalcareous soils in identical bioclimatic belts to those mentioned above there are xerophilous grasslands with a presence of plants such as *Arenaria grandiflora* subsp. *incrasata*, *Koeleria vallesiana*, *Hippocrepis commutata*, *Ononis pusilla*, *Erodium glandulosum*, *Arenaria erinacea* etc. (Koelerio vallesianae-Erodietum glandulosae).

7.5.1.4 Grasslands of *Nardus stricta* (Pastures)

In these Bierzo-Sanabrian territories the most extensive *Nardus* pasture is formed by *Genista anglica*, *Nardus stricta*, *Festuca iberica*, *Potentilla erecta*, *Juncus squarrosus*, *Lotus glareosus* and *Galium saxatile*, among other species, growing on soils with a gleyic character or combined with intense snow cover; that is with a water level close to the surface (Genisto anglicae-Nardetum strictae). Occupying less hydrophilous positions than the pasture described above, although still subject to significant snow cover, there is another type of pasture formed by *Festuca iberica*, *Campanula herminii*, *Gentiana pneumonanthe*, *Juncus squarrosus*, etc. It occupies soils which –due to the higher altitude– have a colder temperature regime (frigid) and are more humid (perudic) than the previous pasture, but it cannot be described as being hydromorphic or having gleyic properties (Campanulo herminii-Festucetum ibericae). With a chionophilous character and located in northern or north-eastern exposures where the snow tends to accumulate and takes longer to melt, there is a presence of grasslands formed by *Poa alpina* subsp. *legionensis*, *Nardus stricta*, *Luzula caespitosa*, *Galium saxatile*, *Jasione laevis* and others (Poo legionensis-Nardetum strictae).

There are other types of pasture in these territories growing on soils with very significant hydromorphia (humic gleysols) and which are in contact with heaths or hydrophilous heathlands and hygro-peaty communities formed by *Luzula caespitosa*, *Nardus stricta*, *Carex nigra*, *Juncus squarrosus*, *Gentiana pneumonanthe* and *Potentilla erecta* (Luzulo carpetanae-Pedicularietum sylvaticae). Finally, pastures of Nardo strictae-Genistetum carpetanae are found on intensely stony soils characterised by the presence of *Genista carpetana, Nardus stricta, Dianthus laricifolius, Juncus squarrosus* and others.

7.5.2 Woodlands in the Galician-Leonese Mountains and the Bierzo Trench

7.5.2.1 *Quercus rotundifolia* Woodlands (Holm-Oakwoods)

The relict forests of *Quercus rotundifolia* and *Genista falcata*, fundamentally in the Bierzo and Caurelian-Navian sector, correspond to what are known as the holm-oakwoods of the Genisto falcatae-Quercetum rotundifoliae. The richest floristically are found in areas of El Bierzo and have a thermophilous nature, enriched by the presence of *Pistacia terebinthus*, *Phillyrea angustifolia*, *Jasminum fruticans*, *Arbutus unedo, Asparagus acutifolius* and *Ruscus aculeatus*. These plants do not accompany these forests in Caurelian territories where these woodlands are more impoverished in vascular flora. In their understorey these woods contain taxa such as *Helleborus foetidus, Glandora diffusa, Daphne gnidium, Lonicera etrusca, Piptatherum paradoxum* and *Carex halleriana*, among others. Holm oakwoods grow on dolostone and crystalline limestone in El Caurel and represent the mature stage on this type of substrates in the aforementioned meso-supramediterranean Bierzo-Sanabrian and supratemperate Caurelian-Navian territories.

In the territories located on the edges of the Bierzo trench, particularly in the mesomediterranean and lower supramediterranean belt, there are forests of *Quercus rotundifolia* that correspond to Bierzo-Sanabrian, Portuguese-Duero and Salamancan holm oakwoods (Genisto hystricis-Quercetum rotundifoliae). These woodland formations grow on base-poor soils with a dry-humid character and have a shrub layer with a notable presence of *Daphne gnidium*, *Ruscus aculeatus*, *Cytisus scoparius, Lavandula sampaiana, Genista hystrix, Crataegus monogyna* and several species of *Cistus* sp. The most frequent herbaceous species include *Rubia peregrina, Arenaria montana, Teucrium scorodonia, Silene nutans* subsp. *nutans*, *Clinopodium vulgare* and *Festuca elegans* subsp. *merinoi*. In mesomediterranean territories these woodlands are enriched by the presence of *Arbutus unedo, Pistacia terebinthus* and *Tamus communis*.

In both cases in the mesomediterranean thermotype in the Bierzo-Sanabrian sector, the first substitution stage of these holm oakwoods consists of phanerophytic scrublands dominated by *Arbutus unedo*, *Lonicera etrusca*, *Phillyrea angustifolia*, *Pistacia terebinthus*, *Rubia peregrina* and *Daphne gnidium*, usually accompanied by *Cytisus scoparius* subsp. *scoparius* and numerous characteristic elements of the class *Rhamno cathartici-Prunetea spinosae* such as *Crataegus monogyna*, *Rosa pouzinii*, *Prunus mahaleb*, *Rosa micrantha*, and others (Lonicero etruscae-Arbutetum unedonis).

7.5.2.2 *Quercus pyrenaica* Woodlands (Pyrenean Oakwoods)

The Pyrenean oakwoods (woods of *Quercus pyrenaica*) in meso-supramediterranean and supratemperate submediterranean Bierzo-Sanabrian territories with subhumid-hyperhumid oceanic ombroclimates grow on humic cambisols, and very occasionally on dystric planosols, and correspond to the association Genisto falcatae-Quercetum pyrenaicae. The most notable phenological aspect of these woods –as is the case of all those in which *Quercus pyrenaica* and *Quercus orocantabrica* are the dominant species– is that during autumn and most of winter, the dry leaves of both trees and their hybrids remain on the trees (marcescent leaves), which gives these woodland formations their highly distinctive appearance. In the shrub layer there is a frequent presence of *Genista falcata*, *Genista florida* subsp. *polygaliphylla*, *Cytisus scoparius*, *Erica arborea* and *Crataegus monogyna*. In the herb stratum there is an abundance of species such as *Holcus mollis*, *Stellaria holostea*, *Physospermum cornubiensis*, *Clinopodium vulgare*, *Teucrium scorodonia*, *Melampyrum pratense*, *Festuca elegans* subsp. *merinoi*, *Doronicum plantagineum*, *Aquilegia vulgaris* subsp. *dichroa*, *Brachypodium sylvaticum*, etc. These woodlands represent the mature stage of a climatophilous Bierzo-Sanabrian silicicolous temperate oceanic submediterranean and Mediterranean pluviseasonal oceanic meso-supramediterranean subhumid-hyperhumid character.

On the western and northern face of the Sanabrian mountains with a temperate oceanic bioclimate with a certain degree of submediterraeneity, there is a presence of Pyrenean oakwoods of the Linario triornithophorae-Quercetum pyrenaicae, whose herb stratum includes *Linaria triornithophora*, *Polygonatum odoratum*, *Erythronium dens-canis*, *Physospermum cornubiense*, *Poa nemoralis*, *Melittis melissophylum*, *Holcus mollis*, *Stellaria holostea*, *Primula vulgaris*, *Pulmonaria longifolia*, *Narcisus triandrus*, etc.

These supratemperate woodlands with a humid to hyperhumid ombroclimate grow equally well on deep soils formed on moraine deposits, on highly pulverized rocks and colluvia (umbric cambisols), and on very shallow soils (leptosols) on granite, gneiss and slate substrates. These are acid soils with a sandy texture and a high content in organic matter that sometimes withstand a certain level of hydromorphia.

7.5.2.3 Riparian Woodlands (Alder, Willow, Ash, Poplar-Willow Woods)

The Bierzo trench, like all those on the Iberian Peninsula, is a flat or slightly deformed area surrounded by mountains. It is also the point of convergence of a series of rivers that flow from these mountains and which have carried the materials that today conform their soils. Notable among these rivers are the Cúa and Burbia, although it is the Sil, which flows from northeast to southwest through the trench and continues until it joins the Miño river –beyond the trench–, which has provided the infill of the trench until its total formation. These detritic soils were early settled and exploited by humans, which has led to the practical annihilation of the pre-existing potential vegetation. Thus, it is today difficult to recognise this potential vegetation due to the use of these territories for growing vegetable crops, livestock farming –decreasingly–, and their exploitation as woodlands and orchards.

However it be said that in the rivers on hard materials such as slate or quartzite there is a presence of alders (Carici broteriani-Alnetum glutinosae), whereas in most of the Bierzo trench and in some sections of the Sil where the materials are finer, and where there are more inputs to form significantly developed soils with sufficient water levels, there are elm or poplar woods of *Ulmus minor* (Aro cylindracei-Ulmetum minoris), willow-poplar woods (Salici neotriochae-Populetum nigrae) and shrubby willows (Salicetum salviifoliae). As we have indicated in this brief introduction, Bierzo-Sanabrian alder woodlands (Carici broteriani-Alnetum glutinosae) grow on luvisol- and fluvisol-type soils on the banks of fast-flowing rivers and streams that do not dry out in summer and run through narrow valleys in the meso- and supramediterranean thermotypes. They are dominated by *Alnus glutinosa*, *Fraxinus angustifolia*, *Ulmus minor* and two western Iberian elements, namely *Carex acuta* subsp. *broteriana* and *Galium broterianum*. Furthermore, given the proximity of the Eurosiberian area there is also a frequent presence of *Fraxinus excelsior*. These woodlands potentially represent the climax of the territory and occupy a large part of the riverbed, with scionitrophilous grasslands in their midst (Geranio robertiani-Cariolophetum sempervirentis), and in contact with permanent water communities formed by *Glyceria declinata*, *Oenanthe crocata*, *Phalaris arundinacea*, etc. (Glycerio declinatae-Oenanthetum crocatae), always located in slower flowing areas or in backwaters.

Poplar-willow woods have a very widespread distribution (Salici neotriochae-Populetum nigrae) and are defined by a floristic combination with a dominance of several tall species in the genus *Salix* (*Salix neotricha*, *Salix rubens*, *Salix* x *erythrocados*, *Salix fragilis*), along with *Populus nigra*, and occasionally *Populus alba*, *Fraxinus angustifolia* and *Ulmus minor*. They occupy broader riverbeds and grow on meadow brown soils originating from silty or clayey and even pebbly sediments, and presenting clearly defined horizons in their final stages. This soil development allows the growth of tree species, unlike in the case of shrubby willows that occupy narrow river beds on soils with a coarser granulometry, which will be discussed below. These now highly altered communities are replaced by spiny scrublands (Rubo ulmifolii-Rosetum corymbiferae), as their protection edge, and also by hay meadows (Festuco rothmaleri-Cynosuretum cristati). In addition, the nitrogenated soil amendments used by humans in these communities encourage the establishment of a broad range of nitrophilous communities with differing requirements in regard to light and soil characteristics. There is a frequent presence of tall hemicryptophyte communities (Galio aparines-Conietum maculati), and scionitrophilous plant communities (Geranio robertiani-Cariolophetum sempervirentis). In areas where these willow-popular woodlands are in contact with free-flowing waters there are common reeds (Scirpo lacustris-

Phagmitetum australis) and bunchgrass (Phalaridetum arundinaceae), with a not infrequent presence of pastures-reed beds (Deschampsio-Juncetum effusi).

The riparian ash woods in the Valdeorras region (Hedero hibernicae-Fraxinetum angustifoliae) constitute the riparian woodlands in the Galician-Portuguese and Galician inland freshwater pluviseasonal Mediterranean oceanic and temperate mesomediterranean oceanic and mesotemperate subhumid-humid submediterranean territories, and have a predominance of *Fraxinus angustifolia* accompanied by *Salix atrocinerea*, *Acer pseudoplatanus*, *Prunus avium* and *Ulmus minor*. These formations have an appearance of closed forests with very little light, and their floristic composition is enriched by plants of the class Querco-Fagetea, such as *Helleborus foetidus*, *Primula acaulis*, *Melica uniflora*, *Crepis lampsanoides*, *Poa nemoralis*, *Dryopteris filix-mas*, *Dryopteris affinis*, *Polystichum setiferum* and *Phyllitis scolopendrium*. In the shrub layer there is a relatively frequent presence of *Prunus spinosa*, *Crataegus monogyna*, *Cornus sanguinea* and *Rosa corymbifera*, among others. Also frequent are creepers such as *Bryonia cretica* subsp. *dioica*, *Tamus communis* and *Hedera hibernica*

In Bierzo-Sanabrian and Queixan territories occupying narrow river beds and on soils with a coarse, occasionally gravelly, sandy and even silty granulometry –but always poor in bases and subjected to a long period of summer drought–, there is a presence of salviifolia willow woodlands (Salicetum salviifoliae), with a predominance of *Salix salviifolia*, *Salix* x *secalliana*, *Salix triandra* subsp. *discolor*, *Salix atrocinerea* and even *Fraxinus angustifolia*. Among the herbs present in these willow woods with a shrubby nature, and distributed throughout the Galician-northern Portuguese, Leonese and Sierra Ibérican sectors, there is a presence of *Brachypodium sylvaticum*, *Saponaria officinalis*, *Potentilla reptans*, etc. These tree formations are frequent in two thermotypes –meso- and supramediterranean–, provided they have acid soils.

When the substrates are rich in bases, there is a presence of shrubby willows with a predominance of *Salix purpurea* subsp. *lambertiana*, *Salix salviifolia*, *Salix* x *legionensis*, *Salix* x *secalliana*, *Salix* x *pseudosalviifolia*, *Salix eleagnos* subsp. *angustifolia* (Salicetum lambertiano-salviifoliae). The herb layer comprises *Brachypodium sylvaticum*, *Solanum dulcamara*, *Geum urbanum*, *Crepis lampsanoides*, *Euphorbia amygdaloides* and *Carex reuteriana*. These willows, as in the case of those mentioned above, grow in the meso- and supramediterranean thermotypes.

7.5.2.4 *Quercus orocantabrica* and *Betula celtiberica* Woodlands (Oro-Cantabrian Oak and Birchwoods)

The oakwoods of *Quercus orocantabrica* (Avenello ibericae-Quercetum orocantabricae), already described above, currently occupy limited areas in the Bierzo-Sanabrian and Queixan territories as a result of their use by humans, who since ancient times have used burning as a means of obtaining grazing land for livestock. These woodlands grow on highly acid dystic cambisols or cambic podsols originating from siliceous parent rock. They are characterised in these areas by the reduced presence of *Quercus orocantabria* and the more widespread presence of its hybrids *Quercus* x *puentei* (*Quercus petraea* x *Quercus orocantabrica*), and the even stronger presence of *Quercus* x *penasi* (*Q. orocantabrica* x *Q. pyrenaica*); the species found in these woodlands include *Vaccinium myrtillus*, *Avenella iberica* and *Melampyrum pratense*. Occasionally these woodlands occupy gelifracts that retain significant levels of water, and it is thus not infrequent to find the presence of plants such as *Festuca paniculata* subsp. *multispiculata*, *Epilobium angustifolium*, *Luzula lactea*, along with others such as *Erica arborea*, or nemoral species with a more widespread distribution such as *Teucrium scorodonia*, etc.

The woodland formations dominated by Celtiberian birch (*Betula celtiberica*), which are fairly frequent in the rainier Cantabrian siliceous territories, constitute the young stages in most of the potential woodlands in these territories. In Galician-Leonese areas they are restricted to riparian environments in the upper valleys of most of the rivers and streams, and succeed in growing at some distance from the river beds in places where the soil moisture continues to be high. It is therefore not infrequent to find these birchwoods in areas near peat bogs or with persistent waterlogging. In these environments, birchwoods (Luzulo henriquesii-Betuletum celtibericae) –which are exclusive to the Cantabrian Range and the Montes de León– generally constitute the upper limit of the forest. As indicated earlier, they grow on oligotrophic nutrient-poor soils of somewhat peaty brown earth, generally with substantial undecomposed organic matter. In the tree layer in the Bierzo-Sanabrian and Queixan territories there is a predominance of *Betula celtiberica*, with the possible presence of *Taxus baccata*, *Sorbus aucuparia*, *Sorbus aria* and *Ilex aquifolium*, among other trees. The woody taxa in the understorey include particularly *Erica arborea*, *Vaccinium myrtillus* and *Corylus avellana*, different species of *Rubus* sp. pl., *Frangula alnus* and –in these territories– *Salix atrocinerea*.

In the herb stratum there is a frequent presence of ferns (*Athyrium filix-femina*, *Dryopteris affinis*, *Blechnum spicant*) and herbs such as *Luzula sylvatica* subsp. *henriquesii*, *Lilium martagon*, *Saxifraga spathularis*, *Holcus mollis*, *Viola riviniana*, *Epilobium montanum*, *Stellaria holostea*, *Poa nemoralis*, *Teucrium scorodonia*, *Ceratocapnos claviculata*, *Crepis lampsanoides*, *Anemone nemorosa*, etc. These woodlands are the mature edaphohygrophilous and chionophilous silicicolous stage of a temperate oceanic supra-orotemperate submediterranean character.

On soils that are cool and rich in organic matter there are birchwoods with a secondary character in parts of the Sanabrian and La Cabrera mountains with a significant presence of *Salix atrocinerea*, *Salix salviifolia*, *Salix caprea*, occupying areas in the headwaters of streams with large blocks in their riverbed, and which, in addition to by the aforementioned species, are formed by *Betula celtiberica*, *Sorbus aucuparia* and *Populus tremula* (Sorbo aucupariae-Betuletum celtibericae).

7.5.3 *Meso-supramediterranean and Supratemperate Submediterranean Scrublands*

7.5.3.1 Broom Communities

The woodland edges growing on base-poor substrates as their first substitution stage are very diverse in Bierzo-Sanabrian territories, and in some cases the communities they constitute also extend throughout specific oro-Cantabrian areas. The Cytisetum scopario-oromediterranei is abundantly represented in the oro-Cantabrian subprovince, although it reaches upper supratemperate submediterranean and orotemperate territories with a Bierzo-Sanabrian and Queixan continental character. These nanophanerophytic scrublands are characterised by the presence of *Cytisus oromediterraneus*, *Genista florida* subsp. *polygaliphylla* and *Cytisus scoparius*, and often accompanied by *Erica arborea*. These communities constitute the edge of the woodlands of *Betula celtiberica*.

More frequent are the also nanophanerophytic communities of the Cytiso scoparii-Genistetum polygaliphyllae with a supramediterranean and supratemperate subhumid to humid character, and a broad distribution in both the oro-Cantabrian and Lugan, Sierra Ibérican, plani-Leonese, upper-Salamancan, Castilian-Cantabrian and Bierzo-Sanabrian territories. They have a peculiarity in the territories in La Cabrera, namely the absence of *Cytisus multiflorus*, which is surprising given the proximity to areas in the Laciana and Ancares sector where this taxon is extremely widespread. These communities represent the first substitution stage of the woodlands dominated by *Quercus pyrenaica*.

As a first substitution stage of the woodlands of *Quercus orocantabrica* (Avenello ibericae-Quercetum orocantabricae), whose distribution (Alto Carrión, southern Picos de Europa-Ubiña and Laciana and Ancares sectors) extends as far as the orotemperate and supratemperate submediterranean Bierzo-Sanabrian territories with a continental character, there are communities with a predominance of *Cytisus oromediterraneus*, *Genista obtusiramea* and *Erica arborea*, and a less frequent presence of other taxa such as *Cytisus soparius* and *Genista florida* subsp. *polygaliphylla*.

In the Valdeorras region and in the foothills of the Peña Trevinca massif (Ortiz et al. 1991), in the meso- and lower supramediterranean thermotypes, there is another woodland edge scrubland characterised by a predominance of *Erica arborea*, *Genista falcata*, *Cytisus scoparius*, with a not infrequent presence of *Cytisus striatus*, *Cytisus multiflorus* and *Genista florida* subsp. *polygaliphylla* (Genisto falcatae-Ericetum arboreae).

7.5.3.2 Spiny Nanoshrublands

As a seral stage of the woodlands of the Pulmonario longifoliae-Quercion roboris, there are the natural shrubby edges of the Tamo communis-Rubetum ulmifolii

(Díaz González and Fernández Prieto 1994b) of Cantabrian-Atlantic distribution with Bierzo-Sanabrian irradiations, occupying humid areas on meso-eutrophic soils. They are formed by *Rubus ulmifolius*, *Tamus communis*, *Crataegus monogyna*, *Prunus spinosa*, *Cornus sanguinea* along with creeping plants such as *Lonicera periclymenum* subsp. *hispanica*, *Climatis vitalba*, *Hedera hibernica* and *Rubia peregrina* and several species in the genus *Rosa* (*Rosa micrantha*, *Rosa pouzinii*, *Rosa squarrosa*).

On moist deep base-poor soils with a widespread western Iberian distribution in the meso- and supramediterranean thermotypes, growing between hay meadows or bordering the remains of elm, poplar and willow woods –and even woods of *Quercus pyrenaica*– there are spiny scrublands characterised by an abundant presence of *Rosa corymbifera* and *Rubus ulmifolius* accompanied by *Prunus mahaleb* and *Rosa stylosa*, among others (Rubo ulmifolii-Rosetum corymbiferae).

7.5.3.3 Heathlands

Chamaephytic scrublands dominated by *Genista sanabrensis*, *Calluna vulgaris* and *Erica umbellata*, sometimes accompanied by *Erica australis* subsp. *aragonensis*, *Pterospartum tridentatum* subsp. *lasianthum*, *Luzula lactea*, etc. (Erico-umbellatae-Genistetum sanabrensis), grow in the orotemperate Bierzo-Sanabrian thermotype on poorly-formed soils; fire plays a significant role in their origins.

In supramediterranean –occasionally mesomediterranean and with penetrations in submediterranean supratemperate and even orotemperate subhumid-hyperhumid– territories, there are dense heathlands on base-poor substrates forming a more regressive substitution stage preferentially of the Pyrenean oakwoods of the Genisto falcatae-Quercetum pyrenaicae. They comprise *Erica australis* subsp. *aragonensis*, *Pterospartum tridentatum* subsp. *lasianthum*, *Luzula lactea*, *Halimium lasianthum* subsp. *alyssoides*, *Halimium umbellatum*, *Calluna vulgaris* and *Vaccinium myrtillus*, and prefer well-drained soils on slopes and crests, which they acidify and tend to podsolize due to the acid material they accumulate (Rivas-Martínez 1979). On acid podsolic and gleyed soils, in trenches with poor drainage so they are subject to hydromorphism almost all year round, occur hygrophilous heathlands with a frequent presence of *Genista anglica*, *Erica tetralix* and *Calluna vulgaris* (Genisto anglicae-Ericetum tetralicis).

There is also another type of community formed by spiny frutescent and semidecumbent chamaephytes accompanied by perennial scrublands of herbaceous species which comprise a heathland growing on moraine substrate with abundant surface stoniness, subjected to a degree of seasonal hydromorphism, and with substantial snow cover. This community includes *Genista carpetana*, *Genista anglica*, *Genista micrantha*, *Thymelaea dendrobryum*, *Erica tetralix*, etc. (Thymelaeo dendrobryi-Genistetum carpetanae).

7.5.3.4 Other Types of Scrubland (Cistus-Lavender-Gorse)

In Bierzo-Sanabrian territories, growing on acid substrates in the upper mesomediterranean and supramediterranean bioclimatic belts, as a result of the maximum degradation of the forests of the Genisto hystricis-Quercetum rotundifoliae, there are scrublands preferentially constituted by chamaephytes and some hemicryptophytes which include *Genista hystrix*, *Lavandula sampaiana*, *Thymus mastichina* and *Cistus salviifolius* (Lavandulo sampaioanae-Genistetum hystricis). They form communities which –depending on the ombroclimate in the territory or whether they are significantly altered by human action– are either enriched or transformed. If the ombroclimate is dry-subhumid there is commonly an addition of *Cistus ladanifer*; in the case of a subhumid upper-humid ombroclimate it includes *Erica australis* subsp. *aragonensis*; and if they are transformed there may be taxa with nitrophilous preferences such as *Santolina semidentata* and *Helichrysum stoechas* subsp. *stoechas*.

The still fairly well-developed holm oakwoods mentioned previously for the mesomediterranean humid thermotype on acid soils, if overmanaged, are substituted by a plant community comprising several species in the genus *Cistus* (*Cistus salviifolius*, *Cistus populifolius*, *Cistus psilosepalus*, *Cistus ladanifer*) and accompanied by *Lavandula sampaiana* and *Thymus mastichina*, (Lavandulo sampaioanae-Cistetum populifolii).

7.5.4 Meso- and Supramediterranean Meadows and Grasslands

Basophilous xerophytic grasslands are not very frequent in these Bierzo territories, due to the scarcity of lower meso- and supramediterranean base-rich substrates. When they occur, they are formed by *Thymus zygis* and accompanied by plants such as *Asperula cynanchica*, *Ruta montana*, *Euphorbia segetalis* and others (Ononido pusillae-Thymetum zygidis).

On base-poor substrates and in the subhumid-hyperhumid meso-supramediterranean and supratemperate bioclimatic belts there is an occasional occurrence –as in the case of oro-Cantabrian territories– of perennial grasslands ("*cerrillares*") that occupy often stony and poorly-developed soils where there is a presence of *Festuca elegans* subsp *merinoi*, *Phalacrocarpum oppositifolium*, *Silene nutans* etc. (Phalacrocarpo oppositifolii-Festucetum merinoi), which occupy verges and clearings in Pyrenean oakwoods, holm oakwoods and broom scrublands. On base-poor soils in oakwood domains (Pyrenean oakwoods) of the Genisto falcatae-Quercetum pyrenaicae and the Linario triornithophorae-Quercetum pyrenaicae, either due to human exploitation or else in forest clearings, there are hay meadows ("*vallicar*") dominated by grasses such as *Festuca rothmaleri*, *Agrostis castellana*, *Trisetum flavescens*, *Holcus lanatus*, *Poa pratensis* and accompanied by *Trifolium pratense*, *Trifolium repens*, *Hypochoeris radicata, Crepis capillaris, Rumex acetosa*, etc. (Festuco rothmaleri-Agrostietum castellanae).

There are also meadows growing on soils with a high water table near river courses and on terraces and alluvial plains which constitute a community formed by *Festuca rothmaleri, Cynosurus cristatus, Holcus lanatus, Agrostis castellana, Anthoxantum odoratum, Arrhenatherum bulbosum, Poa pratensis, Alopecurus pratensis*; leguminous species include *Lathyrus pratensis, Trifolium pratense* and *Trifolium repens*, in addition to other species such as *Rhinantus minor, Prunella grandiflora, Achillea millefolium, Carum verticillatum*, etc. (Festuco rothmaleri-Cynosuretum cristati), which are found in the terrain of the ash and Pyrenean oak community (Fraxino angustifoliae-Quercetum pyrenaicae). These communities may occasionally be very near the pastures-reed beds growing on soils with intense temporary hydromorphia with a presence of *Deschampsia hispanica, Juncus effusus, Scirpus holoschoenus, Juncus acutiflorus*, among other species (Deschampsio hispanicae-Juncetum effusi). On even more hydromorphic soils than those mentioned previously there are formations dominated by *Hypericum undulatum, Juncus acutiflorus, Carum verticillatum, Lotus pedunculatus* and *Molinia caerulea*, among others (Hyperico undulati-Juncetum acutiflori).

7.5.5 Other Types of Vegetation

7.5.5.1 Aquatic Vegetation

In lakes in the Sanabrian mountains and in the Lake of Sanabria itself there are communities in which only *Nitella flexilis* thrives, provided the substrates consist of very fine gravel and sand. If the gravel is coarser they may contact catenally with communities of *Isoetes velata* subsp. *asturicensis*. They can be seen from the banks in shallow waters up to depths of over three or more metres, and in areas on the shores of water bodies where the sunlight penetrates to the bottom.

In substrates with dark sediments that are rich in organic matter and with a sandy matrix or finer, occasionally at depths of over 1.5 m, there are communities dominated by *Potamogeton natans*, sometimes accompanied by *Myriophyllum alternifolium* (Myriophyllo alterniflori-Potametum natantis). There is also an occasional presence of a small community formed by *Potamogeton polygonifolius* that occupies the minor watercourses draining peat-bog and hygrophilous moor communities. It grows in narrow, shallow rivulets carrying little water, which form a stream when they converge. The communities of the Callitricho brutiae-Ranunculetum aquatilis occur in moderately fast-flowing waters, always in mid and high areas above 1500 m, in the source of mountain streams with organic matter in the substrates, near hygro-peaty environments and in contact with other types of communities in the alliance Myosotodion stoloniferae. Communities –in many cases single-species– of *Lemna minor* (Lemnetum minoris) can be found growing in eutrophised stagnant waters in cattle troughs and fountains and in the backwaters of streams.

In river beds and watercourses there are herbaceous communities growing in clumps dominated by *Galium broterianum* and *Carex reuteriana* (Galio broteriani-Caricetum reuterianae). The community Oenantho crocatae-Filipenduletum ulmariae may occur if the watercourses themselves have gravel and pebbles and a low presence of fine sand.

7.5.5.2 Meso-supramediterranean Scree Vegetation

In the meso- and supramediterranean thermotypes in the Bierzo territories, depending on the type of rocks on which they develop, there are chasmophytic plant communities of great interest due to the presence of a large number of endemic taxa in small areas. Fissures on dolomitic and calcareous substrates contain communities with scarce coverage that extend as far as the supramediterranean thermotype in the Bierzo territories, and the meso-supratemperate thermotype in the Sierra del Caurel. These formations are represented by the presence of *Leontodon farinosus*, *Campanula arvatica* subsp. *adsurgens*, *Rhamnus pumila* subsp. *legionensis* (Bierzo-Caurelian endemisms), along with *Saxifraga trifurcata*, *Crepis albida* subsp. *asturica* and others (Saxifragetum trifurcatae). Two different plant communities occur on this same type of substrate: one (Petrocoptidetum grandiflorae) on overhangs and ledges where the endemic *Petrocoptis grandiflora* grows almost exclusively; and the second extending to meso- and lower supramediterranean territories in the Bierzo district, with a very localised presence of communities of the Petrocoptidetum viscosae, preferentially constituted by the endemic *Petrocoptis pyrenaica* subsp. *viscosa*. These last communities are scarce in the territories mentioned, and only four populations are known at present. On this same type of rocky substrates, occupying earth-filled crevices and fissures, there are several chasmophytic communities constituted by *Cheilanthes acrostica*, *Ceterach officinarum* and *Phagnalon saxatile* (Asplenio ceterach-Cheilanthetum acrosticae).

On slate, sandstone and even quartzite in the meso-Mediterranean Bierzo-Sanabrian thermotype, with an occasional presence in the supramediterranean and meso-supratemperate submediterranean thermotypes in the Laciana-Ancares sector, there are communities of chamaephytes and hemicryptophytes colonising fissures of varying widths and earth-cover dominated by *Linaria saxatilis* subsp. *glabrescens*, *Cheilanthes tinaei*, *Cheilanthes hispanica* and others (Linario glabrescentis-Cheilanthetum tinaei).

7.5.5.3 Annual Grasslands

The land use that led to the intense deforestation process of the past has today been halted, and this has given the annual grasslands abundant space to develop. As in the other sections, we should highlight that this type of communities has a very high diversity due to the wide range of soils and bioclimates present on the North Plateau. We will attempt to synthesise them as far as possible.

Typical of ledges in rocky areas in heathlands, also with an early spring phenology, preferentially in the meso-supramediterranean belts on base-rich substrates, is another type of annual grassland whose composition includes particularly *Hornungia petraea, Saxifraga tridactylitis, Campanula erinus* and *Arenaria emarginata*, among others (Saxifrago tridactylites-Hornungietum petraeae).

Acid substrates are home to very different communities formed by *Arenaria leptoclados, Arenaria serpyllifolia, Asterolinum linum-stellatum, Cerastium pumilum, Anthoxanthum aristatum, Campanula lusitanica, Linaria spartea, Crupina vulgaris, Evax pygmaea, Cerastium diffusum, Medicago minima, Molineriella laevis, Logfia minima, Trifolium arvense, Tolpis barbata, Hispidella hispanica* and *Ornithopus compressus*, among many others (Molineriello-Airopsietum tenellae, Hispidello hispanicae-Tuberarietum guttatae, Triseto ovati-Agrostietum truncatulae, Sedetum caespitoso-arenarii, and Anthoxantho-Holcetum setiglumis).

7.5.5.4 Herbaceous Ruderal Vegetation and Others

The long-lasting action of humans on nature is the determining factor in causing the landscape of the North Plateau to have moved so far from its natural potential vegetation. These territories today bear only a remote resemblance to their natural potential. The often sudden alteration of the substrates themselves –particularly in the case of the upper soil horizon– has led to the establishment of communities of invasive plants (although some have been with us for many centuries), whose common denominator is their marked need for ammoniacal substances deriving from the transformation of organic matter.

The denomination of nitrophilous and ruderal vegetation includes plant communities adapted to trampling, and also grasslands typical of roadways and ruderal areas which undoubtedly tend to be seen as weeds in areas of crops, or which are found growing on heaps of debris. They also include scionitrophilous plants; that is, plants that exploit the transformation of organic matter from other plant formations with a tree or shrub character. The former include particularly communities formed by *Poa annua, Polygonum aviculare, Capsella bursa-pastoris, Matricaria matricarioides, Sclerochloa dura, Coronopus didymus, Coronopus squamatus* and *Matricaria discoidea*, among others (Coronopo squamati-Sclerochloetum durae, Polygono arenastri-Matricarietum discoideae), which on numerous occasions coexist with *Plantago major* and *Lolium perenne* (Lolietum perennis).

On disturbed soils, often on the verges of paths and roadways, it is very frequent to find thistle communities formed by tall hemicryptophytes belonging to the subclass Onopordenea acanthii. Also worth mentioning are the communities formed by *Carduus carpetanus, Onopordum acanthium, Cirsium vulgare, Scolymus hispanicus, Centaurea calcitrapa* and *Verbascum pulverulentum*, among others (Carduo carpetani-Onopordetum acanthii). On slightly moist meso-supramediterranean soils there is a presence of formations with no specific soil preferences dominated by *Silybum marianum* and accompanied by *Carduus bourgeanus*, *Carduus pycnocephalus* and *Carduus tenuiflorus* (Carduo bourgeani-Silybetum mariani).

In fields of rain-fed cereal crops growing on soils with a sandy-clay texture, flowering at the start of spring before *Triticum* sp., *Avena* sp. or *Secale cereale* become widely developed, there is a presence of plants typical of the Stellarietea mediae class such as *Veronica persica*, *Veronica triphyllos*, *Veronica hederaefolia* subsp. *triloba*, *Mibora minima*, *Arabidopsis thaliana*, *Senecio vulgaris*, *Erophila verna* and *Cerastium glomeratum*, all with indifferent soil preferences (Miboro minimae-Arabidopsietum thalianae). In horticultural crops growing on temporarily moist soils and in irrigated crops there are communities dominated either by *Echinochloa crus-galli*, accompanied by *Echinochloa colona*, *Digitaria sanguinalis*, *Polygonum lapathifolium*, *Setaria verticillata* and *Bidens tripartita*, among others,

In soils of varying depth on the edges of roadways there is a frequent presence in the North Plateau of grasslands dominated by *Hordeum murinum* or *Hordeum secalinum* (Rivas-Martínez 1977) occasionally accompanied by other plants such as *Sisymbrium irio*, *Sisymbrium officinale*, *Descurainia sophia*, *Hirschfeldia incana* and *Isatis tinctoria*, among others (Papaveri argemone-Sisymbrietum contorti, Bromo scoparii-Hordeetum leporini, Sisymbrio irionis-Malvetum parviflorae, Sisymbrio officinalis-Hordeetum murini).

Among the scionitrophilous communities it is worth noting the community formed by *Galium aparine*, *Conium maculatum*, *Alliaria petiolata*, *Saponaria officinalis* and *Urtica dioica*, among others (Galio aparines-Conietum maculati). It forms grasslands with a substantial biomass which flower in spring and early summer, are parched in summer, and they grow on deep soils with significant contributions of organic matter (humicolous) and almost constant moisture. They are located preferentially on the edges of roadways and verges, and in spite of their heliophilous character they sometimes occur in the shade of deciduous woodlands such as those of the Aro-Ulmetum minoris. On roadsides and verges, on soils rich in organic matter and with high soil moisture, there is a frequent presence of communities, with a summer phenology, and dominated to a large degree by *Sambucus ebulus*, often accompanied –but with less biomass– by *Urtica dioica*, *Dipsacus fullonum*, *Arctium minus*, *Galium aparine* and *Pentaglottis sempervirens* (Urtico dioicae-Sambucetum ebuli). These grasslands are located mainly in areas of ash, elm and poplar woodlands.

References

Alonso JL, Pulgar JA, Pedreira D (2007) *El relieve de la Cordillera Cantábrica*. Enseñanza de las Ciencias de la Tierra 15(2):151–163

Aramburu F, & Bastida C (1995) *Geología de Asturias*. Ed Trea 312. ISBN: 9788489427303

Díaz González TE (2009) Caracterización de los Distritos Biogeográficos del Principado de Asturias (Norte de España). In Llamas F, Acedo C (eds) Botánica Pirenaico-Cantábrica en el siglo XXI. Área Publ. Univ. León. León. pp 423–455

Díaz González TE (2014a) El Paisaje Vegetal de Asturias en el contexto Bioclimático y Biogeográfico de la Península Ibérica. Real Instituto de Estudios Asturianos. Gobierno del Principado de Asturias, p 133

Díaz González TE (2014b) Map of vegetation series, geoseries and geopermaseries from Spain: Asturias 1:250.000. Global Geobotanica 3:1–34

Díaz González TE, Fernández Prieto JA (1988) Caracterización de las unidades fitogeográficas de Asturias. Monografías del Instituto Pirenaico de Ecología, Jaca 4:517–528

Díaz González TE, Fernández Prieto JA (1994a) El paisaje vegetal de Asturias. Itinera Geobot 8:5–242

Díaz González TE, Fernández Prieto JA (1994b) La Vegetación de Asturias. Itinera Geobotanica 8:243–520

Díaz González TE, Fernández Prieto JA (1999) The Vegetal Landscape of the Picos de Europa. Itinera Geobotanica 13:263–290

Díaz González TE, Fernández Prieto JA (2006) Biogeografía de Asturias: Bases para su actualización. Actas I Congreso de Estudios Asturianos. Oviedo, 10–13 mayo 2006. Real Instituto de Estudios Asturianos, Oviedo, p 19

Díaz González TE, Fernández Prieto JA, Bueno Sánchez A, Alonso Felpete JI (2005) Itinerario botánico por el oriente de Asturias. El paisaje vegetal de los Lagos de Covadonga y de los Bufones de Pría. Cuadernos del Jardín Botánico Atlántico (Gijón):1–104

Díaz González TE, Bueno Sánchez A, Alonso Felpete JI (2014) El Paisaje Vegetal de Parque Natural de Ponga (Asturias). Documentos del Jardín Botánico Atlántico (Gijón) 12:1–80

González de Paz L (2012) Flora y Vegetación de la Cabrera baja (León): valoración del estado de conservacion. Tesis Doctoral. Universidad de León

Herrero L, García ME, Díaz González, TE, Penas A, Salegui F (2002) *Euphorbio hybernae-Fraxinetum excelsiois*. In S. Rivas-Martínez & al. Vascular Plant Communities of Spain and Portugal. Addenda to the syntaxonomical checklist of 2001. Itinera Geobotanica 15(1):5–432

Navarro Andrés F, Díaz González TE (1977) Algunas consideraciones acerca de la Provincia corológica Orocantábrica. Anales Instituto Botánico Cavanilles 34(1):219–253

Ortiz S, Amigo J, Izco I (1991) Las orlas forestales fruticosas orensano-sanabrienses: dos nuevas asociaciones del valle del Sil. Lazaroa 12:303–315

Rivas-Martínez S (1977) Datos sobre la vegetación nitrófila española. Acta Bot Malacitana 3:159–167

Rivas-Martínez S (1979) Brezales y jarales de Europa occidental (Revisión fitosociológica de las clases Calluno-Ulicetea y Cisto-Lavanduletea). Lazaroa 1:5–127

Rivas-Martínez S, Coautores (2007) Mapa de Series, Geoseries y Geopermaseries de Vegetación de España. Memoria del mapa de Vegetación Potencial de España. Parte 1ª. Itinera Geobotanica (Nueva Serie) 17:1–436

Rivas-Martínez S, Izco J, Costa M (1971) Sobre la flora y vegetación del Macizo de Peña Ubiña. Trab Dept Bot Fisiol Veg Madrid 3:47–123

Rivas-Martínez S, Díaz González TE, Fernández Prieto JA, Loidi J, Penas A (1984) La vegetación de la alta montaña cantábrica. Los Picos de Europa. Ed. Leonesas, León, p 300

Rivas-Martínez S, Fernández-González F, Loidi J (1999) Checklist of plant communities of Iberian Peninsula, Balearic and Canary Islands to suballiance level. Itinera Geobotanica 13:353–451

Rivas-Martínez S, Fernández-González F, Loidi J, Lousã M, Penas A (2001) Syntaxonomical checklist of vascular plant communities of Spain and Portugal to association level. Itinera Geobotanica 14:5–341

Rivas-Martínez S, Díaz González TE, Fernández-González F, Izco J, Loidi J, Lousã M, Penas A (2002a) Vascular plant communities of Spain & Portugal. Addenda to the syntaxonomical checklist of 2001. Part I Itinera Geobotanica 15(1):5–432

Rivas-Martínez S, Díaz González TE, Fernández-González F, Izco J, Loidi J, Lousã M, Penas A (2002b) Vascular plant communities of Spain & Portugal. Addenda to the syntaxonomical checklist of 2001. Part II Itinera Geobotanica 15(2):433–922

Rivas-Martínez S, Coautores et al (2011a) Mapa de Series, Geoseries y Geopermaseries de la Vegetación de España (Memoria del Mapa de Vegetación potencial de España). Itinera Geobotanica 18(1/2):5–800

Rivas-Martínez S, Rivas-Sáenz S, Penas A (2011b) Worldwide bioclimatic classification system. Global Geobot 1:1–634 + 4 Maps. doi:10.5616/gg:110001

Rivas-Martínez S, Penas A, Díaz González TE, Costa JC, Pinto Gomes C, Herrero L, del Río S (2014) Bioeography of Spain and Portugal. update typology. Int J Geobot Res 4(1):1–64

Vera JA (ed) 2004 *Geología de España*. Ed. Pral. Sociedad geológica de España. Instituto Geológico y Minero de España. ISBN 84-7840-546-1

Chapter 8
The Pyrenees

Josep M. Ninot, Empar Carrillo, and Albert Ferré

Abstract The Pyrenees is a large mountain system stretching over the Iberian isthmus, and thus encompassing three distinct vegetation regions: medio-European, Mediterranean and Alpine. It includes contrasting landscapes in accordance with this large-scale pattern and also with smaller scales related to continentality, bedrock type, landform and ancient anthropic land use. The northern side and the western edge of the Pyrenees are formed of short valleys with steep slopes. Being strongly influenced by Atlantic winds, these areas experience rainy bioclimates across the elevation gradient. Therefore, the vegetation is set according to the altitudinal zonation model found in other medio-European mountains. This starts with a submontane belt supporting mixed deciduous forests, acidophilous oakwoods and heaths, which at 900–1000 m gives way to the montane belt, best characterized by beech forests, and by fir forests, heaths and meadows.

The Iberian side of the Pyrenees extends southwards over secondary ranges—known as the pre-Pyrenees—and internal depressions, and shows a more complex zonation, since it changes from Mediterranean to montane and then to alpine landscapes. This transition is sharper in the central part of the Iberian side, where the submontane belt combines dominant units related to marcescent oakwoods with sclerophyllous oakwoods and xerophilous scrubland on steeper rocky landforms. From 1100 to 1300 m upwards, the montane belt still expresses some continental Mediterranean influence such as bearing widespread Scots pinewoods together with meso-xerophilous grassland and box scrub. Beech and fir forests and other mesophilous units are secluded on moister slopes.

High mountain landscapes develop from 1600 to 1800 m up to the highest summits (peaking at 3404 m with Aneto), with less pronounced differences between the main north and south sides. The potential treeline ecotone, found at 2200–2450 m, makes the transition between the subalpine and the alpine belts; the former bearing forests of mountain pine and related vegetation units, the latter formed by contrasting vegetation mosaics where rock units become dominant towards the highest elevations.

J.M. Ninot (✉) • E. Carrillo • A. Ferré
Institute for Research on Biodiversity (IRBio) & Department of Evolutionary Biology, Ecology and Environmental Sciences, University of Barcelona, Av. Diagonal 643, E-08028 Barcelona, Spain
e-mail: jninot@ub.edu

J. Loidi (ed.), *The Vegetation of the Iberian Peninsula*, Plant and Vegetation 12,
DOI 10.1007/978-3-319-54784-8_8

8.1 Introduction

8.1.1 The Physiographic Framework

The Pyrenees is a mountain system that is more than 400 km long, stretching over the entire Iberian isthmus, and clearly distinct from the adjacent low-lying areas in both geological and biological aspects. According to physiographic and biogeographic criteria, however, the western and eastern edges form part of the Atlantic and Mediterranean low-mountain areas, respectively. Thus, they will not be included in this account, which focuses on the range running from the Roncal Valley in the west to the Canigó massif in the east. It is a complex mountain unit of some 300 km in length and at its central part it is more than 100 km wide. Over the highly diverse physiography and types of bedrock, the main subunits and highest peaks are set roughly along an east–west axis, the Pyrenean axis, which slopes downwards to the main valleys running northwards or southwards. This pattern is less clear in the eastern sector, where the axial unit including the peaks of Puigmal and Canigó is separated from the main axis by the Segre and Tet valleys, oriented NE to SW. Moreover, this eastern Pyrenean sector connects southwards with the northern Catalanidic system through mountains and plateaus that reach 1200–1500 m, all the way to the Montseny massif, which at its highest exceeds 1700 m. These south-eastern mountains, although not belonging to the Pyrenees according to geological and physiographical criteria, may be included in the Pyrenean province due to their biogeographic character (Fig. 8.1).

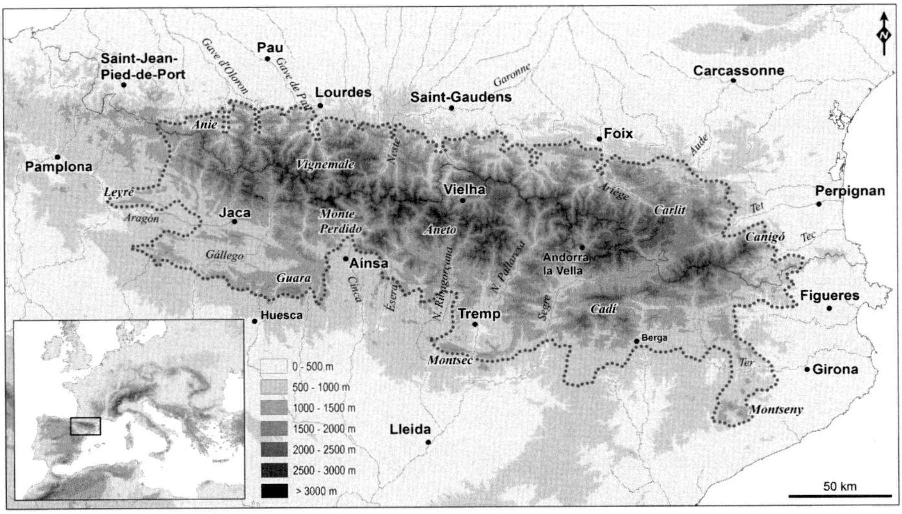

Fig. 8.1 Physiographic map of the Pyrenean area, which includes the Pyrenean mountains from some 500 to 700 m upwards, and also the northern Catalanidic mountains

Fig. 8.2 Head of the Aspe valley in the north-western Pyrenees. The upper part of the montane belt in the foreground (≈ 1400 m) includes acidified Bromion pasture and rich Fagion forests. Above the timberline, the abrupt high mountain hosts contrasting calcicolous vegetation such as *Salix pyrenaica* shrubby carpets and meadows of the Primulion (Photo: A. Ferré)

A remarkable trait of the Pyrenees is the marked asymmetry between the northern side, which is formed of short valleys rapidly decreasing in elevation onto the Aquitanic plain, and its southern side. This latter Iberian part is formed of more complex and extended valleys, which run southwards crossing or surrounding transverse secondary ranges known as the pre-Pyrenees.

The western sector of the main range (from Portalet westwards) is built up of lime-rich sedimentary rocks which have given rise to abrupt relief, including deeply carved valleys, plateaus, and peaks or summits reaching 2000–2800 m (Fig. 8.2). From there to the Noguera Pallaresa river, the central part of the main range includes higher crests and peaks, many of them exceeding 3000 m; with Aneto, the highest Pyrenean summit, peaking at 3404 m. This sector is mainly formed of metamorphic bedrock which produces a highly varied physiography, from the abrupt granite valleys and peaks of Maladeta-Aigüestortes, to more regular slate peaks and summits at the Noguera Pallaresa head. However, it also includes the impressing Vignemale-Monte Perdido massif, with the highest lime peaks in western Europe, and spectacular circs and cliffs, as in Gavarnie. From Andorra towards the east, the main range is mostly made of slate and granite. The relief is less abrupt, since most of the high mountain is formed by regular crests and peaks, as well as noticeable pre-Quaternary erosion plateaus. However, the main elevations exceed 2900 m as far as Madres and Canigó; the easternmost mountains with alpine areas.

Southwards, the pre-Pyrenean area is formed of lower ranges set more or less parallel to the axial range. They are mainly formed of lime bedrock and show contrasting relief, frequently combining sloping areas with abrupt cliffs. A group of higher mountains (San Juan de la Peña, Turbón, Cadí, etc.) reaches 2000–2600 m and is set aside from the main axis; it forms the inner pre-Pyrenees. The other ranges, known as outer pre-Pyrenees, are somewhat detached from the other Pyrenean ranges by noticeable internal depressions. These smaller ranges (Loarre, Guara, Montsec, etc.) hardly exceed 1500 m and mark the end of the Pyrenees towards the wide plains and low mountains of the Ebro depression.

All the streams and rivers that rise on the southern side of the Pyrenees flow to the Mediterranean Sea; most as tributaries of the Ebro (such as the Aragón, Cinca or Segre) though a few directly (the Llobregat and Ter). On the northern side, most rivers (the Saison, Gaves, Garonne, Ariège, etc.) run northwards and eventually empty into the Atlantic Ocean; but those in the eastern sector (the Aude, Tet and Tec) flow eastwards to the Mediterranean Sea. All of them have seasonal regimes, responding to rainfall and snowmelt seasonality, which include strong peak flows from the end of spring to early summer.

The complex hydrological systems have led to dominant fluvial relief in the low and mid elevations, where contrasting altitudes are expressed through extensive slopes forming downwards V-shaped river and stream valleys, occasionally containing narrow terraces. At higher elevations, from 1600 to 1800 m upwards, the dominant glacial relief defines the high mountain. Here, cirques and flat valley bottoms carved by glaciers during the cold Pleistocene periods contrast with steep rocky slopes. In the metamorphic areas, particularly in the central sector, these valley bottoms host a number of high-mountain lakes and mire areas. A few glacier remains, lying on the steep north-facing side of the highest summits (Vignemale, Monte Perdido, Maladeta, etc.) nowadays still show the vanishing nival landscape.

8.1.2 A Complex Climatic Mosaic

Like other large massifs, the Pyrenees shape a complex mosaic of bioclimates primarily related to the contrasting conditions in two distinct zones—Atlantic and Mediterranean—bordering the main range, and also to smaller-scale effects driven by orography (Izard in Dupias 1985; Rivas-Martínez et al. 2007). A strong Atlantic influence penetrates the westernmost Pyrenean edge and into the northern side of the range as far as the Ariège river. This brings a temperate oceanic bioclimate to the lower elevations, involving moderately cold winters and a long growth season with regular rain. In contrast, most of the lower areas along the southern side exhibit a Mediterranean pluviseasonal bioclimate, imposing notable drought conditions on the vegetation in summer. Connecting such distinct bioclimates, the transitional areas found at lower elevations in the eastern Pyrenean edge and in south-western and south-eastern Iberian mountains show a sub-Mediterranean climate; i.e., with moderate drought conditions through summer. In addition, the contrast between the north side and the central part of the Iberian side is enhanced by the orographic effects of the higher Pyrenees through föhn winds. While the low and mid elevations along the south-west, north and south-east Pyrenean areas bear an oceanic influence, the central Iberian sector experiences notable to strong continentality. This latter characteristic involves cold resting winters and at low elevations, dryer Mediterranean summers. These continental effects are most significant in some inner valley areas and depressions, such as Cerdanya, Urgell or Pallars.

From the peripheral, varying outskirts to the core of the Pyrenees, the principal bioclimatic shifts follow ground elevation, mainly centred on decreasing temperatures and increasing rainfall (Ninot et al. 2013a). Mean temperatures decrease quite regularly (some 0.6 °C per 100 m of altitude), which results in a shorter growing period with increasing elevation, as well as lower winter temperatures. Precipitation increases less regularly with altitude and distinctly from one area to another. In the oceanic Pyrenees (i.e., northern side, and most of western and eastern edges) the mid elevations experience a mountain variant of the Atlantic bioclimate, in which shorter growth periods drive shifts in vegetation more than increased rainfall does. In the Iberian valleys, however, altitudinal rainfall increase together with decreasing evapotranspiration make the dry summer season shorter and milder across the mid elevations (Fig. 8.3). At the same time, lower temperatures bring longer winters, and thus shorten the growing season. This creates a sharp environmental gradient in most Iberian valleys, which shift from a subcontinental Mediterranean bioclimate at the lower elevations to a milder sub-Mediterranean bioclimate, and then to conditions very close to those of the Atlantic side of the mountain range at the valley heads. A similar climatic transition is found across lower altitudes in areas connecting Mediterranean and Atlantic areas—thus, from the central outer pre-Pyrenees to around Jaca, leading north-westwards; and from the same area to Berga, leading north-eastwards—where the sub-Mediterranean bioclimate is dominant.

In the high mountain, from 1600 to 1800 m upwards, the primary bioclimatic driver is the decreasing length of the growing period—from 7 months to less than 3 months (Fig. 8.3). Shorter and colder summers mean less structured vegetation and soils with increasing elevation, due to lower primary production and longer freezing periods. Within this altitudinal pattern, from the western to the eastern high Pyrenees, more snowpack accumulates in winter, and summer includes more drought events. Apart from these general trends, the high mountain sees greater bioclimatic differentiation at a small scale; for example, summer water shortages on south-facing slopes and ridges with poor soil, shorter growing period due to north-facing exposure or to local snow accumulation, summer storms due to convective activity, and windy, contrasting conditions at summits and along ridges.

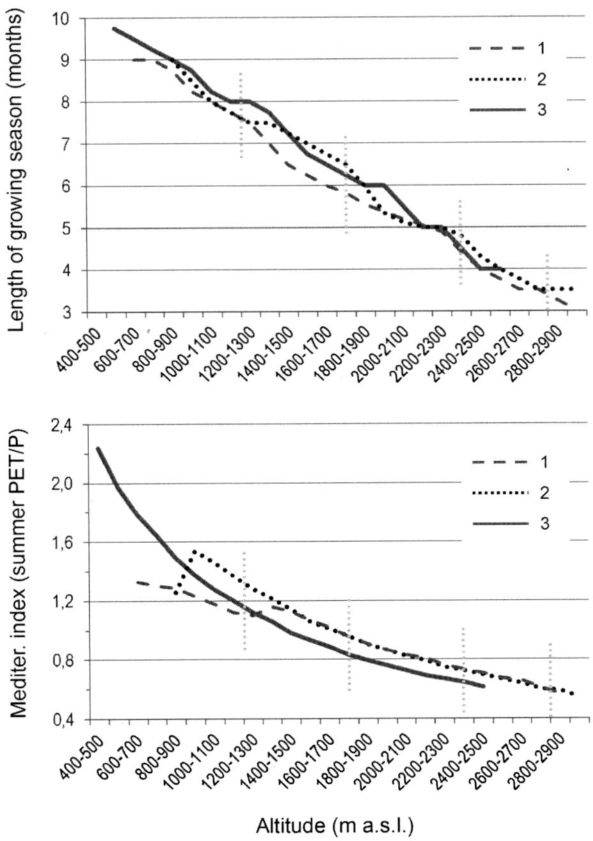

Fig. 8.3 Altitudinal variation of key bioclimatic descriptors in distinct areas of the central Pyrenees: *1*, Atlantic side (Aran valley); *2*, Iberian side (Alta Ribagorça); *3*, pre-Pyrenees (Tremp, Sort). *Vertical dotted lines* indicate the mean altitude for shifts between vegetation belts

8.1.3 The Pyrenean Flora

Although a comprehensive modern Pyrenean flora is still lacking, plant contents and distribution are reasonably well known through partial floras and general atlases, recently compiled in the *Atlas de la Flora de los Pirineos* (http://atlasflorapyrenaea.org/florapyrenaea/index.jsp). A general picture may start from assuming the figure of some 3500 species of vascular plants for the Pyrenean area considered here and following the analysis of chorological and functional groups reported for main sectors (Villar et al. 2001; Ninot et al. 2013a; Loidi et al. 2015).

The most represented chorotype in the Pyrenees corresponds to medio-European species; not only in terms of species richness (27% of the flora), but also as the dominant plants in very general landscapes (such as beech forests, Scots pinewoods or mesophilous grasslands and meadows). This corresponds to both the regional location of the massif, mostly included in the medio-European (= Eurosiberian) region, and the ecological conditions for plant life found in most mid altitudes of the area, which are mostly the same as those of lowlands and hills through the central part of Europe.

Two small collectives related to this temperate flora are the Atlantic and the sub-Mediterranean chorotypes. Atlantic plants, even taken in a broad sense, may represent less than 1.5% of the flora; they are mostly scarce in the landscape, and remain restricted to lower altitudes of the north-western part of the range. The sub-Mediterranean chorotype is also small (around 5%) and is well represented at low and mid altitudes of the southern and eastern Pyrenees, where some notable species (such as *Buxus sempervirens* or *Quercus pubescens*) define well-developed sub-Mediterranean landscapes.

The Mediterranean flora constitutes the second largest collective (19%), which reflects the species richness and ecological diversification of this chorotype. Mediterranean assemblages are particularly notable at low and medium altitudes through the pre-Pyrenees and the south-eastern mountains, mostly on south-facing exposures. But they penetrate the Iberian axial valleys, forming part of xerophilous assemblages, up as far as the beginning of the high mountain, secluded in specific sheltered locations.

Boreo-Alpine plants (7%) are mostly found in the high mountain, where they may be locally dominant in plant assemblages at higher altitudes, such as snowbed carpets or alpine dwarf heaths (e.g., *Salix herbacea*, *Juniperus communis* subsp. *alpina* and *Vaccinium uliginosum* subsp. *microphyllum*). Orophytes specific to south European mountains account for about 17% of the flora; half are Alpine plants. This latter group includes a number of taxa that are as characteristic of high mountain vegetation as *Pinus uncinata*, *Abies alba* or *Rhododendron ferrugineum*. Other orophytes may be related to Mediterranean high mountains (3%–4%), shared with the Cantabrian mountains (1.7%) or exclusive to the Pyrenees (3.5%). These Pyrenean endemics include from very rare plants secluded in particular habitats to such dominant species as *Festuca airoides* or *Echinospartum horridum* (see Box 8.1).

Together with the chorotypes mentioned, the Pyrenees hosts a great number of species found in two or more biogeographic regions, such as the Mediterranean and medio-European regions (pluriregional taxa: 18%). And some alien plants (about 6%) are mainly found at lower elevations. Therefore, the extant Pyrenean flora reveals the functions of this range as a porous barrier for lowland migrating plants, a corridor for Alpine orophytes and other cryophilic plants, refugia for ancient taxa, and an arena for ongoing speciation.

The rich Pyrenean flora is spread across the mountain range into distinct patterns. The variation in the physiographic aspects mentioned above—elevation, landforms, exposure, hydrology—together with the diversification of substrata and bioclimates gives rise to a huge diversity of habitats, which are variously colonized by distinct floral elements. Such diversification in plant assemblages operates from the most comprehensive geographic levels to the finer-grained mosaics of vegetation that respond to intricate local physiography. Due to the sharp differences found at the small scale in soil quality, snow deposition, slope dynamics or thermal inversion, a given landscape may include very contrasting plant assemblages in terms of species composition and ecosystem functioning. Therefore, the main floristic areas distinguished in the Pyrenees always include plants from contrasting chorotypes, as is the case of Mediterranean assemblages that reach the axial high mountain on rocky shelves, or Atlantic taxa secluded along specific rivers of the central Iberian sector.

Box 8.1 Plant Endemics, a Singular Trait of the Pyrenean Vegetation
The endemic species of the Pyrenees have attracted considerable interest from botanists. They have been reported in some specific papers (García and Gómez 2007) and information on their distribution, ecology and biological aspects is compiled in local and regional floras and data banks. Here, we bring this knowledge together to form a general analysis of the endemic Pyrenean flora with regards to vegetation features. We consider strictly endemic taxa (species or subspecies) along with sub-endemics, i.e., taxa mostly found in the Pyrenees, but with a small part of their range in nearby mountains.

All in all, the endemic component represents a significant part of the whole Pyrenean vascular flora, comprising 258 taxa, of which 202 are sexual species or subspecies and 56 are apomictic species. The latter group is mostly formed by 'species' described in the genera *Alchemilla*, *Hieracium* and *Taraxacum*, which although not equivalent to sexual species, reveal the speciation potential of the complex Pyrenean landscape. As for sexual endemics, 128 taxa are strict endemics and 74 taxa are sub-endemics, which represent 3.5% and 2.1% respectively of the entire Pyenean flora—estimated at some 3500 species. These figures are far lower than the percentages given for other mountainous areas: 11.2% for the Alps and 15.2% for southern Iberia (Aeschimann et al. 2011; Melendo et al. 2003). The most likely causes of this notable difference in the proportion of endemic flora are the much smaller size of the Pyrenees and the character of the range as a biogeographical corridor between other nearby mountainous areas.

Endemics are found at similar densities throughout different parts of the Pyrenees; being scarcer only in the southernmost, low-altitude pre-Pyrenean ranges. As for the altitudinal distribution, they are mainly concentred from the mid submontane (≈ 1000 m) to low alpine (≈ 2400 m) belts, with their maximum richness in the subalpine belt (1600–2200 m). This pattern contrasts with that of the entire Pyrenean flora, which shows the maximum

(continued)

Box 8.1 (continued)
species richness at 800–1200 m, and emphasizes the biogeographic isolation and singularity of the Pyrenean high mountain. As for the extent of their range, half of the endemics are absent from more than half of the Pyrenean area, and 25% are narrow endemics found in areas smaller than a quarter of the whole range.

The role of the Pyrenees as a reservoir of ancient taxa is revealed by a few palaeoendemics, such as *Ramonda myconi, Xatardia scabra, Borderea chouardii* and *B. pyrenaica*, some of which are isolated members of tropical families (Fig. 8.4). These and other palaeoendemics must have remained in southern Pyrenean refugia through Pleistocene climatic oscillations. However, the bulk of the Pyrenean endemic flora corresponds to neoendemics, mostly arising during the same changing period. They belong to many plant groups, but are particularly concentred in some genera, chiefly *Androsace, Armeria, Festuca* and *Saxifraga*. This suggests rich *in situ* allopatric speciation occurred within separate glacial refugia; the isolation of which would have been exacerbated at higher altitudes.

Fig. 8.4 *Ramonda myconi* (Gesneriaceae), an emblematic Pyrenean endemic specialist of rock crevices, widespread in limestone areas. Its species name honours both Louis Ramond de Carbonnières, a great French explorer and botanist in the Pyrenees of the early nineteenth century, and Francesc Micó, a Catalan doctor and naturalist of the sixteenth century (Photo: A. Ferré)

(continued)

Box 8.1 (continued)

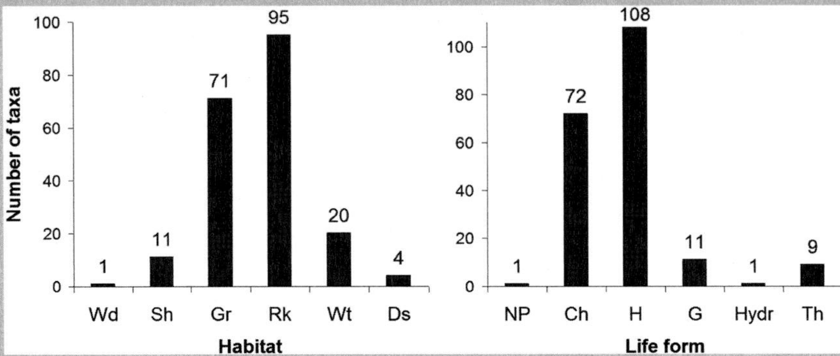

Fig. 8.5 Distribution of Pyrenean endemic flora (202 species and subspecies) in main habitats (*left*) and classification into life forms (*right*). *Wd* woodland, *Sh* shrubland, *Gr* grassland, *Rk* rocky areas, *Wt* water-related habitats, *Ds* disturbed areas, *NP* nanophanerophytes, *Ch* chamaephytes, *H* hemicryptophytes, *G* geophytes, *Hydr* hydrophytes, *Th* therophytes

In comparison with the entire Pyrenean flora, endemics stand out as inhabiting mainly rocky habitats and also grasslands, which host 47% and 35% of the endemics, respectively (Fig. 8.5). The importance of rocky areas for endemics is emphasized if we consider their whole species richness compared to that of pastures or other habitats. For example, in the Catalan Pyrenees, the rocky and the grassland habitats include 12.6% and 29.1% of the entire flora, respectively (Ninot et al. 2013a). This leads to considerable singularity in the vegetation of cliffs, rocky shelves, scree and other habitats with poorly developed soils, where associations and other vegetation units are mostly characterized by narrow endemics. In contrast, endemics are very rare in woodlands and in disturbed areas, where the driving forces are mainly competition and opportunism. Moreover, the relative abundance of chamaephytes and the scarcity of annuals and phanerophytes show even more clearly the relevance of conservative, low-competitive strategies to Pyrenean endemism.

These patterns suggest that, together with ancient historical factors, Pyrenean plant endemics are mainly explained by Quaternary environmental changes. Those changes enhanced the role of the lithological, physiographic and topoclimatic heterogeneity of the mountain system. Thus, the Pyrenean range has played a complex role in the floral history of south-west Europe from Pliocene onwards, through dramatic changes in climate and physiography, and hence landscapes. The present endemic flora singularizes the plant assemblages specific to the habitats where this flora is more frequent (i.e., grassland, scree and rock). This is the basis for the distinction of particular alliances in the Pyrenees, vicariant of those found in other Alpine massifs, and also for to define sectorial associations.

8.1.4 Altitudinal Zonation

Over all the environmental factors discussed, the elevation gradient remains the primary indirect driver of plant distribution for singular species, floral groups and entire chorotypes. A very general pattern across the entire range traces a humped function for species richness with a maximum at mid-low elevations (800–1200 m). Species numbers decrease slowly both towards the lower elevations and upwards, until the beginning of the high mountain, and then decrease regularly with increasing altitude again until the highest peaks (Villar et al. 2001; Grau et al. 2012). The same elevation gradient drives the chorological and life-form spectra and, even more markedly, defines contrasting landscapes based on particular vegetation types. Thus, although the altitudinal gradient produces gradual variations in the main bioclimatic traits, in soil processes, and also in floristic composition, the dominant vegetation units reveal sharper transitions between reasonably distinct main landscapes or vegetation belts (Fig. 8.6).

As mentioned earlier, the Pyrenees straddles two contrasting biogeographic regions: the medio-European and Mediterranean, and thus the altitudinal zonation follows different patterns on the two sides of the range—Atlantic or northern, and

Fig. 8.6 Panoramic view over the northern slopes of the Aran valley. The altitudinal zonation on the Atlantic Pyrenean side is shown by means of dominant vegetation units, such as mixed forests (Isopyro-Quercetum) and seral vegetation in the lowest, submontane belt; beech forests (Scillo-Fagetum) and extensive fir forests (Goodyero-Abietetum, Festuco-Abietetum) in the montane belt; and dominant grassland (Nardion) and heathland (Genisto-Vaccinion) in the subalpine belt (Photo: A. Ferré)

Iberian or southern, respectively—which converge in the transitional areas (south-western, and eastern to south-eastern). The differences are more profound at lower elevations, in the form of more contrast in the landscapes at the two sides (Ninot et al. 2007).

On the Atlantic side, the lower elevations (up to 900–1000 m) correspond to the submontane (or mesotemperate) belt and bear entirely medio-European flora (including the Atlantic chorotype) and vegetation, weakly differenced from those of the neighbouring lowlands. The shift to the montane (or supratemperate) belt is more noticeable, leading to fir and beech forests and associated units, as in most medio-European mountain areas. The following altitudinal change, leading to the subalpine belt, occurs at 1600–1700 m on north-facing slopes and about 100 m higher on south-facing ones; roughly coinciding with the transition to high mountain landforms. From there upwards on both main Pyrenean sides, the medio-European chorotype becomes scarcer, while Boreo-Alpine species and other orophytes increase and even dominate certain vegetation units. For instance, the subalpine belt is defined as a potential woodland level, most typically dominated by the mountain pine, *Pinus uncinata*. The highest pinewoods occur at 2500 m or a bit higher in central Pyrenean granitic areas, such as Néouvielle, Aigüestortes and Cerdanya. More generally, however, the upper subalpine limit is located at 2200–2300 m, and at lower elevations at the oceanic edges (eastern and western) of the range and also on clearly north-facing slopes. The transition to the alpine belt, a naturally treeless zone, is very noticeable where forests remain less depleted; but the floristic shift is very gradual, particularly within non-forest vegetation. Moreover, both the abrupt relief at these elevations and ancient clear-cutting make the transition between subalpine and alpine belts very patchy, and a general boundary less evident.

On the Iberian side, the zonation pattern is more complex, since it encompasses Mediterranean, medio-European and Alpine landscapes. The submontane belt reaches higher elevations (1100–1300 m) and corresponds to larger areas than on the north side; it stretches over diverse pre-Pyrenean and axial Pyrenean slopes and basins. Its general landscape is sub-Mediterranean, as shown through marcescent oakwoods and related scrubland, but it nevertheless includes typical Mediterranean assemblages on south-facing slopes and rocky landforms, according to local drought conditions. The montane belt extends upwards until altitudes similar to those on the Atlantic side, and is characterized by extensive Scots pinewoods. But it also includes noticeable fir and beech forests in milder areas, and thus evidences some Atlantic character. In the subalpine and alpine belts, vegetation units are much the same as on the Atlantic side, although the Mediterranean influence shapes more or less differentiated landscapes.

The following synthesis on the Pyrenean vegetation is based on a great many geobotanic papers, from regional studies to local monographs. Their contents have been accessed through the SIVIM facility (http://www.sivim.info/sivi/) or indirectly by means of regional vegetation maps or surveys (primarily Rivas-Martínez et al. 2011, which moreover is the reference for syntaxa names). Literature citations in the text are then chiefly limited to these general works and to a few particular papers. Following this phytosociological knowledge and the zonation pattern

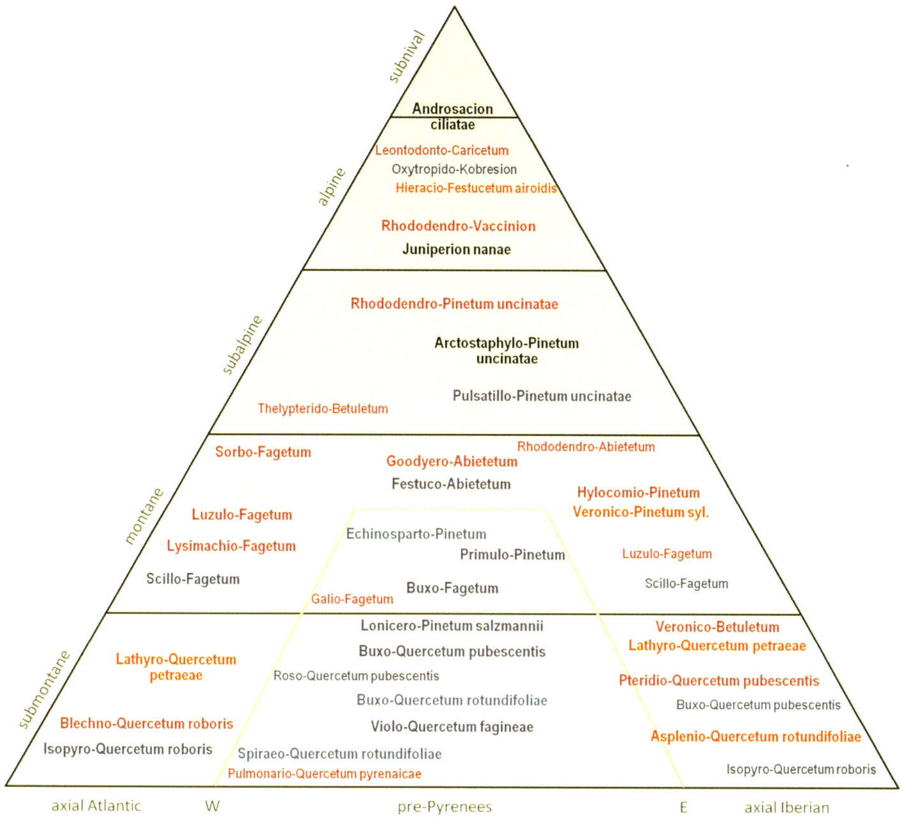

Fig. 8.7 Schematic location of main potential associations (climactic or not) in the Pyrenees, according to the altitudinal belts. The Atlantic and Iberian sides of the axial Pyrenees differ in the lower levels (*left* and *right*) and progressively converge towards the high mountain; the Iberian pre-Pyrenees (*trapezium*) show some distinction between eastern and western parts. Associations in *black* lettering are indifferent to substrate, whereas *blue* means calcicolous (or eutrophic) and *red* means acidophilous units; *light* and *dark* shading within each colour stand for south-facing and north-facing exposures, respectively

outlined above, Fig. 8.7 provides a schematic location in the distinct belts and sectors for the main potential associations discussed in the following sections.

8.2 Mid Elevations

8.2.1 Woodlands at the Atlantic Side

Forests that develop under the Atlantic influence are formed mostly by deciduous meso-hygrophilous trees, where the dominant species—typically oaks or beech—form mixed crown layers together with ash, lime, maple or birch. Fir forests are also notable in the montane belt in subcontinental areas, but otherwise coniferous trees play a very secondary role in the Atlantic woodland landscapes. Generally, these are complex forests including distinct plant strata formed of medio-European sciophytes. Atlantic forests are widespread over the western edge and the northern side of the axial Pyrenees, but they are also found in the transitional areas stretching from the eastern Catalan valleys to Montseny, and from Navarra to western Aragón; or even at the heads of some Iberian valleys and in a few pre-Pyrenean locations (Gaussen 1964; Dupias et al. 1983; Loidi and Báscones 1995; Bolòs et al. 2004).

In the submontane belt, the most widespread climactic forest is the Isopyro thalictroidis-Quercetum roboris: a rich mixed forest including a varied tree layer (formed of *Quercus robur, Fraxinus excelsior, Prunus avium, Tilia cordata, Tilia platyphyllos*, etc.), sparse strata of shrubs and saplings, and a ground layer of vernal geophytes and hemicryptophytes, related to the Fagetalia forests. This association develops in eutrophic, neutral to slightly acidic soils, across the Atlantic side and in particularly humid areas of the oriental Catalan valleys. Where the substrata are more acidic, or on locally dryer slopes, these mixed forests give way to species-poorer Quercion robori-petraeae oakwoods. A common association in milder areas is the Veronico urticifoliae-Betuletum pubescentis, where *Quercus petraea, Quercus robur* or *Betula pendula* may dominate over a light acidophilous understorey. But the Lathyro linifolii-Quercetum petraeae, a meso-xerophilous *Quercus petraea* forest, is a more widespread association, since it is found at locally dry Atlantic sites and also in less oceanic valleys of transitional areas and in most of the Catalan area, from Pallars to Montseny.

The montane belt is characterized by beech (*Fagus sylvatica*) forests, mixed beech and fir (*Abies alba*) forests, and even pure fir forests at its higher levels or in less oceanic areas (Fig. 8.8). The dominance of such contrasting trees depends on interrelated factors, such as elevation, microclimate and soil quality; but it must also respond to ancient forest management. Beech forests are particularly dominant on the Atlantic side, on any exposure and substrate, and reach the lower levels of the high mountain where they may even form the present timberline (Fig. 8.2). Increasing continentality or dryness along some central Pyrenean valleys (such as Aran) or in transitional areas makes beech forests less competitive, compared to fir or even Scots pine forests. Beech forests are highly diversified over the Pyrenees, given the wide environmental scope where they are found and the rich nemoral flora they host. A widespread association is the Scillo liliohyacinthi-Fagetum sylvaticae: a eutrophic forest association rich in bulbous and rhizomatous species that bloom in spring. This beech forest covers a wide scope of gentle slopes and substrata in Atlantic areas, whereas it reaches the Iberian side much scarcely secluded in particularly humid valleys. On acidic substrata, beech forms poorer assemblages, with an acidophilous understorey including sparse shrubs and herbs, partially shared with acidic oak forests. The most general association is the Luzulo niveae-Fagetum sylvaticae, which is found on intermediate or north-facing exposures in south-eastern Pyrenean areas and on dry slopes in Atlantic valleys. In these locations, moister acidic slopes host a richer forest: Lysimachio nemorum-Fagetum sylvaticae.

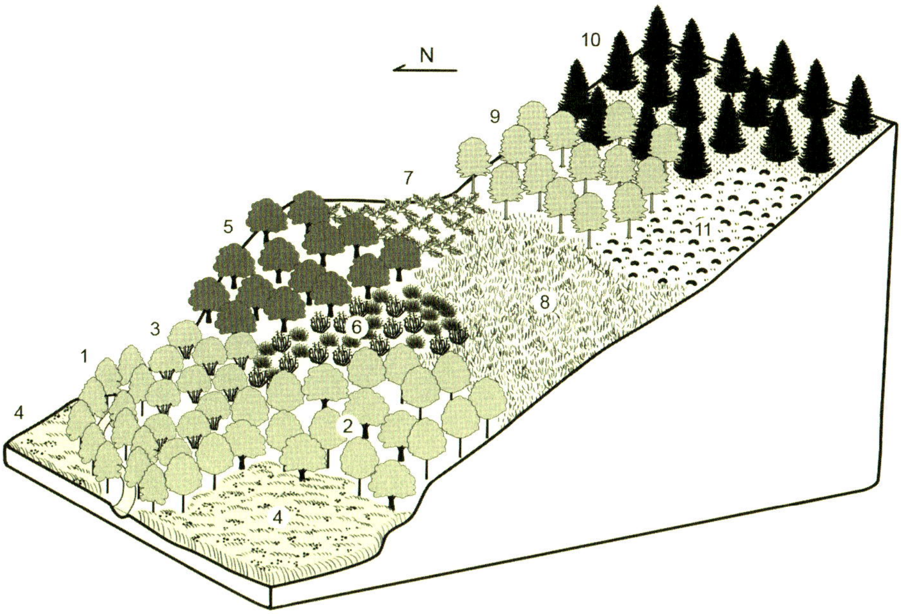

Fig. 8.8 Main potential and seral units in a landscape at mid elevations (submontane and montane belts, ≈ 700–1700 m) representative of central Pyrenees, northern side. *1*, Alnion incanae; *2*, Isopyro-Quercetum roboris; *3*, *Corylus avellana* seral thickets; *4*, Arrhenatherion; *5*, Lathyro-Quercetum petraeae; *6*, Cytision oromediterranei-scoparii; *7*, *Pteridium aquilinum* seral stands; *8*, Bromion; *9*, Scillo-Fagetum; *10*, Goodyero-Abietetum; *11*, Genisto-Vaccinion, Violion caninae (Drawing: Laura Fuentes)

At the higher elevations of the montane belt (from 1200 m upwards) in subcontinental valleys, *Abies alba* woodland becomes very common (Fig. 8.6). This dominant conifer forms dense forests with a sparse sciophilous understorey including Fagetalia character-taxa and acidic humic soil specialists, such as *Orthilia secunda*. The most common type of fir forest is the Goodyero repentis-Abietetum albae; a relatively species-poor assemblage, well represented in the central sector on acidic substrates. Other fir associations are the Festuco altissimae-Abietetum albae, a species-rich forest occupying more eutrophic soils; the Emero majoris-Abietetum albae, secluded in narrow shady lime-rich valleys of Navarra; and fir-dominated forms of the beech associations mentioned above (Scillo liliohyacinthi-Fagetum, Luzulo niveae-Fagetum).

8.2.2 *Woodlands at the Iberian Side*

A particular trait of the Pyrenees that differentiates it from other European Alpine mountains is the noticeable Mediterranean influence in terms of bioclimate, flora and vegetation patterns. Although this influence decreases from the lower Iberian elevations upwards, it is still noticeable up to the high mountain (Bolòs et al. 2004; Rivas-Martínez et al. 2011).

This sub-Mediterranean character is particularly apparent in the submontane belt of the Iberian Pyrenees, through climactic forests formed mainly of marcescent oaks (*Quercus faginea*, *Quercus pubescens*, *Quercus subpyrenaica*), but also pines (*Pinus nigra* subsp. *salzmannii*, *Pinus sylvestris*) and even holm oak (*Quercus rotundifolia*) in dryer areas (Fig. 8.9). The tree layers in these forests are not as dense and complex as in the Atlantic forests, and they give way to a partially heliophilous understorey. Meso-xerophilous oakwoods of *Quercus faginea* (or *Quercus subpyrenaica*) are common in the subcontinental pre-Pyrenees up to some 1000 m, on sloping relief or in depressions affected by thermal inversion, and related to limestone or marls. They belong to the Violo willkommii-Quercetum fagineae association from the Segre river to Navarra, and are a good expression of the transitional character from southern sclerophyllous to northern deciduous forests. On the steeper slopes of harder limestone or conglomerate in the central pre-Pyrenees, under contrasting thermal regimes, Salzmann pine forms a related association: Lonicero xylostei-Pinetum salzmannii. Mediterranean *Quercus*

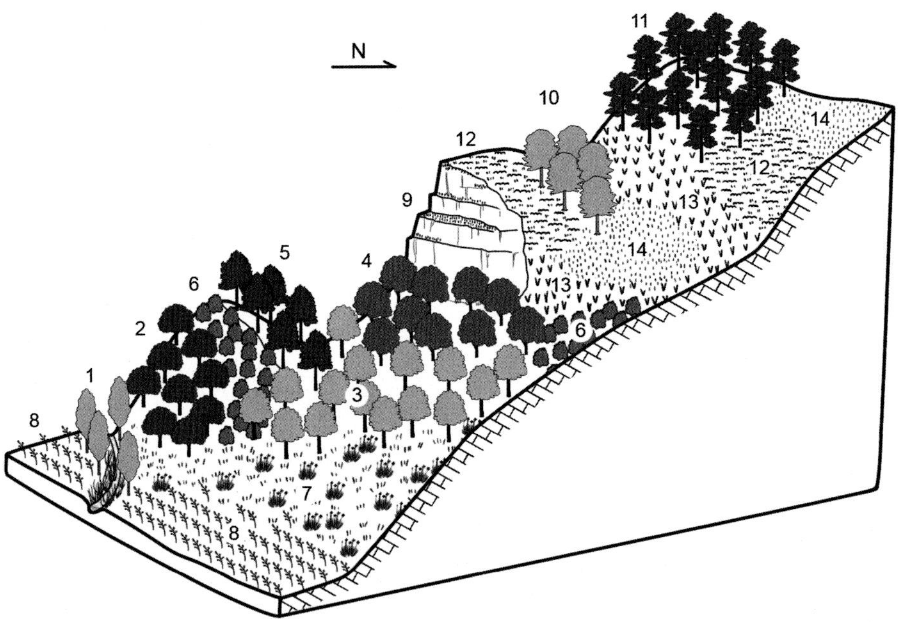

Fig. 8.9 Main potential and seral units in a landscape at mid elevations (submontane and montane belts, ≈ 700–1700 m) representative of central pre-Pyrenees. *1*, Salicion discolori-neotrichae; *2*, Buxo-Quercetum rotundifoliae; *3*, Violo-Quercetum fagineae; *4*, Buxo-Quercetum pubescentis; *5*, Lonicero-Pinetum salzmannii; *6* Amelanchiero-Buxion; *7*, Helianthemo-Aphyllanthion; *8*, sown fields; *9*, rocky units; *10*, Buxo-Fagetum; *11*, Primulo-Pinetum sylvestris; *12*, Ononidion striatae; *13*, Xerobromion; *14*, Bromion (Drawing: Laura Fuentes)

rotundifolia forests and maquis form part of most pre-Pyrenean landscapes, especially in continental valleys. These sclerophyllous formations are common on steep south-facing slopes and, following rocky relief, they can reach the base of the pre-Pyrenean montane belt and penetrate northwards into some axial valleys. These are species-poor assemblages, mostly included in the calcicolous Buxo sempervirentis-Quercetum rotundifoliae—but in the Spiraeo obovatae-Quercetum rotundifoliae in the westernmost pre-Pyrenees—or in the Asplenio adiantinigri-Quercetum rotundifoliae on acidic bedrock in the central Palaeozoic Pyrenean core. Other noticeable Mediterranean elements are the relict stands of *Juniperus sabina* or *Juniperus thurifera*, which form open shrubby communities on particular rocky landforms of axial Pyrenees.

In milder locations within the submontane belt (i.e., at higher elevations, on north-facing areas, and particularly in the eastern and western transitional valleys), the main tree species is *Quercus pubescens*, along with *Quercus subpyrenaica* and *Pinus sylvestris*. The most widespread association is the Buxo sempervirentis-Quercetum pubescentis, covering a wide altitudinal range (600–1500 m) on any lime-rich substrate from the eastern Catalan ranges to Navarra. These forests still reflect some Mediterranean influence, through the abundance of the evergreen *Buxus sempervirens* and the inclusion of some typically Mediterranean taxa in the understorey (more than 10%). In the western transitional valleys of Navarra, the equivalent association is the Roso arvensis-Quercetum pubescentis. In the tree layer of these woodlands, *Pinus sylvestris* is frequent as an occasional tree, but it may also be dominant in locally cold locations, or as a result of forest management, which in places has led to extensive secondary pinewoods. On siliceous bedrock, we find the calcifuge Pteridio aquilini-Quercetum pubescentis (with *Quercus pubescens* or *Quercus petraea*) in the central and eastern sectors; and the Pulmonario longifoliae-Quercetum pyrenaicae at the western edge, in the sole Pyrenean location for *Quercus pyrenaica* (Sierra de Leyre).

A general feature of the landscape of the Iberian montane belt is the abundance of *Pinus sylvestris* forests. The Pyrenean populations of this species have been described as two particular subspecies, *pyrenaica*, in western and central Iberian areas, and *catalaunica*, in Catalonia. However, there is no apparent ecological difference among them; they occur on any exposure and substrate type. This ubiquitous pine species appears to be better adapted to suboptimal conditions—summer drought and thermal continentality— than beech and fir are. On north-facing slopes, an extended association through most Iberian valleys is the acidophilous Hylocomio splendentis-Pinetum catalaunicae; whereas the Polygalo calcareae-Pinetum sylvestris thrives in calcareous soils. These pinewoods are characterized by a notable moss layer and by some boreal and humicolous plants; they are vicariants of similar communities found in the inner continental Alps. On dryer, south-facing slopes, the Scots pine forms xerophilous assemblages with poorer herb and moss layers. On acidic soils, sparse silicicolous herbs and sub-shrubs in the understorey differentiate the Veronico officinalis-Pinetum sylvestris; while in lime-rich areas, pinewoods bear some taxa of the Quercion pubescenti-petraeae or of an oromediterranean character (such as *Viburnum lantana* or *Lavandula angustifolia* subsp. *pyrenaica*). This is the Primulo columnae-Pinetum sylvestris: an association commonly found in the montane belt of the Catalan Pyrenees and pre-Pyrenees. In the central region of Aragón, the endemic *Echinospartum horridum* singularizes the Echinosparto horridi-Pinetum pyrenaicae, which includes open Scots pinewoods with a xerophilous understorey. Following forest succession, it may eventually shift to the climactic Buxo sempervirentis-Quercetum pubescentis, still with dominant *Pinus sylvestris*, or even to the Goodyero repentis-Pinetum sylvestris.

As mentioned earlier, Fagetalia forests progressively become scarcer as we move from the Atlantic to the Iberian areas in the montane belt. However, patchy *Fagus sylvatica* forests and related assemblages thrive in milder locations of the Iberian valleys and even of most pre-Pyrenean ranges. They even reach Montsec and Montseny (see Box 8.2); but they are very scarce or absent from the most continental valleys between Pallars and Cerdanya. In these Iberian beech forests, mostly found on lime bedrock, beech finds suboptimal humidity conditions, and the understorey is generally species-poor. They mostly belong to the Buxo sempervirentis-Fagetum sylvaticae, which is characterized by sub-Mediterranean meso-xerophilous taxa such as *Quercus pubescens* or particularly the sclerophyllous *Buxus sempervirens*. In moister soils, beech may form an association richer in Fagion taxa: Helleboro occidentalis-Fagetum sylvaticae. Other representatives of medio-European vegetation are the *Fraxinus excelsior* stands found near valley bottoms and on humid slopes of most axial Iberian valleys, described as Brachypodio sylvatici-Fraxinetum excelsioris. Given their ecologic location and species composition, which includes subdominant trees such as *Prunus avium* or *Populus tremula* and nemoral herbs, these are the southernmost sub-Atlantic mixed forests in the Pyrenees.

Box 8.2 Montseny: A Pyrenean-Like Mountain Emerging from Mediterranean Lowlands

As the highest and northernmost Catalanidic mountains, the relief, substrate, bioclimate and biota of the Montseny massif resemble those of a Pyrenean mountain. Its summit, at some 1700 m of elevation, abruptly rises almost 1600 m from the south-eastern littoral lowlands and 1200 m from the inland Vic depression. As in most Pyrenean mountains, the Alpine orogeny promoted a steep landscape, which includes high mountain features such as treeless summits, rocky slopes, cliffs and scree. Moreover, and despite the distance from the Pyrenees through low mountains and depressions, much phylogeographic evidence indicates that Montseny has formed part of a corridor for a number of Boreo-Alpine and Alpine elements. Some of these, such as *Abies alba* and a number of species of high mountain grasslands, have their southernmost populations on Montseny.

(continued)

Box 8.2 (continued)

Most of the Montseny bedrock is siliceous (Palaeozoic schist and granite); the same as the lower Mediterranean mountains to the south-east. The associated soils are moderately to strongly acidic; the latter mostly found at higher elevations. Thus, calcifuge vegetation dominates over the whole mountain, except for the lower north-west slopes, where lime-rich bedrock supports calcicolous plant communities.

The bioclimate of the extensive lower slopes is mild Mediterranean: relatively humid and oceanic. Annual precipitation amounts to some 700 mm, which mostly falls in autumn and winter, and defines a clear summer drought period. However, orogenic rains result in a generalized increase in precipitation with increasing elevation, which yields annual values of 1100–1200 mm in the rainiest places. Moreover, although higher areas experience a seasonal regime that is similar to that of Mediterranean lowlands in its irregularity, summer drought is minimized through frequent fog events, visible as clouds covering the mountain from 1000–1200 m upwards on more than 50% of summer days. This elevation roughly coincides with the shift from Mediterranean to medio-European landscapes.

On Montseny, mountain sclerophyllous woodland (Asplenio onopteridis-Quercetum ilicis and its seral vegetation) is dominant across the extensive submontane belt and gives way rather sharply to beech forests along most slopes. This is in contrast to the more gradual transition typical of the eastern Pyrenees (Bolòs 1983). Only in places do marcescent oak forests of *Quercus pubescens* or *Quercus petraea* subsp. *huguetiana* (Quercion pubescenti-petraeae, Quercion robori-petraeae) form an irregular transition to the beech belt. At these intermediate elevations, forests of *Castanea sativa* also cover some slopes irregularly. They do not seem to correspond to a modified pristine chestnut belt, as is the case in similar north-west Mediterranean mountains (Aigoual, Corsica, Apennines), but to ancient plantations traditionally exploited for timber (Bolòs 1983).

In the montane belt, beech forests are extensive and mostly very species-poor, which responds to acidic soils and to the climatically suboptimal conditions for beech forests. They are mostly poor forms of the Luzulo niveae-Fagetum, except for a few richer patches of the Helleboro occidentalis-Fagetum secluded in certain ravines. Also, this deciduous woodland includes small *Abies alba* stands, appearing as spots of subalpine landscape situated in particular north-facing cold areas of the higher forest level (Fig. 8.10). This whole mesophilous landscape also includes small patches of other medio-European vegetation units, such as Brometalia erecti grasslands, Arrhenatherion meadows and riparian alder and ash forests (Alnion incanae).

(continued)

Box 8.2 (continued)

Fig. 8.10 In Montseny, the montane belt is characterized by extensive beech forests (Luzulo-Fagetum) spotted with small stands of *Abies alba*, whereas the supra forest summits (≈ 1600–1700 m) host an alpine-like landscape (foreground) with Antennario-Festucetum short grassland, *Juniperus communis* subsp. *alpina* scattered scrub, and slate rocky units (Photo: A. Ferré)

From 1600 m upwards, hard physiographic and microclimatic conditions hamper forest development. On these summits and ridges, in spite of their moderate elevation, the vegetation mosaic is strongly related to alpine landscapes through a noticeable Boreo-Alpine flora (such as *Juniperus communis* subsp. *alpina*, *Antennaria dioica* or *Vaccinium myrtillus*). The most apparent plant communities are dwarf juniper scrubs (Cytiso oromediterranei-Arctostaphyletum) scattered over short grasslands (Antennario-Festucetum commutatae). Other specific units also related to Pyrenean vegetation are the *Festuca gautieri* grasslands (Luzulo-Festucetum gautieri) found on rocky north-facing slopes, spots of tall meadows (Dryopterido-Allietum victorialis) secluded in fertile mild depressions, and rupicolous vegetation including *Saxifraga vayredana* and *Saxifraga genesiana*, both species exclusive to Montseny. The closeness of these narrow endemics to Pyrenean species highlights the biogeographic interest of Montseny for understanding speciation in mountain flora.

8.2.3 Seral Vegetation and Azonal Units

The mid elevation zone has been strongly anthropized since ancient times, which has led to and maintained secondary vegetation, from fields to scrubland as well as to preclimactic thickets and forests. On the Iberian side of the range, moreover, harsher climatic conditions and erosion slow down secondary succession, which maintains widespread and diversified scrubland, extensive grassland and pinewood—seral or due to reforestation (Fig. 8.11). This includes xerophilous garigue (Quercetum cocciferae) and heaths (Rosmarinion officinalis, Cistion ladaniferi) together with annual-rich pastures (Thero-Brachypodion ramosi) and scrubby pastures (Brachypodion phoenicoidis, Helianthemo italici-Aphyllanthion monspeliensis) on dryer submontane slopes exhibiting a strong Mediterranean influence. At higher elevations or on milder slopes of the Iberian Pyrenees, the landscape mostly consists of box scrubland (Amelanchiero ovalis-Buxion sempervirentis), *Corylus avellana* thickets, *Cytisus oromediterraneus* calcifuge heaths (Cytision oromediterranei-scoparii) and meso-xerophilous pasture diversified into a number of associations included in the Helianthemo-Aphyllanthion (in lime-rich areas under some oceanic influence), the Xerobromion erecti (in subcontinental or montane areas), or the Bromion erecti (in moister locations). Ridges and slopes under contrasting conditions are a place for chamaephytic

Fig. 8.11 Valldarques, in the Segre basin, showing a submontane (≈ 750–950 m) pre-Pyrenean landscape with Salzmann pine forest (Lonicero-Pinetum, *right*) on the north-facing slope and marcescent oak forest and seral communities (Buxo-Quercetum pubescentis, Aphyllanthion, *left*) on the south-facing aspects; fields and remnants of riparian forest (Salicion discolori-neotrichae) at the valley bottom (Photo: A. Ferré)

Ononidion striatae pastures, somewhere dominated by thorny cushions such as *Erinacea anthyllis* (in the central-eastern pre-Pyrenees) or *Echinospartum horridum* (in Aragón).

In the Atlantic valleys, deforestation has produced distinct types of heath, chiefly broom assemblages of *Sarothamnus scoparius* or *Genista hispanica* subsp. *occidentalis* (Cytision oromediterranei-scoparii) and ericaceous heaths of the Genisto-Vaccinion with *Calluna vulgaris* and *Vaccinium myrtillus* in most of the area; or *Erica vagans* and *Erica cinerea* in the western sector. The slopes most suitable for extensive grazing bear mesophilous Violion caninae and meso-xerophilous Bromion erecti grasslands, together with invasive *Pteridium aquilinum* formations, which are expanding nowadays due to rural abandonment. Gently sloping landforms with deep soil have traditionally been used as crop fields and hay meadows. These semi-natural units are decreasing in extension and becoming more intensively managed nowadays, but they are still a typical feature of axial Pyrenean landscapes. They belong to distinct associations of the Arrhenaterion elatioris through most of the area, and to the Polygono bistortae-Trisetion flavescentis in the higher montane belt.

In spite of the strong deforestation of most Pyrenean riversides and the poor development of fluvial terraces, alluvial forests are still found as narrow strips or scattered patches along their margins. The most characteristic are alder assemblages (Alnion incanae), where *Alnus glutinosa* together with other tree species (*Populus nigra, Populus canadensis, Fraxinus excelsior, Salix alba, Betula pendula*) form irregular, dynamic canopies over the tall herbaceous understorey. Wider river beds with irregular water courses and streams with strong seasonal flow variation, typically found in Iberian valleys, host thickets and open forests of the Salicion discolori-neotrichae formed of willows (*Salix elaeagnos, Salix purpurea, Salix cinerea*) which may survive occasional flooding and mitigate mechanical damage.

Secluded margins of carving rivers and humid cliffs are the southernmost place for small patches of the Tilio platyphylli-Acerion pseudoplatani. These rocky, lime-rich, humid locations, with apparent slope dynamics, enhance species-rich forests or *Corylus avellana* thickets with mixed tree strata (of *Tilia platyphyllos, Acer* sp. pl., *Fraxinus excelsior* and *Ulmus glabra*).

Neighbouring the hygrophilous forests just mentioned, or next to springs and other bodies of water, a wide scope of hygrophilous herbage may be found, from reed beds (Molinio-Holoschoenion, Juncion acutiflori) to mixed meadows (Molinion caeruleae, Calthion palustris, Magnocaricion elatae, Filipendulion ulmariae) and to small spring carpets (Glycerio-Sparganion, Cardamino-Montion). On the Atlantic side, even fens and bogs of boreal type (Scheuchzerio palustris-Caricetea nigrae, Oxycocco palustris-Sphagnetea magellanici) occur in the higher montane belt.

Rocky habitats host diverse plant assemblages, characterized at the association level by endemics and other orophytes (see Box 8.1). Lime-rich outcrops, very notable in the pre-Pyrenees and central axial Pyrenees, are habitat for: thermophilous Asplenion petrarchae assemblages at lower elevations on south-facing cliffs;

Saxifragion mediae associations on colder rock surfaces; and Violo biflorae-Cystopteridion alpinae communities on mild sheltered cliffs. Siliceous rocks, more present in the eastern half of the Pyrenees, bear Antirrhinion asarinae vegetation. Scree and gravel slopes in lime-rich areas are colonized by open Achnatherion calamagrostis or by Petasition paradoxi assemblages; while silicate scree hosts Galeopsion pyrenaicae communities.

8.3 High Mountain

8.3.1 Subalpine Woodlands

The subalpine belt is mainly the domain of *Pinus uncinata*, a ubiquitous conifer capable of forming different types of forests, since it can thrive on north-facing to south-facing exposures, in acidic or lime-rich soils, and under a wide range of hydric conditions. Whereas there is notable variation in the species composition of pinewoods in accordance with these environmental factors, regional variation in ecological types along the Pyrenees remains small (Rivas-Martínez 1968; Vigo 1979; Carrillo and Ninot 1992, etc.). However, the role of pinewoods in the present landscape is very irregular across the region. They are a dominant landscape feature in the eastern half of the axial range; but become scarcer and patchy towards western and pre-Pyrenean areas, where they only occur on steep, rocky slopes. Meanwhile, they have vanished from some central Iberian valleys, such as Castanesa or Cabdella.

Another factor in the variation of these *Pinus* forests is elevation. Standing dense and evenly structured in the lower part of the belt (mainly up to 2000 m), they become lighter and more irregular further up or on less favourable substrates, as the pines appear looser and uneven. The highest forest boundary, or timberline, shifts upwards to a contrasting treeline ecotone, where the same species, *Pinus uncinata*, appears in decreasing size and density, and eventually gives way to alpine mosaics (see Box 8.3).

Box 8.3 The Treeline, a Sharp Ecotone Between Mountain Forests and Alpine Mosaics

As in similar mountain systems, the Pyrenean woodlands define an altitudinal boundary with open alpine landscapes: the treeline ecotone. The treeline occurs at varying altitudes, primarily depending on continentality, exposure and landform, and also on the past anthropogenic activities. The uppermost forest remains, which are mostly found between 2200–2450 m, are apparently limited by bioclimatic constraints and therefore are evidence of the potential treeline (Carreras et al. 1996). According to Körner and Paulsen (2004), the most basic

(continued)

Box 8.3 (continued)
ecological factor that defines the potential treeline location is the minimum temperature of the rhizosphere during the growing season (around 7 °C).

Most of the Pyrenean treeline occurs at lower elevations than its potential position, chiefly due to ancient clear-cutting to gain grazing range for domestic cattle, and chiefly on gentler relief and south-facing exposures. The present treelines are thus at much lower altitudes in these areas. Nowadays, the ongoing cessation of traditional land use promotes spontaneous reforestation, which is more marked where the treeline had been lowered more. The chief Pyrenean treeline-forming species is *Pinus uncinata*, although in Atlantic parts of the range, and mainly where the treeline has been lowered the most, it may be based on *Abies alba* or even on deciduous trees. In the Atlantic areas, the present forest belt mostly ends as low, irregular *Fagus sylvatica* stands, at relatively low elevations (1800–2000 m).

Most detailed knowledge of the Pyrenean treeline comes from a number of potential or moderately lowered examples involving *Pinus uncinata*, in eastern and central sectors of the axial chain (Batllori et al. 2009). There, the uppermost forests, or timberline, define the most obvious vegetation border within the ecotone. From there upwards, decreasing pinewood structure promote sharp shifts in different environmental drivers; chiefly the light distribution through the vegetation canopy, the consistency and dynamics of the snowpack, and the aboveground thermal regime. Under these contrasting conditions, isolated trees encounter a better thermal soil regime than in the nearby forest, since aboveground temperatures are more responsive to increased summer irradiation. Thus, most of the ecotone consists of trees scattered over a patchy area formed of pasture and scrub. Among the latter, and depending on the slope exposure and bedrock, alpine dwarf heaths (of *Vaccinium* sp. pl., *Dryas octopetala, Loiseleuria procumbens*) alternate with patches of taller shrubs such as *Rhododendron ferrugineum* or *Juniperus communis* subsp. *alpina* (Fig. 8.12). Pine seedlings occur abundantly in this ecotone, but growing into saplings and trees mostly takes place only near these taller shrubs, which facilitate it through accumulating snow and protecting against herbivores.

Further upwards in the ecotone, *Pinus uncinata* individuals are shorter and have less regular branching and canopies, becoming even shrubby. This is a result of increasing environmental stress and disturbance events, such as winter drought, herbivory and avalanches. These scattered krummholz individuals form the tree limit, or treeline, defined by convenience as the uppermost position of individuals that are 2 m tall. In most cases, the entire treeline ecotone encompassing from regular forest to these last trees occurs within 100–200 ground meters. However, scattered krummholz individuals may be found much further up than the regular treeline; even reaching 2700 m in a few particular locations.

(continued)

Box 8.3 (continued)

Fig. 8.12 An expanded treeline ecotone (including open Rhododendro-Pinetum forest, scattered pines and krummholz) gives way upwards to an alpine mosaic on a gentle slate relief, with low scrub (Saxifrago-Rhododendretum), open grassland (Festucion eskiae) and scree vegetation on most slopes, and patches of short climactic grassland (Hieracio-Festucetum airoidis, foreground). Slopes of the Salòria peak (2789 m, background right) in the high Noguera Pallaresa basin (Photo: E. Carrillo)

Other treeline landscapes are the open *Pinus uncinata* forests combined with xerophilous grasslands and a sub-shrubby understorey that occurs on sunny slopes. In most cases, the creeping *Arctostaphylos uva-ursi* is a key treeline species, since it occurs from encroached pasture to dense carpets hosting sparse pines and to the patchy understorey in open forest. A similar role is performed by *Echinospartum horridum* in low subalpine areas of Aragón, where this thorny cushion plays a very dominant role, favoured by decreasing levels of traditional land use.

The most general type of pinewood is the Rhododendro ferruginei-Pinetum uncinatae, which is associated with acidic, humic, moist soils and near-continuous snow cover in winter. This favours an ericaceous and moss understorey, where dominant *Rhododendron ferrugineum* together with *Vaccinium myrtillus*, *Pyrola minor*, *Homogyne alpina*, *Pleurozium schreberi*, etc. are related to slow litter decomposition and a relatively short growing season. These conditions occur on north-facing slopes and on gentle or flat relief, where this pinewood covers great surfaces, chiefly on siliceous bedrock.

In the lower subalpine belt on north-facing exposures, *Pinus uncinata* shares dominance with *Abies alba* and locally with a few deciduous trees. Particularly, *Abies* outcompetes *Pinus* and may form fir-dominated forests in the best ecological locations. This mostly corresponds to the Rhododendro ferruginei-Abietetum albae, which is found in the eastern and central parts of the axial Pyrenees, but in subcontinental areas of the central Pyrenees (e.g., Aran, high Pallars) the extensive fir forest Galio rotundifolii-Abietetum albae (= Goodyero-Abietetum) protrudes from the montane belt up to 1900 m. Where the Atlantic influence is stronger, thus mostly in western axial areas, even deciduous Fagetalia forests occur somewhere in the lower subalpine belt, in the form of small spots above the fir forests. Beech forests (Sorbo aucupariae-Fagetum sylvaticae) have been reported up to 1850 m, and birch thickets (Thelypterido limbospermae-Betuletum pubescentis) reach 1950 m. Other deciduous forests that occur sparsely in the central Pyrenees (such as the Roso pendulinae-Aceretum platanoidis) seem to be more dependent on slope dynamics and avalanche events, which favour fast-growing resprouters such as *Sorbus aucuparia, Salix caprea* or *Betula pubescens*, and promote diversity in structure and species composition at the local scale.

On south-facing exposures and in steep or windy areas, irregular or variable snow cover in winter together with occasional drought events in summer give rise to contrasting forest communities. This clearly affects the understorey, which is made up of a xerophilous sub-shrub layer (including *Arctostaphylos uva-ursi, Juniperus communis* subsp. *alpina*, *Cytisus oromediterraneus*, etc.) and also the tree layer, where the same *Pinus uncinata* forms more open, irregular populations. These xerophilous pinewoods mostly correspond to the Arctostaphylo uvaeursi-Pinetum uncinatae, both on siliceous bedrocks and on limestone. Other pinewood types are more closely related to ecological conditions, such as the Veronico officinalis-Pinetum sylvestris pinetosum uncinatae, which covers siliceous, dry slopes in the lower subalpine belt, and the Pulsatillo fontquerii-Pinetum uncinatae, which colonizes moist calcareous soils found on steep, north-facing slopes. In lime-rich areas, poorly characterized calcicolous pinewoods (*Pinus uncinata* and *Festuca gautieri* community) develop where dry conditions prevent soil leaching, e.g., on the south-facing, lower hillsides of the axial range and in most of the pre-Pyrenean ranges.

Associated with subalpine forests, dynamic vegetation units appear following spontaneous clearing or anthropogenic clear-cutting. Tall herbaceous stands of the Epilobietum montano-angustifolii spread fast after disturbance, to be progressively substituted by patchy brambles of the Sambuco racemosi-Rubetum idaei. There, fast-growing deciduous trees, such as *Betula pendula, Salix caprea* or *Sorbus aucuparia* form more long-lasting thickets, which eventually become potential pine forest (Fig. 8.13). However, in continental Iberian valleys such as in Pallars Sobirà, widespread *Betula pendula* forests remain as rather persistent vegetation after repeated intense forest exploitation.

Fig. 8.13 Slopes near the Carlit peak, in the north-eastern Pyrenees. The subalpine belt is covered of mountain pine forest (Rhododendro-Pinetum) up to almost 2400 m, and also of seral *Betula pendula* stands and Nardion grasslands (Photo: A. Ferré)

8.3.2 Non-forest Subalpine Vegetation

Although most subalpine landscapes look rather pristine, they are actually the outcome of millennia of anthropic influence. The adequacy of subalpine grasslands for seasonal grazing led to woodland vanishing through massive wild fires and clear-cutting in the most appropriate range areas. In contrast, steeper relief hosts more diverse landscapes including, along with forest patches on regular hillsides, a wide scope of plant communities related to particular biotopes, such as steep slopes, scree, rocky areas and watersides (Figs. 8.13 and 8.14). All in all, the subalpine landscape is very diverse from one valley to another, and also at the landscape scale, as a result of its physiography and traditional land use (Dupias et al. 1983; Rivas-Martínez et al. 1991; Carrillo and Ninot 1992; Benito 2006, etc.).

The main shrub species found in the understorey of subalpine forests are frequently capable of building up distinct scrub communities, in some cases following disturbances (such as an avalanche or anthropogenic clearings) and in others related to poorly developed substrates (Rivas-Martínez 1968). In south-facing areas, *Juniperus communis* subsp. *alpina* and *Arctostaphylos uva-ursi* make Juniperion nanae low heaths, in places extensively covering irregular rocky slopes. In the lower part of the subalpine belt, these low heaths may include notable

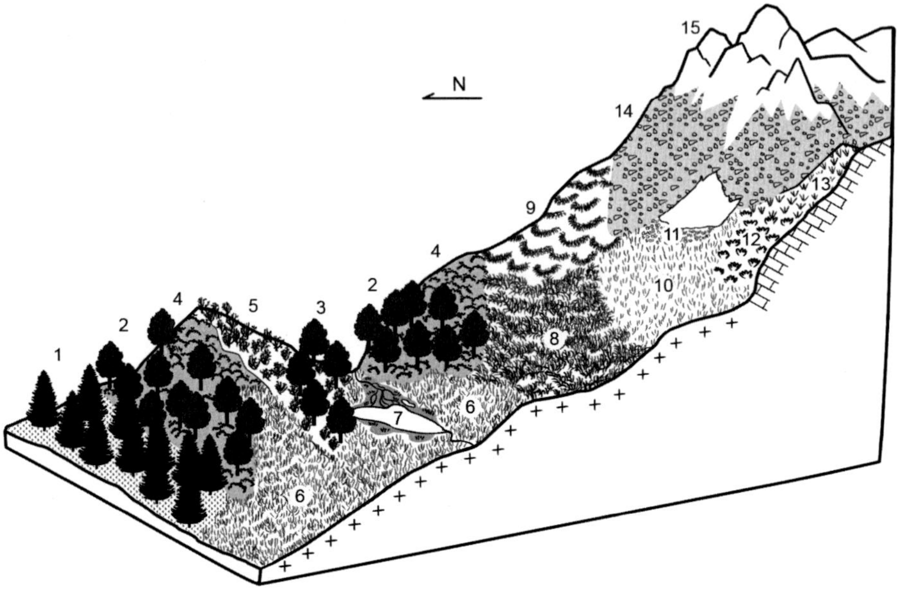

Fig. 8.14 Main potential and seral units in a high mountain landscape (subalpine and alpine belts, ≈ 1700–3000 m) representative of the central Pyrenees. *1*, Rhododendro-Abietetum; *2*, Rhododendro-Pinetum uncinatae; *3*, Arctostaphylo-Pinetum uncinatae; *4*, Rhododendro-Vaccinion; *5*, Juniperion nanae; *6*, Alchemillo-Nardetum; *7*, small lake surrounded of Caricion fuscae; *8*, Selino-Festucetum eskiae; *9*, Festucion eskiae; *10*, Festucion airoidis; *11*, Salicion herbaceae; *12*, Dryado-Salicetum pyrenaicae; *13*, Oxytropido foucadii-Kobresietum myosuroidis; *14*, scree units (Senecionion leucophylli, etc.); *15*, rocky units (Androsacion vandellii, etc.) (Drawing: Laura Fuentes)

calcifuge species (chiefly *Calluna vulgaris* and *Cytisus oromediterraneus*) on siliceous substrates, or shift into *Echinospartum horridum* stands on the extensive limestone of the central Pyrenean and pre-Pyrenean sectors. North-facing scrublands mostly consist of *Rhododendron ferrugineum* stands (Saxifrago geranioidis-Rhododendretum ferruginei), typically related to acidic bedrock, but also found on decalcified soils. Apart from occurring as seral scrub, this association is also widespread on rocky slopes that are inappropriate for forest, and in avalanche corridors.

Most characteristic grasslands that develop on a gentle relief consist of even stands of fine-leaved grass species, capable of dominating through intense tillering, such as *Nardus stricta* and *Festuca nigrescens*. Among these clumps stand sparser populations of other common hemicryptophytes (such as *Alchemilla colorata, Galium verum, Carex caryophyllea* or *Dianthus deltoides*). On acidic bedrock and also on decalcified soils, *Nardus* becomes important and plant assemblages include other calcifuge taxa (such as *Trifolium alpinum, Gentiana acaulis* or *Pilosella lactucella*). This corresponds to the Nardion strictae alliance, which in spite of its ubiquity and moderate species richness from the plot to alliance levels, shows notable regional diversification following ecological variation. In this way, a number of particular associations and subassociations have been described, such as the Alchemillo flabellatae-Nardetum strictae found in most eastern and central sectors, or the Trifolio thalii-Nardetum strictae from western Atlantic valleys.

An interesting Pyrenean feature is the role played by Bromion erecti assemblages in the lower half of the subalpine belt. Mainly on lime-rich substrates and favoured by south-facing mild exposures, a number of generalist Bromion taxa (such as *Plantago media, Cirsium acaule* and *Trifolium montanum*) find there their highest locations. Together with a few altitude differentials (such as *Eryngium bourgatii* or *Daphne cneorum*), they form particular Bromion erecti associations, such as the calcicole Alchemillo flabellatae-Festucetum nigrescentis or the slightly silicicole Chamaespartio sagittalis-Agrostietum tenuis. Higher up, these assemblages make shift on steep slopes to open high mountain grasslands, namely to the calcifuge Festucion eskiae or the calcicolous Festucion scopariae (described in the Alpine Mosaics, below; Fig. 8.14). In the middle of this elevational transition, appropriate locations host more specific subalpine assemblages that stand out due to their species richness and relatively luxuriant aspect. Good examples are the Hieracio hoppeani-Festucetum spadiceae, a tall prairie-like assemblage on rich south-facing acidic soils, and the Dryado-Salicetum pyrenaicae, a sub-shrubby calcicolous grassland from mild north-facing slopes.

The subalpine plant assemblages that colonize cliffs, scree and other rocky areas are roughly the same as those found more extensively at higher elevations, as most plant specialists are widespread through the high mountain. Since these landscape units become more important in the alpine belt, they are better represented in the alpine mosaics. In contrast, water-related areas are here more widespread and diverse than in the alpine belt. Since water courses acquire greater importance as we move further down, hygrophilous vegetation along lower subalpine riversides and on humid foothills may include small willow and birch thickets and noticeable herbaceous communities, such as Juncion acutiflori reed stands or Calthion palustris luxuriant meadows. Following traditional land use, these hygrophilous units have been substituted by Triseto-Polygonion bistortae hay meadows at the most suitable locations. Throughout the subalpine belt and upwards, humid fertile biotopes such as streamsides or wet cliff feet are places for tall herb communities of the Adenostylion alliariae; whereas soils prone to flooding bordering rivulets and lakes bear Scheuchzerio palustris-Caricetea nigrae fens, which in places form complex mire mosaics (see Box 8.4).

Box 8.4 High Mountain Lakes and Mires

Glacial landforms in the high mountain zone include a wide range of biotopes that bear surface waters and promote diverse water-related vegetation, which is more noticeable from mid subalpine to lower alpine elevations. On siliceous bedrock, water bodies are particularly abundant, ranging from small ponds to large alpine lakes. Frequently, their shores and bottom are rocky and abrupt, and thus almost devoid of vegetation. But on gentler relief, the banks of fine-textured substrates bear distinct aquatic or hygrophilous plant communities.

In lakes lying on extensively acidic bedrock, such as the granite at Néouvielle or Maladeta-Aigüestortes or slate in Pallars or Andorra, the water is very poor in mineral contents (Gacia et al. 1994). On sandy and silty bottoms they may support open populations of *Sparganium angustifolium* subsp. *borderei*, in places accompanied by smaller hydrophytes (*Subularia aquatica, Isoetes lacustre, Isoetes echinosporum*; Isoeto-Sparganietum borderei). In not such extremely oligotrophic waters, moderately deep bottoms are a place for the Ranunculo eradicati-Potametum alpinae; an open, partly floating community formed of irregularly branched hydrophytes (*Ranunculus trichophyllus* subsp. *eradicatus, Potamogeton berchtoldii, Potamogeton alpinus, Myriophyllum alterniflorum*).

The gentler lake margins are frequently places for the emergence of *Carex rostrata*, which roots in the shallow bottom and forms almost monospecific communities (Caricetum rostrato-vesicariae, Magnocaricion). These tall, open stands retain incoming sediment, and thus promote primary succession from free water to wetland and to terrestrial vegetation. In a few Pyrenean locations with a particularly humid climate, formations of *Carex rostrata* or *C. lasiocarpa* include a lower carpet of aquatic *Sphagnum*, making semi-floating communities of the Scheuchzeretalia palustris which are the southernmost representations of quaking mires.

The most common Pyrenean mires depend on groundwater: they are geogenous. They can occur as small strips surrounding lakes and ponds on flat shores, but most frequently they border streams. In some flat valley bottoms with braided river systems, they acquire particular importance in the form of complex mosaics of distinct plant communities (Fig. 8.15). These mostly correspond to fens dominated by sedges or other Cyperaceae (*Carex nigra, Carex echinata, Carex davalliana, Eriophorum angustifolium, Scirpus cespitosus*, etc.) and include a lower moss layer particularly well developed in the wettest locations (with *Straminergon stramineum, Warnstorfia exannulata*, etc.). These fens are chiefly formed by the most common taxa of the Scheuchzerio palustris-Caricetea nigrae, where varying conditions in water mineralization and the flooding regime promote noticeable differences in species composition. Thus, contrasting plant communities may be found within areas of a few square meters, depending on the microtopography and water regime (Pérez-Haase 2016).

(continued)

Box 8.4 (continued)

Fig. 8.15 Mire in Aran valley (Pleta de Saboredo, 2180 m). It covers a former glacial lake, and includes *Carex rostrata* stands (*greyish green*) and *Eriophorum angustifolium* and *Carex nigra* formations (*darker*) in the hollows, and extensive *Scirpus cespitosus* fen. In the surrounding granite landscape, irregular patches of *Pinus uncinata* forest, *Rhododendron ferrugineum* scrub and Hieracio-Festucetum spadiceae tall grassland shift upwards to the alpine belt (background), with extensive Festucion eskiae open grassland and rocky vegetation (Photo: A. Ferré)

Where the water is rich in basic ions, mainly calcium, fens correspond to the Caricion davallianae. They are relatively diverse at the community level and between communities occupying distinct environmental locations. *Carex frigida* forms turf fens along rivulets or other flowing water, together with *Saxifraga aizoides* and mosses (*Palustriella falcata, Palustriella commutata*). In sloping soaked soils, *Scirpus cespitosus* forms short stands including calcium indicators such as *Pinguicula grandiflora* or *Primula farinosa*. On the rich soils prone to flooding of the lower subalpine belt, *Carex paniculata* may build up luxuriant fens, in places neighbouring Calthion tall meadows. But the most common calcicolous fen communities are formed by codominant sedges (*Carex davalliana, Carex panicea, Carex lepidocarpa*) and a number of other noteworthy species (*Eriophorum latifolium, Eleocharis quinqueflora, Parnassia palustris, Dactylorhiza majalis*, etc.). They correspond to the Caricetum davallianae, or more rarely to specific related associations.

(continued)

Box 8.4 (continued)

Meanwhile, most fens in oligotrophic acidic waters correspond to the ubiquitous Caricetum fuscae, generally dominated by *Carex nigra* and including sparser *Carex echinata, Eriophorum angustifolium, Juncus filiformis, Viola palustris*, etc. The moss layer is generally noticeable and frequently includes *Sphagnum* populations which promote some peat accumulation. Related communities are some *Scirpus cespitosus* fens (Pediculari mixtae-Scirpetum, Primulo integrifoliae-Scirpetum) which, together with the former fens, may be included in the acidophilous Caricion fuscae alliance.

Under a strong Atlantic influence, *Sphagnum* development may lead to small raised bogs mainly fed by rain water and fog, corresponding to the southernmost ombrogenic Oxycocco palustris-Sphagnetea magellanici (Pérez-Haase et al. 2012). These appear as sparse hummocks, a few decimetres high over flat fens, built up by moss peat and bearing some mesophytes at their top, typically *Vaccinium* sp. pl., *Calluna vulgaris* and poorly developed *Pinus uncinata*, and in a few locations, *Erica tetralix*.

Therefore, the Pyrenean high mountain mire and aquatic plant communities are relatively species poor as peripheral representations of Boreo-Alpine vegetation, and in response to various forms of abiotic stress. But varying ecological conditions at landscape level drive contrasting, well-defined plant assemblages amassing rich hydrophytic flora, which is of great interest in terms of biogeography, ecology and conservation.

8.3.3 Alpine Short Grasslands

Short grassland communities are certainly the most emblematic landscape feature in the alpine belt, and some of these psychro-xerophilous assemblages may be taken as climactic vegetation. They are mainly formed of hemicryptophytes, where one graminoid species generally exerts clear dominance in the form of small detached tussock-like turfs. Typical examples of these engineer species are the calcifuge *Festuca airoides* and *Carex curvula* subsp. *curvula*, and the calcicolous *Kobresia myosuroides* and *Carex curvula* subsp. *rosae*. In between stand other hemicryptophytes (such as the bloomy *Gentiana alpina, Pulsatilla vernalis* or *Aster alpinus*) and small chamaephytes (such as *Minuartia sedoides* or *Silene acaulis*); the latter remaining through winter with very little above-ground structure.

Alpine bioclimate and grassland functioning is related to poorly developed, ranker-like or rendzina-like soils, and is thus mostly reduced to a blackish horizon rich in organic matter. These are relatively nutrient-poor acidic soils on siliceous bedrock and also in some limestone areas, chiefly in the rainiest ones. In other cases, the weakness of leaching and the erosive slope dynamics allow soils originating from lime bedrock to remain calcium-rich and moderately basic (Braun-Blanquet 1948; Nègre 1969).

The species richness is clearly higher on lime-rich soils than on acidic substrates, and characteristic species compositions typify both substrate types. The calcicolous communities correspond to the Carici rupestris-Kobresietea bellardii, whereas calciphobe grasslands are part of the Juncetea trifidi; two vegetation classes that show a Boreo-Alpine distribution. Both grassland groups are singularized in the Pyrenees by notable endemics, such as *Festuca eskia* or *Festuca airoides*, or by Mediterranean orophytes, such as *Festuca gautieri* or *Sideritis hyssopifolia*; the latter are particularly abundant through limestone pre-Pyrenean ranges and in the Iberian axial valleys. These particularities lead to the Pyrenean alpine grasslands being recognized as endemic mid-level vegetation units within the classes mentioned (Braun-Blanquet 1948; Rivas-Martínez et al. 2011, etc.).

Since these alpine grasslands occur neither in areas affected by particular microclimates nor on slopes subject to soil sliding, they cover significant areas only where the alpine landscape includes gentle reliefs, chiefly in the eastern axial sector. From Andorra westwards their landscape occupation decreases as smooth alpine relief becomes scarcer. More abrupt physiography in central and western sectors, with dominant steep slopes and rocky units, limit alpine short grasslands to reduced hilly relief and shelves, thus constraining the theoretical climactic alpine communities to a very secondary role.

In the eastern half of the range, the most extended of these grassland communities on siliceous soils is that formed of *Festuca airoides* (Hieracio breviscapi-Festucetum airoidis), which covers large areas on regular slopes and high periglacial erosion surfaces. It is related to a relatively dry climate through summer, partly due to northern wind events; these are also responsible for winter blizzards making the snowpack irregular, and thus soils prone to low temperatures (Braun-Blanquet 1948; Vigo 1996). The dominant fescue and other typical components such as *Pilosella breviscapa*, *Minuartia recurva* and *Erigeron aragonensis* show some traits related to stress tolerance, such as reduced leaf area or hairy epidermis, which seem coupled with these relatively dry conditions (Fig. 8.16). In the central Pyrenean sector, due to the steeper physiography mentioned and also to a rainier bioclimate, the Hieracio-Festucetum airoidis becomes scarcer compared to the Leontodonto pyrenaicae-Caricetum curvulae. Since this *Carex curvula* subsp. *curvula* association is related to a more regular snowpack, its distribution pattern shows the opposite trend (Gruber 1978; Carrillo and Ninot 1992): it shifts from being more general in the central sector—where, nevertheless, the appropriate relief never covers large areas—to being secluded in small snow-richer patches in the eastern sector. Typical components of this grassland are *Oreochloa disticha* subsp. *blanka*, *Gentiana alpina* and *Leontodon pyrenaicus*, and also a few chionophilous plants such as *Gnaphalium supinum* or *Sibbaldia procumbens*. In spite of these differences, both associations belong to the same Pyrenean acidophilous alliance: Festucion airoidis.

Calcicolous grasslands are frequent in the lime-dominated western half of the main chain, generally covering small areas (Rivas-Martínez et al. 1991; Benito

Fig. 8.16 Extensive alpine grassland of the Hieracio-Festucetum airoidis at the Maià peak ($\approx$ 2500 m), in Andorra (Photo: E. Carrillo)

2006), and are sparser in the eastern Pyrenees. On sloping shelves and other exposed landforms, the stiff turfs of *Kobresia myosuroides* and *Carex curvula* subsp. *rosae* tolerate low temperatures when winter blizzards sweep the snow cover off. They dominate relatively rich plant communities including taxa such as *Antennaria carpatica, Carex atrata* subsp. *nigra*, *Carex rupestris* or *Polygonum viviparum*. This species composition corresponds to the Oxytropido-Kobresion myosuroidis, a clearly defined alliance in which distinct associations have been distinguished, mainly according to the occurrence of particular *Oxytropis* species

(*Oxytropis halleri*, *Oxytropis foucaudii*). In places, *Dryas octopetala* forms notable patches within *Kobresia* communities, apparently related to rocky substrates. Where the snowpack is more persistent and soils more developed, as in gently sloping areas, grasslands become richer and tender, formed by *Festuca nigrescens*, *Trifolium thalii*, *Poa alpina*, *Myosotis alpestris* and others. This is the Festuco-Trifolietum thalii, an association of the Pyrenean endemic alliance Primulion intricatae which shows the highest plant diversity at the plot scale in the Pyrenean alpine belt.

8.3.4 The Alpine Mosaic

As in most mountain systems, growing altitude affords increasing significance to particular relief units and mainly to diversely oriented rocky slopes, compared to flat or gently sloping areas. This section aims to outline the vegetation mosaics that result from such contrasting landforms, which are particularly well-developed in the alpine belt, but also in most higher subalpine areas and in the subnival belt (Fig. 8.14).

Rocky areas become highly important in landscapes made of hard bedrock, such as granite and limestone, and that have been modelled by strong erosive dynamics; foremost in the central and western sectors of the main chain, and in the pre-Pyrenees. Cliffs and other rocky surfaces are primarily places for cryptogamic colonization, and for a few pioneers such as *Poa alpina*, *Helictotrichon sedenense* or *Cystopteris fragilis*, which are found very sparsely in small soil pockets and other interstices. Where outcrop surfaces are more evenly fissured, cliffs host more structured plant assemblages adapted to the particular stressing conditions—chiefly soil scarcity, contrasting temperatures, and drought. The most notable growth-forms are espalier chamaephytes (such as *Rhamnus pumila* or *Globularia repens*) and pulvinular small chamaephytes (such as many species of *Saxifraga*, *Androsace* and *Hieracium*); the latter adopt a highly conservative strategy based on long rosette lifespan and poor vegetative expansion. These biotopes, together with scree habitats, stand out as places for a biogeographically very interesting flora, with intense regional diversification (see Box 8.1). This leads to the distinction of a number of specific fissuricolous plant associations, which are grouped into two ecology-based alliances: the calcicolous Saxifragion mediae and the calciphobe Androsacion vandellii. Important associations in the first group include the Saxifrago iratianae-Potentilletum nivalis (in the western half of the Pyrenees) and the Saxifragetum mediae (in the eastern half); while in the second group, the Androsacetum pyrenaicae and the Artemisio gabriellae-Drabetum subnivalis are also commonly found in western and eastern areas, respectively (Braun-Blanquet 1948; Gruber 1978; Benito 2006, etc.).

Scree and rock fields are well represented in most rocky alpine landscapes, frequently at the foot of cliffs and on steep slopes. These biotopes also pose particular restrictions to plant life, derived from the slope dynamics, scarcity of fine earth, and contrasting microclimate. The vegetation that settles here consists of sparse populations of species with apparent adaptive strategies, such as strong rooting systems, clonality, or greyish leafs—covered by hairs, scales or waxes—tolerant to high radiation. Moreover, in a similar way to the rock vegetation, species show fine-scale geographic diversification, together with ecological specialization into different bedrock types and slope dynamics. Rocky glaciers formed by large pieces are relatively stable, but provide only place for irregular vegetation that can benefit from a few interstices, such as fern communities of the Dryopteridion oreadis found on acidic bedrocks. Finer scree promotes plant assemblages that are more evenly distributed on the sliding substrates, such as the calcicolous Festucetum glaciali-pyrenaicae and Aquilegio pyrenaicae-Bordereetum pyrenaicae in western and central Pyrenees, and Iberidetum spathulatae in the eastern areas; or the calcifuge Galeopsio pyrenaicae-Poetum fontqueri, widespread, and Senecionetum leucophylli (with the endemics *Senecio leucophyllus* and *Cerastium pyrenaicum*), east-Pyrenean.

On slopes with more evolved soils, some tussock-like grass species are particularly successful in building extensive open grasslands, most typically on south-facing exposures. There, the spring thawing cycles lead to soil subtly sliding down the slope which, combined with strong rooting of the grass turfs, results in a combination of small gentle and steeper surfaces over and down from each turf, respectively. From the community to large landscape scales, this slope dynamics produces a staircase-like relief, where turfs of hard spiny leafs acquire wide half-moon shapes, alternating with parallel open ground (Nègre 1969). The most general grass species playing this role are the Pyrenean endemic *Festuca eskia* on acidic materials, and *Festuca gautieri* on lime-rich substrates. Festucion eskiae communities include sparse calcifuge hemicryptophytes, such as *Luzula pediformis, Carex sempervirens* subsp. *pseudotristis* or *Campanula scheuchzeri*, together with occasional opportunists (such as *Pilosella* sp. pl. or *Silene rupestris*). Festucion scopariae grasslands are generally species-richer, including notable orophytes, both Alpine and Mediterranean, such as *Ononis cristata, Scutellaria alpina, Helianthemum oelandicum* subsp. *alpestre* or *Vicia pyrenaica*. These calcicolous grasslands extend downwards through the subalpine belt, and become important in deforested areas of the axial and pre-Pyrenean ranges. This spatial dispersion of the alliance over varying ecological sites, together with high species richness, leads to the distinction of a number of associations, from the Seslerio-Festucetum scopariae in most of the eastern and central axial Pyrenees, to the Oxytropido pyrenaicae-Festucetum scopariae in the westernmost valleys.

In the opposite ecological situation from the sliding slopes, valley bottoms and gently north-facing exposures accumulating fair snowpack support dense grasslands related to stable, moist substrates. In these acidic soils that mostly occur on siliceous bedrock, but also on washed lime-related substrates, *Nardus stricta* and *Festuca eskia* can form continuous swards, together with *Trifolium alpinum, Ranunculus pyrenaeus* and *Crocus vernus*. Most of these communities fit in the Selino pyrenaei-Festucetum eskiae, a Nardion association found throughout the whole of the axial Pyrenees, and particularly important in the landscapes of the central and western sectors, under an Atlantic influence (Nègre 1969; Carrillo and Ninot 1992). Where snow accumulation implies a noticeable delay in the melting calendar, Nardion grasslands appear less dense and host some chionophilous specialists, such as *Alopecurus alpinus*, *Plantago alpina* or *Carex pyrenaica* (Trifolio alpini-Phleetum gerardii), and progressively shift into typical Salicetea herbaceae assemblages towards the snowbeds (see Box 8.5). Other specific vegetation units occur related to surface waters. In a sequence from regular slopes to rivulets or lakes in the valley bottoms, increasing soil moisture promotes hygrophilous Nardion swards (Selino pyrenaei-Nardetum), mire communities (Scheuchzerio-Caricetea) where the water table approaches the soil surface, and eventually aquatic assemblages in standing water (see Box 8.4).

Box 8.5 Dwarf Plant Communities Secluded in Pyrenean Snowbeds

The Pyrenees is one of the southernmost European mountain systems where the snowpack may remain until mid summer in favourable locations of the high mountain: the snowbeds. The long-lasting snow is clearly a limiting factor for vegetation as it shortens the growing period; but it is also advantageous as protection against freezing, since temperatures under the snowpack remain close to zero throughout winter. Moreover, the delayed sprouting of snowbed plants occurs during the summer thermal regime (Fig. 8.17).

The gradual melting of the snowpack imposes a phenological gradient on snowbed vegetation, from the alpine grasslands covering regular slopes to special, chionophilous communities that colonize snowbeds. Specialist plants may accomplish their reproductive cycle there within 2–3 months: from blooming early after sprouting to seed setting between late August and September (Lluent et al. 2013). Incoming radiation during the growing period is also a key driver for snowbed vegetation, given the short time that plants have for growth and reproduction. Snowbeds lying on north-facing exposures, and especially those located under steep slopes or cliffs, are more restrictive than those on sunny exposures.

The same constraints that drive snowbed vegetation also hamper soil processes and evolution. In fact, some of the Pyrenean snowbeds lie on rocky surfaces or block fields, and are thus almost devoid of vegetation. Where the bedrock is more fractured, scree and gravel patches are suitable for species such as *Alchemilla fissa* or *Saxifraga praetermissa*, adapted to short summers and also to unstable, poorly evolved substrates. Even on gentler surfaces, the soils are frequently coarse-grained and poorly developed, although able to sustain typical snowbed communities. In most cases these cover small areas, just a few square metres, and are made up of small plants rarely taller than 2–3 cm, such as the chamaephyte *Salix herbacea* or several hemicryptophytes.

(continued)

Box 8.5 (continued)

Fig. 8.17 Muntanyó d'Àrreu of Pallars Sobirà, in the central Iberian Pyrenees, where the relief of acidic bedrock promotes a diverse alpine mosaic, including one small lake (Pudo, ≈ 2500 m) with hydrophytic flora, short grassland, low scrubs, rocky areas, and a number of snowbeds (Photo: E. Carrillo)

The Pyrenean snowbed communities are closely related to those of other Alpine systems, which altogether are included in the class Salicetea herbaceae, where the main bedrock type conditions calcifuge or calcicolous assemblages: Salicion herbaceae or Arabidion caeruleae, respectively (Braun-Blanquet 1948; Rivas-Martínez 1969).

In siliceous areas, prevalent in the eastern and central Pyrenees, the most typical snowbed communities (Gnaphalio-Sedetum candollei, Carici-Cardaminetum alpinae) are short, open carpets made by *Cardamine bellidifolia* subsp. *alpina, Veronica alpina, Carex pyrenaica, Gnaphalium supinum, Sibbaldia procumbens*, etc. A singularity of this vegetation is the occurrence of a few tiny succulents, mainly *Sedum alpestre* and *Sedum candollei*. The latter is an Iberian orophyte capable of growing in dense populations on sandy soils. Most of these plants show good colonizing capacity and form permanent seed banks, which ensure their persistence in years of very poor seed production (Lluent et al. 2013; Ninot et al. 2013b).

(continued)

Box 8.5 (continued)

Small surfaces of open, wet soil are a place for chionophilous bryophytes (such as *Polytrichum sexangulare, Polytrichum alpinum* or *Brachythecium glaciale*), which form tiny communities (Polytrichetum norvegici) where the snowpack melts later. In contrast, the peripheral part of the snowbeds, with more developed soil, is the optimal place for *Salix herbacea*, which makes dense carpets on fine-textured soils. Together with some of the plants mentioned above and in places with the tiny carpets of the liverwort *Anthelia juratzkana*, it forms one of the most characteristic associations: Anthelio-Salicetum herbaceae.

On lime-rich substrates, the peripheral parts of the snowbeds are often covered by the creeping chamaephytes *Salix reticulata* or *Salix retusa* (or a mixture of them). They colonize irregular, rocky substrates affected by gelifluction and sliding, thanks to strong rhizome and root systems. The small soil pockets in between the willows may bear *Carex atrata* subsp. *nigra, Veronica aphylla, Sedum atratum*, etc., forming the Carici parviflorae-Salicetum retusae. Gentler slopes, with better soils and a more regular snowpack, is the place for the Potentillo dubiae-Gnaphalietum hoppeani. This is found throughout the Pyrenees, but appears scarce, since even on lime-rich bedrock, more developed soils are the place for calcifuge plant assemblages, chiefly the *Salix herbacea* carpets.

Snowbed plant assemblages find in the Pyrenees their southernmost habitats and, facing climate change, they seem certainly doomed to vanish. Recent analyses of their functioning and ongoing detailed monitoring reveal shifts in their microtopographical distribution and species composition. In the case of *Salix herbacea*, the capacity to resprout from woody rhizomes that act as storage systems provides it notable stability in front of interannual climatic variations. But there is still poor knowledge on the capacity of snowbed specialists to overcome environmental changes.

Shrubby units appear in the alpine belt as extreme woody vegetation at particular locations, mainly on irregular rocky relief, somewhere extending upwards from the subalpine belt. In fact, most of these communities belong to the same associations found further downhill, such as the Saxifrago geranioidis-Rhododendretum ferruginei which occurs, further up than the related pinewoods, on acidic north-facing slopes with continuous winter snowpack. In the central Atlantic Pyrenees, this ubiquitous scrub shifts in places to lower heaths with *Empetrum nigrum* subsp. *hermaphroditum* (Empetro-Vaccinietum). Also in north-facing acidic areas, but where winter blizzards sweep the snow cover away, *Loiseleuria procumbens* together with terricolous lichens and other stress-tolerant plants form open dwarf shrub carpets (Cetrario-Loiseleurietum). Lime-rich northern exposures may support low heaths formed of *Dryas octopetala, Salix pyrenaica* or *Arctostaphylos alpinus* (Dryado octopetalae-Salicetum pyrenaicae). In south-facing areas, woody communities are less diversified, although they occur elsewhere and reach 2600 m on protected rocky shelves. They mostly correspond to the Juniperion nanae, frequently dominated by *Arctostaphylos uva-ursi* or by *Juniperus communis* subsp. *alpina*.

8.3.5 The Subnival Belt

On the highest alpine ridges and summits, even the scarce gentle surfaces offer poor conditions for the development of vegetation, thus grasslands shift to open assemblages of plants that tolerate soil poorness, wind erosion, and contrasting temperatures and moisture. The uppermost Festucion airoidis community, the Saxifrago bryoidis-Minuartietum sedoidis, is formed of small creeping and cushion-shaped chamaephytes, and a few clumped hemicryptophytes (such as *Potentilla frigida* or *Festuca borderei*). But from 2800 m upwards, plant assemblages are even looser and may be singularized as an endemic alliance, Androsacion ciliatae, which defines a subnival (or uppermost alpine) belt, as a Pyrenean vicariant of the Androsacion alpine alliance from the Alps (Rivas-Martínez et al. 1991). It includes the calcicole Minuartio cerastiifoliae-Androsacetum ciliatae and the calciphobe Minuartio sedoidis-Androsacetum ciliatae. This very sparse vegetation, together with the predominant rock and scree supporting sparse plant assemblages, occurs next to the last remnants of the Pyrenean glaciers, secluded on the northern sides of the highest massifs of the range (Fig. 8.18).

Fig. 8.18 Gavarnie cirque, on the northern side of central Pyrenees, reaching 3251 m at the Marboré peak. Alpine grasslands are reduced to relatively gentler slopes whereas most of the landscape is place for rock or scree vegetation, including the subnival Minuartio-Androsacetum ciliatae found at the higher elevations, and patches of glacier remnants (Photo: Jordi Carreras)

8.4 Changing Landscapes

Pyrenean vegetation has been experiencing noticeable changes since the mid twentieth century, related to the dramatic cessation of so-called traditional land use, mostly at lower and mid altitudes. These changes are mostly encroachment processes which produce a generalized expansion of scrubland and then young forests, such as dense pinewoods or deciduous coppices (García-Ruiz et al. 1996; Monje 2004; Ferré et al. 2013), and thus illustrate secondary succession along different dynamic series. Meanwhile, the most suitable areas see an intensification of agriculture and livestock.

It has frequently been assumed that the most intense land use occurred between the Middle Ages and the nineteenth century, coinciding with the maximum human habitation of the Pyrenees. However, growing evidence suggests that greater changes in the vegetation occurred much earlier; mostly during the Neolithic period (Pèlachs et al. 2011; Orengo et al. 2014). In the early Neolithic (seven to eight millennia ago) land use was generalized, not only at mid elevations but also up to the lower alpine level. Most of this early occupation apparently consisted of seasonal use of the mountain, which included forest clearing and herding. The activity of middle Neolithic people slowly turned into more permanent occupancy, and they even settled in the subalpine belt, which brought with it greater deforestation, more grazing, and increased cultivation pressure. Between the late Neolithic and Romanization, coinciding with the Holocene thermal optimum, agro-pastoral use became more intensive, causing permanent landscape changes: the upper subalpine areas definitively changed to open habitats and eutrophication of lakes and valley bottoms reached extreme levels. The Middle Ages brought growing intensification linked to the extension of transhumance herding and feudal organization. While the Little Ice Age forced some retreat of agriculture from the subalpine belt downwards, herding and transhumance continued to grow.

In parallel with the landscape changes produced directly by human activity, other major vegetation shifts could be driven by different factors such as climatic oscillations or the slow rate of plant migration and vegetation dynamics in complex mountain areas. One example is the varying role played by *Abies alba* forests, which started a major expansion some 7000 years ago, reached varying degrees of dominance, and has decreased with ups and downs since the end of the Roman Empire. In the Pallars area, *Betula* was the dominant tree in the subalpine belt for millennia, but it declined some 5000 years ago to its present subdominant role (Pèlachs et al. 2011). During such landscape changes, anthropic activity must have influenced the distribution and relevance of generalist species and plant assemblages, such as semi-natural hay meadows or weeds. The introduction or facilitation of alien plants is a more recent and noticeable human influence, particularly expressed at the lower elevations. Therefore, the changes experienced by Pyrenean vegetation over the last millennia are a combined expression of spontaneous processes and of human activity, since both drivers operated at the same time scale.

In the last decades, there has been growing interest in understanding and predicting ongoing changes in vegetation. Some seem to obey complex causes, as in the case of the sharp decay of the *Abies alba* forests in the western Iberian Pyrenees. Nowadays, increased fir biomass after much logging in previous centuries, combined with more important drought events synergistically enhance the vulnerability of fir, and could bring about its decline (Camarero et al. 2011). Other cases of forest decline are due to the growing incidence of pests (such as the pine processionary, *Thaumetopoea pityocampa*, which severely affects *Pinus nigra* and *Pinus sylvestris* in places) or to the combined effects of different components of global change (drought events, high spring temperatures, atmospheric ozone levels, etc.). The response of forests to such decay processes is not easy to predict, as in the case for other spontaneous changes in vegetation such as pasture encroachment by dominant shrubs (e.g., *Juniperus communis, Echinospartum horridum, Buxus sempervirens* or *Cytisus oromediterraneus*) in different Pyrenean landscapes.

The subalpine timberline is expected to move upwards after decades of marked decline in grazing and logging, combined with rising temperatures. However, this response has only been observed in the areas where the subalpine forest had been lowered the most, whereas forests near the potential timberline have experienced only tree densification, with very little or no displacement (Batllori et al. 2009; Ameztegui et al. 2016). Together with the inertia expected in high mountain succession, this seems to be related to climatic irregularity and meteorological events affecting the settlement of young pines further upwards.

Changes in alpine vegetation are expected to be less apparent in general. However, particular units such as mires or snowbed communities may decrease or disappear from their most extreme locations. Although plant species in these habitats show the capacity to overcome years of lower water or snow supply, the response of specialized assemblages to longer or more consistent climate change is unknown. Less specialized vegetation also seems to be subject to variations in its species composition. According to the Global Observation Research Initiative in Alpine Environments, species from lower elevations are slowly infiltrating summit assemblages in most European Alpine mountains, including the Pyrenees (Gottfried et al. 2012). This may be seen as an ongoing plant colonization upwards that eventually reaches the highest summits, which would be mediated by global change drivers.

References

Aeschimann D, Rasolofo N, Theurillat J-P (2011) Analyse de la flore des Alpes. 2: biodiversité et chorologie. Candollea 66(2):225–253

Ameztegui A, Coll L, Brotons L, Ninot JM (2016) Land-use legacies rather than climate change are driving the recent upward shift of the mountain tree line in the Pyrenees. Glob Ecol Biogeogr 25(3):263–273

Batllori E, Blanco-Moreno JM, Ninot JM, Gutiérrez E, Carrillo E (2009) Vegetation patterns at the alpine treeline ecotone: the influence of tree cover on abrupt changes in species composition of alpine communities. J Veg Sci 20:814–825

Benito JL (2006) Vegetación del Parque Nacional de Ordesa y Monte Perdido (Sobrarbe, Pirineo central aragonés). Consejo de Protección de la Naturaleza de Aragón, Zaragoza

Bolòs O (1983) La vegetació del Montseny. Diputació de Barcelona, Servei de Parcs Naturals. Barcelona

Bolòs O, Vigo J, Carreras J (2004) Mapa de la vegetació potencial de Catalunya 1:250.000. I.E.C. & Universitat de Barcelona, Barcelona

Braun-Blanquet J (1948) La végétation alpine des Pyreneés Orientales. Mon Est Est Pir e Inst Esp Edaf Ecol Fisiol Veg 9:1–306

Camarero JJ, Bigler C, Linares JC, Gil-Pelegrín E (2011) Synergistic effects of past historical logging and drought on the decline of Pyrenean silver fir forests. For Ecol Manag 262:759–769

Carreras J, Carrillo E, Masalles RM, Ninot JM, Soriano I, Vigo J (1996) Delimitation of the supra-forest zone in the Catalan Pyrenees. Bull Soc Linn Provence 47:27–36

Carrillo E, Ninot J M (1992) Flora i vegetació de les valls d'Espot i de Boí. Arxius Sec. Cièn. 99 (1 & 2):1–474 & 1–350

Dupias G (1985) Végétation des Pyrénées : notice détaillée de la partie pyrénéenne des feuilles 69 Bayonne – 70 Tarbes – 71 Toulouse – 72 Carcassonne – 76 Luz – 77 Foix – 78 Perpignan. Service de la carte de végétation de la France, Paris

Dupias G, Montserrat P, Izard M (1983) Carte de la Végétation de la France. Feuille n° 76 – Luz. 1: 200.000. C. N. R. S, Toulouse

Ferré A, Ninot JM, Batllori E, Carreras J, Carrillo E, Monje X, Vigo J (2013) From vegetation mapping to a prediction for landscape evolution in the Catalan Pyrenees. Colloq Phytosociologiques 29:227–235

Gacia E, Ballesteros E, Camarero L, Delgado O, Palau A, Riera JL, Catalán J (1994) Macrophytes from lakes in the eastern Pyrenees: community composition and ordination in relation to environmental factors. Freshw Biol 32:73–81

García MB, Gómez D (2007) Flora del Pirineo Aragonés. Patrones espaciales de biodiversidad y su relevancia para la conservación. Pirineos 162:71–88

García-Ruiz JM, Lasanta T, Ruiz-Flano P, Ortigosa L, White S, González C, Martí C (1996) Land-use changes and sustainable development in mountain areas: a case study in the Spanish Pyrenees. Landsc Ecol 11(5):267–277

Gaussen H (1964) Carte de Végétation de la France. N° 77 Foix. 1: 200.000. C.N.R.S, Toulouse

Gottfried M et al (2012) Continent-wide response of mountain vegetation to climate change. Nat Clim Chang 2:111–115

Grau O, Ninot JM, Ferré A, Font X, Grytnes J-A (2012) Altitudinal species richness patterns of vascular plants in the south-eastern Pyrenees and nearby mountains of Catalonia. Plant Ecol Divers 5(1):115–126

Gruber M (1978) La végétation des Pyrénées ariégeoises et catalanes occidentales. Thèse, Fac. Sc. Techn. St. Jerôme. Université Aix-Marseille, III.

Körner C, Paulsen J (2004) A world-wide study of high altitude treeline temperatures. J Biogeogr 31:713–732

Lluent A, Anadon-Rosell A, Ninot JM, Grau O, Carrillo E (2013) Phenology and seed setting success of snowbed plant species in contrasting snowmelt regimes in the Central Pyrenees. Flora 208:220–231

Loidi J, Báscones JC (1995) Memoria del mapa de series de vegetación de Navarra, E. 1:200.000. Gobierno de Navarra, Pamplona

Loidi J, Campos JA, Herrera M, Biurrun I, García-Mijangos I, García-Baquero G (2015) Eco-geographical factors affecting richness and phylogenetic diversity patterns of high-mountain flora in the Iberian Peninsula. Alp Bot 125(2):137–146

Melendo M, Giménez E, Cano E, Gómez-Mercado F, Valle F (2003) The endemic flora in the south of the Iberian Peninsula: taxonomic composition, biological spectrum, pollination, reproductive mode and dispersal. Flora 198:260–276

Monje X (2004) Estudio mediante sistemas de información geográfica (GIS) de la evolución de la vegetación natural en Cataluña. Aproximación al caso del Valle del Freser durante el periodo 1956–2000. Acta Bot Barcin 49:421–438

Nègre R (1969) La végétation du Bassin de l'One (Pyrénées centrales), deuxième note: les pelouses. Portug Acta Biol 10(1–4):1–135
Ninot JM, Carrillo E, Font X, Carreras J, Ferré A, Masalles RM, Soriano I, Vigo J (2007) Altitude zonation in the Pyrenees. A geobotanic interpretation. Phytocoenologia 37:371–398
Ninot JM, Ferré A, Grau O, Font X, Pérez-Haase A, Carrillo E (2013a) Environmental drivers and plant species diversity in the Catalan and Andorran Pyrenees. Lazaroa 34:89–105
Ninot JM, Grau O, Carrillo E, Guàrdia R, Lluent A, Illa E (2013b) Functional plant traits and species assemblage in Pyrenean snowbeds. Folia Geobot 48:23–38
Orengo HA, Palet JM, Ejarque A, Miras H, Riera S (2014) Shifting occupation dynamics in the Madriu–Perafita–Claror valleys (Andorra) from the early Neolithic to the Chalcolithic: the onset of high mountain cultural landscapes. Quat Int 353:140–152
Pèlachs A, Julià R, Pérez-Obiol R, Soriano JM, Bal MC, Cunill R, Catalán J (2011) Potential influence of bond events on mid-Holocene climate and vegetation in southern Pyrenees as assessed from Burg lake LOI and pollen records. The Holocene 21(1):95–104
Pérez-Haase A (2016) Patrons estructurals, ecològics i biogeogràfics en vegetació de molleres i de torberes d'esfagnes. PhD thesis, University of Barcelona
Pérez-Haase A, Carrillo E, Batriu E, Ninot JM (2012) Diversitat de comunitats vegetals a les molleres de la Vall d'Aran (Pirineus centrals). Acta Bot Barc 53:61–112
Rivas-Martínez S (1968) Estudio fitosociológico de los bosques y matorrales pirenaicos del piso subalpino. Publ Inst Biol Aplicada Barcelona 44:5–44
Rivas-Martínez S (1969) Las comunidades de los ventisqueros (Salicetea herbaceae) del Pirineo central. Vegetatio 17:232–250
Rivas-Martínez S, Báscones JC, Díaz González TE, Fernández-González F, Loidi J (1991) La vegetación del Pirineo Occidental y Navarra. Itinera Geobotanica 5:5–456
Rivas-Martínez S et al (2007) Mapa de series, geoseries y geopermaseries de vegetación de España (Memoria del mapa de vegetación potencial de España). Parte I. Itinera Geobotanica 17:5–435
Rivas-Martínez S et al (2011) Mapa de series, geoseries y geopermaseries de vegetación de España (Memoria del mapa de vegetación potencial de España). Parte II. Itinera Geobotanica 18:5–800
Vigo J (1979) Les forêts de conifères des Pyrénées catalans. Essai de révision phytocénologique. Documents Phytosociologiques N.S. 4:929–941
Vigo J (1996) El poblament vegetal de la Vall de Ribes: 19–442. Les comunitats vegetals i el paisatge. Mapa de vegetació 1: 50.000. Institut Cartogràfic de Catalunya, Barcelona
Villar L, Sesé JA, Ferrández JV (1997–2001) Atlas de la flora del Pirineo aragonés, 2 vols. Instituto de Estudios Altoaragoneses & Cons. Protec. Nat. Aragón, Zaragoza

Chapter 9
Trás-os-Montes and Beira Alta

Carlos Aguiar and Carlos Vila-Viçosa

Abstract The Trás-os-Montes and Beira Alta region *grosso modo* includes the northeastern quadrant of Portugal as far as the Mondego River Platform, enlarged further north by neighbouring Spanish territories. Acid and phosphorus-poor palaeozoic schists intruded by variscan ganitoids are the prevailing lithological types. Its relief is characterized by the widespread occurrence of uplifted planation surfaces, the majority of which are between 600 and 900 m in altitude, dissected by deep river valleys, sometimes interrupted by larger tectonic basins. The supratemperate or supramediterranean, mainly sub-humid to humid, plateaus are the climactic domain of climatophilous, mesophilous, deciduous mesoforest of *Quercus pyrenaica*. Old forests are not only species-richer, but share complex, forest-dependent vegetation mosaics. On the Mondego platform prevails a Potential Natural Vegetation of *Q. robur* subsp. *broteroana* forests. The mesomediterranean upper dry to lower sub-humid valleys and basins harbour the most remarkable forests of the study area: one association of semi-deciduous forests of *Q. faginea* subsp. *faginea*, and six associations of perennial oak woodlands, three of them dominated by *Q. suber*, and the other three by *Q. rotundifolia*. Heathlands and gorse heathlands are the most conspicuous substitution stages of *Q. pyrenaica* and *Q. robur* subsp. *broteroana* Querco-Fagetea sylvaticae woodlands; *Cistus* shrublands are seral of Quercetea ilicis evergreen and semi-deciduous woodlands. Also diverse and with a clear-cut bioclimatic control are riparian and tempori-hygrophilous forests. The geobotanical interpretation of regional grassland vegetation complexes requires particular caution. They include several oligotrophic grassland types, dominated by therophytes, *Agrostis* sp. pl. or *Nardus stricta*, among other species. Mesoscale physiography, tree shading, grazing and hay cutting have a strong effect on hay-meadow vegetation complexes. Recent changes in hay-meadow management promote grass abundance and the extension of

C. Aguiar (✉)
CIMO-Centro de Investigação de Montanha, Escola Superior Agrária de Bragança, Aptdo 1172, 5301-855 Bragança, Portugal
e-mail: cfaguiar@ipb.pt

C. Vila-Viçosa
CIBIO – Centro de Investigação em Biodiversidade e Recursos Genéticos/InBIO Laboratório Associado, Universidade do Porto, 4485-661 Vairão, Portugal
e-mail: cvv@cibio.up.pt

J. Loidi (ed.), *The Vegetation of the Iberian Peninsula*, Plant and Vegetation 12,
DOI 10.1007/978-3-319-54784-8_9

Arrhenatherion grasslands. The mafic and ultramafic metamorphic rocks are, simultaneously, the most singular lithologically, and the habitat of the most remarkable phytocoenosis of the study area. Herbaceous weed, ruderal and road vegetation types reflect the widespread oligotrophy of the regional soils. This chapter ends with a brief resumé of the series, and geoseries of the Trás-os-Montes and Beira Alta.

9.1 Introduction

The Trás-os-Montes and Beira Alta region, the study area of this chapter, is located in the NW quadrant of the Iberian Peninsula. *Grosso modo* it covers the territory of two former administrative provinces – Tras-os-Montes and Beira Alta – both situated in the northeastern corner of Portugal, enlarged further north by neighbouring Spanish territories. The southern slopes of the Galician Massif, still in Spanish territory, form its northern border. The canyon of the River Douro is, simultaneously, an outstanding geomorphological feature and an administrative border between the Spanish autonomous community of Castilla y León and the Portuguese Trás-os-Montes region. However, the boundary with the *The Douro Basin* area (see Chap. 10) dissects obliquely the Douro canyon along the border between the climatophilous forests of *Quercus suber* and *Quercus rotundifolia*. Further south, the Trás-os-Montes and Beira Alta regional boundary turns southwest through the eastern hills of the Trancoso Mountain, the eastern mountains of the Beira-Duriese (port. *Beira-Duriense*) Mountain System, as far as the village of Sobral da Serra (in the Guarda administrative unit), at the upstream limit of the Mondego River Platform. From there the frontier runs along the southern margin of the Mondego River Platform near the Serra da Estrêla (see Chap. 13). It crosses the Mondego River valley in the Penacova administrative unit, not far from the city of Coimbra. From the Serra of Buçaco northwards, the study area follows a long frontier with the *lowlands and midlands of Northwestern Atlantic Iberia* (see Chap. 6). After a sharp inflection eastwards through the valley of the Tedo River, and again in the direction of the sea due to the strong Mediterranean influence of the Douro Valley, the two territories are separated by the Galaico-Portuguese Mountains. From south to the north, this mountain system includes the mountains of Aboboreira, Marão, Alvão, Alturas do Barroso and Larouco. The strong impact of the Galaico-Portuguese Mountains on accessibility, climate and landscape explains the designation Trás-os-Montes ('behind the mountains' in English). Regarding the recently published biogeographical typology for the Iberian Peninsula of Rivas-Martínez et al. (2014), most of the study area is part of the Duriese-Lusitanian biogeographical sector, the nortwestern territory of the Mediterranean Region in the Iberian Peninsula. The northeastern plateaus (v.i.), in the Bragança administrative unit, belong to the Bercianese-Sanabrese sector, also in the Mediterranean Region. The Mondego Platform is a penetration into the east of the so-called Montemuro and Estrêla Sierran Sector (Eursosiberian Region) (Fig. 9.1).

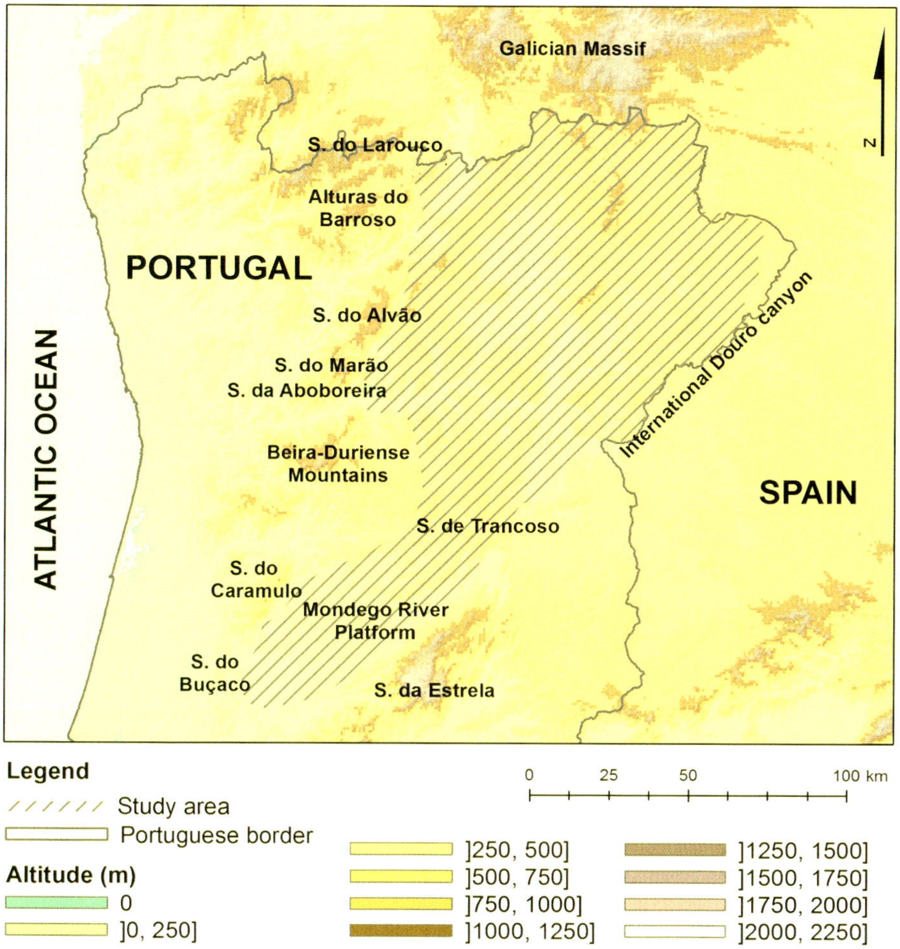

Fig. 9.1 Geographical location of the Trás-os-Montes and Beira Alta region (in the Portuguese Territory)

The study area is positioned at the heart of the Iberian Massif, one of the fragments of the Hercynian Socle. Palaeozoic schists intruded by variscan ganitoids are the prevailing lithological types. Erosion-resistant quartzite outcrops are important landscape features. This monotony – a landscape of siliceous, acid and phosphorus-poor rocks – is interrupted in the Bragança-Vinhais and Morais allochthonous massifs by a complex assortment of felsic, mafic and ultramafic rocks, with varying degrees of metamorphism. As later explained, mafic and ultramafic rocks have a strong impact on plant species assemblage, vegetation dynamics and land use. In the Miranda Plateau, in the northeastern part of the study area, and elsewhere, there are patches of the sediments that fill the Cenozoic Douro Basin. Sometimes hard to find, small pockets of crystalline limestone are the refuge of an interesting chalk flora.

The relief of Trás-os-Montes and Beira Alta is characterized by the widespread occurrence of uplifted planation surfaces dissected by deep river valleys. The majority of these surfaces vary between 600 and 900 m in altitude and are genetically related to the Castilla la Vieja peneplain. As emphasized by Birot (1950), "The Portuguese relief decomposes in more or less large peneplain fragments, more or less high, more or less cut by valleys. In Portugal almost all topographic accidents are scarps that separate peneplains at different altitudes". The genesis of the actual physiography of the Trás-os-Montes and Beira Alta region started with a differential uplift of ancient peneplain blocks through the reactivation of inherited Palaeozoic Variscan faults during the Cenozoic Alpine Orogeny. The same fault system led also to the formation of tectonic depressions (grabens). Since the Pliocene, these primitive peneplain blocks have been dissected by an Atlantic drainage system (Pais 2012). The compressional Alpine Tectonics, due to the N-S convergence between the African and Eurasian Plates, produced the intraplate mountain systems that surround the study area in the north (Galician Massif), west (Galaico-Portuguese Mountains) and south (Central System) (Gutiérrez et al. 2014). Towards the east, the study area touches the border of the Cenozoic Douro Basin. The mountains included in the study area reach just 1486 m in altitude in the Serra of Montesinho, in Portugal, and 1732 m in the Serra de la Tejera in Spain.

Preserved sizable peneplain stretches are correlated with resistance of granitic rocks to erosion, or with the distance to the general basal level, defined by the Douro or Mondego Rivers thalwegs (e.g. plateaus of Miranda do Douro and Beira Alta). Some supratemperate humid plateaus bear a few periglacial terrestrialized lagoons (e.g. Lama Grande, Montesinho Mountain). The drainage networks have a marked structural control in the study area. In the north of the study area they are mostly hierarchized by the Douro River, and further south by the Mondego River, the largest Portuguese-sourced river. The Douro and Mondego Rivers leave the study area at about 50 m above sea level. The majority of the permanent watercourses are turbulent due to marked gradients, and traverse narrow V shaped steep valleys filled with slender longitudinal deposits of colluvial materials. Basins and open valleys (e.g. the Bragança and Mirandela basin and Chaves, Vilariça and Verín valleys) are interpreted as grabens, and are frequently associated with demihorsts (e.g. Nogueira and Bornes Mountains). Basins and open valleys exhibit large sedimentary deposits, with a colluvial or fluvial origin. Some schist areas, for example in the Mirandela Tectonic Basin, show a prematurely aging landscape with scattered hills lowered by erosion, cut through by a chaotic drainage network. The Mondego Platform, at the southern extent of the study area, is a granitic planation surface built in the Mountains of Montemuro and the Estrêla Mountains, crossed by the rivers Mondego and Dão, with a maximum width of 30 km and mean altitudes varying between 200 and 500 m.

The rainfall that irrigates the territory is almost always of frontal origin. The fronts cross the Peninsula from the NW to the SE and are forced to climb the Galaico-Portuguese mountains. A pronounced Foehn effect causes a reduction in rainfall towards the east. In a transect of 100 km the precipitations falls from ca. 3000 mm/year in the Serra do Gerês (ultra-hyperhumid supratemperate bioclimate), in the Galaico Portuguese Sector, to less than 400 mm/year, in the Douro Valley, where the Douro River enters Portuguese territory (lower dry lower mesomediterranean bioclimate). The above-mentioned peneplain stretches are mainly positioned in the lower supratemperate humid or lower supramediterranean humid belts, descending to the sub-humid ombroclimate eastwards in the Planalto de Miranda (Fig. 9.2). The borders of the peneplain stretches usually descend to the mesomediterranean sub-humid belt. The highest mountains of the study area have a distinct hyperhumid supratemperate plain summit, never touching the orotemperate belts. The Duriese-Lusitanian basins and valleys are mesomediterranean, upper dry to sub-humid. The supratemperate and supramediterranean belts are commonly known as 'Cold Land' (port. *Terra Fria*) (Fig. 9.3); the mesomediterranean basins and valleys are called 'Hot Land' (port. *Terra Quente*) (Fig. 9.4). In the Mondego Platform there is a gradual northeast-southwest transition between the upper and the lower mesomediterranean belts. The Mondego Platform is not closed to the sea by high plateaus and for that reason there it rains more than in the Duriese-Lusitanian valleys, and the ombroclimate is lower humid.

9.2 Forests and Woodlands

The idea of a continuous forest cover of *Quercus* species in the first half of the Holocene, before the onset of the anthropic deforestation, is nowadays widely rejected for the Iberian Peninsula. Besides the acceptance of the natural coexistence of diverse seral stages in a "shifting mosaic" (Bormann and Likens 1979), it is also probable that some summital areas in the study region never had a noteworthy forest vegetation cover in the Holocene (Vieira 1995). The floristic structure of the forests also changed during the Holocene. Scattered remains of semi-deciduous *Quercus robur* subsp. *broteroana* forest patches along the Douro valley and tributaries are testimonies to a former Middle Holocene, more benign, climate. Abundant plant macroremains and paleopalynological data prove that *Pinus* species – *Pinus sylvestris* on the higher plateaus and *Pinus pinaster* on steep slopes at lower altitudes – were well represented in the study area till recent historical times (Figueiral 1995). In the *Pinus pinaster* plantations planted since the 1930s the dunar ecotype was extensively used leading to a genetic homogenization of the species in the study area (vd. Ribeiro et al. 2001). Actual forests of indigenous *Pinus* are unknown in the Trás-os-Montes and Beira Alta region. The deep and widespread anthropic deforestation described by the travellers and writers of the nineteenth century, and the accelerated erosion linked with cereal cultivation on marginal soils in the first half of the twentieth century, changed the boundaries between climatophilous and edaphoxerophilous Potential Natural Vegetation (PNV; see concept discussion below). For example, the area potentially covered by edaphoxerophilous forests of *Quercus rotundifolia* became probably larger at the expense of former *Quercus suber* or *Quercus pyrenaica* climatophilous forest

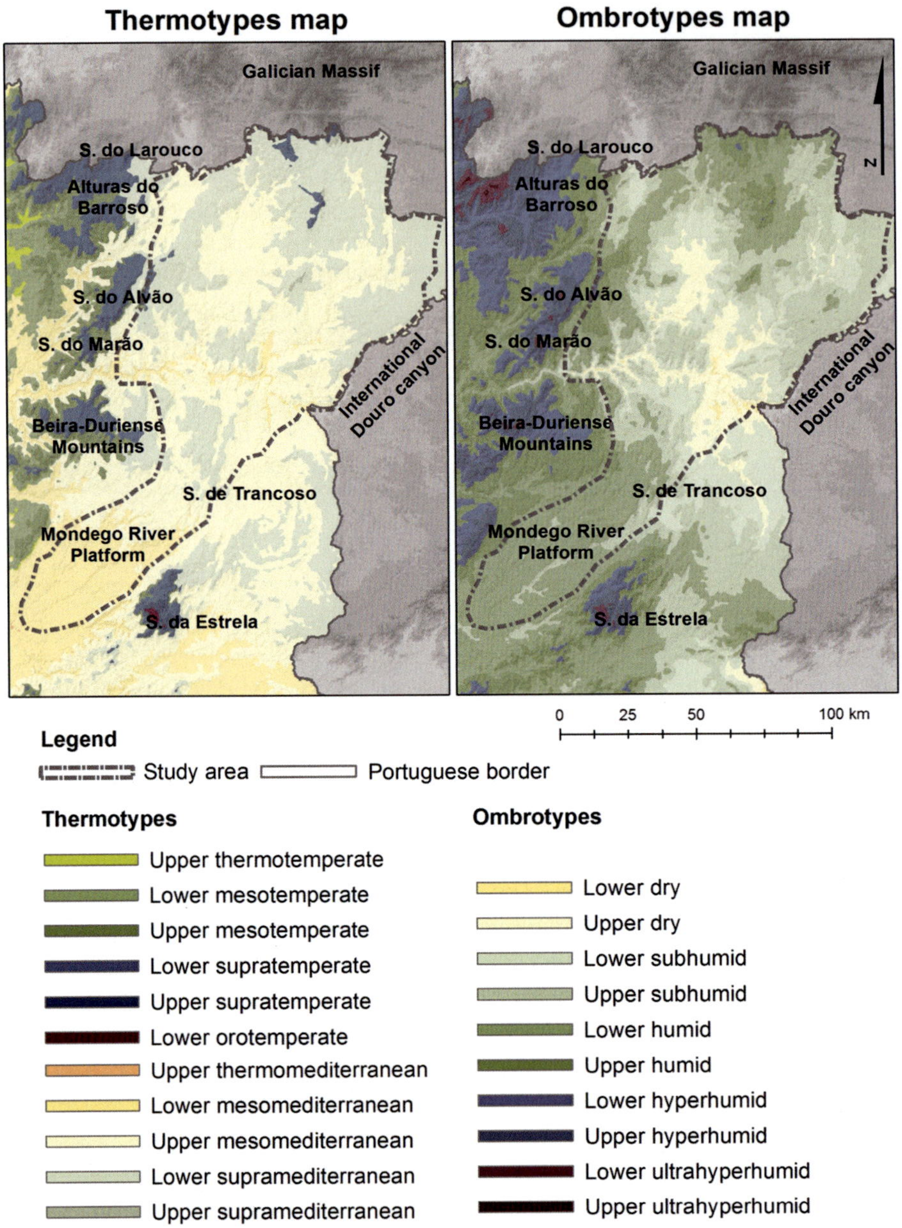

Fig. 9.2 Bioclimatology of the Trás-os-Montes and Beira Alta region (in the Portuguese Territory) according to Rivas-Martínez's bioclimatic system

Fig. 9.3 Landscape of the 'Cold Land' (supratemperate and supramediterranean bioclimatic belts), in the Municipality of Vinhais (northeastern Portugal). The land use of supramediterranean granitic peneplains created islands of fertility around the villages, surrounded by exhausted soils covered by mosaics of *Erica australis* heathlands (Calluno-Ulicetea minoris) and oligothrophic grasslands of *Agrostis* sp. pl. or *Arrhenatherum elatius* subsp. *bulbosum* (Festucion merinoi)

habitats, via soil loss and the mass effect caused by seeding from surviving *Quercus rotundifolia* trees on rock outcrops.

The physiography of the study area creates three distinct equilibrium spaces between pedogenetic and natural erosive processes (Agroconsultores and COBA 1991). On the plateaus pedogenetic processes are preponderant and so are soil catenas with moderate to deep soils (cambisols and regosols) colonized by climatophilous forests of *Q. pyrenaica* or *Q. suber*. The mesomediterranean basins and valleys also develop complex soil catenas that can include luvisols, regosols and eventually fluvisols. These areas share a complex pattern of climatophilous forests mixed with several types of riparian and tempori-hygrophilous (phreatophilous) tall shrublands and forests. On steep slopes the rain/runoff ratio is smaller than on more horizontal landforms, and soil creep and recurrent mass movement events produce a dominance of shallow soils (leptsols). The holm hoak (*Q. rotundifolia*) is the dominant and sometimes exclusive tree of edaphoxerophilous climactic forests on steep valley slopes, but also on stressed, denuded rock outcrops on the plateau and, as will be referred to later, on ultramafic soils.

The plateaus of the study area are the climactic domain of the climatophilous mesophilous deciduous mesoforest of *Quercus pyrenaica* (Quercenion pyrenaicae)

Fig. 9.4 Landscape of the 'Hot Land' (mesomediterranean bioclimatic belt), nearby the village of Algoso (Municipality of Mogadouro, northeastern Portugal). Rock outcrops and former cultivated land mainly covered by patches of edaphoxerophilous *Quercus rotundifolia* forests (Quercetea ilicis), pioneer *Cytisus multiflorus* shrubland (Cytisetea scopario-striati) and *Cistus ladanifer* shrublands (Cisto-Lavanduletea stoechadis)

of the Genisto falcatae-Quercetum pyrenaicae (supramediterranean) and the Holco mollis-Quercetum pyrenaicae (preferentially supratemperate) (Fig. 9.5). The possible past occurrence of climatophilous *Betula celtiberica* forests (Betulion fontqueri-celtibericae) or, at least, mixed forests of *Q. pyrenaica*/*Q. robur* subsp. *broteroana* and *Betula celtiberica* (Quercion pyrenaicae), on the supratemperate hyperhumid summits cannot be excluded (Honrado 2003). The frequent wild fires that cyclically affect the mountains of the study area don't counteract the robust recuperation of *Q. pyrenaica* forests on abandoned agricultural land. Secondary *Q. pyrenaica* forests are species-poor, and contain a significant number of seral shrubs (e.g. *Cytisus* sp.pl.). At the landscape scale, older forests, with a long historical tree presence, are not only species-richer, but share more complex forest-dependent vegetation mosaics. Dense and old ombrophilous *Q. pyrenaica* forest stands, like those found on Nogueira Moutain in Portugal and the Sanabria Valley in Spain, tend to have an understory dominated by esciophilous graminoids and herbs (e.g. *Holcus mollis*, *Poa nemoralis*, *Luzula forsteri*, *Potentilla sterilis* and *Melampyrum pratense*) and geophytes (e.g. *Asphodelus lusitanicus*, *Erythronium dens-canis*, and *Narcissus asturiensis*). In the forest mantle and fringe

Fig. 9.5 Mosaic of climatophilous mesophilous deciduous mesoforest of *Quercus pyrenaica* (Quercenion pyrenaicae) (Trás-os-Montes, Portugal), tall shrub communities of *Erica arborea* (Genisto falcatae-Ericetum arboreae), and *E. australis* heathlands (Pterosparto lasianthi-Ericetum aragonensis)

prosper complex mosaics of tall shrub communities of *Erica arborea* (Genisto falcatae-Ericetum arboreae), lianoid communities rich in endemic *Rubus* (*e.g. Rubus vigoi, Rubus lainzii* or *Rubus vagabundus*) of the Rosenion carioti-pouzinii, species-rich herbaceous perennial communities of the Linarion triornithophorae, and *Festuca elegans* subsp. *merinoi* grasslands (Festucion merinoi). In the north-eastern part of the study area the *F. elegans* subsp. *merinoi* grasslands are the habitat of the endemic *Asteraceae Phalacrocarpum hoffmannseggii*. In contrast, broom communities are heliophilous and functionally independent of the forests. The presence of *Genista florida* segregates the mature broom communities of the plateau from its mesomediterranean counterparts (v.i.).

The distribution of the *Quercus robur* subsp. *broteroana* thermophilous semi-deciduous forests (Viburno tini-Quercetum roboris) encompasses a large area of the northwestern Portuguese lowlands and the Mondego Platform. The Viburno-Quercetum roboris is a mesophyllous mesoforest rich in thermophilous taxa (e.g. *Viburnum tinus* and *Ruscus aculeatus*) and *Quercus* species and hybrids (e.g. *Q. robur* subsp. *broteroana, Q. suber, Q. faginea* subsp. *faginea* and *Quercus x andegavensis* nothosubsp. *henriquesii*. *Quercus robur* subsp. *broteroana* keeps part of its leaves functional till bud burst in the spring: it is a semi-deciduous tree. In the lower section of the Mondego Platform the *Q. robur* subsp. *broteroana* forests are progressively segregated into more ombrophilous habitats and substituted by *Q. suber* forests (Teucrio salviastri-Quercetum suberis, *v.i.*). Nevertheless, the Mondego Platform PNV is difficult to ascertain because it is a crossover area of the Eurosiberian and Mediterranean Phytogeographic Regions, and it has a long history of human use and faces a severe invasion of *Acacia dealbata* and *A. melanoxylon*. The identification of the climactic domains of *Quercus robur* subsp. *broteroana* is further complicated by the millenary-long, anthropic selection that favoured *Q. pyrenaica* because *Q. robur* produces a better wood for carpentry use than *Q. pyrenaica*. Although it is, for these reasons, difficult to ascertain, forests of the Rusco-Quercetum roboris (see Chap. 6) are probably the PNV of the Vila Real Basin (SW of Trás-os-Montes) and elsewhere. The tall shrublands of the Mondego Platform are examined in the next section.

The Duriese-Lusitanian *Quercus suber* forests are one of the most unusual vegetation types of the study area. In the upper sub-humid mesomediterranean peneplain tracts, there develops a climatophilous woodland of *Q. suber* first described from edaphoxerophilous positions in the River Sil Valey (Galicia, Spain), a Mediterranean enclave inside the Galaico-Portuguese biogeographical sector, with the name Physospermo cornubiensis-Quercetum suberis. The same forest community is easy to observe on the slopes and on dry pediment (footslope) deposits in the Chaves-Verín Graben, and in vineyard terraces abandoned during the phylloxera crises (decade of 1870) in the western portion of the Douro Valley and direct tributaries, still inside the Duriese-Lusitanian Sector. The dominant variant of this association, the subas. quercetosum fagineae, is, in the Duriese-Lusitanian Sector, a mixed forest of *Q. suber* and *Q. faginea* subsp. *faginea*, with the frequent presence of *Q. pyrenaica* and *Q. x welwitschii*. The understory exhibits an inconspicuous combination of Quercetea ilicis shrubs and lianas (e.g. *Daphne gnidium* and *Rubia peregrina*) with a varied herb layer in which excel three narrow endemics *Epipactis duriensis*, *Silene coutinhoi* and, on boulders inside the woodland, *Anarrhinum duriminium*.

The eutrophic soils derived from basic rocks of the Morais Massif or from cambric metasediments of the Douro Group are the climactic domain of the Hedero hybernicae-Quercetum fagineae (Aceri granatensis-Quercion fagineae), a mesophilous semi-deciduous woodland of *Q. faginea* subsp. *faginea* with an understory similar to the Physospermo-Quercetum suberis. The rarity of these woodlands in the Duriese-Lusitanian landscape is explained by the fertility of its soils, today mainly occupied by orchards, vineyards, arable land and kitchen gardens. A variant of the broom association Lavandulo sampaioanae-Cytisetum multiflori with an abundant presence of *Cytisus scoparius subsp. scoparius* is the most frequent tall shrub substitution stage of both *Q. faginea* subsp. *faginea* and *Q. suber* forests. The Cytiso grandiflori-Arbutetum unedonis, a strawberry tree-dominated tall shrubland, forms a noticeable forest mantle of the same woodlands in the Douro valley.

The steep slopes of the superior mesomediterranean and inferior supramediterranean valleys that cut through the Duriese-Lusitanian peneplain fragments are the habitat of the Genisto hystricis-Quercetum rotundifoliae forest (Fig. 9.4). The transitional position of this evergreen sclerophyllous arboreal community from steep slopes to euclimatopes (sensu Walter 1985), establishing climatophilous woodlands, is an unmistakable indicator of the transition between Duriese-Lusitanian and Salamancan biogeographic sectors. From the slopes bordering the Mondego Platform two other edaphoxerophilous woodlands associations exclusively dominated by perennial oaks, the Teucrio salviastri-Quercetum rotundifoliae and the Teucrio salviastri-Quercetum suberis, were recorded. The Teucrio-Quercetum rotundifoliae is supramediterranean and contacts the *Q. pyrenaica* climatophilous woodlands (Genisto-Quercetum pyrenaicae); the edaphoxerophilous Teucrio salviastri-Quercetum suberis is mesotemperate and mesomediterranean and develops in the vicinity of the Viburno-Quercetum roboris.

The regosols derived from thick pediment deposits in the dry mesomediterranean belt of the Duriese-Lusitanian Sector are the climactic domain of the Junipero lagunae-Quercetum suberis. It is a common pattern in the Iberian Peninsula that in dry ombroclimate cork oak woodlands there occur deep compensated azonal soils. The Junipero-Quercetum suberis displays an odd combination of trees: three *Quercus* species – *Q. suber, Q. faginea* subsp. *faginea* and *Q. rotundifolia* –, with the gymnosperm *Juniperus oxycedrus*. These woodlands are climatophilous in the sub-humid inferior horizon, *grosso modo* with an Io < 4.5. Nevertheless, an alternative hypothesis shouldn't be discarded: the sampled association individuals of the Junipero-Quercetum suberis in the sub-humid inferior horizon belong in reality to the Physospermo-Quercetum suberis; being immature they are prone to being temporarily colonized by *Juniperus oxycedrus*, a small tree that produces fleshy fructifications with seeds dispersed by birds, and that is easily excluded when the tree canopy closes. In catenas of valley woodland vegetation the Junipero-Quercetum suberis comes into contact with the Rusco aculeati-Juniperetum lagunae, the climatophilous woodland of the Duriese-Lusitanian dry obroclimatic belt. Its dominant trees are *Q. rotundifoliae* and *Juniperus oxycedrus* (Fig. 9.6). The interaction of a low rainfall (dry ombroclimate) and continental climate prevents the development of a continuous tree layer and promotes the penetration of heliophilous bush species characteristic of subseral stages (e.g. *Cistus ladanifer* or *Lavandula pedunculata*) and many other sun-loving species (e.g. *Juniperus oxycedrus*). The tall shrubland mantle in the dry and sub-humid inferior ombroclimate vegetation belts have a constant presence of *Retama sphaerocarpa* (Cytiso multiflori-Retametum sphaerocarpae).

9.3 Shrublands

Secondary succession is slow on the supramediterranean/supratemperate plateaus with a long history of rough grazing with fire, and with scarce sources of diaspores of tree species. Mesotopographic gradients strongly affect species associations under these conditions. In convex areas, with poorer soils (umbric leptosols), there are dry heathlands (Calluno vulgaris-Ulicetea minoris shrublands, port. *urzais*, cast. *brezales*). Thicker soils, mostly regosols, are colonized by broom shrublands (Cytisetea scopario-striati). Heaths can be surprisingly uncommon on

Fig. 9.6 Dry mesomediterranean mixed forests of *Quercus rotundifolia* and *Juniperus oxycedrus* (Rusco aculeati-Juniperetum lagunae)

plateaus with deep granitic regoliths. From a physiognomic and floristic point of view there are two main types of mesic heathlands in the study area: *Erica australis* heathlands (Fig. 9.3), and gorse heathlands of *Erica umbellata*, *Erica cinerea* and *Ulex minor*, eventually with *Ulex micranthus* on schists. The former are dominant on more mediterranean-continental mountains; on temperate mountains they shift to rocky summits and steep rocky slopes. In the upper sub-humid horizon, especially on schists, characteristic species of the Cisto-Lavanduletea stoechadis are abundant in *E. australis* heathlands. The gorse heathlands are restricted to the more oceanic and rainy territories (supratemperate belt). In traditional grazing systems *E. australis* heathlands are roughly grazed by sheep and goats. In Portugal there are four short local cattle breeds – arouquesa, cachena, barrosã and maronesa – adapted to graze *Ulex minor* communities. Without the evidence of forests stands – *Q. pyrenaica* vs. *Q. robur* subsp. *broteroana* climactic forests – the transition between the Galaico-Portuguese Sector (Eurosiberian Region) and the Duriese-Lusitanian Sector (Mediterranean Region) can be mapped with reasonable certainty showing, respectively, the distributions of the heathlands with *Pterospartum tridentatum* subsp. *cantabricum* (Halimio alyssoidis-Pterospartetum cantabrici and Carici asturicae-Ericetum aragonensis) and with *P. tridentatum* subsp. *lasianthum* (Pterosparto lasianthi-Ericetum cinereae and Pterosparto lasianthi-Ericetum aragonensis) (Costa et al. 2008). In northeastern Portugal, mesic (and hygrophilous) heathlands descend into the mesomediterranean bioclimatic belt only in very particular habitats (e.g. quartzitic crests near the Eurosiberian/

Mediterranean border). In contrast, heathlands are common in Southwest Iberian lowlands (see Chap. 2).

In the study area there is a clear correlation between Calluno vulgaris-Ulicetea minoris gorse and/or heath communities with deciduous Querco-Fagetea sylvaticae forests (Fig. 9.3), and *Cistus ladanifer* shrublands of the Cisto-Lavanduletea stoechadis with sclerophyllous evergreen forests and with semi-deciduous forests of the Quercetea ilicis (Fig. 9.4). In Cisto-Lavanduletea stoechadis communities there are common aromatic species like *C. ladanifer*, *Lavandula pedunculata* or *Thymus mastichina*. There are two major associations of the Cisto-Lavanduletea stoechadis vegetation class described for the Duriese-Lusitanian Sector: the Cisto-Genistetum hystricis and the Euphorbio-Cistetum ladanifer. The first is characteristic of colder and more continental habitats than the second. Dominant heathland and Cisto-Lavanduletea stoechadis shrub species not only tolerate and are perpetuated by fire: they increase fire hazards: they are ecosystem builders. In the intermediate stages, the species follow either a 'seeder' (e.g. *Ulex minor* and *Cistus* sp. pl.) or a 'sprouter' fire strategy (e.g. *Erica australis* and *Pterospartum tridentatum*). 'Seeders' are incapable of forming new growth after a fire event, and reproduce from seeds. Sprouters produce stem or root sprouts that enable repeated shoot production despite frequent fire damage (Herrera 1998).

Cytisetea scopario-striati shrublands are dominated by slender (e.g. *Adenocarpus*, *Cytisus* and *Genista florida*) or spiny (e.g. *Ulex europaeus* subsp. *latebracteatus*) tall *Fabaceae* shrubs of the tribe *Cytiseae*. *Cytisus multiflorus* is a pioneer species; soil properties determine the evolution of its basal communities in the direction of broom shrubland or to low *Cistus* shrublands or heathlands (Fig. 9.4). In all the studied area there is a strong association of *Cytisus scoparius* subsp. *scoparius* broom tall shrublands with landscapes with a forest matrix, and of *C. striatus* communities with progressive successions on abandoned agricultural fields or on soils modified by afforestation practices. As mentioned before, in the dry and lower sub-humid mesomediterranean belts, the broom communities are enriched with *Retama sphaerocarpa* (Retamion sphaerocarpae), to the point of being monospecific in pioneer variants. In the supramediterranean and supratemperate belts, *Genista florida* is a newcomer. *G. florida* has a flexible ecology; it appears in subseral tall shrublands but can also colonize deep fissures in granitic boulders or mountain screes, and be a dominant species on temporary watercourse margins in mountains (Fig. 9.7). Wind-exposed supramediterranean and supratemperate granitic surfaces are the habitat of permanent communities of *Echinospartum ibericum* (Echinospartetum ibericum and Teucrio-Echinospartetum pulviniformis). On the Mondego Platform, and in rainy mesomediterranean areas near the Eurosiberian/Mediterranean border, there are common tall shrublands of *Ulex europaeus* subsp. *latebracteatus*, *Cytisus striatus* and *Adenocarpus complicatus* subsp. *complicatus* or subsp. *lainzii*, with the last species frequently being the dominant (Ulici latebracteati-Cytisetum striati). *U. europaeus* subsp. *latebracteatus* is a species that rapidly expands via the new highway network built across the study area.

Fig. 9.7 Riparian woodland of *Betula celtiberica* (Carici reuterianae-Betuletum celtibericae) with patches of *Genista florida*, and under-used Nardus grassland

In tempori-hygrophilous habitats Cytisetea scopario-striati shrublands are replaced by different combinations of spiny tall shrubs and lianas of the *Rosaceae* together with a few plants from other vascular plant families, all with bird-dispersed fleshy fruits. These species are characteristic of the Rhamno-Prunetea class. Among the most common Rhamno-Prunetea shrub species in the study area are *Crataegus monogyna, Prunus spinosa*, *P. institia*, *Pyrus cordata* and *Sambucus nigra*. Almost all the Portuguese species of *Rubus* and *Rosa* lianas are found in the mountains of the Bercianese-Sanabrese sector. The species diversity of Rhamno-Prunetea tall shrublands drops in the mesomediterranean belt, though is partially compensated for by the appearance of a few new species (e.g. *Prunus mahaleb* and *Acer monspessulanum*).

9.4 Riparian and Tempori-Hygrophilous Forests

In the study area, rivers are born on plateaus, experience a rapid descent along deep-carved valleys and reach the Douro and Mondego rivers that flow westerly to the sea. Excluding the plateaus, lentic streams and vegetation are rare. Just a few tens of kilometres from the coast, the Douro and Mondego rivers used to have violent rapids, that today are drowned under hydroelectric dams, giving a habitat that is also inadequate for lentic aquatic vegetation. As later referred to, supratemperate plateaus harbour oligotrophic grassland complexes with remains of fens and hygrophilous heathlands, together with *Nardus* grasslands and other graminoid vegetation types. Scattered *Betula celtiberica* trees prove the presence of a tempori-hygrophilous PNV dominated by this tree (Betulion fontqueri-celtibericae) in these habitats. Fens are uncommon in the supramediterranean belt, and there its PNV usually includes ash trees (Querco pyrenaicae-Fraxinetum angustifoliae). Anthropic plateau drainage has increased the area covered by riparian vegetation. In the supratemperate belt there occurs an interesting riparian woodland of *Betula celtiberica* (Carici reuterianae-Betuletum celtibericae) (Fig. 9.7), substituted below ca. 900 m in altitude by supramediterranean *Alnus glutinosa* riparian woodlands (Galio-Alnetum glutinosae). In the mesomediterranean belt, river margins develop thermophilous alder forests (Scrophulario-Alnetum glutinosae).

In narrow V shaped valleys behind the *Alnus* riparian forests, both in the supra and the mesomediterranean belts, tall shrublands of *Salix salviifolia* (Salicetum salviifoliae) are found, capable of withstanding the turbulent waters of winter floods (Fig. 9.8). In the study area, this is also the vegetation that protects river banks in water rapids and in the outside banks of sharp river bends, and that colonizes many

Fig. 9.8 Cynosurion cristati hay-meadows. The traditional irrigation system enlarged the meadow area up-slope into the climactic domain of the edaphoxerophilous woodlands of *Quercus rotundifolia*. N.B. the greyish canopy of riparian tall shrublands of *Salix salviifolia*

temporary watercourses. In more open valleys, colluvial regosols, and eventually fluvisols, were in the past covered by mixed phreatophilous mesotrophic forests of *Fraxinus angustifolia*, *Prunus avium* and *Q. pyrenaica* (Querco pyrenaicae-Fraxinetum angustifoliae, supramediterranean belt), *Q. faginea* subsp. *faginea* (Fraxino angustifoliae-Aceretum monspessulani, mesomediterranean sub-humid or dry belt), or *Quercus robur* subsp. *broteroana* (Omphalodo nitidae-Fraxinetum angustifoliae, mesomediterranean humid belt). With rare exceptions the *Fraxinus angustifolia* mesotrophic forests were converted into hay-meadows, arable land or kitchen gardens. In forest vegetation catenas in valleys, the mixed ash forests are located between riparian and climatophilous forests, on rarely flooded soils. Rhamno-Prunetea tall thorn shrublands reach their phytosociological optimum – measured by their cover and characteristic species diversity – on the edges and clearings of phreatophilous forests.

In large mesomediterranean basins and valleys, even if profoundly altered by drainage works and farming – e.g. the valley of Chaves –, there still persist small areas of *Salix atrocinerea* and *Alnus glutinosa* dystrophic swamps (Carici lusitanicae-Alnetum glutinosae), invaded by *Rubus* lianas, and other Rhamno-Prunetea species. Sometimes a thick fringe of *Molinia caerulea* grasslands is also present. On the margins of artificially fixed river channels, with a somewhat slow-moving current, emerges a riparian mixed forest of *Salix alba*, *S. atrocinerea*, *F. angustifolia* and *Populus nigra* var. *betulifolia* (Salici neotrichae-Populetum nigrae). On the steep slopes that border the deep mesomediterranean valleys, appear small pockets of humid eutrophic soils due to the emergence of the rainwater from neighbouring plateaus. This is the habitat of *Celtis australis* forests (Clematido campaniflorae-Celtidetum australis), a tree that can also be seen in thermophilous riparian *Alnus glutinosa* forests. The evidence gathered so far doesn't support the presence of phreatophilous *Ulmus minor* forests, a probable archaeophyte in the study area. In the past *U. minor* was widely planted for its leaves, essential for pig and cattle feed during the dry summer months. The Dutch elm disease reduced it to shrub mats with no economic value. *F. angustifolia* leaves were also used to feed domestic herbivores but they don't have the same nutritive value as *U. minor* leaves.

9.5 Grasslands

There has been substantial progress in grassland syntaxonomy in the past few decades in western, acid Iberia. The dichotomy between perennial and annual grasslands, and its successional significance, was established in 1970s by S. Rivas-Martínez. The most recent syntaxonomical synthesis for the Iberian Peninsula (Rivas-Martínez et al. 2001; Costa et al. 2012) split the classification of perennial grasslands in such a way that it implicitly recognizes the subtle relationships between the floristic structure of the perennial grasslands and the characteristics of their thermoclimate, soil texture, chemical fertility and humidity, and biomass disturbance regimes (e.g. hay cutting, mowing and trampling). Grasslands develop an intricate, diverse phytocenotic mosaic in the territories that fringe the Eurosiberian/Mediterranean frontier. Their study demands a consistent syntaxonomy and a prudent and informed interpretation.

The annual grasslands (Tuberarietea guttatae) of the study area reach a species and phytocoenotic diversity maximum in the mesomediterranean thermoclimatic belt (Tuberarion guttatae associations). Therophyty is a widespread adaptation to climates with a long dry hot season. In the Douro Valley these communities are the habitat of the endemic *Holcus annus* subsp. *duriensis*. Although less species-diverse, annual grasslands are also present on supramediterranean and supratemperate plateaus (Molinerion laevis and Thero-Airion associations). The annual grasslands grow in mosaics with low shrubs of the Cisto-Lavanduletea stoechadis and Calluno vulgaris-Ulicetea minoris. Being heliophilous, they are favoured by short cycles of fire. In traditional grazing systems, annual grasslands were roughly grazed by sheep and goats. In the driest areas of the Duriese-Lusitanian Sector, nitrate leaching by rain is less effective and there is a progressive increase in the area occupied by subnitrophilous grasslands of *Taeniatherum caput-medusae* or *Stipa capensis* (Taeniathero-Aegilopion geniculatae), for example.

Heavy grazing by cattle and sheep, but especially by mixed herds – with donkeys eating dry grasses, shrubs and other roughages – depresses Tuberarietea guttatae species and fosters grasslands dominated by *Trifolium subterraneum* subsp. *subterraneum*, called in Spanish *majadales* and in Portuguese *malhadas*. Patches of *malhadas* are very frequent on common land near the villages where the herds finish the grazing day before returning to their shelters by nightfall. That clover is an annual plant adapted to endure heavy grazing, except by sheep during flowering time (sheep graze closer to the ground than cattle and eat the flowers, and the survival rate of seeds passing through the digestive tract is lower). It produces plagiotropic twigs, and cleistogamic flowers close to the ground; after fecundation, the peduncle becomes gravitropic, burying spherical fructifications in the soil with a few seeds protected by sterile flowers (Smetham 2003). Among the most frequent species in *T. subterraneum* subsp. *subterraneum* pastures are *Poa bulbosa* and *Parentucellia latifolia*, and a large group of annual species today frequent in road vegetation (Poetea annuae), e.g. *Plantago coronopus* and *Spergularia rubra*. In rainy years, annual plants adapted to temporary wet soils (Isoeto-Nanojuncetea characteristics) burst in *T. subterraneum* subsp. *subterraneum* grasslands, e.g. *Trifolium cernuum* and *Juncus bufonius*. The *malhadas* have a sub-humid lower supramediterranean and a mesomediterranean optimum.

The higher altitude summits, upper slopes and rocky platforms are often inhabited by mosaics of annual grasslands and *Festuca summilusitana* communities (Hieracio castellani-Plantaginenion radicatae), an Iberian endemic grass with hard blue leaves with silicified epidermal cells. In similar nutrient-exporting habitats, there are also as yet undescribed oligothrophic grasslands of *Agrostis curtisii* (Festucion merinoi), widespread in the understory of *Pinus pinaster* trees, and related to the effect of this tree on soil fertility. A slight improvement in soil chemical fertility enables the differentiation of *Festucion merinoi* grasslands dominated by *Agrostis x fouilladei* (or *Agrostis capillaris*). Although they are very frequent, the syntaxonomy of these grasslands is not yet resolved. In more favourable habitats develops a grassland of *Pseudarrhenatherum longifolium* (mainly supratemperate schists) or *Arrhenatherum elatius* subsp. *bulbosum* (mainly supratemperate and supramediterranean granits), both of the Festucion merinoi alliance. A further additional increase in soil fertility, for example on footslope soils through the repeated depositions of cattle dung, enables the appearance of Molinio-Arrhenatheretea characteristics. Granite soils exposed to cyclical severe fires differentiate into a mobile, superficial soil layer of gravel and coarse sand, that is the habitat of annual/biennial low-productivity grasslands of the Polytricho-Agrostietum truncatulae; it is a phytocoenosis with evident floristic affinities with temperate fossil dunes vegetation (Jasiono sessiliflorae-Koelerietalia crassipedis). All the above mentioned grasslands can cover large tracts of the landscape in mosaic with mesic heathlands, as a substitution stage of *Quercus pyrenaica* forests (Quercenion pyrenaicae).

Downhill, but still on the plateaus, water-accumulating soils develop mosaics of hygrophilous heathlands (Genistion micrantho-anglicae) and *Nardus stricta* grasslands (port. *cervunais*, span. *cervunales*), subseral of phreatophilous forests. Hay cutting and intensive grazing weakens hygrophilous heathlands species (e.g. *Erica tetralix* and *Genista anglica*) to the benefit of Nardetea species (e.g. *Agrostis hesperica*, *N. stricta*, *Juncus squarrosus*, *Festuca nigrescens* and *Danthonia decumbens*). When grazed by cattle *Nardus* grasslands are species-diverse and rich in Molinio-Arrhenatheretea species; sheep grazing positively selects *Nardus stricta*, which can easily exceed a cover of 50%. This effect is obvious when comparing the *cervunais* grazed by cattle in the Trás-os-Montes, with those of the Serra da Estrêla (see Chap. 13) with a long history of use by transhumant sheep herds. Mixed with Nardetea strictae vegetation are frequent communities of *Deschampsia cespitosa* and *Juncus effusus* (Juncion acutiflori), like *Nardus*, two species avoided by grazers. On the western temperate plateaus of the study area Nardetea grasslands frequently come into contact with minerotrophic fens (Scheuchzerio-Caricetea nigrae). Public forest services drained the largest mountain fens in the study area, for example in Padrela and the Montesinho Mountains, in the middle of the twentieth century with the intention of increasing the production of forage or potato seed. Regular drainage ditches and basal communities of *Molinia caerulea* colonizing mineralizing peat are a witness to the existence of former fens.

Hay-meadows with Molinio-Arrhenatheretea grasslands are already embedded in the valley, on footslope regosols or in alluvial terraces. Meadows of plateau headwaters, even when subjected to hay cutting, except when fertilized with phosphorus, are covered by Nardetea vegetation. Plot experiments have proved the role of phosphorus bioavailability in Nardetea strictae grasslands population dynamics (Hejcman et al. 2007). Hay-meadows are by themselves an amalgamation of several grassland communities that fluctuate over time in response to grazing pressure, hay cutting, chemical fertilization or hydraulic works (Aguiar et al. 2000). Water, plant nutrients and grazing pressure tend to concentrate at the bottom of the hay-meadow, creating a strong ecological gradient at hay-meadow scale. Tree shade is another ecological factor controlling the structure of hay-meadow vegetation complexes. A group of specialized tall perennial herbs dwells in the nitrophilic, shady and humid habitats beneath, or in the fringes of riparian hedgerows; e.g. *Geranium robertianum*, *Pentaglottis sempervirens*, *Filipendula ulmaria*, *Silene dioica*, *Lamium maculatum* and *Urtica dioica* (Galio-Alliarietalia petiolatae). Mixed with these hemicryptophytes are annual plants, e.g. several aromatic *Geraniaceae* (Geranio pusilli-Anthriscion caucalidis). Rush communities develop on water-logged, sun-exposed soils (Juncion acutiflori). In heavily grazed and manured hay-meadows, instead of *Juncus effusus* and *J. acutiflorus* rush vegetation, there appear communities of a bluish-green rush, *J. inflexus*, accompanied by other nitrophilous species like *Agrostis stolonifera*, *Potentilla reptans*, *Mentha suaveolens* and *Ranunculus repens* (Potentillion anserinae). The majority of the above listed species are low-productive, avoided by cattle or of poor hay quality. To reduce their presence, traditional hay-meadow management involves river channel deepening, the linearization of riparian forests, tree pollarding (usually on a 7 year cycle), irrigation and weed cutting or pulling. Cynosurion cristati communities are the most productive and palatable (Fig. 9.8). Among its most common species in the study area are *Holcus lanatus*, *Bromus racemosus*, *Cynosurus cristatus*, *Festuca arundinacea* subsp. *arundinacea*, *Plantago lanceolata*, *Trifolium pratense* and *T. repens*. The drier parts of hay-meadows usually have impoverished *Arrhenatherum elatius subsp. bulbosum* grasslands (Arrhenatherion elatioris), or, on even drier soils, grasslands of *Agrostis castellana* (Agrostion castellanae). *A. elatius* subsp. *bulbosum* grasslands of the Agrostio-Arrhenatheretum bulbosi (Arrhenatherion eleatoris) are dominant on leached supratemperate soils. In most situations Cynosurion and Arrhenatherion communities are seral of phreatophilous forests of *Fraxinus angustifolia* (*v.s.*), as is testified to by the systematic presence of Rhamno-Prunetea hedgerows or dispersed trees of this species. The traditional hay-meadow irrigation systems have the objective of enlarging the Cynosurion and Arrhenatherion meadow area up-slope, to soils once occupied by climatophilous or edaphoxerophilous *Quercus* forests (Fig. 9.8).

The integration of autumn grazing and early spring grazing, usually until the first week of April, with one early summer cut, and an efficient irrigation system, favours Cynosurion vegetation and increases productivity and the diversity in hay-meadow species, in particular in the legume component. Today, due to reducing cattle numbers over the past decades, the majority of the hay-meadows is used just to produce hay. Consequently, there is a tangible increase in grass species cover, and, particularly in the western part of the study area, of the area covered by Arrhenatherion grasslands. Simultaneously, the neglect of pollarding practices is facilitating the invasion of annual, unpalatable plants (Geranio pusilli-Anthriscion caucalidis) and excluding the most productive heliophilous species. The expansion of *Brachypodium rupestre* or *Carex paniculata* subsp. *lusitanica*, two plant species avoided by cattle, is the first step in the dynamics following hay-meadow abandonment. The invasion of brooms and *Rosaceae* lianas and tall thorn bushes soon follows (Rhamno-Prunetea). In traditional mountain farming, hay-meadows were the most valued land use type because the cattle number depended on the amount of hay stored for the winter period; today chestnut and olive orchards replace hay-meadows as the most valuable land use.

The Mediterranean communities of the Agrostion castellanae alliance, although imperfectly studied, are pervasive at the mesomediterranean and the lower altitudes of the supramediterranean belts of the study area. They combine Mediterranean perennial grasses with a summer dormancy (e.g. *Agrostis castellana* and *Dactylis glomerata* subsp. *hispanica*), with small chamaephytes (e.g. *Armeria* sp. pl.), hemicryptophytes (e.g. *Sanguisorba verrucosa* and *Centaurea* gr. *paniculata*), geophytes (e.g. *Allium* and *Asphodelus* sp. pl.) and a few tall annuals (e.g. *Gaudinia fragilis*). These grasslands develop as small "fertility islands" near the low shrubs (e.g. *Cistus* and *Erica* species), but are more continuous and frequent in the understory of broom communities. In deeper, tempori-hygrophilous soils there develops a well-known grassland association, the Gaudinio-Agrostietum castellanae, that bears some Molinio-Arrhenateretea species and is already managed with hay mowing and cattle grazing. The Gaudinio-Agrostietum castellanae is frequent in hay-meadow vegetation complexes in supramediterranean territories and is dominant in many mesomediterranean hay-meadow vegetation mosaics.

9.6 Mafic and Ultramafic Metamorphic Rocks Vegetation

The mafic and ultramafic metamorphic rocks are the most singular lithology in the study area. They are dispersed in two massifs: the Bragança-Vinhais and Morais massifs, spread, respectively, in the supramediterranean and in the mesomediterranean belts. In the study area, outside these massifs, excluding a few isolated protrusions of crystalline limestone, *in situ* derived soils are acid and deficient in plant nutrients (specially phosphorus). Mafic rocks (e.g. amphibolites and similar lithologies) produce deep, fertile soils. Although covered by agriculture – the most productive chestnut orchards in northeastern Portugal are planted on soils derived from basic rocks – the Bragança-Vinhais massif holds some of the best preserved *Quercus pyrenaica* forests of the study area and of the aforementioned mesotrophic semi-deciduous forests of *Q. faginea* subsp. *faginea*. It is relevant to report that basic rocks preclude the development of Calluno vulgaris-Ulicetea minoris shrublands, even at ca. 1200 m altitude. Ultramafic soils are, in contrast, intrinsically infertile and highly selective to vascular flora. The main adverse factors conditioning plant life in ultramafic (serpentine) soils – the serpentine effect – are probably the high Mg/Ca quotient (two divalent cations with an antagonistic effect), phytotoxic levels of Ni and the low contents of bioavailable N, P, K and Ca (Kruckeberg 1986). These extreme ecological conditions impose a strong selective effect on the flora, which results in a widespread ecotypic differentiation among generalist plant species and in a high diversity of endemic species and biogeographical disjunctions, some of them with a relict character (Kruckeberg 1986). Currently there are recognized seven narrow endemics on northeastern Portuguese ultramafics: *Antirrhinum rothmaleri*, *Anthyllis sampaioana*, *Arenaria querioides* subsp. *fontiqueri*, *Armeria eriophylla*, *A. langei* subsp. *marizii*, *Avenula pratensis* subsp. *lusitanica* and *Festuca brigantina* subsp. *brigantina*. There is a marked small-scale spatial variation of pH in ultramafic soils. Sequeira and Pinto da Silva (1992) measured, over a distance of a few centimetres, a pH > 8, in fact 8.6, at the weathered surface of stones and gravel, 7 at the fine fraction, and less than 6 at the <0.5 mm fraction. This variation is part of the explanation of the co-existence of acidophilous and neutrophilous flora on ultramafic soils.

The most striking ultramafic effect on the vegetation is the absence of *Quercus* species other than *Q. rotundifolia* (Fig. 9.9). Climactic woodlands on ultramafic outcrops belong to the Genisto hystricis-Quercetum rotundifoliae. Seral shrublands are similar to those seen substituting the same woodlands outside ultramafic rocks: Genisto hystricis-Cytisetum multiflori tall shrublands and Cisto ladaniferi-Genistetum hystricis gum rockrose low shrublands. With the exception of *Avenula pratensis* subsp. *lusitanica*, that colonizes pioneer grasslands of the Agrostion castellanae, the endemic serpentinophytes are characteristic of dwarf-chamaephyte

Fig. 9.9 Ultramafic outcrops of northeastern Portugal in the autumn. There is a clear-cut correlation between ultramafic rocks and *Quercus rotundifolia* woodlands (*greyish green*). In the supramediterranean belt the soils derived from amphibolites are colonized by mesophilous deciduous mesoforest of *Q. pyrenaica* (*yellow* and *reddish leaves*)

Fig. 9.10 The Galician-Lusitanian endemic nickel bioaccumulator *Alyssum serpyllifolium* subsp. *lusitanicum* (*Brassicaceae*)

communities (Jasiono sessiliflorae-Koelerietalia crassipedis). The clustering of endemic species in dwarf-chamaephyte communities, colonizing compacted gravel soils or horizontally fissured rocks, is a common pattern in all the siliceous territories of Mediterranean Iberia. The nitrophilous shrub vegetation class Pegano-Salsoletea is present in the study area with one sole association, the synendemic Alysso lusitanici-Santolinetum semidentatae. This chamaephyte community dominated by the Galician-Lusitanian endemic nickel bioaccumulator *Alyssum serpyllifolium* subsp. *lusitanicum* (*Brassicaceae*), occurs on soils disturbed by freeze-thaw cycles or by human activities (e.g. soil ploughing and machinery movement) (Fig. 9.10).

9.7 Herbaceous Weed, Ruderal and Road Vegetation

Ruderal and road semi-nitrophilous vegetation is a vast subject impossible to summarize in a few lines. The most conspicuous types in the study area are: i) annual vegetation on trampled roads with variable combinations of *Chamaesyce maculata, Crassula tillaea, Herniaria lusitanica, Plantago coronopus, Poa annua, Polygonum aviculare, Sagina apetala, Spergularia purpurea* and *S. rubra*, among others (several associations of the Poetea annuae); ii) annual ruderal communities dominated by early flowering *Brassicaceae*, e.g. *Brassica barrelieri*, *Coincya monensis* subsp. *pl.*, *Diplotaxis catholica*, *Sisymbrium officinale* and *Raphanus raphanistrum* (three associations of the Alysso granatensis-Brassicion barrelieri); iii) ruderal grasslands of *Hordeum murinum* (two associations of the Hordeion leporini); iv) ruderal and fallow mesomediterranean communities, promoted by sheep grazing, dominated by *Galactites tomentosus* (one association of the Echio plantaginei-Galactition tomentosae); and v) thistle communities on disturbed and loose soil with *Carduus carpetanus* (an excellent bioindicator of the supramediterraenan belt), *C. pycnocephalus*, *C. tenuiflorus*, *Centaurea calcitrapa*, *Silybum marianum* and *Onopordum acanthium* among many other species (at least four basal communities and associations of the Onopordetea acanthii). In the ruderal vegetation catenas the effect of soil compaction and, implicitly, of trampling frequency and intensity importantly determines the species associations.

There are three main groups of weed communities in the study area: i) winter weed communities of cereals or perennial crops of oligothrophic sandy siliceous soils (at least five associations of the Arnoseridenion minimae), ii) winter weed communities of footslope soils (two associations of the Scleranthenion annui and one of the Sperguló pentandrae-Arabidopsienion thalianae) and iii) spring summer communities of irrigated crops (or cultivated without irrigation on deep fertile soils) (at least two association of the Digitario ischaemi-Setarienion viridis). Among the most frequent species in the Arnoseridenion minimae weed communities are *Anthemis arvensis*, *Arnoseris minima*, *Eryngium tenue*, *Logfia minima*, *Micropyrum patens*, *Myosotis discolor* subsp. *discolor*, *Scleranthus annuus* and *Spergula arvensis* (Fig. 9.11). In late spring many Scleranthenion annui communities are characterized by a dense blanket of *Stellaria media* from which emerge plants like *Lamium amplexicaule*, *Lamium purpureum* or *Veronica* sp. pl. The white ligulate flowers of *Chamaemelum fuscatum* (Chrysanthemo myconis-Anthemidetum fuscatae) have a strong scenic effect when they bloom in early spring in the olive orchards of the "hot land" of Trás-os-Montes. They indicate acid and nutrient-poor soils, frequently with boron deficiency. In the Digitario-Setarienion viridis weed communities C_4 plants preponderate. Noteworthy are also, in the Douro valley, the citrus orchards and unmanaged kitchen gardens communities of the neophyte *Oxalis pes-caprae*, escorted by *Fumaria capreolata* and *Mercurialis ambigua*, among other species (Fumarion wirtgenii-agrariae).

The ruderal and the weed communities share large distribution areas. Their species find their primary habitats in annual Tuberarietea grasslands (many Arnoseridenion minimae plants), nitrophilous, trampled or disturbed habitats in semi-natural grassland (most of the plants present in the Scleranthenion annui and Poetea annuae communities), or on nutrient-rich river margin sediments (Digitario-Setarienion viridis characteristics). A large number of the Digitario-Setarienion viridis are putative archaeophytes, e.g. *Amaranthus* sp. pl. and *Setaria* sp. pl.

Weed communities, mostly Scleranthenion annui phytocoenoses, have undergone enormous changes in their floristic structure and distribution areas since the decade of the 1990s. The use of herbicides and clean seeds, and the regression of

Fig. 9.11 Oligothrophic weed community in rye. N.B. the iberian endemic *Micropyrum patens* (*Poaceae*)

cereal farming have reduced, to the verge of extinction, at least four weed species in the study area – *Lolium temulentum*, *Buglossoides arvensis*, *Neslia paniculata* subsp. *thracica* and *Avena sterilis* subsp. *ludoviciana* –, and seriously shrank the distribution of *Avena strigosa*, *Agrostemma githago*, *Centaurea cyanus* and *Odontites verna* (Aguiar 2001). A similar trend happened forty years ago with the linicolous weeds, *i.e.* the flax weed flora; *e.g. Cuscuta epilinum*, *Silene linophila* and *Spergula maxima* (M. Sequeira, pers. com.).

9.8 Vegetation Series and Geoseries

Despite its drawbacks the PNV concept is a useful tool for summarizing the knowledge about the vegetation of a territory and for hypothesis generation in vegetation science (Loidi and Fernández-González 2012). The Potential Natural Vegetation (PNV) phytocoenoses of the Trás-os-Montes and Beira Alta region have a forest physiognomy. Using the main reference and update for Portuguese vegetation (Costa et al. 2012), as well as complementary works (Aguiar et al. 2000;

Meireles 2010), a total of 22 PNV associations where identified in the study area. These forests can be displayed in two mains groups:

1. Edaphoxerophilous and climatophilous woodlands
   (a) Semi-deciduous *Quercus robur* subsp. *broteroana* and deciduous *Q. robur* subsp. *broteroana* and/or *Quercus pyrenaica* forests (Quercion pyrenaicae, Querco-Fagetea sylvaticae)
      (i) Rusco aculeati-Quercetum roboris
      (ii) Viburno tini-Quercetum broteroanae
      (iii) Holco mollis-Quercetum pyrenaicae
      (iv) Genisto falcatae-Quercetum pyrenaicae
   (b) Semi-deciduous *Quercus faginea* subsp. *faginea* woodlands (Aceri granatensis-Quercion fagineae, Querco-Fagetea sylvaticae)
      (i) Hedero hibernicae-Quercetum fagineae
   (c) Evergreen *Quercus rotundifolia* and *Quercus suber* woodlands (Quercion broteroi, Quercetea ilicis)
      (i) Physospermo cornubiensis-Quercetum suberis
      (ii) Teucrio salviastri-Quercetum suberis
      (iii) Junipero lagunae-Quercetum suberis
      (iv) Genisto hystricis-Quercetum rotundifoliae
      (v) Teucrio salviastri-Quercetum rotundifoliae
      (vi) Rusco aculeati-Juniperetum lagunae
2. Edapho-hygrophilous woodlands
   (a) Tempori-hygrophilous woodlands (Populion albae, Salici purpureae-Populetea nigrae)
      (i) Querco pyrenaicae-Fraxinetum angustifoliae
      (ii) Omphalodo nitidae-Fraxinetum angustifoliae
      (iii) Clematido campaniflorae-Celtidetum australis
      (iv) Fraxino angustifoliae-Aceretum monspessulani
   (b) Riparian tall shrublands (Salicion salviifoliae, Salici purpureae-Populetea nigrae)
      (i) Salicetum salviifoliae
   (c) Riparian woodlands (Osmundo-Alnion, Salici purpureae-Populetea nigrae)
      (i) Scrophulario scorodoniae-Alnetum glutinosae
      (ii) Galio broteriani-Alnetum glutinosae
      (iii) Carici reuterianae-Betuletum celtibericae
   (d) Dystrophic swamps (Alnetea glutinosae)
      (i) Carici lusitanicae-Alnetum glutinosae

Each PNV forest heads a vegetation series (*sigmetum*). In short, a vegetation series is a complex of plant communities connected by dynamic relationships, segregated from other vegetation series by its mature stage, the PNV. On the large plateaus, profoundly dissected by deep valleys, a feature typical of the Trás-os-Montes and Beira Alta region, the vegetation series are organized in catenas – topographic *geosigmeta* – along a strong regional gradient, inherent to the geomorphological system summit-shoulder-backslope-footslope-toeslope. Five main topographic geoseries are present in the study area, each one presented with the aid of the following scheme:

(a) Climatophilous vegetation series
(b) Edaphoxerophilous vegetation series
    (i) Relictual vegetation series
(c) Riparian geoseries
    (i) Riparian series
    (ii) Tempori-hygrophilous series

1. Supratemperate humid topographic geoseries
    (a) Holco mollis-Querco pyrenaicae Sigmetum
    (b) Edaphoxerophilous vegetation series (usually absent)
    (c) Riparian geoseries
        (i) Carici reuterianae-Betulo celtibericae Sigmetum; Galio broteriani-Alno glutinosae Sigmetum
        (ii) Series headed by a Betula celtiberica woodland: Holco mollis-Betulo celtibericae Sigmetum; Querco pyrenaicae-Fraxino angustifoliae Sigmetum
2. Supramediterranean sub-humid to humid topographic geoseries
    (a) Genisto falcatae-Querco pyrenaicae Sigmetum
    (b) Teucrio salviastri-Querco rotundifoliae Sigmetum; Genisto hystricis-Querco rotundifoliae Sigmetum
    (c) Riparian geoseries
        (i) Galio broteriani-Alno glutinosae Sigmetum
        (ii) Querco pyrenaicae-Fraxino angustifoliae Sigmetum
3. Mesomediterranean humid topographic geoseries
    (a) Viburno tini-Querco broteroanae Sigmetum
    (b) Teucrio salviastri-Querco suberis Sigmetum
    (c) Riparian geoseries
        (i) Scrophulario scorodoniae-Alno glutinosae Sigmetum; Saliceto salviifoliae Sigmetum
        (ii) Omphalodo nitidae-Fraxino angustifoliae Sigmetum

4. Mesomediterranean sub-humid acidophilous topographic geoseries
   (a) Physospermo cornubiensis-Querco suberis Sigmetum
   (b) Genisto hystricis-Querco rotundifoliae Sigmetum
   (c) Riparian geoseries
      (i) Scrophulario scorodoniae-Alno glutinosae Sigmetum; Saliceto salviifoliae Sigmetum
      (ii) Fraxino angustifoliae-Acereto monspessulani Sigmetum; Clematido campaniflorae-Celtideto australis Sigmetum
5. Mesomediterranean dry topographic geoseries
   (a) Rusco aculeati-Junipereto lagunae Sigmetum
   (b) Rusco aculeati-Junipereto lagunae Sigmetum edaphoxerophilous synvariant
   (c) Riparian geoseries
      (i) Scrophulario scorodoniae-Alneto glutinosae Sigmetum; Salici neotrichae-Populetum nigrae; Saliceto salviifoliae Sigmetum
      (ii) Junipero lagunae-Querceto suberis Sigmetum; Fraxino angustifoliae-Acereto monspessulani Sigmetum; Clematido campaniflorae-Celtideto australis Sigmetum

References

Agroconsultores and COBA (1991) Carta dos Solos, Carta do Uso Actual da Terra e Carta da Aptidão da Terra do Nordeste de Portugal. UTAD/PDRITM, Vila Real

Aguiar C (2001) Flora e Vegetação da Serra da Nogueira e do Parque Natural de Montesinho. Dissertação de Doutoramento em Engenharia Agronómica, Universidade Técnica de Lisboa, Lisboa

Aguiar C, Honrado J, Soutinho A (2000) Comunidades e complexos de vegetação pratense do Nordeste de Portugal. *III Reunião Ibérica de Pastagens e Forragens*. Xunta de Galicia, Bragança

Birot P (1950) Le Portugal. Etude de géographie regionale. Librairie Armand Colin, Paris

Bormann FH, Likens GE (1979) Pattern and process in a forested ecosystem: disturbance, development, and the steady state based on the Hubbard Brook ecosystem study. Springer-Verlag, New York

Costa JC, Honrado J, Monteiro Henriques T, Neto C, Aguiar C (2008) Sobre as comunidades de Pterospartum tridentatum sensu lato em Portugal continental. Silva Lusitana 16:123–127

Costa JC, Neto C, Aguiar C, Capelo J, Espírito Santo MD, Honrado J, Pinto Gomes C, Monteiro Henriques T, Sequeira M, Lousã M (2012) Vascular plant communities in Portugal (continental, the Azores and Madeira). Glob Geobot 2:1–180

Figueiral I (1995) Charcoal analysis and the history of *Pinus pinaster* (cluster pine) in Portugal. Rev Palaeobot Palynol 89:441–454

Gutiérrez F, Gutiérrez-Elorza M, Martín-Serrano Á (2014) The geology and geomorphology of Spain: a concise introduction. In: Gutiérrez F, Gutiérrez-Elorza M (eds) Landscapes and landforms of Spain. Springer Science, Dordrecht

Hejcman M, Klaudisova M, Stursa J, Pavlu V, Schellberg J, Hejcmanova P, Hakl J, Rauch O, Vacek S (2007) Revisiting a 37 years abandoned fertilizer experiment on Nardus grassland in the Czech Republic. Agric Ecosyst Environ 118:231–236

Herrera J (1998) Flower and fruit biology in Southern Spanish Mediterranean shrublands. Ann Mo Bot Gard 74:69–78

Honrado J (2003) Flora e Vegetação do Parque Nacional da Peneda-Gerês. Dissertação de doutoramento, Universidade do Porto, Porto

Kruckeberg A (1986) An essay: the stimulus of unusual geologies for plant speciation. Syst Bot 11:455–463

Loidi J, Fernández-González F (2012) Potential natural vegetation: reburying or reboring? J Veg Sci 23:596–604

Meireles C (2010) Flora e vegetação da Serra da Estrêla – aproximação fitossociológica da vertente meridional. Doutoramento Dissertação de Doutoramento, Universidad de Jaén, Jaén

Pais J (2012) The Paleogene and Neogene of western Iberia (Portugal): a Cenozoic record in the European Atlantic domain. Springer, New York

Ribeiro M, Plomion C, Petit R, Vendramin G, Szmidt A (2001) Variation in chloroplast single-sequence repeats in portuguese maritime-pine (*Pinus pinaster* Aiton). Theor Appl Genet 102:97–103

Rivas-Martínez S, Fernández-González F, Loidi J, Lousã M, Penas A (2001) Syntaxonomical checklist of vascular plant communities of Spain and Portugal to association level. Itinera Geobot 14:5–341

Rivas-Martínez S, Penas A, Díaz-González TE, del Río S, Cantó P, Herrero L, Pinto Gomes C, Costa JC (2014) Biogeography of Spain and Portual, preleminary typological synopsis. Int J Geobot Res 4:1–64

Sequeira E, Pinto da Silva A (1992) Ecology of serpentinized areas of north-east Portugal. In: Roberts A, Proctor J (eds) The ecology of areas with serpentized rocks. A world view. Kluwer Academic Publishers, Dordrecht

Smetham ML (2003) A review of subterranean clover (*Trifolium subterraneum* L.): its ecology, and use as a pasture legume in Australasia. Adv Agron 79:303–350

Vieira G (1995) Processos morfogenéticos recentes e actuais na Serra do Gerês. Dissertação de Mestrado em Geografia Física e Ambiente, Faculdade de Letras da Universidade de Lisboa

Walter H (1985) Vegetation of the earth and ecological systems of the geo-biosphere. Springer, New York

Chapter 10
The Duero Basin

Ángel Penas, Sara del Río, Luis Herrero, and Miguel Ladero

Abstract The Spanish North Plateau or the Spanish Duero basin is located in the southern foothills of the Cantabrian Ranges in the north-western quadrant of the Iberian Peninsula. It was formed by the action of the Duero river and its tributaries on both the right and left. The climate can be considered continental between dry and humid. The vegetation in this territory is very varied, and several groups can be differentiated based on their structure and physiognomy, and based on the soil and climate variability in this area. These groups are: (1) Forest and woodlands dominated by *Quercus rotundifolia*, *Quercus pyrenaica*, *Quercus faginea*, *Juniperus thurifera*, *Juniperus oxycedrus*. (2) Riparian ecosystems formed by willows (*Salix neotricha*), poplars (*Populus alba* and *Populus nigra*), ashes (*Fraxinus angustifolia*), alders (*Alnus glutinosa*), elms (*Ulmus minor*) and shrubby willows (*Salix salviifolia*, *Salix elaeagnos* subsp. *angustifolia*). (3) Seral scrub, including spiny nanoshrublands dominated by *Rosa* sp. pl., *Rubus* sp. pl., broom communities formed by several species in the genera *Cytisus* and *Genista*, hygrophilous or non-hygrophilous heathlands, with a predominance of species from the genus *Erica*, secondary scrub communities dominated by *Cistaceae* and *Lamiaceae* or *Thymus* sp. pl. communities, and nitrophilous scrub dominated by *Artemisia* sp. and *Santolina* sp. (4) Gypsophilous vegetation. (5) Grasslands of a wide diversity starting with communities of *Nardus stricta*, and continuing with communities of *Celtica gigantea*, *Poa bulbosa*, *Agrostis castellana*, rushy pastures dominated by *Juncus* sp. pl., meadows, communities of *Festuca elegans* subsp. *merinoi* and *Stipa* sp. pl. (6) Herbaceous ruderal and assorted vegetation with widely differing communities of the *Stellarietea mediae*, with both a roadway ruderal character and also present in rain-fed and irrigated croplands and grasslands produced by human and animal action. (7) Aquatic vegetation, including riparian herbaceous communities, peat bogs, lakes and pools communities, reed communities, reed beds, megaforbic

Á. Penas (✉) • S. del Río • L. Herrero
Department of Biodiversity and Environmental Management (Area of Botany), Mountain Livestock Farming Institute (Join Center CSIC-ULE), Faculty of Biological and Environmental Sciences, University of León, Campus of Vegazana, s/n, León E-24071, Spain
e-mail: apenm@unileon.es

M. Ladero
Department of Botany, Faculty of Pharmacy, University of Salamanca (Spain), Campus Miguel de Unamuno, Salamanca E-37007, Spain

J. Loidi (ed.), *The Vegetation of the Iberian Peninsula*, Plant and Vegetation 12,
DOI 10.1007/978-3-319-54784-8_10

communities and pioneer annual and dwarf perennial isoetid communities on periodically flooded bare soils. (8) Salt marshes. (9) Rock vegetation. (10) Vegetation series.

10.1 Introduction

The mountain chains that almost completely encircle the plains that conform the North Plateau or the Spanish Duero river basin, reaching altitudes of nearly 2600 m in the Sierra de Gredos and in the Picos de Europa, comprise the limits of the territory to which we refer here. These limits are the Cordillera Cantábrica range to the north, the Cordillera Central range to the south and the Cordillera Ibérica range to the east. To the west lie the Montes de León, which in conjunction with the Galaico-Leonese massifs separate these mountain chains from the Sil river basin, and continue southwards along the border with Portugal (Fig. 10.1).

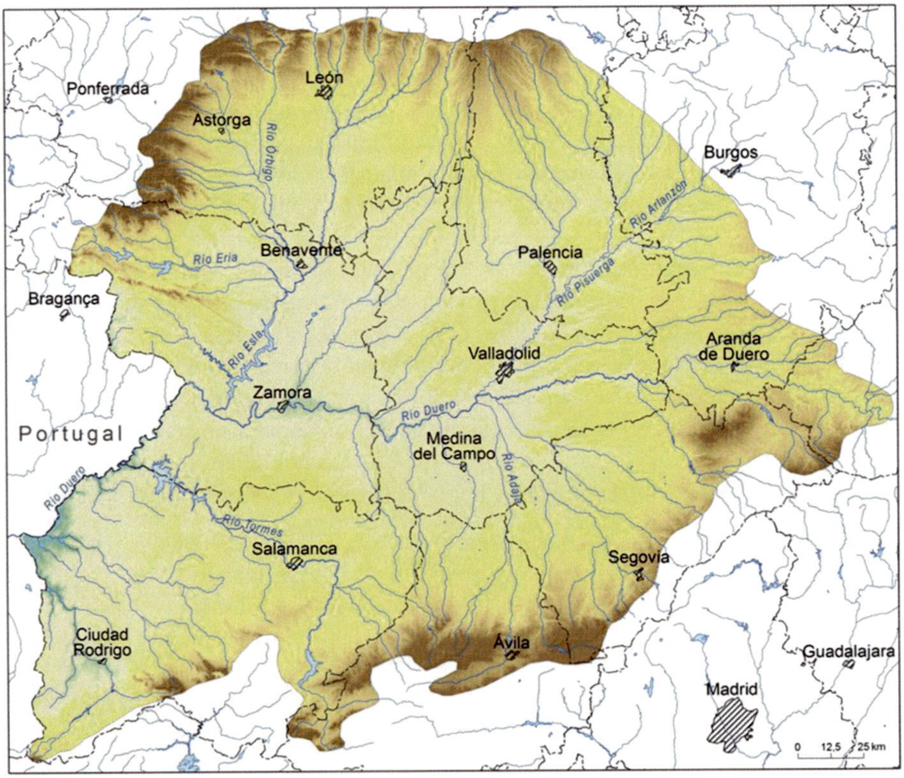

Fig. 10.1 Physical map of the Duero Basin

These plains are almost totally irrigated by the Duero river and can be designated the central basin. Two characteristic spaces can be differentiated here based on the type of surface materials: the western peneplains and the sedimentary basin. The first has a preponderance of granite, slate, quartzite and other materials from the lower Palaeozoic. In the sedimentary basin, which extends through the central and eastern areas of the region, the predominant materials are sand, loam, limestone, clay and other sedimentary materials.

The peneplains developed on Palaeozoic materials are located in the eastern part of the region at heights between 700 and 900 m and in the provinces of Ávila, Salamanca and Zamora. The term peneplain refers to the outcropping of crystalline materials that were dragged and levelled at the end of the Palaeozoic era, giving rise to a flat topography with gentle slopes, only interrupted by certain residual reliefs and by changes in level caused by the subsequent entrenchment of the river system, as occurs in the Arribes del Duero. Two types of peneplain can be differentiated based on their lithology and the different degree of resistance to erosion of their materials:

- **Granite**: on rigid crystalline and resistant rocks, found in the area of Campo de Ledesma, Sayago, Aliste, Carballeda, Vitigudino and Campo de Argañán. It may be intersected or broken by residual quartzite reliefs (Sierra de la Culebra).
- **Slates**: on schist and phyllite, and more undulating due to the lower resistance of the rock. This can be seen in the area of Alcañices, Aliste and south and southwest of Salamanca, between 900 and 1000 m. It may also be intersected by quartzite alignments crossing from northwest to southeast (Sierra de Herreros, Peña Gudiña), or definitively interrupted by the Sierra de Tamames and Las Quilamas.

The sedimentary basin in the centre of the region and in topographic continuity with the peneplains derives from the sinking of the Palaeozoic bedrock beneath the sea level, and is produced by the sedimentation of the materials from the surrounding mountains. This accumulation of materials occurred during the Tertiary Era between the Eocene and the Pliocene, when the segmentation phase terminated and the phase of erosion and entrenchment of the rivers began. The sediments are composed of two types of materials:

- **Detrital**: These are calcareous and clayey materials from the mountains in the north and east.
- **Sedimentary**: Essentially sands from the Cordillera Central range.

In the sedimentary basin, the different consistency of the sediments has caused a differential action of the erosion.

At the lower level there are alternating layers of clay, sand and sandstone interspersed with conglomerates. The mid-level is represented by loam and gypsum, and the upper level comprises several limestone strata of lacustrine origin. Limestone is the most resistant material to erosion in the whole series, as compared to the materials in the first two layers which are more erodable.

On this broad plain, the Duero river and its tributaries have carved out three different topographical spaces that correspond to moorland, green fields and floodplains. Their differentiation is the result of the uneven resistance of the various types of rock to the erosive action of the rivers.

Moorlands can be differentiated from the lower plains and the green fields by virtue of their nature as flat uplands (850–1100 m). They are found in the Montes Torozos mountains, between the provinces of Valladolid and Palencia and in the El Cerrato area in Burgos and Palencia. In the centre and northwest of the region, where the Pontian limestone deposits are more resistant to erosion, there are broad valleys with flat floors and steep sides, between which lie what are known as the calcareous moorlands (850–1000 m).

In the northwest, located at the foot of the Galaico-Leonese mountains, where the Miocene clay is covered with a deposit of the sedimentary formation known as "rañas" (quartzite pebbles in a clay matrix), the rivers from the Cordillera Cantábrica range have carved out deep broad valleys with a flat inter-riverine area and with altitudes of 900–1100 m. These constitute the highest plains in the region, and are known as Leonese, "raña" or detrital moorlands.

Below the level of the moorlands there are extensive landscapes of green fields. These correspond to areas in the centre and south of the region (the areas of Tierra de Campos and La Armuña in Salamanca) where the materials are softer (Miocene clay or sand). Erosion has caused gentle undulations with very broad valleys, slightly sloping out from the central basin (700 m) towards the periphery (800–900 m).

The green fields located to the north of the Duero are formed by materials that are predominantly clayey, forming broad valleys with slight changes in level that conform to a landscape with broad horizons. To the northwest between León and Benavente the green fields are pebbly.

The southern landscapes are characterised by the presence of sedimentary deposits with a sandy texture. These accumulations of sand deriving from the alteration of granite materials in the Sierra de Guadarrama that have been carried by the wind and today form genuine dune accumulations that extend throughout what are known as the "Tierras de Pinares" (pine-forest terrain) in Segovia and Valladolid.

These moorland and green field profiles are interrupted only by the floodplains. Floodplains are extensive discontinuous spaces where materials have been deposited by the river courses as they criss-cross the basin. These floodplains have a considerable breadth in areas where the river courses meander widely, as occurs in the confluence of the Tuerto and Órbigo rivers in León.

The plains of Castile-León are encircled by a belt of mountains of great complexity due to their different origins and the diversity of phenomena and materials involved in modelling them.

In the westernmost end of the system, the aggregate of Peña de Francia (1723 m), Gata (1367 m) and Malvana (1090 m) dominates the plains that extend towards the north. This aggregate represents, through the tectonic trench of Ciudad Rodrigo, a prolongation of the Serra da Estrêla in Portugal, and in the southwest-northeast direction serves as a natural boundary between the provinces of Cáceres and Salamanca. This mountain range is open to the Atlantic influence and receives abundant rainfall to the west, while the areas of Las Hurdes and Las Batuecas to the east receive less rainfall.

The "Béjar corridor" runs between the Sierra de Gata, Peña de Francia and Gredos mountain ranges. The "Béjar corridor" or "Alagón trench" is a major tectonic accident running north-northeast south-southwest near the Sierra de Peña de Francia (1723 m) and the Béjar mountain range (2404 m), along the west and east, respectively. The trench is a sunken block tilting southwards which causes a step of 700 m between the high plains in Salamanca and the low plains in Cáceres. The altitude of the mountains in this area generally remains below 1000 m. This major difference in level means that the river from which it takes its name (Alagón) and its tributaries run through a series of valleys and inter-riverine areas that conform a relief that can be classified as mountainous; this is the reason it is known by the name of Sierra Salmantina. It is also worth noting that in the western part of this step the change in level is enhanced by the Sierra de Tamames (1300–1400 m), although structurally it does not form part of the Cordillera Central range.

The hydrological characteristics of the area derive from its climatic and morphological aspects. Due to the existence of a vast high plain surrounded by a mountain belt, most of the waters that run through the territory flow onto this flat area.

The river system to which we refer here is the Duero river basin, and it is the most important one not only in the region but also in the whole of the Spanish territory. The Duero river is the principal and largest river system in the territory in this study. Its source is the Laguna Negra in the Picos de Urbión mountains in the province of Soria, at an altitude of over 2000 m. Until it enters Portugal, on its long itinerary it flows through the provinces of Soria, Burgos, Valladolid, Zamora and Salamanca, and comprises three sections with different characteristics. The first is the upper or mountain course, where it descends in the form of a torrent until it reaches the city of Soria, but this section will not be discussed in this chapter. It then changes direction before crossing the inner plain from east to west. Once it has passed through the city of Zamora it becomes entrenched in the peneplains, and then regains its turbulence in the area of Los Arribes.

Its main tributary on its right bank is the Pisuerga, one of the longest tributaries on the entire Iberian Peninsula. Its most important tributaries include the Arlanza, which flows down from the Sierra de la Demanda, and the Carrión which flows from the Fuentes Carrionas massif, with an altitude of over 1800 m. The Pisuerga, along with the tributaries mentioned, supplies the Duero with an average of 2100 hm^3/year.

The Esla river is another fast-flowing tributary that enters the Duero on its right bank. Its source is located in the foothills of the central-western massifs of the Cordillera Cantábrica, and it collects the waters of the Porma, Bernesga, Órbigo, Tera and Aliste rivers; it is joined by the Cea river on its left bank. The tributaries on the left bank of the Duero river are slower-flowing as their headwaters are in the

Sistema Central range. These reliefs receive less rainfall and with a more irregular pattern than the Cordillera Cantábrica. The most abundant of these tributaries is the Tormes river, whose source is at an altitude of 2592 m in the Sierra de Gredos. Other notable tributaries on the left bank include the Duratón and the Cega rivers, and the Adaja which collects the waters of the Eresma.

When writing this text we have referred to Cabo Alonso and Manero (1991), Gutiérrez Elorza (coord.) (1994), Julivert et al. (1972), Junta de Castilla y León (1995), Morales et al. (1995), Loidi (1999), Panareda Clopés (coord.) et al. (1996), del Río (2005).

Bioclimatology

Figures 10.2 and 10.3 show the bioclimatology of the territory (thermotypes and ombrotypes respectively). Tables 10.1 and 10.2 establish the climatic parameters and bioclimatic indices for some meteorological stations which served as the basis for the corresponding bioclimatic diagnoses of the territory. Table 10.3 shows the relationships between the vegetation series (sigmeta) and the bioclimatic types and the biogeographic units of the territory.

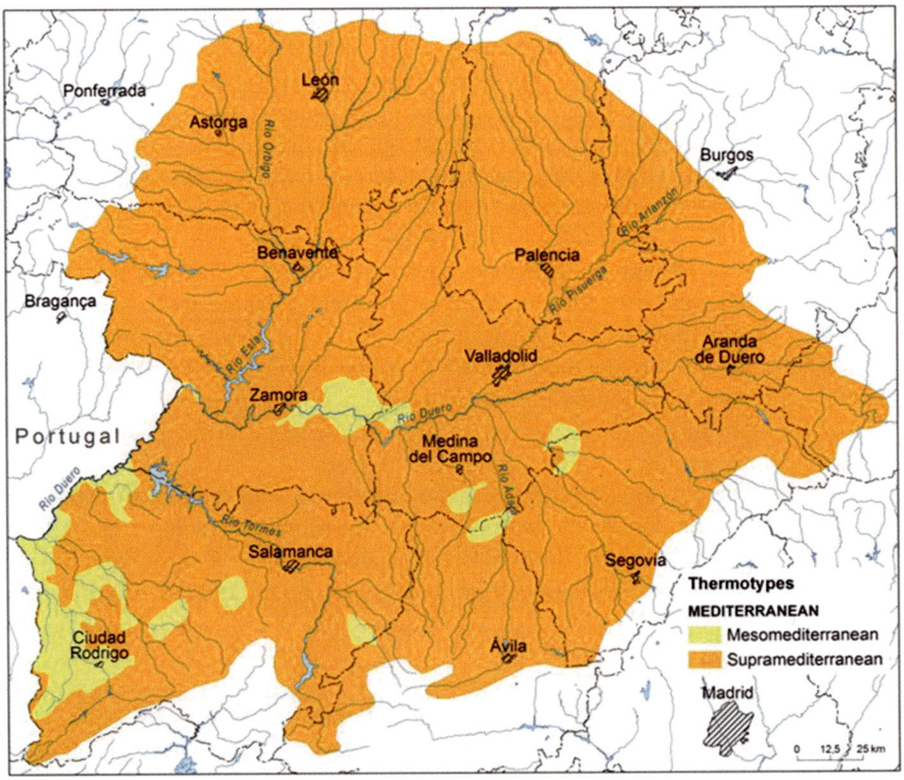

Fig. 10.2 Thermotypes of the Duero Basin

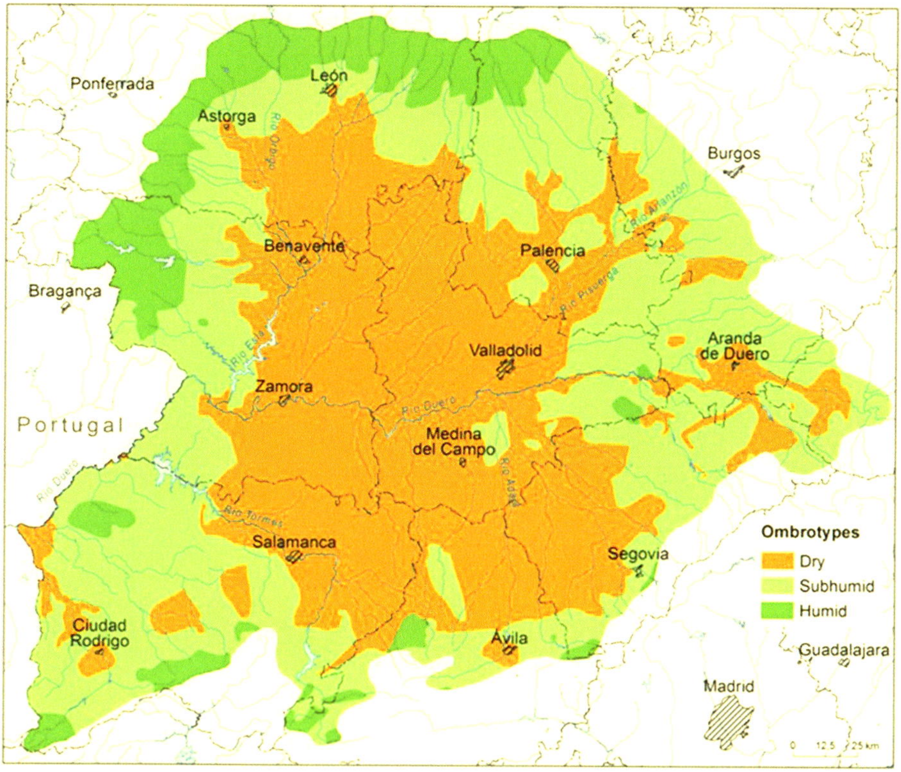

Fig. 10.3 Ombrotypes of the Duero Basin

10.2 Forests and Woodlands

10.2.1 Quercus pyrenaica *Forests*

In Guadarraman territories, in the supramediterranean thermotype and in subhumid-humid ombroclimates, several types of forests of *Quercus pyrenaica* occur. Its mature stage (Luzulo forsteri-Quercetum pyrenaicae) is fairly dense and has frequent spiny plants, an abundant herbaceous layer with a nemoral character and it grows on siliceous substrates on mountainsides and crests at between 1000 and 1300 m.

It occasionally occupies positions on somewhat steeper slopes, particularly those with southern exposures, and on dry crests and promontories in the lower mountain alignments. As the ombrotype changes it gives way to humid forests of Pyrenean oak ("melojares") rich in *Pulmonaria longifolia*, that occupy the length of the river beds and shady slopes (Pulmonario longifoliae-Quercetum pyrenaicae).

Table 10.1 Parameters and bioclimatic indices for some meteorological stations

Locality	Prov	Alt. (m)	Lat.	Long.	POT	POP	T	M	m	P	Itc	Tp	Ic	Io	Ios_2	Ios_3	Ios_4
Almazán	So	938	41° 29′ N	2° 32′ W	38	38	11.8	7.1	–1.8	560	178	1419	19.4	3.95	1.29	1.80	2.26
Arévalo	Av	820	41° 03′ N	4° 42′ W	38	38	12.2	7.8	–2.2	374	190	1465	20.5	2.55	0.60	1.04	1.30
Ávila	Av	1130	40° 39′ N	4° 41′ W	33	33	10.3	7.3	–0.9	368	166	1234	16.7	2.99	0.74	1.20	1.61
Benavente	Za	720	42° 00′ N	5° 40′ W	21	38	13.1	8.7	0.9	434	227	1574	17.5	2.75	0.71	1.09	1.43
Ciudad Rodrigo	Sa	653	40° 36′ N	6° 32′ W	38	38	13.4	9.4	0.9	547	236	1603	17.3	3.41	0,51	0.88	1.38
Linares del Arroyo	Sg	911	41° 31′ N	3° 33′ W	38	38	11.7	8.0	–0.8	461	189	1401	18.1	3.29	0.91	1.30	1.79
León	Le	820	42° 34′ N	1° 58′ W	31	31	10.8	7.4	–1.1	581	171	1300	17.1	4.31	1.24	1.66	2.21
Medina del Campo	Va	721	41° 18′ N	4° 55′ W	33	33	11.6	7.5	–1.4	393	179	1396	18.4	2.82	0.62	0.98	1.31
Palencia	P	750	42° 00′ N	4° 32′W	19	19	11.5	7.6	–1.7	424	174	1381	18.0	3.07	0.91	1.35	1.79
Puebla de Sanabria	Za	960	42° 03′ N	6° 38′ W	38	38	9.6	7.2	–2.4	998	143	1151	15.5	8.67	1.07	1.60	2.64
Salamanca	Sa	797	40° 50′ N	5° 40′ W	38	38	12.2	8.7	–0.9	434	203	1468	18.6	2.95	0.52	0.92	1.38
Segovia	Sg	1005	40° 57′ N	4° 37′W	38	38	11.8	7.5	0.5	459	198	1418	17.7	3.24	0.86	1.29	1.78
Soria	So	1063	41° 46′ N	2° 28′ W	40	40	10.5	6.7	–2.2	572	149	1255	17.3	4.56	1.61	2.12	2.68
Valladolid	Va	700	41° 42′ N	4° 41′ W	24	24	11.9	8.0	–0.4	407	195	1433	17.5	2.84	0.79	1.15	1.56
Venta de Baños	P	720	41° 55′ N	4° 29′ W	38	38	12.0	8.2	–0.1	433	200	1438	17.3	3.01	0.87	1.26	1.64
Zamora	Za	667	41° 29′ N	5° 45′W	38	38	12.6	7.9	0.8	417	212	1510	17.7	2.76	0.66	1.00	1.31

Prov Province, *Alt. (m)* Altitude (m), *Lat.* Latitude, *Long.* Longitude, *POT* Length of the period of temperature measurements (in years), *POP* Length of the period of precipitation measurements (in years), *T* Average temperature, *M* Average maximum temperature of the coldest month, *m* Average minimum temperature of the coldest month, *P* Average precipitation, *Itc* Compensated thermicity index, *Tp* Positive temperature, *Ic* Continentality index, *Io* Annual ombrothermic index, Ios_2 Bimonthly ombrothermic index, Ios_3 Three-monthly ombrothermic index, Ios_4 Four-monthly ombrothermic index

Table 10.2 Bioclimatic diagnosis for some meteorological stations

Locality	Bioclimatic diagnosis
Almazán	Mediterranean pluviseasonal oceanic low supramediterranean low subhumid
Arévalo	Mediterranean pluviseasonal oceanic low supramediterranean low dry
Ávila	Mediterranean pluviseasonal oceanic low supramediterranean upper dry
Benavente	Mediterranean pluviseasonal oceanic upper mesomediterranean low dry
Ciudad Rodrigo	Mediterranean pluviseasonal oceanic upper mesomediterranean upper dry
Linares del Arroyo	Mediterranean pluviseasonal oceanic low supramediterranean upper dry
León	Mediterranean pluviseasonal oceanic low supramediterranean low subhumid
Medina del Campo	Mediterranean pluviseasonal oceanic low supramediterranean upper dry
Palencia	Mediterranean pluviseasonal oceanic low supramediterranean upper dry
Puebla de Sanabria	Mediterranean pluviseasonal oceanic upper supramediterranean low humid
Salamanca	Mediterranean pluviseasonal oceanic low supramediterranean upper dry
Segovia	Mediterranean pluviseasonal oceanic low supramediterranean upper dry
Soria	Mediterranean pluviseasonal oceanic low supramediterranean low subhumid
Valladolid	Mediterranean pluviseasonal oceanic low supramediterranean upper dry
Venta de Baños	Mediterranean pluviseasonal oceanic low supramediterranean upper dry
Zamora	Mediterranean pluviseasonal oceanic low supramediterranean upper dry

The most representative taxa in the community include *Quercus pyrenaica, Crataegus monogyna, Prunus spinosa, Rosa corymbifera, Lavandula pedunculata, Cistus laurifolius, Adenocarpus hispanicus, Genista florida*; and among creepers and herbaceous species *Lonicera periclymenum, Tamus communis, Luzula forsteri, Physospermum cornubiense, Primula canescens, Lathyrus montanus, Lilium martagon, Brachypodium sylvaticum, Melica uniflora, Geum sylvaticum* and *Cruciata glabra*, among others. Its dynamics are shown in Table 10.4.

In the territories between the Duero river to the north and the areas of El Rebollar and Campo de Agadonez in the south, occupying an irregular area in the middle of the provinces of Zamora and Salamanca and with scant presence in the south of the province of León and in adjacent areas in Portugal corresponding to the areas of Miranda do Douro and Almeida, there is a forest of *Quercus pyrenaica* (Pyro bourgaeanae-Quercetum pyrenaicae) in a lower supramediterranean thermotype with an upper dry-upper subhumid character that occupies soils from acid rocks such as granite, sandstone, conglomerate and slate, with a predominance of granite (Ladero et al. 2004). Its floristic composition comprises *Pyrus bourgaeana, Crataegus monogyna, Rosa corymbifera, Rosa micrantha*, with a frequent presence of *Quercus faginea, Daphne gnidium* and *Paeonia broteri*, among others, although in Salamancan territories it is joined by *Genista tournefortii*. Its dynamics are shown in Table 10.5.

Table 10.3 Relationships between the potential natural vegetation and bioclimatology and biogeography

Mediterranean pluviseasonal-oceanic	Thermotypes		Ombrotypes				
Series (S)	mme	sme	dry	shu	hum	hhu	Biogeography in the North Plateau
Climatophilous Series							
Cephalanthero rubrae-Querco fagineae S. (calcicolous)	●	●	●	●			Castilian-Duero Sector
Genisto falcatae-Querco pyrenaicae S. (silicicolous)	●	●		●	●	●	Bierzo and Sanabria Sector
Junipero hemisphaerico-thuriferae S. (calcicolous)		●	●	●			Castilian Subprovince
Pulmonario longifoliae-Querco pyrenaicae S. (silicicolous)		●		●	●		León Plain Sector
Luzulo forsteri-Querco pyrenaicae S. (silicicolous)		●		●	●		Guadarrama Sierran Sector and Oroiberian subprovince
Pyro bourgaeanae-Querco pyrenaicae S. (silicicolous)		●		●	●		Salamancan and Bierzo and Sanabria Sectors
Festuco merinoi-Querco pyrenaicae S. (silicicolous)		●			●	●	Bejar and Gredos Sierras Sector
Climatophilous or edaphoxerophilous Series							
Junipero lagunae-Querco rotundifoliae S. (silicicolous)	●	●	●	●			Guadarrama Sierran and León Plain Sectors
Genisto hystricis-Querco rotundifoliae S. (silicicolous)	●	●	●	●	●		Bierzo and Sanabrian and Salamanca Sectors
Junipero thuriferae-Querco rotundifoliae S. (calcicolous)		●	●	●			Castilian Subprovince
Edaphohygrophilous or climato-temporihygrophilous Series							
Rubio tinctorum-Populo albae S.	●	●	●				South Castilian territory
Carici lusitanicae-Alno glutinosae S.	●	●		●	●		Iberoatlantic territory
Salici salviifoliae S.	●	●		●	●		León Plain Sector and North Oroiberian Sectors
Salici lambertiano-salviifoliae S.	●	●		●	●		Carpetania and León subprovince
Aro cylindracei-Ulmo minoris S.		●	●	●			North Castilian territory and León Plain Sector
Querco pyrenaicae-Fraxino angustifoliae S.		●	●	●	●		Carpetania and León subprovince
Populo nigrae-Salici neotrichae S.		●	●	●			North Castilian territory and León Plain Sector
Galio broteriani-Alno glutinosae S.		●	●	●	●		Carpetania and León subprovince
Salici cantabricae S.		●			●		León Plain Sector

This brief synopsis follows the proposals of Rivas-Martínez et al. (Rivas-Martínez et al. 2011a, b) and Rivas-Martínez et al. (2014) to establish the relationships between the potential natural vegetation at the vegetation series level and the bioclimatic belts (thermotypes and ombrotypes), and the corresponding biogeographic units. *S.* sigmetum, *mme* mesomediterranean, *sme* supramediterranean, *dry* dry, *shu* subhumid, *hum* humid, *hhu* hyperhumid

Table 10.4 Climax and seral stages of Luzulo forsteri-Quercetum pyrenaicae forests

Climax: **Luzulo forsteri-Quercetum pyrenaicae**
Seral stages
Cytiso oromediterranei-Genistetum cinerascentis (in mountain areas in the Sistema Central)
Genisto floridae-Adenocarpetum hispanici (in shady areas)
Thymo mastichinae-Cytisetum multiflori (in areas with high rainfall)
Festuco amplae-Agrostietum castellanae (temporary hydromorphia but with intense summer heat)
Festuco amplae-Cynosuretum cristati (with soil moisture)
Leucanthemopsio pallidae-Festucetum elegantis
Halimio ocymoidis-Cistetum populifolii (on skeletal soils)
Santolino rosmarinifoliae-Cistetum laurifoli (on highly eroded soils)

Table 10.5 Climax and seral stages of Pyro bourgaeanae-Quercetum pyrenaicae forests

Climax: **Pyro bourgaeanae-Quercetum pyrenaicae**
Seral stages
Cytiso scoparii-Genistetum polygaliphyllae
Genisto hystricis-Cytisetum multiflori
Lavandulo sampaianae-Cytisetum multiflori
Echinospartetum lusitanici (permanent community)

Table 10.6 Climax and seral stages of Festuco merinoi-Quercetum pyrenaicae forests

Climax: **Festuco merinoi-Quercetum pyrenaicae**
Seral stages
Thymo mastichinae-Cytisetum multiflori
Leucanthemopsio pallidae-Agrostietum elegantis
Festuco amplae-Cynosuretum cristati (with soil moisture)

In the supramediterranean territories of the eastern Gredos massif there are some Pyrenean oak forests with a marked Atlantic character which –due to their high value for livestock farming– are normally transformed into wooded pastures ("dehesas"), and are characterised by the presence of *Festuca elegans* subsp. *merinoi*, *Leuzea rhaponticoides*, *Nepeta coerulea*, *Euphorbia nevadensis*, *Knautia arvernensis*, *Lilium martagon*, *Melitis melissophyllum*, *Paeonia macrocarpa* and *Pulsatilla apiifolia*, among others (Festuco merinoi-Quercetum pyrenaicae) (Table 10.6).

Elsewhere in Guadarraman-Ayllonensian enclaves, and also in the foothills of the Sistema Ibérico range occupied by the submediterranean supratemperate thermotype, there is another type of forest of *Quercus pyrenaica* (Pulmonario longifoliae-Quercetum pyrenaicae), with abundant representation in the Leonese territories where it grows on siliceous soils in the supramediterranean oceanic pluviseasonal upper subhumid-humid Mediterranean thermotype. In these Leonese

Table 10.7 Climax and seral stages of Pulmonario longifoliae-Quercetum pyrenaicae forests

Climax: **Pulmonario longifoliae-Quercetum pyrenaicae**
Seral stages
Cytiso scoparii-Genistetum polygaliphyllae
Festuco amplae-Agrostietum castellanae (temporary hydromorphia but with intense summer heat)
Festuco amplae-Cynosuretum cristati (with soil moisture)
Pterosparto lasianthi-Ericetum aragonensis (on skeletal soils)

Table 10.8 Climax and seral stages of Genisto falcatae-Quercetum pyrenaicae forests

Climax: **Genisto falcatae-Quercetum pyrenaicae**
Seral stages
Cytiso scoparii-Genistetum polygaliphyllae (with *Cytisus multiflorus*)
Genisto hystricis-Cytisetum multiflori (shallow soils with a granite character)
Festuco amplae-Agrostietum castellanae (temporary hydromorphia but with intense summer heat)
Phalacrocarpo oppositifolii-Festucetum elegantis (deep soils)
Pterosparto lasianthi-Ericetum aragonensis (highly eroded soils)

territories it is enriched with taxa such as *Quercus pauciradiata, Lithodora diffusa* and *Chamaespartium sagittale*; in Iberian territories with *Erica vagans*; and in Ayllonensian territories with *Genista florida*. This community type is characterised by the following taxa: *Quercus pyrenaica, Genista florida* subsp. *polygaliphylla, Adenocarpus complicatus, Crataegus monogyna, Prunus spinosa, Hedera helix, Festuca heterophylla* subsp. *braun-blanquetii, Melampyrum pratense, Teucrium scorodonia, Brachypodium sylvaticum, Holcus mollis, Melica uniflora, Pulmonaria longifolia* and *Serratula tinctoria*, among others. Its dynamics are shown in Table 10.7.

The forests of *Quercus pyrenaica* in the supramediterranean territories with subhumid-humid ombroclimates in La Maragatería grow on humic cambisols and rarely on dystric planosols (Genisto falcatae-Quercetum pyrenaicae). The most notable phenological aspect of these forests, as with all forests which have *Quercus pyrenaica* as their dominant species, is that during autumn and a large part of winter the trees do not shed their dry leaves (marcescent leaves), which gives these forests their highly singular appearance. Frequent in the shrub layer are *Genista falcata, Genista florida* subsp. *polygaliphylla, Cytisus scoparius, Erica arborea* and *Crataegus monogyna*. In the herb layer there is an abundance of species such as *Holcus mollis, Stellaria holostea, Physospermum cornubiensis, Clinopodium vulgare, Teucrium scorodonia, Melampyrum pratense, Festuca elegans* subsp *merinoi, Doronicum plantagineum, Aquilegia vulgaris* subsp. *dichroa, Brachypodium sylvaticum* and others). Its dynamics are shown in Table 10.8.

10.2.2 Quercus rotundifolia *(Holm Oak) Forest and Woodlands (Fig. 10.4)*

In Castilian-Duero and Celtiberian-Alcarreñan territories and in some calcareous Planileonese enclaves with a supramediterranean and occasionally mesomediterranean thermotype and under a dry-subhumid ombrotype there are mixed forests of *Quercus rotundifolia* and *Juniperus thurifera* (sometimes absent) (Junipero thuriferae-Quercetum rotundifoliae). They are poor in shrubs and herbaceous species, but with some orchids such as *Cephalanthera longifolia*, *Ophrys scolopax*, *Ophrys sphegodes* and *Orchis morio*, along with other taxa such as *Osyris alba, Rubia peregrina, Teucrium pinnatifidum* and *Piptatherum paradoxum*, and shrubs such as *Juniperus communis* subsp. *hemisphaerica, Rhamnus infectoria, Jasminum fruticans*. They occupy base-rich soils, and sometimes loams, detrital clays, detrital calcium arkose, and even cohesive limestones, in which case the forest is enriched with the presence of *Quercus faginea*. Its dynamics are shown in Table 10.9.

In the territories located preferentially in the Leonese areas of La Maragatería and Cabrera, in the supramediterranean belt, there are forests of *Quercus rotundifolia* (Genisto hystricis-Quercetum rotundifoliae). These forest formations grow on base-poor soils with a dry-humid character and with the occasional presence –as in Salamancan and Zamoran territories– of *Quercus broteroi*. They have a shrub layer with an abundant presence of *Daphne gnidium*, *Cytisus*

Fig. 10.4 *Quercus rotundifolia* wood near San Justo de la Vega (León) (Photo by Sara del Río)

Table 10.9 Climax and seral stages of Junipero thuriferae-Quercetum rotundifoliae forests

Climax: **Junipero thuriferae-Quercetum rotundifoliae**
Seral stages
Rosetum micrantho-agrestis
Lino differentis-Salvietum lavandulifoliae (Castilian-Duero and Celtiberian-Alcarreñan)
Veronico javalambrensis-Thymetum mastigophori (Castilian-Duero)
Lino differentis-Lepidietum subulati (gypsiferous loams)
Genisto scorpii-Cistetum laurifolii (Castilian-Duero)

Table 10.10 Climax and seral stages of Genisto hystricis-Quercetum rotundifoliae forests

Climax: **Genisto hystricis-Quercetum rotundifoliae**
Seral stages
Genisto hystricis-Cytisetum multiflori (Salamanca with an oceanic influence)
Lavandulo sampaianae-Cytisetum multiflori (western Salamanca)
Genisto hystricis-Echinospartetum lusitanici (lithosols)
Lavandulo sampaianae-Genistetum hystricis (dry lithosols)
Pterosparto lasianthi-Ericetum aragonensis (humid lithosols)
Diantho merinoi-Plantaginetum radicatae (slates)

scoparius, *Lonicera etrusca*, *Lavandula sampaiana*, *Genista hystrix*, *Crataegus monogyna* and various species of *Cistus* sp. pl. The most frequent herb species include *Rubia peregrina, Arenaria montana*, *Teucrium scorodonia*, *Silene nutans* subsp. *nutans*, *Clinopodium vulgare* and sometimes *Festuca elegans* subsp. *merinoi*. In supramediterranean Salamancan territories in the protective edges or first substitution stages, *Genista florida* subsp. *polygaliphylla* (Zamoranan-Maragatan) is replaced with *Genista florida* subsp. *florida*. In contact with the Eurosiberian region they occupy exposed biotopes –particularly south-facing and growing on skeletal soils– and thus function as a permanent community. They are very frequent in the Leonese areas of La Cabrera, La Maragatería, La Cepeda, in areas of El Tera and the western part of Tierra del Pan in Zamora, and in Salamanca, and present a very interesting variation in the area of Los Arribes del Duero, where –given their thermophilous character– they are enriched with *Olea europaea* subsp. *sylvestris*, among others. Their dynamics are shown in Table 10.10.

The forests of *Quercus rotundifolia* (Junipero lagunae-Quercetum rotundifoliae) have their optimum in Guadarraman territories, although they are also frequent in areas of northern Castile, the Leonese plains, Arévalo and Tierra de Pinares. They are eminently supramediterranean, although they appear sporadically in the Guadarraman upper mesomediterranean thermotype developed on solid substrates, which promote the presence of *Retama sphaerocarpa*. They grow under a dry-subhumid ombrotype.

The most characteristic taxa in this community are the following: *Quercus rotundifolia, Juniperus laguna, Cytisus scoparius, Lavandula stoechas* subsp. *pedunculata, Retama sphaerocarpa, Osyris alba, Daphne gnidium, Rubia*

Table 10.11 Climax and seral stages of Junipero lagunae-Quercetum rotundifoliae forests

Climax: **Junipero lagunae-Quercetum rotundifoliae**
Seral stages
Cytiso scoparii-Genistetum polygaliphyllae (with *Lavandula pedunculata*)
Cytiso scoparii-Retametum sphaerocarpae (Salamancan, Guadarraman mesomediterranean)
Festuco amplae-Cynosuretum cristati (subhumid Leonese)
Festuco amplae-Agrostietum castellanae
Arrhenathero baetici-Stipetum giganteae (without hydromorphia)
Centaureo ornatae-Stipetum clausae (with intermediate hydromorphia)
Genisto scorpii-Cistetum laurifolii (Leonese)
Santolino rosmarinifoliae-Cistetum laurifolii (Guadarraman and Bejarano-Gredensian)
Rosmarino-Cistetum ladaniferi (thermic, Guadarraman)

peregrina, *Lonicera etrusca*, *Teucrium scorodonia*, *Agrostis castellana*, *Celtica gigantea*, *Paeonia broteroi* and *Carex distachya*.

On the northern plateau, these forests –depending on the area– are enriched with taxa that are bioindicators of its geographical diversity. In the Leonese areas of El Páramo and Oteros, that is *Brachypodium rupestre*; in Tierra de Pinares, *Adenocarpus aureus* acquires a certain relevance due to the psammophilous character of its soils; and in contact with the Guadarraman oromediterranean thermotype there is a presence of *Cytisus oromediterraneus*. In the Arévalo area, forests can be found growing on sandy-silty detrital sediments. Their dynamics are shown in Table 10.11.

10.2.3 Quercus faginea *Forests*

The forests of *Quercus faginea* (Cephalanthero rubrae-Quercetum fagineae) on the northern plateau occupy both deep clayey-limestone soils and decarbonated red clay. They grow in Castilian-Duero and Celtiberian-Alcarreñan territories in mesomediterranean and supramediterranean thermotypes under a dry-subhumid, and –less frequently– a humid ombrotype.

The mature stage corresponds to a forest dominated by *Quercus faginea*, occasionally with *Sorbus* sp. and *Acer* sp., with an understory with a large number of shrubs, creepers and herbaceous plants with basophilous preferences. The most representative species in the community include the following: *Acer monspessulanus*, *Sorbus aria*, *Jasminum fruticans*, *Euonymus europaeus*, *Crataegus monogyna*, *Ligustrum vulgare*, *Rosa corymbifera*, *Viburnum lantana*, *Lonicera etrusca*, *Lonicera periclymenum*, *Hedera helix*, *Asperula aristata*, *Cephalanthera rubra*, *Cephalanthera longifolia* and *Geum sylvaticum*, among others. In warmer territories there may be a presence of *Quercus coccifera* and *Brachypodium rupestre* in Castilian-Duero areas, indicating its variability. Its dynamics are shown in Table 10.12.

Table 10.12 Climax and seral stages of Cephalanthero rubrae-Quercetum rotundifoliae forests

Climax: **Cephalanthero rubrae-Quercetum fagineae**
Seral stages
Rosetum micrantho-agrestis
Genisto scorpii-Retametum spharocarpae
Mantisalco salmanticae-Brachypodietum phoenicoidis (limestone or loams)
Lino differentis-Lepidietum subulati (Castilian-Duero, on gypsum)
Salvio lavandulifoliae-Linetum appressi (Continental Celtiberian-Alcarreñan areas)
Lino differentis-Salvietum lavandulifoliae (Castilian-Duero and Celtiberian-Alcarreñan)

Table 10.13 Climax and seral stages of Juniperetum hemisphaerico-thuriferae forests

Climax: **Juniperetum hemisphaerico-thuriferae**
Seral stages
Rosetum micrantho-agrestis
Mantisalco salmanticae-Brachypodietum phoenicoides (marly-limestone soils)
Salvio lavandulifoliae-Linetum appressi (continental areas)
Festucetum hystricis (intensively grazed stony calcareous lithosols)
Genisto scorpii-Cistetum laurifolii (in riverbeds with a decalcified substrate)
Lysimachio ephemeri-Holoschoenetum (banks oozing water saturated in CO_3Ca)

10.2.4 Juniperus thurifera *Woodlands*

On the northern plateau, the mature stage of *Juniperus thurifera* woodlands (Juniperetum hemisphaerico-thuriferae) corresponds to a forest with an open structure, dominated by *Juniperus thurifera*, which has a layer of phanerophytes comprising savin junipers with some holm oaks –depending on the area–, and a layer with predominantly junipers, some roses or spiny species, and tree-sized savin junipers. The lower shrubby layer is dominated by fruticose chamaephytes.

It develops on thin, often skeletal, dark-coloured limestone soils, always extremely dry during the summer months and frequently subjected to winter cryoturbation and with the presence of numerous stones. In areas of deeper soils, there may be a presence of *Quercus rotundifolia* or *Quercus faginea*, whereas in decalcified areas it is frequent to find *Cistus laurifolius*. It grows under a Mediterranean pluviseasonal-oceanic supramediterranean dry-subhumid bioclimate. The most characteristic taxa in this community are the following: *Juniperus thurifera, Juniperus communis* subsp. *hemisphaerica, Juniperus sabina, Berberis vulgaris* subsp. *seroi, Rhamnus infectoria, Salvia lavandulifolia, Rosa agrestis, Artemisia lanata, Festuca trichophylla, Festuca hystrix, Poa ligulata* and *Linum appresum*, among others. They sometimes occupy base-poor soils, particularly in Guadarraman territories, in which case it is very frequent to find *Juniperus lagunae*. Its dynamics are shown in Table 10.13.

10.3 Riparian Forests

10.3.1 Ulmus minor *Forest* (Olmedas)

These are Northern Castilian and oroiberian supramediterranean elm forests in dry to humid ombroclimates, with a predominance of *Ulmus minor* and *Fraxinus angustifolia* in the tree layer, and an understory with species such as *Arum cylindraceum, Ligustrum vulgare, Rosa corymbifera* and others. They grow on "vega parda" soils or fluvisols in river valleys, but in a stratum somewhat removed from the watercourse where the river water table is deep and flooding infrequent; that is, on the floodplain (Aro cylindracei-Ulmetum minoris).

10.3.2 Fraxinus angustifolia *Forests* (Fresnedas angustifolias)

Mixed forests of *Fraxinus angustifolia* and *Quercus pyrenaica* (Fraxino angustifoliae-Quercetum pyrenaicae) are very abundant in Carpetan-Leonese territories, preferentially Salamancan and Guadarraman, although we can also find them more specifically in the following areas: the Sistema Ibérico and even in the Bercian-Sanabrian sector in the supramediterranean thermotype under the dry-subhumid ombrotype. Their mature stage corresponds to a forest that occupies riverbeds and valley plains with temporary hydromorphia on deep soils with a sandy texture. The most representative taxa in the community include the following: *Fraxinus angustifolia, Quercus pyrenaica, Acer monspessulanus, Rubus ulmifolius, Rosa canina, Rosa corymbifera, Crataegus monogyna, Arum maculatum, Poa nemoralis, Brachypodium sylvaticum, Viola riviniana, Alliaria petiolata* and *Bryonia cretica* subsp. *dioica*.

In dry-subhumid mesomediterranean territories, preferentially Salamancan and sporadically Guadarraman, there is a presence of forests of *Fraxinus angustifolia* (Ficario ranunculoidis-Fraxinetum angustifoliae) which occupy infrequently flooded sandy siliceous soils, whose most representative taxa are *Fraxinus angustifolia, Populus alba, Populus nigra, Salix atrocinerea, Osmunda regalis, Ranunculus ficaria* and *Arum italicum*, among others. This formation has been highly deforested for use for agriculture and livestock farming, and is thus very scarcely represented, only appearing on certain sites in the province of Salamanca around the Alagón river, and in Los Arribes del Duero.

10.3.3 Populus nigra *Forests* (Choperas-saucedas)

Black poplar forests with a very wide distribution on the North Plateau (Salici neotrichae-Populetum nigrae) colonise the broad riverbeds of the middle and lower

Table 10.14 Climax and seral stages of Salici neotrichae-Populetum nigrae forests

Climax: **Salici neotrichae-Populetum nigrae**
Seral stages
Rubo ulmifolii-Rosetum corymbiferae
Festuco amplae-Agrostietum castellanae

sections of fast-flowing rivers with an insignificant summer low water level. They constitute the tree belt closest to the flowing waters in the broad valleys, and are located near woodlands of tree willows and elms.

They grow on fluvisols and –in spite of their indifference to the substrate–grow better on base-rich clayey biotopes where there is a clear supply of alkaline waters.

The most characteristic taxa in this community are *Populus nigra, Salix neotricha, Salix atrocinerea, Populus alba, Sambucus nigra, Fraxinus angustifolia, Salix* x *erythroclados, Crataegus monogyna, Cornus sanguinea, Rubus ulmifolius, Ranunculus ficaria, Poa nemoralis, Brachypodium sylvaticum, Cucubalus baccifer* and *Bryonia dioica*, among others. In the northern Castilian territories it is enriched with *Salix elaeagnos* subsp. *angustifolia* and in Planileonese areas with *Salix salviifolia*. Its dynamics are shown in Table 10.14.

In mesomediterranean Castilian-Duero territories with a dry ombrotype we find dense gallery forests, preferentially in the middle and lower courses of Mediterranean rivers on floodable eutrophic soils with permanent hydromorphia (Rubio tinctorum-Populetum albae). The following species best define the community: *Populus alba, Populus nigra, Salix alba, Ulmus minor, Tamarix canariensis, Humulus lupulus, Rubia tinctorum, Solanum dulcamara, Saponaria officinalis, Brachypodium sylvaticum* and *Clematis vitalba*, among others.

There are several variants of these poplar woodlands that can be distinguished according to the predominant tree species. The differentiation in their floristic composition is almost imperceptible, although the most common are the woodlands of *Salix neotricha* and *Populus nigra*, which occur depending on either siltier or sandier soils, respectively.

10.3.4 Alnus glutinosa *Forests* (Alisedas)

These are alder forests on hygro-peaty substrates that undergo prolonged flooding, with an iberoatlantic distribution in the subhumid supramediterranean, and more specifically on the Leonese moorland. The community (Carici lusitanicae-Alnetum glutinosae) is dominated by *Alnus glutinosa* and *Salix atrocinerea* in the tree layer, and has an understory with a tuft-like appearance, with species such as *Carex paniculata* subsp. *lusitanica*, together with characteristic plants from the adjacent communities of the Phragmito-Magnocaricetea and Filipenduletalia (*Oenanthe croccata, Iris pseudacorus, Phragmites australis, Galium palustre, Mentha aquatica, Lythrum salicaria*, and others).

The alder woodlands (Carici broteriani-Alnetum glutinosae) grow on luvisol and fluvisol type soils, occupying the banks of fast-flowing rivers and streams which do not dry up in summer and flow through narrow valleys, in meso- and supramediterranean thermotypes. They are dominated by *Alnus glutinosa*, *Fraxinus angustifolia*, *Ulmus minor*, and two western Iberian elements: *Carex reuteriana* and *Galium broterianum*. Given the proximity to Eurosiberian territories it is also frequent to find *Fraxinus excelsior*. The most characteristic taxa in this community are the following: *Alnus glutinosa*, *Ulmus minor*, *Fraxinus angustifolia*, *Salix atrocinerea*, *Populus nigra*, *Frangula alnus*, *Acer pseudoplatanus*, *Crataegus monogyna*, *Ilex aquifolium*, *Carex reuteriana*, *Galium broterianum*, *Paris quadrifolia*, *Galium rotundifolium*, *Poa nemoralis* and *Brachypodium sylvaticum*, among others. On the North Plateau they are present in the upper basin of the Tormes river.

These forests potentially represent the climax of the territory and occupy a large part of the riverbed. They contain scionitrophilous grasslands (Geranio robertiani-Cariolophetum sempervirentis) and fringe permanent water communities formed by *Glyceria declinata*, *Oenanthe crocata*, *Phalaris arundinacea*, and others (Glycerio declinatae-Oenanthetum crocatae). They are always located in areas with slower currents or in backwaters.

10.3.5 **Salix *Woodlands* (Saucedas arbustivas)**

In Leonese and Iberian –and occasionally Castilian-Duero– mountain areas, occupying lesser riverbeds and on soils with a coarse, occasionally gravelly, sandy and even silty granulometry –but always base-poor and subjected to a long period of summer low-water levels– there are formations of *Salix salviifolia* (Salicetum salviifoliae) with a predominance of *Salix salviifolia*, *Salix* x *secalliana*, *Salix triandra* subsp. *discolor*, *Salix atrocinerea* and even *Fraxinus angustifolia* (Fig. 10.5). Among the herbs present in these shrubby willow woodlands distributed throughout Galicia, northern Portugal, León and the Sistema Ibérico, are *Brachypodium sylvaticum*, *Saponaria officinalis*, *Potentilla reptans*, and others. These tree formations are frequent in two meso- and supramediterranean thermotypes, provided their soils have an acid character. Among their companion communities it is worth noting grasslands on pebbles that stand dry during periods of low water, and contain *Lactuca chondrilliflora*, *Andryala ragusina*, *Scrophularia canina* and others (Lactuco chondrilliflorae-Andryaletum ragusinae).

Elsewhere, and even when their character is highly eurosiberian, occasionally in territories in contact with humid Mediterranean supramediterranean thermotypes, it is not uncommon to find formations of *Salix cantabrica* (Salicetum cantabricae) in an isolated representation in some enclaves such as in Planileonese territories and in the westernmost part of Castilian-Cantabrian territories. Its mature stage corresponds to a willow forest with trees colonising the edges of river beds and currents that undergo substantial flooding in spring and do not suffer a prolonged low-water

Fig. 10.5 Riparian vegetation in the Esla river (Photo by Sara del Río)

level, occupying the belt nearest to the floodable areas of the river courses. They thrive on stony, sand and silty substrates, generally rich in calcium carbonate. The most characteristic taxa in this community are *Salix cantabrica, Salix alba, Salix elaeagnos* subsp. *angustifolia, Salix atrocinerea, Salix triandra* subsp. *discolor, Salix* x *expectata* and *Salix* x *legionensis*, the latter in areas with greater Mediterranean influence.

Similar to the previous willow woodlands, but with a preferentially Mediterranean character on base-rich substrates, we find stands of shrubby willows with a predominance of *Salix purpurea* subsp. *lambertiana*, *Salix salviifolia*, *Salix* x *legionensis*, *Salix* x *secalliana*, *Salix* x *pseudosalvifolia* and *Salix eleagnos* subsp. *angustifolia* (Salicetum lambertiano-salviifoliae). The herb layer is constituted by *Brachypodium sylvaticum*, *Solanum dulcamara*, *Geum urbanum*, *Crepis lampsanoides* and *Carex reuteriana*, among others. These willow woodlands grow in the meso- and supramediterranean thermotypes (Díaz González and Penas 1987).

10.4 Seral Scrub

Seral scrublands are highly variable in this broad territory in which there is such a diverse range of climate, soil and land-use factors. It is possible to differentiate the scrub in the first substitution phase with a spiny or aphyllous nanophanerophytic character, and the scrub representing the maximum possible level of seral degradation before encountering pioneer communities with a permanent character. We can thus initially differentiate the spiny nanoshrublands with indifferent soil preferences with a predominance of species belonging to the genera *Rosa* and *Rubus*, together with other species, spiny or not, such as *Prunus* sp. pl., *Crataegus* sp. pl. and others; and on acid substrates the broom communities ("piornales") dominated by nanophanerophytes that may attain heights of between 4 and 6 m, including particularly *Cytisus* sp. pl., *Genista* sp. pl. (in this case preferentially aphyllous), and others. The stages farther removed from the climax and with a chamaephytic character –also on acid soils– include "brezales" or "brezales-jarales"; the first comprising mainly various *Erica* sp. pl. along with *Pterospartum* sp. pl., whereas the "brezales-jarales" –although they also have some *Erica* sp. pl.– have a greater predominance of *Halimium* sp. pl. and *Cistus* sp. pl. Finally there are "jarales", where *Cistus* sp. pl. are dominant, accompanied by *Lavandula* sp. pl.

On base-rich substrates, in addition to spiny shrublands there is presence of gypsophilous scrub –which we will discuss in a specific section–, and "tomillares" or "salviares" with a predominance of *Thymus* sp. pl. or *Salvia* sp. pl. with *Artemisia* sp. pl., respectively.

Seral scrub grows on both skeletal base-rich or base-poor soils (lithosols) and on deep waterlogged soils; in the latter the scrub has a hygrophilous character having also a predominance of *Erica* sp. pl. and spiny *Genista* sp. pl.

10.4.1 Spiny Nanoshrublands

On deep siliceous soils with a Mediterranean optimum (meso- and supramediterranean), with a wide Carpetan-Leonese distribution, there is a community (Rubo ulmifolii-Rosetum corymbiferae) whose composition features a combination of species in the *Rosa* genus (*Rosa micrantha*, *Rosa corymbifera*, *Rosa canina*, etc.), in addition to *Rubus ulmifolius*, *Rubus caesius*, *Tamus communis*, *Hedera helix*, and *Lonicera etrusca*, which form a protective edge and the first seral stage in woodlands of ash, elm, willow and others (Arnaiz and Loidi 1983).

In contrast, on deep base-rich soils and as the first substitution stage of forests of *Quercus rotundifolia*, preferentially in Castilian-Duero territories, there is another community of spiny nanoshrublands (Rosetum micrantho-agrestis), whose floristic composition includes *Rosa agrestis*, *Rosa micrantha*, *Rosa deseglisei*, *Rosa pouzinii* and others.

10.4.2 Broom Communities (Piornales)

We have already noted in the tables showing the natural dynamics of the different types of forests and woodlands that these formations are seral stages, and correspond to the first seral stage. They grow –with exceptions– on base-poor soils, and in this section we will therefore only indicate their floristic composition and their distribution (Penas et al. 1988a; Rivas-Martínez et al. 1987a).

Heathlands dominated by *Cytisus oromediterraneus* and *Genista cinerascens* (Cytiso oromediterranei-Genistetum cinerascentis), accompanied by *Juniperus communis* subsp. *hemisphaerica, Deschampsia flexuosa subsp. iberica, Festuca elegans subsp. merinoi*, grow on deep siliceous soils in subhumid-humid Bejarano-Gredensian supramediterranean territories.

Tall heathlands grow on deep soils with a siliceous nature, characteristic of Zamoran-Maragatan Planileonese supramediterranean territories with a humid-hyperhumid ombroclimate. These communities are dominated by *Genista florida* subsp. *polygaliphylla, Cytisus scoparius, Erica arborea* and *Pteridium aquilinum* (*Cytiso scoparii-Genistetum polygaliphyllae*), and show considerable variability. In the westernmost territories their floristic composition may be enriched with the presence of *Cytisus multiflorus*, and in the Planileonese territories with the presence of *Lavandula pedunculata*.

On siliceous soils formed from slate (and granite in a dry ombroclimate), preferentially Guadarraman –although it can also be found in Salamancan territories in the mesomediterranean belt with a dry ombrotype–, there is a community characterised by the presence of *Cytisus scoparius* and *Retama sphaerocarpa* (Cytiso scoparii-Retametum sphaerocarpae). This community is frequent in Salamancan territories on humic cambisols with a subhumid supramediterranean character, and comprises *Genista cinerascens, Genista florida* subsp. *florida, Cytisus scoparius, Pteridium aquilinum* and *Erica arborea*, among others (Genisto floridae-Cytisetum scoparii).

On dystric cambisols in the subhumid-humid supramediterranean, it is not uncommon to find communities formed by *Genista hystrix, Genista florida* subsp. *polygaliphylla, Cytisus scoparius, Cytisus multiflorus* and *Adenocarpus complicatus*, which in Maragatan-Zamoran territories represent the first seral stage of the forests of the Genisto hystricis-Quercetum rotundifoliae (Genisto hystricis-Cytisetum multiflori).

Occasionally, as the soil horizons are lost due to the effect of forest fires, they become permanent communities, where *Genista hystrix* and *Echinospartum lusitanicum* are exclusive and dominant (Genisto hystricis-Echinospartetum lusitanici).

Also growing on humic cambisols in the subhumid supramediterranean belt is a community formed by *Genista cinerascens, Genista florida* subsp. *florida, Thymus mastichina, Thymus zygis, Cytisus multiflorus, Cytisus scoparius, Cistus ladanifer, Lavandula pedunculata* and *Erica arborea* (Thymo masctichinae-Cytisetum multiflori).

In Zamoran and occasionally Salamancan territories holm-oak woodlands can be found as a substitution stage (Junipero lagunae-Quercetum rotundifoliae, Pyro bourgaeanae-Quercetum rotundifoliae), with a community formed by *Retama sphaerocarpa* and *Cytisus scoparius*, growing on luvisols and acrisols from the edaphisation of pre-Cambrian slates (Cytiso scoparii-Retametum sphaerocarpae).

10.4.3 Heathlands (Brezales)

As in the previous section, we will mention the different communities present, indicating their distribution, bioclimatology, floristic composition and soil classification (Fig. 10.6).

Chamaephytic heathlands grow on dystric lithosols formed from siliceous rocks, and in humid supramediterranean territories; they are characterised by the presence of *Halimium ocymoides* and *Erica australis* subsp. *aragonensis*, together with *Calluna vulgaris* (Halimio ocymoidis-Ericetum aragonensis). In dry meso-supramediterranean areas, however, the community is formed by *Erica australis* subsp. *australis Erica australis subsp. australis*, along with *Erica scoparia* and *Halimium ocymoides, Erica umbellata, Pterospartum lasianthum, Cistus ladanifer, Tuberaria vulgaris* and *Polygala microphylla* (Polygalo microphyllae-Cistetum populifolii). Both formations are widespread through the western part of the North Plateau.

In these western territories there are other degraded heathlands typical of lithosols or skeletal soils, also with a Mediterranean character with a high coverage (Pterosparto lasianthi-Ericetum aragonensis). Floristically they show a predominance of *Erica australis* subsp. *aragonensis* and *Halimium umbellatum*, along with *Pterospartum tridentatum* subsp. *lasianthum, Erica umbellata, Erica cinerea* and *Calluna vulgaris*, and have a Zamoran, Sanabrian and Planileonese distribution.

Fig. 10.6 Heathland with pine plantation in the surroundings of Astorga (León) (Photo by Sara del Río)

10.4.4 Hygrophilous Heathlands (Brezales higrófilos)

These formations also include heathlands developed on hydromorphic pseudogley podzols. They grow in areas with a tendency to waterlogging as they have very little gradient, such as foothills, depressions and inter-riverine areas. They are characterised floristically by the presence of *Erica tetralix*, *Genista anglica* and *Genista micrantha* (*Genisto anglicae-Ericetum tetralicis*), sometimes enriched by *Erica vagans* (Genisto anglicae-Ericetum vagantis), in both cases in subhumid-humid supramediterranean territories.

10.4.5 Secondary Scrub Communities Dominated by Cistaceae *and* Lamiaceae (Jarales)

On fairly decarbonated Miocene loams, as a substitution stage amidst the basophilous holm-oak forests (Junipero thuriferae-Quercetum rotundifoliae), there are formations dominated by *Genista scopius* accompanied by a large number of acidophilous species such as *Cistus laurifolius*, *Lavandula pedunculata* and *Halimium viscosum*, along with other basophilous species such as *Stahelina dubia* (Genisto scorpii-Cistetum laurifolii) (Díaz González et al. 1988c).

In Guadarraman supramediterranean territories, preferentially at shady sites and on lithosols, we find communities formed by *Cistus laurifolius*, *Santolina rosmarinifolia*, *Lavandula pedunculata* and *Thymus mastichina*, and others (Santolino rosmarinifoliae-Cistetum laurifolii) (Díaz González et al. 1989).

In these same territories but with a more continental character than those mentioned previously there are "jarales" dominated by *Cistus ladanifer* along with *Rosmarinus officinalis*, and other species (Rosmarino officinalis-Cistetum ladaniferi) (Rivas-Martínez 1968).

As a more regressive stage of the dry-subhumid supramediterranean forests of the Genisto hystricis-Quercetum rotundifoliae in La Maragatería there is a community on slate lithosols formed by *Lavandula sampaioana*, *Genista hystrix*, *Erica umbellata* and *Cistus ladanifer*, and others (Lavandulo sampaianae-Genistetum hystricis).

10.4.6 Thymus *Communities* (Tomillares)

These are communities dominated by hemicryptophytes and decumbent chamaephytes such as *Veronica jabalambrensis, Thymus mastigophorus, Fumana procumbens, Linum appressum, Teucrium expansum, Koeleria vallesiana* and *Coronilla minima*, among others (Veronico jabalambrensis-Thymetum mastigophori), with a fundamentally subhumid supramediterranean northern

Castilian-Duero and western Castilian-Cantabrian distribution. They grow on more or less decapitated marly soils.

On acid substrates we find communities growing on shallow lithosols or ranker soils with a frequent presence of *Thymus zygis*, *Plantago radicata*, *Corynephorus canescens*, *Koeleria crassipes* and *Hieracium castellanum*, which in Maragatan territories are joined by the endemic *Dianthus merinoi* (Diantho merinoi-Plantaginetum radicatae, Thymo zygidis-Plantaginetum radicatae).

10.4.7 Nitrophilous Scrub (Bolinares)

Communities dominated by *Santolina rosmarinifolia*, *Artemisia campestris* subsp. *glutinosa* and *Helichrysum serotinum* among others (Artemisio glutinosae-Santolinetum rosmarinifoliae) attain their optimum in Guadarraman territory. These are the most widely distributed communities and they appear on dystric or chromic cambisols or arenosols, linked to series of evergreen oak and deciduous oak forests. In the Zamoran-Sanabrian and Planileonese territories (Rivas-Martínez et al. 1986), on slopes, verges and sides of roadways or on forest tracks, and in other places where the soil has been disturbed, the dominant taxon is *Santolina rosmarinifolia* subsp. *semidentata* (Artemisio glutinosae-Santolinetum semidentatae).

10.5 Gypsophylous Vegetation

The central territory in the Duero basin is formed by sedimentary materials from the Miocene era, largely gypsiferous loams which often form gypsum. In these strata there is an abundance of gypsum crystals, occasionally mixed with arrowhead or desert rose forms, all deriving from continental freshwater lacustrine evaporitic sedimentation. The soils are basic and slightly saline, and contain calcium; they have a pH between 7.7 and 8.4 (Loidi 2011).

These gypsiferous outcrops occur preferentially in the central region of the basin, particularly on the moorlands of Esgueva, in the areas of Peñafiel and Medina del Campo in the province of Valladolid, and in the area of El Cerrato in the province of Palencia. In the east they reach the area of the lower Arlanzón valley in Burgos and the uplands of Torresandino; in the south they extend as far as the areas around Cuéllar in the province of Segovia; and westwards as far as Toro (Zamora).

These outcrops have an abundance of Miocene reliefs carved out by the rivers Arlanza, Arlanzón, Pisuerga, Duero, Esgueva and Cega. They occupy 5179 km^2 and their average altitude is around 850 m (Loidi 2011).

The main potential natural vegetation consists of forests of holm oaks and Phoenician juniper (Junipero thuriferae-Quercetum rotundifoliae), although there are also areas with Portuguese oak woodlands (Cephalanthero-Quercetum fagineae), or even more reduced juniper woodlands (Juniperetum hemisphaerico-thuriferae).

The management of this territory by humans has transformed its potential natural variation into areas of cereal cultivation, leaving uncultivated areas or seral waves in which the abandoned croplands have led to their colonisation by a large number of gypsophilous species, including particularly *Lepidium subulatum*, *Ononis tridentata*, *Reseda stricta* subsp. *stricta*, *Hedysarum boveanum* subsp. *palentinum*, *Gypsophila bermejoi* and *Campanula fastigiata*; which along with *Salvia lavandulifolia*, *Linum appresum*, *Helianthemum cinereum* subsp. *rotundifolium* and *Fumana procumbens*, and others typical of the class Rosmarinetea officinalis, form the communities of the Lino differentis-Lepidietum subulati. This contrasts with the area of La Bureba in Burgos, where there are formations of the Sideritido linearifoliae-Gypsophletum hispanicae and a frequent presence of species such as *Gypsophila hispanica*, *Sideritis linearifolia* and *Helianthemum incanum*, among others.

Along with these heathlands there are the nitrophilous formations in the class *Pegano-Salsoletea*, corresponding to the Artemisio herba-albae-Santolinetum squarrosae, and representing the nitrophilous chamaephyte vegetation typical of marly-gypsiferous and marly-limestone substrates in the areas of Tierra de Campos and the valleys of El Cerrato (Ladero et al. 1994). They include *Artemisia herba-alba*, *Santolina squarrosa*, *Kochia prostrata* and occasionally *Camphorosma monspeliaca*.

There are also "hargamales" of the Pegano harmalae-Salsoletum vermiculatae in these territories. They reach their western extension in the Tierra del Pan and in the basin of the river Guareña (province of Zamora), and in Castronuño (Valladolid) where *Peganum harmala*, *Salsola vermiculata* and *Mercurialis tomentosa* occur.

10.6 Grasslands

Grasslands occur very widespread throughout the length and breadth of the territory, depending on its climate and soil characteristics, although the differences in floristic composition basically depend on three factors: the level of soil moisture, its depth, and the altitude.

The floristic composition can differ drastically from the natural state, owing to the actions of humans and livestock, as overgrazing can drastically transform the species composition of the grasslands. These differences in plant composition of these grasslands (pastures and meadows) have given rise to a series of popular denominations that clearly denote their different composition. Thus, on the North Plateau the potential plant communities that formerly occupied these territories have today been destroyed –through natural processes, but more often due to human action– and are replaced by plant communities used as grazing for livestock.

Here we include the "cervunales" (dominated by *Nardus stricta*), "berceales" (dominated by *Celtica gigantea*), "majadales" (dominated by *Poa bulbosa*), "vallicares" (dominated by *Agrostis castellana*), "rushy pastures" (dominated among others by *Deschampsia flexuosa*, *Juncus effusus* and *Cirsium pyrenaicum*), "juncales" (dominated by *Juncus acutiflorus*), "fenalares" (dominated by *Brachypodium phoenicoides*), meadows (dominated by *Cynosurus cristatus*), "cerrillares" (dominated by *Festuca elegans* subsp. *merinoi*), "lastonares" (dominated by *Stipa clausa*), and a series of annual grasslands which due to their great diversity we will discuss later.

10.6.1 Nardus stricta *Communities* (Cervunales)

The fundamental character of the soils on which the "cervunales" grow is their almost permanent moisture, which enables the development of matgrass (*Nardus stricta*), which together with other species such as *Campanula herminii*, *Juncus squarrosus*, *Merendera pyrenaica*, *Plantago media*, *Genista anglica* and *Gentiana pneumonanthe*, produce hard grasses for use by livestock, and which in many very dry territories serve as general reservoirs of cattle feed in the summer months.

10.6.2 Celtica gigantea *Communities* (Berceales)

Extremely tall in height due to the dominance of *Celtica gigantea* and *Arrhenatherum baeticum* (Arrhenathero baetici-Stipetum giganteae), they grow in dry supramediterranean territories on deep cambisols with well-structured organic horizons, not subjected to temporary hydromorphia. These grasslands have a Guadarraman optimum, but extend into Planileonese areas (Rivas-Martínez et al. 1987b).

10.6.3 Poa bulbosa *Communities* (Majadales *and* gramales)

With an extensive dry-subhumid supramediterranean Carpetan-Leonese distribution, these are densely caespitose grasslands on siliceous sandy soils, formed by *Poa bulbosa*, *Parentucellia latifolia* and *Trifolium subterraneum*, in addition to *Trifolium filiforme*, *Plantago lanceolata*, *Chamaemelum nobile* and *Festuca ampla* (Festuco amplae-Poetum bulbosae, or in drier areas Poo bulbosae-Trifolietum subterranei). These grasslands have developed in some intensively grazed areas near towns and villages, and persist in certain parts of the cattle migration network used in the movement of livestock. In mesomediterranean areas we find "gramales" with varying degrees of density that grow on humid sandy eutrophicated soils that are fairly compacted by trampling, in river courses or in shallow depressions that may become temporarily waterlogged in winter. Dominated by *Carex chaetophylla* and accompanied by *Trifolium resupinatum* and *Cynodon dactylon* (Trifolio resupinati-Caricetum chaetophyllae), they are frequent in the western part of the North Plateau. These grasslands are very productive, and in general are the result of intense grazing of other rushy pastures.

The plant communities growing on slightly basic and strongly nitrophilous soils are commonly known as "gramales", and have a predominance, among other species, of *Cynodon dactylon* (Trifolio-Cynodontetum dactyli). On base-rich substrates, these formations of *Poa bulbosa* are enriched with species such as *Astragalus sesameus*, *Astragalus stella*, *Convolvulus lineatus*, *Plantago loeflingi*, *Trifolium scabrum* and *Trigonella gladiata*, and others (Astragalo sesamei-Poion bulbosae) (Cano et al. 2007).

10.6.4 Agrostis castellana *Communities* (Vallicares)

On siliceous oligotrophic soils with moderate temporary hydromorphia, that suffer intense summer heat and drought, used for haymaking and occupying spaces potentially belonging to the forests of *Quercus rotundifolia* and *Quercus pyrenaica*, we find hemicryptophyte communities dominated by *Festuca ampla*, *Agrostis castellana* and *Holcus lanatus* among others (Festuco amplae-Agrostietum castellanae), used for grazing. If this occurs on an intensive scale, they become transformed into "majadales" (Festuco amplae-Poetum bulbosae).

10.6.5 Rushy Pastures (Praderas juncales)

Growing in the areas near source waters and the banks of streams, on deep acid soils on valley floors and in river beds, and used for grazing, are communities of *Juncus effusus*, *Deschampsia hispanica*, along with other hygrophilous plants of the Molinietalia coeruleae such as *Juncus acutiflorus*, *Carum verticillatum* and *Lotus pedunculatus* and others (Deschampsio hispanicae-Juncetum effusi). They serve as a source of water for livestock, and are transformed by the enrichment in nitrogen into communities with a predominance of *Juncus inflexus*.

On soils previously occupied by riverbank forests (Aro cylindracei-Ulmetum minoris) and used for intensive grazing, these rushy pastures are transformed into "majadales", plant formations dominated by hemicryptophytes, containing *Scirpus holoschoenus*, *Juncus inflexus*, *Juncus effusus*, *Cirsium pyrenaicum* and *Althaea officinalis* among others (Cirsio monspesulani-Holoschoenetum vulgaris).

There are also dense hygrophilous and somewhat eutrophic rush beds, growing on soils compacted by trampling and with moisture most of the year. Characteristic species are *Mentha suaveolens, Mentha longifolia, Mentha* x *rotundifolia* and *Juncus inflexus*, among others. They are distributed throughout Eurosiberian and Mediterranean territories (Mentho suaveolentis-Juncetum inflexi).

Other rush beds present in the meso-supramediterranean belts of this territory occupying oligotrophic soils with a sandy or sandy-silty texture with a pseudogley horizon are those formed by *Hypericum undulatum*, *Juncus acutiflorus*, *Lotus pedunculatus*, *Carum verticillatum*, *Holcus lanatus*, *Trifolium pratense* and *Anthoxantum odoratum*, and others (Hyperico undulati-Juncetum acutiflori).

10.6.6 Meadows (Prados de siega)

These are medium or low-growing subhumid-humid supratemperate Carpetan-Leonese meadows. The most important species are *Cynosurus cristatus, Festuca ampla, Trifolium repens, Lolium perenne, Bellis perennis, Festuca rothmaleri, Plantago lanceolata, Holcus lanatus* and *Trifolium pratense*. They grow on soils with temporary hydromorphia, are used for livestock grazing or, intermittently, for grazing and as meadows, and possibly altered by fertiliser. They tend to remain green and productive all year round. They are usually linked to the edaphohygrophilous series of ash woodlands in inter-riverine areas and on valley floors (Festuco amplae-Cynosuretum cristati).

10.6.7 Festuca elegans *subsp.* merinoi *Communities* (Cerrillares)

These are communities formed by hemicryptophytes that occupy deep ranker-type soils in humid supramediterranean territories, distributed throughout the whole area. These include *Festuca elegans* subsp. *merinoi*, *Phalacrocarpum oppositifolium*, *Saxifraga fragosoi*, *Festuca summilusitana*, *Silene legionensis* and *Leucanthemopsis pallida*, among others (Phalacrocarpo oppositifoliae-Festucetum elegantis, Leucanthemopsio pallidae-Festucetum elegantis).

10.6.8 Stipa sp. pl. *Communities* (Lastonares)

Dry supramediterranean “lastonares” dominated by *Stipa clausa* subsp. *clausa* frequently occur. They thrive on relatively deep cambisols with well-structured organic horizons and tend to colonise flat biotopes in situations with an intermediate hydromorphia, between the “berceales” (*Celtica gigantea*) and the “vallicares” (*Agrostis castellana*). Their biogeographical optimum is Carpetan-Leonese. They constitute seral stages of dry supramediterranean “carrascales” (*Quercus rotundifolia*) (Centaureo ornatae-Stipetum clausae). Another type of frequent “lastonar” grows on deep base-rich soils where *Stipa celakovskyi* has its optimum, along with *Dactylis hispanica* (Dactylo hispanicae-Stipetum celakovskyi), and which forms part of the seral stages of the holm-oak forests of the Junipero thuriferae-Quercetum rotundifoliae.

10.6.9 Annual Grasslands

The land use which caused the intense process of deforestation in the past has today been halted, and this has provided the annual grasslands with abundant space to develop. As in the other sections, we should highlight that this type of communities has a very high diversity due to the diversity of soils and bioclimates on the North Plateau. We will attempt to synthesise them as far as possible.

On base-rich substrates there are communities of ephemeral therophytes characteristic of calcareous terrain and clay (Velezio rigidae-Astericetum aquatici) with an early spring phenology, dominated by *Asteriscus aquaticus* and *Velezia rigida*. Typical of ledges in rocky areas in heathlands, also with an early spring phenology, preferentially in the Castilian-Duero meso- and supramediterranean belts, we see another type of annual grassland, whose composition includes particularly *Hornungia petraea, Saxifraga tridactylitis, Campanula erinus* and *Arenaria emarginata*, and others (Saxifrago tridactylites-Hornungietum petraeae).

There are also communities of small crassifoliar therophytes dominated by *Crassula tillaea* and *Sedum caespitosum* that occupy small sites (around 1 m^2 or less) adjacent to rocky outcrops on somewhat compacted sandy soils (Crassulo tillaeae-Sedetum caespitosi).

Annual ephemeral grasslands characterised by the Mediterranean western Iberian endemic *Ctenopsis delicatula* grow on shallow siliceous soils that undergo short spring episodes of hydromorphia. They bloom in late spring-early summer and are distributed in the meso- and supramediterranean belts of the inner sectors of the Mediterranean western Iberian biogeographic province (Ctenopsietum delicatulae).

On acid substrates we find very different communities formed by *Arenaria leptoclados, Arenaria serpyllifolia*, *Asterolinum linum-stellatum*, *Cerastium pumilum*, *Anthoxanthum aristatum*, *Campanula lusitanica*, *Eryngium tenue*, *Linaria spartea*, *Crupina vulgaris*, *Evax pygmaea*, *Cerastium diffusum*, *Medicago minima*, *Pistorinia hispanica*, *Molineriella laevis*, *Logfia minima*, *Trifolium arvense*, *Tolpis barbata*, *Hispidella hispanica* and *Ornithopus compressus*, among many others (Molineriello-Airopsietum tenellae, Hispidello hispanicae-Tuberarietum guttatae, Triseto ovati-Agrostietum truncatulae, Sedetum caespitoso-arenarii, Trifolio cherleri-Plantaginetum bellardi, Agrostio truncatulae-Sedetum lusitanici, Anthoxantho-Holcetum setiglumis).

The Crassulo-Saginetum apetalae is annual and nitrophilous and is a roadway community typically found in cracks between paving stones on streets, paths and roads subjected to frequent waterlogging and trampling, and in which *Crassula tillaea* and *Sagina apetala* are dominant. There are also annual and sub-nitrophilous communities such as the Medicagini rigidulae-Aegylopetum geniculatae on acid soils and the Trifolio cherleri-Taeniatheretum caput-medusae on basic soils (Rivas-Martínez and Izco 1977).

10.7 Herbaceous Ruderal and Similar Vegetation

The action of humans on nature since time immemorial is the determining factor in causing the landscape of the North Plateau to have shifted so far from its natural potential vegetation. It can be said that today these territories bear only a remote resemblance to their natural potential (Díaz González and Penas 1984).

The often sudden alteration of the substrates themselves –particularly with regard to the upper soil horizon– has led to the establishment of communities of invasive plants (although some have been with us for many centuries), whose common denominator often is a marked need for ammoniacal substances deriving from the transformation of organic matter.

The denomination of nitrophilous and ruderal vegetation includes plant communities adapted to trampling, and also grasslands typical of roadways and ruderal areas which undoubtedly tend to be seen as weeds in areas of crops, or which grow on heaps of debris. They also include scionitrophilous plants; that is, plants that exploit the transformation of organic matter from other plant formations with a tree or shrub character. Among the first we should mention communities formed by *Poa annua*, *Polygonum aviculare*, *Capsella bursa-pastoris*, *Matricaria matricarioides*, *Sclerochloa dura*, *Coronopus didimus*, *Coronopus squamatus* and *Matricaria discoidea*, among others (Coronopo squamati-Sclerochloetum durae, Polygono arenastri-Matricarietum discoideae), which on numerous occasions coexist with *Plantago major* and *Lolium perenne* (Lolietum perennis) (Díaz González et al. 1988a, b).

On disturbed soils, often on the verges of paths and roadways, it is very frequent to find communities formed by tall hemicryptophytes belonging to the subclass Onopordenea acanthii. It is also worth mentioning the communities formed by *Carduus carpetanus*, *Onopordum acanthium*, *Cirsium vulgare*, *Scolymus hispanicus*, *Centaurea calcitrapa* and *Verbascum pulverulentum*, and others (Carduo carpetani-Onopordetum acanthii). On slightly moist meso-supramediterranean soils there is a presence of formations with no specific soil preferences dominated by *Silybum marianum* and accompanied by *Carduus bourgeanus*, *C. pycnocephalus* and *C. tenuiflorus* (Carduo bourgeani-Silybetum mariani). On base-rich substrates, as found in Castilian-Duero territories, these communities are preferentially dominated by *Onopordum acanthium* and *Onopordum nervosum* subsp. *castellanum* (Onopordetum acantho-castellani). There is another community with similar characteristics but with a summer phenology growing on permeable siliceous or sandy soils that have their optimum in the mesomediterranean belt, but penetrating into the supramediterranean area, and whose characteristic plants are *Carthamus lanatus*, *Carlina corymbosa*, *Scolymus*

Fig. 10.7 Grain crop fields in Lédigos (Palencia) (Photo by Sara del Río)

hispanicus and *Centaurea ornata*, among others (Carlino hispanicae-Carthametum lanati) (Penas et al. 1988b).

In fields of rain-fed cereal crops (Fig. 10.7) growing on soils with a sandy-clay texture, flowering at the start of spring before *Triticum* sp., *Avena* sp. or *Secale cereale* become widely developed, one finds plants typical of the *Stellarietea mediae* such as *Veronica persica*, *Veronica triphyllos*, *Veronica hederifolia* subsp. *triloba*, *Mibora minima*, *Arabidopsis thaliana*, *Senecio vulgaris*, *Erophila verna* and *Cerastium glomeratum*, all with indifferent soil preferences (Miboro minimae-Arabidopsietum thalianae). On base-rich soils these species are accompanied by *Ceratocephala falcata* and *Androsace maxima* (Ceratocephalo falcatae-Androsacetum maximae). As the cereals achieve their optimum size it is very frequent to find plants such as *Centaurea cyanus*, *Brassica nigra*, *Lolium temulentum*, *Raphanus raphanistrum*, *Papaver rhoeas*, *Agrostemma githago*, *Ranunculus arvensis*, *Galium tricornutum*, *Lupinus angustifolius*, *Camelina microcarpa*, *Buglossoides arvensis*, *Biscutella auriculata*, *Euphorbia serrata*, *Anchusa undulata*, *Linaria spartea*, *Anchusa azurea* and *Salvia verbenaca*, among others.

If these communities grow on base-rich substrates they are joined by more exclusive plants such as *Vaccaria piramidata*, *Platicapnos spicata*, *Cnicus benedictus*, *Roemeria hybrida*, *Hypecoum pendulum* and *Hypecoum imberbe*, among others (Roemerio hybridae-Hypecoetum penduli). In croplands that have already been harvested or are lying fallow it is frequent in autumn to find plants such as *Xanthium strumarium*, *Heliotropium europaeum*, *Xanthium spinosum*, *Amaranthus albus*, *Solanum nigrum*, *Datura stramonium* and *Tribulus terrestris* (Heliotropio europaei-Amaranthetum albi).

In horticultural crops growing on temporarily moist soils and in irrigated crops there are communities presided either by *Echinochloa crus-galli*, accompanied by *Echinochloa collina*, *Digitaria sanguinalis*, *Polygonum lapathifolium*, *Setaria verticillata*, *Galinsoga parviflora* and *Bidens tripartita*, among others, or else by *Portulaca oleracea*, *Amaranthus retroflexus*, *Amaranthus hybridus*, *Eragrostis cilianensis* and more.

On soils of varying depth on the edges of roadways there is a frequent presence in the North Plateau of grasslands dominated by *Hordeum murinum* or *Hordeum secalinum*, occasionally accompanied by other plants such as *Sisymbrium irio*, *Sisymbrium officinale*, *Descurainia sophia*, *Hirschfeldia incana* and *Isatis tinctoria*, among others (Papaveri argemone-Sisymbrietum contorti, Bromo scoparii-Hordeetum leporini, Sisymbrio irionis-Malvetum parviflorae, Sisymbrio officinalis-Hordeetum murini).

Among the scionitrophilous communities we should mention the community formed by *Galium aparine*, *Conium maculatum*, *Alliaria petiolata*, *Saponaria officinalis* and *Urtica dioica* among others (Galio aparines-Conietum maculati). It forms grasslands with a substantial biomass which flower in spring and early summer, are parched in summer, growing on deep soils with significant contributions of organic matter (humicolous) and almost constant moisture. The community is located preferentially on the edges of roadways and verges, and in spite of their heliophilous character its species sometimes seek the shade of deciduous woodlands such as the Aro-Ulmetum minoris or Fraxino angustifoliae-Quercetum pyrenaicae. With a summer phenology, growing on roadsides and verges on soils rich in organic matter and with high soil moisture, there is a frequent presence of communities dominated to a large degree by *Sambucus ebulus*, often accompanied –but with less biomass– by *Urtica dioica*, *Dipsacus fullonum*, *Arctium minus*, *Galium aparine* and *Pentaglottis sempervirens* (Urtico dioicae-Sambucetum ebuli). These grasslands are located mainly in the areas of ash, elm and poplar woodlands.

10.8 Aquatic Vegetation

Wetlands are places in which to a greater or lesser degree the presence of water can be seen in its liquid state for all or most of the year. This definition includes some highly heterogeneous formations, and several different levels can be distinguished. Some of these formations comprise hygrophytic species, understood as species that are either wholly or partially submerged, or that are found on sites with high soil moisture. These plants live in lakes, ponds, pools, rivers, streams and springs.

Among hygrophytic plants we can distinguish plants that are submerged and thus have (nearly) the whole of their body below water. Floating plants may have their whole body floating or else be rooted to the bottom with only one part floating–frequently the leaves and flowers. There are also amphibious plants which have one part submerged and another above water; we can differentiate those with a short growing period which generally develop in one season and are found in sites that are periodically flooded, from others that live in areas in which the level and the permanence of the water is relatively constant and that are rooted in aquatic soils, generally with a silty character.

There are other plants that need large quantities of water to live, giving rise to formations such as the hygrophilous heathlands of which we have already spoken, or megaforb communities, peat marshes and bogs, and salt marsh communities.

10.8.1 Riparian Communities

All along rivers with permanent water all year round –although subjected to low-water levels in summer–, it is frequent to find herbaceous communities of different sizes near formations of shrubby willow woodlands, as is the case of the Galio broterianae-Caricetum reuterianae, which is very frequent in the western part of the plateau. This community forms on large rocks standing in the river itself and comprises, among other species, *Carex reuteriana*, *Galium palustre* and *Oenanthe crocata*. Sometimes, if the soil in the river bed is formed by gravel and pebbles and has a scant presence of fine materials, this community fringes on another one comprising *Phalaris arundinacea* (Phalaridetum arundinaceae); and on other occasions on rushy pastures of the Oenantho crocatae-Filipenduletum ulmariae subjected to intense summer low-water levels.

It is also frequent to find helophytic grasslands on the edges of rivers and streams with slow-flowing water, dominated by *Oenanthe crocata* and *Glyceria declinata* (Glycerio declinatae-Oenanthetum crocatae), accompanied by other taxa such as *Phalaris arundinacea*, *Galium palustre*, *Sparganium erectum* subsp. *neglectum* and *Lythrum salicaria.* It is present in steep gorges with fast-flowing water in the midst of forests of the Galio broterianae-Alnetum glutinosae, Populo nigrae-Salicetum neotrichae, Salicetum angustifolio-salviifoliae and Salicetum lambertiano-salviifoliae.

10.8.2 Peat Bogs (Turberas)

In Sanabrian territories on typic haplofibrist (fibric histosols) we can find some relict communities in the supramediterranean belt with a hygro-peaty character comprising *Rhynchospora alba*, *Eleocharis multiculmis*, *Juncus squarrosus*, *Drosera rotundifolia*, *Parnassia palustris* and *Carex echinata*, accompanied by *Sphagnum* sp. pl. (Eleocharito multicaulis-Rhynchosporetum albae), e.g. at Aveseda, Lago de Sanabria at 1040 m.

In these same territories at a higher altitude in the nearby Galaico-Leonese mountains there are other peat-loving communities that are described or discussed in another chapter of the book.

10.8.3 Lakes and Pools (Lagos y charcas)

In these environments there is a presence of a number of very different communities, ranging from the aquatic carpets of nymphoides (*Potamogeton natans*) and myriophylloides (*Myriophyllum alterniflorum*), often accompanied by batrachiids (*Ranunculus pseudofluitans*) and small elodeids growing in slow-flowing or standing mesotrophic and moderately deep (a maximum of 50-200 cm in depth) waters associated to river courses or pools subject to acute changes in water levels, but which rarely dry out completely (Myriophyllo alterniflori-Potametum natantis). There we also find formations of small or medium non-rooted plants that float on the surface of fresh waters in pools, lakes or water courses with a slow current, and that are typical of polluted waters and backwaters (rivers, canals, water deposits, pools and so on) that are eutrophicated and rich in soluble ions such as phosphates, nitrates and others, and characterised by the presence of *Lemna gibba*. (*Lemnetum gibbae*); as well as communities in shallow meso-eutrophic –including oligohaline– waters in areas that totally dry out, such as water deposits, the edges of pools, temporarily flooded dry river beds and others, which have their optimum in spring. They are characterised by the presence of *Callitriche butria* and *Ranunculus baudotii*, accompanied by *Ranunculus trichophyllus* in environments with more pronounced flooding (Callitricho butriae-Ranunculetum baudotii), and helophytic associations (Glycerio declinatae-Eleocharidetum palustris) typical of base-poor shallow freshwater lakes where it is common to find both the cypergrass *Scirpus palustris* and the western decumbent grass *Glyceria declinata*, and other low-growing helophytes such as *Mentha aquatica*, *Galium palustre*, *Equisetum palustre* and others. This association widely occurs on the Iberian Peninsula in the Mediterranean-Iberoatlantic superprovince, the Orocantabrian province and the Cantabrian-Atlantic province. It has numerous variants, of which the most typical are the one with *Scirpus palustris*, another with *Alopecurus aequalis* in pioneer communities, *Alopecurus geniculatus* on intensively grazed and nitrified sites, and *Paspalum distichum* on substrates flooded by strongly eutrophicated waters.

Another very particular community is the Subulario aquaticae-Isoetetum asturicensis which in Sanabria reaches the lake of the same name at an altitude of 995 m, although it also grows at altitudes of over 1600 m, not only in these territories but also in the Sistema Central (Sierra de Candelario, Salamanca). In this association it is not infrequent to find *Subularia aquatica*, *Isoetes velatum* subsp. *asturicensis*, *Fontinalis antipyretica* and *Myriophyllum alternifolium*.

10.8.4 Reed Communities (Carrizales)

These communities are also frequent in ponds, lakes and on riverbanks and receive the name of *Phragmites australis*, a species which is sometimes dominant in these communities and is accompanied by *Scirpus tabernaemontani* or by *Schoenoplectus littoralis*, *Scutellaria galericulata*, *Bolboschoenus compactus* and others (Bolboschoeno compacti-Phragmitetum australis, Bolboschoeno compacti-Scirpetum tabernaemontani or Bolboschoeno compacti-Schoenoplectetum littoralis).

10.8.5 Reed Beds (Cañaverales)

These are communities of amphibious plants, low- or medium-high, rooted in the bottom and constituting the outermost vegetation belt in ponds and lakes, and the belt farthest from the water courses, almost always in contact with humid pastures. There is a frequent presence of *Typha latifolia*, *Typha angustifolia*, *Typha dominguensis*, *Iris pseudacorus*, *Alisma plantago-aquatica*, *Carex acuta* and *Glyceria declinata*, and others (Typho-Scirpetum tabernaemontani, Irido pseudacori-Caricetum lusitanici).

10.8.6 Megaforbic Communities

These herbaceous perennial communities with a great height and a megaforbic and edaphophilous character are physiognomically dominated by *Filipendula ulmaria*, and also have a frequent presence of *Angelica sylvestris, Centaurea carpetana, Epilobium hirsutum, Heracleum montanum, Lysimachia vulgaris, Scrophularia lyrata, Scrophularia reuteri* and *Scrophularia scorodonia*, and others. They have a limited distribution in the supramediterranean belt. They develop on permanently moist gleyed soils on verges and on the boundaries of hygrophilous pastures, the banks of permanent stream beds or rivers, or as perennial edge vegetation in woodlands of alder and weeping willow in supramediterranean Bejarano-Gredensian territories (Angelico sylvestris-Filipenduletum ulmariae).

10.8.7 Pioneer Annual and Dwarf Perennial Ephemeral Isoetid Communities on Periodically Flooded Bare Soils

In small shallow ponds or temporary lakes that dry up every summer and fill with water in winter and spring there are communities formed by small plants that grow in rings on the dry soil when evaporation begins and until the soil becomes totally dry. The species include *Mentha pulegium*, *Mentha cervina*, *Polypogon maritimus*, *Myosurus minimus*, *Juncus tenageia*, *Juncus capitatus*, *Juncus bufonius*, *Eryngium viviparum*, *Eryngium galioides*, *Eryngium cornucopioides*, *Damasonium alisma*, *Pulicaria uliginosa*, *Molineriella minuta*, *Lytrum hyssopifolium* and *Cicendia filiformis*, along with *Elatine macropoda* and the false four-leafed clover, *Marsilea strigosa*; or the more summer-autumnal species such as *Cyperus fuscus*, *Crypsis schoenoides* and *Heliotropium supinum*, among others (Isoetion, Menthion cervinae, Agrostion pourretii, Cicendion, Nanocyperion, Verbenion supinae and Lythrion tribracteati).

10.8.8 Other Aquatic Communities

These include amphibious communities composed of species with both submerged and floating leaves that are characterised by their temporality, as their biotopes are regulated by temporary waters. Here the dominant and characteristic species is *Ranunculus peltatus*, accompanied by *Callitriche brutia*, *Fontinalis antipyretica* and *Myriophyllum alterniflorum*, which grow in ponds, springs and lakes with shallow, cold and neutral or acid waters (which may dry up) (Callitricho brutiae-Ranunculetum peltati).

This vegetation is widely distributed throughout the Iberian Peninsula along slow-flowing stream beds or in pools which may have low-water levels in summer and my even totally dry up. They comprise low-growing helophytic herbs covering a wide area, including those dominated by *Apium nodiflorum*, *Veronica anagallis-aquatica*, *Rorippa nasturtium-aquaticum* and *Veronica beccabunga* (Helosciadetum nodiflori).

10.9 Salt Marshes (*Saladares*)

On the higher plateau in several parts of the middle course of the Duero river valley, (Ladero et al. 1984), located in the basins of its left-bank tributaries –the Cega, Eresma, Adaja, Zapardiel, Trabancios and Guareña–, and its right-bank tributaries – the Valderaduey and Sequillo– there are several places with saline and saline-alkaline soils, rich in salts formed from sandy-clayey Miocene and Quaternary sediments deriving from the alteration of sandstone and granite and which may include calcium solonetzs or solonchak with a pH of over 7. These soils tend to be waterlogged in winter and spring, and when they dry out in summer they leave a saline efflorescence.

In these conditions, in our territory there are halophilous communities of pioneer crassicaulous plants with aphyllous therophytes and a summer phenology, as occurs in some localities where *Microcnemum coralloides* is dominant (Microcnemetum coralloidis); or with an autumn phenology with a presence of *Salicornia ramosissima*.

On saline soils, with ephemeral nanotherophytes that have their physiological optimum in mid spring in territories in the regions of Ávila, Salamanca and Valladolid on soils with a sandy-silty or sandy-clayey texture we find communities formed by *Sphenopus divaricatus*, *Frankenia pulverulenta*, *Parapholis incurva*, *Hymenolobus procumbens* and others (Parapholi incurvae-Frankenietum pulverulentae). On soft soils with a nitrophilous character we find *Cressa cretica* forming almost single-species communities, sometimes accompanied by *Salicornia ramosissima*.

With an autumnal character on disturbed soils in dry watercourses which receive run-off waters that are chloride-rich and somewhat nitrophilous there is a community formed mainly by *Salsola soda* and *Atriplex patula*.

Perennial communities include the halophilous rushes characteristic of the edges of temporary pools and puddles, pastures of *Schoenus nigricans*, and hemicryptophytes on temporarily flooded saline soils.

Among the halophilous rushes occupying the banks of irrigation channels we find *Juncus maritimus*, *Elymus curvifolius*, *Sonchus maritimus* and occasionally *Sonchus crassifolius* (Soncho crassifolii-Juncetum maritimae). On other occasions on soils with a sandy-clayey texture there is a presence of formations dominated by *Schoenus nigricans*, accompanied by *Linum maritimum*, *Plantago maritima*, *Tetragonolobus maritimus*, and others.

In the Villafáfila depression (Zamora), there are some dense salt-water formations developing in spring which require very moist and waterlogged soils for most of the year. In years of scarce precipitation the populations of these formations are very reduced. They are characterised by the presence of *Juncus subulatus*, *Aeluropus littoralis* and occasionally *Juncus gerardi* (Aeluropodo littoralis-Juncetum subulati). These formations sometimes fringe on rushy pastures with dense coverage and low-growing taxa on saline soils that are persistently waterlogged after the rainy period. They are characterised by *Juncus gerardi*, *Bupleurum tenuissimum* and *Aeluropus littoralis* (Bupleuro tenuissimi-Juncetum gerardii).

On bare soils subject to intense summer drought, poor in organic material and waterlogged in spring there are practically pure communities of *Puccinellia fasciculata*, accompanied on occasions by *Frankenia laevis*. On non-marly Miocene substrates with less flooding we can find a community formed by *Plantago maritima* and *Camphorosma monspeliaca*, among others, and on Miocene and gypsiferous substrates there are communities formed by *Artemisia gallica*, *Limonium costae*, *Gypsophila tomentosa*, and others.

Finally, with regard to this type of communities, we should mention the hyperhalophilous fruticose communities characteristic of salt marshes, salt lakes and on the banks of endorreic lakes, dominated by *Suaeda vera*.

10.10 Rock Vegetation

In the western territories with a mesomediterranean optimum, and characterised above all by the presence of *Rumex induratus* and *Phagnalon saxatile*, there are some formations on acid rock beds which also include *Dianthus lusitanus*, *Umbilicus rupestris*, *Matthiola fruticulosa* and others (Rumici indurati-Dianthion lusitani). There are also communities of chasmo-comophytes with nitrophilous preferences such as *Parietaria judaica*, *Ceterach officinarum*, *Cymbalaria muralis* and *Phagnalon saxatile*, among others, which give rise to different communities (Parietarietum judaicae, Cymbalarietum muralis).

10.11 Vegetation Series

The vegetation series present on the North Plateau, following the proposals of Rivas-Martínez et al. (2011a) are as follows:

Climatophilous Bercian-Sanabrian silicicolous temperate oceanic supratemperate submediterranean and Mediterranean pluviseasonal oceanic meso-supramediterranean subhumid-hyperhumid series of forests of *Quercus pyrenaica* and *Genista falcata* with *Festuca braun-blanquetii* and *Prunella hastifolia*: **Genisto falcatae-Querco-pyrenaicae sigmetum**

Climatophilous Guadarraman Celtiberian-Alcarreñan and Oroiberian silicicolous Mediterranean pluviseasonal oceanic supramediterranean subhumid-humid series of forests of *Quercus pyrenaica* and *Luzula forsteri* with *Milium vernale* and *Cistus laurifolius*: **Luzulo forsteri-Querco pyrenaicae sigmetum**

Climatophilous Salamancan and Zamoran-Maragatan silicicolous Mediterranean pluviseasonal oceanic supramediterranean subhumid-humid series of forests of *Quercus pyrenaica* and *Pyrus bourgaeana* with *Paeonia broteroi* and *Crataegus monogyna*: **Pyro bourgaeanae-Querco pyrenaicae sigmetum**

Climatophilous oroiberian Ayllonensian and Planileonese silicicolous temperate oceanic supratemperate and Mediterranean pluviseasonal oceanic supramediterranean subhumid-humid submediterranean series of forests of *Quercus pyrenaica* and *Pumonaria longifolia* with *Teucrium scorodonia* and *Hypericum pulchrum*: **Pulmonario longifoliae-Querco pyrenaicae sigmetum**

Climatophilous Bejarano-Gredensian silicicolous Mediterranean pluviseasonal oceanic supramediterranean humid-hyperhumid series of forests of *Quercus pyrenaica* and *Festuca merinoi* with *Cytisus multiflorus* and *Leuzea rhaponticoides*: **Festuco merinoi-Querco pyrenaicae sigmetum**

Climatophilous Guadarraman Celtiberian-Alcarreñan and Oroiberian silicicolous Mediterranean pluviseasonal oceanic supramediterranean subhumid-humid submediterranean series of forests of *Quercus pyrenaica* and *Luzula forsteri* with *Milium vernale* and *Cistus laurifolius*: **Luzulo forsteri-Querco pyrenaicae sigmetum**

Climatophilous Castilian calcicolous Mediterranean pluviseasonal oceanic meso-supramediterranean dry-subhumid series of forests of *Quercus faginea* and *Cephalanthera rubra* with *Lonicera hispanica* and *Paeonia humilis*: **Cephalanthero rubrae-Querco fagineae sigmetum**

Climatophilous and edaphoxerophilous Lusitanian Duero-Bercian-Sanabrian and Salamancan silicicolous Mediterranean pluviseasonal oceanic meso-supramediterranean dry-humid series of forests of *Quercus rotundifolia* and *Genista hystrix* with *Cytisus multiflorus* and *Euphorbia broteroi*: **Genisto hystricis-Querco rotundifoliae sigmetum**

Climatophilous and edaphoxerophilous Castilian and Iberian maestracensean calcicolous Mediterranean pluviseasonal oceanic supramediterranean dry-subhumid series of forests of *Quercus rotundifolia* and *Juniperus thurifera* with *Rhamnus infectoria* and *Satureja gracilis*: **Junipero thuriferae-Querco rotundifoliae sigmetum**

Climatophilous and edaphoxerophilous Guadarraman Castilian northern and Planileonese silicicolous Mediterranean pluviseasonal oceanic meso-supramediterranean dry-subhumid series of forests of *Quercus rotundifolia* and *Juniperus lagunae* with *Carex distachya* and *Lavandula pedunculata*. **Junipero lagunae-Querco rotundifoliae sigmetum**

Climatophilous Castilian and oroiberian Mediterranean pluviseasonal and temperate oceanic supramediterranean and supratemperate dry-subhumid submediterranean series of forests of *Juniperus thurifera* and *Juniperus hemisphaerica* with *Artemisia assoana* and *Festuca asperifolia*: **Junipero hemisphaerico-thuriferae sigmetum**

Fluvial Carpetan-Leonese soft freshwater Mediterranean pluviseasonal oceanic supramediterranean dry-humid series of the forests of *Fraxinus angustifolia* with *Quercus pyrenaica* and *Salix salviifolia*: **Querco pyrenaicae-Fraxino angustifoliae sigmetum**

Fluvial-lake iberoatlantic freshwater lentic dystrophic temperate hyperoceanic and Mediterranean pluviseasonal oceanic thermo-mesotemperate and mesomediterranean subhumid-humid submediterranean series of forests of *Alnus glutinosa* and *Carex lusitanica* with *Salix atrocinerea* and *Betula celtibé rica*. **Carici lusitanicae-Alno glutinosae sigmetum**

Fluvial-alvear and fluvial Carpetan-Leonese soft freshwater Mediterranean pluviseasonal oceanic supramediterranean dry-humid series of forests of *Alnus glutinosa* and *Galium broterianum* with *Betula celtiberica* and *Caltha palustris*: **Galio broteriani-Alnetum glutinosae sigmetum**

Fluvial-riverine northern Castilian and oroiberian hard or slightly hard freshwater Mediterranean pluviseasonal oceanic supramediterranean dry-subhumid series of forests of *Ulmus minor* and *Arum cylindraceum* with *Ligustrum vulgare* and *Rosa corymbifera*: **Aro cylindracei-Ulmo minoris sigmetum**

Fluvial-alvear and fluvial lower-Aragonese and Castilian southern Mediterranean hard or very hard freshwater pluviseasonal and xeric oceanic lower meso-supramediterranean semiarid-dry series of forests of *Populus alba* and *Rubia tinctorum* with *Salix neotricha* and *Tamarix canariensis*: **Rubio tinctorum-Populo albae sigmetum**

Fluvial-alvear and fluvial northern Castilian and oroiberian slightly hard or hard freshwater Mediterranean pluviseasonal oceanic supramediterranean dry-subhumid series of forests of *Populus nigra* and *Salix neotricha* with *Populus alba* and *Salix salviifolia*: **Populo nigrae-Salico neotrichae sigmetum**

Fluvial-alvear internal and rivular Galician-north Portuguese Leonese Sistema Ibérico soft freshwater temperate oceanic and Mediterranean pluviseasonal meso-supratemperate and meso-supramediterranean subhumid-humid submediterranean series of high shrublands of *Salix salviifolia* and *Salix atrocinerea* with *Fraxinus angustifolia* and *Alnus glutinosa*: **Salico salviifoliae sigmetum**

Fluvial-alvear internal and rivular Carpetan and Montes de Toledo-Ciudad Real soft and slightly hard freshwater Mediterranean pluviseasonal and temperate oceanic meso-supramediterranean and supratemperate subhumid-humid submediterranean series of high shrublands of *Salix salviifolia* and *Salix lambertiana* with *Salix atrocinerea* and *Salix neotricha*: **Salico lambertiano-salviifoliae sigmetum**

References

Arnaiz C, Loidi J (1983) Sintaxonomía del Pruno-Rubion ulmifolii (Prunetalia) en España. Lazaroa 4:17–22

Cabo Alonso A, Manero Miguel F (1991) Castilla y León. In: Maurel J (dir.) Geografía de España vol. 6: 255–589. Ed. Planeta. Barcelona

Cano E, Ladero M, García-Fuentes A, Pinto-Gomes C, Cano-Ortíz A (2007) Estado actual de la clase Poetea bulbosae en la Peninsula Iberica. Phytocoenologia 37(3–4):645–661

del Río S (2005) El cambio climático y su influencia en la vegetación de Castilla y León (España). Itinera Geobotanica 16:1–534

Díaz González TE, Penas A (1984) Datos sobre la vegetación terofítica y nitrófila leonesa. Acta Botanica Malacitana 9:233–254

Díaz González TE, Penas A (1987) Estudio de las saucedas mediterráneas de la provincia de León. Publ Univ La Laguna, Ser Informes 22:87–120

Díaz González TE, Penas A, Puente E, Herrero L, García ME, López Pacheco MJ, Llamas F (1988a) Síntesis de la vegetación nitrófila de la provincia de León (NW España). Acta Bot Barc 37:133–148

Díaz González TE, Penas A, Herrero L, Pérez Morales C, Llamas F, Terrón A (1988b) Estudio de los herbazales nitrófilos vivaces y comunidades ruderal viarias de la provincia de León (NW España). Acta Bot Barc 37:113–131

Díaz González TE, Penas A, López Pacheco MJ, Puente E, Andrés J (1988c) Datos sobre los matorrales de la Ononido-Rosmarinetea en la provincia de León. Studia Botanica 7:147–157

Díaz González TE, Penas A, López Pacheco MJ, Morales C, Llamas F (1989) La clase Cisto-Lavanduletea en la provincia de León. Acta Bot Malacitana 14:226–230

Gutiérrez Elorza M (coord.) (1994) Geomorfología de España. Rueda, Madrid

Julivert M, Fonboté J, Ribeiro A, Conde L (1972) Mapa Tectónico de la Península Ibérica y Baleares. Escala: 1:1.000.000. ITGE.

Junta de Castilla y León. Dirección General de Urbanismo y Calidad Ambiental (ed.) (1995) Atlas del Territorio de Castilla y León. Consejería de Medio Ambiente y Ordenación del Territorio

Ladero M, Navarro F, Valle C, Marcos B, Ruiz Téllez T, Santos MT (1984) Vegetación de los saladares castellano-leoneses. Studia Bot 3:17–62

Ladero M, Valle C, Gutiérrez Balbás A (1994) On the class Pegano harmalae-Salsoletea vermiculatae Br.-Bl. & Bolòs 1957 in the Duero basin (Spain). Candollea 49:499–507

Ladero M, Amor A, Santos MT, Sánchez ME, Ferro G (2004) Robledales supramediterraneos de ombroclima subhúmedo en los sectores Salamantino y Lusitano-Duriense. Quercetea 4:5–11

Loidi J (1999) General description of the Iberian Peninsula: substrate and relief. Itinera Geobotanica 13:17–40

Loidi J (2011) Los yesos del valle del Duero. In: Mota JF, Sánchez-Gómez P, Guirado JS (eds) Diversidad vegetal de las yeseras ibéricas. El reto de los archipiélagos edáficos para la biología de la conservación. ADIF-Mediterráneo Asesores consultores, Almería, pp 349–351

Morales G, Méndez B, Reques P, López L (1995) Asturias, Cantabria, Castilla y León. In: Guisán S (ed) Geografía de España, vol VII. Instituto Gallach, Barcelona, pp 1347–1550

Panareda Clopés JM (coord.), et al. (1996) Atlas de Castilla y León. El Mundo de Castilla y León. Valladolid.

Penas A, Díaz González TE, García ME, Herrero L, Puente E (1988a) Aportaciones al conocimiento de los piornales (Cytisetea scorpario-striati) en la provincia de León. Monogr Inst Piren Ecol 4:687–694

Penas A, Díaz González TE, García ME, López Pacheco MJ, Puente E, Herrero L (1988b) Datos sobre los cardales y tobales (Onopordetea acanthii) en la provincia de León. Lazaroa 10:65–79

Rivas-Martínez S (1968) Los jarales de la Cordillera Central. Collect Bot 78(2):1033–1082

Rivas-Martínez S & coauthors: Andalucía (Alfredo Asensi, Blanca Díez-Garretas, Joaquín Molero, Francisco Valle & Eusebio Cano), Aragón (Manuel Costa & Luis Villar), Asturias (Tomás E. Díaz & José Antonio F. Prieto), Baleares (Leonardo Llorens), Canarias (Marcelino del Arco, W. Wildpret, P.L. Pérez de Paz, O. Rodríguez, J.R. Acebes, A. García, V. E. Martín, J. A. Reyes Betancort, M. Salas, M.A. Díaz, J.A. Bermejo, R. González, M.V. Cabrera y S. García), Cantabria (José Antonio F. Prieto & Tomás E. Díaz), Castilla-La Mancha (Federico Fernández & Daniel Sánchez-Mata), Castilla y León (Ángel Penas, Luis Herrero & Sara del Río), Cataluña (Ramón Masalles & Manuel Costa), Ceuta y Melilla (Alfredo Asensi & Blanca Díez-Garretas), Extremadura (Miguel Ladero & Ángel Amor), Galicia (Jesús Izco & Javier Amigo), La Rioja (Javier Loidi & Gonzalo Navarro), Madrid (Paloma Cantó), Murcia (Francisco Alcaraz), Navarra (Javier Loidi & Juan Carlos Báscones), País Valenciano (Manuel Costa & Pilar Soriano), País Vasco (Javier Loidi). 2011a Mapa de series, geoseries y geopermaseries de vegetación de España (Memoria del mapa de vegetación potencial de España). Parte II. Itinera Geobotanica 18 (1): 5–424

Rivas-Martínez S & coauthors: Andalucía (Alfredo Asensi, Blanca Díez-Garretas, Joaquín Molero, Francisco Valle & Eusebio Cano), Aragón (Manuel Costa & Luis Villar), Asturias (Tomás E. Díaz & José Antonio F. Prieto), Baleares (Leonardo Llorens), Canarias (Marcelino del Arco, W. Wildpret, P.L. Pérez de Paz, O. Rodríguez, J.R. Acebes, A. García, V. E. Martín, J. A. Reyes Betancort, M. Salas, M.A. Díaz, J.A. Bermejo, R. González, M.V. Cabrera y S. García), Cantabria (José Antonio F. Prieto & Tomás E. Díaz), Castilla-La Mancha (Federico Fernández & Daniel Sánchez-Mata), Castilla y León (Ángel Penas, Luis Herrero & Sara del Río), Cataluña (Ramón Masalles & Manuel Costa), Ceuta y Melilla (Alfredo Asensi & Blanca Díez-Garretas), Extremadura (Miguel Ladero & Ángel Amor), Galicia (Jesús Izco & Javier

Amigo), La Rioja (Javier Loidi & Gonzalo Navarro), Madrid (Paloma Cantó), Murcia (Francisco Alcaraz), Navarra (Javier Loidi & Juan Carlos Báscones), País Valenciano (Manuel Costa & Pilar Soriano), País Vasco (Javier Loidi). 2011b Mapa de series, geoseries y geopermaseries de vegetación de España (Memoria del mapa de vegetación potencial de España). Parte II. Itinera Geobotanica 18 (2): 425–800

Rivas-Martínez S, Izco J (1977) Sobre la vegetación terofítica subnitrófila mediterranean (Brometalia rubenti-tectori). Anales Inst Bot Cavanilles 34(1):355–381

Rivas-Martínez S, Penas A, Díaz González TE (1986) Datos sobre la vegetación nitrófila y terofítica leonesa. Nota II Acta Bot Malacitana 11:273–288

Rivas-Martínez S, Belmonte MD, Cantó P, Fernández-González F, Fuente V, Moreno JM, Sánchez-Mata D, Sancho LG (1987a) Piornales, enebrales y pinares oromediterráneos (Pino-Cytision oromediterranei) en el Sistema Central. Lazaroa 7:93–124

Rivas-Martínez S, Fernández González F, Sánchez-Mata D (1987b) El sistema Central español: De la Sierra de Ayllón a Serra da Estrêla. In: Peinado M, Rivas-Martínez S (eds) La vegetación de España. Serv. Publ. Univ, Alcalá de Henares, pp 419–451

Rivas-Martínez S, Penas A, Díaz González TE, del Río S, Cantó P, Herrero L, Pinto Gomes C, Costa JC (2014) Biogeography of Spain and Portugal. Preliminary typological sinopsis. Int J Geobot Res 4(1):1–64

Chapter 11
The Iberian Ranges and Highlands

Manuel Peinado, Juan Luis Aguirre, and Alejandro Aparicio

Abstract One of Spain's main mountain ranges, the Iberian System, separates the Ebro basin and Meseta Central (Inner Plateau) and runs northwest-southeast from the La Bureba corridor in Burgos to the Mediterranean coastal ranges in Castellón and Valencia. The barren, rugged slopes of this range cover an area close to 61,140 km^2. Each mountain chain has different geological characteristics depending on the outcropping materials of the Alpine orogeny, either those of the Hercynian basement or Mesozoic materials. The former give rise to the acid soils of the Sierras of Demanda or Moncayo and Peaks of Urbión, and the latter to the basic substrates of the Sierras of Gúdar, Javalambre, the Iberian High Plateau ("parameras" or "páramos") and the serranías of Cuenca and Albarracín.

Vegetation varies according to both geological/edaphic factors and a complex orogeny. The prevailing ombroclimate is subhumid, reaching humid to hyperhumid, and the predominant thermoclimate is supramediterranean. Potential vegetation mainly comprises forests. Rocky areas, steep slopes or southern exposures of lower mountain zones sustain large patches of supramediterranean *Quercus rotundifolia* forests. Marcescent oak forests, both of *Quercus pyrenaica* and *Quercus faginea* subsp. *faginea* (*Q. faginea* hereafter), dominate large areas of the gentle relief of the *sierras'* foothills though, in many zones, they have been repopulated with pines or given over to grazing. In more moist areas, generally of northern or eastern exposures, we find beech forests dominated by *Fagus silvatica*. In the north-west zone of the Iberian System, these occupy extensive areas.

Large expanses of conifers from plantations of the twentieth century exist, though in some zones we find natural pine forests dominated by *Pinus sylvestris* var. *iberica*, *P. nigra* subsp. *salzmannii* and *P. nigra* subsp. *latisquama* (*P. iberica*, *P. salzmannii* and *P. latisquama*, respectively, hereafter). Also appearing are some relict groves dominated by *P. uncinata*. The dominance of coniferous formations is magnified by the large expanses of *Juniperus thurifera* woodlands of the Iberian parameras and of creeping *Juniperus sabina* in some summit areas.

The online version of this chapter (doi:10.1007/978-3-319-54784-8_11) contains supplementary material, which is available to authorized users.

M. Peinado (✉) • J.L. Aguirre • A. Aparicio
Department of Life Sciences (Plant Biology), University of Alcalá, 28871-Alcalá de Henares, Madrid, Spain
e-mail: manuel.lorca@uah.es

J. Loidi (ed.), *The Vegetation of the Iberian Peninsula*, Plant and Vegetation 12,
DOI 10.1007/978-3-319-54784-8_11

Scrublands range from heaths dominated by *Cytisus* and *Erica* species in the rainiest zones, to scrub of the laurel-leaf rock rose *Cistus laurifolius* ("jarales") and *Genista scorpius* ("ahulagares") in drier or more exposed areas. In some shady zones appear formations of *Buxus sempervirens* or *Berberis hispanica* and in windier zones spiny formations of *Genista rigidissima* or *Erinacea anthyllis*. In the northern region occur mixed *Ilex aquifolium* woodland formations comprising several deciduous trees (*Acer monspeliensis, A. campestre, Corylus avellana*, etc.).

Riverbank formations include those typical of the interior peninsula with a dominance of series of forest, woodlands and thickets dominated by willows (*Salix purpurea, S. alba, S. atrocinerea*, etc.), poplars (*Populus alba, P. nigra*) and elms (*Ulmus minor*). The sierras comprise large rocky habitats that sustain the growth of an interesting rupicolous flora with several significant endemisms. Habitats of basic substrate on rocky, steep slopes are dominated by open thickets of *Juniperus phoenicea* ("sabina negral").

The Iberian System is one of the most under-populated regions of Spain with large expanses of Teruel, Soria, Burgos, Cuenca and Guadalajara being home to fewer than 10 inhabitants/km^2. Its setting is therefore one of immense, well-preserved zones. The area comprises mountains and sierras that are largely protected, sometimes because they form part of the Red Natura 2000 or because they harbour a protected feature such as the Natural Parks of the Alto Tajo, Moncayo, Sierra de Cebollera, Laguna Negra and Circos Glaciares de Urbión.

11.1 Introduction

Limits Running for over 500 km NW-SE between the Ebro basin and Meseta Central, the Iberian System is one of Spain's principal mountain ranges and the main forest region of Mediterranean Spain. The system's haphazard and motley series of mountain chains ("sierras" or "serranías"), massifs, high plateaus and interior depressions extends southeastwards from the Corredor de La Bureba in Burgos, close to the Cantabrian Ranges, to the Mediterranean coastal ranges in Castellón and Valencia, covering an area approaching 61,140 km^2 (see Figs. 11.1, and S.11.1, S.11.2 as electronic supplementary material).

Geological and Orographical Features (Figs. S.11.3, S.11.4, S.11.5, S.11.6, S.11.7 and S.11.8) After the Variscan or Hercynian orogeny, a gigantic mountain range formed and crossed in a NW-SE direction all the current land occupied by the Spanish Meseta Central. This huge Palaeozoic massif was so intensely and persistently eroded that, before the start of the Mesozoic era, a large portion had developed into a peneplain sloped towards its eastern flank. This flank was to be shaped by several Mesozoic marine transgressions and regressions such that from the Triassic (which saw the origin of the spectacular red sandstones known as "rodenos"), to the Cretaceous, deposits were left by invading marine waters and by continental sediments arising from erosion of the old Palaeozoic massif (Pérez-Estaún et al. 2004). This was followed in the Late Mesozoic by fracturing due to

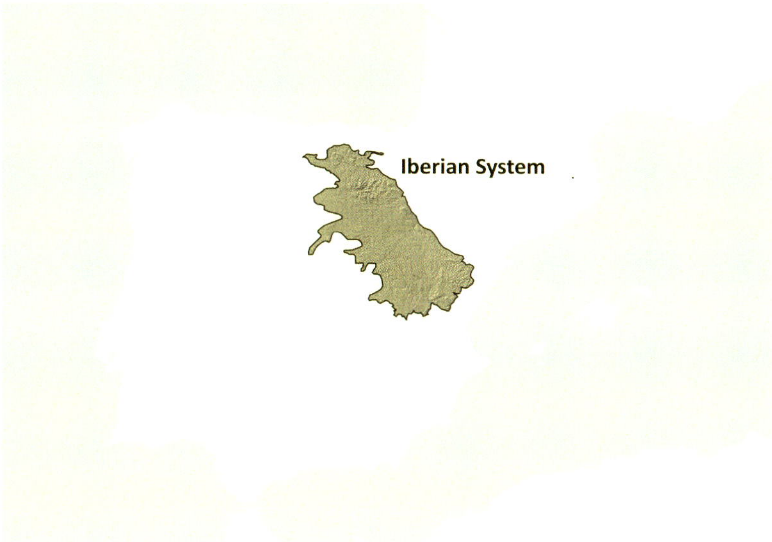

Fig. 11.1 Study area

Alpine orogeny pressures of the hardened Palaeozoic basement and folding of Mesozoic deposits to induce anticlines (horsts) and synclines. The latter comprised depressions of varying size that filled with continental sediments throughout the Tertiary and Quaternary (Fig. 11.2).

Besides folded Mesozoic rocks, the Alpine orogeny gave rise to different Iberian System relief features: on one hand to the Palaeozoic outcrops (quartzite, schist, Cambrian limestone and dolomite) of La Demanda, and on the other hand to the Tertiary materials that filled interior depressions. The Palaeozoic basement of La Demanda forms the series of mountains of the North Oroiberian Sierran biogeographical sector, divided into two well-differentiated parts by the narrow depression of the river Najerilla, a tributary of the Ebro. To the north of the Najerilla appear the sierras of San Millán (2131 m) and San Lorenzo (2262), formed mainly by Silurian quartzite. To the south of this river, that basement, more reduced, sinks below the Mesozoic mantle, which forms an impressive mass of Jurassic and Cretaceous limestones comprising the sierras Urbión (2228) and Cebollera (2147). The landscape of these mountains is abrupt, especially in the narrow valleys of the northern slope, and summits show a glacier erosion model. Most of the rocks are siliceous (except for the Cambrian dolomites of Mansilla) and belong to the Palaeozoic core as well as to the Mesozoic mantle: quartzite, slate, sandstone or conglomerate. Large islands of base-rich materials, limestone and marl of the Jurassic and Cretaceous, are also present. Some of these materials, such as the siliceous conglomerates of the Lower Cretaceous, are highly resistant to erosion and form marked shelves and ridges, among which the most outstanding peaks are those of the Urbión (Urbión, Castillo de Vinuesa, etc.).

Fig. 11.2 Evergreen oak forest (Teucrio scorodoniae-Quercetum rotundifoliae) in the Moncayo foothills (Soria)

The sierras making up the massif are arranged as three differentiated complexes (see orographical profiles as electronic supplementary material S.11.4, S.11.5, S.11.6, S.11.7 and S.11.8): the more north-western complex occurs around the sierra La Demanda; to the south of which and separated by the Canales depression appears the Neila-Urbión-Cebollera complex, the longest and largest of the three, which with the Alba sierra, forms the source of the Duero. From here, now in a southeast direction, continue the small sierras of Almuerzo and Madero in Soria, which connect with the Sierra del Moncayo, where the peak summit of the Iberian System is found at 2316 m. From Moncayo southeastwards arise a series of modest alignments that extend to the headwaters of the river Huerva, where they join with the Sierra de Gúdar. The Moncayo is a half raised fault block (semihorst), with a Palaeozoic core and Triassic cover reworked by glacial and periglacial erosion.

Except in the Moncayo and its south-eastern irradiations, whose waters all flow into the Ebro both directly or via its tributaries Jalón, Jiloca and Huerva, the northern massifs of the Iberian system constitute the divide between the waters of the Ebro and Duero. The northern slope drains towards the Ebro via a series of rivers (Tirón, Oja, Najerilla, Iregua, Leza and Cidacos) that carve narrow valleys ending in the Rioja plains of the Ebro depression (Fig. 11.3). The northern slope continues southeastwards through the valleys of the river Alhama and Tierra de Ágreda until the foot of the Moncayo. The southern slope of the Neila-Urbión-Cebollera alignment rests on the Meseta Central, which in this region of Soria

Fig. 11.3 Cliffs and bluffs of conglomerates usually mark the transition between the Iberian System and the Ebro valley. Iregua river gorge at Islallana, La Rioja

reaches heights close to 1000 m. A glacis of sandy substrates fed by the river Duero confers a gentle landscape rising towards the summit of the Picos de Urbión.

The central zone of the Iberian System is formed by a great longitudinal depression –Depresión Ibérica– whose most distinctive feature is the Calatayud-Teruel trough, which is traced from north to south by the rivers Turia and Jiloca, and crosswise by the river Jalón (Fig. 11.4). The Teruel trough is entrenched between the massifs of Albarracín, Javalambre and Palomera. The Tertiary and Quaternary sediments that fill it have been modelled, as in Calatayud, into "muelas" (steep flat-topped mounds), hills and small depressions. The Jiloca trough is clearly traced between the horst of Albarracín and the Palomera sierra. Tertiary sediments have been overlain by large well-preserved glacis that connect with the foothills of the sierra. From the Jalón, the System divides into two mountain alignments, one running parallel to the Meseta and the other parallel to the Ebro.

To the east of Soria and La Rioja, the alignment that parallels the Ebro mostly commences on the eastern margin of Zaragoza, where there is a massive presence of siliceous rocks of the Palaeozoic core. Its Mesozoic sedimentary cover is reduced to narrow bands, some appearing on slopes and others in the tectonic troughs. The Palaeozoic zone is split by NW-SE longitudinal troughs, among which we should mention those of Calatayud-Daroca and Almazán-Ariza, whose dimensions have

Fig. 11.4 Typical upper-mesomediterranean landscape of the Jalon river depression near Arcos del Jalón, Soria. PNV: ecotone between Asparago acutifolii-Quercetum rotundifoliae and Junipero thuriferae-Quercetum rotundifoliae

Fig. 11.5 Sabinares albares (juniper woodlands: Juniperetum hemisphaerico-thuriferae) on the high plateaus (páramos) around the Turia river valley, Teruel

allowed for extensive Miocene infill under a lacustrine regimen marked by gypsum, which supports a similar plant cover as in the Ebro depression.

The branch of the Iberian System in Aragón, rather than a compact, continuous alignment, is fragmented into smaller units, apparently unconnected and disorganized, separated by páramos, often at heights of more than 1000 m (Fig. 11.5).

There is only some orographic continuity in the central-western zone where the summit line, trending NW-SE, ranges from 1200 m to 1600 m, with valleys and depressions that outline the sierras of Tabuenca and de la Virgen (1433 m), Vicort (1420 m), Algairén (1279 m), Modorra (1299 m), Peca (1279 m), Herrera (1348 m), Culalón (1478 m), San Just (1613 m) and Ejulve (1528 m). In the south-eastern corner, more defringed and of lower altitude, arise the small sierras of Calanda (791 m), Arcos (978 m), Alcorisa (879 m), Alcaine (854 m) and the alignments of Gargallo-Los Molinos-Castellote in the low Ebro depression (Bajo Aragón). Towards the south, the flank that parallels the Ebro continues along the sierras of the Gúdar massif (2024 m), in Teruel; between Teruel and Castellón rises the Maestrazgo. The final foothills are those of Peñagolosa (1813 m), which join up with the Catalan ranges, Javalambre (2020 m), Espina, Espadán and Sabinar.

South of the Jalón, the interior alignment parallel to the Meseta commences with the Sierra de Solorio, continues with Sierra Ministra, which merges with the Central Range, and runs on southwards via the parameras of Sigüenza, Atienza and Molina, the Alto Tajo, the sierras Menera, Pardos (1264), Santa Cruz (1423), Albarracín (Pico Caimodorro, 1935 m) and Montes Universales (Muela de San Juan, 1814 m). From the latter, starts the Serranía de Cuenca east of the river Júcar's basin, whose foothills spread in the south until the Sierra de Mira. Finally, this alignment concludes south of the Cuenca province in La Mancha and the sierras of Martés and Aledua, whose flora has affinities with the coastal Levantine flora.

Biogeographical Sketch The map in Fig. 11.6 is a biogeographical sketch to the sector level. A more detailed map is provided as supplementary material (S.11.9). According to the biogeographic classification scheme described in Chap. 5, the Iberian System encompasses the following biogeographical units:

Mediterranean region: West Mediterranean subregion. Central Iberian Mediterranean province. A Low Aragonese subprovince: 25 Rioja and Estella sector (330,415). B Oroiberian subprovince: 26 South Oroiberian sector (1,819,160) (Fig. 11.7), 27 North Oroiberian Sierran sector (868,416), 28 Cantabrian Castilian sector (644,838); IIbc, Castillian subprovince: 30 Celtiberia and Alcarria sector (2,451,243). Figures in brackets indicate the surface area (hectares) occupied by each sector in the study area.

Although from an orographic and geologic point of view they belong to the Iberian System, we have not included in this section the lowlands of La Alcarria de Guadalajara and Cuenca, the mountains that isolatedly splash the La Mancha Sector (described in Chap. 3 in volume 2) or the mountains south of Cuenca and Teruel, along with most Iberian mountains of Valencia (Pina, Espadán, Calderona, Negrete, Palomera, Martés, Cavall Bernat de Corbera, Montdúver, described in Chap. 14).

Climate Features (Figs. S.11.10, S.11.11 and S.11.12) To summarize the climate trends of the Iberian System, we compiled data for 99 weather stations within the study zone cited by Rivas-Martínez (2007). The Iberian system features a markedly diverse climate due to two main factors: its geographic position and its orography. Owing to the System's central position in the Iberian Peninsula, it experiences the confluence or transition of the conditions of both of its eastern and western halves.

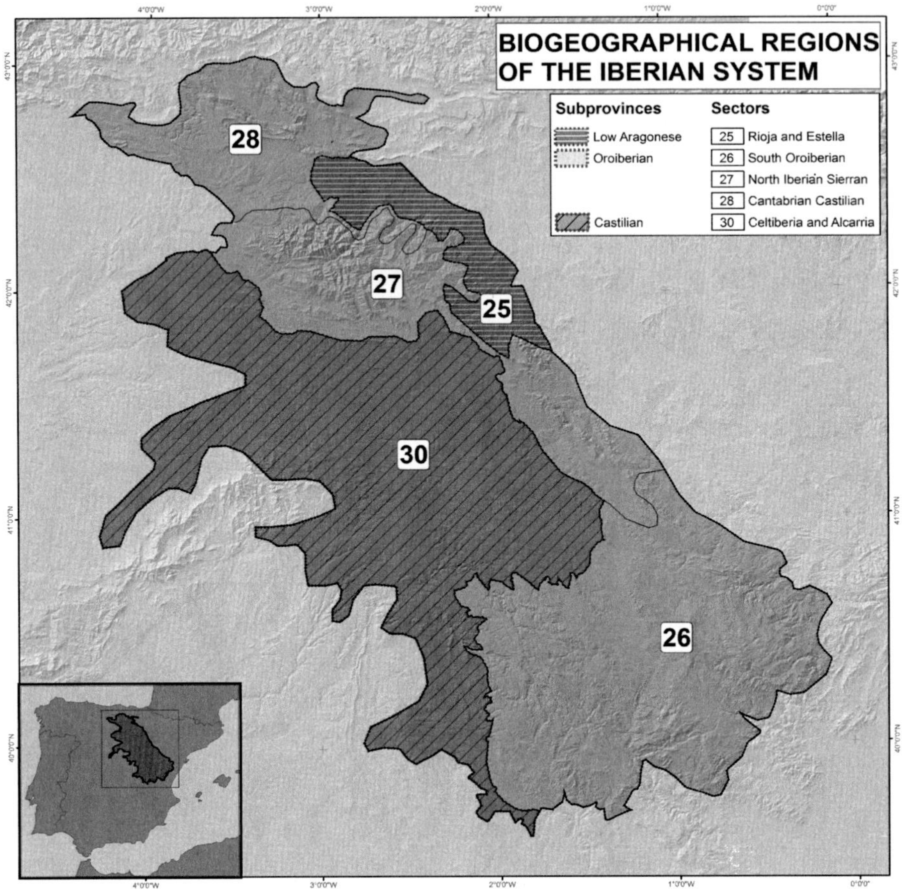

Fig. 11.6 Biogeographical map of the Iberian ranges and highlands

This means that its precipitation regimen is influenced both by the Atlantic slope and summer cyclogenic activity from the east. Naturally, the Atlantic regimen is all the greater the more western the position we adopt, while the Mediterranean influence increases in the opposite direction. A marked climate diversity is obvious if we consider we are in a territory of elevated mean altitude (1057 m compared to 660 m for the peninsular mean) and of complex orography. These conditions induce strong phenomena of cold air damming, whereby low-level cold air masses are trapped topographically in depressions and föhn effects are produced in the lee of the mountain ranges, exaggerating thermopluvimetric gradients typical of the regional climate and determining sharp climate contrasts at the provincial or local level.

Temperature (Fig. S.11.10) As a consequence of polar air damming, some zones of the Iberian System, especially Guadalajara, Soria, Cuenca and Teruel, endure the

Fig. 11.7 Supratemperate submediterranean landscape of the South Oroiberian sector. Junipero thuriferae-Pinetum latisquamae on deep soils and Rhamno lycioidis-Juniperetum phoeniceae on lithosols with *Quercus rotundifolia*. Rincón de Ademuz, Valencia

harshest temperatures of Spain (Fig. 11.8). Thus, zones of the Montes Universales and Maestrazgo have mean annual temperatures under 8–9 °C. In some of these zones, as also in the Serranía de Cuenca and Páramos de Molina, more than 120 days of frost are recorded per year. This situation is only comparable to that of some high Pyrenean zones since although mean temperatures may be lower there, the greater presence of clouds prevents night temperatures from falling as much as in the Iberian System.

The Jiloca valley holds the record for cold temperatures if we exclude the high mountain zone, and figures of −30 °C have been registered in Calamocha; other towns such as Molina de Aragón and Monreal del Campo have seen −28 °C on several occasions. Teruel and Calamocha form, with Molina de Aragón, the triangle considered as the "Spanish cold pole" owing to the frequency of minimum temperatures of the order of −20 °C or lower (Aupí 2005).

Besides the Teruel-Calamocha-Molina de Aragón triangle, other zones that are notorious for their low temperatures are the Serranía de Cuenca, Montes Universales de Teruel and the area around the Picos de Urbión in Soria. In the coldest regions of Soria, night frosts are sometimes recorded in mid-summer, mostly in the second half of August when nights start to lengthen after the summer solstice. In the Soria locations of Castejón, Vovaleda and Lubia, and in the

Fig. 11.8 Winter in the Junipero sabinae-Pinetum ibericae, summits of Sierra de Javalambre (Teruel), February, 19, 2015

surroundings of the Picos de Urbión, this is not exceptional but rather often occurs in July and August. In the first of these towns, a temperature of −6 °C is on record for August 20th of 1961, and in the latter, −6 °C was recorded the following day. In both areas, minimums of −2 to −5 °C were reached in the summers of 1957 and 1958. These data indicate that in many zones of the Iberian System, freezing temperatures may be recorded at any time of the year.

Rainfall (Fig. S.11.11) The source of most rainfall is the North Atlantic, where cyclonic perturbations are the consequence of the circulation of air masses predominating at these latitudes. These air masses along with frontal systems enter the Peninsula via the west or northeast in an alternating sequence of periods of instability and weather improvement coinciding with the passage of fronts and the influence of high pressures respectively. South of the train of perturbations, lies the band of high subtropical pressures, including the Azores anticyclone which deviates towards the north Atlantic storms on their way towards the European continent. These high pressures are the culprits of the dry summers of the Mediterranean climate that prevails in the Peninsula complex and Iberian System.

Bioclimatology (Fig. S.11.12) The bioclimatic classification of the 99 weather stations examined can be found in Table S.11.1 (Supplementary Material). In Table 11.1, we summarize this classification and relate bioclimates to the potential natural vegetation (PNV). In principle, the PNV follows the Vegetation Series Map by Rivas-Martínez (1987). Following our field studies, this PNV was updated according to Rivas-Martínez et al. (2011a, b) and each weather station was initially assigned to a particular PNV, defined as the plant community that would become established if all successional sequences were completed without interference by man under the present climate and edaphic conditions (Ricotta et al. 2002).

Table 11.1 Summary of the relationships among bioclimates, physiognomical formations and associations in the Iberian System

Mediterranean	(10) Mesomediterranean Dry (9) and Subhumid (1)	(10) Evergreen oak forests	(7) *Asparago acutifolii-Quercetum rotundifoliae*	C1
			(2) *Junipero thuriferae-Quercetum rotundifoliae*	
			(1) *Teucrio scorodoniae-Quercetum rotundifoliae*	
	(23) Supramediterranean Dry	(18) Evergreen oak forests	(13) *Junipero thuriferae-Quercetum rotundifoliae*	C1
			(2) *Asparago acutifolii-Quercetum rotundifoliae*	
			(2) *Junipero lagunae-Quercetum rotundifoliae*	
			(1) *Teucrio scorodoniae-Quercetum rotundifoliae*	
		(5) Marcescent oak forests	(3) *Cephalanthero rubrae-Quercetum fagineae*	C2
			(1) *Luzulo forsteri-Quercetum pyrenaicae*	
			(1) *Sileno melliferae-Quercetum fagineae*	
	(37) Supramediterranean Subhumid	(22) Marcescent oak forests	(19) *Cephalanthero rubrae-Quercetum fagineae*	C1
			(2) *Sileno melliferae-Quercetum fagineae*	
			(4) *Spiraeo obovatae-Quercetum fagineae*	
		(10) Evergreen oak forests	(7) *Junipero thuriferae-Quercetum rotundifoliae*	EX
			(1) *Junipero lagunae-Quercetum rotundifoliae*	
			(2) *Asparago acutifolii-Quercetum rotundifoliae*	
		(5) Juniper woodlands	(5) *Juniperetum hemisphaerico-thuriferae*	C3
	(7) Supramediterranean Humid	(4) Marcescent oak forests	(2) *Cephalanthero rubrae-Quercetum fagineae*	C1
			(1) *Pulmonario longifoliae-Quercetum pyrenaicae*	
			(1) *Spiraeo obovatae-Quercetum fagineae*	
		(3) Juniper woodlands	(3) *Juniperetum hemisphaericae-thuriferae*	C3

(continued)

Table 11.1 (continued)

Temperate	(1) Supratemperate Dry (Xeric)	(1) Evergreen oak forests	(1) *Junipero thuriferae-Quercetum rotundifoliae*	C1
	(4) Supratemperate Subhumid (Steppic)	(4) Marcescent oak forests	(4) *Sileno melliferae-Quercetum fagineae*	C1
	(12) Supratemperate Subhumid (Submediterranean)	(11) Marcescent oak forests	(2) *Cephalanthero rubrae-Quercetum fagineae*	C1
			(2) *Spiraeo obovatae-Quercetum fagineae*	
			(1) *Corylo avellanae-Quercetum fagineae*	
			(1)*Luzulo forsteri-Quercetum pyrenaicae*	
			(5) *Sileno melliferae-Quercetum fagineae*	
		(1) Pine forests	(1) *Junipero sabinae-Pinetum ibericae*	C3
	(9) Supratemperate Humid (Submediterranean)	(6) Marcescent oak forests	(2) *Cephalanthero rubrae-Quercetum fagineae*	C1
			(1) *Corylo avellanae-Quercetum fagineae*	
			(2) *Pulmonario longifoliae-Quercetum pyrenaicae*	
			(1) *Spiraeo obovatae-Quercetum fagineae*	
		(2) Beech forests	(2) *Galio rotundifolii-Fagetum sylvaticae*	C2
		(1) Pine forests	(1) *Junipero sabinae-Pinetum ibericae*	C3

Numbers in brackets are the number of weather stations. C1 Climatophilous, C2 Climatophilous on northern expositions and deeper soils, C3 Climatophilous on more continental areas and mountain summits, EX Edaphoxerophilous.

Of these 99 weather stations, 81 feature a Mediterranean macrobioclimate. The remaining stations show a Temperate macrobioclimate, which in this terrain of marked Mediterraneanity is determined essentially by three factors. The first of these is that high mountains offset summer aridity owing to rainfall falling on their slopes, such that a Temperate climate can be assigned to their middle and high zones as islets within a Mediterranean climate. Accordingly, this type of climate may be found at the stations at higher altitudes such as those of La Poveda-Puerto de Piqueras (1709 m), Gúdar (1587 m), Guadalaviar (1519 m), Fortanete (1408 m), Alustante (1404 m) and Tragacete (1342 m); this last site holds the precipitation record for the whole Iberian System (Fig. 11.9). Notwithstanding, a Temperate climate and high altitude do not necessarily mean that highest rainfall values will be recorded. Paradigms of this are the stations of Guadalaviar and Gúdar, both in Teruel and Tragacete (Cuenca), and at the same latitude and belonging to the same biogeographic unit (South Oroiberian sector). In Guadalaviar (1007 mm) and Tragacete (1019 mm) rainfall is 2 or 3 times higher than that recorded at Gúdar (470 mm). The difference lies in the fact that the two former stations occur upwind from the Atlantic fronts that empty on the western sierras of Cuenca and Teruel (Serranía de Cuenca, Albarracín) rainshadowing both the Jiloca-Teruel depression and Gúdar-Javalambre massif, while Gúdar occurs on the other side of these

Fig. 11.9 Supratemperate submediterranean landscape near Mosqueruela, Teruel. PNV: Festuco gautieri-Pinetum salzmannii. *Juniperus hemisphaerica, Erinacea anthyllis* and and grasslands of Festucetum hystricis dominate open spaces

Fig. 11.10 Supratemperate submediterranean landscape near Losilla, Valencia. Climatophilous black pine forests (Junipero thuriferae-Pinetum latisquamae) interspersed with edaphoxerohilous juniper woodlands (Rhamno lycioidis-Juniperetum phoeniceae)

depressions, further away from the already debilitated Atlantic fronts and in the Gúdar-Javalambre massif's rainshadow. Effectively, in rainshadows, we also find the station at Alfambra (1047 m; P = 416 mm), also in Teruel, as the only station featuring a Temperate Xeric bioclimate (Steppic variant) (Fig. 11.10).

The second broad factor that determines the appearance of the Temperate macrobioclimate is latitude, or in other words, closeness to the Cantabrian sea. The peninsular north, or northeast Portugal, Galicia, the Cantabrian Cornisa (or rim) and Cordillera, and Pyrenees, is affected in different measure by the tails of the cyclone fronts that circulate in summer at high latitudes, making the summer period sufficiently rainy for its adscription to the limits of the Temperate macrobioclimate. Consequently, within the Iberian System, the more northern lands of the Cantabrian Castilian sector are influenced by ocean disturbances in the form of summer rains, which -without resolving the summer aridity of the Mediterranean climate- reduce it to the minimum required for their inclusion in the Submediterranean variant of the Temperate macrobioclimate. Along the north-south transect of the mountain systems of the Iberian System, we may observe a Mediterraneanization hastened by the rainshadowing effects of each alignment on its southern foothills. The northern part of the Iberian System, included in the North Oroiberian Sierran biogeographical sector, collects sufficient summer rains on its main mountain core (Demanda, Cebollera-Urbión) to also be considered Temperate Submediterranean.

The third factor accounting for a Temperate macrobioclimate is proximity to the Mediterranean coast. This peninsular zone suffers the summer and autumn cyclogenic activities of the Golfo de León that marginally reach the Spanish east, along with the torrential rains produced by the phenomenon of "cut-off low", or "gota fría", caused by reheating of Mediterranean waters in the summer. The influence of these summer disturbances is felt in the rhythm of precipitations at the weather stations with greater rainfall in late summer and/or autumn and at the stations with least rainfall in winter. This determines a different regimen to the typical Mediterranean rhythm in which summer (followed by autumn) is always the driest season. This precipitation rhythm and a Temperate Oceanic bioclimate (Steppic variant) is shown by four stations of Teruel (La Puebla de Valverde, Torre Los Negros, Vivel del Río Martín and Sarrión-La Escaruela) at the headwaters of the river Palancia, whose corridor connects the Teruel meseta with Valencia.

Finally, given the relatively low altitudes of the analyzed weather stations (the highest occurs some 300 m below the 2000 m summit line), stations with an Oromediterranean thermotype have not been included in Table S.11.1. However, certain vegetation types, in particular the pine forests of *Pinus iberica*, are indicators of the broad presence of this thermotype in the System's high mountains.

11.2 Potential Vegetation

Using the Vegetation Series map (Rivas-Martínez 1987) as the data source, Table 11.2 outlines the surface areas occupied by the different PNV types in the study area. A more detailed table with the corresponding vegetation series is provided as supplementary material (Table S.11.2). Across most of the Iberian System (46.4%), the PNV comprises evergreen oak forests. This is followed by marcescent *Quercus faginea* forests (26.4%), though the area occupied by these is overestimated, because in Rivas-Martínez's map, the forests of *Pinus nigra* are

Table 11.2 Surface areas occupied by the different PNV types in the study area

PNV	Surface (hectares)	%
Grassland	30	0.00
Coniferous	807,788	13.21
Deciduous	154,289	2.52
Marcescent	2,289,539	37.45
Evergreen	2,777,778	45.43
Riparian	65,976	1.08
Others	18,671	0.31
Total	6,114,072	100.00

For detailed data see Table S.11.2 as supplementary material ascribed to the *Q. faginea* series. This means that the area occupied by pine forests (2.8%) is smaller than the surface area they cover in the territory examined. The land sustaining the dominant evergreen formations –*Quercus rotundifolia* forests– coincides with the dominant Mediterranean macrobioclimate; the marcescent formations dominated by *Quercus faginea* or *Quercus pyrenaica*, second in extension (45.4%), coincide with the Mediterranean-Temperate transition zones, i.e., submediterranean territories.

Nervertheless, reality wins over potentiality. The long cultural history of humans occupying this land has marked this landscape. Presently, the Third National Forest Survey (Tercer Inventario Forestal Nacional: http://www.magrama.gob.es) is underway. Of the provinces covered by the Iberian System, only data for four are avalable at this portal. The information provided, however, does reflect the impacts suffered by the PNV in these provinces. Hence, of Cuenca province's 1,714,112 ha, more than 900,000 ha have been deforested. This situation is repeated in the other three provinces: Guadalajara (1,221,209 vs. 500,000 ha), Soria (1,030,642 vs. 600,000 ha), and Teruel (1,480,956 vs. 900,000 ha).

Forests and Woodlands

Forests and woodlands make up most of the Iberian System's PNV. Figures S.11.3, S.11.4, S.11.5, S.11.6, S.11.7 and S.11.8 show the distributions of the major forest and woodland types in three latitudinal transects of the study area (Acronyms for the different types of vegetation in these profiles correspond to map series in Fig. S.11.12). Besides types linked to given biotopes with special edaphic characteristics such as hygrophilous forests, the most extensive of these in decreasing order are: evergreen oak forests and tall shrublands (forests dominated by *Quercus rotundifolia* and macchias or "coscojares" of *Quercus coccifera*), marcescent oak forests ("quejigares" dominated by *Quercus faginea*, and "melojares" dominated by *Quercus pyrenaica*), pine forests, juniper woodlands (dominated by *Juniperus thurifera*) and deciduous forests (of which the beech forests dominated by *Fagus sylvatica* are the most extensive, yet with the occasional presence of groves dominated by the deciduous oak *Quercus petraea*.

Evergreen Oak Forests and Tall Shrublands Forests and woodlands dominated by the sclerophyllous evergreen oak *Quercus rotundifolia* ("encina" or "carrasca") known as "encinares" or "carrascales" at one time covered over three million hectares of the Iberian System (Table S.11.2). The following key (Table 11.3) serves to identify these encinares of the Iberian System:

Most encinares appear in the supramediterranean belt and comprise four described associations. Growing on calcareous substrates, the most extended are the supramediterranean Junipero thuriferae-Quercetum rotundifoliae and the mesomediterranean Asparago acutifolii-Quercetum rotundifoliae. In mesomediterranean areas and on basic soils, often brown rendzina soils, encinares of the association Asparago acutifolii-Quercetum rotundifoliae dominate. Precipitation ranges from 350 to 550 mm per year; if precipitation exceeds 550 mm (subhumid and upper dry ombroclimates), encinares usually give way to the quejigares of the Cephalanthero rubrae-Quercetum fagineae. The forest mantle of that

Table 11.3 Key to evergreen oak forests of the Iberian System

1a	Mesomediterranean evergreen oak forests and woodlands, usually with *Quercus coccifera* as seral shrub: ***Asparago acutifolii-Quercetum rotundifoliae***.	
1b	Supramediterranean evergreen oak forests and woodlands	**2**
2a	Evergreen oak forests and woodlands growing on calcareous soils	**3**
2b	Evergreen oak forests and woodlands growing on acidic soils	**4**
3a	With *Juniperus thurifera* as co-dominant: ***Junipero thuriferae-Quercetum rotundifoliae***.	
3b	Without J. thurifera but with *Quercus ilex* subsp. *gracilis* and *Spiraea obovata*; evergreen oak forests and woodlands thriving on the Cantabrian Castilian sector and on calcareous outcrops of the North Oroiberian Sierran sector (Moncayo area): ***Spiraeo obovatae-Quercetum rotundifoliae***.	
4a	Submediterranean forests of the Ibérico-Serrano sector with *Teucrium scorodonia* and *Quercus ilex* subsp. *gracilis*: ***Teucrio scorodoniae-Quercetum rotundifoliae***.	
4b	Mediterranean forests and woodlands of the Celtiberico-Alcarreño sector with *Juniperus oxycedrus* subsp. *lagunae*, *Cytisus scoparius* and *Lavandula pedunculata*: ***Junipero lagunae-Quercetum rotundifoliae***.	

mesomediterranean evergreen association is formed by either brooms or "retamares" (Genisto scorpii-Retametum sphaerocarpae) on deep soils or coscojares (Daphno gnidii-Quercetum cocciferae) in the more rugged and drier biotopes, though characteristic of lithosol ridges are the edaphoxerophilous "sabinares negrales" (Rhamno lycioidis-Juniperetum phoeniceae). Other regressive stages of the series are scrub formations (Lino differentis-Salvietum lavandulifoliae, Paronychio-Astragaletum tumidi, Salvio lavandulifoliae-Erinaceetum anthyllidis) and some grasslands (Stipo offneri-Helictotrichetum filifolii, Phlomido lychnitis-Brachypodietum ramosi, Saxifrago tridactylitae-Hornungietum petraeae, Astragalo sesamei-Poetum bulbosae).

In shaded zones or in ravines or rifts with some soil, mesomediterranean encinares contain plants with greater water demands such as *Quercus faginea, Acer monspessulanum, Geum sylvaticum, Colutea hispanica* or *Ligustrum vulgare*. The appearance of these species determines the subassociation quercetosum fagineae, as the transit to the quejigares of the Cephalanthero rubrae-Quercetum fagineae. Today, the natural area of these encinares coincides with those of characteristic dryland crops such as cereal and grape. Vast areas occupied by this series have been repopulated with "carrascos" (*Pinus halepensis*) or "pinos negros" (*Pinus nigra*). Their richness in aromatic species such as rosemary, sage, lavender, or thyme has been exploited by bee keepers.

The largest encinares of the Iberian System are represented by the association Junipero thuriferae-Quercetum rotundifoliae that inhabits the supramediterranean belt of the sectors Celtiberian and Alcarrian, North Oroiberian Sierran and South Oroiberian, at altitudes of 1000 m to 1400 m (Fig. 11.11). This association alternates topographically with the quejigares of the Cephalanthero-Quercetum fagineae that establish themselves in the more shady places, while on the windy exposed slopes of the parameras it is replaced by the supramediterranean juniper woodlands

Fig. 11.11 Evergreen oak forest (Junipero thuriferae-Quercetum rotundifoliae) interspersed with *Pinus nigra* subsp. *salzmannii* on the Alto Tajo, Cuenca

(Juniperetum hemisphaericae-thuriferae), locally called "sabinares albares", from the "sabina albar" *Juniperus thurifera*. The mature stage of the series is an encinar with *J. thurifera*. In the zones where they make contact, these supramediterranean forests can be mistaken for mesomediterranean encinares (Asparago acutifolii-Quercetum rotundifoliae) that have *J. thurifera* in the upper horizon of the mesomediterranean belt. Notwithstanding, both are clearly distinguishable by their replacement stages.

Microclimate plays an essential role in the development of one or another forest. Over distances of only a few metres, factors such as orientation, soil, wind exposure, relief shadows causing local shade etc., will provoke a temperature variation determining a change from a meso- to supramediterranean habitat. Linked to the mesomediterranean encinares are the communities Daphno gnidii-Quercetum cocciferae, Arrhenathero erianthi-Stipetum tenacissimae, Genisto scorpii-Retametum and Cisto clusii-Rosmarinetum, always absent in supramediterranean zones, in which it is common to find spiny mantles (Rosetum micrantho-agrestis), grasslands (Festucetum hystricis, Paronychio capitatae-Artemisietum lanatae) and scrublands dominated by thorny and cushion dwarf and spiny scrub such as the shepherd's seat *Erinacea anthyllis* and/or *Genista pumila* subsp. *rigidissima* (Lino appressi-Genistetum rigidissimae, Scabioso turolensis-Erinaceetum anthyllidis). In springtime, the study of ephemeral therophyte communities will also reveal the

Fig. 11.12 Dense evergreen oak forest (Teucrio scorodoniae-Quercetum rotundifoliae) near Magaña, Soria

presence of one or other potential vegetation: the association Saxifrago tridactylitae-Hornungietum petraeae in mesomediterranean zones, while the Bupleuro baldensis-Arenarietum ciliaris is confined to supramediterranean areas.

In the rainiest supramediterranean Cantabrian-Castilian territories, on lithosols associated with cretaceous hard limestones, we find the large encinares of the Spiraeo obovatae-Quercetum rotundifoliae (Fig. 11.12), described by Loidi & Fernández-Prieto (1986) and Loidi et al. (1997) and also mentioned for the North Oroiberian Sierran sector (Navarro 1986; Medrano 1994). These are basiphilous forests of *Quercus rotundifolia* and hybrid (*Quercus ilex* subsp *gracilis*) evergreen oaks, in which the former predominate, that thrive under a subhumid ombrotype, sometimes reaching mesomediterranean areas if the ombrotype persists. These closed woods are usually comprised of low, somewhat stunted trees, under which thrive shrubs such as *Spiraea hypericifolia* subsp. *obovata*, *Lonicera etrusca*, *Amelanchier ovalis*, *Rhamnus alaternus* or *Rosa agrestis* and a few replacement stage bushes such as *Genista scorpius*, *Genista occidentalis*, *Erica vagans* or *Arctostaphylos uva-ursi* subsp. *crassifolia*. Other tree or bushy plants that coexist with the evergreen oak are *Juniperus phoenicea* and *J. oxycedrus* subsp. *lagunae*. Also, in the more mesophytic enclaves of these woods such as stream or ravine beds, the presence of the hybrid oak is not rare along with *Arbutus unedo*, *Pistacia terebinthus* or *Viburnum tinus*. On sandy soils of some zones of the Castilian-

Fig. 11.13 Evergreen oak and marcescent oak forests (Junipero lagunae-Quercetum rotundifoliae and Luzulo forsteri-Quercetum pyrenaicae, respectively) on the Moncayo foothills

Cantabrian sector we find the cork oak, or "alcornoque", *Quercus suber* (García-Mijangos 1995).

Two associations growing on siliceous soils have been described, the Junipero lagunae-Quercetum rotundifoliae (Fig. 11.13), and Teucrio scorodoniae-Quercetum rotundifoliae (Fig. 11.2). The first of these is sometimes observed in some Palaeozoic outcrops of the Celtiberian and Alcarrian sector, from where it has been described by Peinado et al. (2008). In the zones with most precipitation in the Cameros sierra (North Oroiberian Sierran sector), there are the oaks of the association Teucrio scorodoniae-Quercetum rotundifoliae, in which some species with a temperate optimum occur that are absent in the association Junipero lagunae-Quercetum rotundifoliae, e.g. *Erica vagans, Erica cinerea, Calluna vulgaris* or *Brachypodium sylvaticum* (Loidi et al. 1997: 299).

Marcescent Forests The character marcescence refers to the retention of dead plant organs that are normally shed and may be viewed as intermediate between evergreen and deciduous. Thus, while deciduous trees lose their leaves at the beginning of winter, a marcescent tree will keep most of its leaves (lacking chlorophyll until) the new leaves appear in spring, when burgeoning buds displace these withering leaves. Examples are the melojo (*Quercus pyrenaica*), the Portuguese oaks or quejigos (*Q. faginea, Q. broteroi* and *Q. canariensis*) and downy oaks (*Q. pubescens*). Only *Q. pyrenaica* and *Q. faginea* inhabit the Iberian System.

Table 11.4 Key to marcescent oak forests of the Iberian System

1a	*Quercus pyrenaica* is the dominant tree on acidic substrates	**2**
1b	*Quercus faginea* dominates usually on limestones and marls	**4**
2a	With *Pulmonaria longifolia*; *Q. pyrenaica* forests on humid-hyperhumid areas of the North Oroiberian Sierran sector: ***Pulmonario longifoliae-Quercetum pyrenaicae.***	
2b	Other *Q. pyrenaica* forests and woodlands	**3**
3a	Forests and woodlands of the Celtiberian and Alcarria sector: ***Luzulo forsteri-Quercetum pyrenaicae***.	
3b	*Q. pyrenaica* forests and woodlands of the South Oroiberian sector (Peñagolosa massif and neighbouring areas): ***Cephalanthero rubrae-Quercetum pyrenaicae***.	
4a	With *Corylus avellana*; supratemperate *Q. faginea* forests of the South Oroiberian sector: ***Corylo avellanae-Quercetum fagineae.***	
4b	Other *Q. faginea* forests and woodlands	**5**
5a	With *Spiraea obovata*; *Q. faginea* forests thriving in the Cantabrian Castilian and Rioja and Estella sectors and on calcareous outcrops of the North Oroiberian Sierran sector (Moncayo area): ***Spiraeo obovatae-Quercetum fagineae.***	
5b	Other *Q. faginea* forests and woodlands	**6**
6a	With *Silene mellifera, Teline patens, Genista valentina* and *Stachys heraclea* subsp. *valentina*; *Q. faginea* forests from the eastern parts of the South Oroiberian sector (Gúdar, Albarracín and Javalambre mountain ranges: ***Sileno melliferae-Quercetum fagineae.***	
6b	Without those taxa; *Q. faginea* forests and woodlands of the Castilian subprovince: ***Cephalanthero rubrae-Quercetum fagineae.***	

The oak forests of the Iberian System dominated by *Q. pyrenaica* and *Q. faginea* may be distinguished according to the following key (Table 11.4).

Forests of *Quercus pyrenaica*, as those of other deciduous trees, differ clearly from encinares both in their structure and physiognomy (Fig. 11.14). The canopy is composed of a tree layer of *Q. pyrenaica* (usually some 10 m tall, though some specimens are taller than 20 m), with some trees of overall Atlantic character (*Sorbus torminalis, Sorbus aria, Taxus baccata, Ilex aquifolium, Frangula alnus*, etc.), which give rise to an open understorey. This undergrowth would be easily penetrable due to the lack of a thick shrub layer, but this is complicated by the melojo's capacity to produce seedlings, runners and stolons, which are scarce when the forest is mature but highly abundant when its above-ground mass has been decimated by the axe and underground mass increases. In contrast, a relatively well-developed herbaceous stratum exists, dominated by evergreen bulbous grasses, which flower extremely early to make good use of the light conditions prevailing in the forest interior before the dominant trees leaf out. Shade tolerant herbaceous plants are also present (sciophilous), adapted to the nemoral conditions of the wood, in which sunlight is a limiting factor. The shrub layer is well-developed but only at a forest border. In this manner, the surroundings of the melojares are protected by microphyllous brushwoods ("piornales") in which legumes such as *Cytisus scoparius* subsp. *scoparius*, *Genista florida*, *Genista cinerascens*, *Adenocarpus complicatus*, *Adenocarpus hispanicus*, etc., prosper alongside heathers (*Erica arborea*) and various species of roses.

Fig. 11.14 Marcescent oak forest or "melojar" (Pulmonario longifoliae-Quercetum pyrenaicae) with plantations of *Pinus sylvestris* on Montes de Oca, Burgos

Across the Iberian System's landscape, melojares can be easily distinguished at the start of spring through the reddish hues of their young leaves, and then through the ash green of their lobulated leaves that turn yellow in autumn, but above all by the brown colour of the withering leaves that remain on the tree through most of the winter. The melojares, though often victims of abusive fells, are still common in some siliceous sierras of the Iberian System. Their wood has been used to make railway lines, staves for barrels, floorings and charcoal. Forest regeneration is slow (up to 150 years), such that today it is common to observe patches of melojar in shrub state used as browsing or cleared grazing land. In the sandstone outcrops of the Iberian System (rodenos), melojos coexist with cluster pines (*P. pinaster*), whose spread has been undoubtedly prompted by man, yet it cannot be fully ruled out that in the original forests, pines occupied edaphoxerophilous positions within the stands of the melojares that occupy the most favourable ecological positions.

The melojares constitute the PNV of large supramediterranean and supratemperate (submediterranean) zones of the Iberian System that fulfil at least three conditions: acidic substrates and deep soils, and subhumid to hyperhumid ombroclimates. Their broadest distribution area is in the Cantabrian Castilian and North Oroiberian Sierran sectors, where they generally occupy a band between the driest encinares of the Junipero lagunae-Quercetum rotundifoliae and the beech forests of the Galio rotundifolii-Fagetum sylvaticae, which are usually more demanding of precipitation. In good measure, the melojares have been repopulated with *Pinus pinaster* and *Pinus sylvestris*. Notwithstanding, although these pine forests may be dominant in the landscape, the abundant regrowth of *Quercus pyrenaica* in the understorey clearly indicates the potentiality of the territory. Moreover, the melojares are mostly young forests coppicing after the traditional felling for firewood. Only in some areas do well-developed forests exist including specimens of a certain size, though usually they appear as very dense formations with numerous seedlings and a large density of young specimens, often sprouts, per hectare. This juvenile state of the ancient forest masses has the added consequence of a scarcity of accompanying nemoral species that are common in mature, closed-canopy forests.

The melojares of the Pulmonario longifoliae-Quercetum pyrenaicae (Fig. 11.15) are found in the rainier zones (humid to hyperhumid ombroclimates) of the Cantabrian Castilian and North Oroiberian Sierran sectors. Here, they usually alternate with those of the Luzulo forsteri-Quercetum pyrenaicae, which replace them in zones where precipitation is less (subhumid ombroclimate) or in sunny places where evapotranspiration is greater. A third association, Cephalanthero rubrae-Quercetum pyrenaicae, appears in some enclaves of the Peñagolosa massif (South Oroiberian sector).

Although these three forest types have more floristic similarities than differences, they may be distinguished by the presence or absence of certain plants and communities associated with some or others. The following are differential species of the association Pulmonario longifoliae-Quercetum pyrenaicae: *Ilex aquifolium*, *Viola reichenbachiana*, *Pulmonaria longifolia*, *Melampyrum pratense*, *Fagus sylvatica*, *Festuca braun-blanquetii*, *Hypericum pulchrum*, *Lonicera periclymenum*, *Rosa arvensis*, *Erica vagans*, *Erica arborea*, *Erica australis*,

Fig. 11.15 Humid "melojares" (Pulmonario longifoliae-Quercetum pyrenaicae) with hollies (*Ilex aquifolium*). Sierra Cebollera, Soria

Ribes alpinum, Genista occidentalis and *Melica uniflora*. When the forest is in a relatively well-preserved state, *Quercus petraea, Fagus sylvatica* (sometimes forming mixed woods), *Acer monspesulanum, Acer opalus, Acer campestre, Sorbus aucuparia, Sorbus aria, Sorbus torminalis, Pinus iberica*, etc. appear in the canopy. The common holly or "acebo" (*Ilex aquifolium*), which can be frequent, is a good differential species against the melojares of the Luzulo forsteri-Quercetum pyrenaicae. Dense holly populations ("acebedas") are a common landscape feature in the northern Iberian System. These populations seems to be associated with felling and grazing in melojares and beechwoods (Oria de Rueda 1992, 2003) or with soils showing evidence of clay illuviation (argillic horizon) in which the association Saniculo europaeae-Ilicetum aquifolii (Rivas-Martínez et al. 2011b: 469) thrives.

On wet sites, the forest mantle of the association Pulmonario longifoliae-Quercetum pyrenaicae is a dense bramble of *Rubus ulmifolius*, accompanied by *Rosa corymbifera, R. canina, Crataegus monogyna, Ligustrum vulgare, Lonicera periclymenum* subsp. *hispanica* and *Prunus spinosa*. In drier places, the forest mantle comprises the Cytiso scoparii-Genistetum polygaliphyllae. Heathlands linked to this association are ascribed to the associations Ericetum scopario-vagantis (on sandy soils), Arctostaphylo crassifoliae-Daboecietum cantabricae and Calluno-Genistetum occidentalis (the latter restricted to Moncayo).

In a subhumid ombroclimate in the North Oroiberian Sierran and South Oroiberian sectors, Luzulo forsteri-Quercetum pyrenaicae forests dominate. In its forest optimum, the most moisture-demanding elements of the Pulmonario-Quercetum pyrenaicae such as *Fagus sylvatica* or *Quercus petraea* are now lacking. Heathlands disappear and are replaced by "jarales" or thickets of *Cistus laurifolius, Cistus ladanifer, Cistus salvifolius, Lavandula pedunculata, Halimium ocymoides, H. umbellatum* subsp. *viscosum* and *Thymus mastichina*, corresponding to the associations Santolino rosmarinifoliae-Cistetum laurifoli and Erico arboreae-Arctostaphylletum crassifoliae (see below in the section Scrublands).

However, these thickets have been extensively ploughed and replaced with repopulations of *Pinus sylvestris* and *Pinus pinaster* (Fig. 11.16). Unlike natural pinewoods, the structure of these pine forests usually follows perfectly traced lines, individuals belong to few age classes and there are few accompanying species. When they are not managed, the regrowth of jarales and *Q. pyrenaica* takes over the understorey and initiates the recovery of the regional climax. It should be noted that very few pine seedlings manage to thrive in the undergrowth of the melojares, and this reflects the favouring of pine forests since ancient times. While melojos (*Quercus pyrenaica*) do sprout and prosper under the pines, it is difficult for them to survive the nemoral conditions of the melojares' understorey. Modern forest management of these repopulated pine forests has focused on selective clearing to preserve the most vigorous melojo stumps. Over time, this will generate mixed forests of *Quercus* and *Pinus* rich in flora and fauna. The abandoning of lands across large zones of the provinces of Teruel, Soria, La Rioja and Burgos is provoking a natural succession in these potential areas that starts with the occupation of

Fig. 11.16 Many pine forests of the northern part of the Iberian System coming from forestry practices. Climatophilous "melojares" (Luzulo forsteri-Quercetum pyrenaicae) and *Pinus pinaster* plantations near Zorraquín, Soria

wastelands by scrub followed by a slow steady process of regeneration, sometimes interrupted by forest fires.

Finally, in the eastern corner of the South Oroiberian sector and restricted to scarce sandstone outcrops appear mixed forests of *Quercus pyrenaica* and *Pinus iberica*, described in Peñagolosa as Cephalanthero rubrae-Quercetum pyrenaicae pinetosum sylvestris (Vigo 1968: 139). Floristically, the most significant feature of these forests is the presence of shrubs of the association Erico arboreae-Cistetum populifolii, which are rather scarce in the rest of the Iberian System.

In general terms, *Quercus faginea* forests (quejigares) may be said to ecologically substitute melojares on limestones and marls. In effect, quejigares occupy areas of subhumid ombroclimate or fresh shady sites (north-facing slopes, ravines, wet valleys, etc.), where soils rich in bases are deep and humic, even reflecting periods of transient hydromorphism which are better withstood by the quejigo (*Q. faginea*) than encina (*Q. rotundifolia*). Often, these series take refuge in shady sites, though in zones receiving sufficient rainfall (more than 550 mm) they may expand across large areas, alternating with different forests of evergreen oaks, black pines, or with juniper woodlands, at altitudes from 500 to 1500 m (Fig. 11.17). As in *Q. pyrenaica*, the leaves of *Q. faginea* are marcescent making the masses of quejigos interspersed with encinares or sabinares (juniper

Fig. 11.17 Slope, exposition and soil gradients near Zorraquín, Soria. Evergreen oak forests (Junipero thuriferae-Quercetum rotundifoliae) on rocky and undeveloped soils, and marcescent oak forests (Cephalanthero rubrae-Quercetum fagineae) on deep, well developed soils

groves) easily distinguishable; their natural colouring being highly characteristic of the Castilian landscape. Marcescence determines the combination of two physiological-ecological strategies. On one hand, the sclerophyllic character of the quejigo, though more moderate than in the case of encinas, allows it to survive the long summer drought, yet it makes it less competitive than the encina on sunny slopes. On the other hand, the strategy of annual leaf fall means that the quejigo can make better use of the sun's rays during summer; in other words its photosynthesis rate is higher than that of sclerophyllous plants. The consequence of all this is that, in conditions of adequate moisture, the quejigo will eventually replace the encina.

Although some specimens may be 20 m tall, most of Spain's quejigos are of modest height especially since most are young trees grown from seedlings. Frequently, the quejigo coexists with specimens of encinas and black pines, or *pinos salgareños*; the former are sometimes codominants in the Celtiberian and Alcarrian sectors, and the latter in the serranias of the South Oroiberian sector. In shadier places, they are enriched by broad-leaved hard wood species such as maples (*Acer monspessulanum, Acer granatense*) and rowans (*Sorbus aria, Sorbus domestica, Sorbus torminalis*), while at their margins, spiny shrubs and lianas are common. As in other deciduous forests, both the herbaceous stratum and that of lichens and mosses are poorly developed though there is no lack of herbaceous species, especially of nemoral grasses, peonies, violets and orchids.

Forests of the Cephalanthero-Quercetum fagineae are mainly spread across the Celtiberian and Alcarria sector and Serranía de Cuenca (López 1978), in the western South Oroiberian sector. They grow on deep soils under subhumid and humid ombroclimates, in positions facing north and east, on sunny sites alternating with evergreen oak forests of *Quercus rotundifolia*. Thus, ecotones show an abundance of mixed forests of marcescents and evergreen oaks, and small changes in slope exposures will prompt the transition of a pure forest of quejigos, to first a mixed forest and then a forest of encinas. Occasionally, in the Iberian parameras, quejigares alternate with juniper woodlands (Juniperetum hemisphaericae-thuriferae), forming different mosaics according to microtopography and

Fig. 11.18 Cephalanthero rubrae-Quercetum fagineae with *Pinus nigra* subsp. *salzmannii* in Serranía de Cuenca (Guadalajara)

orientation (Fig. 11.18). The ecological behaviour of *Quercus faginea* and *Pinus nigra*, and of mixed formations of these in the southern Iberian System is described below in the section "Pine forests".

Well-preserved Cephalanthero-Quercetum fagineae quejigares form a closed canopy in which normally appear *Acer monspesulanum*, *Sorbus torminalis*, *Sorbus aria* and, towards the forest's borders, tree-like shrubs such as *Viburnum lantana*, *Amelanchier ovalis*, *Ligustrum vulgare*, *Prunus spinosa*, *Cornus sanguinea* and others. Beneath the canopy, junipers are common (*Juniperus hemisphaerica*, *Juniperus phoenicea* and *Juniperus oxycedrus* subsp. *oxycedrus*), and in the more continental zones, especially in the parameras, we find the groves of sabina albar (*Juniperus thurifera*). The presence in the understorey of these species is a consequence of the behaviour of frugivore birds that find shelter at night and excrete the juniper seeds at the foot of the quejigo trunks. Sciophytes are also frequent such as *Bupleurum rigidum*, *Saponaria ocymoides*, *Cephalanthera longifolia*, *Cephalanthera rubra*, *Epipactis helleborine* and other orchids. In the Serranía de Cuenca and the Cantabrian Castilian sector, the presence of dogwoods (*Buxus sempervirens*) is frequent, while bearberries (*Arctosthaphylos uva-ursi*) are very common on slopes. *Buxus* thickets, or "bojedas" (Berberido-Buxetum sempervirentis), occupy large expanses of the Serranía de Cuenca, as the consequence of degradation of the quejigares of the Cephalanthero-Quercetum fagineae and Corylo avellanae-Quercetum fagineae. The bojedas show a preference for drier sites. In the more humid zones appear communities dominated by *Ligustrum vulgare* and *Berberis hispanica* subsp. *seroi* (Ligustro vulgaris-Berberidetum seroi).

The forest mantle comprises brambles belonging to the Rosetum micranto-agrestis. The main replacement stages are scrubs of the associations Lino differentis-Salvietum lavandulifoliae, Genisto scorpii-Cistetum laurifolii (on lixiviated luvisols) and, on lithosols exposed to the wind, we find the Lino appressi-Genistetum rigidissimae. When these scrubs are subject to intense grazing, woodier species vanish giving way to grasslands of the Festuco hystricis-Poetalia ligulatae.

Quejigares of the Corylo avellanae-Quercetum fagineae were described from the massifs of Gúdar and Javalambre, where summer rains are more intense due to their closeness to the cyclogenic area of the northeast Mediterranean. In the eastern mountains of the South Oroiberian sector, hazels (*Corylus avellana*) are common in the quejigar, though they become less frequent as we move westwards such that in the Serranía de Cuenca they take refuge as hydrophiles in the subhydrophytic association Astrantio-Coryletum avellanae, locally known as "avellanares" from the hazel tree, or "avellano". Quejigares of Corylo avellanae-Quercetum fagineae differ from the neighbouring Cephalanthero-Quercetum fagineae in the trees that accompany the quejigo, the composition of the canopy and in the cortege of nemoral grasses. The avellanares and quejigares are populated by a heterogenous group of trees and nemoral or border shrubs including *Acer monspessulanum, Betula fontqueri* subsp. *fontqueri, Corylus avellana, Fraxinus excelsior, Ilex aquifolium, Populus tremula, Prunus mahaleb, Pinus iberica, Pinus latisquama, Rhamnus alpina, Rhamnus cathartica, Ribes alpinum, Sorbus aria, Sorbus latifolia, Sorbus torminalis, Taxus baccata, Tilia platyphyllos, Ulmus glabra* and *Viburnum opulus*.

Interspersed with these nemoral species, some plants find refuge in the shady understorey of these closed woods (see Peinado et al. 2008: 509–510). In the easternmost zone of the South Oroiberian sector (Morella and Peñagolosa comarque) along boundaries with the Valencia biogeographical subprovince, quejigares show the floristic influence of the coastal mountains of the Valencia-Provençe and Balearic province and have been included in the association Sileno melliferae-Quercetum fagineae. These quejigares always appear dotted depending on the microclimate at shady sites and never occupy large expanses of the landscape dominated by the forests of the Festuco gautieri-Pinetum salzmannii (see section Pine forests for the ecological relationships of the black pines –*Pinus salzmannii* and *Pinus latisquama*– and quejigos). A complete description of this association (as Violo willkommii-Quercetumn fagineae) and of its dynamic relationship with black pinewoods may be found in Roselló (1994: 431–436).

In the Cantabrian Castilian, Rioja and Estella and North Oroiberian Sierran sectors, *Quercus faginea* forests of the association Spiraeo obovatae-Quercetum fagineae grow on soils overlying limestones and marls of greater water storage capacity (Fig. 11.19). These quejigares are the dominant forests of the Castilian-Cantabrian sector, though they are today practically absent from their potential area due to agricultural and sometimes livestock land uses. They share numerous species with other quejigares of the Iberian System yet the presence of northerly distributed species such as *Erica vagans, Brachypodium pinnatum* subsp. *rupestre, Spiraea hypericifolia* subsp. *obovata* or *Genista occidentalis* serves to differentiate them from these. Such differences between one quejigar and another also apply to

Fig. 11.19 The Spiraeo obovatae-Quercetum fagineae with dogwoods (*Buxus sempervirens*) in Iregüa canyons (La Rioja)

communities associated with these northern quejigares. The forest mantle belongs to the association *Lonicero etruscae-Rosetum agrestis*, and as serial stages, we often find bearberry communities of the Arctostaphylo crassifoliae-Genistetum occidentalis.

Deciduous Forests The most widespread deciduous forests of the Iberian System are beech forests or "hayedos", from the word for beech "haya" (*Fagus sylvatica*). In the northern hemisphere, there are ten species of beech, of which –*F. sylvatica*– is endemic to the European continent, mainly its western half, being replaced eastwards by *F. orientalis*, another European species. Beech forests of *F. sylvatica* are among the most widespread and characteristic temperate broad-leaf forests of Europe, which span from the Carpathians and Balkans in the east to the southern English and southern Scandinavian beech forests. The fossil record for only 9000 years ago reveals a total lack of beech forests across Central Europe. However, in the Carpathians there are records of beech pollen. The westward expansion of the beech is observed as recently as four thousand years ago, passing through Central Europe to reach the Atlantic front displacing, as it advanced, other deciduous forests, mainly of oaks, but also evergreen or mixed forests with pines and firs. In Spain, the fossil record dates the expansion of beech at some three thousand years ago meaning that the beech forests, or hayedos, we find in the

Iberian System represent the last expansions of European beech forests that took place towards the west and south.

The Iberian System's N-S trending ranges played an important role in the migration of beech from Eurosiberian Spain to the Ayllón massif in Guadalajara, where we can observe the southernmost Spanish beech forests. Hence, in times that were favourable for deciduous species, the beech was able to migrate from the north of Spain passing through the sierras of Cameros, Piqueras, Demanda and Moncayo –where we can still find good beech forests– to the Central System, where they became established in Ayllón. According to the monograph by Hernández Bermejo and Sáinz Ollero (1984), Spain's hayedos have been totally cleared on several occasions such that current populations are young and coeval. These populations have been grown from stump shoots, though in some rugged, steep escarpments escaping the axe, ancient beeches coexist with yews or "tejos" (*Taxus baccata*). Nothwithstanding, given that they reach reproductive maturity between the ages of 50 and 120 years, beeches are rarely over 300 years of age.

The summer aridity that characterizes the Mediterranean climate is the southern limiting factor for beech and other deciduous species that are not adapted to the long Mediterranean dry summer period. This explains why their ancient distribution ranges practically completely disappeared and they persisted only in a few micro-climatic enclaves which, owing to their particular topographic position on shady, cool slopes, enjoy more rainfall, condensation of summer rains and diminished evapotranspiration. Thus, while beech forests are common in hyperhumid and humid areas, they may also appear in zones of subhumid ombroclimate, in places where the potential vegetation comprises forests of *Quercus pyrenaica*, but there, they occupy sheltered positions. The hayedos of the Iberian System inhabit extremely high altitudes in the Sierra de la Demanda, reaching 1850 m on the Pico de San Lorenzo (García-Baquero 2005), an altitude only surpassed in Spain by the beech forests of the Parque Natural de Tejera Negra, Guadalajara, lying at 1950 m (Peinado et al. 2008).

In some zones of the Sierra de la Demanda and Sierra Cebollera, along with Ayllón in the Central System, we may find mixed formations of beech and *Quercus pyrenaica*. Under these conditions, beech trees end up displacing these melojos because beech plantlets are good at withstanding the sciophilous conditions of the understorey and will eventually replace the oaks (Aguirre et al. 1991a, b, 1994). The Iberian System is home to three types of beech forests: Galio rotundifolii-Fagetum sylvaticae, on acid or decarbonated soils (Fig. 11.20); Epipactido helleborines-Fagetum sylvaticae, growing on carbonate soils, and Carici sylvaticae-Fagetum sylvaticae, in shady places with some degree of soil moisture.

The silicicolous and oligotrophic, supratemperate, humid-hyperhumid association Galio rotundifolii-Fagetum sylvaticae, corresponds to the beech forests of the North Oroiberian Sierran sector, flourishing on the northern and western slopes of the sierra de la Demanda, Moncayo, Cebollera and neighbouring areas. Its differential species is the nemoral *Galium rotundifolium*. The first replacement stage comprises small woodlands of trees or shrubs such as *Ilex aquifolium, Sorbus aria, Sorbus aucuparia, Corylus avellana, Acer monspessulanum* and *Acer campestre*,

Fig. 11.20 Beech forest (Galio rotundifolii-Fagetum sylvaticae) near Enciso, La Rioja

which make use of the lack of beech trees to develop (Fig. 11.21). Other replacement shrubs are the broom formations, or "piornales" (Cytiso scoparii-Genistetum polygaliphyllae) and heath formations, or "brezales" (Calluno vulgaris-Genistetum occidentalis and Cytiso scoparii-Ericetum aragonensis). Some associated grasslands are pastures formed by intensive grazing (Festuco amplae-Cynosuretum cristati) and swards of the Luzulo carpetanae-Nardetum strictae. The association Epipactido helleborines-Fagetum sylvaticae has its ecological optimum in the Eurosiberian region. In the Iberian System, however, it flourishes as submesophytic on calcareous deep soils of the Cantabrian Castilian sector (Sierra de Tesla and Montes Obarenes) and in the limestone outcrops of the North Oroiberian Sierran sector. In the herbaceous stratum appear species such as *Primula columnae, Melica uniflora* or even *Lilium martagon*. The forest mantle is composed of diverse species of *Rosa* in addition to *Viburnum lantana, Crataegus monogyna, Lonicera etrusca* and *Ligustrum vulgare*. Associated with the potential area of these beech forests is the dwarf cushion scrub of the Arctostaphylo crassifoliae-Genistetum occidentalis.

In humid microclimates at some limestone outcrops of the North Oroiberian Sierran sector (sierras of Camero Nuevo, Canales de la Sierra, Montenegro de Cameros and the Mesozoic outcrops of La Demanda) and other Cantabrian Castilian enclaves, we find beech forests of the Carici sylvaticae-Fagetum syvaticae, also with an Eurosiberian optimum. Apart from the lack of Mediterranean species, which are common in the former association, differential species of these hayedos are *Ajuga reptans, Allium ursinum, Carex sylvatica, Daphne laureola, Euphorbia amygdaloides, Galium odoratum, Helleborus occidentalis, Galeobdolon luteum, Lilium martagon, Mercurialis perennis, Polystichum setiferum, Potentilla sterilis,*

Fig. 11.21 Mixed populations of *Ilex aquifolium, Crataegus monogyna, Sorbus aria, S. aucuparia, Acer campestre* and *A. monspesulanum* constitute the first seral stage to the Galio rotundifolii-Fagetum sylvaticae in northern areas of the Iberian System

Primula acaulis, Sanicula europaea, Scilla lilio-hyacinthus, Scrophularia alpestris, Veronica montana, etc. In forest clearings, patches of *Prunus avium* or even *Fraxinus excelsior* arise. Along woodland edges grow the subnitrophilous herbaceous communities of the Geranietum robertiano-lucidi. The most widespread grasslands are those of the Potentillo montanae-Brachypodion rupestris; as a result of intense grazing and disturbance communities of thistles appear (Carduo nutantis-Cirsietum richterani). Transiently waterlogged sites are inhabited by communities of the Junco inflexi-Menthetum longifoliae and Glycerio declinatae-Catabrosetum aquaticae.

Beech forests are relatively well preserved since they occupy steep, rocky zones and thus their agricultural or farming use has been less intense than that of other forest types. Even so, hayedos are exploited for firewood and logging and are also used as a refuge for intensive farming, especially cattle breeding. Some areas have been cleared, sometimes to make charcoal, and these are currently used as pastures for intensive farming. From the 1950s to the 1970s, these zones were repopulated with *Pinus sylvestris*, and currently, they form large tree formations, some of mixed pines and beeches. A large proportion of the Iberian System's beech formations are protected by the Red Natura, as natural parks, LICS ("Lugares de Importancia Comunitaria") or ZEPAS ("Zonas de Especial Protección para las Aves).

In the Llanada Alavesa, i.e., in the plains and valley floors of the foothills of mountains bordering the Mediterranean and Cantabrian basins of the Rioja and Estella sector, forests dominated by pedunculate oaks (*Quercus robur*) occur. These forests, belonging to the association Crataego laevigatae-Quercetum roboris, grow in areas of high rainfall (humid ombroclimate) and moderate temperatures (upper mesotemperate thermoclimate), but what seems to determine their appearance is edaphic moisture as given by the topographical position and soil texture. Soils are gently sloped (planisols), of loamy texture and contain many fine elements and mineral salts. This makes them good agricultural soils and they have been exploited as such. In structural terms, they are forests whose tree cover is mainly composed of *Q. robur* which coexists with *Quercus x coutinhoi* (*Q. robur* x *Q. faginea*), especially towards the west. Other trees may appear such as *Fraxinus excelsior, Fagus sylvatica, Ilex aquifolium* or *Acer campestre*, but almost everywhere they look like large oakwoods. Because of the richness of the soil on which the Crataego-Quercetum roboris develops, dense shrub and herbaceous strata exist with a large variety of species. Clinging vines, mainly *Hedera helix, Tamus communis* and *Lonicera xylosteum* are important structural elements of these forests. The shrub and herbaceous layers of the forests are well-developed and comprise, among many other species, *Cornus sanguinea, Corylus avellana, Crataegus laevigata, Ligustrum vulgare, Prunus spinosa, Pteridium aquilinum, Rosa arvensis, Viburnum lantana* and, sometimes, *Viburnum opulus*. Among the herbaceous plants, the most common are *Agrostis capillaris, Potentilla sterilis, Primula veris, Pulmonaria longifolia, Ranunculus nemorosus, Stellaria holostea* and *Viola reichenbachiana*. Variants growing on slightly hydromorphic soils usually feature plants such as *Solanum dulcamara, Lycopus europaeus* or *Scutellaria minor*, indicators of greater soil moisture, while those of sloping soils that make contact with quejigares of the Spiraeo obovatae-Quercetum fagineae often incorporate plants of these climatophilous forests.

In the gorges of the river Tajo, between the confluences of the rivers Cabrillas and Hoz Seca (districts of Beteta, Tragacete and Checa), there are populations of mountain ash (*Fraxinus excelsior*), which is a common species of Central Europe that flourishes in wetter Spain from Galicia to the Pyrenees. Its southern Iberian limit occurs at these gorges of the South Oroiberian sector, and its microclimatic position reveals the relict nature of these ashwoods described in detail by Herranz (1995). The mountain ash appears in two habitats: in edapho-hydrophilic gallery forests, as occurs in the North Oroiberian Sierran sector, and also it may be a component of quejigares on shady slopes with basic deep soils. The floristic composition of quejigares with mountain ash of the South Oroiberian sector, which contain a good part of the trees, shrubs and the herbaceous cortege of *Quercus faginea* forests (Herranz l.c: Table 2), reveals that mountain ashwoods, or "fresnedas", are a variant of the Corylo avellanae-Quercetum fagineae.

Nutrient-rich, sheltered, moist and shady slopes and gully floors lying between 850 and 1200 m in the supratemperate submediterranean belt of the northern slopes of La Demanda (Alto Oja and Cárdenas valley in Rioja, Alto Tirón valley in Burgos and northern slope of Moncayo), have as the potential vegetation a type of mixed forest or ashwood (Aceri campestris-Fraxinetum excelsioris) composed of mountain ash (*Fraxinus excelsior*), field maple (*Acer campestre*), wild cherry (*Prunus avium*) and, sometimes, large-leaved lindens (*Tilia platyphyllos*). In the shrub layer exist *Corylus avellana, Euonymus europaeus, Rosa arvensis, Crataegus monogyna* and *Prunus spinosa*; in the climbing layer *Hedera helix, Tamus communis* and *Lonicera periclymenum*, while in the herbaceous layer species characteristic of the order Fagetalia prevail (see García-Baquero 2005: Table 56). Currently there are very few well-preserved woods left and their habitats are normally used for intensive farming and agriculture. In addition, these woods have been the areas traditionally taken for the development of many towns and villages. The first succession stage corresponds to communities of hazels, or “avellanos” (Laserpitio eliasii-Coryletum avellanae) and the forest mantle to brambles (Rubo ulmifolii-Tametum communis). The main type of herbaceous community is that of pastures of the Malvo moschatae-Arrhenatheretum bulbosi. In these valleys, rushes of the Junco inflexi-Menthetum longifoliae, thickets dominated by *Sambucus ebulus* (Urtico dioicae-Sambucetum ebuli) and apple trees (*Malus domestica*) are common. From the northern slope of Moncayo, in similar ecological positions and of similar floristic composition, other ashwoods assigned to the Pyrenean association Brachypodio sylvatici-Fraxinetum excelsioris have been described (see Navarro 1989: 25).

In the same sierras and in the close-by Cebollera sierra, in the scree of large siliceous block slopes forest communities of *Quercus huguetiana* developed accompanied by *Sorbus aucuparia, Teucrium scorodonia, Betula celtiberica, Physospermum cornubiense* and *Veronica officinalis* (see García-Baquero 2005: 184). From the Monte de La Mata of Moncayo, edaphoxerophilous oak forest dominated by *Quercus petraea* subsp. *petraea* and the hybrid *Quercus x trabutii* (*Q. petraea* x *Q. pyrenaica*) have been described. In these oak forests (Galio odorati-Quercetum petraeae) common species are *Ilex aquifolium, Sorbus torminalis, Fagus sylvatica* and *Acer monspesulanum*, alongside an array of herbaceous nemorals that also appear in the undergrowth of the Pulmonario longifoliae-Quercetum pyrenaicae (see Navarro 1989: Table 7).

Coniferous Forests, Woodlands and Scrublands Along with oak forests, “sabinares” (juniper woodlands and scrublands) and “pinares” (pine forests and woodlands) are the most representative formations of Mediterranean Spain. However, while encinares usually occupy our mountainsides, both sabinares and pinares are generally orophilous plant formations that take up higher positions forming the timberline (oromediterranean and supramediterranean belts) or adopt edaphoxerophilous positions at lower altitudes. These forests or woodlands are structured as a tree canopy of gymnosperms (*Juniperus thurifera, J. phoenicea, Pinus sylvestris, P. nigra, P. halepensis*) to which is subordinated a shrubby stratum of junipers, creeping junipers (*Juniperus alpina* subsp. *alpina, Juniperus sabina, Juniperus hemisphaerica*) or “piornos” (*Fabaceae* brooms of the genera *Cytisus, Genista* or *Echinospartum*) which, together with diverse types of pasture, especially those that thrive in high-mountain frost-churned soils, make up an open formation in which tree canopies rarely touch. This peculiar arrangement of climactic or edaphoxerophilous woods of Iberian conifers is determined by two essential climate factors. One of these is a prolonged period of severe cold as well as persisting low temperatures and snow. These conditions are obviously met at other latitudes and because of this we find conifer woods in the subalpine and high mountain Eurosiberian belts. Notwithstanding the decisive factor that confers our sabinares and pinares, their characteristic feature is summer drought. Thus, it is only in the Iberian Mediterranean highlands that winter cold and summer aridity combine to give rise to the harsh environment that has evidently conditioned and still conditions the particular physiognomy of these juniper groves and pinewoods.

It is said of these forests, or more appropriately creeping shrublands splashed with trees, that they are clearly residual or relict. These are forests linked to the glaciations of the Quaternary, when they likely appeared close to zones of permanent ice cover, in areas experiencing a dry period. With the retreat of the ice, other plant formations had the opportunity to colonize lands that were once frozen. Thus, alternating periods of progressive and retrogressive vegetation development resulted in the pattern observed today. In the periods of expanding ice cover, periglacial conifer forests only found shelter on mountain massifs, far from present-day zones of high constant summer rainfall as occurs in the high Mediterranean mountains, where they persist but with a marked residual character.

Despite these adaptations, one has only to visit a juniper forest of the páramos of Guadalajara or Teruel, or a high Mediterranean mountain pine forest, to realise the precarious conditions they have to withstand (Fig. 11.22). Wind punishes the trees making them take on fastidious forms: bushes stick to the ground seeking soil warmth; weighed down by snow, many plants produce creeping forms; dominant trees form pyramids to easily keep off snow, etc. In summer, the drying effect of the

Fig. 11.22 Juniper woodland (Juniperetum hemisphaericae-thuriferae) on the Sierra de Javalambre, Teruel

wind is intensified by the lack of rain. Excessive grazing and fire, soil degradation and the slow growth of savin junipers suffering acute water deficit are the factors that have led to the present precarious nature of these so typically Mediterranean ecosystems. To the desolate appearance of the landscape contributes the lack of good pasture lands. This can be attributed to the leaf litter produced by junipers, savines and pines which poorly moistens and acidifies soils preventing the germination of seeds of many species.

Pine Forests One of the most famous botanists to cross Spain in the nineteenth century, the German Mauritius Willkomm, wrote: «In the eastern zone [of the Iberian Peninsula], conifers dominate while in the western zone, fagaceae do so. Conifer forests, and more specifically, pinewoods partially cover the mountains of Alto Aragón, Serranía de Cuenca, Segura sierra and limiting mountains [...]» (Willkomm 1852). The Iberian System is no exception in the context of the Mediterranean basin, where pinewoods are one of the most characteristic and diversified plant formations. Along with *Quercus*, the genus *Pinus* features the greatest number of tree species, subspecies and varieties of not only all circum-Mediterranean zones but throughout the entire Holarctic realm. Among the species of Mediterranean pine, four —*Pinus nigra, Pinus sylvestris, Pinus uncinata* and *Pinus halepensis*— appear naturally in the Iberian System as dominants in pure woods ("pinares") or accompanied by other trees, especially of the genus *Quercus*.

Generally speaking, in the Iberian System climatophilous pinewoods are those dominated by *Pinus sylvestris* var. *iberica* (*P. iberica* hereafter), which occupy the supramediterranean (upper horizon) and oromediterranean belts in all the System's sierras. Accompanying this pine, relict enclaves very occasionally occur in which *P. uncinata* may be dominant or codominant. The peculiar feature of the Iberian System is that in its southern region—the sectors Celtiberia and Alcarria and South Oroiberian—extensive *P. nigra* forests form on dolomite substrates in the supramediterranean belt and this circumstance is not repeated in northern territories (Cantabrian Castilian and North Oroiberian Sierran sectors). This is most likely because these rocks are scarce there or because the more abundant rainfall favours forests of *Quercus* at the expense of pines. Lastly, in the Iberian System's mesomediterranean belt, *P. halepensis* replaces *P. nigra* on similar dolomite substrates, but also on limestones, marls and even gypsum sediments. Such Aleppo pines also exist in other areas of the Iberian Peninsula such as those described in Chap. 14 and Chap. 3 in volume 2 of this book.

The pine forests of the Iberian System examined in this chapter may be differentiated using the following key (Table 11.5).

Pinus mugo, or dwarf mountain pine, is a conifer species native to high altitude habitats from southwest to central Europe. In Spain, *P. mugo* is represented by subsp. *uncinata* (*P. uncinata* hereafter), a larger, usually single-stemmed tree 20 m tall that bears asymmetrical cones. Across the Iberian system, *P. uncinata* never forms large forests; it rather appears as isolated groves in the middle of the extensive Oromediterranean forests dominated by *P. iberica*. These groves belong

Table 11.5 Key to pine forests of the Iberian System

1a	Black pine (*Pinus uncinata*) dominant or co-dominant	**2**
1b	Other pine forests	**3**
2a	Calcicolous relict forests thriving on the mountain summits of the South Oroiberian sector. With *J. sabina, Berberis seroi* and *Veronica javalambrensis*: ***Junipero sabinae-Pinetum uncinatae.***	
2b	Orotemperate relict forests thriving on acidic soils (Picos de Urbion): ***Avenello ibericae-Pinetum uncinatae.***	
3a	*Pinus nigra* as dominant or co-dominant in the forest canopy. Edaphoxerophilous forest thriving on limestones and dolomites.	**4**
3b	Conditions not met. *P. iberica* is usually the dominant tree	**5**
4a	*P. salzmannii* is the dominant tree. South Oroiberian sector (Javalambre Sierran District): ***Festuco gautieri-Pinetum salzmannii.***	
4b	*P. latisquama* dominant. Oreoconquense subsector (Alto Tajo, Albarracín and Serranía de Cuenca): ***Junipero thuriferae-Pinetum latisquamae***.	
5a	*P. iberica* forests on acidic soils	**6**
5b	*P. iberica* forests on calcareous soils	**7**
6a	Pine forests of the North Oroiberian Sierran sector: ***Vaccinio myrtilli-Pinetum ibericae***.	
6b	Pine forests of the South Oroiberian sector: ***Calluno vulgaris-Pinetum ibericae***.	
7a	Orotemperate *P. iberica* forests of the South Oroiberian sector; *Juiperus sabina* is the dominant shrub in the understorey: ***Junipero sabinae-Pinetum ibericae.***	
7b	Supratemperate *P. iberica* forests of the South Oroiberian sector; with *Ononis aragonensis; J. sabina* usually lacks: ***Ononido aragonensis-Pinetum ibericae.***	

to either of two associations. The Avenello ibericae-Pinetum uncinatae includes orotemperate submediterranean hyperhumid micro-mesoforests, growing on strongly acid tangel rankers developed on siliceous rocks on the summits (1850–2050 m) of the Reserva Nacional de Urbión, around the Castillo de Vinuesa peak (North Oroiberian Sierran sector, Urbión Sierran district). Besides the dominants *P. uncinata* and *P. uncinata* x *P. iberica*, common species in the underwood are *Juniperus alpina*, *Vaccinium myrtillus*, *Calluna vulgaris, Nardus stricta, Avenella iberica* and *Erica arborea*. In the high peaks of the sierras Gúdar and Javalambre (South Oroiberian sector), on calcareous substrates, these relict groves have been asigned to the endemic association Junipero sabinae-Pinetum uncinatae, whose differential species are *Prunus prostrata, Berberis seroi, Galium idubadae* and *Veronica javalambrensis*.

P. sylvestris (Scots pine or "pino albar" in Spanish), a species of pine readily identified by its fairly small, blue-green needles and orange-red bark, is native to Europe and Asia, ranging from western Europe to eastern Siberia, south to the Caucasus Mountains and Anatolia and north to well inside the Arctic Circle in Scandinavia. In the south of its range, it is a high altitude mountain tree growing at 1200–2600 m. Over one hundred *P. sylvestris* varieties have been described in the literature, differing only minimally in morphology, but with more pronounced differences in genetic factors and resin composition (Farjon 2005). Except for

Fig. 11.23 Magnificent specimen of Iberian Scots pine (*Pinus sylvestris* var. *iberica*) in the Maestrazgo region, Teruel. Shrubs are boxwoods (*Buxus sempervirens*)

populations inhabiting the Baetic mountains, in Spain they have been designated as the variety *iberica* (Fig. 11.23), which in the Iberian System is the dominant tree in four associations. The calcicolous association Junipero sabinae-Pinetum ibericae in the east occupies the high sierras of Maestrazgo (Gúdar, Javalambre), in the west it reaches the Alto Tajo de Guadalajara, while southwards it spreads to the Serranía de Cuenca (Sierra de San Felipe and Montes Universales) at altitudes from 1500 to 1800 m (Fig. 11.24). Its mature stage corresponds to a creeping carpeted scrubland almost exclusively made up of *Juniperus sabina*, over which emerge pines forming a tree layer of scarce cover. *J. sabina* plays a pioneering or colonizing role on leptosols; the invasion of new substrates by savines gives rise to regosols on which the mature stage may become established, prospering both on rendzinas and calcareous kastanozems. At sites sheltered from the wind, a closed forest may form, though generally, the association shows a woodland structure in whose clearings it is common to find *Rosa sicula, Berberis seroi*, *Ribes uva-crispa, Juniperus hemisphaerica*, *Satureja intrincata* subsp. *gracilis, Erinacea anthyllis* and perennial grasslands (Cirsio microcephalae-Onobrychidetum hispanicae, Festucetum hystricis and Paronychio capitatae-Artemisietum lanatae).

In the foothills of the Sierra de Pela, mainly between Somolinos and Valvenedizo, in the Muela de Somolinos, in the parameras of Galve de Sorbe and in the small valleys, or "navas", of Cantalojas, patches of Scots pine appear whose

Fig. 11.24 Junipero sabinae-Pinetum ibericae, Valdelinares, Sierra de Gúdar, Teruel

physiognomy and floristic composition distance them from the previous association. They have been included in the association Ononido aragonensis-Pinetum ibericae. Apart from the absence of *Juniperus sabina*, the presence of *P. iberica* and *J. hemisphaerica* highlight the floristic composition of these pinares. To distinguish them from the abundant repopulated pinewoods found in adjacent areas whose potential vegetation are quejigares (Cephalanthero-Quercetum fagineae) or encinares (Junipero thuriferae-Quercetum rotundifoliae), *Pulsatilla rubra, Vicia pyrenaica, Galium idubedae, Carex humilis, Rosa pimpinellifolia* and *Ononis aragonensis* serve as differentials. In zones where they coexist with encinares and quejigares, these pinares occupy biotopes in which hardwood species are at a competitive disadvantage owing to the substrate (lithosols on limestones, dolomites and occasionally Cretaceous sandstones) or to the topography (steep slopes or "navas" showing temperature inversion).

Throughout the potential zone of the pinares of the Ononido aragonesis-Pinetum ibericae there is evidence of strong human influence and great pressure from sheep and goats. This determines that the original structure of the pinewood cannot be recreated, although based on its similarity with that of the Junipero sabinae-Pinetum sylvestris, we can imagine an open population of pines above a lower canopy of *Juniperus hemisphaerica*, superpositioning thickets and thyme pastures harbouring *Ononis aragonensis* and *Festuca hystrix*. Currently, as seral stages, one finds communities of *J. hemisphaerica* forming a mosaic with shrublands of the

Saturejo-Erinaceetum anthyllidis, fragments of rose gardens including *Rosa pimpinellifolia, Rhamnus alpina* and *Ononis aragonensis* (Ononido aragonensis-Berberidetum seroi), and grasslands of the Festucetum hystricis.

The *Pinus iberica* pine forests of the Iberian System growing on acidic soils have been assigned to two associations, namely the Vaccinio myrtilli-Pinetum ibericae (North Oroiberian Sierran sector) and the Calluno vulgaris-Pinetum ibericae (South Oroiberian sector). The first association corresponds to orotemperate humid-hyperhumid micro-mesoforests, growing on shallow acid tangel cambisols on acidic bedrocks with a deep winter snow cover (1700–2100 m) in the Urbión Sierran district (Fig. 11.25). Common species of the association's underwood are *Avenella iberica, Calluna vulgaris, Juniperus alpina* and *Vaccinium myrtillus.* In the South Oroiberian sector, it is replaced by its counterpart the Calluno vulgaris-Pinetum ibericae, which comprises the forests of *P. iberica* in the upper-supratemperate and lower-orotemperate submediterranean Maestrazgo and Montes Universales. These grow on siliceous rocks on strongly acid rankers, cambisols or planosols, often with local and limited stagnant properties. Characteristic species (territorials) are *Avenella iberica, Calluna vulgaris, Nardus stricta, Juniperus communis* var. *intermedia* and *Vaccinium myrtillus.*

Pinus nigra (black pine) is a large coniferous evergreen tree occurring across central-southern Europe from Spain to the eastern Mediterranean, including the

Fig. 11.25 Open climatophilous pine forests (Vaccinio myrtylli-Pinetum ibericae) with *Calluna vulgaris*. Puerto de Piqueras, Soria

Anatolian peninsula, Corsica/Cyprus, the Crimea and the high mountains of the Maghreb in North Africa (Farjon 2005). Mature *P. nigra* forests can develop a high and closed tree canopy and have several age classes and trees well over 30 m tall, as well as a shaded understorey. These forests often occupy substrates rich in magnesium (dolomites) and even though some populations may thrive on deep soils, in most cases the soil is superficial and not mature.

P. nigra is a collective species that is usually divided into the two subspecies: *Pinus nigra* subsp. *nigra* in the east of its distribution range and *Pinus nigra* subsp. *salzmannii* in the west of its range, from southern Italy to southern France, Spain and North Africa (Christensen 1993). In the mid-nineteenth century, Willkomm (*l. c.*) described the pinares of Cuenca, in those times practically virgin, and within them identified pine specimens older than 1000 years (Willkomm 1844, 1896). Willkomm & Lange (1870: 18) described the pines of the Serranía de Cuenca as an endemic variety —var. *latisquama*— which Rivas-Martínez (2011b: 487) combined as the subspecies *Pinus nigra* subsp. *latisquama* (*P. latisquama* hereafter).

P. salzmannii and *P. latisquama* make up about half a million hectares of pure stands in eastern Spain and southern France (Tíscar & Linares 2011). In Spain, where they are known as "pinos salgareños", they have their central distribution areas at heights in the range 1000–1600 m in the dolomite-limestone mountain ranges of the Iberian System, and account for two-thirds of the Iberian Peninsula's black pine formations (Regato-Pajares & Elena-Rosselló 1995). The Iberian System's pinewoods have suffered a long history of felling, fires and forest management, including massive timber production for the ship industry (Bauer 1991), though they preserve both their impressive appearance and complex ecological position. *P. salzmannii* and *P. latisquama* are well adapted to rocky substrates and have a main root that is much less developed than their secondary roots, which vigorously develop to become firmly fixed to the rock's fissures.

Hence, there we have an indication of the ecological behaviour of both pines: a better tolerance of rocky substrates and poorly-developed slope soils than their main competitor in their bioclimatic belt, which is none other than the quejigo *Quercus faginea*. There are conspicuous differences in trunk development between these species: while *Q. faginea* splits close to the ground to open out into a broad canopy, both *Pinus salzmannii* and *Pinus latisquama* produce a strong, straight trunk that reaches heights of up to 30 m, surpassing five times the height of the largest quejigos.

This erect stature is typical of heliophilous, fast-growing trees that successfully manage to emerge into the sun-lit canopy in competition with leafy trees. Perhaps these simple observations are key to the original ecological role played by quejigos and salgareño pines in the Iberian System. Thus it could be that salgareño pines were codominants in the climatophilous quejigar (Cephalanthero rubrae-Quercetum fagineae), which would have been structured as a forest with two canopies: an emerging canopy of salgareño pine and a subordinate canopy formed by shade trees (Fig. 11.26). The dominant in this latter canopy would be *Quercus faginea*, but it would also include some of the species mentioned when describing

Fig. 11.26 Pine forest (Festuco gautieri-Pinetum salzmannii with *Juniperus sabina*) and marcescent oak forest (Sileno melliferae-Quercetum fagineae), near Orihuela del Tremedal, Teruel

the *Q. faginea* forests: *Acer monspessulanum, Prunus mahaleb, Sorbus aria, S. latifolia, S. torminalis, Taxus baccata, Tilia platyphyllos* and *Ulmus glabra*.

Q. faginea is a tree of more mesophytic tendency; it requires deep soils and does not reject transiently waterlogged soils. Accordingly, the Iberian System's quejigares are considered climatophilous and temporarily hydrophilous; to emerge from seedlings into juveniles and mature trees they tolerate neither direct, continuous sunlight nor the lack of a moderately evolved soil. With such requirements, in the sheer, rugged sierras of the Iberian System with its vast rocky outcrops, steep ledges and stony slopes, many biotopes remain to be colonized by *P. salzmannii*. Thus, the pinewood dominated almost exclusively by *P. salzmannii* would act as edaphoxerophilous across all the highlands. In addition, unlike *Q. faginea*, the salgareño pine grows easily on dolomite rocks and soils generated from that. This will also prompt the establishment of that pinewood on deeper soils which, if overlying another bedrock, could have been occupied by the quejigar.

In the Iberian System's southern sierras, *P. salzmannii* and *P. latisquama* appear dispersed among the pinewoods of pino albar, or Scots pine (Junipero sabinae-Pinetum ibericae), and even manage to become dominants when substrates are massive dolomites, relinquishing their vocation as edaphoxerophilous dolomiticolous. Towards lower altitudes, the pinewoods of salgareño pines start to alternate with pinares of *P. iberica*. However, while these preferentially occupy shady sites, the forests of *P. salzmannii* establish at sunny places. With the gradual drop in altitude and coinciding with more rough and rockier zones, *P. salzmannii* and *P. latisquama* seem to better tolerate heat and drought than *P. iberica* and as a consequence they eventually occupy about half the mountain. In gorges and valley floors quejigos manage to abide but the foothills of the highlands are transformed into mixed pine-marcescent oak forests.

In the Iberian System, the forests of salgareño pines have been ascribed to two associations: the Festuco gautieri-Pinetum salzmannii (Fig. 11.27), and the Junipero thuriferae-Pinetum latisquamae (this one includes the association Berberido seroi-Pinetum latisquamae described by Peinado et al. 2008). In their natural state, the pinewoods of both associations show a relatively simple structure, with the pine forming a monostratum and very open canopy at edaphoxerophilous positions and on massive dolomites. However, this canopy gradually closes to create a shady underwood where it acts as climatophilous. *P. iberica* is usually a constant companion in the tree stratum, while *Juniperus thurifera* normally appears as a companion in sunny positions in the case of the association Junipero thuriferae-Pinetum latisquamae. At more open places (Fig. 11.28), the second stratum is formed by *Juniperus phoenicea*, *J. hemisphaerica*, *Berberis seroi*, *Buxus sempervirens* and small roses especially *Rosa pimpinellifolia*, alongside

Fig. 11.27 Festuco gautieri-Pinetum salzmannii growing on dolomites. Sierra de Gúdar (Teruel)

Fig. 11.28 Junipero thuriferae-Pinetum latisquamae on the Serranía de Cuenca (Cañizares, Cuenca)

grasses and shrubs that are common in the replacement stages of the Siderito incanae-Salvion lavandulifoliae and Festucetea indigestae.

When the forest further closes, what normally occurs at mid mountain heights (1100–1500 msl) as the understorey becomes denser, *J. phoenicea* practically vanishes, giving way to *Prunus mahaleb*, *Acer monspessulanum, Ligustrum vulgare*, *Lonicera etrusca, Sorbus aria*, *S. domestica*, *S. torminalis*, *Cornus sanguinea*, *Rhamnus saxatilis* and *Q. faginea*. Under these conditions, the herbaceous layer is rich in nemoral species: *Brachypodium sylvaticum, Geum sylvaticum, Hedera helix, Hepatica nobilis, Primula veris* subsp. *columnae, Tanacetum corymbosum* and *Viola reichenbachiana.*

Due to their rapid growth and better use for timber, mixed woods of *Q. faginea* and salgareño pines have been managed to the benefit of the conifers, favoured by their massive rejuvenation. In consequence, today we find either extensive coeval homogenous pinewoods or relatively heterogeneous woods, yet lacking or with scarce trees and leafy saplings of the herbaceous cortege that are typical of the subordinate stratum when the wood shows a mature structure. Willkomm was perhaps the last botanist to describe these woods in their virgin state.

Juniper Formations Communities dominated by species of the genus *Juniperus* are highly characteristic of the Mediterranean basin's landscape. These dominant

Table 11.6 Key to juniper woodlands and shrublands of the Iberian System

1a	Open woodlands dominated by *J. thurifera* and *J. hemisphaerica*: ***Juniperetum hemisphaerico-thuriferae.***	
1b	Scrublands dominated by *Juniperus phoenicea*, *J. sabina* or *J. alpina* subsp. *alpina*	**2**
2a	*J. phoenicea* open rupestrian shrublands thriving on limestones and dolomites	**3**
2b	Other shrublands	**4**
3a	Usually with *Rhamnus lycioides* subsp. *lycioides* and *Pistacia terebinthus*. Celtiberia and Alcarria, and South Oroiberian sector: ***Rhamno lycioidis-Juniperetum phoeniceae***.	
3b	*Buxus sempervirens* codominant; *R. lycioides* subsp. *lycioides* and *Pistacia terebinthus* usually lacking. Cantabrian Castilian and Rioja and Estella sectors: ***Buxo sempervirentis-Juniperetum phoeniceae***.	
4a	*Juniperus sabina* creeping shrublands with *Ephedra nebrodensis*. (Moncayo summits): ***Ephedro nebrodensis-Juniperetum sabinae.***	
4b	*Juniperus alpina* creeping shrublands with *Vaccinium myrtillus* (acidic rocks of the North Iberian sector): ***Vaccinio myrtilli-Juniperetum nanae.***	

species in the Iberian System are known as "sabinas" when their leaves are scale-like, as in the case of "sabina albar" (*J. thurifera*), "sabina rastrera" (*J. sabina*) and "sabina negral" (*J. phoenicea*), or as "enebros", when their leaves are needle-like as in the case of *Juniperus communis*. The five juniper associations in the Iberian System can be distinguished according to this key (Table 11.6).

The association Juniperetum hemisphaerico-thuriferae corresponds to the typical woodlands of the parameras of Guadalajara, Cuenca and Teruel (sectors Celtiberian and Alcarria, and South Oroiberian). These woods always colonise flat high mesetas (1000–1400 m) and exposed slopes, where the climate conditions are harsher and they meet with less competition. The mature woodland stage, preferentially growing on rendzinas, is a sabinar whose structure is one of an open, non-dense, heliophilous wood comprised of two woody strata. In the upper tree layer, *J. thurifera* dominates in the form of isolated individuals that are usually not taller than 6–8 m (Fig. 11.29), while *J. hemisphaerica* dominates in the shrubby stratum. Another stratum of small creeping bushes and xerophytic grasses may develop, but these woodlands are generally very poor in species due to their ancient relict nature. Further, other plants find it difficult to grow in the more calcic humus (tangel) generated by the fallen leaves of enebros and sabinas (Fig. 11.30).

Owing to their altitude, these woodlands represent the transition between supramediterranean evergreen oak forests (Junipero thuriferae-Quercetum rotundifoliae) and submediterranean pine forests (Junipero sabinae-Pinetum ibericae, Ononido aragonensis-Pinetum ibericae), and alternate in shady zones with deep soils with marcescent oak forests (Cephalanthero-Quercetum fagineae). In the higher areas, there is a subassociation with creeping juniper (Juniperetum hemisphaerico-thuriferae juniperetosum sabinae) and that is the ecotone between typical *J. thurifera* woodlands and pine forests (Junipero sabinae-Pinetum ibericae). The ecotone with the quejigares is indicated by the occurrence of species typical of deep soils (*Brachypodium phoenicoides*, *Thalictrum tuberosum*,

Fig. 11.29 Giant *Juniperus thurifera* tree at Puebla de San Miguel Natural Park, Rincón de Ademuz, Valencia. PNV: Juniperetum hemisphaerico-thuriferae

Teucrium chamaedrys subsp. *pinnatifidum*), while at the lower limits, oaks gradually appear representing the transition towards the encinares of the Junipero thuriferae-Quercetum rotundifoliae. The juniper grove, or sabinar, may also develop on siliceous substrates, as for example, occurs in the sabinares of Tamajón (Guadalajara). These sabinares growing on slates and gneisses have been described as a variant with enebros, the Juniperetum hemisphaerico-thuriferae juniperetosum lagunae.

At the core of the high plateaus of the Celtiberia and Alcarria sector (Fig. 11.31), the climate is continental and summers are rain-free, and there, in a dry ombroclimate, develops the typical variant of the association. Further eastwards, both total rainfall (subhumid ombroclimate) and summer rainfall increase along with the submediterranean influence, such that the sabinar gradually becomes enriched with salgareño pines (*Pinus latisquama*). In these eastern parameras there appears a juniper-pine mosaic with pines generally occupying the more mesophytic positions. While pinewoods, or pinares, have been promoted for wood production, the farming vocation of the paramera has prompted the development of meadowed sabinares to the detriment of pines. In the past years, with the abandoning of villages and farming activities, a clear recovery of pinewoods is

Fig. 11.30 Juniper woodland (Juniperetum hemisphaerico-thuriferae) on the Sierra de Javalambre (Teruel)

Fig. 11.31 Typical structure of the Juniperetum hemisphaerico-thuriferae. Canyon of Anchuela del Pedregal (Guadalajara)

observed and it is ever more common to see what were probably the original structures: dispersed pyramidal junipers under an emerging canopy of salgareño pines.

Human degradation of sabinares of the páramos has led to the spread of a spiny, cushioned scrub dominated by *Genista pumila* subsp. *rigidissima* (Lino appressi-Genistetum rigidissimae), which also occurs characteristically on the wind-exposed crests. In the clearings of this community and in those of the sabinares the "tomillares" or fields of thyme occur, rich in gramineae such as *Festucetum hystricis*. When soils are thinner, both because of surface outcrops of rocks disaggregated by cryoturbation and the constant stomping of sheep, the association Paronychio capitatae-Artemisietum lanatae develops. In spring, therophyte communities (Bupleuro baldensis-Arenarietum ciliaris) arise on cryoturbated soils. Areas occupied by this series are farming lands and areas of dryland crops that are harvested late (by mid August) such that risks of drought are high. Salgareño pines are used for reforestation.

The "sabina negral", or "sabina mora" (*J. phoenicea*), is a true specialist in biotopes where rock makes up a large part of the landscape. It is a usual component of some scrublands that colonise small ledges, cracks and fissures of the limestone, and dolomite walls, cliffs and crestones all over Mediterranean Spain (Fig. 11.32). These microforests develop as permanent edaphoxerophilous communities on lithic

Fig. 11.32 Landscape with Rhamno lycioidis-Juniperetum phoeniceae. *J. phoenicea* and *Erinacea anthyllis* growing on limestones (Sierras de Préjano, La Rioja)

lithosols, alternating with encinares, quejigares, sabinares albares and pinares, which replace the sabinares negrales as soon as soils are more developed. Sabinares negrales or sabinares moros are of great palaeoecological and palaeobotanical value, since they are relicts of an ancient xerothermic vegetation that during periods of glaciations sought refuge in these warm dry environments.

The Rhamno lycioidis-Juniperetum phoeniceae shows a simple structure that is physiognomically dominated by the pyramidal silhouettes of *J. phoenicea*, accompanied by a heterogeneous display of shrub and herbaceous species which thrive in the clearings left by juniper (Fig. 11.33). Among the larger shrubs it is usual to find the thorny "espino negro" (*Rhamnus lycioides* subsp. *lycioides*), buckthorn or "aladierno" (*Rhamnus alaternus*), jazmin or "jazmín" (*Jasminun fruticans*), terebinth or "cornicabra" (*Pistacia terebinthus*), and on stony ground, the odd shadbush or "guillomo" (*Amelanchier rotundifolia*), all of which gain in abundance towards zones of accumulated fine materials where soils better develop. Here, it is also possible to find young specimens of *Quercus rotundifolia*, *Quercus faginea* or *Juniperus thurifera* and this indicates that the vegetation develops towards climatophilous stages. Accompanying the above-mentioned species are other shrubs and herbs such as *Linum differens*, *Salvia lavandulifolia*, *Lavandula latifolia*, *Erinacea anthyllis*, *Genista rigidissima*, *Satureja gracilis*, *Teucrium expansum*, *Sideritis incana* or *Scabiosa turolensis*. These are the habitual

Fig. 11.33 Rhamno lycioidis-Juniperetum phoeniceae on rocky outcrops near Préjano, La Rioja

components of the successional shrublands of the climatophilous woodlands that surround rocky outcrops.

In northern zones of the Iberian System, the "sabinar negral" corresponds to the association Buxo sempervirentis-Juniperetum phoeniceae, an especially dominant community in the gorges of the river Ebro, but that encroaches upon all bounding sierras. The great physiognomic difference with the Rhamno lycioidis-Juniperetum phoeniceae is the codominance of *Buxus sempervirens* and presence of rupicolous and subrupicolous chamaephytes and hemicryptophytes such as *Hormatophylla spinosa, H. lapeyrousiana, Iberis saxatilis* or *Aethionema saxatile*.

The association Vaccinio myrtilli-Juniperetum nanae corresponds to the silicicolous, orotemperate, submediterranean creeping "enebrales" that thrive in the North Oroiberian Sierran Sector. Besides *Juniperus alpina* subsp. *alpina*, we find *Calluna vulgaris*, *Vaccinium myrtillus* and *Cytisus oromediterraneus*, along with hemicryptophytes, low chamaephytes and grasses such as *Agrostis tenuis*, *Arrhenatherum elatius* subsp. *bulbosum*, *Avenella iberica, Avenula lodunensis, Koeleria crassipes, Nardus stricta* or *Festuca aragonensis*. These dwarf creeping shrubs grow on wind-exposed slopes, walls of glacial cirques, moraines and stony outcrops stabilized and fixed by the vegetation such that they never appear in snowbeds or temporarily swamped areas (Loidi et al. 1997: 211).

Sabinares of *J. sabina*, Ephedro nebrodensis-Juniperetum sabinae, are the climatophilous vegetation in topographically-exposed biotopes of the calcareous high mounds ("muelas") of eastern Moncayo, where they topographically alternate with the quejigares of the cooler and wetter valley floors and slopes with deeper soils. The mature sabinar is dominated by nanophanerophytes (*J. sabina, J. alpina, J. hemisphaerica, Ephedra nebrodensis, Rhamnus saxatilis* and *Erinacea anthyllis*) that leave gaps in which many therophytes, chamaephytes and hemicryptophytes belonging to the successional scrublands and grasslands prosper (Lonicero pyrenaicae-Rhamnetum alpinae, Iberidi ibericae-Erinaceetum anthyllidis, Androsaco villosae-Festucetum hystricis, Bupleuro baldensis-Arenarietum ciliaris). If the substrates are mobile limestone screes of steep slopes, pioneer communities of the association Linario badalii-Cochlearietum aragonensis develop. In the more or less vertical crags-cliffs-rocks occur three rupicolous communities: the Saxifragetum segurae-moncayensis, Chaenorrhino semiglabri-Asplenietum celtiberici and Globulario repentis-Saxifragetum longifoliae, rich in local and regional endemisms (see Navarro 1989).

Riparian Forests and Willow Thickets Riparian wet deciduous woodland and willow communities are included in the Salici purpureae-Populetea *nigrae* class. Character species are *Alnus glutinosa, Brachypodium sylvaticum, Elymus caninus, Equisetum hyemale, E. telmateia, Frangula alnus* subsp. *alnus, Humulus lupulus, Populus nigra, Saponaria officinalis, Solanum dulcamara* and *Vitis vinifera* subsp. *sylvestris*. Two orders are recognized in the class: Populetalia albae (riparian meso-macroforests growing on wet fluvisols with a high water-table) and Salicetalia purpureae (willow forests and shrubby pioneer thickets growing in low river and streambed zones that are frequently flooded). In the Iberian System, the first order is

Table 11.7 Key to riparian forests and willow thickets of the Iberian System

1a	*Populetalia albae* forests	**2**
1b	*Salicetalia purpureae* forests and thickets	**7**
2a	*Osmundo-Alnion* microforests dominated by *Salix atrocinerea:* ***Rubo corylifolii-Salicetum atrocinereae.***	
2b	Other *Populetalia albae* forests	**3**
3a	*Fraxinus angustifolia* and *Quercus pyrenaica* forests growing on valley bottoms: ***Fraxino angustifoliae-Quercetum pyrenaicae.***	
3b	Other *Populetalia albae* forests	**4**
4a	Alder forests from the Cantabrian Castilian sector: ***Humulo lupuli-Alnetum glutinosae.***	
4b	Other *Populetalia albae* forests	**5**
5a	Elm forests. *Ulmus minor* dominant	**6**
5b	Poplar forests dominated by *Populus alba* and/or *Populus nigra*, along with *Fraxinus angustifolia, Salix neotricha, Ulmus minor* and, sometimes on hard-water river beds, with *Tamarix gallica:* ***Rubio tinctorum-Populetum albae.***	**8**
6a	Cantabrian Castilian elm forests: ***Viburno lantanae-Ulmetum minoris.***	
6b	Celtiberian and Alcarria elm forests: ***Opopanaco chironii-Ulmetum minoris.***	**7**
7a	*Salicion albae* forests	**8**
7b	*Salicion salviifoliae* thickets	**9**
8a	*Salix alba* forests: ***Salicetum lambertiano-albae.***	
8b	*S. fragilis* subsp. *neotricha* dominates; with *Populus nigra, Salix triandra* subsp. *discolor* and *Tamarix gallica:* ***Salicetum neotrichae.***	
9a	*Salix* thickets in hard-water river areas. With *S. eleagnos* subsp. *angustifolia*, and usually with *S. triandra* subsp. *discolor*: ***Salicetum discoloro-angustifoliae.***	
9b	*Salix* thickets in soft-water river areas. With *S. salviifolia, S. fragilis* subsp. *fragilis, S. atrocinerea* and *S. fragilis* subsp. *neotricha*: ***Salicetum salviifoliae.***	

represented by the alliances Osmundo-Alnion (growing on poor sandy or siliceous rock soils in river beds with running oligotrophic water) and Populion albae, including two suballiances: Populenion albae (in low, sometimes flooded, river bed zones) and Fraxino angustifoliae-Ulmenion minoris (in high scarcely flooded river bottom zones). The order Salicetalia purpureae is represented by the alliances *Salicion albae* (meso-supratemperate wooded willow growing in flooded deep river beds) and Salicion salviifoliae (wooded willow growing in zones of sandy soils and oligotrophic water). Riparian associations flourishing in the Iberian System can be differentiated according to the key provided in Table 11.7.

Forest Mantles and Herbaceous Fringes The class Rhamno-Prunetea encompasses the locally called "espinales", "zarzales" or "setos", i.e., deciduous, shrubby, mesophytic, or xerophytic mantle and seral communities mostly of Querco-Fagetea and Salici-Populetea nigrae forest vegetation. These are dominated by microphanerophytes, mainly thorny, prickly scrambling shrubs developing on nutrient-rich humic soils at forest edges close to grasslands or river-banks. They also appear as permanent communities on shallow stony soils in cliffs, steep slopes or mountain gullies. In the study area and linked to the climactic territories of the

Aceri-Quercion fagineae and Junipero sabinae-Pinion ibericae, the Rhamno-Prunetea is represented by the meso-supratemperate and submediterranean order Prunetalia spinosae and its alliance Berberidion vulgaris. Common taxa are *Amelanchier ovalis, Berberis seroi, Clematis vitalba, Cornus sanguinea, Crataegus laevigata, C. monogyna, Ligustrum vulgare, Lonicera xylosteum, Prunus insititia, P. spinosa, Rosa agrestis, R. canina, R. corymbifera, R. elliptica, R. nitidula, R. pimpinellifolia, R. squarrosa, R. stylosa, R. villosa, Rubus bifrons, R. canescens, R. corylifolius, R. caesius, R. ulmifolius* and *Sambucus nigra*.

In the Cantabrian Castilian sector, the class is represented by two associations: *Amelanchiero ovalis-Spiraeetum obovatae*, including *Amelanchier ovalis* and *Spiraea obovata* growing on the edges of forests of Spiraeo obovatae-Quercetum rotundifoliae and Lonicero etruscae-Rosetum agrestis, brambles dominated by roses (*Rosa agrestis, Rosa micrantha*) and blackberries (*Rubus ulmifolius*) around forests of Spiraeo obovatae-Quercetum fagineae.

From the South Oroiberian sector, four associations have been described: *Berberido seroi-Juniperetum sabinae*, in which *Berberis seroi* emerges from carpets of *Juniperus sabina* in the vicinity of pine forests of Junipero sabinae-Pinetum ibericae; Ononido aragonensis-Berberidetum seroi, with *Ononis aragonensis* and *Berberis seroi* growing around *Pinus salzmannii* forests; xerophytic communities of the Berberido seroi-Buxetum sempervirentis, locally called "bujedas" after the boxwood *Buxus sempervirens*, flourishing on the eroded soils of steep slopes of the Serranía de Cuenca; and, finally, the mesophytic association Ligustro vulgaris-Berberidetum seroi, which is the deciduous shrubby mantle of Salici-Populetea nigrae forest vegetation inhabiting deep fluvisols on valley floors.

In the North Oroiberian Sierran sector, the most widespread associations are the silicicolous Rubo ulmifolii-Rosetum corymbiferae, appearing close to marcescent forests of the Pulmonario longifoliae-Quercetum pyrenaicae and the above mentioned Lonicero etruscae-Rosetum agrestis in calcareous soils and linked to marcescent forests of Spiraeo obovatae-Quercetum fagineae. Communities growing on calcareous soils of the Moncayo belong to Lonicero pyrenaicae-Rhamnetum alpinae, in rock crevices and screes, and Roso pendulinae-Loniceretum nigrae, including *Lonicera nigra, Ribes alpinum, R. petraeum* and *Rosa pendulina*, which thrive on the northwestern slope of the Moncayo summit. Finally, brambles of the Celtiberian and Alcarria sector are assigned to Rosetum micrantho-agrestis, in which it is common to find *Crataegus monogy*na, several species of blackberry (*Rubus*), *Prunus spinosa, Ligustrum vulgaris, Lonicera etrusca, L. xylosteum, Berberis hispanica*, along with diverse species of the genus *Rosa*, among which *R. agrestis* is the character species. These mantles encircle both Cephalanthero rubrae-Quercetum fagineae and Junipero thuriferae-Quercetum rotundifoliae forests.

The class Cytisetea scopario-striatae groups the locally called "retamares" or "piornales", i.e., forest mantles, seral or permanent edaphoxerophilous shrubby communities dominated by brooms (*Leguminosae* scrub, nano- and microphanerophytes with long flexible, slender green stems and branches, mainly of the genera *Genista* and *Cytisus*) flourishing on humic, non-eroded soils without gleyic or stagnant properties as climactic or mantle formations, and a secondary vegetation of sclerophyllous or deciduous climax forests. In the Iberian System, character species are *Adenocarpus complicatus* subsp. *complicatus*, *Cytisus scoparius* subsp. *scoparius*, *Avenella iberica*, *Erica arborea*, *Orobanche rapum-genistae* and *Pteridium aquilinum* var. *aquilinum*. The class is represented in the study area by the silicicolous order Cytisetalia scopario-striati and two of its alliances: Genistion floridae and Genistion polygaliphyllae. The supramediterranean Genistion floridae, characterized by *G. florida* subsp. *florida*, is the mantle of Luzulo forsteri-Quercetum pyrenaicae forests and mainly spreads across the Celtiberian and Alcarria sector, though it has also been mentioned for some zones of Moncayo (Navarro 1989). The widespread supratemperate Genistion polygaliphyllae, characterized by *G. florida* subsp. *polygaliphylla*, is represented by two associations: Cytiso scoparii-Genistetum polygaliphyllae, a dense brushland dominated by *C. scoparius* subsp. *scoparius*, *G. polygaliphylla*, *Erica arborea* and *Pteridium aquilinum* var. *aquilinum* (Fig. 11.34). In the North Oroiberian Sierran sector, it constitutes the mantle of beech and marcescent forests (Galio rotundifolii-Fagetum sylvaticae and Pulmonario longifoliae-Quercetum pyrenaicae); and Genistetum polygaliphyllo-obtusirameae, in which *Genista obtusiramea* is the

Fig. 11.34 Forest mantle (Cytiso scoparii-Genistetum polygaliphyllae) seral to Galio rotundifolii-Fagetum sylvaticae on the Sierra de la Demanda (La Rioja)

character species, described for the San Lorenzo mountain of the Sierra de la Demanda (Loidi et al. 1997: 315).

Semi-shaded perennial herb communities of scarce moisture requirements growing as external fringe woodlands (Trifolio-Geranietea, Origanetalia vulgaris) are represented in the Iberian System by the following alliances and associations: Trifolion medii (growing at the edges of mesophilous deciduous forests with the association Trifolio medii-Lithospermetum officinalis); Geranion sanguinei (growing at the edges of mesoxerophilous forests, mostly on calcareous soils, with the association Sileno nutantis-Geranietum sanguinei); and Linarion triornithophorae (growing on acidic soils with the association Trifolio medii-Lathyretum nigri). The Cardamino hirsutae-Geranietea purpurei and its only order Cardamino-Geranietalia purpurei encompasses annual, slightly-nitrified, semi-shade communities growing on organic nutrient-rich soils, which are represented in the Iberian System by the alliance Geranio-Anthriscion caucalidis and its association Anthrisco caucalidis-Geranietum lucidi.

11.3 Scrublands

One of the most characteristic features of the landscape of the study area are the Iberian System's rich, diversified scrublands, or "matorrales", dominated by nanophanerophytes, chamaephytes and other dwarf woody plants. These, often spiny and/or aromatic plants, with small leaves, hairs and glands, represent a large proportion of the indicator and endemic taxa of the Iberian System. They are usually secondary communities arising from the destruction of the potential natural vegetation of woodlands or forests due to fires, or grazing or agricultural uses and subsequent erosion of the upper soil layer. Alternatively, "matorral" communities may flourish on steep, dry, sunny slopes, usually facing south or west.

The floristic composition of these matorrales is firstly determined by the nature of the substrate: on substrates rich in bases, such as Mesozoic and Tertiary limestones and dolomites or gypsums of the Triassic or Miocene, appear communities of the class Rosmarinetea officinalis; while on the siliceous substrates of the Palaeozoic core, and depending on climate conditions, either Cisto-Lavanduletea thickets in zones of lower rainfall (dry and subhumid ombroclimates) or Calluno-Ulicetea heathers in rainier zones (subhumid to hyperhumid ombroclimates) occur.

Basophilous Scrublands Calcicolous, gypseous or dolomiticolous scrublands in thermo- to oromediterranean arid to subhumid areas of the western Mediterranean belong to Rosmarinetea officinalis. Two main types of basophilous scrublands are recognized: communities of Sideritido incanae-Salvion lavandulifoliae (Rosmarinetalia officinalis) flourish on rendzinas and regosols lacking gypsum, and Lepidion subulati (Gypsophiletalia) dominate on Triassic or Miocene gypsisols.

Communities of Sideritido incanae-Salvion lavandulifoliae are the most widespread in the Iberian System. In scrublands of the supramediterranean and lower-oromediterranean belts, thorny and cushion dwarf and spiny shrubs such as *Erinacea anthyllis* and/or *Genista pumila* subsp. *rigidissima* dominate along with various composites and labiates, many of which are local endemics. On wind-exposed lithosols of the Celtiberian plateaus, *G. pumila* subsp. *rigidissima* is the dominant cushion shrub, characterizing the association Lino appressi-Genistetum rigidissimae. In the North Oroiberian Sierran sector, *Iberis saxatilis, Serratula nudicaulis, Ononis striata, Carduncellus pseudomitissimus* and *Androsace villosa* are the characteristic species of the association Iberido saxatilis-Erinaceetum anthyllidis.

From the South Oroiberian sector, three associations have been described. *Satureja intrincata* subsp. *gracilis, Thymus leptophyllus* and *Euphorbia mariolensis* characterize Saturejo gracilis-Erinaceetum anthyllidis, which is the sector's most widespread association. At windy sites, it is replaced by *Genisto* hispanicae-Erinaceetum anthyllidis, whose characteristic species are *Genista hispanica* and *Dianthus algetanus* subsp. *turolensis* (Fig. 11.35). These two species are absent in the eastern Maestrazgo association, Salvio-Aphyllanthetum monspeliensis. The latter is a succession stage of marcescent forests (Sileno melliferae-Quercetum fagineae), while the other two associations replace sabinares (Juniperetum

Fig. 11.35 Genisto hispanicae-Erinaceetum anthyllidis near Allepuz, Teruel

hemisphaericae-thuriferae) and pinares (Junipero sabinae-Pinetum ibericae, Ononido aragonensis-Pinetum ibericae, and Festuco gautieri-Pinetum salzmannii).

In the upper-mesomediterranean and lower-supramediterranean belts, cushion plants are rare or only occur on northern slopes, and thyme and sage scrubs thrive, among which aromatic species dominate (*Thymus mastichina, T. vulgaris, T. zygis, Lavandula latifolia, Rosmarinus officinalis, Salvia lavandulifolia*, etc). Locally given the name "romerales" (after rosemary, or "romero", *Rosmarinus officinalis*), "salviares" (after species of the genus *Salvia*, or sage), "esplegares" (after "espliego", or lavender, *Lavandula latifolia*) or "tomillares" (after "tomillos", species of the genus *Thymus*, or thyme), these matorrales have been ascribed to various associations.

Chamaephytes such as *Linum suffruticosum* subsp. *differens, Fumana ericoides, Stipa offneri, Rosmarinus officinalis, Teucrium capitatum* and *T. gnaphalodes* are typical and dominant in Lino differentis-Salvietum lavandulifoliae, a widespread association in the Celtiberia and Alcarria sector. *Sideritis linearifolia* and *Teucrium expansum* are the characteristic species of the association Siderito linearifoliae-Teucrietum expansi, thriving in the lowlands of Gúdar and Javalambre. Eastwards, in the Maestrazgo mountains, the endemic *Lavandula angustifolia* subsp. *turolensis* characterizes the association Salvio lavandulifoliae-Lavanduletum turolensis. From limestone outcrops of the North Oroiberian Sierran sector two associations are described: Siderito spinulosae-Lavanduletum latifoliae, endemic to the sunny, dry slopes of the Moncayo, is characterized by *Sideritis spinulosa* and *Centaurea linifolia*, while the most widespread sage scrub formation in that sector correspond to the association Salvio lavandulifoliae-Linetum appressi. Lastly, in many zones of the Iberian System, *Astragalus granatensis*, the characteristic species of Santolino-Astragaletum boissieri, is dominant along the margins of sheep tracks.

Gypsum substrates are frequent in several sedimentary basins of the Iberian System. Some of these substrates are evaporite rocks showing flat stratifications and are formed as a result of lacustrine sedimentation during the Miocene, when the Ebro basin and surrounding areas were an interior lake that was not connected to the Mediterranean sea. In addition, other gypsum outcrops featuring folded stratifications of earlier origin formed in the continental Tertiary-Oligocene and early Miocene. This abundance of gypsum confers the landscape of much of the Iberian System a whitish colour which enhances its arid, steppe-like appearance. Soils formed on such bedrocks are gypsisols, which, under the Mediterranean severe summer drought, determine the existence of a very specialized flora of gypsophytes, almost all Iberian endemics confined to those soils, which constitute the core of the order Gypsophiletalia. In the Iberian System, *Gypsophila hispanica* is the main characteristic species. In the Cantabrian Castillian sector (gypsum soils of La Bureba, Burgos), *G. hispanica* occurs along with *Sideritis linearifolia*, characterizing the association Siderito linearifoliae-Gypsophiletum hispanicae. In the South Oroiberian sector, *Jurinea pinnata* and *Teucrium expansum* are the characteristic species of the endemic association Teucrio expansi-Gypsophiletum hispanicae. In the Miocene deposits of the interior depressions, *J. pinnata* and *T. expansum* are lacking and some species such as *Boleum asperum*, *Helianthemum syriacum*, *Launaea fragilis*, *L. pumila* and *Reseda stricta* are typical components of the association Helianthemo thibaudii-Gypsophiletum hispanicae.

Acidophilous Scrublands ("Brezales" and "Jarales") Heathlands and moorlands ("brezales") are seral dwarf shrub communities of deciduous or sclerophyllous potential natural forests degraded by fires, grazing or wood-cutting. These communities occur on poor soils (eroded dystric cambisols or ferric podzols with acid raw humus and occasionally gleyic or stagnant properties) in supratemperate and supramediterranean subhumid to hyperhumid oceanic and hyperoceanic bioclimates. Common taxa in the Iberian System are *Agrostis curtisii*, *Allium ericetorum*, *Calluna vulgaris* f. *vulgaris*, *Daboecia cantabrica*, *Erica australis* subsp. *aragonensis*, *E. scoparia*, *E. tetralix*, *E. vagans*, *Genista anglica*, *G. pilosa*, *Halimium umbellatum*, *Scorzonera humilis*, *Simethis mattiazzi* and *Tuberaria lignosa*. According to soil moisture regime, wet and dry heathlands are distinguished.

Wet heathlands with gleyic or stagnant soils are represented in the Iberian System by the alliance *Genistion micrantho-anglicae*, and two associations: Genisto anglicae-Daboecietum cantabricae (with *Daboecia cantabrica* and without *Erica tetralix*, *E. vagans* and *Genista micrantha*) occurring in some areas of the Cantabrian Castilian sector such as Montes de Oca and the comarques Tozo and Valdelucio in Burgos province; and Genisto anglicae-Ericetum vagantis in the North Oroiberian Sierran and Cantabrian Castilian sectors, in which *D. cantabrica* is lacking and *E. vagans* is dominant.

Dry heathlands occur on eroded dystric cambisols or ferric podzols with acid raw humus contents but without gleyic or stagnant properties. Their communities belong to the alliances Daboecion cantabricae and Ericion umbellatae. The former comprises heathlands developed on rankers or podsols and are succession stages of beech forests and deciduous and marcescent oak forests of the Cantabrian Castilian sector. In the Iberian System, the alliance is represented by two associations: Ericetum scopario-*vagantis*, occurring on sandy soils and dominated by *Erica scoparia* and *E. vagans*; and Arctostaphylo crassifoliae-Daboecietum cantabricae, in which *Arctostaphylos uva-ursi* subsp. *crassifolia*, *Genista occidentalis* and *Thymelaea ruizii* grow on less sandy, but permeable soils, whose potential vegetation comprises marcescent forests (Spiraeo obovatae-Quercetum fagineae).

Ericion umbellatae groups western Iberian and North African associations, and is restricted, in the Iberian System, to North Oroiberian Sierran territories and the Serranía de Cuenca. Characteristic species are *Cistus crispus*, *C. psilosepalus*, *Erica australis* subsp. *aragonensis*, *Halimium ocymoides*, *H. umbellatum* and *Luzula lactea*. From the North Oroiberian Sierran sector, two associations have been described: Calluno vulgaris-Genistetum occidentalis, endemic to the Moncayo (Fig. 11.36); and Cytiso scoparii-Ericetum aragonensis, thriving in

Fig. 11.36 Scrub dominated by the endemic *Genista occidentalis* (Calluno vulgaris-Genistetum occidentalis) near Puerto de Oncala, Soria

the heathlands of Picos de Urbión and la Demanda, an association dominated by *Erica australis* subsp. *aragonensis, Calluna vulgaris* and *E. cinerea*, in which *Genista pilosa* replaces *G. occidentalis*. In the sandstones of the South Oroiberian sector, *Thymelaea subrepens* and *Erica umbellata* characterize the endemic association Thymelaeo subrepentis-Ericetum aragonensis.

In continental areas with less rainfall, heathers are replaced with "jarales" ("jara" is the Spanish name for *Cistus* species) belonging to the class Cisto-Lavanduletea, which groups thermomediterranean to supramediterranean dry and semiarid, scarcely subhumid, silicicolous or calcifugous Western Mediterranean secondary scrub communities, mainly *Cistaceae* and *Labiatae*. These heaths arise from the destruction of potential natural sclerophyllous forests by fire or due to wood production and subsequent erosion of the upper soil layers. Six associations have there been described in the Iberian System. Erico scopariae-Cistetum populifolii is an association dominated by *Cistus populifolius* and *Erica scoparia* growing in ravines and shady slopes. The five remaining associations are dominated by *Cistus laurifolius* in upper mesomediterranean and supramediterranean areas: Genisto scorpii-Cistetum laurifolii has *Genista scorpius* as a codominant and differential species and forms a calcifugous scrubland on lixiviated, carbonate-poor, luvisols originating from Mesozoic limestones; Erico scopariae-Arctostaphyletum crassifoliae in the South Oroiberian sector, includes *Erica scoparia* and *Arctostaphylos uva-ursi* subsp. *crassifolia* as dominants along with *Cistus laurifolius;* and the three North Oroiberian Sierran sector associations are: Halimio ocymoidis-Cistetum laurifolii, including *Halimium ocymoides, Erica cinerea* and

Calluna vulgaris, described for Puerto del Mojón Pardo, Soria; Erico arboreae-Arctostaphyletum crassifoliae, a succession stage of suhumid oak forests (Luzulo forsteri-Quercetum pyrenaicae) endemic to the supramediterranean belt of Moncayo; and Santolino rosmarinifoliae-Cistetum laurifolii, which corresponds to the "jarales de estepa" dominated by *Cistus laurifolius* and *Lavandula pedunculata*, which are seral stages of the same subhumid oak forests.

11.4 Grasslands

Grasslands are habitats dominated by graminoids (*Poaceae*, *Cyperaceae* and *Juncaceae*). They are among the most widely distributed vegetation types of the world's landscapes. Besides natural grasslands, large areas covered by grassland are a product of prolonged human influence. Grasslands thrive under varying ecological conditions (dry, mesic, wet) and under different management regimens (mown, grazed, fertilized), resulting in different biological patterns. In this section, we focus on dry and mesic grasslands. The wetter tall grasslands (Phragmito-Magnocaricetea) of the Iberian System are dealt with in the following section on amphibious vegetation.

Dry Grasslands In the Iberian System, perennial grasslands belong to the classes Lygeo-Stipetea, Poetea bulbosae, Festucetea indigestae and Sedo-Scleranthetea. Annual grasslands are included in the class Tuberarietea guttatae. Both types of dry grassland appear on the dry, well-drained slopes, and contrast with the tender evergreen mesic grasslands of humid riparian basins and wet meadows, which benefit from a water table near the surface.

Perennial Dry Grasslands The class Lygeo-Stipetea groups together Mediterranean perennial, basophile, xerophytic, tall and bunchy or short and dense open grasslands. In the ecological succession, these grasslands are one of the first subseral stages of the climactic sclerophyllous forest of Quercetea ilicis thriving on deep, rich calcareous soils. The Iberian system features grasslands of the order Lygeo-Stipetalia and two of its alliances: Thero-Brachypodion retusi (basophile, rather hard short-grass dominated by *Brachypodium retusum*) and Stipion tenacissimae (medium to tall grassland communities dominated by species of *Stipa* and *Helictotrichon*). Communities belonging to Thero-Brachypodion retusi are usually found in areas of long summer drought. Under semiarid conditions, the grass community usually "hides" under a shrub cover. The habitat is found mostly on eutrophic, and less frequently neutral siliceous soils, often stony and degraded by erosion. When this habitat is in good ecological condition, ground cover is often high, sometimes complete. Since Thero-Brachypodion retusi is considered to be the last substitution stage of Mediterranean sclerophyllous forests and pine forests and shrublands, it usually occurs in association with these woody communities, both in small- to medium-sized gaps or clearings and under their often light canopies. Its presence is usually linked to extensive grazing (sheep and goats) and frequent fires.

The only association included in Thero-Brachypodion retusi recognized in the Iberian System is the widespread Phlomidio lychnitis-Brachypodietum ramosi. This association takes the form of short grasslands dominated by *B. retusum* growing on calcareous regosols, often with a stony or gravelly upper layer, produced by the erosion of fine grained mineral elements. In the succession, this association represents a more degraded stage than the grasslands of Stipion tenacissimae, whose bunchy grass communities grow on deep calcareous and clayey soils (rendzinas and luvisols). In the Iberian System, we find two associations of Stipion tenacissimae: Stipo offneri-Helictotrichetum filifolii, appearing as medium-sized grasslands dominated by *Helictotrichon filifolium* and *Stipa offneri* in the Celtiberian and Alcarria and South Oroiberian sectors; and the Gúdar grassland Helictotricho bromoidis-Stipetum pauneroanae, which lacks *H. filifolium* and *S. offneri* and commonly features *Helictotrichon bromoides* and *Stipa pauneroana* (as characteristic species).

Grasslands of Poetea bulbosae can be regarded as a "cultural" climax that includes dense, short but highly productive, Mediterranean perennial swards dominated by the bulbous meadow grass *Poa bulbosa*, along with many annual species. When the habitats of this class are in good ecological and management conditions, ground cover is complete, woody vegetation is usually absent or scarce (due to intensive browsing in times of summer and winter green grass shortage) and legumes are abundant. The grassland is created and maintained by the intense and persistent activity of livestock (usually sheep) on both oligotrophic and eutrophic soils. Thus, grazing and species selection along with the fertilization and accelerated nutrient cycling as affected by faeces give rise to the optimum Mediterranean sward grassland for livestock grazing. The ecological requirements of this class are a thermo- to supra-Mediterranean thermoclimate, a dry to subhumid ombroclimate and a summer drought period longer than two months, at least for the upper soil horizon, where most grass roots are found. The vegetation of this class grows on acidic (Periballio-Trifolion subterranei) or basic (Astragalo-Poion bulbosae) soils. Oligotrophic substrates are not so well suited for agriculture and more appropriate for livestock grazing. Periballio-Trifolion subterranei is, by far, the alliance that covers the most extensive area. Whatever the substrate, the upper soil horizon is always rich in organic matter and nutrients as a result of intense and persistent livestock activity. Another essential feature is the high level of activity of soil meso- and microfauna, e.g. worms, ants, bacteria. The oligotrophic alliance Periballio-Trifolion subterranei is represented in the Iberian System by the association Festuco amplae-Poetum bulbosae. The eutrophic swards of Astragalo-Poion bulbosae are scarce in the Iberian System. When they do occur, their floristic composition is typical of the widespread association Astragalo sesamei-Poetum bulbosae.

Festucetea indigestae encompasses silicicolous communities dominated by caespitose, tufted grasses of small size (>25 cm), mostly gramineae or graminoid, and small decumbent or pulviniform dwarf-chamaephytes, growing in silicate substrates (granite, gneiss, slate, quartzite, ultrabasic rocks, etc.) on geliturbate rankers or cambisols without histic or gleyic properties. These supra- to cryoromediterranean and submediterranean grasslands are zonal at supratimberline altitudes and secondary at lower altitudes. The class is represented in the Iberian System by two orders. Festucetalia indigestae, groups together zonal associations of high altitudes. Its alliance Minuartio-Festucion curvifoliae encompasses oro- and cryorotemperate associations of the Central and Oroiberian ranges. The second order, Jasiono sessiliflorae-Koelerietalia crassipedis and its alliance Hieracio castellani-Plantaginion radicatae, comprise supra- to lower oromediterranean dry to humid, secondary, dwarf-chamaephyte, grassland-like communities growing on shallow siliceous, geliturbate soils of the Iberian System and related Baetic and Mediterranean Central Iberian high plateaus and mountains. Three associations have been described from the Iberian System. The first is Antennario dioicae-Festucetum curvifoliae, a silicicolous orotemperate association endemic to the North Oroiberian Sierran sector (Picos de Urbión, Sierra de la Demanda and Moncayo). The remaining two are supra- to lower oromediterranean, secondary, dwarf-chamaephyte, grassland-like associations thriving on shallow, siliceous and geliturbate soils (Hieracio castellani-Plantaginion radicatae): Potentillo velutinae-Leucanthemopsietum assoi, flourishing on sandstones of the Serranía de Cuenca (with *Potentilla velutina* and *Leucanthemopsis assoi* the absence of *Corynephorus canescens* and *Sedum amplexicaule* is a characteristic feature versus the next association), and Scleranthо perennis-Corynephoretum canescentis, thriving on sandy soils of the Gúdar and Javalambre sierras (including *C. canescens, S. amplexicaule* and *Potentilla velutina*, and lacking *Leucanthemopsis assoi*).

Sedo-Scleranthetea includes open communities with perennial succulent Crassulaceae (*Sedum, Sempervivum*) and other dwarf chamaephytes and geophytes, accompanied by ephemeral therophytes, growing on lithosols and rock surfaces of cohesive siliceous or calcareous substrata. On siliceous cohesive rocks of the Iberian System, the class is represented by the oceanic supra- to lower-orotemperate and submediterranean alliance Sedion pyrenaici, whose most common species is *Sedum anglicum* subsp. *pyrenaicum*. The latter is dominant in two associations described from the Iberian System: Sedetum brevifolio-pyrenaici (with *Sedum brevifolium*, but without *Rumex angiocarpus*; North Oroiberian Sierran sector) and Sedetum micrantho-pyrenaici (with *R. angiocarpus* but without S. *brevifolium*; Cantabrian Castilian sector).

Alpine, subalpine and supra-oromediterranean calcareous grasslands occur above the treeline on base-rich soils in the high mountains of Europe. Harsh climate conditions (i.e., low temperatures, cryoturbation, prolonged frost, heavy snow accumulation), which limit the vegetative period to a few months, characterize their habitats. These grasslands harbour several plant communities, mainly of the phytosociological class Festuco hystricis-Ononidetea striatae, of dry, basophilous grasslands, rich in hemicryptophytes, pulvinate chamaephytes or dwarf cushion shrubs growing in soils with ephemeral snow cover, often with cryoturbation phenomena and without gleyic features. This class is represented by the two orders: Ononidetalia striatae and Festuco hystricis-Poetalia ligulatae.

Submediterranean grasslands of the Cantabrian Castilian sector on deep or rendzina soils and rich in dwarf cushion shrubs belong to the alliance Genistion occidentalis (Ononidetalia striatae), in which only one association has been described for the Iberian System: Arctostaphylo crassifoliae-Genistetum occidentalis, a grassland rich in dwarf cushion shrubs dominated by the bright yellow inflorescence of *Genista occidentalis* that flourishes on deep rendzina soils. The association is also known from the North Oroiberian Sierran sector, though there it lacks *Arctostaphylos uva-ursi* subsp. *crassifolia*.

Festuco hystricis-Poetalia ligulatae comprises intensely cryoturbated communities flourishing in Mediterranean high mountain zones and is represented in the Iberian System by two short grass alliances: Sideritido fontquerianae-Arenarion aggregatae and Plantagini discoloris-Thymion mastigophori. The latter groups together supramediterranean and submediterranean grasslands rich in dwarf chamaephytes on calcareous clayey soils of the Cantabrian Castilian sector. In this alliance, whose characteristic species are *Arenaria vitoriana, Onobrychis reuteri, Plantago discolor* and *Thymus mastigophorus*, three associations have been described: Veronico javalabrensis-Thymetum mastigophori, a chamaephytic grassland characterized by *Veronica javalambrensis* colonizing the western part of the Cantabrian Castilian sector; Koelerio vallesianae-Thymetum mastigophori, thriving in the eastern part of that sector and lacking *V. javalambrensis*; and Festuco hystricis-Genistetum eliassennenii, a creeping, chamaephytic shrubland dominated by the microendemism *Genista pumila* subsp. *elias-sennenii*, flourishing on wind exposed slopes with skeletal soils in some northern sierras of the Cantabrian Castilian sector (Montes Obarenes, Sierra de Arcamo and Sierra de Codés).

The Sideritido fontquerianae-Arenarion aggregatae alliance is represented in the Iberian System by seven associations. The main four of which are: Androsaco villosae-Festucetum hystricis, harbouring *Androsace villosa* and *Iberis saxatilis* as characteristic species, comprising supra-oromediterranean grasslands of the Moncayo; Paronychio capitatae-Artemisietum lanatae, which include typical biotopes of *Artemisia pedemontana* subsp. *assoana*, and dominates in euthrophic grasslands overgrazed by sheep; the widespread *Festucetum histricis*, which replaces the previous one in places where grazing is moderate, with an abundance of *Poa ligulata* and *Festuca hystrix*; and Drabo estevei-Ononidetum striatae on Jurassic hard limestones of the Picos de Urbión and Sierra de la Demanda, including *Draba aizoides* subsp. *estevei* and *Ononis striata* as dominant species. The other three associations occur in the summit areas of the sierras of Gúdar and Javalambre and have *Arenaria aggregata* subsp. *erinacea, Astragalus muticus, Ononis cristata* and *Sideritis fontqueriana* as typical species of the alliance: Sideritido fontquerianae-Arenarietum erinaceae, flourishing on windy sites with skeletal soils; Astragalo austriaci-Ononidetum cenisiae, with *Astragalus austriacus* and *Ononis cristata* dominating in grazed pastures. Lastly, the presence of *Erodium celtibericum* characterizes the association Erodio celtiberici-Erinaceetum anthyllidis, endemic to Peñagolosa and surrounding areas.

Annual Dry Grasslands Pioneer spring and early summer ephemeral, annual grasslands, usually growing on lithosols, are assigned to Tuberarietea guttatae. Given their colonizing nature, these therophytic communities occupy gaps between bushes and in perennial grasslands in xeric, non-nitrified biotopes. Species typical of Tuberarietea pastures possess the R-strategy, involving intense seed production and high germination rates. This strategy is observed when a community shows a strongly seasonal development which implies that its survival is ensured through seed banks and depends on whether or not ideal conditions for germination are met every autumn-winter.

The acidophilic or calcifugous order Tuberarietalia guttatae is represented in the Iberian System by two alliances: Tuberarion guttatae and Molineriellion laevis. Tuberarion guttatae, grouping thermo- to lower supramediterranean associations growing on sandy-loamy soils, is represented in the study area by only one association, Sedo caespitosi-Tillaeetum muscosae; it comprises communities occupying small plots (*c.* 1 m^2 or less) that are dominated by succulent microtherophytes (*Crassula tillaea*, *Sedum caespitosum*) colonizing sandy soils arising from the erosion of Palaeozoic rocks in the North Oroiberian Sierran sector. At higher altitudes, communities of Tuberarion guttatae are replaced by supra- to lower oromediterranean communities of Molineriellion laevis, represented in the same sector, but also by the association Triseto ovati-Agrostietum truncatulae on sandstones of the South Oroiberian sector.

Pioneer and ephemeral basophilous annual grasslands, usually growing on calcium-carbonated, clayey, gypseous, dolomite, serpentinic or mafic soils in Mediterranean areas, belong to Brachypodietalia distachyi. Along with perennial grasses of Lygeo-Stipetea, the communities of Brachypodietalia distachyi configure the European priority habitat 6220* (pseudo-steppe with grasses and annuals), which includes a variety of xeric, thermophilous and mostly-open Mediterranean perennial and annual grasslands growing on, usually eutrophic, but also oligotrophic, soils. In the Iberian System they are found on eutrophic soils, but which are usually degraded or poorly developed (lithosols).

Brachypodietalia distachyi is represented in the Iberian System by the alliance Brachypodion distachyi and three associations. At low altitudes, *Hornungia petraea* dominates two pioneer associations developing on calcareous lithosols after the first spring rains: Saxifrago tridactylitae-Hornungietum petraeae, with *Saxifraga tridactylites* as characteristic species, growing in the western Iberian System (Celtiberian and Alcarria sector and Serranía de Cuenca), and Erophilo spathulatae-Hornungietum petraeae (Gúdar and Javalambre), lacking *S. trydactilites* and with *Erophila verna* subsp. *spathulata* as its differential species.

Mesic Grasslands Molinio-Arrhenatheretea includes mown meadows and mesic pastures of temperate regions of Europe and adjacent regions of Asia. Mesic meadows and pastures are widespread, diverse semi-natural habitats of temperate Europe that have their geographic distribution limit in the Iberian Peninsula. Their biogeographic optimum extends across the temperate Eurosiberian region, while favourable conditions in the Mediterranean region are restricted to valley floors and foothills in mountain areas with fine-textured soils and sufficient moisture (gleysols and fluvisols). The class is represented in the Iberian System by four orders.

Molinietalia coeruleae encompasses pastures and meadows on permanently moist, often peaty, occasionally-mowed soils and has two alliances, Molinion caeruleae and Juncion acutiflori. In the Iberian System, the oligo-mesotrophic, unmanured, moist pastures of the alliance Molinion caeruleae are represented by the supratemperate association Deschampsio refractae-Molinietum caeruleae. This association comprises oligotrophic, moist pastures on sandstones and its main species are *Molinia caerulea, Deschampsia caespitosa* subsp. *subtriflora* and *Carum verticillatum*. The oceanic alliance Juncion acutiflori of unmanured, rushy pastures on marshy or mineral peaty soils, includes two associations: Hyperico undulati-Juncetum acutiflori, dominated by *Juncus acutiflorus* at ungrazed, oligotrophic sites and Deschampsio hispanicae-Juncetum effusi, dominated by *Juncus effusus* in places that are overgrazed and soaked.

Arrhenatheretalia elatioris encompasses mesophytic, eutrophic, manured meadows appearing on well-drained, fertile, mineral soils. Three alliances of meadows and pastures belonging to this order have been described for the Iberian Peninsula: Arrhenatherion elatioris, Cynosurion cristati and Triseto flavescentis-Polygonion bistortae. The last alliance only occurs in the Pyrenean mountains while the former two are restricted to mountains and valleys in the northern half of the Peninsula, including the Iberian System. Arrhenatherion elatioris, comprising manured hay meadows on well-drained, fertile, mineral soils that are mowed at least twice a year, is represented in the Iberian System by the eutrophic association Galio veri-Arrhenatheretum bulbosi, in which *Arrhenatherum bulbosum* dominates. Cynosurion cristati contains pastures generated by intense grazing on nutrient-rich, damp soils; its dominant species is *Cynosurus cristatus*. Pastures of *C. cristatus* in the North Oroiberian Sierran and Castillian Cantabrian sectors belong to the association Festuco amplae-Cynosuretum cristati, whereas those of the South Oroiberian sector are ascribed to Cynosuro cristati-Trifolietum repentis.

The wettest meadows of the class are included in the orders Holoschoenetalia vulgaris and Plantaginetalia majoris. The former groups Mediterranean summer green pastures and grasslands, appearing on deep, damp soils and is represented in the Iberian System by the alliances Molinio-Holoschoenion and Deschampsion mediae. In the study area, Molinio-Holoschoenion encompasses two associations, Lysimachio ephemeri-Holoschoenetum, grouping club-rushy pastures on sites with a phreatic layer dominated by *Scirpus holoschoenus* ("junco churrero"), and Inulo viscosae-Schoenetum nigricantis, comprising rushy pastures dominated by *Schoenus nigricans* and *Molinia caerulea* subsp. *arundinacea* on sites soaked by dripping calcareous waters. Deschampsion mediae defines grasslands on temporary wet clayey, often expanded, base-rich soils and is represented in the Iberian System by two associations: Prunello hyssopifoliae-Plantaginetum serpentinae (Cantabrian Castilian sector) and Sanguisorbo lateriflorae-Deschampsietum hispanicae (South Oroiberian sector).

Plantaginetalia majoris groups together widespread, trampled, grazed, hygrophilous, perennial, creeping pastures with low forbs, often transiently flooded, on moist to wet eutrophic soils, rich in organic and mineral nutrients. In the Iberian System, this order includes the alliances Trifolio fragiferi-Cynodontion dactyli, Mentho-Juncion inflexi and Potentillion anserinae. The first alliance encompasses grazed, trampled, summer swards growing on moist, nutrient-rich soils, usually dominated by creeping sedges, grasses and clovers (*Carex divisa* subsp. *divisa, Cynodon dactylon*, *Cyperus laevigatus* subsp. *distachyos*, *Trifolium fragiferum* or *T. repens*, among others). The creeping sedge *Carex chaetophylla* and *Trifolium resupinatum* dominate the Trifolio resupinati-Caricetum chaetophyllae colonizing eutrophized sandy soils, and *Cynodon dactylon* dominates the Trifolio fragiferi-Cynodontetum dactyli, growing in clayey soils, where *C. chaetophylla* does not thrive.

Mentho-Juncion inflexi comprises rushy, heavily-grazed pastures on temporarily waterlogged but summer-moist, mineral and organic nutrient-rich soils. Two associations have been identified in the Iberian System: Festuco fenas-Caricetum hirtae, flourishing at dam margins, and Junco inflexi-Menthetum longifoliae, including *Juncus inflexus* dominating rushy, heavily-grazed pastures on temporary swamped but summer moist, mineral and organic nutrient-rich soils.

Heavily trampled, overgrazed, occasionally flooded, wide-spread, pioneer, perennial grass-like and low forb communities that grow on rich, summer-moist soils with an excess of manure and nutrients correspond to Potentillion anserinae. There is only a single association in the Iberian System, Lolio perennis-Plantaginetum majoris, a kind of pioneer, perennial, grass-like, low forb community dominated by Plantago major.

Anthropogenic, intensely-grazed, acidophilic, climactic grassland swards of the upper treeline zones of high mountainswith a long-lasting snow cover, are assigned to the class Nardetea strictae. They grow on summer-moist, deep, strongly-acidic, organic mineral or peaty mineral soils and are assigned to the unique order Nardetalia strictae, which is represented in the Iberian System by the alliance Campanulo herminii-Nardion. Five associations have been recognized in the Iberian System: Luzulo carpetanae-Pedicularietum sylvaticae, grasslands of the North Oroiberian Sierran sector dominated by *Juncus squarrosus* and *Nardus stricta* occupying wet soils with a histic horizon; Plantago penyalarensis-Festucetum ibericae in the same sector, but on well-drained soils, with *Plantago alpina* subsp. *penyalarensis, Nardus stricta* and *Festuca iberica* as dominants; Luzulo carpetanae-Nardetum strictae, endemic to Picos de Urbión (lacking *Festuca iberica* and *Plantago alpina* subsp. *penyalarensis*); Galio idubaeae-Nardetum strictae, with *Galium idubaeae* in the Gúdar and Javalambre high mountains; and Carici pallescentis-Luzuletum multiflorae including *Genista anglica* in the Serranía de Cuenca.

The Festuco-Brometea class comprises anthropogenic, grazed, baso-neutrophilous or slightly acidophilic, mesophytic or slightly xerophytic, nutrient-rich pastures, largely covered by perennial grasses growing on deep, rich, well-drained soils without gleyic properties. The class is represented in the Iberian

System by the order Brometalia erecti, which clusters closed, short-grass pastures, and also by its mesophytic, baso-neutrophilous, calcicolous alliance Potentillo montanae-Brachypodion rupestris. The two recognized associations are Cirsio acaulis-Onobrychidetum hispanicae, comprising mesophytic grasslands dominated by *Onobrychis hispanica* that flourish in the supramediterranean belt of the Celtiberian and Alcarria and Ibero-Maestracense sectors; and Brachypodio rupestris-Seselietum cantabrici, which are mesophytic grasslands endemic to the Sierra de la Demanda, dominated by *Brachypodium rupestre* and accompanied by *Carduncellus mitissimus, Helianthemum nummularium, Lotus corniculatus* and *Seseli cantabricum.*

Tall grass pastures ("fenalares"), in which *Brachypodium phoenicoides* is the main species, occurring on basophilous to neutrophilous deep soils, characterize the Mediterranean order Brachypodietalia phoenicoidis. Its communities usually are found in areas of higher moisture because of greater rainfall or a greater soil water retention capacity where the potential natural vegetation corresponds to subhumid-humid *Quercus faginea* forests or to *Populus alba* forests in valley bottoms. Two associations have been described for the Iberian System: Mantisalco salmanticae-Brachypodietum phoenicoidis, grouping pastures abundant in *Elytrigia intermedia, Medicago sativa, Centaurea aspera, Foeniculum piperitum* and other subnitrophilous plants living on roadsides, in fallow fields and other places with human influence of the Celtiberian and Alcarria and South Oroiberian sectors, and the tall grass pastures of Festuco andres-molinae-Brachypodietum phoenicoidis, with *Bromus erectus* as co-dominant species, flourishing on deep clayey soils in the North Oroiberian Sierran sector.

The class Stipo giganteae-Agrostietea castellanae and its unique Iberian order Agrostietalia castellanae comprises silicicolous, perennial grasslands ("vallicares") on deep cambisols, and is seral to natural potential forests of *Q. rotundifolia, Q. faginea* and *Q. pyrenaica*. Only the meadow-like and usually heavily-grazed association Festuco amplae-Agrostietum castellanae occurs on fresh soils (gleyic cambisols) in the study area. The order's main species are *Agrostis castellana, Festuca ampla* subsp. *ampla*, *Trifolium cernuum* and *T. retusum.*

11.5 Other Vegetation Types

For floating or rooted submerged aquatic vegetation (Potametea, Lemnetea), amphibious vegetation of fresh-waters, springs and fens (Phragmito-Magnocaricetea, Bidentetea tripartitae, Isoeto-Littorelletea, Isoeto-Nanojuncetea, Montio-Cardaminetea, and Scheuchzerio-Caricetea fuscae), chasmophytic and scree vegetation (Asplenietea trichomanis, Adiantetea, Parietarietea, Petrocoptido pyrenaicae-Sarcocapnetea enneaphyllae, Thlaspietea rotundifolii and Phagnalo-Rumicetea indurati), and synanthropic vegetation (Artemisietea vulgaris,

Stellarietea mediae, Polygono-Poetea annuae, Pegano-Salsoletea, Galio-Urticetea, Epilobietea angustifolii) see "Other vegetation types" as Supplementary electronical material.

11.6 The Iberian System's Vegetation and Human Impacts Suffered: From the Palaeolithic to the Present

The history of the Iberian System's deforestation is as old as that affecting the rest of the Iberian Peninsula, for which we have records of very ancient human settlements. The artificialization of nature is coeval with these first settlements rather than with the origin of Man. According to anthropologists and prehistory experts, the population of the Iberian Peninsula in the Palaeolithic was limited to a few thousand inhabitants living in small communities. In those times, forests must have afforded these hunters/gatherers protection and we could argue that these inhabitants, being dependent on forests, were very much interested in their conservation.

In the Sierra de Atapuerca (Burgos) or the north-western corner of the Iberian System, we find records of the most ancient human activity in the Iberian Peninsula. Dated at older than 700,000 years, these records indicate communities of hunters-fishermen and collectors of seeds, fruits and roots which, presumably, had little impact on the country's plant formations. However, large animal populations eliminated through the actions of hunters may have induced some of the extinctions produced during the Pleistocene and Holocene. Thus, changes in vegetation were likely discrete barring those attributed to the large-scale use of fire, and in any case, any deforestation would have been limited to zones close to seasonal settlements of human communities that were essentially nomadic.

Since a review of the complete landscape history of the Iberian System is beyond the scope of this chapter, we here focus on the Middle Ages, as the period that saw the start of the most intense deforestation. It was in the Middles Ages and early Modern Age when the causes of deforestation or deficient management became appreciable in Spain. Three main historic factors contributed to this: the Spanish reconquest of land occupied by Arabs (known as the Reconquista), an increase and promotion of land devoted to grazing, and finally, an increasing demand for wood for industrial use, especially for shipbuilding. Unfortunately, Spanish-Arab wars and the rise in population that took place from the eighth to the fourteenth century were extremely detrimental to the country's vegetation, which was regularly burned or felled by alternating waves of invaders.

The peculiar war-faring strategies of the eight centuries of Reconquista, including large-scale raids (razzia), land burning to avoid ambushes and limit the enemy's resources, massive destruction of large expanses of plant cover and the transfer of conquered lands to new settlers, had detrimental impacts on forest regions despite royal decrees dictating norms for their conservation and rational exploitation. Thus, the tenth to fourteenth centuries saw the advent of crop cultivation in response to the rise in population, to the Reconquista's repopulating surge and to the demands of numerous refugees of the Duero meseta and Ebro valley, whose diet was based on cereals. This event threatened every forest in the kingdom of Navarra in the eleventh century and of the Montes Burgaleses of Castilla in the twelfth century and prompted restrictions to protect forests as a necessary resource for purposes of construction and heating. Some historians consider the powers given by the Catholic Monarchs (Reyes Católicos) to the transhumance farmers of the "Honorable Mesta Council" a political impulse for social cohesion of the recently created central monarchy. Others relate the southwards transfer of the Reconquista to a need for new lands for cultivation and grazing with as a consequence the replacement of large forest areas by extensive cereal crops and vine groves. This is supported by references in ancient texts, such as the Libro de la Montería (a hunter's guide) written by the Monteros de Castilla in the fourteenth century at the request of king Alfonso XI, which indicate still the persistence of large expanses of forests. The Mesta and sheep and goat grazing under a transhumance regimen were not of recent appearance. Rather, this was a common nomad tradition that arose in very remote times (documented as the vegetation patterns of the Liber Iodiciorum or Visigothic code Fuero Juzgo, compiled by Recesvinto in 654) and whose origins lay in the land use cycling that was necessary where, during dry season a grass deficit enforces livestock migration to mountain or northern zones, where rains or the phenological delay in species maturity allows for summer grazing.

The economic benefits of livestock rearing were high during times in which wool was in great demand, and was thus an important source of income for medieval monarchies that were trying to recover from the economic burden of constant battles amongst each other and against Arabs to reconquer lands lost in earlier times. The monetary returns of transhumance elevated by the quality of merino wool reached extraordinary proportions and gave rise to large expanses of land given over to this activity either as large estates or joined communities. The success of this industry was such that, in 1273, king Alfonso X the Wise created the Mesta. Among the tasks of this council were the ordering and protection of transhumance and the creation of paths for herds to follow, mainly the so-called "cañadas reales" and their branches or "veredas" and "cordeles". The "Concejo de la Mesta", a lobby of those times in the manner of today's electricity or fuel companies, built up political and economic power. In the times of the Catholic Kings (Reyes Católicos) Elisabeth and Ferdinand, when money was needed for warfare and political cohesion between unified kingdoms, the council imposed a series of laws that were severely detrimental to agricultural practices and forests. These privileges of the council must have lasted several centuries, because the political levies of farmers were not abolished until the merino wool market declined and the Mesta was abolished in 1836.

This priority given to sheep rearing led to the disappearance of a large proportion of the underwood, which hindered natural regeneration. Shepherds, authorized by royal orders to benefit from mountain and forest regions with practically no restrictions whatsoever, burnt down vast areas of forest to expand their grazing lands, left their herds to browse impairing the recovery of shrublands, built cabins, pens, and sheepfolds, and obtained warmth from forest products depriving many lands of trees and exposing them to erosion. The importance of the Mesta in the Iberian System, where sheep concentrated during the summer months, is reflected by the fact that out of the nine large pathways ("cañadas reales") that crossed the Peninsula in every direction, five commenced at, or crossed, lands of the Iberian System. Only within the comarques of Cameros there were, in the eighteenth century, a million head of wool sheep that wintered on the pastures of Extremadura and sparked an active textile industry in the lands of Cameros. The end of the Mesta in 1836 brought the sudden demise of this activity and, although livestock rearing continued, the textile industry collapsed, unable to compete with modern industry.

The final factor to affect the exploitation of woods and forests at the beginning of modern times was the progressive use of timber for industrial and shipbuilding purposes. The Middle Ages was a period of intense metal use. Manufacturing tools and armour required large amounts of fuel to supply the smelting furnaces, a fuel that until the nineteenth century was exclusively provided by the forest. It is calculated that an iron smelting furnace requires three tons of coal per ton of iron produced, which would mean the clearing of up of a forest patch with a diameter of up to 1.5 km per 40 days of iron casting. The metal industry was especially important in Spain given its mining resources and intensive mining practices since ancient times.

In the times of the Reyes Católicos, ships of 500–1000 tons were constructed. The capacity of shipbuilders in terms of the ratio of raw wood to used wood was 10:1, meaning that for a small ship, let us say a 500 ton caravel, some 5000 m^3 cubic metres of good quality timber would be needed. Before offering more data, the reader should consider that although agriculture and husbandry had mainly affected the more accessible, productive lands or lands closer to markets, the demand for good quality wood for shipbuilding was covered by terrains that were earlier protected, mountain areas sustaining broad-leaved and conifer forests that had been so far well maintained and preserved. The reign of the Reyes Católicos saw the discovery of America, while their successor Charles V witnessed the first trip around the world started by Magallanes and, after his death, completed by Juan Sebastián Elcano in 1522. This feat was of such magnitude that it is well documented in the Spanish navy's archives at Viso del Marqués. The ship used by Magallanes and Elcano, La Victoria, with a crew of 60 sailors, weighed 300 tons and its wood (*Quercus robur*) came from the oak forests of Irisasi-Zarauz in

Guipúzcoa province, while its masts were provided by the straight poles of fir trees (*Abies alba*) from the Pyrenees.

Admiring the Invincible Armada with which King Phillip II planned to invade England, the poet Lope de Vega referred to the vessel as «a floating forest of the sea and jungle». Leaving aside naval disasters such as that of the ephemeral Armada, according to Bauer (1991: 170), if we consider that the mean life of a vessel in those times was around 25 years and that a rough record of the Spanish fleet in 1550 amounted to some 300,000 tons, we are left with a fairly good idea of the cost of Spain's naval monopoly in terms of the loss of its best forests: six million trees are needed to build a fleet like that, requiring the felling of some 1.200 km^2 of dense forests.

If there is cause for optimism regarding the conservation of Spain's mountain systems when reading the Libro de la Montería of the times of Alphons XI, the same cannot be said of subsequent publications such as the Historic-Geographic Relations of the era of Philip II written only two centuries later in which a forest landscape is described that should seriously be worried about. However, it was not until the Bourbons' monarchy and Age of Enlightenment that institutional measures were adopted for the conservation and expansion of Spain's forests. These measures, nevertheless, were always designed with a vision of their future exploitation and forests were not viewed as assets contributing to biological balance. Indeed, this concept was still not nourished by generations that considered Nature as something to benefit from rather than resources that need to be conserved and that are not inexhaustible. Concern for the state of conservation is documented in the detailed forest censuses undertaken in the eighteenth century, with the definition of the "Montes de Marina", as any mountains less than 25 leagues away from the coast or from large sailable rivers whose organizing, conservation and, above all, exploitation regimen for shipbuilding was entrusted to the three large naval institutions: Cádiz, Ferrol and Cartagena.

Only towards the end of the eighteenth century did ideas of agricultural reform gradually start to infiltrate, influenced by the European cultural movement of enlightenment, though the measures arising were often based on concepts that were too naive. We could definitely say that the nineteenth century was characterized by the industrial revolution, fossil fuels, modern agricultural technologies including specialized machinery and the use of fertilizers for intensive cultivation designed to feed a rapidly growing population. This of course meant the transformation by deforestation and ploughing of vast areas that up until then had remained uncultivated. Although the industrial revolution arrived later than in other European countries, the clearing of Spanish forests was promoted by the confiscation laws of Mendizábal and his liberals and of Madoz and his conservatives. This enabled the low-cost privatization of extensive natural areas that had until then been mortmain, i.e. in "unproductive hands", which were none other than those of the State itself, the Church, or the lands of village communities. It is also no surprise that such lands acquired for an absurdly low price were immediately subjected to irrational and abusive use with solely economic ends (agriculture, farming, wood, coal, etc). This determined that, while between 1876 and 1880 Spain could boast of having 48,000 km^2 of mountains and meadows, by 1896 this area was reduced to 7000 km^2.

Among the human pressures that have had most impacts on primery forests are the ploughing of lands for agricultural use and forest fires. Not to mention the old tale that a squirrel could cross the whole of Spain from tree to tree without touching the ground, before any post-Palaeolithic deforestations, 90% of the Spanish territory was covered by forests of the various oaks, beech, conifers and other characteristic trees of the Mediterranean and Temperate floras, which at a conservative estimate amounted to some 40 million wooded hectares. According to the Spanish Forestry's Third National Forest Inventory (1997–2000), Spain's forested surface area is today 26,273,235 ha, representing 51.93% of the territory. This percentage is higher than that of most European countries, though unlike these, a large part of Spain's forested area suffers from a low density of trees and only 56% (14,732,247 hectares) of this area can be considered true forest. The data for the 1997–2000 inventory indicate an increase of 827,588 ha (or almost 6%) as compared to the area given in the second inventory (1986–1995). The increase in forested surface area per inhabitant has been even greater (8.7%) due to a stabilized Spanish population in the past 10 years (2005–2015). Of note, even more so than the forested area, is the improved tree density between the two surveys, suggesting a mean increase of more than 12 trees per hectare (3.2%).

This expansion of forested surface area has been the outcome of several factors. The first of these is the large scale reforestation effort over the past 60 years and the Agricultural Land Forestation programme. However, by far the strongest contributing factor has been the progressive succession in our mountains as the consequence of the profound transformation experienced by the Spanish rural society in the past 50 years. Large zones of the mountainous Iberian System have undergone heavy depopulation since the early twentieth century. There are many ghost towns and abandoned villages scattered across different parts of the Iberian System, especially in Teruel Province. Currently a great number of surviving towns have only a residual population.

The exodus from rural mountainous Spain was the consequence of General Franco's stabilization programme of 1959. The population declined steeply as people emigrated towards large cities and coastal towns where the tourism industry was growing. Other causes of high emigration have been the abandoning by local youth of traditional agricultural practices that were the mainstay of village economies, such as sheep and goat rearing, as well as the lifestyle changes that have swept across rural Spain during the second half of the twentieth century.

The distribution of the described Vegetation Series and Geoseries in the territory is shown in Fig. 11.37.

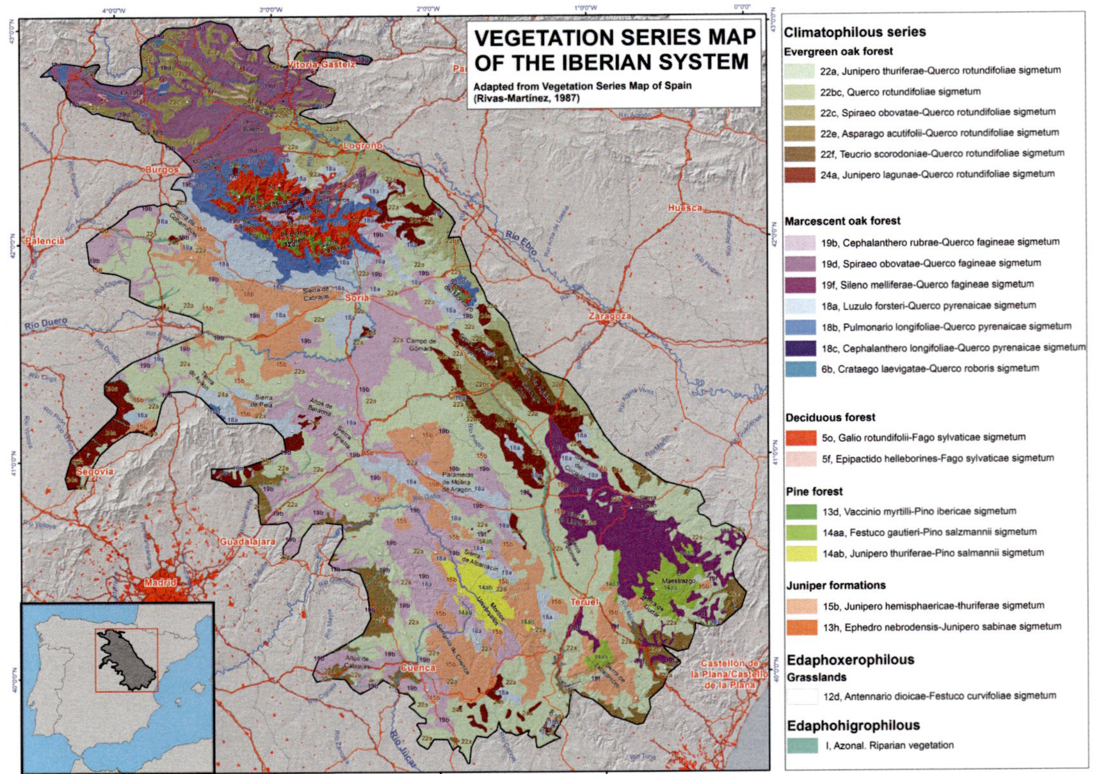

Fig. 11.37 Vegetation Series and Geoseries distribution in the Iberian ranges and highlands

References

Aguirre JL, Álvarez J, Bartolomé C et al (1991a) Crecimiento de plántulas de haya (*Fagus sylvatica* L.) en el Sistema Central. Stvdia Oecologica 8:9–22

Aguirre JL, Álvarez J, Bartolomé C et al (1991b) Mortalidad primaveral de plántulas de haya (Fagus sylvatica L.) durante los meses de verano en el Sistema Central (España). Stvdia Oecologica 8:127–138

Aguirre JL, Álvarez J, Bartolomé C et al (1994) Producción depredación de hayucos y polinización cruzada en un hayedo mediterranean en dos años con cosecha masiva. Stvdia Oecologica 10:339–349

Aupí V (2005) Guía del clima de España. Omega, Barcelona

Bauer E (1991) Los montes de España en la historia. Ministerio de Agricultura Pesca y Alimentación. Fundación Conde del Valle de Salazar, Madrid

Christensen KI (1993) Comments on the earliest validly published varietal name for the Corsican Pine. Taxon 42:649–653

Farjon A (2005) Pines drawings and descriptions of the Genus Pinus, 2nd edn. Brill, Leiden

García-Baquero G (2005) Flora y vegetación del Alto Oja (Sierra de La Demanda, La Rioja, España). Guineana 11:1–249

García-Mijangos I (1995) Los alcornocales del sector Castellano-Cantábrico. Lazaroa 15:241–244

Hernández Bermejo JE, Sáinz Ollero H (1984) El análisis de semejanza aplicado al estudio de barreras y fronteras fitogeográficas: su aplicación a la corología y endemoflora ibéricas. Anales Jard Bot Madrid 40(2):421–432

Herranz JM (1995) Fraxinus excelsior L. en el Alto Tajo, límite meridional ibérico. Ecología 9:191–200

Loidi J, Fernández-Prieto JA (1986) Datos sobre la biogeografía y la vegetación del sector CastellanoCantábrico (España). Documents Phytosociologiques 10(1):323–362

Loidi J, Biurrun I, Herrera M (1997) La vegetación del centro–septentrional de España. Itinera Geobot 9:161–618

López G (1978) Contribución al conocimiento fitsociológico de la Serranía de Cuenca, II. Anal Inst Bot Cavanilles 34(2):597–702

Medrano LM (1994) Flora y vegetación de las sierras de La Demanda y Cameros (La Rioja). PhD dissertation. Universidad de Navarra.

Navarro G (1986) Vegetación y flora de las sierras de Urbión, Neila y Cabrejas. PhD Universidad Complutense, Madrid

Navarro G (1989) Contribución al conocimiento de la vegetación del Moncayo. Opuscula Bot Pharm Complutensis 5:5–64

Oria de Rueda JA (1992) Las acebedas de Castilla-León y La Rioja: Origen, composición y dinámica. Ecología 6:79–91

Oria de Rueda JA (2003) Las acebedas de España. Aproximación al origen, dinámica y conservación de las mismas. PhD dissertation, Universidad Politécnica de Madrid

Peinado M, Monje L, Martínez-Parras JM (2008) El paisaje vegetal de Castilla–La Mancha. Manual de Geobotánica. Editorial Cuarto Centenario, Madrid

Pérez-Estaún A, Bea F, Bastida F et al (2004) La cordillera varisca europea: El Macizo Ibérico. In: Vera Torres JA (ed) Geología de España. Sociedad Geológica de España e Instituto Geológico y Minero de España, Madrid, pp 17–25

Regato-Pajares P, Elena-Rosselló R (1995) Natural black pine (Pinus nigra subsp salzmannii) forests of the Iberian eastern mountains: development of the phytoecological basis for their site evaluation. Ann Sci Forest 52(6):589–606

Ricotta C, Carranza ML, Avena G et al (2002) Are potential vegetation maps a meaningful alternative to neutral landscape models? Appl Veg Sci 5:271–275

Rivas-Martínez S (1987) Mapa de series de vegetación de España. ICONA/Serie Técnica, Madrid

Rivas-Martínez S (2007) Mapa de series, geoseries y geopermaseries de vegetación de España (Memoria del Mapa de Vegetación Potencial de España). Parte I. Itinera Geobot 17:55–435

Rivas-Martínez S et al (2011a) Mapa de series, geoseries y geopermaseries de vegetación de España (Memoria del mapa de vegetación potencial de España). Parte II Itinera Geobotanica 18(1):5–424

Rivas-Martínez S et al (2011b) Mapa de series, geoseries y geopermaseries de vegetación de España (Memoria del mapa de vegetación potencial de España). Parte II. Itinera Geobotanica 18(2):425–800

Roselló R (1994) Catálogo florístico y vegetación de la comarca natural del Alto Mijares (Castellón). Servei de Publicacions, Diputació de Castelló

Tíscar PA, Linares JC (2011) Pinus nigra subsp. salzmannii forests from Southeast Spain: using structure and process information to guide management. In: Frisiras CT (ed) Pine forests: types, threats and management. Nova Science Publishers, Hauppauge, pp 1–27

Vigo J (1968) La vegetació del Massís de Penyagolosa. Institut D'Estudis Catalans, Barcelona

Willkomm M (1844) Botanische Berichte aus Spanien. Botanische Zeitung 21(8):721–728

Willkomm M (1852) Wanderungen durch die nordöstlichen und centralen Provinzen Spaniens: Reiseerinnerungen aus dem Jahre 1850. Urnold Buchhandlung, Leizpig

Willkomm M (1896) Grundzüge der Pflanzenverbreitung auf der Iberischen Halbinsel. Wilhelm Engelmann Verlag, Leipzig

Willkomm M, Lange J (1870) Prodromus florae hispanicae, seu synopsis methodica omnium plantarum in Hispania, vol 1. E. Schweizerbart, Stuttgart

Chapter 12
The Ebro Basin

Javier Loidi

Abstract The Ebro Basin is located in the northeastern quadrant of the Iberian Peninsula, in the southern foothills of the Pyrenees, and excavated by its homonymous river. It is one of the largest catchments areas in the Iberian Peninsula. Its climate is Mediterranean with both a continental and semi-arid tendency due to the rain shadow effect of the surrounding mountain ranges. The vegetation in dry areas is made up by several groups: (1) Forests and woodlands dominated by *Quercus rotundifolia*, *Q. coccifera*, *Pinus halepensis*, *Juniperus phoenicea* and *J. thurifera*, and represented in small patches due to deforestation; (2) Garrigue on limestone or on gypsum substrates; on the latter, a group of gypsophytes constitutes the core of the communities. (3) Grasslands, which can be divided into: non-steppic, dominated by *Brachypodium phoenicoides* and *B. retusum;* steppic, dominated by *Lygeum spartum* and *Stipa* species; heavily grazed, dominated by *Poa bulbosa;* and annual grasslands of plants with ephemeral life cycles. There are two edaphically exceptional ecosystems: salty depressions and riparian ecosystems. The first are populated by a set of communities of succulent *Chenopodioideae*, such as *Microcnemum coralloides*, *Salicornia patula* and, mostly *Suaeda braun-blanquetii* which plays the most relevant role, together with *Limonium latebracteatum*; and woodland communities with *Tamarix canariensis* and *T. boveana;* reed communities with *Juncus maritimus*, *J. gerardii* and *J. subulatus*.; and annual communities dominated by *Aeluropus littoralis* and *Sphenopus divaricatus*. The riparian ecosystems are formed by forests, with willows (*Salix neotricha*), poplars (*Populus alba, P. nigra*), ashes (*Fraxinus angustifolia*), alders (*Alnus glutinosa*) and elms (*Ulmus minor*). In this environment *Scirpoides holoschoenus* communities are frequent, as well as *Cynodon dactylon* grasslands. Some comments about land use, concerning deforestation and irrigation are made, highlighting the historically intense human pressure on this territory.

J. Loidi (✉)
Department of Plant Biology and Ecology, University of the Basque Country (UPV/EHU), Bilbao, Spain
e-mail: javier.loidi@ehu.eus

J. Loidi (ed.), *The Vegetation of the Iberian Peninsula*, Plant and Vegetation 12,
DOI 10.1007/978-3-319-54784-8_12

12.1 Introduction

The Ebro catchment area, located in the northeastern quadrant of the Iberian Peninsula (Fig. 12.1), lies at the southern foothills of the Pyrenees. It is excavated by its homonymous river, forming one of the largest hydrographic catchment areas in the Peninsula with an extent of 85,000 km^2. The abundant waters, drained by the Ebro and its hierarchical fluvial network, flow into the Mediterranean Sea.

The notion of the Ebro Basin, or Ebro Depression, is not coincident with that of the Ebro Valley as the former only occupies the central portion of the Ebro catchment area (Fig. 12.1). The Basin is a triangular depressed geomorphological unit, at an average altitude of 200 m, concealed among the three main mountain ranges which surround it: the Pyrenees in the north, the Iberian Range in the south and southwest, and the Catalan Coastal Ranges in the east. The Ebro Depression was formed during the mid-Tertiary as a result of the orogenic activity of that period, which gave rise to a tectonic depression. This formed an interior sea that was disconnected from all of the surrounding oceans and filled with sediments from the eroded materials of the surrounding ranges. In this endorheic sedimentary basin intense lacustrine sedimentation took place during the Tertiary. This sedimentary regime determines that in the centre of the Depression there are materials of chemical sedimentation (evaporites) such as gypsum or salt, with limestone in the upper strata of the hills. In the peripheral areas detrital materials such as

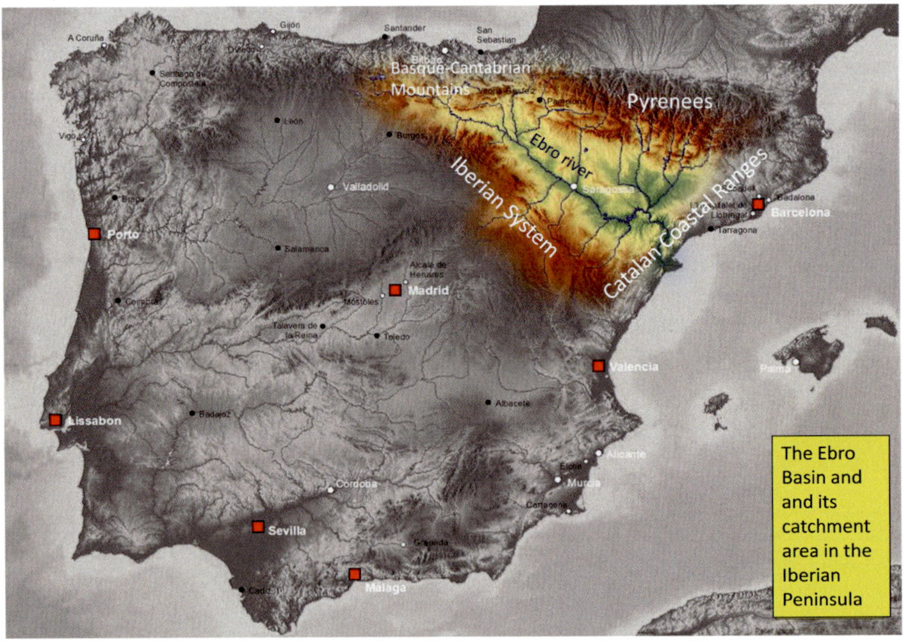

Fig. 12.1 The Ebro Basin and its catchment area in the Iberian Peninsula

Fig. 12.2 Monadnocks in the central portion of the Ebro Basin as those of the Bardenas Reales in southern Navarra

conglomerates (beside the mountain foothills), sandstone or marl dominate. These Tertiary materials are arranged in horizontal layers as no relevant tectonic activity has folded them. As a result of this sedimentary endorheic regime, the interior lake of the Ebro Basin reached an elevation up to 200 m above the level of the open sea, enforced by accumulation of sediments in the lake. This rise in the level of the waters facilitated the change into an exorheic regime at the end of the Tertiary as a connexion with the Mediterranean Sea opened and the interior lake emptied into it. At that time the Ebro fluvial network was established as we know it today, and the sediments started to erode producing the characteristic relief of this territory. Such relief is typical of a sedimentary basin, having the following structural elements: platforms, glacis, terraces and alluvial plains. The former are flat hilltops called *muelas*, which are monadnocks or non-eroded remains of the top of the original sedimentary formation. These high platforms dominate the landscape in the central portion of the Depression, causing quite an abrupt relief (Fig. 12.2). Some examples are the Sierra de Alcubierre (822 m), Montes de Castejón (744), El Castellar (742), La Muela de Borja (700), La Plana de la Negra (646) and La Muela (627). Conservation of those summits is due to their having upper calcareous hard strata which prevent them from being easily eroded.

The glacis are the result of the accumulation of materials dug from the *muelas* and deposited on their slopes. These glacis were formed by an extensive network of small valleys (*vales*) that resulted from the erosion by temporary streams in a spasmodic regime, characteristic of a highly seasonal climate.

The alluvial plains are often concealed by terraces and correspond to different morphoclimatic events of the Pleistocene.

The fluvial terraces occupy the lower stretches of the valleys, relative to the main rivers which cross the Depression: the Ebro and its tributaries Ega, Arga, Aragón, Gállego, Cinca and Segre on the left, and Oja, Najerilla, Iregua, Alhama, Jalón, Huerva, Martín and Guadalope on the right.

Such morphology is produced under conditions of low vegetation cover, which favours the laminar flux of the runoff water, and is probably related to a severe climate, i.e. low rainfall regime (arid or semiarid) and low temperatures. Such conditions are consistent with those that are supposed to have reigned in southern Europe during the Pleistocene ice-ages.

For a long time it was accepted that several territories in central Iberia have a steppic character, an idea that was launched by Willkomm (1852, 1896) and later sustained by the German plant geographic tradition (Jäger 1971). It was speculated that some plants, such as *Krascheninnikovia ceratoides* and others, reached the Iberian Peninsula from Central Asia during the Messinian crisis in the upper Miocene. This idea was progressively discarded as detailed studies were made during the twentieth century (Huguet del Villar 1925; Terradas 1986) and today it is generally accepted that the Ebro Basin is a Mediterranean territory in terms of its current climate, flora and vegetation, although it has some features of the central Asian steppes, with which it was probably connected in some periods during the Tertiary. Looking at the current situation, we can say that the conditions of the Ebro Basin can be qualified as totally Mediterranean. This does not exclude the possibility that during the cold and dry episodes of the Pleistocene some regions in central Iberia could have been covered by a wooded steppe or even by a treeless steppe. Summarizing, the current landscape of the Ebro Basin, with abundant gypsiferous substrates, seems to have its origin in a steppic period under an arid and cold Mediterranean climate. Later, after several fluctuations, the climate has evolved to the current state, also dry and quite continental, but with not such cold winters. If we accept this interpretation, we can compare some vegetation patterns existing today, such as the hard grasslands of the Stipo parviflorae-Lygeetum sparti in mosaic with juniper woodland of the central basin, with that in other steppic areas in Turkestan and other Mediterranean territories.

12.1.1 Climatic Features

As pointed out above, the Ebro Basin is completely immersed in the Mediterranean climatic area of southern Europe, with a dry summer. More precisely, we can establish that the whole basin is in the meso-mediterranean thermotype and the ombrotypes vary between subhumid-dry in the periphery and semi-arid in the core. In some areas of the central districts, the steppic variant can be recognized, which probably was much more widespread in the cold phases of the Pleistocene. Within this, we can highlight some features which can distinguish this area from the neighbouring ones and make the Ebro Depression a climatically well characterized territory.

Rain Shadow from the Surrounding Mountain Ranges This is a common feature of all valley bottoms in the world, and the Ebro valley is not an exception; it is caused by the Pyrenees, the Iberian System and the Catalan Coastal Ranges. All of them, but particularly the Pyrenees, are responsible for a huge rain shadow effect which is at a maximum in the central core of the depression. This is the main reason for the enhanced drought of this central sector (Bardenas-Monegros) which also experiences low levels of relative humidity and a high continentality, i. e. a large temperature oscillation between summer and winter (cold winters and torrid summers).

Thermal Inversion Another feature of the climate of the Ebro Basin is the frequency and intensity of episodes of thermal inversion, which affect to a large extent the lowlands of the central basin. It produces a cooling effect in these areas, increasing their climatic continentality, thus explaining the occurrence there of species typical of higher altitudes, such as *Juniperus thurifera*, and the absence of some more thermophilous ones, such as *Quercus rotundifolia*.

Wind Finally, another unpleasant feature of the climate in this region is the frequency and intensity of the northwestern wind, called *cierzo*, which blows frequently and is enhanced by the relief and the lack of barriers in the landscape. When this wind blows hard, it is one of the most uncomfortable events happening during winter.

Following human and geomorphological criteria, the Ebro basin can be divided into several districts as shown in Fig. 12.3.

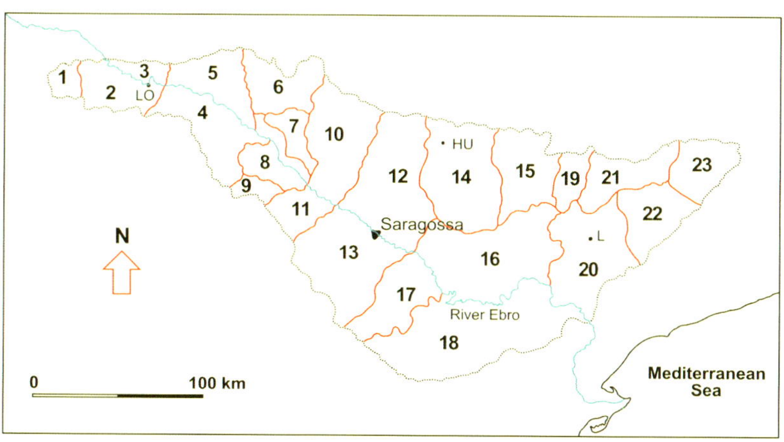

Fig. 12.3 Geographical districts of the Ebro Basin. (*1*) La Riojilla; (*2*) Rioja Alta; (*3*) Sonsierra; (*4*) Rioja Baja; (*5*) Ribera Estellesa; (*6*) Navarra Media Oriental; (*7*) Las Bardenas; (*8*) Ribera Tudelana; (*9*) Tarazona; (*10*) Cinco Villas; (*11*) Campo de Borja; (*12*) Zaragoza; (*13*) Campo de Cariñena; (*14*) Hoya de Huesca; (*15*) Somontano; (*16*) Los Monegros; (*17*) Campo de Belchite; (*18*) Bajo Aragón; (*19*) Litera; (*20*) Segriá; (*21*) Noguera; (*22*) Pla d'Urgell; (*23*) Segarra. *LO* Logroño, *HU* Huesca, *L* Lleida

12.2 Forests and Woodlands

The Ebro Basin is an area subjected to severe summer drought and continentality, conditions that determine the dominance of evergreen sclerophyllous forests and shrublands, which share the territory with semideciduous marcescent forests in the most humid areas and with juniper coniferous forests in the areas with more extreme temperatures. Furthermore, due fundamentally to the human factor, the landscape offers a semblance in which forests and woodlands are rare, almost absent, maximizing the appearance of aridity and extreme conditions of the area. Forests are supposed to have been cut down by humans in historic times (Terradas 1986) and the current state is supposedly mostly the result of human action. Moreover, not only the ancient history of human settlement but also the slowness of the regeneration capacity of the forest ecosystems of the area share responsibility for the current state, showing a picture of a devastated treeless country. However, most of the area of this territory is considered to have a potential for forest, woodland or tall shrub vegetation. Concerning this, some remnants of natural populations of native trees and certain patches which have survived till now, permit us to reconstruct a situation with the forests occupying their potential area and depict its diversity, structure and composition. We will classify the different types of forests which can be found in the Ebro Basin by their dominant trees.

12.2.1 Quercus rotundifolia *(Holm Oak) Forests and Woodlands*

They occupy the peripheral area of the basin, where rainfall is above 350–400 mm and conditions permit growing holm oak forests. The general association for the majority of holm oak forests in the Ebro Basin is the Quercetum rotundifoliae, which was first described by Braun-Blanquet and Bolòs (1958). The remnants of this forest type are irregularly distributed in the territory being mostly represented by isolated trees, but there are some examples in the Sierra de Alcubierre, the Montes de Castejón and in La Plana de la Negra (Bardenas Reales). All the stands and individual trees present mutilations from being severely pruned for firewood, as a signal of the strong human pressure upon this vegetation. The surviving stands are young and of low height, being now favoured by the replacement of firewood by fossil fuels in domestic uses. The diversity of vascular plants in these forests is quite low due to the severe climate. Together with *Quercus rotundifolia*, often accompanied by *Pinus halepensis*, there are some tall shrubs such as *Lonicera implexa, Quercus coccifera, Rhamnus alaternus, Prunus spinosa* and *Juniperus oxycedrus*. Chamaephyte communities are favoured by the high light conditions created by the gaps in the canopy of the taller vegetation. They are formed by the surrounding scrub species such as: *Centaurea linifolia, Thymus vulgaris, Genista scorpius, Coronilla minima, Bupleurum rigidum, Rosa pimpinellifolia*, etc. The lianoid habit is only represented by *Rubia peregrina* and the herb layer is formed by *Teucrium chamaedrys*, *Brachypodium retusum* and *Carex hallerana*. There is a colder variant on the northern slopes of the higher elevations with *Arctostaphylos uva-ursi* subsp. *crassifolia*. The main sources of data are those of Braun-Blanquet and Bolòs (1958), Bolòs (1961, 1996), Molero et al. (1998) and Ursúa (1986). In the northern part of the depression, in the transition to the Pyrenees, those forests show an important increase in more moisture demanding plants, such as *Buxus sempervirens*, *Brachypodium phoenicoides*, *Emerus major*, *Helleborus foetidus* and even *Quercus pubescens*, *Quercus faginea*, *Pinus nigra* subsp. *salzmanii*, and *Lavandula angustifolia* subsp. *pyrenaica*. This floristic change, indicative of the rainfall gradient towards the foothills of the Pyrenees, permits the recognition of the association Buxo sempervirentis-Quercetum rotundifoliae.

12.2.2 **Quercus faginea *Forests***

This marcescent or semideciduous oak is endemic to the Iberian Peninsula, where it occupies submediterranean areas, usually on marly base-rich substrates. It is thus distributed in the central north-eastern quadrant of Iberia and, within the Ebro Basin, it occurs only on its northern fringe, in the transition to the Pyrenean foothills contacting the *Quercus pubescens* dominated forests. There it occupies a strip between *Quercus rotundifolia* forests (Buxo-Quercetum rotundifoliae) and the *Quercus pubescens* groves of the Pyrenean lowlands (Buxo-Quercetum pubescentis). In such a situation, hybridization between both oaks, i.e. *Quercus x allorgeana* (*Quercus subpyrenaica*) is frequent. The association for those forests is the Violo willkommi-Quercetum fagineae, and it is represented by scattered patches in the northern sector of the territory, always constituted by young stands of low height and high stems density. As usual in submediterranean semideciduous forests, the abundance of light inside the forest, combined with the richness of nutrients, produces a dense cover in the understory vegetation, particularly in the shrub layer, which is formed by many spiny species, making it practically impassable. The tree layer is formed mostly by *Quercus faginea* and hybrids, but often individuals of *Sorbus domestica*, *Acer campstre*, *Acer monspessulanum* and *Quercus rotundifolia* share the canopy. The shrub layer is species-rich and dense, and formed by typical species of European submediterranean forests, such as *Amelanchier ovalis*, *Viburnum lantana*, *Buxus sempervirens*, *Cornus sanguinea*, *Crataegus monogyna*, *Emerus major*, *Ligustrum vulgare*, *Lonicera etrusca*, *Juniperus communis*, *Prunus mahaleb*, *Prunus spinosa*, *Rhamnus saxatilis*, etc. The herb layer is formed by *Brachypodium phoenicoides*, *Helleborus foetidus*, *Paeonia officinalis* subsp. *microcarpa*, *Primula veris* subsp. *columnae*, *Tanacetum corymbosum*, *Viola willkommii*, and others. The Mediterranean influence is conspicuous and represented by *Carex hallerana*, *Juniperus oxycedrus*, *Quercus coccifera* and *Teucrium chamaedrys*. Information about this forest type is available in Rivas-Martínez et al. (1991).

12.2.3 **Juniperus thurifera** *Woodlands*

The natural vegetation in the central section of the Ebro Basin, in the district called *Los Monegros*, is a juniper woodland. The main species are *Juniperus thurifera* and *Juniperus phoenicea;* they form an open woodland of steppic appearance, similar in physiognomy to those of southwestern North America (Utah, Nevada and Arizona) or some north African and western and central Asian areas (Morocco, Cyrenaica, Anatolia, Transcaucasia, southern Turkestan and the upper Indus Valley). This vegetation is an open woodland of *Juniperus thurifera*, a juniper tree of medium height, distributed in several western Mediterranean countries (western Alps, French Massif Central, Spain, and Moroccan Grand Atlas) where it is found in areas under continental climatic conditions. Its distribution in central Iberia is concentrated in the central-eastern midlands, usually at altitudes above 800 m, but in the Ebro Basin it is found at significantly lower altitudes in the central core area of the basin, where continentality is highest. The existence of this woodland in this area is attributed to the high frequency of thermal inversion episodes, since *Juniperus thurifera* commonly occurs at substantially higher elevations (Terradas 1986). It also prefers deep soils, avoiding the drier shallow rocky soils on the steep slopes and rims. The association described for this vegetation is the Juniperetum phoeniceo-thuriferae. It has a low number of forest plants, such as *Asparagus acutifolius, Ephedra nebrodensis, Juniperus phoenicea, Juniperus thurifera, Pinus halepensis, Quercus coccifera, Rhamnus alaternus, Rhamnus lycioides*, etc. In the broad gaps which remain open between the trees, the corresponding scrub develops, almost always dominated by *Genista scorpius, Rosmarinus officinalis* and *Thymus vulgaris*. There is not one stand in which the forest closes its canopy and casts a continuous shade on the understory. The particular growth and reproductive strategy of junipers and the extreme conditions of drought and continentality combined with human action are responsible for the few remaining examples of this forest type presenting the appearance of an open wooded scrub more than of a true forest. Moreover, most of the original juniper forests have disappeared and only a few examples remain: Bujaraloz-Retuerta de Pina, Castejón de Monegros, Pallaruelo, Monegrillo (Fig. 12.4), Leciñena and few others. Those remnants are found in the district known as Los Monegros, a name clearly indicating that the area was densely covered by a dark tall vegetation (*monte negro*), very probably corresponding largely to this juniper woodland (Fig. 12.4). Information about this woodland is available in Braun-Blanquet and Bolòs (1958).

12.2.4 *Tall* **Quercus coccifera** *Shrubland (Macchia) and* **Pinus halepensis** *Woodlands*

This ligneous sclerophyllous shrubby vegetation is extensively represented in the Ebro Basin, where it can be found scattered on the hilltops and slopes where

Fig. 12.4 A fragment of the Juniper woodland with *Juniperus thurifera* near Monegrillo

agriculture is not possible. It forms a mosaic with scrub of *Rosmarinus officinalis*, *Genista scorpius*, *Dorycnium pentaphyllum* and *Thymus vulgaris* and usually shows a dominance of *Quercus coccifera*, which is almost always accompanied by *Rhamnus lycioides*, *Rhamnus alaternus*, *Juniperus oxycedrus*, *Juniperus phoenicea* and *Pinus halepensis*. The physiognomy of this vegetation is that of a tall and dense shrubland in which pines are often leaning out, producing the aspect of a wooded shrubland or even a true pine forest. *Rubia peregrina* is a prevailing/predominant vine which climbs across the dense shrubs, and the herb layer is dominated by *Brachypodium retusum*. The association encompassing those stands is the Rhamno lycioidis-Quercetum cocciferae, distributed in the base-rich central territories of the Iberian Peninsula. It constitutes the potential natural vegetation of the central part of the Ebro Basin (Bardenas and Monegros sector). In this area, continentality and summer drought are more intense and this shrubland occupies an altitudinal range between 400 and 700 m. Substrates are mostly of marl, limestone and gypsum, in any case basophilous and lime-rich. In areas where temperatures are higher in winter –southern slopes, gorges, low altitude areas, etc.– this shrubland is enriched with *Pistacia lentiscus*, a thermophilous species indicative of a lower incidence of frosts and low temperatures. This variant is the dominant one in the low altitude area surrounding Caspe. In contrast, at higher elevations (between 550 and 700 m) *Carex humilis* indicates a colder and moister variant. Examples of this vegetation are abundant all over the Ebro Basin, partially covering the hills which were traditionally used for grazing. In recent times, rural abandonment and the decrease of grazing pressure have caused a recovery of this vegetation all over the territory. Information about this woodland is available in Braun-Blanquet and Bolòs (1958). Along the northern fringe of the Ebro Basin, in the transition to the foothills of the

Pyrenees, *Quercus coccifera* shrublands are enriched with boxwood, *Buxus esmpervirens*, enabling one to distinguish two communities: the first one with *Phillyrea angustifolia, Lonicera implexa* and *Pistacia therebinthus*, plants not frequent in the central districts of the basin, which form the Quercetum cocciferae buxetosum sempervirentis, related to the *Quercus rotundifolia* forests; and the second one, more humid and expanding northwards to the pre-Pyrenees, the Ononido fruticosae-Buxetum sempervirentis, with *Coriaria myrtifolia, Hepatica nobilis, Ononis fruticosa* and *Thalictrum tuberosum*, which is related to the *Quercus faginea* forests of the area. Further information about these *Buxus sempervirens*-rich shrublands is available in Bolòs (1961) and in Braun-Blanquet and Bolòs (1958).

12.2.5 Rocky Woodland of Boxwood and Juniper

This type of woodland is formed by *Juniperus phoenicea* and *Buxus sempervirens*, together with other shrubs such as *Rhamnus alaternus, Amelanchier ovalis, Asparagus acutifolius, Helichrysum stoechas*, etc. It occupies the calcareous rocky rims, boulders and outcrops of the narrow valleys surrounding the basin, in steep rocky slopes almost without soil. The association for these woodlands is the Buxo sempervirentis-Juniperetum phoeniceae, common in all the ragged districts of the eastern Iberian Peninsula and southern France.

12.3 Scrub

Ligneous vegetation dominated by chamaephytes and other dwarf woody plants, often spiny and/or aromatic, with small leaves, hairs and glands, constitute one of the most characteristic features of the landscape in Mediterranean countries. This vegetation, called *matorral* in Spanish, contains an important proportion of the indicator and endemic taxa of each of the provinces of that region. This abundance of stenochorous plants is probably due to the origin of this flora and vegetation in the late Tertiary as an adaptation to the increasing aridity of the climate. This process took place in the absence of human disturbance when landscapes were covered mostly by mature vegetation and degraded secondary habitats were isolated, occupying small patches scattered like islands, thus enhancing the speciation of its flora. When humans started to destroy mature vegetation during the Neolithic, particularly forests and tall shrublands, the situation reversed and the scrub expanded as a result of human land use, particularly animal husbandry. This has resulted in the current situation, in which these *matorrales* have become the vegetation type covering the majority of the area not used for agriculture. They contain numerous taxa, among which there are many endemics. In the Ebro Basin substrates are almost exclusively basophilous and thus the scrub is clearly responding to that condition. However, the mineralogical composition of the bedrock, specifically the proportion of calcium sulphate (gypsum), makes it possible to distinguish two main groups in the Ebro Basin scrub.

12.3.1 Scrub on Limestone and Marl

We can differentiate several geographically distributed associations (Fig. 12.5), sorted by temperature and moisture gradients. The broad territories of the central districts of the Ebro Basin are dominated by the most extended association: the Rosmarino officinalis-Linetum suffruticosi. This vegetation type contains a combination of thermophilous plants of coastal distribution, such as *Helianthemum marifolium*, *Linum suffruticosum* and *Thymelaea tinctoria*, with some regional endemics, such as *Centaurea linifolia*, and a large group of species of wide distribution, such as *Bupleurum fruticescens*, *Genista scorpius*, *Fumana ericifolia*, *Fumana thymifolia*, *Helianthemum violaceum*, *Koeleria vallesiana*, *Rosmarinus officinalis*, *Teucrium capitatum*, *Teucrium gnaphalodes*, *Thymus vulgaris* and many others. The alfa diversity of vascular plants is high, with the average number of species per relevé being 25. It is a scrub of low density, usually covering around 30 to 40% of the total area, and a height of 20–60 cm. The abundant aromatic species confer a typically scented character to this vegetation. Information about this association is abundant: Braun-Blanquet and Bolòs 1958, Loidi et al. 1997, Molina et al. (1993).

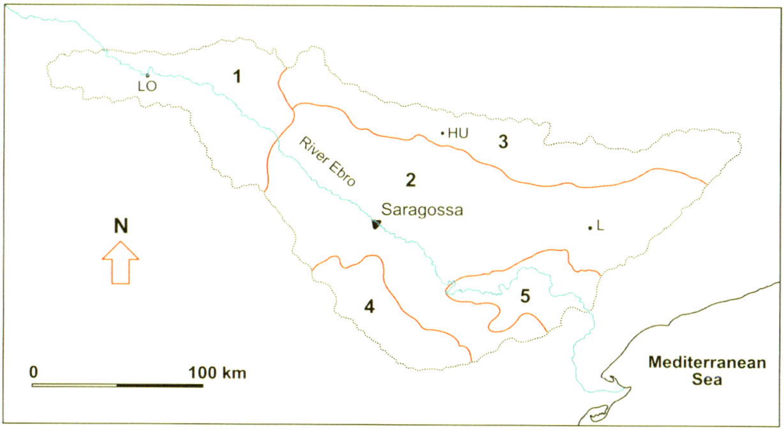

Fig. 12.5 Areas of the five main limestone scrub associations of the Ebro Basin: (*1*) Salvio lavandulifoliae-Ononidetum fruticosae; (*2*) Rosmarino officinalis-Linetum suffruticosi; (*3*) Teucrio aragonensis-Thymetum fontqueri; (*4*) Sideritido spinulosae-Lavanduletum latifoliae; (*5*) Cytiso fontanesii-Cistetum clusii

In addition to the above mentioned association, a number of associations occupy the transition to the outer territories in the periphery of the basin (Fig. 12.5):

1. In the eastern part of the depression, where altitudes are lower and some thermophilous plants enter into the depression from the coastal Mediterranean area following the Ebro river up, the dominant association is the Cytiso fontanesii-Cistetum clusii, with *Cistus clusii, Coronilla minima* subsp. *lotoides, Cytisus fontanesii, Fumana hispidula* and *Globularia alypum*. Further information in Bolòs (1973), Braun-Blanquet and Bolòs (1958), Molina et al. (1993)
2. The transition to the Iberian central high plateau (meseta) on the southern fringe of the basin is occupied by the association Sideritido spinulosae-Lavanduletum latifoliae, which is characterized by plants of the central highlands of the Iberian Peninsula, such as *Helianthemum cinereum* subsp. *rotundifolium, Lavandula latifolia, Salvia lavandulifolia* and *Teucrium expassum*, mixed with plants of the central basin lowlands. Information in Molina et al. (1993).
3. The transition to the northern pre-Pyrenean fringe is occupied by the association Teucrio aragonensis-Thymetum fontqueri, with *Teucrium aragonense, Thymus fontqueri, Carex humilis, Convolvulus cantabrica, Genista hispanica, Helianthemum oelandicum* subsp. *italicum, Satureja montana* and others which indicate cooler and more humid conditions. Information in Bolòs 1961, Loidi et al. (1997), Molero and Vigo (1981).
4. In the northwestern corner of the basin, the Rioja region, the association is the Salvio lavandulifoliae-Ononidetum fruticosae with *Ononis fruticosa, Salvia lavandulifolia* and *Helianthemum cinereum* subsp. *rotundifolium* which replaces *Helianthemum marifolium*, common in the central part of the basin. Other diagnostic species are *Euphorbia minuta, Thymus loscosii* and *Helianthemum apenninum* subsp. *stoechadifolium*. Information available in Fernández-González et al. (1986) and in Loidi et al. (1997)

12.3.2 Scrub on Gypsum

Gypsum substrates are frequent in several of the sedimentary basins of the Iberian Peninsula and very common in the Ebro Basin (Fig. 12.6). In this territory most of them are evaporitic rocks formed as a result of lacustrine sedimentation during the Miocene, when the Ebro Basin was an interior lake without connection to the Mediterranean Sea. They are easily recognized by the white flat strata appearing in the hills. There are also others group of gypsum outcrops of earlier origin formed in the continental Tertiary -Oligocene and early Miocene- which show a folded stratification and are abundant in the Ribera Estellesa and Rioja Baja districts. Such abundance of gypsum outcrops confers the landscape of many areas of the Ebro Basin a whitish colour, enhancing its steppic semblance and the impression of aridity to the observer. The soils found on such bedrocks are gypsisols, which under

Fig. 12.6 Gypsum scrub near Nuez de Ebro

Mediterranean climatic conditions (severe summer drought) determine the existence of a very particular flora, specialized in inhabiting those gypsisols and often confined to them; these plants are called gypsophytes and almost all of them are Iberian endemics. This flora is distributed in other gypsum areas in the Iberian Peninsula too, and this results in a particular type of scrub in which generalist species mix with a number of gypsophytes. This is a special case of edaphic habitat, particularly widespread in the Iberian Peninsula, which constitutes one of its most valuable and original vegetation types in the European context (Fig. 12.6).

The main species found in that scrub are the gypsophytes: *Boleum asperum*, *Gypsophila struthium* subsp. *hispanica*, *Helianthemum squamatum*, *Herniaria fruticosa*, *Launaea fragilis*, *Launaea pumila*, *Lepidium subulatum*, *Ononis tridentata* and *Reseda stricta*. They combine with a group of basophilous scrub plants of wider ecological scope such as *Genista scorpius*, *Linum suffruticosum*, *Rosmarinus officinalis*, *Thymus vulgaris*, *Helianthemum syriacum*, *Fumana ericifolia*, etc. The annual community inhabiting the gaps in between the scrub has two other gypsophytes: *Chaenorhinum reyesii* and *Campanula fastigiata*. The association grouping all the Ebro Basin gypsophilous scrub is the Helianthemo thibaudii-Gypsophiletum hispanicae, which has several variants depending on ecological conditions or territory (Loidi and Fernández-González 1994). There is abundant information about these communities in Báscones et al. (2011), Bolòs (1996), Braun-Blanquet and Bolòs (1958), Font (2011), Guardia and Ninot (2000), Loidi and Fernández-González (1994), Loidi et al. (1997), Molina et al. (1993), Rivas Goday et al. (1956), Villar et al. (2011).

A schematic view of the relationships between climatophilous forest types and scrub types sorted out according to a moisture gradient is shown in Fig. 12.7.

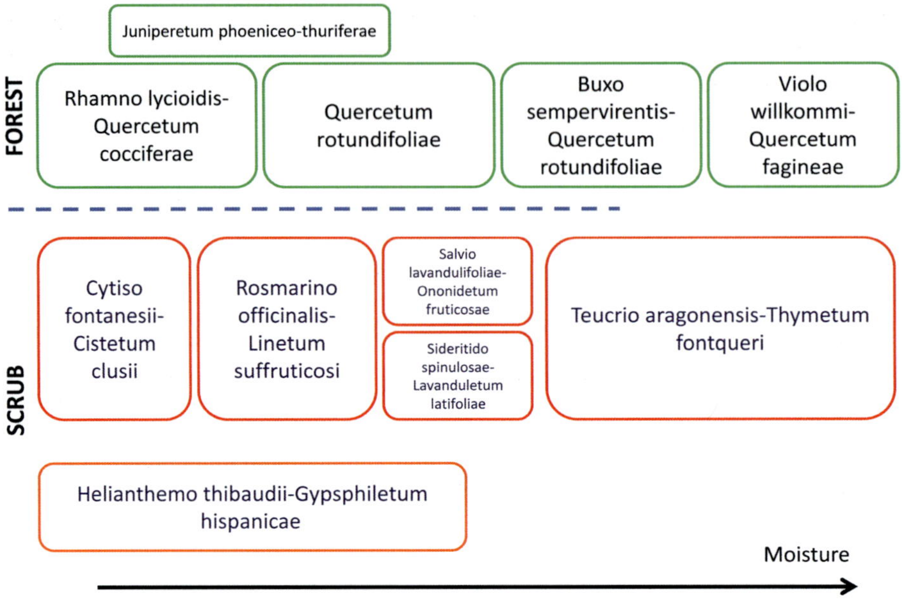

Fig. 12.7 Scheme of the forest and scrub associations of the Ebro Depression in relation with the moisture gradient

12.4 Dry Grasslands

Herbaceous vegetation in Mediterranean countries presents adaptations to endure the severe summer drought. In the case of the Ebro Basin, summer aridity favors the accumulation of soluble ions in the upper layers of the soil, affecting sometimes the osmotic potential of the soil solution. Under such conditions, herbs, and particularly grasses, develop hard tissues in their leaves and stems, mostly sclerenchyma bundles, hairs and silicified epidermal cells, which prevent water loss and predation by herbivores. Another adaptation of herbs to a dry season is becoming annual and spending the unfavourable season as seeds. Both types of strategies are found in the arid hills of the Ebro Basin, contrasting with the tender evergreen grasslands of the humid alluvial plains which can benefit from the shallow water table. We will focus on the dry grasslands as the humid grasslands will be treated as part of the riparian system.

12.4.1 Non-steppic Dry Grasslands

In the peripheral areas of the Ebro Basin, under conditions of lower aridity and continentality, these dry grasslands are widespread on potential areas of *Quercus rotundifolia* and *Quercus faginea*. The main association in the mentioned areas is the Ruto angustifoliae-Brachypodietum retusi, in which *Brachypodium retusum* is the dominant species. They form dense communities in areas where the soil is rich in clay, usually in foothills, often in combination with scrub patches. *Avenula bromoides, Ruta angustifolia, Phlomis lychnitis, Convolvulus lineatus, Plantago albicans* etc., are the most frequent plants, together with the inevitable high number of annuals visible only in spring, which intermingle in this perennial grassland occupying its gaps. In areas with higher moisture due to more rainfall or to edaphic accumulation, where the potential natural vegetation corresponds to *Quercus faginea* forest or to *Populus alba* in alluvial foothill positions, the dry grassland association is the Elytrigio campestris-Brachypodietum phoenicoidis, with *Brachypodium phoenicoides* and *Elymus pungens* subsp. *campestris* as the main species. Other perennial herbs are *Festuca capillifolia, Hypericum perforatum, Medicago sativa, Festuca interrupta, Poa pratensis, Lotus corniculatus, Galium verum, Trifolium campestre, Daucus carota, Dactylis glomerata* subsp. *hispanica*, etc.

12.4.2 Steppic Grasslands in the Central Basin

Perennial grasslands dominated by *Lygeum spartum, Elymus pungens* subsp. *campestris* and several *Stipa* species, all of them with strong root systems, are distributed in the central territories of the Ebro Basin, mostly in the Bardenas-Monegros sector, where aridity and continental conditions are extreme. These grass species, due to the relatively large size of their aerial parts as well as their rhizomatous systems, produce a high amount of organic matter. Thus, this vegetation enhances the development of the soils, producing fertile horizons which have traditionally been used for agriculture. They settle on gentle slopes or even in small depressions where soils can easily develop and fine materials accumulate (Fig. 12.8). The physiognomy given by such grasses, particularly *Stipa*, has induced these grasslands to be considered as steppic, due to its similitude with the central and eastern European steppic grasslands. These grasslands are locally called *albardinares*, as the most common grass, *Lygeum spartum*, is called *albardín* in Spanish, a name of Arabic origin. This grass was commonly used in ancient times to make ropes, shoes (espadrilles) and baskets because of its strong fibers. Other common species found in these grasslands are *Brachypodium ramosum, Dactylis glomerata* subsp. *hispanica, Koeleria vallesiana* and *Sedum sediforme*. Very often, these grasses do not cover the entire surface but tufts of them are spaced out leaving gaps in which many annuals get established. Such annual communities form a mosaic with the tufts of the steppic grassland, in a characteristic composition of both entities, already highlighted by Braun-Blanquet and Bolòs (1958). There are two associations in the Ebro Basin: firstly, on calcareous soils, with no salinity and flat terrain or on very gentle slopes the Stipo parviflorae-Lygeetum sparti, with dominance of *Stipa parviflora* and *Lygeum spartum* and sometimes with *Stipa*

Fig. 12.8 "Albardinar" or *Lygeum spartum* grassland in a small depression between the hills in the central Ebro Basin

lagascae. Secondly, on gypsiferous soils, where a certain salinity is also recorded, and on slightly steeper slopes, there is the association Agropyro pectinati-Lygeetum sparti, with *Agropyron cristatum* subsp. *pectinatum*, and dominated by *Lygeum spartum*. At the foot of the gypsiferous hills, the soils often become saline due to the accumulation of soluble ions in their upper layers. This circumstance is indicated by the presence of the succulent *Aizoon hispanicum*. These steppic grasslands are found in territories where juniper woodland and *Quercus coccifera* shrubland are the potential natural vegetation. Information about these communities can be found in Berastegi (2013), Braun-Blanquet and Bolòs (1958) and Ursúa (1986).

12.4.3 Annual Grasslands

The Mediterranean Region is well known for having a high diversity in annual plants. Annuality can be considered as one of the adaptations to the severe summer drought in which plants avoid the unfavorable season by spending it in the state of seed. The plants germinate in autumn with the first rains and stay in their seedling stage during the whole winter. In early spring they start to grow rapidly and they flower and fruit in a very short time in a manner that by late spring almost all have already sown their numerous and little seeds and are wilting before the first strong heat waves of the summer arrive. This strategy is performed by a large number of species belonging to a variety of families but that have some traits in common: small size, fast growth and explosive flowering and seed production. These annual grasslands are usually very rich in species number and there is a long list of plants which are found in them:

Alyssum granatense, Alyssum simplex, Anagallis arvensis, Arenaria serpyllifolia, Asterolinon linum-stellatum, Astragalus stella, Atractylis cancellata, Bombycilaena erecta, Brachypodium dystachyon, Bupleurum baldense, Bupleurum semicompositum, Campanula erinus, Cerástium pumilum, Clypeola jonthlaspi, Crucianella angustifolia, Desmazeria rigida, Echinaria capitata, Euphorbia exigua, Euphorbia falcata, Euphorbia sulcata, Filago gallica, Filago pyramidata, Galium parisiense, Hedypnois rhagadioloides, Helianthemum ledifolium, Helianthemum salicifolium, Hippocrepis multiliquosa, Holosteum umbellatum, Hornungia petraea, Hymenolobus procumbens, Linum strictum, Lomelosia simplex, Malva aegyptia, Malva trifida, Medicago minima, Medicago orbicularis, Neatostema apulum, Paronychia capitata, Plantago afra, Platycapnos spicata, Polygala monspeliaca, Senecio gallicus, Sherardia arvensis, Sideritis montana, Trifolium scabrum, Trigonella monspeliaca, Vulpia ciliata, Vulpia myuros, Wangenheimia lima, Xeranthemum inapertum, Ziziphora hispanica, etc.

There is a broad association gathering most of these communities in the Mediterranean Iberian Peninsula which is well represented in the Ebro Basin on limestone and marl substrates: the Saxifrago tridactylitae-Hornungietum petraeae with *Saxifraga tridactylites* as indicator species. On gypsum substrates there is another association: the Chaenorhino reyesii-Campanuletum fastigiatae with the annual gypsophytes *Campanula fastigiata*, *Chaenorhinum exile* and *Chaenorhinum reyesii*. Information about these grasslands is available in: Báscones and Ursúa (1986), Berastegi (2013), Loidi et al. (1997).

12.4.4 Grasslands Subjected to Heavy Grazing

Some areas, such as paths, livestock (mainly sheep) trails and places where animals graze intensively, are covered by special grasslands dominated by *Poa bulbosa* with some other species adapted to grazing, such as *Astragalus sesameus*, *Convolvulus lineatus* or *Plantago albicans*, and many other therophytes. These grasslands are adapted to intense grazing during the humid season, i.e. from autumn to spring, and wilt (*agostan*) in summer. For that reason they cannot be grazed in summer and the sheep have to find other ways of nourishment, i.e. wet grasslands nearby or move on to find grasslands which are green in summer, usually in mountain ranges. This causes a regular seasonal migration called transhumance, which in this case is due to seasonal drought (xerofugous), in contrast with the transhumance done by flocks in other regions where migration is prompted by seasonal cold, in a summer-winter pattern between highlands and lowlands (cryofugous). These grasslands are found in areas where traditional management of sheep has been performed for centuries, particularly along the tracks used by the animals. Some of the paths have been used since the track network system of the transhumance between the Pyrenees and the lowlands of the Ebro Basin, within the kingdoms of Aragón and Navarra, was established in the Middle Ages. In this territory there is the basophilous association Astragalo sesamei-Poetum bulbosae. Information about this grassland type is available in Berastegi (2013).

12.5 Nitrophilous Vegetation in Dry Environments

The Ebro Basin has been heavily populated since ancient times and it has been subjected to intense land use: agriculture and husbandry in different forms. This has produced a high variety of environmental situations in terms of deposition of nitrogen and phosphorus compounds, soil humidity, regularity and intensity of disturbance, etc. not only in cultivated and grazed areas but also in urban and peri-urban areas. Those different biotopes are occupied by a diverse set of plant communities, commonly called nitrophilous, because all of them require a high amount of nitrogen compounds, mostly resulting from the decomposition of proteins from accumulated organic matter. These environments are also known as hemerobic, as they are strongly linked to human activity. Other terms used to refer to them are *ruderal* (from *ruderis* = rubble), indicating that they are near human buildings, or *viarious* (from *via* = road) as many of them are found at the edges of roads and paths. Summarizing, three factors condition this vegetation and its diversity: abundance of nitrogenous and phosphate compounds of organic origin, intensity and type of disturbance (soil digging or trampling) and moisture of the soil (Fig. 12.9).

12.5.1 Low Soil-Disturbance, Shrubby Nitrophilous Vegetation

In semi-arid to arid territories, like the central districts of the Ebro Basin, nitrified sites which suffer low levels of disturbance can develop a shrubby community in which ligneous *Chenopodioideae* are particularly frequent. This also occurs in all the Mediterranean, Irano-Turanian and Saharo-Arabian regions, and it seems to be caused by the long permanence of soluble nitrates and phosphates which are slowly leached by the low rainfall or even tend to accumulate in the soil. The most widespread association is the Salsolo vermiculatae-Peganetum harmalae, with *Artemisia campestris* subsp. *glutinosa, Camphorosma monspeliaca, Marrubium alysson, Peganum harmala, Salsola vermiculata* (*sisallo*), and other therophytes such as O*nopordum nervosum, Bromus rubens* and many others. Those *sisallares* are very abundant in the lower parts of the depression, dominating field edges, road and track borders and any marginal land left after human exploitation. As a result of the greyish colour of the *silsallo*, this vegetation is responsible for a significant increase in the arid look of the cultural landscape in the territory. In areas where soil

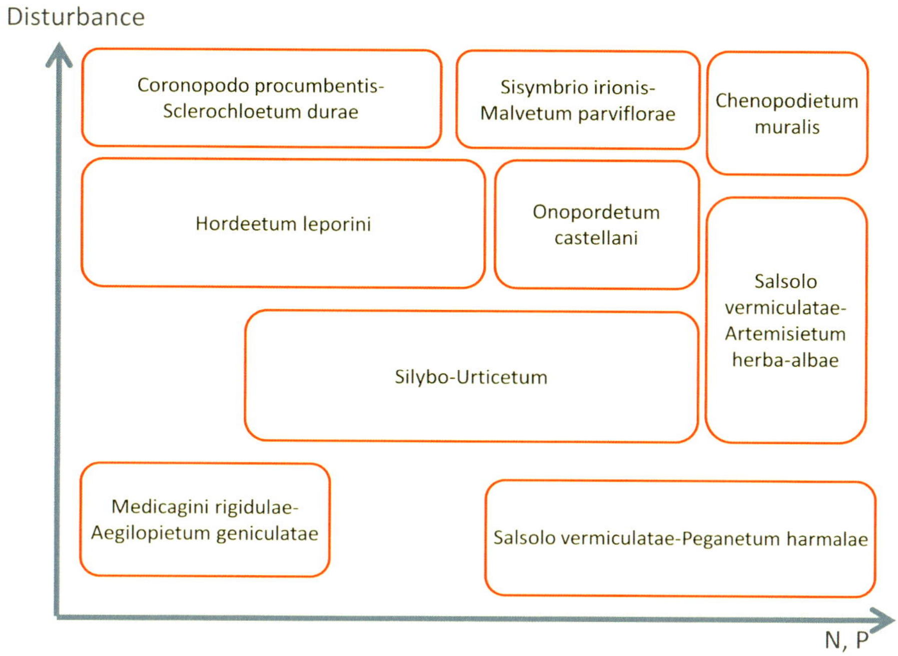

Fig. 12.9 Scheme of the dry nitrophilous associations of the Ebro Depression in relation to disturbance (soil digging, trampling, etc.) and abundance of N and P

humidity is higher, and often also salinity, the species *Atriplex halimus* appears dominating the communities and gives way to the association Salsolo vermiculatae-Atriplicetum halimi. Under certain grazing pressure, *Artemisia herba-alba* dominates and these communities can cover wide areas in abandoned fields. They are gathered under the association Salsolo vermiculatae-Artemisietum herba-albae.

12.5.2 Herbaceous Ruderal and Viarious Vegetation

When disturbance, i. e. soil digging, removal of biomass, trampling, etc. is heavy, the response of the nitrophilous vegetation is to build herbaceous communities. Depending on the disturbance regime and intensity, amount of nitrogen-rich matter added, and other soil traits constraints, different community types result. Under extremely high nitrification conditions, the association is the Chenopodietum muralis, an open community type which is unstable as it varies its floristic composition easily due to its dependence on the high input of ammonia compounds. It is formed by a high number of *Chenopodioideae*, such as *Chenopodium album, Chenopodium ambrosioides, Chenopodium murale, Chenopodium opulifolium, Chenopodium urbicum, Chenopodium vulvaria*, as well as *Amaranthus blitoides,*

Amaranthus deflexus, Amaranthus retroflexus, Sisymbrium irio, Urtica urens, Xanthium spinosum, etc. When the nitrification intensity decreases, many *Chenopodium* species disappear and the association Sisymbrio irionis-Malvetum parviflorae is established. It is dominated by colourful flowering species such as *Anacyclus clavatus, Bromus rubens, Descurainia sophia, Erodium cicutarium, Erodium malacoïdes, Eruca vesicaria, Lamium amplexicaule, Malva parviflora, Papaver hybridum, Papaver rhoeas, Roemeria hybrida, Sisymbrium irio, Sisymbrium orientale, Sisymbrium runcinatum, Urtica urens*, etc., which explode in colours in spring and decorate ephemerally the human-transformed areas, particularly the rural ones. If the nitrification is lower, a spiny vegetation of high biomass formed by annual and biannual thistles of the genera *Carduus, Cirsium* (*cardos*), *Carthamus, Centaurea*, and *Onopordum* (*tobas*) is established, where some specimens of the tallest species clearly stand out. The widespread association inhabiting plowed soils where organic matter, either of animal or of vegetal origin, such as straw from cereal harvesting, accumulates is the Onopordetum castellani and the most common species are *Carduus bourgeanus, Carduus pycnocephalus, Carduus tenuiflorus, Carthamus lanatus, Centaurea calcitrapa, Cirsium vulgare, Onopordum acanthium, Onopordum corymbosum, Onopordum nervosum* subsp. *castellanum, Silybum eburneum* and others. If soil humidity is a bit higher because the terrain is flat or a small depression, *Silybum marianum* dominates the thistle communities accompanied by *Arctium minus, Ballota nigra* and *Marrubium vulgare*. This is the association Silybo-Urticetum.

The viarious vegetation, i. e. that which is linked to roads, trails and pathways covering the moderately trampled areas, such as the central lane of lightly trampled paths or the edges of trails submitted to more intense traffic, are also part of the nitrophilous vegetation of any region. The areas which suffer the heaviest trampling intensity endurable by vascular plants are inhabited by the association Coronopodo procumbentis-Sclerochloetum durae, a vegetation of small prostrate therophytes covering a low proportion of the floor area. This is a community type with few species specialized in this highly disturbed habitat: *Coronopus didymus, Coronopus squamatus, Poa annua, Polygonum aviculare, Sagina apetala*, etc. Along the edges of roads, where trampling is lower, the vegetation increases dramatically in cover, biomass and height, being dominated by grasses (*Hordeum, Bromus*). These grasslands often form a continuous strip along the edges of the trails and fill, like a matrix, all the urban and peri-urban spaces in villages and towns. The association is the Hordeetum leporini and the main species are *Hordeum murinum* subsp. *leporinum, Avena barbata, Anacyclus clavatus, Bromus rubens, Bromus diandrus, Bromus hordeaceous, Bromus madritensis, Bromus tectorum, Capsella bursa-pastoris, Cardaria draba, Erodium ciconium, Erodium malacoides, Lolium rigidum, Malva sylvestris, Rostraria cristata, Sisymbrium irio*, etc. In areas of lower nitrification, like abandoned crop fields and edges in remote, rather scarcely walked tracks, edges of pastures and so on, the vegetation becomes dominated by *Aegilops* species, particularly *Aegilops geniculata* and *Aegilops triuncialis*, which are accompanied by *Convolvulus arvensis, Medicago minima, Medicago rigidula,*

Medicago sativa, *Trifolium angustifolium*, *Trifolium campestre*, *Trifiolium scabrum* and others. This is the association Medicagini rigidulae-Aegilopietum geniculatae.

Concerning the weed communities that were used to be found on crop fields mixed with cultivated plants, the current intensive use of herbicides in agricultural practices has caused their practical disappearance, with the consequent loss of plant diversity. For that reason, it has been considered pointless to mention and comment on them.

Nitrophilous communities contribute substantially to the plant diversity of an area, and they should be considered as part of the biological diversity patrimony in any territory. Thus, any comprehensive conservation policy should take into account this vegetation and try to preserve it from disappearing/becoming extinct. Animal populations in rural areas are decreasing and the use of concrete and asphalt is increasing, encroaching on all patches that used to be occupied by this ruderal vegetation. Data about this vegetation for the Ebro Basin are found in Braun-Blanquet and Bolòs (1958), Loidi et al. (1997), Ursúa (1986).

12.6 Salt Marshes

The history of the sedimentary processes in the Ebro Basin, which over a long period in the Miocene was under endorheic conditions, has resulted in a high proportion of the sedimentary materials deposited being evaporitic, i.e. resulting from the deposition by chemical precipitation of soluble compounds, usually $CaSO_4$, $CaCO_3$ and NaCl. Those salts were dissolved in the lake water covering the depression during the Tertiary and precipitated when oversaturation happened under high evaporation conditions with high aridity. As a result, the Ebro Basin is rich in materials with a high content of gypsum and salt, which have a tendency to accumulate in the depressions where drainage has a low efficiency or is non-existent (arheic and endorheic). A few of these depressions have water permanently (Salada de Alcañiz), but most of them dry out in early summer, as soon as seasonal drought and high temperatures start, forming a white pan with salt micro-crystals on the surface. Those depressions (*cubetas*) have different sizes and forms depending on topography, and are locally called *salitrales* or *saladas* (Fig. 12.10). In the small draining channels of the drier districts, such as the Bardenas for example, an accumulation of salt on the soil also occurs because water flows only after the rain; in such places (*barrancos*) the ecosystem of the salty depressions (Fig. 12.11) is found. Those salt marshes can reach very high salinity values, particularly in summer when the concentration in the soil water is maximal. As a result, there is a decreasing salinity gradient from the central core of the depression towards the periphery.

The starting point of the flood and salinity gradient is in the centre of the depressions, where regular annual floods last longer in the wet season and dry out in spring or early summer. It is occupied by a set of highly specific annual communities dominated by succulent *Chenopodioideae* such as *Halopeplis*

Fig. 12.10 Salt marsh of Laguna La Playa in the central portion of the Ebro basin at Monegros district

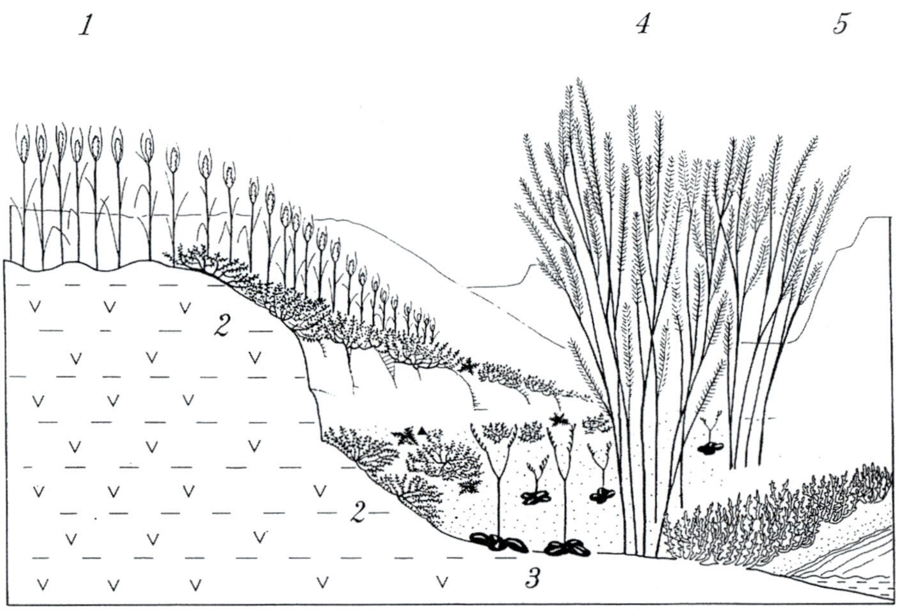

Fig. 12.11 A *barranco* in the Bardenas district. (*1*) Field of barley; (*2*) Pegano harmalae-Salsoletum vermiculatae; (*3*) Limonietum latebracteati; (*4*) Agrostio stoloniferae-Tamaricetum canariensis; (*5*) Suaedetum braun-blanquetii (After Loidi and Báscones 1995 with some alterations)

amplexicaulis, *Microcnemum coralloides*, *Suaeda spicata*, *Suaeda splendens* and *Salicornia patula*. These annual communities flower in late summer and autumn, and present no signs of nitrophily. The association Suaedo braun-blanquetii-Salicornietum patulae is found in sites where the flood lasts longer and the Microcnemetum coralloidis where the flooding is shallower and shorter. Halopeplidetum amplexicaulis is the association for the southeastern part of the area, in the Bajo Aragón district.

Most of the surface on these *saladas* under high salinity conditions is covered by a vegetation type dominated by perennial succulents. It also contains a considerable number of annuals which are visible only in spring, when the floor dries up but still keeps some humidity before the high temperatures of the summer arrive. Associated with the formation of these white pans of crystallized salt is the most frequent association in the central core of the salt marshes, the Suaedetum braun-blanquetii (*almajal*), a succulent scrub constituted mainly by *Suaeda braun-blanquetii* (*almajo*), an Iberian endemic *Chenopodioideae* of interior salt marshes, which often coexists with *Limbarda crithmoides* and *Limonium costae*. In the central districts of the basin (Los Monegros) the above mentioned association is replaced by the Suaedo braun-blanquetii-Arthrocnemetum macrostachyi with *Arthrocnemum macrostachyum* which occurs in addition to the former species and which endures higher salinity levels due to the more extreme climatic conditions. In the northeastern district of the basin, in the Pla d'Urgell district, under more moderate conditions, the central part of the salt marshes are occupied by the Gypsophiletum tomentosae, with the Iberian endemic halophyte *Gypsophila tomentosa*.

Slightly higher elevations in the microrelief determine a more ephemeral flood during the rainy season and a higher salinity in the soil during the dry season. These areas are occupied by the association Limonietum latebracteati, with *Limonium latebracteatum*, *Limonium catalaunicum*, *Limonium costae*, *Limonium supinum*, *Suaeda braun-blanquetii*, *Limbarda crithmoides* and a bulk of annuals. When flooding decreases together with salinity, *Schoenus nigricans* encroaches the area and the association Schoeno nigricantis-Plantaginetum maritimae is found, with *Althaea officinalis*, *Elymus pungens* subsp. *campestris*, *Linum maritimum*, *Plantago maritima* and even *Phragmites australis*.

This perennial vegetation, rich in succulents and rosettes, occupies the highly halophytic areas in the core of most of the salt marshes, especially when they dry up, but the gaps between the plants are colonized by annual communities which develop in spring and provide an important resource to the grazing sheep flocks, a reason why this halophytic vegetation has a certain nitrophilous character (halo-nitrophilous). The most widespread association is the Parapholido incurvae-Frankenietum pulverulentae, which has a high number of halophytic therophytes, such as *Aeluropus littoralis*, *Aizoon hispanicum*, *Frankenia pulverulenta*, *Hordeum marinum*, *Hymenolobus procumbens*, *Limonium echioides*, *Parapholis incurva*, *Spergularia diandra*, *Spergularia marina*, *Spergularia media*, *Sphenopus divaricatus*, etc., which mix with other non-halophytic annuals and constitute the species-rich ephemeral communities typical of these salty depressions. Areas that are higher up the slopes of hills experience a slightly lower salinity. There the dry grasslands of *Lygeum spartum* (*albardín*) dominate on soils with a lower salinity and less moisture, and *albardín* encroachment causes a decrease in the abundance of hyperhalophytic plants such as *Limonium* and *Suaeda* species. Other plants common in these communities are *Plantago maritima, Dorycnium gracile* and the Ebro Basin endemic *Limonium viciosoi*. The association for those communities is the Limonio viciosoi-Lygeetum sparti, known mostly from La Rioja, in the north-western districts of the territory.

Along the gradient towards lower salinity but with a water table near to the soil surface, in the transition to fens and non-halophilous wetlands, some communities, mostly dominated by different species of reeds, can be distinguished. In areas of higher salinity there is the Soncho crasifolii-Juncetum maritimi, with *Juncus maritimus, Elymus curvifolius, Limbarda crithmoides, Plantago maritima, Puccinellia fasciculata, Sonchus crassifolius* and *Sonchus maritimus*. If the soil is flooded for long periods with salty or brackish water, we have the Aleuropo littoralis-Juncetum subulati, with *Juncus subulatus, Aleuropus littoralis* and *Lactuca saligna*. In the next step in this decreasing salinity gradient we can find the Bupeuro tenuissimi-Juncetum gerardii, with *Bupleurum tenuissimum, Hordeum marinum, Juncus gerardii* and *Juncus maritimus*, which grows on wet soils of brackish water.

In this complex of communities, where strong nitrification and soil disturbance happens, some of the above mentioned communities evolve towards particular halo-nitrophilous associations, which are also linked to the salty depressions and can be considered as part of the complex of communities typical of them. In abandoned fields after being tilled and the soil turned over, the annual plants association Aizoo hispanici-Suaedetum splendentis establishes almost immediately, with *Aizoon hispanicum, Frankenia thymifolia, Sphenopus divaricatus* and *Suaeda splendens*. In pathways and fallows another halo-nitrophilous community develops, in areas with medium salinity, dominated by grasses and with a denser cover, the Polypogono maritime-Hordeetum marini, with *Hordeum marinum, Polypogon maritimus, Polypogon monspeliensis, Puccinellia fasciculata*, etc.

Finally, we mention the woodland vegetation of those *saladas*, which can somehow be considered as the potential natural vegetation given that all these ecosystems have been intensely used by humans as rangeland. This vegetation scarcely covers small patches and is almost exclusively made up of species of the genus *Tamarix* (*tarayales* or *tamarizales*). There are two associations recognized: the Agrostio stoloniferae-Tamaricetum canariensis, with *Tamarix canariensis* and *Tamarix gallica*, and in more arid areas at lower altitudes in the eastern half of the Ebro Basin the Suaedo braun-blanquetii-Tamaricetum boveanae, dominated by *Tamarix boveana*.

The salt marshes are one of the most characteristic and valuable ecosystems of the Ebro Basin, contributing a great deal to its floristic richness; a schematic diagram is shown in Fig. 12.12. Their conservation has to be considered a priority, not only by regional, national and international Conservation Agencies, but also by the stakeholders dealing with agriculture and resource exploitation.

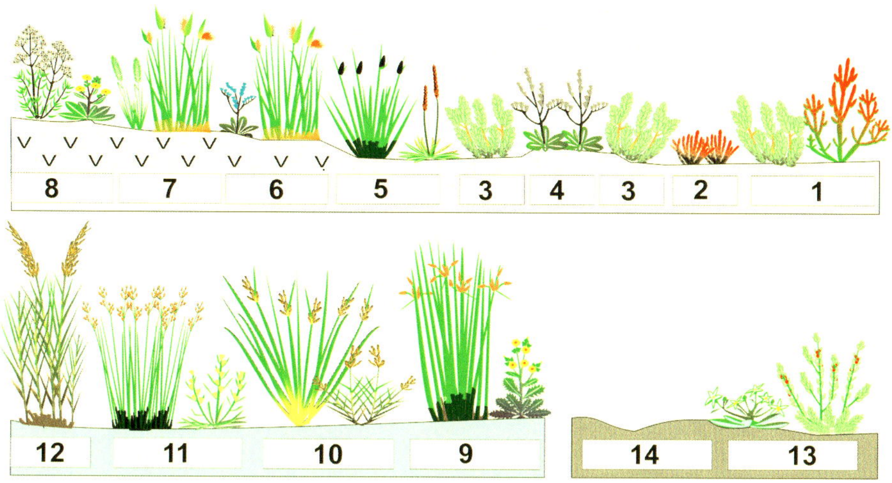

Fig. 12.12 Scheme of an inland salt marsh in the Ebro Basin. *Above*: gradient of saline concentration. (*1*) Suaedo braun-blanquetii-Salicornietum patulae; (*2*) Microcnemetum coralloidis; (*3*) Suaedetum braun-blanquetii; (*4*) Limonietum latebracteati, (*5*) Schoeno nigricantis-Plantaginetum maritimae; (*6*) Limonio viciosoi-Lygeetum spartii; (*7*) Agropyro pectinati-Lygeetum spartii; (*8*) Helianthemo thibaudii-Gypsophiletum hispanicae. *Below*: gradient on humid soils along decreasing salinity. (*9*) Soncho crassifolii-Juncetum maritimi; (*10*) Aleuropo littoralis-Juncetum subulati; (*11*) Bupleuro tenuissimi-Juncetum gerardii; (*12*) Phragmites; Gradient of salinity in disturbed soils: (*13*) Aizoo hispanici-Suaedetum splendentis; (*14*) Polypogono maritimi-Hordeetum marini (Drawn by José Pizarro)

There are numerous works dealing with this vegetation and providing data; we mention: Aguilella and Riera (1997), Belmonte and Laorga (1987), Biurrun (1999), Blanché and Molero (1986), Braun-Blanquet and Bolòs (1958), Conesa (1991), Fernández-González et al. (1990), Loidi et al. (1999), Longares (1997), Ursúa (1986), Ursúa and Báscones (2000).

12.7 Riparian Vegetation

Riverbeds, edges and surrounding flood plains constitute the riparian ecosystem, which occurs all over the world. In the Ebro Basin those environments contrast a lot with the surrounding territories, mainly due to the water availability in the soil, but also due to their geomorphologic and climatic particularities. In countries with severe water shortage, as in our case, riparian environments have always been considered as an oasis in the middle of a harsh semi-desert, and this has driven the distribution of human population, economy and even mentality. Historically an intensive land use has been developed in these riparian areas, exploiting the gleyic soils which have a near-surface water table. The natural vegetation of these areas has thus been extremely altered, having practically disappeared in many cases.

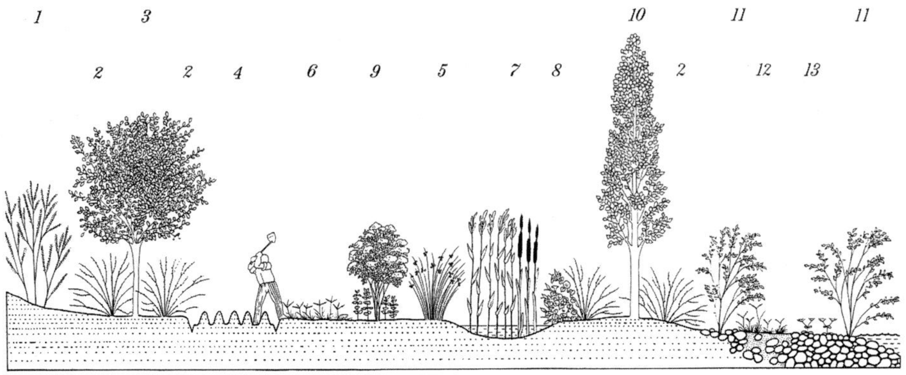

Fig. 12.13 Representation of the riverine and alluvial plain vegetation mosaic in the middle Ebro river. (*1*) Tamaricetum gallicae; (*2*) Pruno-Rubion ulmifolii; (*3*) Elm forest (Aro italici-Ulmetum minoris or Rubio tinctorum-Populetum albae); (*4*) Orchard; (*5*) Cirsio monspessulani-Holoschoenetum vulgaris; (*6*) Trifolio fragiferi-Cynodontetum dactyli; (*7*) Typho-Scirpetum tabernamontani; (*8*) *Calystegia sepium* comm. (*9*) Urtico dioicae-Sambucetum ebuli; (*10*) Poplar forest (Aro italici-Ulmetum minoris or Rubio tinctorum-Populetum albae; (*11*) Salicetum neotrichae; (*12*) Pasapalo distichi-Agrostietum verticillatae; (*13*) Equiseto ramosissimi-Erianthetum ravennae (After Loidi and Báscones 1995 with some alterations)

Human pressure has been so intense and goes back so long in history that almost all the examples of plant communities existing are semi-natural or entirely artificial. Nevertheless, excluding the urban areas, gardens, orchards and other crop lands, there is an important set of semi-natural plant communities which contribute greatly to the biodiversity of the area (Fig. 12.13). They are explained in the following paragraphs, together with the remnants of the fragmented natural vegetation.

12.7.1 Forests and Woodlands

The condition of the riparian forests is critical in the Ebro Basin. There are very few remnants in the territory, only at the edges of the rivers where agriculture is not possible due to regular flooding and where the land is public property (*dominio público hidráulico*) since it is considered part of the riverbed. With reference to the big rivers, i. e. the Ebro and its main tributaries, linear forests of willows and ashes, corresponding mostly to the association Salicetum neotrichae, can be found, with *Fraxinus angustifolia*, *Salix neotricha*, *Salix purpurea* subsp. *lambertiana, Salix triandra* subsp. *discolor*, *Populus nigra* and a long list of herbs living in the riparian environments. On soils with a high proportion of sand or gravel, under these climatic conditions in which summer aridity causes a deep decrease in the water level, the willow forest common along the river is replaced by a forest dominated by *Tamarix gallica* with *Tamarix africana* in the association *Tamaricetum gallicae*. Frequently found and characteristic accompanying plants are *Cynanchum acutum* and *Glycyrrhiza glabra*, as well as some herbs common at river edges. In a higher position, where flooding is less frequent and intense, the natural forests become dominated by poplars and elms, in the associations Aro italici-Ulmetum minoris in the upper stretches of the Ebro and tributaries, and Rubio tinctorum-Populetum albae in the lower ones under more arid conditions. These associations highly contribute to the plant diversity of the territory, as they include many riparian tree species: *Alnus glutinosa, Populus alba, Populus nigra, Salix neotricha, Fraxinus angustifolia, Ulmus minor* and *Tamarix gallica*. The shrub layer is also rich, with *Cornus sanguinea, Crataegus monogyna, Ligustrum vulgare, Prunus spinosa, Rosa micrantha, Rosa sempervirens, Rubus ulmifolius*. Lianas are numerous: *Bryonia dioica, Clematis vitalba, Hedera helix, Humulus lupulus, Rubia tinctorum* and *Vitis vinifera* subsp. *sylvestris*. The herb layer is formed by a group of nemoral species also found in the European temperate riparian forests, such as *Brachypodium sylvaticum, Chamaeiris foetidissima, Cucubalus baccifer, Elymus caninus, Solanum dulcamara*, and a high number of nitrophilous herbs. The variant of this forest type with *Ulmus minor* encompasses the broader part of the flood plain and makes contact with the climatophilous Rhamno lycioidis-Quercetum cocciferae territory, on dry substrates without a near-surface water table.

12.8 Grasslands and Herb Vegetation

12.8.1 Herb Communities in the Riverbeds Subject to Regular Annual Flooding

These herb communities colonize the riverbeds subject to regular flooding with running water and to a long drought period typical of the Mediterranean summer. In these areas the stream removes the organic matter and remixes the topsoil materials with every flood. On beds formed by gravel, sand or silt, a community of annual plants that develop in summer and early autumn, reaching maturity and sowing their seeds just before the rise of the water level with the start of the rainy season, gets established. This vegetation is represented by the species-rich association Xanthio-Polygonetum persicariae, with numerous allochthonous species. The most common are *Bidens tripartitus, Echinochloa crus-galli, Paspalum paspalodes, Lythrum salicaria, Polygonum persicaria, Polygonum lapathifolium, Symphyotrichum squamatum, Xanthium strumarium* and a high number of opportunistic nitrophilous species. The species composition varies depending on the materials of the riverbed. Some species, such as *Atriplex prostrata* and *Datura stramonium*, indicate gravely substrates within this association. On sandy or gravely substrates, in contact with the *Tamarix gallica* and *Tamarix africana* riparian forests, there is a tall grassland community type with Tropical-Mediterranean species, such as *Saccharum ravennae, Imperata cilindrica* and *Arundo micrantha*; other plants are *Cynanchum acutum, Equisetum ramosissimum*,

Dittrichia viscosa and *Glycyrrhiza glabra*. This is the association Equiseto ramosissimi-Erianthetum ravennae, occurring in the central sectors of the Ebro Basin. On similar substrates but out of the core area of the depression, mostly along the tributaries coming from the Pyrenees, the association Rorippo sylvestris-Cyperetum longi is found, with *Rorippa sylvestris* as the dominant species, as well as *Agrostis stolonifera*, *Paspalum paspaloides*, *Rumex conglomeratus* and some others. On the silt and fine sand banks of the riverbeds, a dense grassland covers the site and forms a thick carpet of the stoloniferous prostrate grass *Paspalum paspaloides*. It is the association Pasapalo distichi-Agrostietum verticillatae, with *Cyperus longus*, *Lythrum salicaria*, *Polygonum lapathifolium*, *Polypogon monspeliensis*, *Polypogon viridis*, *Rorippa sylvestris*, *Rumex obtusifolius*, *Xanthium strumarium* and other species common in these environments. At the edges of the ponds under certain salinity conditions and submitted to eventual flooding when the water level rises, a community characterized by *Teucrium scordium* subsp. *scordioides* establishes; it is the association Mentho aquaticae-Teucrietum scordioidis, with *Althaea officinalis*, *Mentha pulegium*, *Plantago major*, *Sonchus maritimus*, *Symphyotrichum squamatum*, etc.

12.8.2 Grasslands Never or Rarely Subject to Floods from the River

Above the frequently flooded riverbed, in the alluvial plain where floods from the river waters occur with low frequency, humidity can be very high, generally forming gleyic soils. Conditions permit the formation of long lasting ponds, where aquatic vegetation can become established, but waters are still. In these sites there are some herbaceous communities which replace or are related to the above mentioned potential forests, and their species composition depends on several environmental constraints, mostly related to human influence. If human pressure is low or lacking, a herbaceous evergreen vegetation dominated by *Scirpoides holoschoenus* gets established: the species-rich association Cirsio monspessulani-Holoschoenetum vulgaris. It develops very frequently in this riverbed environment and is made up by a high number of plants: *Agrostis stolonifera*, *Althaea officinalis*, *Asparagus officinalis*, *Brachypodium phoenicoides*, *Cirsium monspessulanum*, *Cirsium pyenaicum*, *Centaurea jacea*, *Elymus pungens* subsp. *campestris*, *Festuca interrupta*, *Juncus maritimus*, *Lotus glaber*, *Oenanthe lachenalii*, *Phragmites australis*, *Potentilla reptans* and many others. In frequently flooded strongly nitrified puddles, such as troughs for livestock and similar biotopes, a reedbed vegetation type dominated by *Juncus inflexus* is established, the association Mentho suaveolentis-Juncetum inflexi. Other frequent species are *Agrostis stolonifera*, *Carex cuprina*, *Centaurea jacea*, *Daucus carota*, *Festuca arundinacea*, *Holcus lanatus*, *Lotus glaber*, *Mentha suaveolens*, *Potentilla reptans*, *Ranunculus repens*, *Rumex conglomeratus*, *Trifolium fragiferum*, *Trifolium pratense*, *Verbena officinalis* and many others. If the water of the puddle is salty, the community is dominated by *Carex divisa* and *Juncus compressus*: the Junco compressi-Caricetum divisae, with *Agrostis stolonifera*, *Juncus inflexus*, *Hordeum marinum*, *Hordeum secalinum*, *Lotus glaber*, *Puccinellia fasciculata*, *Trifolium fragiferum*, etc. Under heavy grazing conditions, the grassland becomes dense and carpet-like, due mainly to the presence of *Cynodon dactylon* (*grama*), the dominant species. In this case the association is the Trifolio fragiferi-Cynodontetum dactyli, a valuable grassland used not only by sheep but also by other domestic animals such as donkeys and mules, animals that were important in earlier days. Common species are *Agrostis stolonifera*, *Plantago coronopus*, *Plantago lanceolata*, *Lolium perenne*, *Lotus glaber*, *Trifolium fragiferum*, *Trifolium pratense*, *Trifolium repens*, etc.

12.8.3 Nitrophilous Megaforbic Vegetation

This vegetation type can be found in the surroundings of human and domestic animal dwellings, in disturbed sites without grazing pressure. On moist hydromorphic soils a noteworthy community type dominated by *Conium maculatum* becomes established, in the association Galio aparines-Conietum maculati, with *Arctium minus*, *Calystegia sepium*, *Dipsacus fullonum*, *Galium aparine*, *Lactuca serriola*, *Rumex crispus*, *Rumex conglomeratus*, *Rumex obtusifolius*, *Urtica dioica*, etc. If the soil is not so moist, the association is the Urtico dioicae-Sambucetum ebuli, with *Cirsium vulgare*, *Sambucus ebulus*, *Verbena officinalis* and without *Conium*.

Information about this riparian vegetation is available in: Báscones and Ursúa (1986), Berastegi (2013), Biurrun (1999), Braun-Blanquet and Bolòs (1958), Fernández-González et al. (1986), Loidi et al. (1997), Ursúa (1986).

12.9 Vegetation Sigmeta and Geosigmeta

This approach attempts to sort out the vegetation units into groups according to their dynamic relations, i. e. in the absence of disturbances the different plant communities of each group have the natural tendency to evolve towards the Potential Natural Vegetation, considered the most mature vegetation type within each group. In this approach, aside from the general documents for the Iberian Peninsula, there are some regional works which have established the basic criteria and the cartography of these Vegetation Series in the Ebro Basin area: Loidi and Báscones (1995) and Rivas-Martínez et al. (2000). There are seven main units in which the whole territory of the Ebro Basin can be divided, five of them are climatophilous vegetation series or sigmeta, and the other two are geoseries or geosigmeta, i. e, complexes of communities conditioned by a dominant

Fig. 12.14 Mesomediterranean semiarid series of *Quercus coccifera* with *Pinus halepensis*: Rhamno lycioidis-Querco cocciferae sigmetum. (*1*) Rhamno lycioidis-Quercetum cocciferae; (*1′*) The same in a *Pinus halepensis* facies; (*2*) Rosmarino officinalis-Linetum suffruticosi; (*3*) Saxifrago tridactylitae-Hornungietum petraeae; (*4*) Ruto angustifolii-Brachypodietum retusi; (*5*) Agropyro pectinati-Lygeetum sparti; (*6*) Helianthemo thibaudii-Gypsophiletum hispanicae (a *Ononis tridentata* facies, b typical facies); (*7*) Chaenorhino reyesii-Campanuletum fastigiatae (After Loidi and Báscones 1995 with some alterations)

non-climatic factor (salinity, ground water, etc.), which are sorted after their gradient. They are:

1. Mesomediterranean semiarid series of *Quercus coccifera*with *Pinus halepensis*: Rhamno lycioidis-Querco cocciferae sigmetum (Fig. 12.14)
2. Mesomediterranean continental dry-semiarid series of *Juniperus thurifera*and *Juniperus phoenicea*: Junipero phoeniceo-thuriferae sigmetum
3. Mesomediterranean dry series of *Quercus rotundifolia*: Querco rotundifoliae sigmetum (Fig. 12.15).
4. Supramediterranean dry series of *Quercus rotundifolia*: Buxo sempervirentis-Querco rotundifoliae sigmetum
5. Supra-mesomediterranean subhumid series of *Quercus faginea*: Violo willkommii-Querco fagineae sigmetum.
6. Halophilous Aragonese inland geosigmetum
7. Hygrophilous Aragonese riparian geosigmetum

The plant communities associated to each of these units are shown in Table 12.1.

Fig. 12.15 Mesomediterranean dry series of *Quercus rotundifolia*: Querco rotundifoliae sigmetum. (*1*) Quercetum rotundifoliae; (*2*) Rhamno lycioidis-Quercetum cocciferae; (*3*) Salvio lavandulifoliae-Ononidetum fruticosae; (*4*) Ruto angustifolii-Brachypodietum retusi; (*5*) Saxifrago tridactylitae-Hornungietum petraeae; (*6*) Agropyro pectinati-Lygeetum sparti; (*7*) *Pinus pinaster* plantation; (*8*) Olive grove; (*9*) Vineyard; (*10*) Cereal crop (After Loidi and Báscones 1995 with some alterations)

12.10 Land Use

The Ebro Valley has a human population of less than three million inhabitants, a population density of below 35 persons/km^2, quite low even for Iberian standards. Such a low figure is the result of the natural conditions of the territory, which determine a strong concentration of human activities in the alluvial plains due to irrigation agriculture. The modern industrial and post-industrial economy has caused rural abandonment and migration of the population to the big cities, leaving many villages and marginal districts practically deserted. From a historical perspective, human pressure has been intense in the Ebro Basin since antiquity. Archaeological sites of the Neolithic and Iron Age periods reveal a relatively dense human population and later, during the Roman Period, the territory experienced an intense demographic and socioeconomic development. Important cities

Table 12.1 The main associations of the Ebro Basin related with the main types of Sigmeta and Geosigmeta: red, inside or matching with the unit, yellow, also present. Sigmeta and geosigmeta names in the text

	Climatophilous sigmeta					Edaphohygrophilous geosigmeta	
Sigmeta/Geosigmeta	**1**	**2**	**3**	**4**	**5**	**6**	**7**
Forests and woodlands							
Juniperetum phoeniceo-thuriferae	red						
Rhamno lycioidis-Quercetum cocciferae	yellow	red	yellow				
Quercetum rotundifoliae			red				
Buxo sempervirentis-Quercetum rotundifoliae				red			
Violo willkommi-Quercetum fagineae					red		
Suaedo braun-blanquetii-Tamaricetum boveanae						red	
Agrostio stoloniferae-Tamaricetum canariensis						red	
Salicetum neotrichae							red
Rubio tinctori Populetum albae							red
Scrub							
Quercetum cocciferae buxetosum sempervirentis				red	red		
Rosmarino officinalis-Linetum suffruticosi	red	red	yellow				
Cytiso fontanesii-Cistetum clusii							
Sideritido ilicifoliae-Thymetum loscosii			red				
Sideritido spinulosae-Lavanduletum latifoliae			red				
Teucrio aragonensis-Thymetum fontqueri				red	red		
Salvio lavandulifoliae-Ononidetum fruticosae			red				
Helianthemo thibaudii-Gypsphiletum hispanicae	red	red	yellow				
Grassland							
Ruto angustifolii-Brachypodietum retusi	red	red	red				
Elytrigio campestris-Brachypodietum phoenicoidis	red	red					
Stipo parviflorae-Lygeetum sparti	red	red	red				
Agropyro pectinati-Lygeetum sparti	red	red					
Saxifrago tridactylitae-Hornungietum petraeae	red	red	red	red	red		
Chaenorhino reyesii-Campanuletum fastigiatae	red	red					
Astragalo sesamei-Poetum bulbosae	red	red	red	red	red		
Nitropilous vegetation							
Salsolo vermiculatae-Peganetum harmalae	red	red	yellow				
Salsolo vermiculatae-Artemisietum herba-albae	red	red	yellow				
Sisymbrio irionis-Malvetum parviflorae	red	red	red	yellow	yellow		
Onopordetum castellani	red	red	red	yellow	yellow		
Silybo-Urticetum	red	red	red				
Hordeetum leporini	red	red	red	yellow	yellow		
Medicagini rigidulae-Aegilopietum geniculatae	red	red	red	red	red		
Halophytic vegetation							
Suaedo braun-blanquetii-Salicornietum patulae						red	
Microcnemetum coralloidis						red	
Halopeplidetum amplexicaulis						red	
Suaedo braun-blanquetii-Arthrocnemetum macrostachyi						red	
Gypsophiletum tomentosae						red	
Limonietum latebracteati						red	
Schoeno nigricantis-Plantaginetum maritimae						red	

(continued)

Table 12.1 (continued)

Limonio viciosoi-Lygeetum sparti							
Parapholido incurvae-Frankenietum pulverulentae							
Soncho crasifolii-Juncetum maritimi							
Aleuropo littoralis-Juncetum subulati							
Bupeuro tenuissimi-Juncetum gerardii							
Aizoo hispanici-Suaedetum splendentis							
Hygrophilous vegetation							
Xanthio-Polygonetum persicariae							
Equiseto ramosissimi-Erianthetum ravennae							
Rorippo sylvestris-Cyperetum longi							
Pasapalo distichi-Agrostietum verticillatae							
Mentho aquaticae-Teucrietum scordioidis							
Cirsio monspessulani-Holoschoenetum vulgaris							
Mentho suaveolentis-Juncetum inflexi							
Junco compressi-Caricetum divisae							
Trifolio fragiferi-Cynodontetum dactyli							
Galio aparines-Conietum maculati							
Urtico dioicae-Sambucetum ebuli							

were founded, such as Caesar Augusta (Saragossa), Osca (Huesca), Bilbilis (Calatayud), Calagurris (Calahorra), Ilerda (Lleida), etc. and an efficient network of roads were built to connect them (the North Via). During this period a substansive migration to the natural areas supposedly happened as the Roman society was organised in such a way that agrarian activities were oriented to supplying a broad market beyond merely a subsistence production. This required more land for cultivation, more wood for fire and for buildings, and more pastures for livestock, all of them demanded by a growing human population. Thus, large scale deforestation took place during this period. This was the beginning of a process which has resulted in a modern landscape that seems really devastated. The image of the Ebro Basin is really that of a semi-desert if it were not for a generous and dense network of large rivers that permits the life and prosperity of the human population.

During the Middle Ages the situation kept oscillating but from the 8th century onwards, the Ebro Basin was occupied by the Arabs for several hundreds of years (as other parts of the Iberian Peninsula) and they introduced new technologies and ways of exploiting the territory. Of particular relevance is the irrigation, which expanded dramatically in the Ebro Basin during their domination thanks to their construction of important hydraulic works. That heritage has resulted in the current importance of irrigation in many riverbank areas (called *riberas*, like the Ribera de Navarra) and the abundance of dams and channels currently present in the Ebro catchment area (224 altogether); most of which have been recently built to enlarge the irrigated area.

Concerning the restoration efforts to re-establish the vanished greenness of the landscape, the plantation of artificial forests using several tree species, mostly pines, can be mentioned. The species preferably selected in the Ebro Basin is *Pinus halepensis*, which is a natural species of the regional flora and this can be considered a convenient circumstance. Nevertheless, these plantations have yielded poor results due to the hard climate and the eroded soils.

Acknowledgments I am grateful to my friend José Pizarro for his extraordinary drawings which illustrate this chapter. The funds of project IT299-10 for research groups of the Basque Government have supported this research.

References

Aguilella A, Riera J (1997) Estudios sobre la vegetación y flora halófilas de las saladas de 'El Plano' (Alcañiz-Calanda, Aragón, España). Publicaciones del Consejo de Protección de la Naturaleza de Aragón. Serie de Investigación, p 43

Báscones JC, Ursúa C (1986) Estudio fitosociológico de los pastos de la Ribera Tudelana. Príncipe de Viana Supl Ciencias 6:101–140

Báscones JC, Medrano LM, Ursúa C (2011) Yesos del distrito Bardenero y territorios aledaños. In: Mota JF, Sánchez-Gómez P, Guirado JS (eds) Diversidad vegetal de las yeseras ibéricas. El reto de los archipiélagos edáficos para la biología de la conservación. ADIF-Mediterráneo Asesores Consultores, Almería, pp 357–360

Belmonte MD, Laorga S (1987) Estudio de la flora y vegetación de los ecosistemas halófilos de la Rioja logroñesa (Logroño-España). Zubia 5:63–125

Berastegi A (2013) Prados y pastizales de Navarra: descripción, tipificación y ecología. Guineana 19:1–505

Biurrun I (1999) Flora y vegetación de los ríos y humedales de Navarra. Guineana 5:1–338

Blanché C, Molero J (1986) Las cubetas arreicas al sur de Bujaraloz. Contribución al estudio fitocenológico Lazaroa 9:277–299

Braun-Blanquet J, de Bolòs O (1958) Les groupements végétaux du bassin moyen de l'Ebre et leur dynamisme. Anales Estación Experimental Aula Dei 5(1–4):1–266

Conesa JA (1991) Comunitats vegetals de l'embassament d'Utxesa-Secà i rodalies, I: Comunitats aquàtiques, halòfiles, rupícoles, ruderals, nitro-halòfiles i arvenses. Ilerda (Ciències) 49:7–55

de Bolòs O (1961) La transición entre la Depresión del Ebro y los Pirineos en el aspecto geobotánico. Anales Inst Bot Cavanilles 18:199–254

de Bolòs O (1973) La vegetación de la Serreta Negra de Fraga. Mem Real Acad Ci Artes Barcelona 42(6):269–313

de Bolòs O (1996) Contribució al coneixement de la vegetació del territorio Auso-Segarric. Mem Real Acad Ci Artes Barcelona 55(4):2–272

Fernández-González F, Loidi J, Molina A (1986) Contribución al estudio de los matorrales aragoneses: los salviares riojano-estelleses. Anales Jard Bot Madrid 42(2):451–459

Fernández-González F, Molina A, Loidi J (1990) Los tarayales de la Depresión del Ebro. Acta Botanica Malacitana 15:311–322

Font X (2011) Yesos del anticlinal Barbastro-Balaguer a los Pirineos. In: Mota JF, Sánchez-Gómez P, Guirado JS (eds) Diversidad vegetal de las yeseras ibéricas. El reto de los archipiélagos edáficos para la biología de la conservación. ADIF-Mediterráneo Asesores Consultores, Almería, pp 367–372

Guardia R, Ninot JM (2000) Sobre la ecología de las comunidades gipsícolas en la comarca de la Noguera. Congreso de Botánica en homenaje a Francisco Loscos. Teruel, pp 797–809

Huguet del Villar E (1925) Avance geobotánico sobre la pretendida estepa central de España. Ibérica (Tortosa) 13 (576): 281–283; 13 (577): 297–302; 13 (579): 328–333; 13 (580): 344–350

Jäger EJ (1971) Die Pflanzengeographische Stellung der "Steppen" der Iberischen Halbinsel. Flora 160:217–256

Loidi J, Báscones JC (1995) Mapa de series de vegetación de Navarra. Memoria y mapa. Gobierno de Navarra, Pamplona, p 9
Loidi J, Fernández-González F (1994) The gypsophilous scrub communities of the Ebro Valley (Spain). Phytocoenologia 24:383–399
Loidi J, Biurrun I, Herrera M (1997) La vegetación del centro-septentrional de España. Itinera Geobotanica 9:161–618
Loidi J, Herrera M, Biurrun I, García-Mijangos I (1999) Relationships between syntaxonomy of *Thero-Salicornietea* and taxonomy of the genera *Salicornia* and *Suaeda* in the Iberian Peninsula. Folia Geobot 34:97–114
Longares LA (1997) El paisaje vegetal en el entorno de la reserva ornitológica 'El Planerón' (Belchite-Zaragoza). Consejo de Protección de la naturaleza de Aragón, p 193
Molero J, Vigo J (1981) Aportació al coneixement florístic i geobotànic de la Serra d'Aubenç. Treb Inst Bot Barcelona 6:82
Molero J, Sáez L, Villar L (1998) Interés florístico y geobotánico de la Sierra de Alcubierre (Monegros, Aragón). Acta Bot Barcinonensia 45:363–390
Molina A, Loidi J, Fernández-González F (1993) Sobre las comunidades de matorral de la Depresión del Ebro (España). Botanica Complutensis 18:11–50
Rivas Goday S, Borja J, Esteve F, Fernández-Galiano E, Rigual A, Rivas-Martínez S (1956) Aportaciones a la fitosociología hispánica (proyectos de comunidades hispánicas) Nota II (Comunidades gipsófilas fruticosas del centro y sudeste de España) Anales Inst. Bot Cavanilles 14:433–500
Rivas-Martínez S, Báscones JC, Díaz TE, Fernández-González F, Loidi J (1991) La vegetación del Pirineo occidental y Navarra. Itinera Gobotanica 5:5–456
Rivas-Martínez S, Pizarro J, Sánchez-Mata D (2000) Series de vegetación del valle medio del río Ebro. Congreso de Botánica en homenaje a Francisco Loscos. Teruel, pp 641–652
Terradas J (1986) El paisatge vegetal dels Monegros: assaig d'interpretació. Orsis 2:71–95
Ursúa C (1986) Flora y vegetación de la Ribera Tudelana. PhD Universidad de Navarra, Pamplona, p 640
Ursúa C, Bascones JC (2000) Vegetación de las lagunas endorréicas de Navarra. Congreso de Botánica en homenaje a Francisco Loscos. Teruel, pp 687–701
Villar L, Segarra-Moragues JG, Minaya MA, Pérez-Collazo E, Virual J, Catalán P (2011) El valle del Ebro (yesos monegrinos) y territorios colindantes. In: Mota JF, Sánchez-Gómez P, Guirado JS (eds) Diversidad vegetal de las yeseras ibéricas. El reto de los archipiélagos edáficos para la biología de la conservación. ADIF-Mediterráneo Asesores Consultores, Almería, pp 361–366
Willkomm M (1852) Strand- und Steppengebiete der Iberischen Halbinsel und deren Vegetation. Ein Beitrag zur physikalischen Geographie, Geognosie und Botanik. Friedrich Fleischer, Leipzig
Willkomm M (1896) Grundzüge der Pflanzenverbreitung auf der Iberischen Halbinsel. In: Engler A, Drude O (eds) Die Vegetation der Erde. Wilhelm Engelmann Verlag, Leipzig, p 395

Chapter 13
The Sistema Central (Central Range)

Daniel Sánchez-Mata, Rosario G. Gavilán, and Vicenta de la Fuente

Abstract The Sistema Central or Central Range has an overall length of 600 km and is the most extensive mountain range on the Iberian Peninsula. It has an alpine orogeny and is located in the centre of the peninsula running northeast-southwest in its eastern part (Ayllón and Guadarrama ranges), and east-west in its western part (Guadarrama, Gredos and Estrêla ranges). The whole system comprises a continuous series of smaller mountain ranges, including valleys and various natural areas. It occupies a vast territory from the middle of Portugal (Serra da Estrêla) to the Sierra de Ayllón range in the east, and acts as a natural division of the Castilian central plateau, separating the northern plateau (Castile and León) from the southern plateau (Madrid, Castile-La Mancha and Extremadura). It generally represents the watershed between two of the most important Iberian rivers: the Duero and the Tagus. The highest point of the whole Iberian Central Range is the Almanzor Peak (2592 m) in the central massif of the Gredos mountains. Its characterisation and location on the Iberian Peninsula have interesting phytogeographical and geobotanical implications. These mountains are noteworthy for their bioclimatic, geomorphological and lithological features. Their remarkable bioclimatic complexity is a result of the dissymmetry between the northern and southern slopes (with average values of 1200 vs. 500 m), and their lithological homogeneity, dominated by plutonic igneous and –to a lesser extent– metamorphic rocks. All these factors condition the landscape in these territories. The natural vegetation in these mountains has a woodland character, except in the summit areas above the tree line, which are characterised by Mediterranean alpine grasslands. However other types of vegetation grow in these siliceous high-mountain areas, in the rocky places, peatland, meadows and shrubland that conform the landscape here. Natural Iberian

D. Sánchez-Mata (✉) • R.G. Gavilán
Department of Plant Biology II, Faculty of Pharmacy, Universidad Complutense, Ciudad Universitaria, 28040, Madrid, Spain
e-mail: dsmata@ucm.es

V. de la Fuente
Department of Biology, Faculty of Sciences, Universidad Autónoma de Madrid, Campus de Cantoblanco, Calle Darwin 2, 28049, Madrid, Spain
e-mail: vicenta.fuente@uam.es

J. Loidi (ed.), *The Vegetation of the Iberian Peninsula*, Plant and Vegetation 12,
DOI 10.1007/978-3-319-54784-8_13

Scots pine (*Pinus sylvestris* var. *iberica*) and Spanish black pine (*Pinus nigra* subsp. *salzmannii*) forests are distributed throughout the high- and mid-mountain territories respectively in spite of extensive deforestation –more pronounced in the westernmost mountain areas–, in addition to oak forests (*Quercus pyrenaica* and *Quercus rotundifolia*) in the rest of the mountain landscape, where natural forests have often been eliminated in order to cultivate several strains of pines, and to encourage the spread of meadows and grasslands which support an important livestock population. Some interesting small relict forests such as beech (*Fagus sylvatica*), birch (*Betula celtiberica*) and juniper woodlands (*Juniperus thurifera* and *Juniperus oxycedrus* subsp. *lagunae*) are scattered within the mountain areas. Also significant are the riparian and hygrophilous forests and other vegetation, and the seral vegetation complexes –both forest fringe vegetation and shrub communities– which often predominate throughout vast mountain areas in the landscape.

13.1 Introduction. Environmental Features

The Iberian Sistema Central, also known as the Iberian Central Range, is a long mountain chain stretching for around 600 km which divides the Iberian plateau into two units, the northern sub-plateau and the southern sub-plateau. Its structure, location and crosswise orientation (ENE-WSW) are the result of Hercynian movements, subsequently shaped by the alpine orogeny that affected the Iberian central zone and produced the current configuration of the mountain chain (Fig. 13.1).

Throughout the process of emergence, vast areas of the Sistema Central underwent the intense and active erosive action of glaciers, which left their characteristic imprint on the summit areas of the highest mountains, and can still be recognised today (Mayo 2002) (Fig. 13.2).

The lithology of the mountain range consists predominantly of Palaeozoic siliceous materials such as granites, gneisses, slates, schist, quartzite and others. In the foothills these materials are covered by Mesozoic and Cenozoic sediments with a diverse lithology, and particularly by sandy and silty Pliocene-Quaternary glacis and certain Jurassic and Cretaceous fragmentary limestones in the easternmost massifs of the range.

In biogeographical terms, the Sistema Central is considered to begin in the eastern spurs of the Sierra de Ayllón, whose highest summit is the Pico del Lobo (2273 m). Towards the east, Somosierra (Pico de las Tres Provincias, 2129 m) connects with the mountain alignments of the Sierra de Guadarrama, whose highest peaks are El Nevero (2209 m), Peñalara (2430 m) and Cabezas de Hierro (2383 m). The tectonic trench of the headwaters of the Lozoya river extends between these last two, forming the valley of El Paular. The southern branch of the Guadarrama

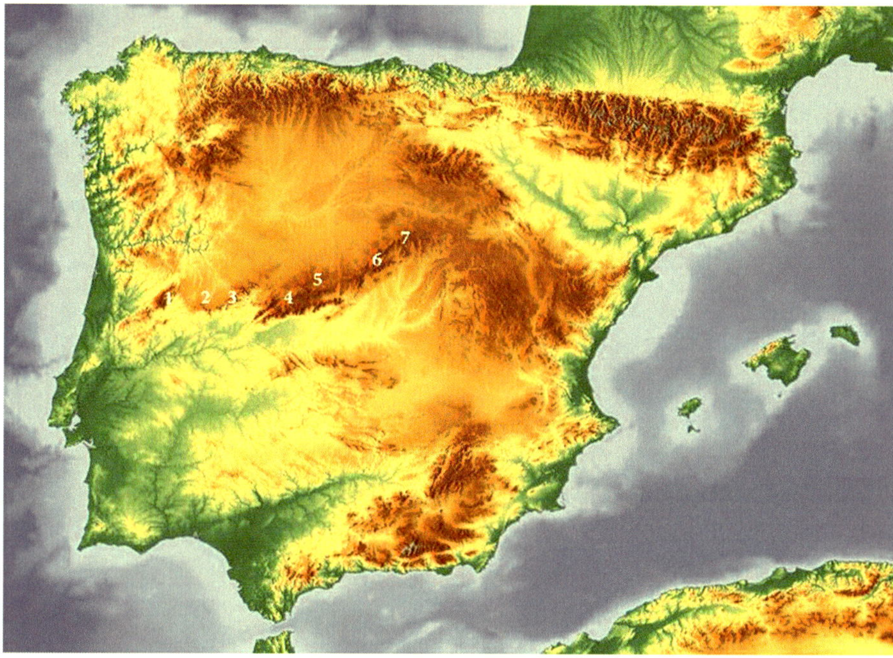

Fig. 13.1 Map of the Iberian Peninsula showing the main mountain ranges. Sistema Central is the numbered range (1, Portugal; 2–7, Spain); from W to E: 1. Serra da Estrêla, 2. Sierra de la Peña de Francia, 3. Sierra de Gata, 4. Sierra de Gredos (highest summit of the Sistema Central: Almanzor Peak, 2592 m), 5. Sierra de la Paramera, 6. Sierra de Guadarrama, and 7. Sierra de Ayllón

mountains continues towards La Almenara (1260 m), or –according to some authors– until the peak of Cadalso de los Vidrios. The northern branch, which starts in the Sierra de Malagón (Cueva Valiente, 1902 m), extends through the Paramera de Ávila (Zapatero Peak, 2158 m), La Serrota (Serrota, 2294 m) and the Sierra de Villafranca (Cerro de Moros, 2059 m) in Piedrahita.

The trench of the Tiétar river delimits the Sierra de Gredos in the south (Fig. 13.3), which is the highest massif in the whole mountain range (Almanzor, 2592 m), and also extends northwards along the broad Alberche and Tormes river basins. The vast fracture of the Jerte and Aravalle rivers separates the Sierra de Tormantos (La Covacha, 2398 m) to the east from the Sierra de Béjar (La Ceja, 2425 m) to the west.

There is a sharp drop in altitude to the west of the Sierra de Béjar produced by the deep depression of the Alberche river, which then rises again in the mountain ranges of the Peña de Francia (1723 m) and Sierra de Gata (1363 m) in Spanish territory. Finally the vast massif of Serra da Estrêla (1993 m) –in Portugal– is separated by the tectonic trench of Ciudad Rodrigo.

Fig. 13.2 Glacier lakes are frequent throughout the Sistema Central high-mountain landscape. 1. Glacier cirque and lake in the Gredos mountains (Laguna Grande, central massif, 1950 m). 2. Glacier lake in the Béjar mountains (2100 m). 3. Glacier lake in the Guadarrama mountains (Peñalara Lake, Peñalara massif, 2017 m) (Photos by Enrique Luengo)

13.2 Bioclimatology

The location and character of the mountain chains in the Sistema Central produce a sharply divergent bioclimatology. These contrasts can be seen both in an east-west transect and in the acute altitudinal dissymmetry between the northern and southern faces. The drainage system is conditioned by the relief of the different mountain massifs in the Sistema Central, and affects the bioclimates, thermotypes and ombrotypes present in their territories.

Following the latest bioclimatic proposals of Rivas-Martínez, the bioclimatic belts that can be recognized in the Sistema Central are: meso- and supramediterranean and oro- and cryorosubmediterranean (oro- and cryorotemperate submediterranean), although the mesomediterranean belt is present in only its continental territories (Sierra de Guadarrama biogeographical sector). In most western sectors the mesomediterranean territories correspond to the Luso-Extremaduran biogeographical subprovince. The change in biogeographical aspect and vegetation series occurs on the edge of the meso-supramediterranean bioclimatic belt. The cryorosubmediterranean belt can only be recognized in the higher parts of the Sistema Central, in the Béjar, Gredos, Guadarrama and Ayllón ranges, over 2200 m. Its lower limit drops significantly among the large massifs shaped by glaciers (Peñalara lake, central Gredos massif and others), due to the effect of the local formation of large masses of cold air occurring in glacier cirques. Recently (Rivas-Martínez et al. 2007) the oro- and cryoromediterranean belts in the Iberian Peninsula have been restricted to the Sierra Nevada, while the rest of the Iberian Mediterranean high mountains are considered submediterranean, a type of variant of the temperate bioclimate; these belts should hereafter be designated orosubmediterranean and cryorosubmediterranean.

From east to west along the Sistema Central there is an inverse relationship between continentality and the mean minimum temperature of the coldest month (m); thus continentality decreases when there is an increase in winter temperature, and vice versa (Gavilán et al. 1998; Gavilán 2005).

Throughout the Sistema Central the ombroclimate is conditioned by the pattern of western winds, which –because of their key components, NE and SW– cause a significant rise in precipitation in the eastern- and westernmost areas of the range (Ayllón, Gredos, Béjar, Peña de Francia, and Estrêla mountain territories). There is also a marked difference between the north and south faces. Most of the rainfall recorded in winter and spring comes from the southwest, and predominantly affects the southern slopes of the western mountains reaching often the humid type. As a result, the supramediterranean belt has a subhumid ombroclimate (P: 600–1000 mm annually) in the territories in Paramera-Serrota and Guadarrama, which are in the rain shadow due to the effect of the Gredos central massif. Humid and hyperhumid ombroclimates are typical of the Ayllón, Gredos, Béjar-Tormantos, Peña de Francia and Estrêla mountain territories.

East of the Somosierra Pass (Sierra de Ayllón) and west of the Sierra de Malagón (Paramera-Serrotensa ranges) there appears to be a total absence of pine forests. This can be attributed both to the impact of management practices and to the vegetation's response to the decrease in continentality; precipitation in the form of snow is much more significant than in the Guadarramean sector. However, some *Pinus* forests have very recently been recognized as relicts in the Ayllón Range.

The shortage of meteorological stations at high altitudes in the Sistema Central, in conjunction with the particular mountain geomorphology, make it difficult to obtain substantial bioclimatic series to facilitate the study of vegetation along bioclimatic belts. We can, however, highlight a number of general aspects that contribute to an understanding of the plant communities:

- A very short period of plant activity –less than 5 months– conditioned by low temperatures.
- A long period of frost risk throughout the year. The frost-free period is slightly over 3 months in the lower part of the belt (1700–1750 m, with a mean annual temperature of approximately 8 °C).

Fig. 13.3 A view of the southern slope of the Gredos Range from the Tiétar river valley. Foreground, managed holm-oak woodlands ("encinares") growing in territories with a mesomediterranean thermotype (Pyro bourgaeanae-Quercetum rotundifoliae) (Photo by Enrique Luengo)

- The presence of three types of ombroclimate: subhumid, humid and hyperhumid. A very marked rainfall gradient. The lowest rainfall recorded in the mountains of Ávila is found in the upper sections of the Alberche and Tormes rivers, which are protected from the influence of westerly winds due to the proximity of the highest peaks in the Sistema Central in the Sierra de Gredos. Precipitation increases gradually to the east, with a peak in the Ayllón Range, which has a highly humid ombroclimate. A sharp rise in precipitation can be seen in the west and south of the Sistema Central. The maximum recorded is in the Béjar Range, where estimated rainfall for the mountain summits is over 3000 mm annually.

The vegetation in those oro- and cryorosubmediterranean territories, with the highest precipitation in the form of snow, is dominated by grasslands rich in *Nardus stricta*, locally known as "cervunales" (Campanulo herminii-Nardion strictae) (Fig. 13.4), with a lower frequency of psychroxerophilous grasslands (Hieracio castellani-Plantaginion radicatae, Minuartio biguerrensis-Festucion carpetanae), which tend to predominate in the central areas of the massif (Sierra de Guadarrama).

In summer, the subtropical jet shifts north, and one of its warm-air masses –the Azores anticyclone– brings warm weather to the Iberian Peninsula. The territories in the western and southwestern parts of the Sistema Central are the most affected

Fig. 13.4 *Nardus stricta* grasslands ("cervunales") on the highest summits of the Béjar mountains. The vegetation complex includes aquatic, spring, peatland and grassland communities (Photo by Enrique Luengo)

by these phenomena, resulting in very dry summers and very humid winters, as explained above.

The geographic units of the ranges and mountains in the Sistema Central, from the rank of district through to sub-province, are justified by the geographic characterisation and geolithological and bioclimatic diversity of the various mountain massifs it comprises.

13.3 Vegetation

13.3.1 Orotemperate and Orosubmediterranean Vegetation (Submediterranean Orotemperate)

The orotemperate (Estrêla Range) and orosubmediterranean belts are only present in the highest massifs of the Iberian Peninsula, and can be recognised at between approximately 1600 and 2000 m. The altitudinal range varies depending on exposure, latitude and the greater or lesser influence of the Atlantic Ocean. In the Guadarrama and Paramera ranges they can be found at around 1650–1700 m, and in the Gredos and Béjar ranges at over 1750–1800 m. The natural vegetation forms dense thickets, with or without a canopy of Scots pine (Pino-Juniperetea class).

Fig. 13.5 *Cytisus oromediterraneus* seral broom communities (Senecioni carpetani-Cytisetum oromediterranei) developed after fires in the Guadarrama mountains (Peñalara massif in the background). Iberian Scots pine forest vegetation series (Avenello ibericae-Pino ibericae sigmetum) (Photo by Enrique Luengo)

Throughout the Sistema Central these plant communities are established on siliceous substrates and are represented by pines, junipers, brooms and other woody creeping vegetation.

The abusive practice of using fire to promote grazing has trivialised the floristic composition of the diverse vegetation units. *Cytisus oromediterraneus* is favoured by fires (Fig. 13.5), while creeping juniper (*Juniperus communis* subsp. *alpina*) has been relegated to hard rock sites which are less accessible to flames. This has had a significant impact on the landscape, which is now almost entirely dominated by *Cytisus oromediterraneus*. This situation is currently changing, as these areas have now been protected and fires are less common.

13.3.1.1 Iberian Scots Pine Forests and Creeping-Juniper Bushlands

The consideration and interpretation of Iberian Scots pine forests (*Pinus sylvestris* var. *iberica*), also called Valsaín pine, within the whole of the Sistema Central has been the object of treatment and study by numerous geobotanists who disagree in their recognition of its natural existence and potential and current distribution (Rivas-Martínez 1964; Fuente 1986; Peinado and Rivas-Martínez 1987; Rivas-Martínez and Cantó 1987; Rivas-Martínez et al. 1987, 2001, 2002; Sánchez-Mata

1989, 1999, 2015; Fernández-González 1991; Blanco et al. 1997; Sardinero, 2004; López-Sáez et al. 2013, 2016).

The latest geobotanical contributions recognise Iberian Scots pine forests as potential vegetation in the Sistema Central, generally at altitudinal levels above the vegetation belt corresponding to Iberian white oak forests of *Quercus pyrenaica* ("melojares"). This implies that Iberian Scots pine forests constitute the upper vegetation belt in the corresponding altitudinal catena, and that *Pinus sylvestris* var. *iberica* constitutes the tree-line environment in our territory. However this widely-accepted fact in current Iberian geobotany does not correspond to the actual vegetation present in a large part of the territory in the Sistema Central. This discrepancy can be attributed to two causes: (1) the extensive deforestation that has taken place throughout all these mountain territories to provide wood for construction and fuel since historic times, and the need to create large wide-open spaces for the development and management of meadows and grasslands to allow the maintenance of livestock (mainly bovine and equine) which is of such crucial importance to the economy of the rural communities in the mountain areas; and (2) Scots pine cultivation by humans, who have extended it into broad mountain territories to favour its management and exploitation, and its indiscriminate use (with different cultivars) for plantations and reforestation. Thus Iberian Scots pine (potential in mid- to high-mountain areas with humid-upper subhumid ombrotypes and with substantial winter snow cover) is recognised as the potential vegetation in areas with a supramediterranean (upper) and oromediterranean thermotype, and in high-mountain territories that can be included in the temperate macrobioclimate: supra- and orosubmediterranean (supra- and orotemperate with a submediterranean character).

Two types of genuinely orophilous vegetation types can be recognised throughout the Sistema Central: Iberian Scots pine forests (Pino-Juniperetalia: Avenello ibericae-Pinion ibericae) and creeping juniper woodlands (Juniperetalia hemisphaericae: Cytision oromediterranei). These formations constitute the genuine potential orotemperate vegetation with a submediterranean character (orosubmediterrarean) or genuinely orotemperate vegetation (Estrêla Range) found in territories with humid-hyperhumid ombroclimates in the Carpetan and Leonese siliceous high mountains sometimes as relictic vegetation. Deforestation, livestock farming practices, and the cultivation of allochthonous forest species often degrade the natural landscape in high-mountain areas of the Sistema Central; however the widespread seral stages, favoured since historic times by traditional livestock farming uses, form a characteristic landscape in these mountains.

A total of two woodland associations are currently recognised as potential natural vegetation within the territorial scope of the Sistema Central, corresponding to mature stages of Iberian Scots pine forest series (Avenello ibericae-Pinetum ibericae and Echinosparto barnadesii-Pinetum ibericae), and two associations – (climatophilous or edaphoxerophilous, and with a clearly chionophobous and rupestrian character): one represents creeping juniper woodlands in areas with an upper orosubmediterranean thermotype and humid-hyperhumid ombrotypes,

Fig. 13.6 Iberian Scots pine forests (Avenello ibericae-Pinetum ibericae) growing in the Guadarrama mountains. In the background, the Peñalara massif (Photo by Enrique Luengo)

widely distributed throughout the high mountains in the Sistema Central and containing a range of recognisable biogeographical strains (Avenello ibericae-Juniperetum alpinae s.l.); and a second, restricted to the orotemperate territories of the Serra da Estrêla (Lycopodio clavati-Juniperetum alpinae). As indicated, these types of mature vegetation are either partially or totally substituted by the set of seral stages that constitute the successional dynamics of the corresponding vegetation series.

The presence of natural Iberian Scots pine forests in the Sistema Central is restricted to small stands with a relict character that can be clearly recognised in the central (Sierra de Guadarrama in the broadest sense) and western massifs (Sierra de Gredos mountains in the broadest sense). They do not exist today in an evident and natural form in either the Sierra de Ayllón and the surrounding mountains, nor in the ranges of Béjar-Tormantos, Peña de Francia or in high-mountain areas in Portugal.

The natural orosubmediterranean Iberian Scots pine forests in the Sierra de Guadarrama (Avenello ibericae-Pinetum ibericae) grow at between 1700–2100 m in humid-hyperhumid areas, and constitute the mature stage of the corresponding vegetation series (Avenello ibericae-Pino ibericae sigmetum) (Fig. 13.6).

In the Gredos Range, orosubmediterranean Iberian Scots pine forests occupy an altitudinal band between 1600–1800 (1900) m and grow in humid-hyperhumid

Fig. 13.7 Small stands of Iberian Scots pine forests near the summit of the Gredos mountains (Echinosparto barnadesii-Pinetum ibericae, eastern massif, 1800 m) (Photo by Enrique Luengo)

territories (Echinosparto barnadesii-Pino ibericae sigmetum); the mature stage corresponds to an Iberian Scots pine forest with a relict character (Echinosparto barnadesii-Pinetum ibericae), where species such as *Echinospartum barnadesii*, *Erica arborea*, *Festuca gredensis* and others serve as differentials compared to the Guadarramean orosubmediterranean pine forests (Fig. 13.7).

Unlike the Iberian Scots pine forests, the orosubmediterranean and orotemperate climactic creeping juniper woodlands (Estrêla Range) are widely distributed throughout cacuminal areas in the highest mountains in the Sistema Central (Avenello ibericae-Juniperetum alpinae, Lycopodio clavati-Juniperetum alpinae), where they represent the mature stages of the potential vegetation with a chionophobous rupestrian character in territories with a humid-hyperhumid ombrotype between altitudes of (1900) 2000–2450 m. The most frequent landscape (orosubmediterranean and orotemperate areas) comprises vegetation mosaic complexes of both seral shrubby plant communities (broom) with a pyrophilous nature and the typical azonal vegetation on relict reliefs with a glacial and periglacial origin (grassland and peatland vegetation colonising glacial moraines, rock and scree beds and others). These genuinely orophilous communities are of special interest from the biogeographical point of view; they are the ongoing object of research into the ecological interactions in high-mountain habitats (Gavilán et al. 2002), and are also used to monitor the parameters affected by climate change at the global level (Global Observation Research Initiative in Alpine Environments – GLORIA– Project) (Fig. 13.8).

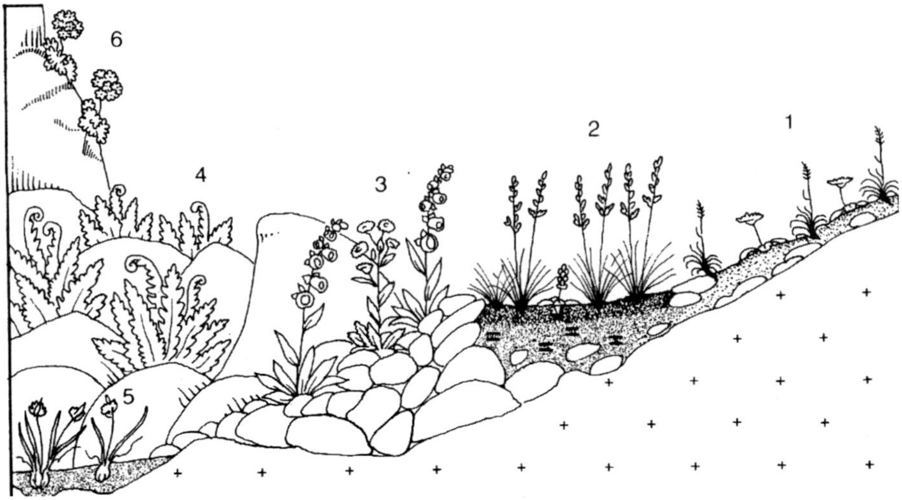

Fig. 13.8 Catena of the vegetation complex at the Peñalara summit (2492 m, Guadarrama mountains). 1. Psychroxerophilous cryorosubmediterranean grasslands (Hieracio myriadeni-Festucetum carpetanae); 2. Chionophile *Nardus stricta* grasslands or "cervunales" (Campanulo herminii-Festucetum ibericae); 3. Stabilized scree communities (Digitali carpetani-Senecionetum carpetani); 4. Scree fern communities (Cryptogrammo crispae-Dryopteridetum oreadis); 5. Wet rupicolous communities (Allietum latiorifiori); 6. Rock crevices communities (Saxifragetum willkommianae) (Adapted from Rivas-Martínez et al. 1990 by J.M. Pizarro)

13.3.1.2 Seral Broom Communities and Orophilous Vegetation

As seral stages, shrublands dominated by *Genisteae* (brooms) are widely predominant in the landscape; they have a marked pyrophilous character and an anthropic origin, and are structured by mountain broom (*Cytisus oromediterraneus*) accompanied by other *Genisteae* which account for their physiognomy (Cytision oromediterranei). We recognise the associations Senecioni carpetanae-Cytisetum oromediterranei (Sierra de Guadarrama), Festuco gredensis-Cytisetum oromediterranei (Paramera and Serrota ranges), Cytiso oromediterranei-Echinospartetum barnadesii (Sierra de Gredos), Echinosparto pulviniformis-Cytisetum oromediterranei (Béjar, Tormantos and Peña de Francia ranges), and Teucrio salviastri-Echinospartetum pulviniformis (Serra da Estrêla). Also significant among the seral vegetation are the xerophilous chamaephytic and hemicryptophytic communities (Hieracio castellani-Plantaginion radicatae), with a large number of endemics such as *Anthemis alpestris*, *Arenaria querioides*, *Bufonia macropetala*, *Centaurea amblensis*, *Centaurea janerii*, *Festuca carpetana*, *F. gredensis*, *F. rivas-martinezii*, *F. vettonica*, *Hieracium castellanum*, *Jasione sessiliflora*, *Jurinea humilis*, and *Koeleria crassipes*, among others; and therophytic grasslands with a pioneering character in the colonisation of granitic lithosols (Triseto ovati-Agrostion truncatulae), along

Fig. 13.9 Vegetation complexes: aquatic, spring, peatland and grassland communities grow in this territory as specialised plant communities. Béjar mountains, 1900 m (Photo by Enrique Luengo)

with the more ephemeral grasslands in the Sedion pedicellato-andegavensis alliance.

In soils with a favourable water balance these perennial pastures are replaced by *Nardus stricta* grasslands (Campanulo herminii-Nardion strictae), which may be hygrophilous (Festuco rothmaleri-Juncetum squarrosi) or influenced by snow (Campanulo herminii-Festucetum ibericae) (Fig. 13.9). If broom or pine forests are cut or burned, pyrophyte communities of the Linarietum niveae become established. Rock, crevice and cliff vegetation is a complex of communities rich in endemics (Figs. 13.10 and 13.11); communities of foxgloves (Digitali carpetanae-Senecietum carpetani) occur on large rocks that may be covered by snow. Vigorous pteridophyte communities (Cryptogrammo crispae-Dryopteridetum oreadis) appear on shady scree formed by large outcrops. Vegetation in the form of chasmophytic communities (Saxifragion willkommianae) appears in narrow crevices and on rocky cliffs. Finally, sedge communities (Caricion nigrae) are common oligotrophic peatland vegetation ("trampales").

Another broom community appears in the Gredos Range, rich in what are locally known as "cambriones" (*Echinospartum barnadesii*), Gredos chamomile (*Santolina oblongifolia*) and certain hemicryptophytes forming the association Cytiso oromediterranei-Echinospartetum barnadesii, which is physiognomically different from its Guadarramean counterpart. They are found at 1800 m, reaching roughly the eastern and central massif areas of the Sierra de Gredos. The use of fire and the

Fig. 13.10 Some notable endemics from the Sistema Central Range: left, *Antirrhinum grosii* from the Gredos central massif; right, *Misopates* (*Pseudomisopates*) *rivas-martinezii* from the La Serrota massif (Photos by Enrique Luengo)

Fig. 13.11 Some notable endemics from the Sistema Central Range (Guadarrama mountains): left, *Senecio boissieri* (Peñalara massif); right, *Erysimum penyalarense* (Cuerda Larga)

cutting of broom favour the development of pyrophyte plant communities such as the Linarietum niveae and extensions of "joragales" (Arenario querioidis-Festucetum gredensis), which give way to *Nardus stricta* grasslands (Campanulo herminii-Nardion strictae) when the soil-water balance is favourable.

Nardus stricta grasslands under long-lasting snow cover are common in the high altitude areas (Poo legionensis-Nardetum strictae), while hygrophilous grasslands can reach lower altitudes in the supramediterranean belt (Genisto anglicae-Nardetum strictae); rock communities are enriched with *Biscutella valentina* subsp. *pyrenaica* and *Scrophularia herminii*; chasmophytic communities are represented by the Hieracio amplexicaulis-Saxifragetum gredensis, and bogs are also common, although found very rarely in situations in the Guadarramean sector (Sedo lagascae-Eriophoretum latifolii).

Scree and rock vegetation are frequently rich in Gredos chamomile (Santolinetum oblongifoliae) and foxgloves (Digitali carpetanae-Senecietum carpetani) or orophilous ferns (Cryptogrammo crispae-Dryopteridetum oreadis). Highly specialised chasmophytic vegetation is represented by a set of diverse communities (Antirrhinetum grosii, Valerianetum tripteridis, Hieracio amplexicaulis-Saxifragetum gredensis). Therophytic grasslands develop during late spring and summer, and are considered a type of pioneer vegetation (Triseto ovati-Agrostion truncatulae), together with the more ephemeral Sedion pedicellato-andegavensis (Agrostio truncatulae-Sedetum lusitanici).

Box: The Genus *Festuca* and Its Significance Throughout the Iberian High Mountains and the Sistema Central

The genus *Festuca* L. is one of the most complex of the vascular plant genera due to its diversity and phenotypical plasticity (Clayton and Renvoize 1986). This is an orophilous genus and its taxa structure the mountain landscapes in regions with a temperate and Mediterranean climate throughout the world (where it reaches its maximum diversity), and on the summits of tropical high mountains. Practically three quarters of the known taxa are holartic. The Eurasian part alone holds almost 70% of all the *Festuca* taxa in the world (Ferrero and Fuente 1996; Ferrero 1999).

The Mediterranean region is home to a considerable percentage of these taxa, causing this biogeographic region to be considered one of the most important centres of diversity for the genus in the world. The high percentage of diploids and the presence of all levels of ploidy –from 2× to 12×– over broad territories suggest the importance of the Mediterranean in the evolution of the genus. The areas of greatest diversity coincide with the mountain massifs, particularly those oriented in an east-west direction, pointing to the significance of glaciations in the evolution and current distribution of *Festuca*.

Studies on the phylogenetic relationships between species of *Festuca* and other genera –while interesting– do not as yet provide sufficient data to enable new proposals to be made for the taxonomic treatment of the whole genus to replace the traditional taxonomic schemes.

Three subgenera are recognised for the Iberian Peninsula within those proposed by Clayton and Renvoize (1986): *Festuca*, which includes the

(continued)

greatest number of taxa; *Drymanthele* Krecz. & Bobr.; and *Schoedonorus* (P. Beauv.) Peterm. Of the 100 or so recognised Iberian taxa, at least 40% are endemic, and most are included in the subgenus *Festuca* and its respective sections.

Within the subgenus *Festuca*, the section *Festuca* includes the group with the highest taxonomic complexity worldwide; this is the group of species with a leaf section with a continuous schlerenchymatic ring. Markgraff-Dannenberg (1980) separated the Linnean taxon *Festuca ovina*, thus facilitating the subsequent taxonomic study of these groups. This complex is formed by the following taxa: *Festuca airoides*, *F. altopyrenaica*, *F. niphobia*, *F. ochroleuca* subsp. *bigorrensis*, *F. ovina* and *F. yvesii*, preferentially in the Pyrenees; *F. aragonensis* and *F. valentina* in the Iberian System; *F. hystrix*, *F. indigesta*, *F. longiauriculata* and *F. segimonensis* in the Cazorla-Segura and Sierra Nevada ranges; *F. carpetana*, *F. gredensis*, *F. summilusitana* and *F. vettonica* in the Sistema Central, Montes de León and Iberian System (Fig. 13.12). This complex includes *F. brigantina* on serpentinised peridotites in northeast Portugal (Tras-os-Montes and Alto Douro province). In caryological terms these are polyploid complexes, and in most mountain massifs there is a coexistence of taxa with different ploidy levels; hexaploids are particularly abundant, as evidenced by their success in colonising the orophilous areas of the Mediterranean biogeographic region, where they constitute the dominant vegetation in uplands and high summits. There are diploids with a relict character in a restricted area in the Sistema Central (*Festuca henriquesii*, *F. vettonica*) and Baetic ranges (*F. longiauriculata*); the recognised tetraploids are located in the Pyrenees (*F. niphobia*) and Iberian System –Moncayo– (*F. aragonensis*). The subordination of some of these taxa to *Festuca indigesta* (hexaploid), originating in the Sierra Nevada, cannot in our opinion be justified by either the caryological or morphological data. The current data suggest a speciation of hexaploid polyploids (and sometimes octoploids and decaploids) in each mountain massif in an isolated and independent way from different and related ancestors. The various taxa of *Festuca* have successfully colonised a wide range of high-mountain biotopes on the Iberian Peninsula; the extreme ecological conditions to which they are subjected (long lasting snow cover, intense radiation due to summer insolation, wind and so on) have led to a high degree of specialisation, and adaptation to those harsh environmental factors has resulted in the emergence and development of some very characteristic morphological structures. Some of the most noteworthy include tunicate and intravaginal foliar innovations, leaf folding to protect the stomata, thick structures in the subepidermal schlerenchyma, and the presence of siliceous cells, thorns, trichomes and waxes that cover the epidermal surfaces of leaves and flowers (Fig. 13.13).

(continued)

Fig. 13.12 Distribution map of some orophilous taxa of *Festuca* in the mountains of the Iberian Peninsula: *F. aragonensis*, *F. carpetana*, *F. gredensis*, *F. henriquesii*, *F. summilusitana* and *F. vettonica*

In the Sistema Central we basically recognise the following taxa:

- Subgenus *Festuca*
  - Section *Festuca*:
    - *Festuca ampla* Hack., *Cat. Rais. Gramin. Portugal*: 26. 1880
    - *Festuca carpetana* Fuente, Sánchez Mata & Rivas Martinez. *Lazaroa* 35: 133. 2014
    - *Festuca gredensis* Fuente & Ortúñez in *Lazaroa* 21: 4. 2000
    - *Festuca henriquesii* Hack., *Monogr. Festuc. Eur.*: 126. 1882
    - *Festuca hystrix* Boiss., *Elench. Pl. Nov.*: 89. 1838
    - *Festuca rivas-martinezii* Fuente & Ortúñez in *Bot. J. Linn. Soc.* 114: 25. 1994

(continued)

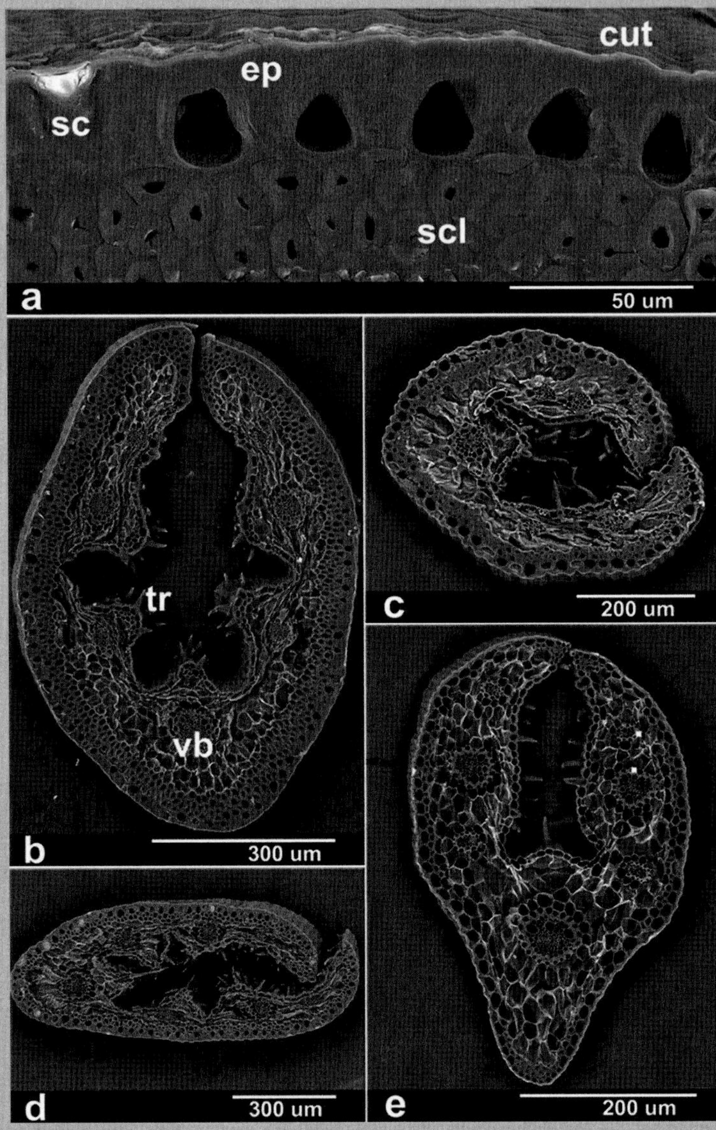

Fig. 13.13 Comparative scanning electron micrographs of blade cross section of *Festuca* species from the Sistema Central Range and *Festuca ovina* from Norway: (**a**), image showing the thickness of the cuticles, epidermal cells and the presence of several rows of sclerenchymatous cells of *F. vettonica* (La Serrota, 2000 m); (**b**), *F. gredensis* (Sierra de Gredos, 2150 m); (**c**), *F. carpetana* (Sierra de Guadarrama, 2100 m); (**d**), *F. summilusitana* (Serra da Estrêla); (**e**), *F. ovina* (Jotunheimen National Park, Norway, 1400 m, in creeping hemicryptophytic and dwarf-shrub tundra). Abbreviations: cut, epidermis cuticle of leaf blade; ep, foliar epidermis; sc, silica cell; scl, sclerenchyma cells; vb, vascular bundle; tr, trichome

(continued)

  - *Festuca summilusitana* Franco & Rocha Afonso in *Bol. Soc. Brot.* ser. 2, 54: 94. 1980
  - *Festuca vettonica* Fuente, Ortúñez & Ferrero in *Lazaroa* 20: 5. 1999

- Section *Aulaxyper* Dumort.

  - *Festuca braun-blanquetii* (Fuente, Ortúñez & Ferrero) Rivas Mart. & al. in *Itinera Geobot.* 13: 417. 1999
  - *Festuca duriotagana* Franco & Rocha Afonso in *Bol. Soc. Brot. ser.* 2, 54: 91. 1980
  - *Festuca iberica* (Hack.) K. Richt., *Pl. Eur.* 1: 99. 1890
  - *Festuca rivularis* Boiss., *Elench. Pl. Nov.*: 90. 1838
  - *Festuca rothmaleri* (Litard.) Markgr.-Dann. in *Bot. J. Linn. Soc.* 76: 325. 1978

- Section *Eskia* Willk.

  - *Festuca merinoi* Pau in *Bol. Soc. Aragonesa Ci. Nat.* 1(3): 51. 1902

- Section *Subbulbosae* Nyman ex Hack.

  - *Festuca durandoi* Clauson in Billot, *Annot. Fl. France Allemagne*: 163. 1859
  - *Festuca paniculata* subsp. *multispiculata* Cebolla & Rivas Ponce in *Lagascalia* 15 (extra): 408. 1988

- Subgenus *Schoedonorus* (P. Beauv.) Peterm.

  - Section *Schoedonorus*

    - *Festuca arundinacea* Schreb., *Spic. Fl. Lips.*: 57. 1771
    - *Festuca fenas* Lag., Nov. Gen. Sp.: 4. 1816
    - *Festuca mediterranea* (Hack.) Rouy ex Prain, *Index Kew, Suppl.* 5: 106. 1921

In the Sistema Central there are two easily recognisable species of *Festuca* (section *Festuca*), whose leaves have discontinuous and generally decurrent sclerenchyma: *F. henriquesii* and *F. rivas-martinezii*. *F. henriquesii* has leaves which all have a flat section, and this species appears to be distant from the rest of the intravaginal taxa in the *Festuca* section. Its geographic isolation and its primitive and exclusive morphological features point to its paleoendemic character, although in caryological terms its symmetry is more typical of a young taxon than a primitive ancestor. The explanation for this paradox may perhaps lie in the orophilous character of *F. henriquesii* and in the direct action of glaciation in the Serra da Estrêla Range in Portugal.

Festuca carpetana is also an important element in the psychroxerophilous hemicryptophyte grasslands developed throughout the summit areas of the

(continued)

Guadarrama, Ayllón, Neila, Urbión and Demanda ranges growing on siliceous substrates. Its communities are widespread throughout supramediterranean, submediterranean supratemperate and submediterranean orotemperate high-mountain landscapes in central and central-north-eastern territories of the Iberian Peninsula, and framed within the alliance Minuartio-Festucion carpetanae.

Festuca vettonica (single diploid species) grows in La Serrota and the eastern Gredos above 2000 m, in clear regression compared to *F. gredensis* (hexaploid), which extends densely to the summits of all the massifs in the eastern, central and western Gredos, forming part of the orophilous silicicolous vegetation (Fig. 13.14). Similarly, *F. summilusitana*, whose populations have the highest level of ploidy (10x, decaploid) in the group, grows on the summits of the Serra da Estrêla in Portugal and in the massifs of Los Ancares and Montes de León. *F. summilusitana*'s populations with a decaploid ploidy level (10x) grow in the massifs of Los Ancares and Montes de León, whereas populations with decaploid and dodecaploid (12x) specimens –the highest level known in this group (Loureiro et al. 2007)– are found coexisting at the summits of the Serra da Estrêla in Portugal.

Fig. 13.14 *Festuca gredensis*. Gredos mountains (central massif, 2100 m) (Photo by Enrique Luengo)

13.3.2 *Supramediterranean Vegetation*

13.3.2.1 Mountain Forests: Iberian Scots Pine (*Pinus sylvestris* var. *iberica*) Forests

Below the orosubmediterranean level, the extension of natural Iberian Scots pine forest is currently reduced in the Sistema Central, unfortunately; in many areas – particularly in the westernmost mountains in the range– it has completely disappeared or been relegated to small forest enclaves that are protected from the continuous forest fires to which they are submitted, and frequently associated to extremely scarce populations of a particular strain of Spanish black pine (*Pinus* aggr. *nigra*) we now assume as *Pinus nigra* subsp. *salzmannii* in a recent study (López-Sáez et al. 2016). We can recognise their potential evidences in these deforested areas today by studying the corresponding vegetation series dynamics through the seral stages.

The diversity of mountain landscapes in the Sistema Central contains several vegetation series that have Iberian Scots pine forests as their mature forest stage. In some mountain massifs we can still recognise forests with an acceptable state of conservation, whereas in others (the westernmost massifs) they are today non-existent, or reduced to token stands with a relict nature. It should also be noted that Scots pine forests may evolve an edaphoxerophilous vegetation character in potential areas of Iberian white oak, beech and birch on sites that are unfavourable for the development of forests of *Fagaceae* and *Betulaceae* due to their topographic and soil conditions.

From east to west in the Sistema Central we can mention particularly the Iberian Scots pine forests in the Ayllón (Alto Rey) and Pela, Guadarrama, Gredos, Béjar and Tormantos ranges. There are natural Iberian Scots pine forests in the easternmost parts of the mountains in the Sistema Central and in the Pela and Alto Rey ranges. The Iberian Scots pine forests in the Sierra de Pela are located on carbonated substrates at an altitudinal interval of between 1370–1450 m due to the location of these mountains. These are supramediterranean (locally growing in lower oromediterranean areas) basiphilous relict mesoforests (Ononido aragonensis-Pinetum ibericae), where perennating shrubby and herbaceous species such as *Bupleurum ranunculoides*, *Juniperus communis* subsp. *hemisphaerica*, *Ononis aragonensis*, *Pulsatilla rubra* and *Vicia pyrenaica* characterise the floristic spectrum of this peculiar forest vegetation series (Ononido aragonensis-Pino ibericae sigmetum). However at altitudes of between 1520–1800 m in the Alto Rey mountains, the Iberian Scots pine forest series is characterised by siliceous soils (developed on quartzites, slates and sandstones). These supramediterranean, suprasubmediterranean and orosubmediterranean (orotemperate with submediterranean character) subcontinental Iberian Scots pine forests are present throughout territories with a humid-hyperhumid ombroclimate (Erico aragonensis-Pinetum ibericae). The species in the communities in this vegetation series (Erico aragonensis-Pino ibericae sigmetum) include particularly the common juniper

(*Juniperus communis* subsp. *hemisphaerica*), several *Ericaceae* (*Calluna vulgaris*, *Erica arborea*, *Erica australis* subsp. *aragonensis*), laurel-leaf cistus (*Cistus laurifolius*), bear berry (*Arctostaphylos uva-ursi*) and perennating herbaceous species such as *Avenella flexuosa* subsp. *iberica*. Another noteworthy aspect of the whole floristic spectrum of the communities in this vegetation series is the negligible presence of *Genisteae* species.

Iberian Scots pine forests have their altitudinal optimum between 1450–1700 m in the mountain massifs of the Sierra de Guadarrama (territories in the central and central-western biogeographical sector of the Sierra de Guadarrama), where they develop as potential vegetation in supramediterranean subhumid-humid or suprasubmediterranean territories (supratemperate with a submediterranean character); on occasion these pine forests (Pteridio aquilini-Pinetum ibericae) can behave like vegetation with an edaphoxerophilous character in an environment of Iberian white oak, birch and beech. The characteristic taxa in the supramediterranean Guadarramean Scots pine forest series (Pteridio aquilini-Pino ibericae sigmetum), in addition to *Pinus sylvestris* var. *iberica*, include particularly *Avenella flexuosa* subsp. *iberica*, *Conopodium pyrenaeum*, *Cytisus oromediterraneus*, *Festuca braun-blanquetii*, *Galium rotundifolium*, *Genista florida*, *Juniperus communis* subsp. *hemisphaerica*, *Luzula fosteri*, *Pteridium aquilinum*, *Sanicula europaea*, *Veronica officinalis*, *Viola riviniana* and *Viscum album* subsp. *austriacum*, among others.

Finally, in the westernmost territories of the Sistema Central (Gredos, Béjar and Tormantos ranges) the potential area of Scots pine forests extends throughout the supramediterranean and suprasubmediterranean humid mountain areas at an altitudinal range of between 1450–1600 m. These Scots pine forests (Festuco merinoi-Pinetum ibericae) (Fig. 13.15) often incorporate, as mentioned earlier, some individuals from a particular type of Spanish black pine we now assume as *Pinus nigra* subsp. *salzmannii* (also found structured some scattered and reduced relict forests recently studied by López-Sáez et al. 2016). Both the forest stage of this vegetation series (Festuco merinoi-Pino ibericae sigmetum) and the seral stages (grasslands, scrub broom and scrub chamaephytic communities) have a sharply differential characteristic floristic spectrum compared to the rest of the Scots pine series in the Sistema Central. The most notable taxa include *Anthemis alpestris*, *Arenaria querioides*, *Centaurea amblensis*, *Cytisus multiflorus*, *Erysimum merxmuelleri*, *Festuca gredensis*, *F. merinoi*, *F. paniculata* subsp. *multispiculata*, *Genista cinerascens*, *Genista florida*, *Ornithogalum concinnum*, *Santolina oblongifolia*, among others.

13.3.2.2 Mountain Forests: Relict Birch (*Betula celtiberica*) and Beech (*Fagus sylvatica*) Forests

Birch and beech forests have an eurosiberian optimum, with a clear affinity for more northerly latitudes. Their presence in reduced and disjointed enclaves in the Sistema Central highlights their relict character from recent Quaternary eras, in

Fig. 13.15 Gredos Iberian Scots pine woodland (Festuco merinoi-Pinetum ibericae) in the high area environment of La Rubia (Gredos mountains, 1500 m) (Photo by Enrique Luengo)

which the prevailing cooler and rainier climate favoured their expansion towards the south. Their current distribution, limited mainly to opposing extremes with the highest rainfall in the whole mountain range, also illustrates the dual route of penetration in the Sistema Central of elements with a Euro-Siberian and Atlantic optimum: the Sistema Ibérico and its connection with the Sierra de Ayllón on the one hand; and the western Cantabrian Galician-Portuguese and Orensian-Sanabrian ranges in the western Carpetan ranges on the other.

In concordance with their relict character, the Carpetan beech and birch forests are located particularly in upper and rainier levels (humid and hyperhumid ombroclimates) in the supramediterranean belt, generally in refugia at the head of north-facing valleys, open to humid winds and/or subjected to intense snowfall, and even on soils with a degree of water stagnation.

The flora in these forests partly coincides with that of other Carpetan deciduous and silicicolous flora –Iberian white oak (*Quercus pyrenaica*)–, but is distinguished by having –sometimes with an exclusive character– a large number of demanding ombrophilous taxa, rare in the mountain range and with a northern optimum. The arboreal elements include the sessile oak (*Quercus petraea*), common ash (*Fraxinus excelsior*), holly (*Ilex aquifolium*) and yew (*Taxus baccata*), in addition to the birch itself (*Betula celtiberica*) and the beech (*Fagus sylvatica*); and the herbaceous species *Asperula odorata*, *Blechnum spicant*, *Ceratocapnos claviculata*, *Corydalis cava*, *Dryopteris expansa*, *Epilobium montanum*, *Galium rotundifolium*, *Luzula sylvatica* subsp. *henriquesii*, *Paris quadrifolia*, *Saxifraga spathularis*, *Scrophularia alpestris*, *Stellaria nemorum*, *Stellaria holostea* and *Veronica officinalis*, among others.

Within the Carpetan and Leonese biogeographical subprovince, beech is only present at the forest scale in the easternmost end (Ayllonensian subsector), where its most famous examples are the beech forests of Montejo de la Sierra, Puerto de la Quesera and the head of the Lillas river (Cantalojas). Outside these enclaves, the presence of some beech is known from the Sierra de la Peña de Francia. The beech forests in the Ayllonensian subsector constitute a particular series (Galio rotundifolii-Fago sylvaticae sigmetum) representing an impoverished and relict southern irradiation of the homologous oroiberian series (Ilici aquifolii-Fago sylvaticae sigmetum). Its seral stages are similar to those of the ombrophilous Ayllonensian Iberian white oak woodlands: shrubby, broom-rich fringes in tree heaths, and heathlands of the Halimio ocymoidis-Ericetum aragonensis, among others.

Two associations of birch woodland are known in the Sistema Central. The continental association (Melico uniflorae-Betuletum celtibericae), represented particularly in the Sierra de Guadarrama biogeographical sector (birch woodlands in the Somosierrra, Canencia and Peñalara), occupies reduced enclaves in some shady areas and rainy valley heads, with intense snow coverage, often on soils with water stagnation. The oceanic birch woodlands in the association Saxifrago spathularis-Betuletum celtibericae have their optimum in the Orensian-Sanabriense and Galician-Portuguese mountains, with minor representations in the Serra da Estrêla biogeographical sector. Certain nemoral elements of a western character (*Saxifraga spathularis*), shrubby fringes with *Genista florida* subsp. *polygaliphylla* (Cytiso striati-Genistetum polygaliphyllae), and the heathlands of Junipero alpinae-Ericetum aragonensis, are characteristic of these birch woodlands. In addition to this type of birch woodland there are relict populations of *Betula fontqueri* as pioneering colonisers of scree beds with large blocks in high-mountain areas with periglacial geomorphological reliefs in the massifs in the Béjar-Tormantos ranges, and in all probability, in similar habitats in the Guadarrama mountains.

The relict character of these Carpetan woodlands implies that their destruction is generally irreversible. For this reason, and considering the importance of the flora they contain, they are covered by maximum legal protection measures.

13.3.2.3 Mountain Forests: Juniper Woodlands of *Juniperus thurifera* or "Sabina Albar"

Juniper (*Juniperus thurifera*) woodlands, one of the most singular forest communities in the western Mediterranean, are very open woodlands typical of extreme continental climates that have their optimum in very cold, dry periods of the recent Quaternary era. They currently have a fragmentary and disjunct distribution that shows symptoms of regression compared to the forests of oak species, competitively favoured by recent climate change. On the peninsula, their distribution has its main focus in the Alcarreñan-Turolensian and Sorian uplands, with irradiations that extend throughout the vast Iberian continental arc: the Castilian face of the Cantabrian range, depression of the Ebro river and the eastern Manchegan edge. They also survive within the Sistema Central in certain calcareous corridors –such as those of Segovia, Tamajón and Torrelaguna– that border on or penetrate into the Sierra de Guadarrama biogeographical sector, the most continental of the Carpetan ranges. Currently at least three enclaves of incense juniper woodlands are conserved growing on siliceous substrate: Tamajón, Lozoya and Prádena.

In spite of the significant and successional floristic divergences compared to the basophilous Alcarreñan and Maestracensian incense juniper woodlands in the association]uniperetum hemisphaerico-thuriferae, the biogeographical relation and the relict character of these Guadarramean silicicolous juniper woodlands suggests they should be treated as a subassociation of this (juniperetosum lagunae). The floristic composition and the seral communities have close affinities with those of the supramediterranean silicicolous holm oak woodlands (Junipero lagunae-Quercetum rotundifoliae), with introgressions between both types of forests. The dominance of juniper woodlands accompanied by common juniper (*Juniperus oxycedrus* subsp. *lagunae*, *Juniperus communis* subsp. *hemisphaerica*) and often also with holm oak, together with other shrubs in the seral stages, constitute the best differential feature of these relict woods deserving of protection.

The vegetation of other basophilous and meso- (Asparago acutifolii-Querco rotundifoliae sigmetum) and supramediterranean (Junipero thuriferae-Querco rotundifoliae sigmetum, Cephalanthero-Querco fagineae sigmetum) Manchegan and Celtiberian-Alcarreñan series penetrate to varying degrees in the Sierra de Guadarrama biogeographical sector through these calcareous alignments. In some cases such as in the sedimentary limestone outcrops of the El Paular valley or those of Soto del Real, these basophilous vegetation series constitute islands within the aforementioned biogeographic sector.

13.3.2.4 Mountain Forests: Iberian White Oak (*Quercus pyrenaica*) Forests ("Melojares")

The forests of Iberian white oak (*Quercus pyrenaica*) (Fig. 13.16) constitute the most widespread potential vegetation type in the supramediterranean belt of the Carpetan and Leonese biogeographical subprovince. Strictly silicicolous and with greater rainfall requirements than the holm oak (*Quercus rotundifolia*), they alternate with them –particularly in the more continental sectors– according to orientation under subhumid ombroclimates, and displace them completely in humid regimes (Gavilán et al. 2007). Their altitudinal interval increases in direct relation to the precipitation gradients in the mountain range, and thus in western sectors they ultimately occupy the whole breadth of the supramediterranean belt and may even descend to the mesomediterranean belt in the southern valleys directly exposed to

Fig. 13.16 Detail of a mature branch of Iberian white oak (*Quercus pyrenaica*) with fruits (acorns) (Photo by Enrique Luengo)

south-westerly winds, and which are extremely rainy as a result (humid-hyperhumid). A particular mesomediterranean series is distinguished (Arbuto unedonis-Querco pyrenaicae sigmetum) in these areas, with distribution in the Luso-Extremaduran biogeographic subprovince.

The territorial and bioclimatic breadth of the Carpetan-Leonese Iberian white oak woods is a source of great diversity, in which their behaviour and dynamism depend on the rainfall and continentality gradients (López-Saez et al. 2015). Based on these factors, two series can be distinguished with an optimum humid-hyperhumid ombroclimate distributed in the extremes of the range: the biogeographical sectors of the Salamanca and Montemuro-Estrêla ranges; and the territories in the Ayllon range, both separated by the vast Guadarramean-Gredensian massifs. Its greater floristic richness in nemoral elements and its seral stages of heathlands are differential compared to the subhumid series, of which three can be distinguished: one central-eastern Guadarramean; another central-western Gredensian; and the westernmost and with the greatest oceanic influence – Salamancan–, all with cistus shrublands as a second regressive woody stage.

The dynamic patterns of these series are closely related. The shrubby fringes, predominant as a result of extensive grazing systems and forestry management in the form of thinning, correspond to shrubby broom formations structured by *Genisteae*, which in the westernmost series are enriched with broom and "cambriones" (spiny *Genisteae*). Only in some Iberian white oak woods on somewhat hygrophilous soils and in ecotone areas with riparian forests is the fringe represented by brambles in the alliance Pruno-Rubion ulmifolii. The scrub layer on soils decapitated by erosion corresponds to shrublands of *Cistus* and Cistion laurifolii in the subhumid series, and heathlands (Ericenion aragonensis) in the humid-hyperhumid series.

Other notable woodland communities are the perennial herbaceous fringes growing in the subsciophilous environment of the forest boundaries and cleared stages, on soils enriched with nitrogen, phosphorus and other nutrients due to the input from the litterfall from the tree canopy. Frequent plants in these communities, which are phytosociologically part of the alliance *Linarion triornithophorae* (Trifolio-Geranienea, Trifolio-Geranietea), are *Centaurea lingulata*, *Clinopodium vulgare*, *Geranium sanguineum*, *Lathyrus niger*, *Origanum virens*, *Silene nutans*, *Silene latifolia*, *Teucrium scorodonia*, *Trifolium ochroleucon*, *Vicia tenuifolia*, and others, in addition to *Physospermum cornubiense*, *Rhaponticum exaltatum*, and others, in the western sectors of the mountain range. The seral stages of grasslands are habitats with great diversity.

Among them it is worth mentioning the bentgrass meadows (Agrostio castellanae-Stipion giganteae) which have their optimum in the subhumid series, and the *Festuca merinoi* grasslands (Festucion elegantis), widespread in less humid climates in Bejaran-Gredensian territories on soils with developed humic horizons. Diverse chamaephytic communities developed on lithosols (Hieracio castellani-Plantaginion radicatae) are also frequent in the more eroded biotopes; particularly noteworthy are the pyschroxerophilous grasslands of *Festuca carpetana* (Sierra de Guadarrama biogeographical sector) and *F. gredensis* -western sectors-, characteristic of the upper levels of the supramediterranean belt. Pioneering therophytic grasslands correspond to several associations in the alliances Triseto ovati-Agrostion truncatulae and Sedion pedicellato -andegavensis. The grasslands of Agrostion castellanae are established on deeper and cooler soils; and the supramediterranean in the alliance Campanulo herminii-Nardion strictae, particularly in the ombrophilous series, may be widely developed. In managed grasslands we can highlight the meadows of false oat grass (Arrhenatherion elatioris, Cynosurion cristati), which are fairly widespread in mountain areas (Sánchez-Mata and Rodríguez-Rojo 2006).

The subhumid Iberian white oak woods reveal a relative floristic impoverishment compared to the more ombrophilous oak woods. Their tree layer frequently includes ivies (*Hedera helix*), honeysuckle (*Lonicera periclymenum* subsp. *hispanica*) and other elements that have their optimum in forest fringes. The nemoral herbaceous layer comprises a wide range of species, particularly several prevernal flowering geophytes and hemicryptophytes typical of deciduous forests (*Hepatica nobilis*, *Hyacinthoides non-scripta*, *Narcissus triandrus* s.l., *Primula veris*, *Viola riviniana*, certain orchids and others), and other plants such as *Arenaria montana*, *Aquilegia vulgaris* subsp. *hispanica*, *Aristolochia paucinervis*, *Brachypodium sylvaticum*, *Centaurea lingulata*, *Conopodium pyrenaeum*, *Epilobium lanceolatum*, *Geum sylvaticum*, *Holcus mollis*, *Lathyrus montanus*, *Luzula forsteri*, *Melica uniflora*, *Milium vernale*, *Moehringia trinervia*, *Poa trivialis*, and others.

The Ayllonensian humid-hyperhumid series (Festuco braun-blanquetii-Querco pyrenaicae sigmetum) has a parallel significance to that of other irradiations of vegetation types with an Atlantic character in the easternmost part of the Sistema Central range, and of which the most illustrative examples –discussed below– are the beech forests of the Galio rotundifolii-Fagetum sylvaticae. This oroiberian and Ayllonensian series (Fig. 13.17) occupies an intermediate bioclimatic space between the aforementioned relict beech forests and the subhumid Iberian white

Fig. 13.17 Ayllonensian "melojares" (Festuco braun-blanquetii-Quercetum pyrenaicae). These marcescent forests grow on the north slopes of the Ayllón mountains (territories with a supramediterranean thermotype) (Photo by Jorge Martínez)

oak woods of the Luzulo forsteri-Querco pyrenaicae sigmetum, also present in Ayllonensian territories. Differential with these Iberian white oak woods are *Asperula odorata*, *Helleborus foetidus*, *Pulmonaria longifolia* and *Stellaria holostea*, among other nemoral elements; and also seral heathlands of Halimio ocymoidis-Ericetum aragonensis. The shrubby fringes are similar, but with a greater preponderance of tree heath (*Erica arborea)*, a common occurrence under these ombroclimatic conditions. Finally, there are significant extensions of subhygrophilic grasslands –"cervunales" and "vallicares"–, most certainly linked to the same factor.

The Guadarramean series, which reaches the most continental Bejaran-Gredensian territories (Luzulo forsteri-Querco pyrenaicae sigmetum), is the most continental and impoverished series, in spite of which it has an important territorial and altitudinal diversity. Thus in addition to the mature stage of the typical subassociation (Luzulo forsteri-Quercetum pyrenaicae subass. typicum), we can distinguish several forest subassociations with a supramediterranean optimum throughout the Guadarramean territories, which should perhaps constitute a specific series: subass. paeonietosum broteroi (warm and somewhat drier levels of the supramediterranean belt), subass. deschampsietosum ibericae (upper-supramediterranean Iberian white oak woods), subass. galietosum rotundifolii (ecotone and influence of birch), subass. aretosum maculati (soils with a degree of hydromorphia and contacts with supramediterranean riparian ash woods). The shrubby fringes correspond to several associations in the alliance Genistion floridae: Cytiso scoparii-Genistetum floridae and Genisto floridae-Adenocarpetum hispanici in the Sierra de Guadarrama biogeographical sector, Thymo mastichinae-Cytisetum multiflori in the Béjar and Sierra de Guadarrama biogeographical sector, and Genisto cinerascentis-Cytisetum oromediterranei in the higher supramediterranean levels of both sectors. The seral scrubland consists of cistus and Spanish lavender belonging to various associations: Santolino rosmarinifoliae-Cistetum laurifolii, Erico arboreae-Arctostaphyletum crassifoliae (Guadarramean), Halimio ocymoidis-Cistetum laurifolii (Ayllonensian) and Rosmarino officinalis-Cistetum ladaniferi (restricted to the warmer supramediterranean levels). These communities are absent from most of the territories in the Bejaran-Gredensian sector.

The Bejaran-Gredensian subhumid-humid Iberian white oak woods (Festuco merinoi-Querco pyrenaicae sigmetum) develop in the supramediterranean belt of the westernmost districts of the Bejaran-Gredensian mountain sector (Fig. 13.18). The mature stage of the series (Festuco merinoi-Quercetum pyrenaicae) corresponds to a forest rich in geophytes and nemoral species such as *Arabis stenocarpa*, *Allium massaesylum*, *Anthericum liliago*, *Arenaria montana*, *Lilium martagon*, *Luzula forsteri*, *Ornithogalum pyrenaicum*, *Paeonia broteroi*, *Paeonia officinalis* subsp. *microcarpa*, *Polygonatum odoratum*, *Vicia sepium* and others. In some enclaves in these Iberian white oak woods there are some endemics or species with biogeographical importance such as *Actaea spicata*, *Convallaria majalis*, *Genista falcata*, *Pulsatilla alpina* subsp. *apiifolia*, *Rhaponticum exaltatum*, *Senecio coincyi* (Hoyocasero, Ávila) and *Trollius europaeus* (Sierra de Villafranca mountains in Piedrahita), among others. The shrubby seral stages of these Iberian white oak woods correspond either to the Cistion laurifolii, or to shrublands of white broom (Thymo mastichinae-Cytisetum multiflori), which at high altitudes are replaced by mountain broom, here in transition towards potential Scots pine forests (Genisto cinerascentis-Cytisetum oromediterranei). The perennial graminoid communities basically correspond to "cerrillares" (Leucanthemopsio pallidae-Festucetum merinoi), whereas thyme scrublands of hemicryptophytes and chamaephytes (Hieracio castellani-Plantaginion radicatae) are predominant in clearings. The Bejaran-Gredensian woods of *Juniperus oxycedrus* subsp. *lagunae* (Festuco merinoi-Juniperetum lagunae) have their optimum in the climactic environment of these Bejaran-Gredensian Iberian white oak woods on lithosols, with negligible edaphogenesis. This is a relict series with an edaphoxerophilous character, very widespread in the granitic formations in the Bejaran-Gredensian mountains.

Finally, we should comment briefly on the Iberian white oak woods developing in the westernmost territories of the Iberian Sistema Central: the Salamancan-Estrelensian humid-hyperhumid series (Holco mollis-Querco pyrenaicae sigmetum) grows on the Peña de Francia massif above the subhumid Iberian white oak woods (Genisto falcatae-Querco pyrenaicae sigmetum), and spreads throughout the Gata mountain ranges and the surrounding areas until the Serra da

Fig. 13.18 Gredensian "melojares" (Festuco merinoi-Quercetum pyrenaicae). These marcescent forests grow in the Gredos mountains (territories with a supramediterranean thermotype). The pictured territory shows a big deforested area by human activities covered by seral vegetation communities

Estrêla in Portugal, where it reaches its greatest altitudinal scope. Elements with a Western optimum such as *Linaria triornithophora*, *Erythronium dens-canis*, *Physospermum cornubiense* and *Omphalodes nitida* characterise the nemoral herbaceous layer of these Iberian white oak woods, where there is a frequent presence of *Galium rotundifolium*, *Potentilla montana*, *Melittis melissophyllum* and even –more sporadically– *Quercus robur*. The shrubby fringes dominated by *Genista florida* subsp. *polygaliphylla* (Genistion polygaliphyllae) and *Cytisus striatus* correspond mainly to the associations Cytiso striati-Genistetum polygaliphyllae (Genistion polygaliphyllae), or –on soils with less capacity for water retention– Cytiso multiflori-Echinospartetum iberici (Cytisenion multiflori). Seral heathlands belong to the association Junipero alpinae-Ericetum aragonensis, which is clearly differentiated from the Ayllonensian association due to the presence of western elements such as *Erica umbellata* and *Halimium alyssoides*.

13.3.3 Meso-Supramediterranean Vegetation

13.3.3.1 Foothill Woodlands: Iberian Evergreen, or Holm oak ("Encinares") Woods

Woodlands of holm oak (*Quercus rotundifolia*) (Fig. 13.19) have a basal or foothill significance in the Sistema Central. They are widely extended as potential vegetation on both Iberian plateaus and barely venture up the lowest foothills of the mountains, except in the most inland and continental ranges, where the summer drought favours their altitudinal expansion to the detriment of other competing forests with greater moisture requirements, such as Iberian white oak woods. Both types of *Quercus* forests constitute the potential vegetation of broad areas of the Sistema Central and extend throughout territories with meso- and supramediterranean thermotypes (Gavilán et Fernández-González 1997).

In general, the holm oak woodlands in the Sistema Central (Carpetan) are distributed throughout territories with meso- and supramediterranean thermotypes. They represent the exclusive potential vegetation in dry ombroclimates, alternating –based on their exposure and soil depth– with Iberian white oak woods in subhumid ombroclimates. When they are well structured they constitute forests with a dense perennial-leafed tree layer dominated by oak, and with juniper (*Juniperus oxycedrus* subsp. *lagunae*) as a frequent secondary tree; in these situations there is a frequent presence of juniper stands ("enebrales") growing on areas with hard substrates (lithosols). As is habitual in continentalised climates, the understorey is particularly poor in perennial-leafed and creeper-type shrubby components (Fig. 13.20). Shrubs such as *Daphne gnidium* and *Ruscus aculeatus*, creepers such as *Lonicera etrusca*, *Rubia peregrina* and some sedges (*Carex distachya*, *Carex hallerana*) are the scarce representatives of the class; *Doronicum plantagineum* and *Paeonia broteroi* are mediterranean-iberoatlantic elements of the alliance; wild asparagus (*Asparagus acutifolius*) and *Phillyrea angustifolia* differentiate the versions in areas with mesomediterranean thermotypes.

We can differentiate two types of Carpetan holm oak woodlands based on the continentality gradient. The most continental series (Junipero lagunae-Querco rotundifoliae sigmetum) extends widely throughout the Sierra de Guadarrama biogeographical sector and also reaches as far as some areas in the Sierra de Béjar and Gredos biogeographical sector (Paramera-Serrota mountains); their greatest altitudinal scope is on the southern slope of the Sierra de Guadarrama, where they can reach heights of 1200–1400 m on steep and sunny rock faces, and in some extremely continental mountains in the Ávila biogeographical district where they practically contact with some types of oromediterranean vegetation. In contrast, on the northern faces they rarely exceed 1000–1100 m. Notable effects of altitudinal inversion with regard to Iberian white oak woods can be seen in some valleys with an orientation parallel to the mountain range.

The floristic poverty of the Guadarramean holm oak woodlands with junipers is further accentuated in its supramediterranean versions, which are considered

Fig. 13.19 Detail of a mature branch of holm oak (*Quercus rotundifolia*) with fruits (acorns)

Fig. 13.20 Holm-oak woodlands ("encinares") in the Guadarrama mountains (Junipero lagunae-Quercetum rotundifoliae) (Photo by Enrique Luengo)

independent subassociations. Other noteworthy floristic variations are the presence of Portuguese oak (*Quercus faginea*) in some shady sites, the presence of cork oaks (*Quercus suber*) in small valleys, or terebinths (*Pistacia terebinthus*) in the western mesomediterranean parts of the Sierra de Guadarrama biogeographical sector, anticipating the transition towards the Iberian oak forests with Iberian pear (Pyro bourgaeanae-Quercetum rotundifoliae) from the Toledo and Tagus biogeographical sector (Luso-Extremaduran biogeographical subprovince).

13.3.3.2 Seral Communities (Bush, Shrub and Grassland Vegetation)

The dynamic patterns of the Carpetan holm oak woodlands are very homogeneous. The shrubby form of this oak itself tends to constitute the first forest layer or the pre-forest states of garrigue. In the Guadarramean holm oak woodland, when the soil conserves its forest structure, the elimination of holm oak favours the development of broom-type shrubby stands (Cytisetea striato-scoparii), including broom (*Retama sphaerocarpa*) in the mesomediterranean layer (Cytiso scoparii-Retametum sphaerocarpae) and flowering broom (*Genista florida*) in the supramediterranean layer (Cytiso scoparii-Genistetum floridae). In the detritic sandy substrates of the Arévalo district –massively repopulated with maritime and stone pines– there is a common presence of "cambrionales" of *Adenocarpus aureus* (Lavandulo pedunculatae-Adenocarpetum aurei). These broom communities show a wide distribution throughout the Sistema Central range landscapes (Gavilán et al. 2011).

Shrublands of gum cistus and Spanish lavender (Rosmarino officinalis-Cistetum ladaniferi) grow on decapitated soils and steppe rockrose (Santolino rosmarinifoliae-Cistetum laurifolii) in some subhumid supramediterranean zones. Perennial meadows in the series correspond above all to "lastonares" (Centaureo ornatae-Stipetum lagascae, with a mesomediterranean and lower supramediterranean optimum) and bentgrass (Arrhenathero baetici-Stipetum giganteae, upper mesomediterranean and supramediterranean), growing on relatively deep soils with well-developed organic horizons. On very degraded lithosols there is an extensive presence of xerophilous perennial meadows and thyme scrublands in the alliance Hieracio castellani-Plantaginion radicatae. The "vallicar" of *Agrostis castellana* (Agrostion castellanae) has its optimal development in the adjoining edaphohygrophilous series, and may only secondarily occupy significant extensions in the holm oak forest series. The most productive meadows in terms of livestock farming are the "majadales" structured by *Poa bulbosa* (Periballio involucratae-Trifolion subterranei: Trifolio subterranei-Poetum bulbosae (mesomediterranean), and Festuco amplae-Poetum bulbosae (supramediterranean). The development of the serial stages of the holm oak forests, in conjunction with anthropic activity, favours the establishment and development of populations of allochthonous species with a potentially invasive character (Gavilán et al. 2016); environments such as motorway verges, the edges of paths, central verges and embankments in road infrastructures, walls and other elements are some of the preferred habitats for their expansion, on many occasions inadvertently caused by humans through their cultivation.

The western Carpetan series of the holm oak (Genisto hystricis-Querco rotundifoliae sigmetum) begins when the head of the Adaja river and the basin of the Zapardiel river are transposed towards the west to enter the valleys of Corneja and Tarabancos, respectively, and there is a frequent presence of some elements with a western optimum such as *Euphorbia oxyphylla*, *Genista hystrix* and *Genista tournefortii*. The series extends across the Salamancan plain and the northern foothills of the Béjar, Peña de Francia, Gata and Malvana mountain ranges. The greater spring rainfall, the accentuation of the summer drought and the moderation of the thermal contrast are differential climate features of this series, as opposed to the homologous Guadarramean series.

Apart from the diagnostic forest elements indicated earlier, in this series the juniper is practically non-existent –except in certain mesomediterranean facies and on lithosols– and the seral woody stages are different: white broom (Genisto hystricis-Cytisetum multiflori, Lavandulo sampaianae-Cytisetum multiflori) and gum cistus with "aulaga" (Genisto hystricis-Cistetum ladaniferi) tend to form extensive mosaics in deforested areas. With regard to the meadows, there is a greater extension of "vallicares" –both annual (Agrostion salmanticae) and perennial (Agrostion castellanae)– promoted by the increased summer rainfall, and by the evenness of the topography. Its altitudinal range is less than that of the Guadarramean series, as above 400–800 m in its westernmost parts and 1000–1200 m in the east it is displaced by the more ombrophilous Iberian white oak forests of the Genisto falcatae-Quercetum pyrenaicae.

In the westernmost territories, frequently alternating with Iberian white oak woods or in favoured exposures with steep supramediterranean sunny sites, there are also juniper woodlands with a relict nature (Festuco merinoi-Juniperetum lagunae) with an occasional presence of some specific populations of Spanish black pine (*Pinus nigra* subsp. *salzmannii*). This edaphoxerophilous vegetation is common on lithosols on humid-hyperhumid southern slopes of the Sierras de Gredos and Peña de Francia.

13.3.4 Edapho-Hygrophilous Vegetation

The vegetation growing on river beds and adjacent river banks, developed on soils with permanent or temporary hydromorphia, constitutes one of the most singular sets of ecosystems in the Sistema Central. Their dependence on soils that do not totally dry out allows the development of plant communities often formed by plants with a cosmopolitan or multiregional distribution. Leaving aside purely aquatic vegetation such as the communities of batrachids, elodeids, nymphaeids and others, and herbaceous vegetation types with a broad helophytic character such as "carrizales", "espadañares", "cañaverales" and others, deciduous copses and woods generally constitute the mature stages of these riparian or riverbank ecosystems.

Based on the geomorphology and the hydrological regime of streams, rivers and riverbanks conditioned by water flows throughout the year and the existence of aquifers, it can generally be said that in flood paths and low-water channels there are some highly specialised woodland formations due to the abrasion of floodwaters; these include willows and alder woods. In the broader riverbeds and riverbanks, which are not directly submerged every year by waters from irrigation ditches but conserve the coolness of their soils until the arrival of summer, there is development of hygrophilous ash, elm and oak woodlands. All this synecosystem, conditioned and defined by a permanent availability of water, constitutes the edapho-hygrophilous geoseries.

In the whole of the Sistema Central we can recognise four riparian series: ash and alder woodlands, grey willow (*Salix atrocinerea*) woodlands belonging to the order Populetalia albae, and willow woodlands of *Salix salviifolia* that can be included in the order Salicetalia purpureae. All of these grow on base-poor substrates, and are very fragile ecological environments covered by certain types of protection or effective conservation measures.

13.3.4.1 Ash (*Fraxinus angustifolia*) Woods

The supramediterranean ash woodlands are possibly the riparian woodlands with the greatest territorial extension throughout the Sistema Central, particularly in the eastern and central zones. Alder woodlands are however very frequent and extensive in the westernmost sections of the mountain range due to the great complexity and diversification of the hydrographic network in these areas, a somewhat different fluvial regime, and to the scarcity of broad riverbeds and terraces that allow the establishment and optimum development of ash woodlands.

The Carpetan ash woodlands occupy valley floors, broad plains and sloping valleys where there is a variable water table (pseudogley-ised soils). Physiognomically these are open woodlands dominated by the narrow-leafed ash (*Fraxinus angustifolia*), and usually accompanied by other trees such as the Iberian white oak (*Quercus pyrenaica*), poplar (*Populus nigra*), grey willow (*Salix atrocinerea*), Montpellier maple (*Acer monspessulanum*), elm (*Ulmus minor*) and others. The thermoclimatic breadth of these ash woods is circumscribed to the meso- and supramediterranean belts. Mesomediterranean ash woodlands (Ficario ranunculoidis-Fraxinetum angustifoliae) grow on sandy Miocene soils in the westernmost territories belonging to the Luso-Extremaduran chorological province. Supramediterranean ash woodlands (Querco pyrenaicae-Fraxinetum angustifoliae), with a broad territorial extension in the Sistema Central, have a Carpetan and Leonese chorological optimum. Unlike the western mesomediterranean ash woodlands, they occupy large areas on ramps and in inland valleys in the Sierra de Salamanca and Guadarramean biogeographical sectors; the special geomorphology of the territories in the Bejarano-Gredensian biogeographical sector means that in this sector supramediterranean ash woodlands are much scarcer, and occupy almost negligible spaces. This is not the case in the mesomediterranean belt in the westernmost territories in the Sistema Central (with a more oceanic climate and here clearly Luso-Extremaduran), where the Iberian white oak has a greater altitudinal breadth and may be found in mesomediterranean ash woodlands. Although the ash tends to be the dominant tree in this type of riparian vegetation, selective cutting has led to ash woodlands assuming the appearance of a "dehesa" or wooded pasture, where (in the case of supramediterranean territories) the Iberian white oak is practically non-existent or has been relegated to a few copses or boundaries. The practice of mass cutting and clearing to which this type of woodland –and in general all riparian arboreal vegetation– has traditionally been subjected in order to favour the development of highly productive perennial grasslands explains the frequency and abundance of the characteristic elements of seral stages.

The important economic value of the soils colonised by ash woodlands has been the cause of their extensive exploitation by man since historic times. The profound transformation of the primary forests owing to their use for livestock farming, and to a lesser degree for agriculture, has led to their current state, where they are today transformed into the wooded pastures known as "dehesas", extensive grazing areas, and particularly into systems of mesh-type "espinales" (small plots of meadow or horticultural cultivation delimited by hedges comprising woodland trees and brambles from their fringes). These formations are characteristic elements denoting the traditional uses of these woodlands, and contain a wide variety of fauna; they are often protected landscapes and habitats.

The non-forest perennial vegetation in the ash woodland series in the Sistema Central is constituted by fringes of spiny shrubs (Prunetalia spinosae), perennial meadow communities (Agrostion castellanae, Cynosurion cristati), rush meadows (Juncion acutiflori), herbaceous communities on forest fringes (Origanietalia vulgaris) and scionitrophilous species (Alliarienion petiolatae, Convolvulion sepium, etc.). In mesomediterranean territories there is a frequent presence of perennial grasslands, in addition to "vallicares" (Agrostion castellanae, Agrostion salmanticae), "gramales" (Trifolio resupinati-Caricetum chaetophyllae) and rush meadows, especially of the Trifolio-Holoschoenetum.

13.3.4.2 Alder (*Alnus glutinosa*) Woods

Alder woods constitute the most characteristic example of flood-prone woodlands. Together with woodlands of grey willow (*Salix atrocinerea*), they grow on alluvial soils that conserve moisture throughout the whole year, unlike the rest of the riparian woodlands.

In the Sistema Central the alder woods (supra- and mesomediterranean) are relegated to certain narrow gorges, river headwaters, enclosed watercourses, narrow valleys and others, and together with the grey willow are the tree formation most adjacent to the river course. Due to the particular structures in the relief and to the distribution of the hydrographic network, they are frequent in the easternmost and westernmost territories of the mountain range, with very isolated examples in its central parts.

We can still find good examples of these genuine gallery woodlands along the length of permanent river courses, primarily in the westernmost areas: the upper areas of the Alberche, Tormes, Tiétar and Alagón rivers, among others. In the rest of the mountain range the alder woods are represented essentially in the basins of the Jarama and Lozoya rivers.

The mature stages of these vegetation series correspond to dense and shady woodlands dominated by the black alder (*Alnus glutinosa*), in which there are also other trees such as ash (*Fraxinus angustifolia*), woodlands of grey willow, *Salix salviifolia*, hazel (*Corylus avellana*) and European nettle tree (*Celtis australis*), among others. In alder woods in the supramediterranean belt (Galio broteriani-Alnetum glutinosae), certain elements with an optimum in the north-western part of the Iberian peninsula are common and are found as refugees in these woodlands; these include birch (*Betula celtiberica*), holly (*Ilex aquifolium*), quaking aspen (*Populus tremula*), and numerous nemoral taxa such as *Luzula sylvatica* subsp. *henriquesii*, *Paradisea lusitanicum*, *Paris quadrifolia* and others. In contrast, in mesomediterranean alder woods in the westernmost territories of the mountain range –here Luso-Extremaduran (Scrophulario scorodoniae-Alnetum glutinosae)– there are certain nemoral taxa typical of less contrasting climates and sensitive to the cold such as *Clematis campaniflora*, *Dryopteris affinis* subsp. *borreri*, *Osmunda regalis*, *Scrophularia scorodonia* and others, while the previously mentioned taxa are absent.

Microphanerophytes and creepers tend to be scarce; they are basically hawthorn (*Crataegus monogyna*), common hawthorn (*Prunus spinosa*), brambles (*Rubus corylifolius*, *Rubus ulmifolius*), honeysuckles (*Lonicera periclymenum* subsp. *hispanica*), ivy (*Hedera helix*), and wild vines (*Vitis vinifera* subsp. *sylvestris*) among others. Common in the herbaceous layer are numerous nemoral ferns (*Athyrium filix-femina*, *Dryopteris filix-mas*) and other taxa that tend to acquire considerable biomass (*Carex broteriana*, *C. vesicaria*, *Galium broterianum*, etc.).

The helophytic vegetation developing on potential areas of alder woodlands mainly corresponds to communities of large sedges (Caricenion broterianae), perennial herbaceous communities on forest fringes (Vicio sepium-Centaureetum carpetanae) and megaphorbic communities (Filipendulion ulmariae). These territories tend to be devoted to sustained horticultural cultivation and to the creation of leisure and recreation areas, particularly in summer, an aspect that represents a serious threat of alteration to the complex riparian ecosystem.

From the dynamic point of view the alder woods occur adjacent either to shrublands (Salicion salvifoliae) or with ash woodlands (Fraxino angustifoliae-Ulmenion minoris), and in more isolated cases directly with climatophilous series corresponding to the holm oak or Iberian white oak.

13.3.4.3 Grey Willow (*Salix atrocinerea*) Woodlands

Tree-type willows, physiognomically dominated by the grey willow (*Salix atrocinerea*) grow on ammoriform soils of gley developed in waterlogged depressions or on the banks of river courses with very slow drainage, frequent throughout the whole of the Sistema Central. They constitute a particular vegetation series, whose mature stage or series head corresponds to shady woodlands (Rubo corylifolii -Salicetum atrocinereae) dominated by willows, and generally with a limited extension. In these woodlands there is also a proliferation of other trees and microphanerophytes such as ash, *Salix salviifolia* willow, buckthorn (*Frangula alnus*), common hawthorn (*Crataegus monogyna*), hawthorn (*Prunus spinosa*), in addition to certain nemoral creepers such as honeysuckle (*Lonicera periclymenum* subsp. *hispanica*), ivy (*Hedera helix*) and wild vine (*Vitis vinifera* subsp. *sylvestris*), among others. In the lower layers there is a frequent presence of brambles (*Rubus corylifolius*, *Rubus ulmifolius*, *Rubus caesius*), nemoral ferns (*Athyrium filix-femina*, *Dryopteris filix-mas*) and others.

The subseral stages or permanent communities of this series (Rubo corylifolii-Salico atrocinereae sigmetum) correspond to hygrophilous brambles (Rubo ulmifolii-Rosetum corymbiferae franguletosum alni), rush meadows that are waterlogged until late summer (Deschampsio hispanicae-Juncetum effusi, Hyperico undulati-Juncetum acutiflori), hygrophilous communities with a large biomass and an Atlantic-Central-European optimum (Galio palustri-Caricetum lusitanicae, Filipendulion ulmariae) and others. The potential territories occupied by the grey willow have the same socio-economic value as those occupied by the alders.

13.3.4.4 Willow Woodlands of *Salix salviifolia*

The microphanerophytic willow woodlands dominated by the *Salix salviifolia* willow grow on riverbanks as primary shrubby vegetation, unlike the vegetation types described earlier, which grow on stable soils with a pioneering and colonising character on alluvial riverbeds subject to the periodic flooding typical of certain river regimes. They may constitute the seral vegetation in potential areas of alder woodlands when these have been radically altered. They have their optimum in rivers with sandy riverbeds or with siliceous pebbles frequent in the Sistema Central.

The mature stage of this series (Salico lambertiano-salvifoliae sigmetum) corresponds to small copses formed by numerous willows (*Salix salviifolia*, *Salix purpurea* subsp. *lambertiana*, *Salix neotricha*, *Salix triandra* subsp. *discolor*, etc.) and some creepers and shrubs typical of adjacent riparian woodlands, under which a very species-poor nitrophilous herbaceous layer may occur.

As seral or permanent communities we can highlight certain helophytic formations such as the rush meadows of the Glycerio declinatae-Eleocharidetum palustris, communities of cress (Helosciadietum nodiflori), communities presided by the hemlock water dropwort (Glycerio declinatae-Oenanthetum crocatae), and others.

Early colonising willow woodlands play a key role in moderating fluvial erosion and in the retention and mineralisation of the materials carried by the rivers. The potential areas occupied by this type of riparian vegetation generally have little economic value due to the highly fluctuating and sometimes devastating water regime to which they are subjected. However they have a very significant ecological value.

References

Blanco E, Casado MA, Costa M, Escribano R, García M, Génova M, Gómez A, Gómez F, Moreno JC, Morla C, Regato P, Sainz H (1997) *Los bosques ibéricos. Una interpretación geobotánica.* Ed. Planeta, Barcelona, p 572

Clayton SD, Renvoize SA (1986) *Genera Graminun. Grasses of the World.* Kew Bulletin Additional Series XIII. Royal Botanic Gardens, Kew, p 389

de la Fuente V (1986) Vegetación orófila del occidente de la provincia de Guadalajara (España). Lazaroa 8:123–219

Fernández-González F (1991) La vegetación del Valle del Paular (Sierra de Guadarrama, Madrid), I. Lazaroa 12:153–272

Ferrero Lomas LM (1999) Citogenética, evolución y biogeografía de las secciones *Festuca* L. (intravaginales) y *Eskia* Willk. del género *Festuca* L. (Poaceae) en la Península Ibérica. Memoria doctoral. Universidad Autónoma, Madrid, p 327

Ferrero Lomas LM, Fuente, V de la (1996) Aportaciones al estudio cariológico de algunas especies del género *Festuca* L. endémicas del Mediterráneo Occidental. Boletin Sociedade Broteriana, sér. 2, 67:303–308

Fuente V de la (1986) Vegetación orófila del occidente de la provincia de Guadalajara (España). Lazaroa 8:123–219

Gavilán RG (2005) The use of climatic parameters and indices in vegetation distribution: a case study in the Spanish Sistema Central. Int J Biometeorol 50(2):111–120

Gavilán RG, Fernández-González F (1997) Climatic discrimination of Mediterranean broad-leaved sclerophyllous and deciduous forests in central Spain. J Veg Sci 8:377–386

Gavilán RG, Fernández-González F, Blasi C (1998) Climatic classification and ordination of the Spanish Sistema Central: relationships with potential vegetation. Plant Ecol 139(1):1–11

Gavilán RG, Sánchez-Mata D, Escudero A, Rubio A (2002) Interspecific interactions in Mediterranean high mountain vegetation (Sistema Central, Spain). Israel J Plant Sci 50(3):217–228

Gavilán RG, Sánchez Mata D, Vilches B, Entrocassi GS (2007) Modelling current distribution of Spanish *Quercus pyrenaica* forest using climatic parameteres. Phytocoenologia 37:561–581

Gavilán RG, Vilches B, Fernández-González F (2011) Syntaxonomical review of Cytisetea scopario-striati communities in central Spain. Lazaroa 32:29–72

Gavilán RG, Sánchez-Mata D, Gaudencio M, Gutiérrez-Girón A, Vilches B (2016) Impact of the non-indigenous shrub species *Spartium junceum* (Fabaceae) on native vegetation in central Spain. J Plant Ecol 9(2):132–143

López-Sáez JA, Sánchez-Mata D, Alba-Sánchez F, Abel-Schaad D, Gavilán RG, Pérez-Díaz S (2013) Discrimination of Scots pine forests in the Iberian central system (*Pinus sylvestris* var. *iberica*) by means of pollen analysis. Phytosociological considerations. Lazaroa 34:191–208

López-Sáez JA, Alba-Sánchez F, Sánchez-Mata D, Abel-Schaad D, Gavilán RG, Pérez-Díaz S (2015) A palynological approach to the study of *Quercus pyrenaica* forest communities in the Spanish central system. Phytocoenologia. doi:10.1127/0340-269X/2014/0044-0572

López-Sáez JA, Sánchez-Mata D, Gavilán RG (2016) Syntaxonomical update on the relict groves of Scots pine (*Pinus sylvestris* var. *iberica*) and Spanish black pine (*Pinus nigra* subsp. *salzmannii*) in the Gredos Range (central Spain). Lazaroa 37:153–172

Loureiro J, Kopecký D, Castro S, Santos C, Silveira P (2007) Flow cytometric and cytogenetic analyses of Iberian Peninsula *Festuca* spp. Plant Syst Evol 269:89–105

Markgraff-Dannenberg I (1980) Festuca L. In: Tutin et al. *Flora Europaea* 5:125–153. Cambridge University Press, Cambridge, p 476

Mayo Arlanzón C (2002) La historia geológica de las Sierras de Gredos. In: Corrales Bermejo L (coord.) *Recursos Naturales de la Sierras de Gredos*: 23–29. Publ. Institución "Gran Duque de Alba", Monografías n. 62. 2ª ed. Diputación provincial de Ávila. Vicolozano (Ávila), p 375

Peinado Lorca M, Rivas-Martínez S (eds) (1987) *La vegetación de España.* Publicaciones Universidad de Alcalá de Henares (col. Aula Abierta, 3), Alcalá de Henares, p 544

Rivas-Martínez S (1964) Estudio de la vegetación y flora de las sierras de Guadarrama y Gredos. Anales Instituto Botánico Cavanilles 21(1):1–325

Rivas-Martínez S, Cantó P (1987) Datos sobre la vegetación de las Sierras de Guadarrama y Malagón. Lazaroa 7:235–257

Rivas-Martínez S, coauthors (2007) Mapa de series, geoseries y geopermaseries de vegetación de España. Itinera Geobotanica 17:5–436

Rivas-Martínez S, Belmonte D, Cantó P, Fernández-González F, de la Fuente V, Moreno JM, Sánchez-Mata D, Sancho LG (1987) Piornales, enebrales y pinares oromediterráneos (Pino-Cytision oromediterranei) en el Sistema Central. Lazaroa 7:93–124

Rivas-Martínez S, Fernández-González F, Sánchez-Mata D, Pizarro JM (1990) Vegetación de la Sierra de Guadarrama. Itinera Geobotanica 4:3–132

Rivas-Martínez S, Fernández-González F, Loidi J, Lousã M, Penas A (2001) Syntaxonomical checklist of vascular plant communities of Spain and Portugal to association level. Itinera Geobotanica 14:5–341

Rivas-Martínez S, Díaz TE, Fernández-González F, Izco J, Loidi J, Lousã M, Penas A (2002) Vascular plant communities of Spain and Portugal. Addenda to the Syntaxonomical checklist of 2001. Itinera Geobotanica 15(1–2):1–922

Sánchez Mata D, Rodríguez-Rojo MP (2006) Hay meadow communities (Arrhenatherion) Koch 1926 throughout the Iberian central range (Spain). Phytocoenologia 36(2):231–246

Sánchez-Mata D (1989) *Flora y vegetación del macizo oriental de la Sierra de Gredos (Ávila).* Publicaciones Institución Gran Duque de Alba n. 25. Diputación Provincial, Ávila, p 440

Sánchez-Mata D (1999) Bioclimatología: una ciencia avanzada para la caracterización del medio natural. *Discursos de Entrada* 1998: 95–112. Publicaciones Institución Gran Duque de Alba. Diputación Provincial, Ávila, p 156

Sánchez-Mata D (2015) *Hábitats y vegetación natural en la alta montaña del Parque Regional de la Sierra de Gredos (Castilla y León Ávila).* Discurso de ingreso en la Academia de Farmacia de Castilla y León. Ed. CERSA, Las Rozas (Madrid), p 54

Sardinero S (2004) Flora y vegetación del macizo occidental de la Sierra de Gredos (Sistema Central, España). Guineana 10:1–474

Chapter 14
The Coastal Levantine Area

Pilar Soriano and Manuel Costa

Abstract The Coastal Levantine area is a very diverse territory situated between the Cap de Creus and the Cap de la Nau. The wide variety of environments and ecological conditions that occur in this area determine its great diversity and richness from the vegetation point of view. Forest composition depends mainly on climatic factors, geomorphology and soil features, and are currently dominated by evergreen sclerophyllous (*Quercus ilex*, *Q. rotundifolia*, *Q. suber*), conifer (*Pinus halepensis*, *P. pinaster*, *P. pinea*, *P. salzmannii*, *P. sylvestris*), and deciduous (*Fagus sylvatica*, *Quercus petraea*) and semi-deciduous (*Quercus faginea*, *Q. canariensis*, *Q. pubescens*, *Q. pyrenaica*) species in inland areas with heavier rainfall or humidity. As a result of land use and human disturbances, shrubby vegetation, scrub, grasslands, along with thermophilic natural or reforested pine forests, are the most widespread vegetation formations, especially at lower altitudes. Scrub formations spread on both siliceous and calcareous substrates, particularly those that grow on thermophile calcareous ones are species-rich communities with numerous endemic plants. Coastal vegetation, including dunes, marshes, saltmarshes and cliffs, is well represented, with slight variations in some parameters, like salt concentration, water availability, soil features, etc., responsible for the floristic composition of the communities growing in these environments. As in the rest of the Iberian Peninsula, vegetation, especially forests and coastal areas, is affected by a long-standing tradition of human use and transformations (fires, massive urbanisation, grazing, felling, etc.), and well-structured formations currently have become quite scarce.

14.1 Introduction

The Coastal Levantine area is a highly diverse territory situated between the Cap de Creus and the Cap de la Nau. Variations in the orography, geology and climatic parameters throughout the territory have led to diverse ecosystems that contain a wide variety of plant life and vegetation. As a whole, the territory is crossed by

P. Soriano (✉) • M. Costa
ICBiBE-Botanical Garden, Universitat de València, Quart 80, 46008, València, Spain
e-mail: Pilar.Soriano@uv.es

J. Loidi (ed.), *The Vegetation of the Iberian Peninsula*, Plant and Vegetation 12,
DOI 10.1007/978-3-319-54784-8_14

large mountains criss-crossed by channels and gullies, surrounded by sedimentary plains. Northwards, the scenery is dominated by the Catalonian coastal mountain ranges (Serralades Costaneres), which link up to the NW-SE running Iberian Mountain Range to the south, the southernmost mountains of the region lie within the Baetic Range, which runs NE-SW.

North of the Ebro spreads a series of plains and mountain stretches, among which the Serralades Costaneres stand out, which are formed by two chains of mountains, the Serralada Litoral, which is closer to the sea and is lower (Montnegre 752 m, Agulles 652 m, Corredor 632 m, etc.), and the longer, more extensive Serralada Prelitoral, which stands more inland and is higher (Turo del l'Home, 1712 m, Les Agudes 1706 m, Caro 1447 m, etc.). Both run NE-SW and more or less parallel to the sea. The Serralada Prelitoral finishes northward in the Serralada Transversal, which comes into contact with the Pyrenees and the Pre-Pyrenean Range. In the northern part of this land, the littoral presents sedimentary coastal stretches, with sandy beaches, marshes and lagoons (Aiguamolls de l'Empordà, Delta del Llobregat, Prat de Cabanes-Torreblanca, etc.) and cliffs. Alongside the low sedimentary coastal areas, there are erosional coasts formed by very abrupt cliffs where a dynamic sea pressure predominates, and between which small "bay-type" beaches have formed (Cap de Creus, Cap Gros, Cap Norfeu, Cap Begur, etc.). Some coastal deltaic areas stand out, of which the most important is the large Ebre Delta between the Baix Ebre and Montsiá, formed by materials that the Ebro River brought, with a surface area covering over 320 km^2 which penetrates into the sea over a distance of some 22 km. The Iberian mountains are at some distance from the sea, which has long, sandy or pebbly beaches, marshlands or lagoons (El Saler-Albufera, Canet-Sagunt, El Brosquil, Marjal de Almenara, Marjal del Moro, etc.), the only stretches of eroding coastline being found in the Irta mountain range and at Cap de Cullera. The Baetic Mountain Range is predominantly calcareous rock, particularly in the northernmost prolongation, which is formed by the Serra Grossa, Agullent and Benicadell (1104 m). Southwards, there is a system of mountain ranges, the most important of which are the Mariola Range, which contains the Montcabrer peak (1390 m), and Aitana (1558 m). The Baetic Mountains sweep down to the sea to form a spectacular eroding coastline that includes the Cap de Sant Antoni and the Cap de La Nau.

Siliceous rocks are more abundant in the north where, apart from the volcanic area of Olot and the Columbrets Islands, we find several major nuclei aside the Pyrenean axis. One of them comprises granitic batholiths that stretch between Girona and Barcelona, as well as outcrops of Paleozoic schists and Tertiary sandstones, among which those from Montseny stand out. Another outstanding nucleus of siliceous, schist and sandstone materials is found in the Prades and Priorat mountains. Other small siliceous patches are located in the Serra d'Espadà or the Calderona Mountain Range. These elements are formed by reddish rock (Bunter sandstone and argillites Buntsandstein) from the Lower Triassic period, with some intercalated shell limestone (Muschelkalk).

Hydrology The northern territory is furrowed by a complex hydrographic network formed by rivers that flow into the Mediterranean Sea. They correspond to two large basins, the Ebro and the so-called inner basin of Catalunya. The Catalonian Ebro Basin begins at the confluence with the river Segre. The inner basin is formed by other rivers which rise in the Pyrenees (Llobregat, Ter, Fluvià and Muga) and the shorter and less fast-flowing rivers that rise in the Serralada Litoral, Serralada Prelitoral and in the Plana de l'Empordà (Daró, Tordera, Besòs, Foix, Gaià, Francolí and Sènia). In the southern territories, a number of rivers and watercourses flow through the region, always in a W-E direction. These include permanent waterbodies, i.e. major rivers, the source of which usually lies in inland mountain ranges (Xúquer, Túria, Mijares, Serpis) and other small rivers, mainly in the south. Temporary watercourses, or "ramblas", characterised by torrential flows during the rainy season and by dry, gravel- and sand-filled beds in summer, are particularly frequent in central and southern areas.

Climatology and Bioclimatology The climate in this Coastal Levantine area is typical Mediterranean, characterised by a summer drought period. It is a transition area with Atlantic influences from the SW–W during the cooler period of the year and with Mediterranean conditions from the E–NE. This situation conditions the limited influence of westerly fronts. Torrential rainfall in autumn, which affects mainly the southern coastal zones, is also important. In this area precipitation levels vastly increase from south to north and particularly from the coastal parts to higher elevations. In coastal areas annual rainfall is around 400–700 mm. Two nuclei are remarkable where rainfall usually exceeds 800 mm, in the area south of València and north of the Alicante coasts (Pego, 956 mm), and in different mountain areas where higher precipitation values allow the occurrence of deciduous and semi-deciduous forests. The influence of the sea leads to decreasing average temperatures from the coast inland, and also from south to north. The highest average temperatures are found on coastal plains (17 °C) and the lowest (below 10 °C) in higher inland zones, and they fall even below 4 °C e.g. in the inland zones of the Maestrazgo or in some other higher areas.

From the bioclimatological point of view, the territory is comprised into the Mediterranean macro-bioclimate zone mostly with a pluviseasonal-oceanic bioclimate. There are some submediterranean areas (Morella, Vistabella del Maestrat) and others with even a Temperate macrobioclimate (Montseny). The thermotypes range from thermomediterranean to oromediterranean, and ombrotypes also vastly vary and range from semiarid (Camp de Túria) to hyperhumid (Montseny).

14.2 Forests

This Mediterranean area, which mainly occupies coastal lands, is primarily influenced by two determinant factors: the effect of the sea, almost all of which is restricted to less than 100 km off the coast; and the effect of demographic pressure due to a large concentration of urban settlements. Forests are currently dominated by evergreen sclerophyllous, coniferous,and, in areas with higher rainfall or humidity, deciduous and semi-deciduous species. Their structure and species composition depend primarily on climatic factors, geomorphology and soil features.

14.2.1 Evergreen Forests

Holm Oak Forests *Quercus rotundifolia* forests are sclerophyllous formations that appear well-adapted to the harsh Mediterranean climate with its lack of water availability in summer. These forests mostly have been profoundly modified by human activities and exploitation and this resulted in a simplification of their structural complexity. The habitat of these forests is in the southern territories, at the lowest altitudes and near the coast, occupying highly anthropic territories, where environmental conditions make rapid regeneration difficult. As in the rest of the Iberian Peninsula, these holm oak forests are the result of long-standing human use. Because of various human-induced transformations, like grazing, felling, urban development, fire, etc., they have become quite scarce currently, and it is difficult to find well-structured ones. Indeed, most of these forests grow on rocky or degraded soils that do not allow the development of trees, and the majority remain as open tree-bearing formations. Two different associations have been described according to their thermotype and distribution area.

Rubio longifoliae-Quercetum rotundifoliae grows principally on calcareous soils in thermomediterranean and mesomediterranean thermotypes, and in areas with rainfall between 400 and 800 mm. They constitute low forests dominated by *Quercus rotundifolia* and a large number of sclerophyllous shrubs, like *Quercus coccifera*, *Phillyrea angustifolia*, *Pistacia lentiscus*, *Chamaerops humilis* or *Rhamnus alaternus*, lianas like *Asparagus acutifolius*, *Rubia peregrina*, *Lonicera implexa*, *Smilax aspera*, etc., and a herbaceous layer that consists of plants typical of degradation stages. Currently, there are very few areas, especially at lower altitudes, where these forests have withstood high anthropic pressure and remain well-structured. In those scarce cases, the forests are dense and may be impenetrable due to their understory composition with its high density of sclerophyllous shrubs. It is noteworthy that they are rare in Catalan territories and become scarcer going northwards, where *Quercus rotundifolia* is gradually being replaced by *Quercus ilex*. In the wetter parts of their distribution range, a variant with several moist deciduous elements occurs, including *Fraxinus ornus*, *Pistacia terebinthus* or *Crataegus monogyna*, and other moisture-demanding species like *Arbutus unedo* or *Viburnum tinus*.

Hedero helicis-Quercetum rotundifoliae In higher parts, with mesomediterranean, supramediterranean and supratemperate thermotypes, and mainly on calcareous substrates, a different species combination, including *Hedera helix*, *Teline patens*, *Genista hispanica*, etc., typify this different association. Geographical variation from north to south along its distribution range, based on soil features, altitude and amount of rainfall gives rise to variants in floristic composition. At supramediterranean and supratemperate levels, some key species of these thermotypes, like *Festuca capillifolia*, *Galium maritimum*, *Cytisus reverchonii* or *Juniperus communis*, occur in this association. Where rainfall is higher, or at submediterranean sites where the rainfall pattern almost allows the summer drought period to be eliminated, mesophyllous species like *Acer granatense*, *Lonicera etrusca*, *Viola sp. pl.*, *Quercus faginea*, etc., or in the north *Quercus ilex*, are components of these forests. Well-preserved stands can be found in the Benifassar and Vallibana mountains and, in the south, in the Font Roja Natural Park (Alcoi) in the province of Alicante, an area that has been protected since 1332.

Quercus ilex forests are dense formations with a well-developed shrubby layer whose physiognomy is determined by the dominance of *Quercus ilex*. They occur in the mediterranean and submediterranean areas within the distribution range of *Quercus ilex*, on deep, well-structured soils, primarily in the northern Catalan areas close to the coast, and extend southward to the subhumid territories of the Valencian Community (Els Ports-Maestrat), where this species gets scarce and restricted to shady sites on north-facing slopes or valley floors. Here it is generally associated with other tree species like *Quercus rotundifolia* or *Quercus faginea*, to give rise to mixed forests. As in *Quercus rotundifolia* forests, only a few stands still shelter the original structure and species composition because of exploitation and continuous human disturbance. However in the last few decades, abandonment of part of these anthropic practices has led to the regeneration of some of these forests, especially in the most protected situations. Soil composition is one of the main differentiating factors determining the two different associations growing in this territory:

Viburno tini-Quercetum ilicis This community, first described by Braun-Blanquet in 1915, occurs on calcareous substrates with a Mediterranean bioclimate, from sea level to 800–1000 m. Its physiognomy arises from *Quercus ilex* and a dense coverage of shrubs like *Viburnum tinus*, *Phillyrea latifolia*, *Arbutus unedo*, etc., lianas like *Clematis flammula*, *Rubia peregrina*, *Lonicera implexa* or *Clematis vitalba*, and some ferns like *Polypodium cambricum* or *Asplenium onopteris*, which form part of lower, shaded layers (Els Ports, Montsià, etc.). In more humid situations, variants with different species of the genus *Quercus* occur, which results in higher species richness. In these mixed forests, *Quercus pubescens* and *Quercus x cerrioides* (Collserola, Montseny, etc.) or *Quercus canariensis* (especially in Collserola Garraf) dominate in the north, while *Quercus faginea* is the principal tree south of Garraf. In the Serra de Prades this type of forest contains *Quercus pyrenaica*.

Asplenio onopteridis-Quercetum ilicis In Mediterranean and submediterranean areas, at higher elevations than the above-mentioned forests (600–1200 m), and primarily on siliceous or decarbonated soils, the floristic composition of the holm oak forests is different. This is apparent in the replacement of a significant number of thermophilous species by species from deciduous forests, including *Sorbus torminalis*, *Sorbus aria*, *Lonicera xylosteum*, *Hepatica nobilis*, *Fragaria vesca* and *Brachypodium sylvaticum*. This association is not that common in the littoral area, where it only occurs in the more humid and colder parts of the Prelitoral and Litoral Serralades (Montnegre and Corredor, Montseny, Guilleries, etc.).

Presently, the last two types of forests described above, both growing on calcareous and siliceous substrates, are mostly found as small patches, and mainly as low or even shrubby forests. Their actual, detailed floristic composition depends on local environmental factors and disturbances.

Cork Oak Forests These are forests with *Quercus suber* as the dominant species in the tree layer and a dense understory composed of sclerophyllous shrubs and lianas. They grow on acid soils or with a significant level of decarbonisation (sandstone, granite and sandy soils), under moist conditions, and mostly between 100 and 600 m. They are more abundant in the northern part of the territory, because acid soils are more common there, vanish in the central part and then reappear, although much scarcer, on siliceous patches in the southern part of the territory. These forests have been subject to frequent wildfires characteristic of the Mediterranean areas, and to sustainable cork extraction (stripping) for millennia, and this has favoured their expansion. Both factors make it nowadays difficult to delimit their original area. Two different associations are distinguished: one in the north and the other in the south.

Carici basilaris-Quercetum suberis In mesomediterranean and supramediterranean thermotypes, with rainfall usually exceeding 600 mm, *Quercus suber* is accompanied by sclerophyllous species, and also climbing plants like *Lonicera implexa*, *Lonicera periclymenum*, *Rubia peregrina*, etc. It is frequently found as shrubby formations dominated by acidophilous shrubs, like *Erica scoparia* or *Erica arborea*, with an arboreal canopy (Cytiso villosi-Ericetum arboreae) and, in more disturbed situations that result in open structures, heliophilous species like *Cistus salviifolius*, *Cistus crispus*, *Lavandula stoechas*, *Calluna vulgaris*, etc., take advantage. This association occurs in the north east littoral area (Maresme, la Selva, Gironès, l'Empordà), where the most extensive granitic outcrops are located.

Asplenio onopteridis-Quercetum suberis This association, first described from the Espadà Mountains, grows on the more humid slopes and valleys in deep and moist Triassic soils. Under natural conditions, in its usual floristic combination, *Quercus suber* is accompanied by *Crataegus monogyna*, *Arbutus unedo*, *Ruscus aculeatus*, *Daphne gnidium*, *Osyris alba*, *Pistacia terebinthus* etc., and has a herbaceous layer with *Galium maritimum*, *Brachypodium sylvaticum*, *Carex sp. pl.* and shade-demanding species like *Asplenium onopteris* or *Polypodium cambricum*. In heliophilous situations, *Quercus suber* frequently forms mixed forests with *Quercus rotundifolia*. This community, which is characteristic of Valencian territories, is not common: it occupies about only 700 ha because calcareous substrates are dominant in this area. Currently it inhabits isolated microsites in ravines and shaded areas like the Mosquera valley (Fig. 14.1), this being the best-developed forest, or in other nearby areas (Serra Calderona, Desert

Fig. 14.1 *Quercus suber* forest. Mosquera Valley (Photo Pilar Soriano)

de les Palmes) and some isolated *Quercus suber* formations in the decarbonated sandy soils of the rainiest regions in the south (Barx and Pinet mountains).

From the same situations as the above-cited association Cytiso-Ericeum arboreae, a vicariant community has been described as part of the succession dynamics of these cork oak woods: the Hedero helicis-Ericetum arboreae.

14.2.2 Decidous and Semidecidous Forests

The physiognomy of these forest communities is determined by different dominant tree species, e.g. *Quercus* (*Quercus petraea*, *Quercus faginea*, *Quercus canariensis*, *Quercus pyrenaica* or *Quercus pubescens*), *Fagus*, *Castanea*, and a floristic combination that varies depending on environmental conditions.

Beech Forests In these forests *Fagus sylvatica* may be joined by other trees like *Quercus pubescens*, *Abies alba*, etc.; they have sparse understories owing to the dense shadow cast by beech. They mainly occur in the more humid territories (Montseny, Guilleries) from 1000–1600 m. Different associations have been identified, which principally differ in terms of soil features and moisture requirements. From acid soils we mention the Luzulo niveae-Fagetum sylvaticae (most

Fig. 14.2 Beech forests in autumn. Montseny (Photo Manuel Costa)

widespread on silicic soils, Montseny) and Helleboro occidentalis-Fagetum sylvaticae, and on calcareous substrates and with less humid situations the Buxo sempervirentis-Fagetum sylvaticae and Primulo acaulis-Fagetum sylvaticae (Ports de Beceit where beech reaches its southern limit) (Fig. 14.2).

***Quercus petraea* Forests** Lathyro linifolii-Quercetum petraeae, are found in humid northern areas on acid substrates (Montseny, Montnegre, Guilleries) which usually occur as open forests with a dense herbaceous layer that is rich in acidophilous plants (*Lathyrus linifolius*, *Teucrium scorodonia*, *Luzula forsteri*, *Deschampsia flexuosa*, etc.).

***Quercus pubescens* Forests** They are more widespread than the previous formations, and are characterised by a tree layer of *Quercus pubescens* accompanied by a rich shrub layer. These forests have been heavily exploited for timber and firewood, so it is difficult to find well-structured examples. Two associations have been described: one on acid soils, the Pteridio aquilini-Quercetum pubescentis (Guilleries, Montseny, etc.), and a vicariant mainly on calcareous substrates, the Buxo sempervirentis-Quercetum pubescentis (Prades, Montsant, etc.).

***Quercus faginea* Forests** *Quercus faginea* subsp. *faginea* tends to grow at altitudes from 500 to 1500 m, but can be found at lower altitudes in some places, even almost at sea level, but always on northern slopes or in valley bottoms. This species does not usually constitute monospecific communities in those littoral areas, but generally joins other species of the same genus, which leads to mixed-oak forests. Indeed it is often associated with *Quercus ilex*, or even with *Quercus pubescens* and *Quercus x cerrioides*, where the summer drought is neither long nor very intense (e.g. Montsant), and also with *Quercus rotundifolia* in southern areas to form the more humid, shady aspect of *Quercus rotundifolia* forests. In wetter northern areas, mesophilous plants are frequent (*Acer sp. pl.*, *Sorbus sp. pl.*, *Rosa sp. pl.*, *Primula columnae*, *Geum sylvaticum*, *Viola sp. pl. Brachypodium phoenicoides*, etc.), and they gradually change towards the drier southern lands, where Mediterranean species are more abundant. Three associations have been described for these marcescent forests. As stated above, *Quercus faginea* does not usually form monospecific forests in this territory and these communities are better represented in nearby areas in more inland territories.

Violo willkommii-Quercetum fagineae This association grows in northern territories in Mediterranean and submediterranean areas in a subhumid ombrotype. There, *Quercus faginea* is often accompanied by *Quercus ilex*, *Quercus pubescens* or *Pinus salzmanii*, depending on the thermotype and other ecological variations. These differences also influence changes in the floristic composition of lower strata.

Telino patentis-Quercetum fagineae It grows in southern territories and comes into contact with the previous association. At drier Mediterranean locations, *Quercus rotundifolia* replaces *Quercus ilex* in the floristic composition of the tree layer to form mixed forests (Baix Maestrat, Alt Millars, Els Ports). In supramediterranean and oromediterranean situations, and always on north-facing slopes, *Pinus salzmannii* or *Pinus sylvestris* can even participate.

Fraxino orni-Quercetum fagineae This association was described in 1960 from southern setabense areas as a mixed forest of *Quercus faginea* and *Fraxinus ornus*, accompanied by mesophilous species like *Acer granatense*, *Sorbus aria*, *Teline patens*, *Pistacia terebinthus*, *Crataegus monogyna*, etc. Currently, it is difficult to find well-structured *Quercus faginea* forests in the southern territory; only small patches can be observed in the rainiest mountains in the south (Font Roja, Mariola, Aitana, etc.), and *Quercus rotundifolia* plays a significant role in the floristic composition of most of them (Fig. 14.3). This species becomes gradually more abundant in drier areas and forms a transitional floristic combination towards the more widespread association Hedero-Quercetum rotundifoliae.

***Quercus pyrenaica* Forests (Cephalanthero rubrae-Quercetum pyrenaicae)** The presence of *Quercus pyrenaica* in the area under discussion is most interesting as this species only forms small nuclei in isolated areas (Penyagolosa, Sierra de Pina, Serra d'Espadà and Serra de Prades) at altitudes from 1000–1200 m. It is often mixed with other species of *Quercus* or *Pinus*, like *Pinus sylvestris*, in higher areas, and with *Pinus pinaster* in mesomediterranean thermotypes.

***Quercus canariensis* Forests (Carici depressae-Quercetum canariensis)** Considering the whole Iberian Peninsula, *Quercus canariensis* forests have a restricted

Fig. 14.3 Fraxino orni-Quercetum fagineae. Font Roja Natural Park (Photo Pilar Soriano)

area. In that specific area, this tree forms small fragmented communities located in the Serralada Litoral and Prelitoral (Montseny, Montnegre, La Selva), where it generally occupies transitional areas between holm oak forests and riparian ones. Regarding its climatic requirements, it grows in thermophilous areas at moderate altitudes (100–1000 m) and with sizeable rainfall values, but always on moist, deep, well-drained, acid substrates, usually as mixed forests with other tree species involved, such as *Quercus ilex*, *Sorbus sp. pl.*, *Populus tremula*, *Castanea sativa*, *Quercus pubescens*, or even *Quercus petraea*, and different hybrid taxa, depending on the environmental conditions. Their undergrowth is characterised by the presence of mesophilous species, principally in the herbaceous layer. *Crataegus monogyna*, *Rubus ulmifolius*, *Ilex aquifolium*, etc., several lianas like *Lonicera periclymenum*, *Hedera helix* or *Rubia peregrina*, and different acidophilous species like *Calluna vulgaris*, *Erica arborea* or *Erica scoparia*, are typical components of the upper understory, while *Asplenium onopteris*, *Carex depressa*, *Luzula forsteri*, *Teucrium scorodonia*, *Brachypodium sylvaticum*, *Prunella grandiflora*, *Fragaria vesca*, etc., often participate in low strata.

Laurus nobilis **and** ***Fraxinus ornus*** **Forests (Viburno tini-Fraxinetum orni)**
This community is restricted to small patches in ravines and on north-facing slopes of the small mountains located in the warm and rainiest parts of the Setabense sector (Barranc de l'Infern, La Murta and La Safor). It occurs as a combination of

Fraxinus ornus, *Laurus nobilis* and some sciophilous species like *Viburnum tinus*, *Myrtus communis*, *Teline patens*, etc.

In Catalan territories, these formations generally appear dominated by *Laurus nobilis*, accompanied by an understory composed of ferns and other shade-demanding plants. They have been assigned to two different associations, the thermophilous Rusco-Lauretum nobilis (Montnegre and Corredor, Collserola) and the Osmundo-Lauretum nobilis, with a tree layer dominated by *Alnus glutinosa* or *Laurus nobilis*, and often related with riparian areas, as evidenced by the participation of species like *Alnus glutinosa*, *Salix atrocinerea*, *Ulmus minor*, *Populus nigra*, etc. This association thrives in the warmest northern areas in Cap de Creus and Montseny and in the low lands of Massis de Cadiretes, where the best preserved stand is found.

Chestnut Forests *Castanea sativa*, cultivated since ancient times, currently lays dense formations on acid soils developed from igneous rocks in the potential territories of deciduous forests, and even in those of *Quercus ilex* (Montnegre, Montseny, Guilleries).

Pine Forests These forests are physiognomically dominated by different species of *Pinus* (*Pinus pinaster*, *Pinus halepensis*, *Pinus salzmannii*, *Pinus sylvestris*, *Pinus pinea*); they grow within a wide range of ecological conditions and depict the most widespread arboreal vegetation in large areas, particularly at the lowest altitudes. Frequently, they grow as monospecific formations as a result of human activities, like reforestation, management or man-induced landscape changes.

Pinus halepensis is the main pine species to grow in this territory, from northern to southern areas, especially on the coast. It usually occurs on calcareous soils, but also on siliceous ones, as part of all types of successional stages. The general species combination corresponds to a monospecific tree layer with *Pinus halepensis*, surrounded by sclerophyllous species as part of a dense shrub layer. The species combination changes from north to south. Plants like *Chamaerops humilis* or *Olea sylvestris* form part of the habitual floristic combination, occurring as from the southern coastal lands of Catalunya. This has led to the description of the association Pistacio lentisci-Pinetum halepensis in which *Pinus halepensis* is accompanied by a shrub understory composed of plants like *Asparagus horridus*, *Ceratonia siliqua*, *Osyris lanceolata*, *Pistacia lentiscus*, *Rhamnus angustifolia*, *Juniperus oxycedrus*, etc. This association occurs on calcareous soils in warm, dry and semiarid areas of the thermomediteranean and mesomediterranean thermotypes. Depending on the occurrence of fires it can act as primary or secondary vegetation, mainly in the Rubio-Quercetum rotundifoliae series. In the drier areas *Pinus halepensis* often appears as monospecific communities due to deforestation and numerous episodes of wildfires, resulting in a low arboreal layer and a poorly developed understory.

Pinus pinaster formations are mainly restricted to the siliceous sites situated in northern granitic areas and a small patch in the Triassic areas from the Espadà and Calderona mountains. This species is often a component of scrub communities, and co-occurs with different chamaephytes from acid soils like *Cistus monspeliensis*,

Cistus salviifolius, *Cistus crispus*, *Lavandula stoechas*, etc. From the phytosociological point of view, these communities have been assigned to the alliance Cistion ladaniferi. In the most favoured situations, sclerophyllous shrubs, and even diverse *Quercus* species (*Quercus suber*, *Quercus rotundifolia* and *Quercus ilex*), can form part of the high understory layers.

As with *Pinus halepensis*, *Pinus pinaster* forms similar monospecific communities in similar situations, but on acid or sandy substrates.

Pinus salzmannii forests (Telino patentis-Pinetum salzmannii) grow in wet areas from Prades, Montsant, Montsiá, Tinença de Benifassa, Ports, etc. When mature they are dense with a relatively closed tree layer of *Pinus salzmannii* and a typically submediterranean sciophilous flora like *Sorbus sp. pl.*, *Acer sp.pl.*, *Lonicera sp.pl.*, *Buxus sempervirens*, *Teline patens*, etc., and a herbaceous layer with *Hepatica nobilis*, *Brachypodium sylvaticum*, *Fragaria vesca*, *Primula columnae*, *Thalictrum tuberosum*, etc.

Pinus sylvestris forests. These forests reach their southern limit on the high mountains of the Iberian Peninsula where they occur in the coldest environments on calcareous or acid substrates. In this territory different floristic combinations result from various ecological conditions, which have led to the description of three different associations: Arctostaphylo crassifoliae-Pinetum catalaunicae. These forests in which *Pinus sylvestris* var. *catalaunica* is often accompanied by *Quercus pyrenaica*, and with a characteristic underlayer of *Arctostaphylos crassifolia*, occur on acid substrates (Serra de Prades). Buxo sempervirentis-Pinetum catalaunicae, characterised by the same *P. catalaunica* and an understory composed of *Buxus sempervirens*, *Amelanchier ovalis*, *Acer granatense*, etc., grows on the shady slopes located in the southernmost areas (Ports de Beceit, Morella); finally, Junipero sabinae-Pinetum ibericae, a widespread association in neighbouring areas (Javalambre, Rincón de Ademuz) that reaches high altitudes in Penyagolosa, where it appears on calcareous soils as an open forest of *Pinus sylvestris* var. *iberica*, with a shrub layer in which *Juniperus sabina*, *Berberis seroi* and *Prunus prostrata* are the usual components. Information about forest is available in Bolòs O de (1959, 1983b, 1988), Bolòs et al. (1993), Carreras et al. (2005), Costa et al. (1995, 1985), Folch Guillén (1981), Garcia Pausas (1989), Hernández Cardona (1999), Molero Briones (1976), Nuet-Badia et al. (2010), Salvá-Catarineu et al. (2012), Vigo (1968), Vilar et al. (1989).

14.3 Riparian Vegetation

This concept embraces several environments, especially in this large area that includes many rivers with different hydrologic regimes. The communities to which this section refers grow in riverbeds, intermittent watercourses ("ramblas"), and also in river mouths. Generally, this set of communities is located in areas strongly influenced by human activities, mainly agriculture and plantations for wood supplies; therefore it is difficult to find well preserved stands. The first group refers to the vegetation of riverbeds with permanent water courses. In these permanent water courses, changes in flow erode, transport and deposit sediments affecting the stability of the substrate. As a result, the riparian zone shows clear variation in vegetation depending on the length and intensity of inundation.

The alluvial *Alnus glutinosa* forests develop in the rainiest zones on permanently moist soils and occur as small areas of deciduous forests with a shrub layer in which plants like *Corylus avellana*, *Prunus avium*, *Sambucus nigra*, etc. are common, and with a diverse herbaceous layer. These formations mainly correspond to two associations, both growing in the northernmost areas: the Equiseto hyemalis-Alnetum glutinosae, at altitudes above 800 m and the Lamio flexuosi-Alnetum glutinosae, which is more widespread and occupies lower stations than the previous one.

Soils with significant seasonal fluctuations in water level host formations characterised by *Fraxinus sp. pl.*, Doronico pardalianchis-Fraxinetum excelsioris has its optimum on the rainiest siliceous mountains in the potential areas of deciduous forests (Guilleries), to become more fragmented and scarce in Montseny, and reach some more Mediterranean and lower areas where they occupy north-facing slopes, valley bottoms or edges of watercourses (Collserola, Montnegre, Corredor, etc.). Moreover an association dominated by *Fraxinus angustifolia* (Rusco aculeati-Fraxinetum angustifoliae), growing on edges of streams with a low water table in summer, has been described from l'Empordà territories. In shady areas and places with high atmospheric humidity, but with less moist soils, a community (Polysticho setiferi-Coryletum avellanae), dominated by *Corylus avellana* and other moisture-demanding species like *Prenanthes purpurea*, *Stachys sylvatica* or *Melampyrum catalaunicum*, develops. This community occupies small areas (Montnegre and Corredor, Collserola, Sant Lorenç de Munt, etc.) and often acts as a shelter of mesophilous species in Mediterranean areas.

The most active portions of riverbeds, gravel bars and islands subject to periodic flooding are the habitat of willows formations. Generally they are bushy communities of *Salix* species with different adaptations, like branch flexibility to reduce their flow resistance, or other features to overcome the effects of flooding. This shrubby vegetation is represented mainly by three associations from north to south: the Carici pendulae-Salicetum atrocinereae, with *Salix atrocinerea* as the key species, in northern siliceous territories; the Saponario officinalis-Salicetum lambertianae, with a more extended distribution area, particularly in the central part, and growing as a combination of *Salix* species (*Salix purpurea*, *Salix eleagnos*, *Salix alba*, etc.); and finally, the Coriario myrtifoliae-Salicetum angustifoliae, in the warmer areas in southern Valencian territories.

In the first lines along temporary and permanent watercourses, where the water table is not too deep, and in the space between *Salix* and *Ulmus* formations, *Populus alba* forests occur. Two associations have been distinguished in this forest formation: the Populetum albae in the cooler areas of the north, which hosts a significant number of mesophilous species in its herbaceous layer, and the Vinco difformis-Populetum albae, which is more widespread, covering littoral lands and reaching the southernmost areas of La Marina. The more distant and driest areas along watercourses, with deeper water tables, are the potential sites of *Ulmus* forests. As these positions are commonly used for agriculture, it is difficult to find extensive, well-structured formations. The most widespread association is the Lithospermo purpureocaerulei-Ulmetum minoris, while in the driest and thermic areas, the Hedero helicis-Ulmetum minoris thrives, a less moisture-demanding community, as reflected in the differences in the floristic composition of the understory; in the first association this is richer in species typical of the Querco-Fagetea class. Small patches of *Ulmus minor* on quaternary low lands in the south of the area have been ascribed to the association Acantho-Ulmetum minoris, which is rich in termophilous species like *Vinca difformis*, *Celtis australis*, etc.

On brackish soils, usually located in river mouths (Delta de L'Ebre, Aiguamolls de L'Empordà, and small rivers in southern areas), shrub formations of different species of *Tamarix* (*Tamarix gallica*, *Tamarix canariensis*) replace the above-mentioned *Salix* communities. These communities have been classified in three associations, the Tamaricetum canariensis and Saccharo ravennae-Tamaricetum canariensis, which represent the communities that grow in the riverbeds with a low salinity level, and the Inulo crithmoidis-Tamaricetum boveanae, which thrives in coastal saltmarshes and other places, but always on saline soils.

In the southern dry areas, which usually correspond to *Quercus rotundifolia* territories, intermittent watercourses are widespread. Under these environmental conditions, many of the above-described formations are replaced with a *Nerium oleander* community, named locally "adelfar" or "baladrar", the Rubo ulmifolii-Nerietum oleandri, which practically finds its northern limit at the Ebro River. These same habitats, but north of this river, are colonised by *Vitex agnus-castus* formations, the Vinco majoris-Viticetum agni-casti. In sandy soils located in riverbeds, and also in inter-dune depressions, occurs a scarce community dominated by *Saccharum ravennae*, the Equiseto ramosissimi-Saccharetum ravennae.

A special group of communities, which is also related with riverbeds and ravines, are the shrubby formations dominated by *Osyris alba* or *Myrtus communis*: the Clematido flammulae-Osyrietum albae, low shrubland formed mainly by *Osyris alba* and situated in cool open environments, like borders of ravines and streams, and which mainly radiate into the humid areas of the Catalunya littoral, from l'Empordà to the wet areas of La Safor and La Marina; the Calicotomo-Myrtetum, a dense maquis of *Myrtus communis* that occurs in warm wet areas near rivers and ravines, and is scarce in the north of this territory. Information concerning this type of vegetation is obtainable from Ballesteros (1981), Bolòs (1959, 1962, 1979a, b, 1980), Curcó (2008), Gesti et al. (2003), Royo (2006).

14.4 Shrublands

Shrublands, commonly called "maquia", are stiff-leaved evergreen formations whose usual combination involves plants like *Quercus coccifera*, *Pistacia lentiscus*, *Olea europaea*, *Myrtus communis*, *Osyris alba*, etc., and they almost always occur on calcareous soils in the coastal area, and they mostly result from the impact of human activities. These formations can act as successional stages of evergreen forests, or as potential vegetation or permanent communities when soil conditions do not allow forests to develop. They are distributed over almost all the territory, with various associations, and often share space with scrublands.

The Querco cocciferae-Pistacietum lentisci occurs along a broad stretch of the coast, with, southwards, a gradual increase in thermophilic species like *Chamaerops humilis* or *Asparagus horridus*. This vegetation generally has a layer of *Pinus halepensis*, which makes it look like a pine forest. On the shady slopes of El Montgó, a special combination occurs with *Buxus sempervirens*, *Ruscus hypophyllum*, *Teucrium glaucum*, and other mesophytic species. A scarce community dominated by *Euphorbia dendroides*, a summer deciduous species of the Mediterranean basin, ascribed to the association Myrto communis-Pistacietum lentisci, mainly grows on rocky outcrops of calcareous material on sunny slopes in the Cap de Creus Peninsula. The Hedero helicis-Telinetum patentis, dominated by *Teline patens* and other moist-demanding plants like *Pistacia terebinthus* or *Jasminum fruticans*, occurs from Els Ports to the humid areas of La Safor, La Marina Alta and La Marina Baixa. the Junipero oxycedri-Quercetum cocciferae (Fig. 14.4), first described from the River Mijares valley, is a medium-high shrub community with a particular physiognomy that arises due to the combination of *Juniperus oxycedrus*, often *Juniperus phoenicea*, and occasionally *Jasminum fruticans*, and with *Quercus coccifera* also participating at lower altitudes. On

Fig. 14.4 Junipero oxycedri-Quercetum cocciferae and *Juniperus phoenicea* community in rock faces (Maestrat territory) (Photo Edoardo Biondi)

siliceous rocks two associations occur, the Erico arboreae-Arbutetum unedonis (Cap de Creus and northern areas) and the Cytiso sessilifolii-Bupleuretum fruticosi first decribed by Rivas-Martínez from the Montserrat Mountain.

In this group we include the edafoxerophilous communities that grow on limestone rock faces, where *Juniperus phoenicea* is the key species: the Rhamno infectoriae-Juniperetum phoeniceae for the supramediterranean territories of the Mijares River basin, the Rhamno lycioidis-Juniperetum phoeniceae for the mesomediterranean areas and the Chamaeropo humilis-Juniperetum phoeniceae for the thermomediterranean southern areas (La Marina, La Safor, etc.).

A special group of shrublands, generally known as "bardisa", is dominated by thorny plants like *Rubus sp.pl.* or *Rosa sp.pl.* These communities have its optimum at the borders of both deciduous and riparian forests, where they occupy vast areas because both of the potential formations, that are part of the dynamic mesophilous series of the forests in this environment, have been degraded. Among them, the Rubo ulmifolii-Coriarietum myrtifoliae thrives in the moist areas from the north to the Ebro River and in the driest areas sheltered in ravines and humid depressions. A less moist-demanding community is the Pyro amygdaliformis-Paliuretum spinae-christi; it is characterised by *Paliurus spina-christi* as a dominant, and also by the absence of *Coriaria myrtifolia*. Southwards these communities come into contact with a poor formation which is almost monospecifically made up of *Rubus ulmifolius*, sometimes accompanied by plants of the Quercetea ilicis class, the Rubo ulmifolii-Crataegetum monogynae. Other communities that are less represented are the Lamio maculati-Rubetum (Montseny), the Buxo sempervirentis-Rubetum ulmifolii in the coldest humid situations in Els Ports, Tinença de Benifassa and northern areas; and in Penyagolosa and also on nearby mountains the Rubo-Rosetum micranthae and Genisto scorpii-Berberidetum seroi. Information is available in Carreras et al. (2006), Fayos (2004), Franquesa (1995), Pérez Badia (1997), Roselló (1994), Vigo (1968).

14.5 Scrub

Scrublands, also called "garrigues", along with thermophilic natural or reforested pine forests, are the most widespread vegetation, especially in the eastern coastal areas of the Iberian Peninsula. In most cases, this extensive and diverse vegetation is the result of years of human intervention (land use, fires, etc.). The commonest scrublands in this coastal area are of the thermophilic type, dominated by chamaephytes and nanophanerophytes, which developed different morphological and physiological strategies to overcome or avoid the effects of significant drought in summer. They occur on both siliceous and calcareous substrates, and particularly those growing on calcareous substrates constitute species-rich communities with numerous endemic plants and are considered a centre of speciation for some families such as *Labiatae* and *Cistaceae*. This group of communities can be subdivided, depending on the bedrock, into calcareous and siliceous scrub.

Limestone substrates. This heading embraces two groups of scrub; those that grow at low altitudes and shape the landscape of vast areas of this territory; and, scarcer, those that grow at higher supramediterranean levels. Among the first group, the Rosmarino officinalis-Lithospermetum fruticosi colonises the northernmost areas, and also reaches the coast of south France. The Thymelaeo tinctoriae-Ericetum multiflorae is most widespread in the northern calcareous areas and occurs principally in areas that potentially support *Quercus ilex*; the Helianthemo mollis-Ulicetum parviflori, a shrubby community dominated mainly by *Ulex parviflorus* is found in the warm areas from the Baix Maestrat to the Mijares River basin. The Thymo piperellae-Helianthemetum marifolii, the most widespread scrub community in southern areas, is characterised by *Thymus piperella*, "pebrella", with a significant participation of other thermophilous elements. On the littoral cliffs of the Dianic territories we find the Erico multiflorae-Lavanduletum dentatae, mainly characterised by *Lavandula dentata* and *Elaeoselinum asclepium*, and in situations closer to the sea and on loamy-sandy substrates the Erico multiflorae-Saturejetum fontanesii, with a particular floristic combination that includes *Helianthemum caput-felis* on the more compact and saline substrates (cliffs of Moraira and Penyal d'Ifac).

At higher elevations than the above-described communities (mesomediterranean and supramediterranean levels), and mostly in the central and southern parts of the area, several associations composed of small and medium-sized shrubs are found. Among them, we mention the Genisto hispanicae-Erinaceetum anthyllidis (Alt Millars, Maestrat); the Salvio lavandulifoliae-Aphyllanthetum monspeliensis (Montsant, Penyagolosa, Els Ports); the Brachypodio phoenicoidis-Aphyllanthetum monspeliensis (from the north to Els Ports); and the Armerio alliaceae-Salvietum mariolensis, an endemic association of the Dianic mountains, characterised by *Salvia lavandulifolia* subsp. *mariolensis* (Font Roja, El Cocoll, Serra Xortà, Serrella, Mariola, Aitana, etc.) The highest areas are occupied by thickets, dominated mainly by thorny cushion plants such as *Erinacea anthyllis*, adapted to the harsh conditions of the high mountains and the Mediterranean summer drought: the Erodio celtiberici-Erinaceetum anthyllidis (Alt Millars, Maestrat) and Erinaceo anthyllidis-Genistetum longipedis, only growing at the highest altitudes of the Serra de Aitana and characterised by *Genista longipes* and *Vella spinosa*.

Other associations, whose distribution area may be limited by the amount of rainfall, soil composition or moisture availability, are the Anthyllido cytisoidis-Cistetum clusii, physiognomically characterised by *Anthyllis cytisoides* over an extensive potential area, and always on marl substrates; the Genisto hispanicae-Anthyllidetum onobrychioidis which grows only in wetter mountains in the La Safor region, where rainfall usually exceeds 800 mm; the Hippocrepido fruticescentis-Anthyllidetum lagascanae (Fig. 14.5), typical of a small territory on the right margin of the Turia River in a dry or semiarid ombrotype. The Centaureo segariensis-Cistetum albidi is an association characterised by the endemic *Centaurea rouyi*, and is restricted to inland subhumid Dianic areas. The Carici humilis-Ericetum terminalis, contains a combination of *Erica multiflora*, *Erica terminalis* and different species of the genus *Carex* in the mountains further south, and on soils

Fig. 14.5 *Anthyllis lagascana* (Photo Josefa Prieto)

that are moist because of runoff water. The Teucrio edetani-Anthyllidetum terniflorae, an endemic association of semiarid territories of the Camp de Turia (province of València), is characterised by *Anthyllis terniflora*, the endemic *Teucrium edetanum*, and other species from semiarid southern territories such as *Onobrychis stenorhiza*, *Cistus carthaginensis*, etc. The Gypsophilo struthium-Ononidetum edentulae, described from gypsum soils in Cofrentes and the Xúquer valley, is characterised by the presence of *Ononis tridentata* subsp. *angustifolia*. Cracks in lithosols, where water accumulates, are the habitat of communities characterised by species such as *Hypericum ericoides*, *Chiliadenus glutinosus*, etc. (Fumano ericoidis-Hypericetum ericoidis, Helianthemo mollis-Hypericetum ericoidis, Sideritido juryi-Hypericetum ericoidis, Thymo piperellae-Hypericetum ericoidis, Jasonio saxatilis-Chaenorhinetum cadevallii or Melico minutae-Saturejetum fruticosae).

The siliceous scrublands are constrained to northern territories and some small patches of the Triassic or decarbonated southern areas. Of the described associations, the Cytiso villosi-Ericetum arboreae is the most widespread association of littoral areas from the north to the Priorat; it comprises a combination of the species of *Cistus sp.pl.*, *Erica arborea*, *Calicotome spinosa*, *Calluna vulgaris*, *Cytisus arboreus* subsp. *catalaunicus*, etc. The Cisto crispi-Calicotometum spinosae occurs in the siliceous areas of La Selva, l'Empordá, Barcelones and the Espadà mountains; a variant characterised by a tree layer of *Pinus pinaster* gives it the physiognomy of a *Pinus* forest. Other, less common associations are the Erico scopariae-Lavanduletum stoechadis, physiognomically dominated by *Erica scoparia*, together with some elements from the northern territories which occur on deep soils, mainly in La Selva; the Ulici parviflori-Cistetum laurifolii, an heliophilous community that grows on stony soils (Sierra de Pina), the Hedero helicis-Ericetum arboreae on Triassic sandstones in the Serra d'Espadà, which is part of the successional series towards *Quercus suber* forests; the Thymo piperellae-Cistetum crispi, which grows on sandy soils mainly in the decarbonated territories of La Safor; and the Ericetum arboreo-cinereae, a scarce community that develops in small patches in the subhumid territories of La Selva and neighbouring areas. Three more associations have been described for the higher areas: the Prunello hastifoliae-Cytisetum scoparii (NE Catalunya), Erico scopariae-Cistetum populifolii, and Pteridio aquilini-Lavanduletum pedunculatae (Penyagolosa). Information is available in Álvarez de la Campa (2004), Bolòs (1957, 1967, 1983a), Crespo (1989, 1999), Folch Guillén (1981), Roselló (1994), Solanas (1996), Soriano (1995), Stübing et al. (1989), Vigo (1968).

14.6 Grasslands and Herbaceous Vegetation

Annual grasslands. These are characterised by small plants like *Cerastium sp. pl.*, *Galium sp. pl.*, *Asterolion linum-stellatum*, *Erophila verna*, *Hornungia petraea*, *Arabis auriculata*, etc., which usually grow in the lower layers of forests, scrub or shrublands, that provide the shade and the soil moisture they require. Two associations, on different soils, are the main communities: the Saxifrago tridactylitae-Hornungietum petraeae, common on limestone in the Iberian Peninsula and, hence, in this Levantine area, and the Helianthemetum guttati on siliceous soils of the northern areas and, in small stands, on acid or decarbonated soils further south (Serra d'Espadà, La Safor, etc.). Other common annual grasslands in silicic, and often sandy, substrates at low altitudes in northern areas are the Corynephoretum catalaunicum, Scillo-Ophioglossetum lusitanici, Tillaeetum muscosae, Polycarpo alsinifolii-Crassuletum campestris, Sedetum caespitoso-andegavense, Trifolio cherleri-Plantaginetum bellardii, Moenchio erecti-Vulpietum muralis, Airo-Crassuletum tillaeae and Lino bienne-Gaudinietum fragilis.

Perennial grasslands. The more extensive perennial grasslands are those of *Brachypodium retusum* ("lastonares"), which cover large areas on relatively deep soils; they develop as a result of frequent wildfires and other disturbances, especially under the driest conditions. These grass communities comprise various associations which correlate with differences in climate and substrate composition. In northern lands, at the areas that potentially support *Quercus ilex*, the Irido chamaeirido-Brachypodietum retusi occurs, and from l'Horta territory to the south this association is being replaced by the Teucrio pseudochamaepityos-Brachypodietum retusi, a more thermophilic formation, characterised by meridional species like *Teucrium pseudochamaepitys*. The Trifolio-Brachypodietum retusi also is found in the potential areas of *Q. ilex*, but on siliceous soils. In the driest areas, located mostly in the south, a set formations of high grasses is common, dominated physiognomically by *Stipa tenacissima* ("espartales"), and accompanied by other hemicryptophytes and geophytes (*Avenula bromoides*, *Brachypodium retusum*, *Heteropogon contortus*, etc): the Heteropogono contorti-Stipetum tenacissimae, Sedo dianii-Stipetum tenacissimae and Helictotricho filifolii-Stipetum tenacissimae. The last one is scarce and grows in the central, inland territories close towards the La Mancha region. On warm and sunny slopes, a kind of slightly nitrophilous grasslands, with *Hyparrhenia hirta* and *Hyparrhenia sinaica* thrives, distributed from north to south (Hyparrhenietum hirto-sinaicae and Heteropogono contorti-Hyparrhenietum sinaicae).

Another species of *Brachypodium*, *Brachypodium phoenicoides*, characterises a group of grasslands mainly growing in the rainiest areas along borders of water-courses or, in dry areas, in soils which remain moist all year long. Four associations have been described: three for calcareous soils, the Brachypodietum phoenicoidis and Mantisalco salmanticae-Brachypodietum phoenicoidis for northern and central areas, and the Lathyro tremolsiani-Brachypodietum phoenicoidis for areas farther south; plus the Polygalo gerundensis-Ononidetum spinosae that thrives on siliceous soils in the rainiest areas.

In the high Dianic mountains, three communities occur in which various grasses, e.g. *Helictotrichon filifolium*, *Avenula bromoides*, *Festuca capillifolia*, *Festuca gautieri*, *Arrhenatherum sardoum*, etc., are frequent. At higher levels, above 1200 m, the Centaureo mariolensis-Festucetum scopariae grows on deep, moist soils, while the Daphno hispanicae-Festucetum capillifoliae and Festuco hystricis-Helictotrichetum filifolii are found on rocky slopes. A special, scarcely represented group of grassland communities restricted to high elevations, and located mainly in Penyagolosa and Montseny, is characterised by different species of *Festuca*: the Conopodio arvensis-Festucetum scopariae (Montsant, Els Ports, Penyagolosa), Conopodio arvensis-Seslerietum elegantissimae (Montserrat), Helianthemo tomentosi-Festucetum ovinae, Luzulo montsignaticae-Festucetum gautieri, and Antennario dioicae-Festucetum commutatae (Montseny). In the highest areas there are communities of short grasses, the Galio idubedae-Nardetum strictae and Sclerantho polycarpi-Corynephoretum canescentis, on acid substrates, and the Festucetum hystricis on calcareous ones. More information is obtainable in Alvarez de la Campa (2004), Bolòs (1967), De la Torre et al. (1997), Folch Guillén (1981), Molero Briones (1976), Royo Pla (2009), Solanas (1996), Soriano (1995) and Vigo (1968).

14.7 Coastal Vegetation

This coastal environment is widely developed in the Levantine area, as it comprises most of the east coast of the Iberian Peninsula. Like most Mediterranean coastal ecosystems, including dunes, salt-marshes, cliffs, etc., it has a long-standing history of intense human impacts, especially massive urbanisation along almost the entire Mediterranean coastline, and it is considered one of the most severely degraded areas.

14.7.1 Sand Dunes Complex

Nowadays, sand dunes are particularly sensitive systems and are strongly influenced by both natural and man-induced dynamics. So, although sandy coasts are widespread in this territory, there are very few localities in the littoral areas where psammophilous communities remain well conserved, and only some fragmented sandy areas are still natural. These areas are restricted to mainly two protected areas: the Delta de l'Ebre Natural Park (Tarragona) and l'Albufera Natural Park (València).

Under such restricted ecological conditions, a complex group of communities grow, which depend on slight variations in some parameters like soil composition, structure and texture, soil salinity, water availability, nutrients etc. Among other factors, tolerance to salt and hydric stress, the ability to overcome burial by sand, and darkness during the germination period, reduce the competitiveness with other species, and are key factors in controlling the floristic composition of the communities growing in these environments. This results in vegetation of different combinations in terms of structure and floristic composition, along the gradient from embryonic to wooded dunes. Several associations have been described in relation to different ecological conditions:

- Cypero mucronati-Elytrigietum junceae. A narrow band of vegetation with low coverage situated on small mounds of mobile sand close to the sea, and colonised by pioneer plants like *Elymus farctus*, a salt-tolerant species whose structure can retain sand brought by the wind. Other regular components, which are less abundant, like *Calystegia soldanella*, *Cyperus capitatus* or *Sporobolus pungens*, can form part of the floristic composition of this community.
- Medicagini marinae-Ammophiletum australis. This first dune ridge can be several metres wide and usually develops behind the previous association in situations that are more protected from both the sea and the wind. In these situations, diversity is enhanced and a well-established microhabitat is generated for many other plants. *Ammophila arundinacea* is the dominant species from the physiognomical point of view, which usually grows together with more associated species than the previous association. This rhizomatous plant plays a fundamental role in dune establishment and stabilisation, and enables the participation of other species like *Medicago marina*, *Otanthus maritimus*, *Echinophora spinosa*, *Lotus creticus*, etc.
- Crucianelletum maritimae. Further away from the sea, the wind effect is attenuated and allows the stabilisation of substrate and the accumulation of a larger proportion of organic matter, which results in more fertile growth conditions,

and thus in increased species richness. Under these conditions a psammophilous scrub grows, dominated by chamaephytes, which are possibly less tolerant to the harsh environmental conditions prevailing in the areas closest to the sea, like *Crucianella maritima*, *Pancratium maritimum*, *Ononis ramosissima*, *Helichrysum stoechas*, *Malcolmia littorea*, *Launaea fragilis*, etc., and some other dune species.

These three associations are found, apart from El Saler and Delta de l'Ebre, in small areas located at different places (Roses, Aiguamolls de l'Emporda and Palls in Girona; Castelldefels in Barcelona; Torredembarra in Tarragona; Torre la Sal and Moncofa-Almenara in Castelló; Canet, Marenys de Cullera, Xeraco, Oliva in València and Pego, Denia, Xavia and Calp in Alacant).

- Teucrio belionis-Halimietum halimifolii. This is a coastal sclerophyllous formation that grows in fixed dunes and clearings of juniper communities or dune pine forests. These plants form a chamaephytic scrub, which provides medium vegetation coverage, of varying sizes and with a higher level of diversity than the communities that colonise the rest of the dune strip. This is a remarkable association, in which plants of the Rosmarinetea like *Teucrium dunense*, *Helianthemum syriacum*, *Halimium halimifolium*, *Ephedra distachya*, *Cistus clusii*, *Helichrysum stoechas*, *Coris monspeliensis*, *Anthyllis cytisoides*, *Erica multiflora*, etc., are frequent (Fig. 14.6). Currently, this formation is no longer common along the coastline because of the transformations that have taken place in what would be its potential space.

Fig. 14.6 Teucrio-Halimietum halimifolii. El Saler (Photo Pilar Soriano)

Shrubby and arboreal vegetation

- Phillyreo-Rhamnetum angustifoliae. Nowadays, a relevant example of plant life of the sand ecosystems of this Levantine coastal area is the tree vegetation that grows in El Saler (València). Located on stabilised soils, its structure is completely closed, and almost impenetrable, and it is predominated by nanophanerophytes (*Rhamnus angustifolia*, *Phillyrea angustifolia*, *Asparagus officinalis*, *Quercus coccifera*, *Rhamnus alaternus*, *Chamaerops humilis*, etc.). Above these there is a tree canopy of *Pinus halepensis*, which becomes increasingly more abundant with the distance from the sea and it also becomes increasingly scarce, and takes on a bushy appearance or can even disappear when approaching the sea.
- The vegetation characterised by *Juniperus macrocarpa* has been ascribed in the Levantine littoral to the association Asparago-Juniperetum macrocarpae. This species, which is more widespread in southwestern areas of the Iberian Peninsula and the Balearic Islands, appears in this territory in small nuclei, especially on the northern coast of the Valencian Community (Prat de Cabanes-Torreblanca) and in l'Albufera (València). At the first locality, it grows on gravelly seashores and is accompanied by scrub plants belonging to the Rosmarinetea class, while it grows at the second locality on sand substrates as a component of the sclerophyllous vegetation growing on stabilised dunes.
- Along the coast, seminatural pine-forests with *Pinus halepensis*, *Pinus pinea* or *Pinus pinaster* can also be found. These formations are scarce and are generally quite degraded because of anthropic activities.

Therophytic Communities

These annual communities are widespread mainly on pioneer dunes, mobile dunes, and also on fixed and semi-fixed dunes, or in situations where perennial vegetation has disappeared due to different disturbances. They occur in early spring and occupy inter-dune areas and the open spaces in sabulicolous vegetation. In this territory, these communities belong mainly to two associations, the Sileno ramosissimae-Cutandietum maritimae and the Erodio laciniati-Maresietum nanae, with *Silene sp. pl.*, *Vulpia fasciculata*, *Vulpia ciliata*, *Pseudorlaya pumila*, *Cutandia maritima*, *Maresia nana* and *Erodium laciniatum* among their typical constituents.

Nitrophilous Vegetation

Nitrophilous vegetation on sand beaches is represented by various associations: the Salsolo kali-Cakiletum aegyptiacae, comprising mainly annual plants (*Cakile maritima*, *Salsola kali*, *Polygonum maritimum*, etc.) that can resist the direct effects of strong winds and high solar radiation levels; the Centaureo maritimae-Echietum sabulicolae, which grows on slightly disturbed sandy soils in zones that are better protected from sea wind, and includes plants such as *Centaurea seridis*, *Echium sabulicola*, *Centaurea aspera* subsp. *stenophylla*, *Scabiosa atropurpurea*, *Sonchus tenerrimus*, etc.; and the Atriplicetum hastato-tornabenei, a halonitrophilous community, characterised by *Atriplex tornabenei*, and described from deep, sandy

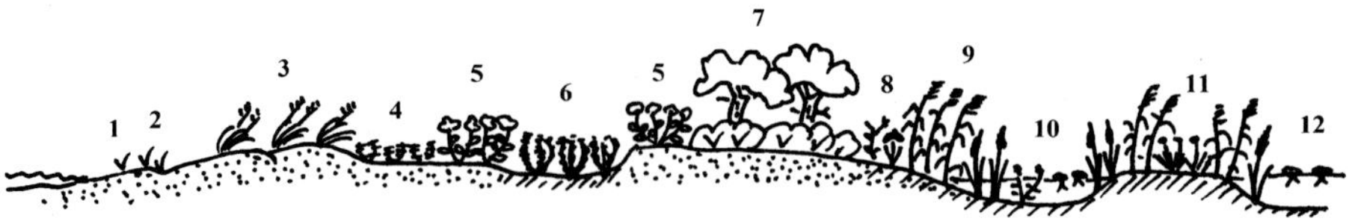

Fig. 14.7 Idealised catenal scheme of coastal vegetation (El Saler-l'Albufera). *1*-Salsolo kali-Cakiletum aegyptiacae; *2*-Cypero mucronati-Elytrigietum junceae; *3*-Medicagini marinae-Ammophiletum australis; *4*-Crucianelletum maritimae; *5*-Teucrio-Halimietum halimifolii; *6*-Puccinellio maritimae-Sarcocornietum fruticosae; *7*-Phillyreo-Rhamnetum angustifoliae; *8*-Irido pseudacori-Polygonetum salicifolii; *9*-Typho-Schoenoplectetum tabernaemontani; *10*-Potametum denso-nodosi; *11*-Hydrocotylo-Mariscetum serrati; *12*-Lemnetum gibbae

beaches of the Catalan littoral. On gravelly beaches, plants that have adapted to unstable substrates, like *Glaucium flavum*, *Hypochoeris radicata*, *Matthiola sinuata*, etc., characterise the association Hypochoerido-Glaucietum flavi (Fig. 14.7).

14.7.2 Coastal Lagoons and Marshes

This habitat type is widely represented in this coastal Levantine area, as there is a large complex of ponds and marshes all the way from north to south (Aiguamolls de l'Empordà, Torredembarra. Aiguamolls del Llobregat, Delta de l'Ebre, Marjal d'Almenara, Prat de Cabanes-Torreblanca, Peñíscola, Marjal del Moro, Marjal de Rafalell y Vistabella. Albufera, El Estany, Brosquil, Marjal de la Safor, Marjal de Pego-Oliva). There are two types of coastal lagoons in this area: those closely related with the sea, and often separated from it by a sandbar (Albufera de València); those directly connected to the sea as a result of a river mouth, like the Delta de l'Ebre (Fig. 14.8). Usually, both types are surrounded by a more or less extended area of marshes and channels with marsh vegetation. In these ecosystems the mixture of both saline and fresh water provides a variable salinity level, which acts as a key factor in the control of the floristic composition, and thus in vegetation. In this complex environment vegetation differs in dependence of salinity and depth and duration of waterlogging, and this has resulted in the description of many associations, most of which are included in two phytosociological classes: the Magnocarici elatae-Phragmitetea australis and the Potametea.

The first class includes helophytic communities, which are involved mainly in two groups known colloquially as "carrizales" and "masegares". Carrizales are communities whose physiognomy is determined by species of reeds of the genera *Phragmites* and *Typha*. In this coastal Levantine area, they occupy large tracts on the edges of marshes and channels. The two widespread associations in this territory correspond to the Typho-Schoenoplectetum tabernaemontani with *Phragmites australis* as the characteristic species, and to the Typho domingensis-Phragmitetum maximi (Fig. 14.9) with *Phragmites altissimus*. Masegares are dense formations

Fig. 14.8 Delta de l'Ebre Natural Park (Photo Ana M[a] Ibars)

dominated by *Cladium mariscus* which usually occupy edges of watercourses and small peat islands where water is almost always present. In some littoral places, they act as a shelter for protected species such as *Kosteletzkya pentacarpos* (Albufera de València). These communities are often managed as a refuge for breeding birds and for hunting. Two associations have been described: the Hydrocotylo-Mariscetum serrati for southern areas; the Soncho maritimi-Cladietum marisci, which is more widespread, even in river courses in areas farther from the sea. On moist saline soils, which may dry up during the dry season, and on edges of ponds, channels of calm water or river mouths, a community occurs that is dominated by *Bolboschoenus maritimus*, the Bolboschoeno compacti-Schoenoplectetum.

Small eutrophic channels with shallow, calm water frequently hold a community of small widespread helophytes (Helosciadietum nodiflori). To the same class belongs a group of communities that are more or less abundant in these environments but differ in their ecological conditions, e.g. in degree of salinity, or soil moisture, etc. They include the Lythro salicariae-Caricetum ripariae, Calliergonello cuspidatae-Eleocharitetum palustris, Irido pseudacori-Polygonetum salicifolii, etc. In the same environment as many of the aforementioned communities, on moist soils which are more or less waterlogged and sometimes anthropogenic, some generally short-stature communities occur, which are included in the Molinio-Arrhenatheretea class: (Lippio nodiflorae-Panicetum repentis, Trifolio fragiferi-Cynodontetum dactyli, Panico-Paspaletum vaginati, Cirsio monspessulani-Holoschoenetum vulgaris, Parietario judaicae-Equisetetum ramosissimae, etc.).

Fig. 14.9 Helophytic vegetation in floating islands "mates" (l'Albufera Natural Park) (Photo Manuel Costa)

The second class (Potametea) comprises the communities that grow in irrigation channels, streams and shallow ponds. The plants of these communities root in the soil and part of their leaves float on the water surface. Most of these communities are dominated by species of the genera *Potamogeton* (*Potamogeton pectinatus*, *Potamogeton natans*, *Potamogeton coloratus*, *Potamogeton nodosus*, etc.), *Ceratophyllum* (*Ceratophyllum demersum*, *Ceratophyllum submersum*), *Myriophyllum* (*Myriophyllum spicatum, Myriophyllum verticillatum*), etc. The more common of these communities are the Myriophyllo verticillati-Potametum pectinati, Potamo pectinati-Myriophylletum spicati and Potametum denso-nodosi, and less frequently the Potametum pectinati (primarily Delta de l'Ebre and Aiguamolls de l'Empordá), and in rice fields and shallow ponds the Callitricho-Ranunculetum baudotii. The following communities (characterized by *Nymphaea alba*, *Zannichellia palustris*, *Utricularia sp.pl.* or *Najas marina*) are scarce due to their requirements of oligotrophic water and to the anthropic pressure faced in some of their potential habitats: the Nymphaeo albae-Nupharetum luteae, Potamo-Utricularietum, Potamo-Najadetum marinae.

In standing waters of small ponds and irrigation channels almost monospecific communities of small free-floating plants, included in the class Lemnetea, grow. Several associations have been described, correlating with the eutrophication level of the water. The most widespread is the Lemnetum gibbae, a nitrophilous association that is frequently encountered on the edges of rice fields and channels between crops. Other associations only cover small areas: the Riccietum fluitantis and Lemno-Azolletum (Delta de l'Ebre), and the Ricciocarpetum natantis (Marjal de Xeraco). In brackish lagoons or river mouths with a high salt concentration in summer, we occasionally find populations of *Ruppia maritima* (Enteromorpho intestinalidis-Ruppietum maritimae).

Moreover, monospecific formations of the alien species of the genus *Ludwigia* are often observed, which colonize backwaters in shallow lagoons, edges of ponds, irrigation ditches or rice fields throughout the territory.

Lagoons and marshes are widespread along the south east Iberian Peninsula, but their conservation level is often unsatisfactory due to the proximity of tourism or crop areas, and to a generally lacking environmental awareness.

14.7.3 Salt Marshes

The halophilic vegetation growing in these environments faces two major challenges: on the one hand, the high salt concentration in the soils (ion toxicity); on the other, the difficulty of water absorption. Salt tolerance in the plants living in these communities depends on a range of ecological factors, like edaphic factors (soil texture, cations and anions concentrations, saline and moisture fluctuations, saline water table and salt water flooding) and adaptations that embrace many aspects of plant physiology. These halophytic communities usually grow on the edges of salt lagoons or in inter-dune depressions, known as "mallades", with seasonally fluctuating water levels. For this reason, these habitats are subject to significant annual changes in salt concentrations and the level of the water table, and that allows plants with different salt tolerances to grow together. This co-occurrence has led to the inclusion of the plant communities of these ecosystems in different associations.

Perennial Communities Some of these communities are dominated by perennial shrub species of the Salicornioideae and form associations that are very poor in species, often monotypic. They occur at lower sites that are subject to periodic flooding and a high salt concentration during drought periods. Among them, *Sarcocornia fruticosa*, occupies the lowest sites, where the salt concentration reaches maximum values, while *Arthrocnemum macrostachyum* is situated at places of slightly lower salinity, and both also grow in inter-dune depressions. From north to south, different associations have been described (Puccinellio maritimae-Sarcocornietum fruticosae, Limonio bellidifolii-Sarcocornietum fruticosae; Sphenopo divaricati-Arthrocnemetum macrostachyi) and on permanently or periodically flooded lands the Halimiono portulacoidis-Sarcocornietum alpini.

At higher sites in these littoral Mediterranean salt marshes different salt steppes are frequently, characterised by species of the genus *Limonium* (*Limonium virgatum*, *Limonium ferulaceum*, *Limonium cossonianum*, *Limonium angustebracteatum*, *Limonium narbonense*, *Limonium bellidifolium*, etc.). Many of them are endemic or have a very restricted distribution range. These communities, of poor diversity, occur on temporarily moist soils, and are subjected to an arid period with a significant salt concentration. These salt steppes mainly belong to three associations in this coastal Levantine area: the Artemisio gallicae-Limonietum angustebracteati, Artemisio-Limonietum virgati, and Zygophyllo albi-Limonietum latebracteati. The last one is an association with a restricted distribution area in the Delta de l'Ebre, located on small sandy mounds that are protected from flooding. In the same environment as the above described communities, but on more or less nitrified soils that remain humid for longer periods, bush communities characterised by Suaedoideae (*Suaeda splendens*, *Suaeda vera*, *Suaeda maritima*, *Salsola soda*, etc.) develop: the Suaedo-Salsoletum sodae, Atriplici-Suaedetum maritimae, Suaedetum verae and Lavatero davaei-Suaedetum verae, an exclusive association of the Columbrets islands characterised by *Suaeda vera* and the endemic *Lobularia maritima* subsp. *columbretensis*).

An endemic association growing in sandy mounds on halophilous soils in the Delta de l'Ebre is the Limonio densissimi-Limoniastretum monopetali, characterised by another Plumbaginaceae, *Limoniastrum monopetalum*.

Annual Communities Two different kinds of communities occur in dependence of soil features. On seasonally flooded soils, mainly at sites within shrub formations, communities constituted by annual pioneer swards belonging to the genus *Salicornia* grow (Suaedo spicatae-Salicornietum patulae, Salicornietum emerici, Suaedo-Salicornietum europaeae), while on drier soils, without a period of waterlogging, ephemeral coenoses grow, which are composed of small therophytes (Parapholido incurvae-Frankenietum, Junco minutuli-Parapholidetum filiformis, Catapodio spicati-Saginetum maritimae and Spergularietum marinae).

Saline Reedbeds This vegetation is composed of different species of Juncaceae, mostly *Juncus acutus* and *Juncus maritimus*. These reedbeds grow on saline-sandy soils in saltmarshes or margins of coastal lagoons. Their structure and zonation are the result of soil features and level of the watertable, which are factors that control flooding episodes, soil moisture and salt concentration. Several associations have been described from these territories that are mainly represented in some protected areas, like Aiguamolls de l'Empordà, Delta del Llobregat, Delta de l'Ebre, Prat de Cabanes-Torreblanca and El Saler. On saline soils that remain flooded for most of the year, the association Juncetum maritimo-subulati, characterised by *Juncus subulatus* and other species of *Juncus*, occurs. At sites with a moderate salt concentration, which also remain inundated during long periods, and on sandy-clay soils, the association Spartino-Juncetum maritimi thrives, characterised by *Spartina versicolor* and some Juncaceae, especially *Juncus maritimus*, which confer large size and extensive coverage. This association is scarce and is mainly found in Aiguamolls de l'Empordà, Delta del Llobregat and Delta de l'Ebre. On slightly saline soils at somewhat elevated sites, with a flooding level that depends on topography and the dry summer period, and on sandy-clay substrates that remain temporarily inundated, the Junco gerardii-Triglochinetum maritimae is found, a sparse, short community with good coverage, dominated by *Triglochin maritima*. On drier and moderately saline soils, which are rarely flooded, the Junco-Iridetum spuriae thrives.

Juncus acutus inhabits sandy saline soils located in drier situations. Two different associations have been described to comprise the vegetation characterised by this species, the Juncetum acuti for northern territories, and the Artemisio gallicae-Juncetum acuti described for southernmost localities (Prat de Cabanes-Torreblanca).

Occupying the inter-dune depressions and outer areas of salt marshes under moderately saline conditions, *Plantago crassifolia* and *Schoenus nigricans*, among others, are characteristic of the association Schoeno nigricantis-Plantaginetum crassifoliae. This association is widespread in the coastal areas of this Mediterranean territory.

14.7.4 Coastal Cliffs

Rocky shores are hostile environments where plants must adapt to lack of soil, strong winds and salinity. This type of habitat is widespread in this fragment of the Mediterranean coast (from Cap de Creus to Cap de la Nau) and shelters many endangered endemic species, especially those belonging to the genus *Limonium*. If we move from north to south we find several associations and almost each one with a different *Limonium* as characteristic species. In the northern areas of Catalunya (Cap de Creus), we find the Thymelaeo hirsutae-Plantaginetum subulatae and Armerietum ruscinonensis, the latter being an endemic association that shelters endemic species like *Armeria ruscinonensis*, *Limonium tremolsii* and *Limonium geronense*; the Dauco hispanici-Crithmetum maritimi, described from the Medes islands that spreads to the northern Catalan coast; the Reichardio-Crithmetum maritimi and Crithmo-Limonietum gibertii, characterised by the endemic *Limonium gibertii* that is shared with the Balearic Islands, (central and southern Catalan coasts); the Crithmo-Limonietum girardiani (rocky cliffs from northern Valencian territories, Oropesa to the Irta mountains); the Dauco commutati-Sonchetum dianii, an endemic association from the Columbrets Islands; the Crithmo-Limonietum dufourii, characterised by the endemic *Limonium dufourii* that grows at two separated localities, Cap deOropesa (province of Castelló) and Cap deCullera (province of València); and the Crithmo maritimi-Limonietum rigualii with a restricted area almost in dianic territories, characterised by the endemic *Limonium rigualii* and other plants like *Daucus hispanicus*, or occasionally *Diplotaxis ibicensis* (Fig. 14.10). On the rocky coast of Cap de Creus, positioned farther from the influence of the sea, and adjacent to the aforementioned communities, there are two associations characterised by the endemic *Astragalus tragacantha*: one in siliceous substrates (Cisto repentis-Astragaletum tragacanthae) and the other (Senecioni cinerariae-Astragaletum tragacanthae) on calcareous substrates in southern areas. Likewise, in the windy areas of Cap de Creus, an endemic

Fig. 14.10 Cap de Sant Antoni (Photo Pilar Soriano)

shrub community occurs on schist substrates, dominated by *Juniperus oxycedrus* and also containing other species, such as *Pinus halepensis*, *Myrtus communis*, *Pistacia lentiscus*, etc. (Myrto communis-Juniperetum oxycedri) (Fig. 14.11). Moreover in the calcareous cliffs of the more southern areas (Oropesa del Mar and the rocky cliffs of la Marina Alta), *Juniperus macrocarpa* occupies the same situations and forms part of shrub communities (Querco-Pistacietum lentisci, Chamaeropo humilis-Juniperetum phoeniceae). The gaps and crevices of all these rocky substrates are colonised by annual halophytic grasslands (Catapodio marini-Frankenietum pulverulentae and Triplachno nitentis-Desmazerietum marinae) (Fig. 14.12).

Where a certain amount of organic matter, often generated by birds, accumulates in these saline soils, nitrophilous communities like the Medicagini citrinae-Lavateretum arboreae and the Lavatero davaei-Suaedetum verae (Columbrets Islands and southernmost territories), the Carduo tenuiflori-Lavateretum arboreae (islets off the northern Catalan coast), the Gasouletum cristalino-nodiflori or the more widespread Soncho-Salsoletum vermiculatae may emerge.

Another group of communities are those that live below the sea surface and are often threatened by mechanical damage from trawlers and boats, coastal development and eutrophication. Different communities exist in relation to soil texture and the eutrophication level: the Posidonietum oceanicae, characterised by *Posidonia oceanica*, endemic to the Mediterranean Sea and the most-widespread seagrass

Fig. 14.11 Myrto communis-Juniperetum oxycedri. Cap de Creus (Photo Edoardo Biondi)

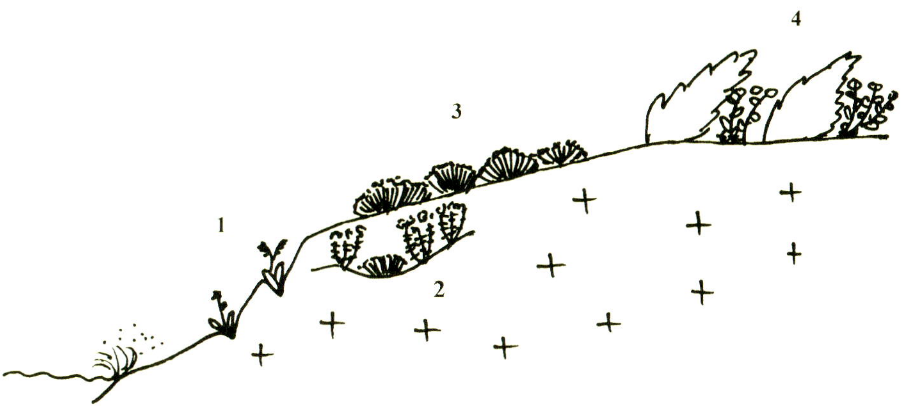

Fig. 14.12 Idealised catenal scheme of cliff communities (Cap de Creus). *1*-Armerietum ruscinonensis; *2*-Thymelaeo-Plantaginetum subulatae; *3*-Cisto repentis-Astragaletum tragacanthae; *4*-Myrto communis-Juniperetum oxycedri

species; the Cymodoceetum nodosae in muddy substrates and relatively eutrophic waters, and the Giraudio-Zosteretum noltii in bays near Cap de Creus and the Delta de l'Ebre. There are numerous references about coastal vegetation, e.g. Boira (1992), Costa (1982), Alonso and de la Torre (2002), Costa and Boira (1981), Crespo et al. (2003), Curcó (1990, 1996a, b, 2001), Curcó et al. (2008), Farràs and

Velasco (1994), Franquesa (1995), Ninot et al. (2000), Peric and Vilar Sais (2002), Royo (2006), Soriano (1995).

14.8 Rupicolous Vegetation

This kind of vegetation embraces a large diversity of communities that often grow under very different ecological conditions (altitude, slope, water availability, substrate, etc.), but always related to rocky substrates. Because of this, we attempt to divide these communities into different groups, mainly according to their habitat, and after considering only the commonest associations and those which, for some reason, are of special interest. In nicks and shelves of rocks in northern areas we find the Anomodonto viticulosi-Moehringietum catalaunicae (Montseny) and Polypodio cambrici-Saxifragetum fragilis (Montsant, Els Ports). A widespread group is formed by chasmophytic communities, mostly dominated by *Sarcocapnos* species that live in limestone caves or overhangs: the Antirrhinetum pertegasii, Chaenorhino crassifolii-Sarcocapnetum enneaphyllae or Sarcocapno saetabensis-Chaenorhinetum tenelli. In oozing rocks and walls, communities characterised by *Adiantum capillus-veneris* include the Eucladio verticillati-Adiantetum capilli-veneris and Trachelio coerulei-Adiantetum capilli-veneris. A heterogeneous group is formed by communities with diverse ecological conditions, all of which are characterised by different species of ferns: the Polypodio cambrici-Saxifragetum fragilis on carbonated nutrient-rich substrates (Prades, Montsant, Els Ports); Arenario intricatae-Polypodietum cambrici (southern areas); Sedo cepaeae-Polypodietum cambrici on silicic soils in northern areas; Centaureo paui-Cheilanthetum hispanicae, first described from cracks of siliceous rocks in sunny areas of the Serra d'Espadà; Cheilantho maderensis-Notholaenetum marantae, in southern territories. The communities that grow in dry sunny limestone crags include the Jasonio saxatilis-Chaenorhinetum cadevallii (a widespread association in central and northern territories); Phagnalo sordidi-Jasonietum mansanetianae, characterised by the endemic *Jasione mansanetiana* (in shaded thermic areas of the Serra d'Espadà) and Melico minutae-Saturejetum fruticosae, on sunny slopes at higher altitudes. On shaded stony slopes we find the Biscutello stenophyllae-Scrophularietum sciophilae, first described from the Serra d'Espadà, Selaginello-Anogrammetum leptophyllae, Scrophulario tanacetifoliae-Arenarietum intricatae and Resedetum paui (growing in southern territories).

Another fairly abundant group is composed of sub-nitrophilous chasmophytic communities characterised by *Parietaria judaica*, *Umbilicus rupestris*, etc.: Parietarietum judaicae, Parietario-Matthioletum incanae, Capparietum rupestris, Cymbalarietum muralis, Umbilico rupestris-Cheilanthetum maderensis, Hylotelephio maximi-Umbilicetum rupestris, Umbilicetum gaditani and Centranthetum rubri.

A set of associations that grows only in the southern parts of this territory comprises: the Jasionetum foliosae, described from shady supramediterranean areas of the Serra Mariola and characterised by *Jasione foliosa*, *Campanula hispanica*, *Saxifraga longifolia*, etc. (Serra de Bernia, Xortá, Montgó, Aitana, etc.); the Saxifragetum cossonianae, described from the same areas and characterised by the endemic *Saxifraga corsica* subsp. *cossoniana*; the Hippocrepido valentinae-Scabiosetum saxatilis, which occurs in the Dianic areas and is characterised by various endemic plants such as *Teucrium hifacense*, *Hippocrepis valentina* and *Sanguisorba ancistroides*; or the Antirrhino valentini-Trisetetum cavanillesiani characterized by the endemic *Antirrhinum valentinum* and with a restricted distribution area (Serra de Corbera, Buixcarró and Montduver).

Finally, there is a series of sparse communities, with only small distribution areas in this territory: the Ramondo myconi-Asplenietum fontani (Montsià); Sempervivo tectorum-Sedetum rupestris, a pioneer vegetation that thrives on rock ledges and crevices (Montseny); Saxifragetum vayredanae of silicic areas of Montseny, characterised by the endemic *Saxifraga vayredana*; Antirrhinetum asarinae on silicic rocks of Cap de Creus and characterised by *Asarina procumbens*; and the Minuartietum valentinae of shaded, stony slopes in the Serra d'Espadà and Calderona and characterised by the endemic *Minuartia valentina*. More information is available in Bolòs (1975), Molero Briones (1976), Royo (2006), Roselló (1994).

14.9 Nitrophilous Vegetation

Nitrophilous vegetation constitutes one of commonest and most diverse vegetation types in the Levantine coastal area. This group comprises a large number of communities that occur under a wide range of ecological conditions, but are always linked to human activity. Here, we consider only the most widespread associations and those of special interest given their floristic composition, ecological conditions, etc. A large, varied group is formed by the communities associated with different types of crops, composed of annual plants that relate closely to agricultural activities. Among them we find those that are established in cereal fields: the Centaureo collinae-Galietum valantiae and Roemerio hybridae-Hypecoetum penduli (calcareous soils) and the Airo cupanianae-Papaveretum rhoeadis (northern siliceous soils). In woody crop fields (olive, almond, carob, vineyards, etc.), the Diplotaxietum erucoidis occurs, an association that is widespread almost throughout the territory, the Eragrostio majoris-Chenopodietum botryos in northern and central areas, and the Calendulo-Chrysanthemetum paludosi in the southern Dianic zones. In irrigated citrus orchards in warm areas we find the association dominated by the introduced *Oxalis pes-caprae*, the Citro-Oxalidetum pedis-caprae and the Poo annuae-Arabidopsietum thalianae (on decarbonated soils), the Setario glaucae-Echinochloetum colonae in irrigated crops throughout the territory. In rice fields we find the Oryzo sativae-Echinochloetum cruris-galli, a weed community formed by plants of tropical origin, like *Cyperus difformis*, *Ammannia coccinea* and *Echinochloa crus-galli*. In strongly ruderalised areas, generally next to buildings, paths or fields, there are optimal summer communities dominated by Chenopodioideae and Amaranthaceae (*Chenopodium album*, *Chenopodium ambrosioides*, *Amaranthus muricatus*, *Amaranthus blitoides*, *Amaranthus retroflexus*), the Chenopodietum muralis or Amarantho-Chenopodietum ambrosioidis.

A large group of varied communities are those of path margins or fallows, some of which are characterised by different Gramineae, the Asphodelo-Hordeetum leporini, Hordeetum leporini, Medicagini rigidulae-Aegilopetum geniculatae, Bromo madritensis-Galactitetum tomentosae, Medicagini littoralis-Stipetum capensis, Reichardio picroidis-Stipetum capensis, Dittrichio viscosae-Piptatheretum miliacei, Sisymbrio irionis-Malvetum parviflorae, Salvio verbenacae-Plantaginetum albicantis; Trifolietum angustifolio-campestris in subhumid areas and Moricandio arvensis-Carrichteretum annuae in the driest ones.

The perennial vegetation rich in thistles that corresponds to the genera *Carduus*, *Silybum*, *Onopordon*, *Galactites*, *Carthamus*, etc., thrives on roadsides, embankments and abandoned crop lands. The numerous described associations include the Urtico piluliferae-Silybetum mariani, Verbasco montani-Onopordetum acanthi, Dipsaco fullonum-Cirsietum criniti, Cirsio echinati-Galactitetum duriaei, Centaureo pullatae-Cynaretum cardunculi, etc. The last two are characteristic of areas located further south.

In urban areas, on trampled and dry soils, mainly between paving slabs on streets and road verges, we find two associations characterised by small species of the genus *Euphorbia*, the Euphorbio-Eleusinetum geminatae and Euphorbio serpentis-Alternantheretum caracasanae.

Two special communities grow on small islands, the Euphorbio terracinae-Lobularietum columbretensis on the Columbrets Islands and the Alysso maritimi-Verbascetum boerhavii on the Medes Islands.

A group of nitrophilous communities, with some human influence, is constituted by hay meadows and formations grazed by different types of livestock. Among the first, there are those dominated by *Arrhenatherum* species (Arrhenatheretum elatioris and Galio veri-Arrhenatheretum bulbosi). The second group involves a heterogeneous set of communities which belong to different vegetation classes (Cynosuro cristati-Trifolietum repentis, Lolietum perennis and Astragalo sesamei-Poetum bulbosae). More particulars on this type of vegetation can be found in Bolòs (1957, 1962, 1967, 1975), Carretero (1989), Carretero and Aguilella (1995a, b), Curcó (2000), Ninot et al. (2010–2011), Royo (2008).

References

Alonso MA, de la Torre A (2002) Las comunidades fruticosas de Suaeda Forsskål ex J. F. Gmelin en la Península Ibérica e islas Baleares. Lazaroa 23:95–105

Álvarez de la Campa J (2004) Vegetació del massís del Port. Collecció Pius Font i Quer 3. Institut d'Estudis Ilerdencs

Ballesteros E (1981) Dues noves comunitats forestals al Massís de Cadiretes (la Selva). Folia Bot Barc 2:9–13

Boira H (1992) La vegetación del marjal de Torreblanca-Ribera de Cabanes. Actes del Simposi Internacional de Botànica 'Pius Font i Quer' 2 (Fanerogàmia): 233–239
Carreras J, Carrillo E, Ferré A, Masalles RM (2005) Manual dels hàbitats de Catalunya VI (4 Boscos). Departament de Medi Ambient i Habitatge, Generalitat de Catalunya, Barcelona
Carreras J, Carrillo E, Ferré A, Font J, Font X, Gesti J, Ninot JM, Vilar L (2006) Manual dels hàbitats de Catalunya IV. 3 Vegetació arbustiva i herbàcia (vegetació arbustiva). Departament de Medi Ambient i Habitatge, Generalitat de Catalunya, Barcelona
Carretero JL (1989) La vegetación emergente de los arrozales europeos. An Biol 15(4):135–141
Carretero JL, Aguilella A (1995a) Flora y vegetación nitrófilas del término municipal de la ciudad de València. Ajuntament de València, València
Carretero JL, Aguilella A (1995b) Algunas comunidades arvenses sabulícolas de la provincia de Valencia. Lazaroa 4:23–35
Costa M (1982) La vegetación costera valenciana: los cabos. Doc Phytosoc NS 6:354–364
Costa M, Boira H (1981) La vegetación costera valenciana: los saladares. An Jard Bot Madr 38 (1):233–244
Costa M, Peris JB, Figuerola R, Stübing G (1985) Los alcornocales valencianos. Doc Phytosoc 9:301–318
Costa M, Perez-Badia R, Soriano P (1995) Acerca de algunos bosques relictuales en el Mediterraneo occidental, Viburno tini-Fraxinetum orni ass. nova. Fitosociologia 29:181–185
Crespo MB (1989) Contribución al estudio florístico, fitosociológico y fitogeográfico de la Serra Calderona (Valencia-Castellón). Colección Tesis Doctorales Servicio de Publicaciones. Universidad de Valencia
Crespo MB (1999) Novedades sintaxonómicas y nomenclaturales en matorrales seriales valencianos del orden Rosmarinetalia Br.-Bl. ex Mounter 1934. Acta Bot Malacitana 24:208–220
Crespo MB, de la Torre A, Costa M (2003) Las comunidades de Crithmo-Staticion Molinier 1934 (clase Crithmo-Staticetea Br.-Bl. in Br.-Bl., Roussine & Nègre 1952) en la Península Ibérica e Islas Baleares. Phytocoenologia 33(2):527–559
Curcó A (1990) La vegetació del delta de l'Ebre (I): les comunitats dunars (Classe Ammophiletea Br.-Bl. et Tx. 1943). Butll Parc Nat Delta de l'Ebre 5:9–18
Curcó A (1996a) La vegetació del delta de l'Ebre (II): les comunitats halòfiles i halo-nitròfiles (Classes Puccinellio-Salicornietea i Cakiletea maritimae). Folia Bot Misc 10:113–139
Curcó A (1996b) La vegetación del delta del Ebro (III): las comunidades acuáticas de macrófitos (Clases Lemnetea minoris y Potametea). Doc Phytosoc 16:273–291
Curcó A (2000) La vegetació del delta de l'Ebre (IV): Les comunitats nitròfiles (classes Asplenietea trichomanis i Ruderali-Secalietea). Acta Bot Barc 46:143–178
Curcó A (2001) La vegetación del delta del Ebro (V): las comunidades helofíticas e higrófilas (Clases Phragmiti-Magnocaricetea y Molinio-Arrhenatheretea). Lazaroa 22:67–81
Curcó A (2008) La vegetació del Delta de l'Ebre (VII): els Tamarigars (Classe Nerio-Tamaricetea Br.-Bl. et O. Bolòs 1958). Revisió sintaxonòmica dels Tamarigars de Catalunya Acta Bot Barc 51: 79–91
Curcó A, Ferré A, Font J, Gesti J, Vilar L, Ballesteros E (2008) Manual dels hàbitats de Catalunya 2(1). Departament de Medi Ambient i Habitatge, Generalitat de Catalunya, Barcelona
de Bolòs O (1957) De vegetatione valentina, I. Collect Bot Barcelona 5:528–596
de Bolòs O (1959) El paisatge vegetal de dues comarques naturals: la Selva i la Plana de Vic, I.E.C. Arx. Secc. Cienc, vol 26, pp 1–175
de Bolòs O (1962) El paisaje vegetal barcelonés. Facultad de Filosofía y Letras, Universidad de Barcelona, Barcelona
de Bolòs O (1967) Comunidades vegetales de las comarcas próximas al litoral situadas entre los ríos Llobregat y Segura. Mem R Acad Cienc Art Barcelona 38(1):1–269
de Bolòs O (1975) De vegetatione valentina, II. Anales Inst Bot Cavanilles 32:477–488
de Bolòs O (1979a) Hedero-Ulmetum (l'omeda amb heura) i Lithospermo-Ulmetum (l'omeda amb mill gruà). Folia Bot Misc (Barc) 1:13–17

de Bolòs O (1979b) De vegetatione valentina, III. Butll Inst Catalana d'Hist Nat Sec Bot 44:65–76
de Bolòs O (1980) Les aulnaies (Alno-Padion) du Montseny en Catalogne. Coll Phytosociol 9:131–141
de Bolòs O (1983a) La brolla calcífuga (Cisto-Lavanduletea) als Països Catalans. Mem R Acad Cienc Art Barcelona 814
de Bolòs O (1983b) La vegetació del Montseny. Servei de Parcs Naturals. Diputació de Barcelona, Barcelona
de Bolòs O (1988) La roureda acidofila (Quercion robori-petraeae) a Catalunya. Monogr Inst Pir Ecol (Hom a P Montserrat) 4:447–453
de Bolòs O, Montserrat JM, Romo À (1993) El bosc mesòfil a les Muntanyes Catalanídiques septentrionals. Collect Bot (Barc) 22:55–71
De la Torre A, Crespo MB, Solanas JL (1997) Aportación al conocimiento de los espartales ibéricos (All. Stipion tenacissimae). Lazaroa 18:173–187
Farràs A, Velasco E (1994) Les comunitats vegetals de les zones humides de l'Alt Empordà. In: Gosálbez J, Serra J, Velasco E (eds) Els sistemes naturals dels aiguamolls de l'Empordà, Treb Inst Catalana Hist Nat, vol 13, pp 167–228
Fayos JM (2004) Vegetació del massís del Port. Col.lecció Pius Font i Quer, 3. Institut d'Estudis Ilerdenc, p 459
Folch Guillén R (1981) La vegetació dels Països Catalans. Ketres, Barcelona
Franquesa T (1995) El paisatge vegetal de la península del cap de Creus. Institut d'Estudis Catalans, Barcelona
Garcia Pausas J (1989) La lloreda (Osmundo-Lauretum), un bosc de ribera del Montseny. II Trobada d'Estudiosos del Montseny: 109–111
Gesti J, Font J, Vilar L (2003) Rusco aculeati-Fraxinetum angustifoliae, una nova associació forestal de ribera del territori ruscínic. Acta Bot Barc (Barc) 48:57–66
Hernández Cardona AM (1999) Les lloredes del Montnegre i el Corredor, fragments de la laurisilva mediterrània. II trobada d'estudiosos del Montnegre i el Corredor. Monografies n° 28. Diputació de Barcelona
Molero Briones J (1976) Estudio florístico y fitogeográfico de la sierra de Montsant y su área de influencia. PhD. Universitat de Barcelona
Ninot JM, Carreras J, Carrillo E, Vigo J (2000) Syntaxonomic conspectus of the vegetation of Catalonia and Andorra. I Hygrophilous herbaceous communities. Acta Bot Barc 46:191–237
Ninot JM, Font X, Masalles RM, Vigo J (2010–2011) Syntaxonomic conspectus of the vegetation of Catalonia and Andorra. II: Ruderal communities. Acta Bot Barc. 53: 113–189.
Nuet-Badia J, Romo-Díez A, Salvà-Catarineu M, Salvador-Franch F (2010) Les fagedes dels turons de Vilarmau i de la Tremoleda, al Montseny. VII Monografies del Montseny: 167–176
Pérez Badia R (1997) Flora vascular y vegetación de la comarca de la Marina Alta (Alicante). Colección Técnica 16. Instituto de Cultura Juan Gil Albert, Alicante
Peric JC, Vilar Sais J (2002) La vegetació halòfila dels aiguamolls de l'Empordà. Butll Inst Cat Hist Nat 70:21–40
Roselló R (1994) Catálogo florístico y vegetación de la comarca natural del Alto Mijares. Diputació de Castelló, Castelló
Royo Pla F (2006) Flora i vegetació de les planes i serres litorals compreses entre el riu Ebro i la serra d'Irta. PhD Dissertation. Universistat de Barcelona
Royo Pla F (2008) Comunitats vegetals de les planes i serres litorals del Baix Ebre, Montsià i Baix Maestrat, I: l'aliança Diplotaxion erucoidis Br.-Bl. in Br.-Bl., Gajewski, Wraber & Walas 1936. Toll Negre 10:21–42
Royo Pla F (2009) Comunitats vegetals de les planes i serres litorals del Baix Ebre, Montsià i Baix Maestrat, II: l'ordre Thero-Brachypodietalia Br.-Bl. (1931) Molinier 193. Toll Negre 11:46–65
Salvá-Catarineu M, Nuet-Badia J, Romo A, Salvador-Franch F (2012) Los hayedos marginales de La Calma (Montseny, cordillera prelitoral catalana) 396(401):396–401. Universitat Autónoma de Barcelona

Solanas JL (1996) Flora, vegetació i fitogeografía de La Marina Baixa. PhD Dissertation. Universidad de Alicante.
Soriano P (1995) Flora y vegetación de la comarca de la Safor (València). PhD Dissertation. Universitat de València.
Stübing G, Peris JB, Costa M (1989) Los matorrales seriales termófilos valencianos. Phytocoenologia 17(1):1–69
Vigo J (1968) La vegetació del massís de Penyagolosa, I.E.C., Arx. Secc. Cièn, vol 37, pp 1–247
Vilar L, Polo L, Domínguez-Planella A (1989) Los alcornocales de la provincia de Girona. Sci Gerundensis 15:143–151

Index of Geographic Names

J. Loidi (ed.), *The Vegetation of the Iberian Peninsula*, Plant and Vegetation 12,
DOI 10.1007/978-3-319-54784-8

V

X

Z

Index of Plant Taxa

J. Loidi (ed.), *The Vegetation of the Iberian Peninsula*, Plant and Vegetation 12,
DOI 10.1007/978-3-319-54784-8

C

F

M

T

Index of Syntaxa

J. Loidi (ed.), *The Vegetation of the Iberian Peninsula*, Plant and Vegetation 12,
DOI 10.1007/978-3-319-54784-8

B

D

E

F

I

J

Q

R

S

Printed in the United States
By Bookmasters